The CASSELL

DICTIONARY of SCIENCE

The CASSELL

DICTIONARY of SCIENCE

PERCY HARRISON
and
GILLIAN WAITES

A CASSELL BOOK

First published in the UK in 1997 by

Cassell
Wellington House
125 Strand
London WC2R 0BB

British Library Cataloguing-in-Publication Data
A catalogue record for this book is available from the British Library

ISBN 0-304-34483-4

Designed and typeset by Book Creation Services, London

Printed and bound in Great Britain by
Mackays of Chatham PLC, Chatham, Kent

Contents

How to use *The Cassell Dictionary of Science*

Arrangement of the dictionary
Entries are arranged alphabetically on a letter by letter basis, ignoring hyphens and spaces between words. Headwords – or main entries – are shown in **bold** type; ***bold italics*** are used to indicate an alternative form of the main headword.

Cross-references
Words that appear in SMALL CAPITALS in articles have their own entries elsewhere in the dictionary. Certain very common scientific words, such as 'element' or 'atom', are not automatically cross-referenced each time they are mentioned in the text.

See denotes a direct cross-reference to another article. *See also* indicates related articles or entries that contain more information about a particular subject.

Units
SI and metric units are used throughout the dictionary.

Abbreviations
In those cases where the part of speech of a headword is specified, the abbreviations used are as follows:

adj. adjective
n. noun
vb. verb

A

abdomen In vertebrates, the part of the body below the THORAX containing the intestines, liver, kidneys and other organs except the heart and lungs. In mammals it is separated from the thorax by a muscular DIAPHRAGM. The lower region of the abdomen is called the PELVIS, which is bounded by a set of bones called the PELVIC GIRDLE, to which the lower limbs are attached.

In invertebrates the abdomen is the hind part of the body. In insects and spiders the abdomen is characterized by the absence of limbs.

aberration Any DISTORTION in an optical image, in particular, one formed as a result of imperfections in a LENS or MIRROR, which mean that rays passing though different parts of the lens or mirror, or which have different WAVELENGTHS, are not all brought to a focus at a single point.

The two commonest forms of aberration are chromatic aberration and spherical aberration. In chromatic aberration, different colours of light are refracted to differing degrees by a lens, resulting in coloured fringes. The effect can be corrected by using an ACHROMATIC LENS – two lenses made of different types of glass. Spherical aberration results from the fact that a spherical surface will not bring parallel light to a perfect focus. Aberration can often be reduced by minimizing the diameter of the lens or mirror.

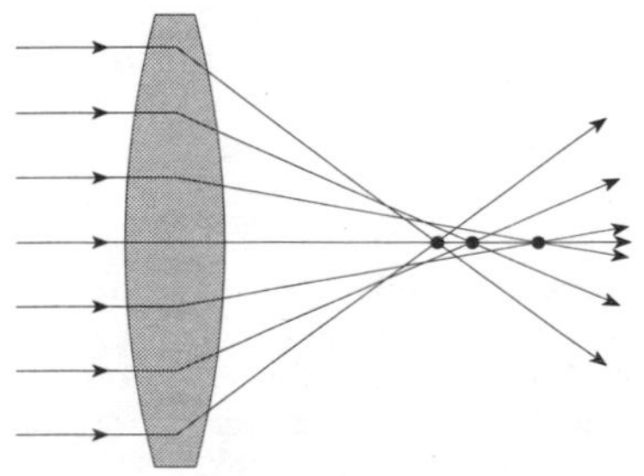

One form of aberration: spherical aberration.

See also ASTIGMATISM, CAUSTIC, COMA, REFRACTION.

abiotic (*adj.*) Describing the non-living or non-organic elements in an ECOSYSTEM, such as light, temperature, soil factors and rainfall.

ABO system *See* BLOOD GROUP SYSTEM.

ABS *See* ACRYLONITRILE-BUTADIENE-STYRENE.

abscissic acid A PLANT GROWTH SUBSTANCE that inhibits growth, in contrast to the other plant growth substances AUXIN, GIBBERELLIN and CYTOKININ. Abscissic acid inhibits NUCLEIC ACID and PROTEIN SYNTHESIS. Its main effect is on leaf or fruit drop (ABSCISSION); it causes the formation of an abscission layer between the main body of the plant and the part that is to fall. Auxin inhibits this process, so it is a balance of auxin and abscissic acid that eventually causes abscission. Auxin can be applied to a crop to prevent premature fruit drop. Abscissic acid can be applied to a crop to regulate the fruit drop.

abscission The shedding of a leaf, fruit, or other part from a plant. Abscission is controlled by PLANT GROWTH SUBSTANCES, particularly ABSCISSIC ACID.

abscission layer *See* ABSCISSIC ACID.

absolute temperature A TEMPERATURE SCALE in which the temperature is proportional to the energy of the random thermal motion of the molecules. This is the same as a temperature scale based on the PRESSURE LAW, which states that temperature is directly proportional to the pressure exerted by an IDEAL GAS held in a fixed volume. The SI UNIT of absolute temperature is the KELVIN. *See also* ABSOLUTE ZERO, INTERNATIONAL PRACTICAL TEMPERATURE SCALE.

absolute value 1. *See* MODULUS

2. A measurement of any physical quantity that is independent of any arbitrary or variable standard. An example is a pressure measured as the total force exerted by a gas, rather than in terms of a pressure difference relative to ATMOSPHERIC PRESSURE.

absolute zero The lowest temperature that is theoretically attainable; zero on the ABSOLUTE TEMPERATURE scale (0 K). Absolute zero is the temperature at which molecules would stop moving and an IDEAL GAS would produce no pressure. It is equivalent to –273.15°C. It is impossible to achieve absolute zero, but temperatures as low as 2 x 10^{-9} K have been reached.

absorptance, *absorptivity* (α) The fraction of ELECTROMAGNETIC RADIATION falling on a body that is absorbed by that body. The absorptance of any body is equal to its EMISSIVITY. *See also* PRÉVOST'S THEORY OF EXCHANGES.

absorption 1. (*chemistry*) The taking up of a gas by a solid or a liquid, or of a liquid by a solid. The molecules of the absorbed substance penetrate throughout the whole of the absorbing substance. *Compare* ADSORPTION.

2. (*physics*) The removal of energy from a wave, or particles from a particle beam, as the wave or beam passes through a material. The energy of the wave or beam is usually converted into heat within the absorbing material. *See also* FILTER.

3. (*biology*) The passage of substances such as nutrients or drugs into and through body tissues and blood vessels.

absorption spectrum A SPECTRUM formed when a sample absorbs certain wavelengths from a continuous background of ELECTROMAGNETIC RADIATION. *See also* ATOMIC ABSORPTION SPECTROSCOPY, SPECTROSCOPY.

absorptivity *See* ABSORPTANCE.

abyssal zone The deepest part of the ocean (below about 2,000 metres), where very little light penetrates.

a.c. *See* ALTERNATING CURRENT.

acceleration (a) The change in the velocity of a body divided by the time over which that change takes place.

In physics, acceleration is usually measured in metres per second per second, also called metres per second squared (ms^{-2}). However, the acceleration figure of a car is normally quoted as the time to reach a given speed, usually 60 miles per hour, from rest. Thus an acceleration time of 10 s from 0–60 mph represents an average acceleration of 10 mph per second.

acceleration = change in velocity/time

See also ACCELEROMETER.

acceleration due to gravity (g) The acceleration experienced by an object in FREE FALL, a measure of the strength of the GRAVITATIONAL FIELD. On Earth, g is about 9.8 ms^{-2}, but it varies from place to place, being greater at the poles as these are closer to the centre of the Earth than the equator. *See also* GRAVITATIONAL FIELD STRENGTH.

accelerator A catalyst that increases the rate of a chemical reaction. In particular, an accelerator is a catalyst that increases the rate of a POLYMERIZATION leading to the formation of a hard material, such as the setting of certain glues and RESINS.

accelerometer A device for the measurement of acceleration. Accelerometers range from simple devices measuring the extension of a spring produced by an accelerating mass and used to check the efficiency of brakes on motor vehicles, to sophisticated devices based on lasers and used in INERTIAL NAVIGATION SYSTEMS.

acceptor atom 1. (*electronics*) An atom that accepts an electron to produce a HOLE in a SEMICONDUCTOR, to form a P-TYPE SEMICONDUCTOR.

2. (*chemistry*) An atom that receives a share in a pair of electrons in the formation of a CO-ORDINATE BOND.

acceptor impurity An element added to a SEMICONDUCTOR in a DOPING process that accepts an electron, thereby producing a HOLE.

access time The time taken for a computer to write to, or read from, MEMORY once an instruction has been given.

accommodation *See* EYE.

accumulator 1. (*computing*) The SHIFT REGISTER in which the output from the ARITHMETIC LOGIC UNIT of a MICROPROCESSOR is stored.

2. (*physics*) An obsolete term for a BATTERY of LEAD-ACID CELLS.

acellular (*adj.*) Describing organisms or their parts that have no real cellular structure, for example tissues forming a SYNCYTIUM and many fungi. *Compare* MULTICELLULAR, UNICELLULAR.

acetaldehyde *See* ETHANAL.

4-acetamidophenol Systematic name for PARACETAMOL.

acetate Common name for ETHANOATE.

acetic acid Common name for ETHANOIC ACID.

acetic ester *See* ETHYL ETHANOATE.

acetic ether *See* ETHYL ETHANOATE.

acetone Common name for PROPANONE.

acetyl CoA *See* ACETYL COENZYME A.

acetyl coenzyme A, *acetyl CoA* COENZYME A that is carrying an ACYL GROUP.

acetylation *See* ETHANOYLATION.

acetylbenzoic acid *See* ASPIRIN.

acetylcholine A NEUROTRANSMITTER that is mostly concerned with the transmission of NERVE IMPULSES across NEUROMUSCULAR JUNCTIONS, resulting in MUSCULAR CONTRACTION.

acetylene Common name for ETHYNE.

acetylide *See* DICARBIDE.

acetylsalicylic acid *See* ASPIRIN.

achene A dry, one-seeded fruit, for example the fruit of the buttercup, dandelion, grasses and daisies. Achenes do not split open to release their seeds (they are INDEHISCENT), and develop from a single CARPEL. The fruit wall can be winged, as in ash and sycamore, or can have hairs attached, as in the dandelion. Both features help DISPERSAL.

achromatic lens A lens designed to correct chromatic aberration (*see* ABERRATION). The simplest type of achromatic lens is a combination of two lenses made of different types of glass, such that their DISPERSIONS neutralize each other.

acid (i) Any compound that contains hydrogen and liberates hydrogen ions when dissolved in water. Acids release hydrogen gas when they react with metals, and they react with BASES to form a SALT plus water. For example, sulphuric acid, H_2SO_4, is IONIZED in water:

$$H_2SO_4 \Leftrightarrow 2H^+ + SO_4^{2-}$$

Dilute sulphuric acid reacts with magnesium to release hydrogen

$$H_2SO_4 + Mg \rightarrow MgSO_4 + H_2$$

and with sodium hydroxide to form sodium sulphate plus water

$$H_2SO_4 + 2NaOH \rightarrow Na_2SO_4 + 2H_2O$$

(ii) In the Lowry–Brønsted theory of acids and bases, an acid is any compound that gives up a proton to some other substance. For example, when sodium chloride is dissolved in ammonia,

$$NH_3 + NaCl \Leftrightarrow NaNH_2 + HCl$$

the ammonia donates a proton and is thus regarded as an acid.

(iii) In the Lewis theory of acids and bases, an acid is defined as any substance that accepts electrons. This encompasses the traditional definition, as in reactions with metals and bases, hydrogen ions in the acid are neutralized to form hydrogen or water molecules. However, it also extends to REDOX REACTIONS and to the formation of CO-ORDINATE BONDS, for example

$$H_3N: + BCl_3 \rightarrow H_3NBCl_3$$

where the ammonia is a LEWIS ACID because it donates a pair of electrons to form the bond with the boron.

See also ACIDIC, ACIDIC HYDROGEN, BASE, PH, PK, SALT, STRONG ACID, WEAK ACID, and under individual acids.

acid anhydride Any chemical compound obtained by the removal of water from another compound, usually an acid. The resulting compound is a dehydrated acid, for example sulphur trioxide, SO_3, is the anhydride of sulphuric acid, H_2SO_4.

In organic chemistry an acid anhydride is formed by the removal of one molecule of water from two molecules of a CARBOXYLIC ACID, for example:

$$2CH_3COOH - H_2O \rightarrow CH_3COOCOCH_3$$

in which ethanoic acid is converted into ethanoic anhydride.

The general formula for an organic acid anhydride is RCOOCOR, where R is an ALKYL GROUP. The chemistry of acid anhydrides is similar to the ACID CHLORIDES except they are less reactive.

acid chloride, *acyl chloride* A CARBOXYLIC ACID in which the –OH HYDROXYL GROUP has been chlorinated, leaving an ACYL GROUP (RCO–, where R is an ALKYL GROUP) attached to chlorine. Thus the general formula for acid chlorides is RCOCl. An example is ethanoyl chloride, CH_3COCl, derived from ethanoic acid, CH_3COOH. Acid chlorides are named by replacing the suffix *-ic* of the carboxylic acid from which it is derived with *-yl*.

Acid chlorides are more reactive than acids since the chlorine atom is easily replaced by other NUCLEOPHILES. They are readily hydrolysed (*see* HYDROLYSIS) in cold water to form hydrochloric acid and induce tears if the vapour is close to the eyes. Acid chlorides are termed acylating agents since they are able to add an acyl group to a molecule.

acid dye *See* DYE.

acidic (*adj.*) Describing an ACID, a solution with a high concentration of hydrogen ions, or a material that produces an acid when dissolved in water. Thus sulphur dioxide, SO_2, which forms sulphurous acid, H_2SO_3, when dissolved in water, may be termed acidic. *See also* AMPHOTERIC, BASIC.

acidic hydrogen The portion of hydrogen in an ACID that will be replaced by a metal to form a SALT. The acidic hydrogen will form positive ions if the acid DISSOCIATES in water. *See also* ACIDIC SALT.

acidic salt Any salt in which not all of the ACIDIC HYDROGEN has been replaced by a metal. The hydrogen will be released as positive ions, forming an acidic solution, if the salt is soluble. An example is the HYDROGENCARBONATES, which contain the HCO_3^- ion. These are formed when only one of the hydrogen atoms in carbonic acid, H_2CO_3, a DIBASIC ACID, are replaced by a metal ion. *See also* BASIC SALT, NORMAL SALT.

acid rain Rain with a high acidity, caused mainly by sulphur dioxide (from the burning of FOSSIL FUELS) dissolving in water to form sulphuric and sulphurous acids. Nitrogen oxides (from industry and car exhausts) also contribute to acid rain. Acid rain causes damage in particular to coniferous forest species and some aquatic species, either directly due to the acidity or indirectly when the rain leaches toxic aluminium from soils (*see* LEACHING).

Levels of sulphur dioxide in the air can be measured by the use of INDICATOR SPECIES, such as LICHENS and MOSSES, which have variable tolerances to sulphur dioxide levels so their survival is indicative of the sulphur dioxide concentration.

See also POLLUTION.

acinus In certain glands, such as the liver or pancreas, the functional part of any of the lobules that make up the gland.

acoustics The study of sound. The term also refers to the behaviour of sound in a particular room, such as a concert hall. The acoustics of a room may be modified by introducing sound-absorbing materials to reduce the REVERBERATION TIME.

acquired immune deficiency syndrome *See* AIDS.

acquired immunity *See* IMMUNITY.

acridine ($C_{13}H_9N$) An organic compound used to make DYES.

acrosome A specialized structure at the tip of a SPERM that is formed from the GOLGI APPARATUS. It contains enzymes that are involved in the ACROSOME REACTION, which enables the sperm to penetrate the OVUM. *See also* FERTILIZATION.

acrosome reaction The process by which a SPERM is able to penetrate an OVUM. Enzymes are released from the ACROSOME at the tip of the head of the sperm, and soften the outer membrane of the ovum. The acrosome then inverts and a fine filament develops at the tip of the sperm that is able to pierce the ovum, allowing the sperm to enter the ovum. *See also* FERTILIZATION.

acrylic A synthetic wool-like fabric made from the COPOLYMERIZATION of PROPENOIC ACID derivatives.

acrylic acid *See* PROPENOIC ACID.

acrylonitrile *See* PROPENENITRILE.

acrylonitrile-butadiene-styrene (ABS) A synthetic RUBBER consisting of a copolymer (*see* COPOLYMERIZATION) of BUTADIENE, PHENYLETHENE and PROPENENITRILE (acrylonitrile). ABS is rigid and tough and used in telephone receivers and suitcases.

ACTH *See* ADRENOCORTICOTROPHIC HORMONE.

actin A eukaryotic (*see* EUKARYOTE) protein that is a major constituent of muscle fibres and plays a vital role in a cell's CYTOSKELETON. Globular protein MONOMERS of actin (G-actin) polymerize (*see* POLYMERIZATION) to form long fibrous molecules of filamentous actin (F-actin). Two of these then twist around one another to form the thin filaments characteristic of muscle MYOFIBRILS. Filaments of actin and MYOSIN can contract together as ACTOMYOSIN, which is vital for MUSCULAR CONTRACTION.

actinide, *actinoid* Any one of the series of elements with ATOMIC NUMBERS from 89 (ACTINIUM) to 103 (LAWRENCIUM). The actinides all have two electrons in the 7s ORBITAL. Increasing atomic number corresponds to filling the 5f (and sometimes 6d) orbitals. All are radioactive and have similar chemical properties, which differ only slightly with atomic number.

actinium (Ac) The element with ATOMIC NUMBER 89; melting point 1,050°C (approx.). It is a white metal, the first in the ACTINIDE series of elements. All known ISOTOPES of the element are radioactive with HALF-LIVES of 21.7 years (actinium–227) or less. Despite this, actinium occurs in nature in very small quantities, being produced in the decay of heavier nuclei with longer half-lives. Actinium is used as a source of high-energy ALPHA PARTICLES.

actinoid *See* ACTINIDE.

actinomorphic *See* FLOWER.

Actinozoa, *Anthozoa* A CLASS of the PHYLUM CNIDARIA, including corals and sea anemones. Actinozoans are mostly marine animals.

action and reaction Newton's third law states that for every force causing an action, there will be an equal and opposite force causing a reaction. *See* NEWTON'S LAWS OF MOTION.

action potential A change in the voltage across the membrane of a NEURONE or muscle fibre that occurs when a NERVE IMPULSE travels along it. This is accompanied by the passage of sodium and potassium ions across the membrane. *See also* MUSCULAR CONTRACTION.

activation analysis A technique for detecting small quantities of an element present in a sample. The sample is bombarded with neutrons from a NUCLEAR REACTOR, forming unstable ISOTOPES of the elements present. These decay, emitting GAMMA RADIATION with energies characteristic of the elements present.

activation energy (E_a) The minimum amount of energy required before a particular process can take place. The term is usually applied to chemical reactions, where the activation energy is the ENERGY BARRIER that must be overcome for the reaction to occur. On the atomic scale, the process may be a NUCLEAR FISSION event or the release of an electron or a photon of light, such as in the THERMOELECTRIC EFFECT or the PHOTOELECTRIC EFFECT. *See* ACTIVATION PROCESS. *See also* CATALYST.

activation process Any process in which the particles involved can only take part in the process if they have more than a specified amount of energy, known as the ACTIVATION ENERGY. This acts as an ENERGY BARRIER, which must be overcome for the process to take place. To do this, energy may be supplied externally (for example by light in the case of some chemical reactions). The energy barrier may also be overcome by the action of a catalyst that allows the reaction to proceed via some intermediate state that requires less energy. Another possibility is to raise the temperature of the material so that more molecules have sufficient energy. Many chemical reactions, the evaporation of liquids and CREEP are all examples of activation processes. *See also* MAXWELL–BOLTZMANN DISTRIBUTION, REACTION PROFILE.

activator In biochemistry, a substance that enables an enzyme to bind to a SUBSTRATE. *See also* COFACTOR.

active anode An ANODE that is chemically involved in an ELECTROLYSIS process.

active device **1.** An artificial satellite that receives and retransmits signals after amplification (*see* AMPLIFIER).

2. A RADAR that emits MICROWAVES and gathers information on an object from the reflected radiation.

3. An electronic device that can act as an AMPLIFIER or SWITCH. The behaviour of active devices, in contrast to devices such as RESISTORS or CAPACITORS, cannot be described in a simple mathematical way. Most active devices are based on semiconductors. A common example is the TRANSISTOR, which can use a small current to control a larger one. Modern active devices are often manufactured as INTEGRATED CIRCUITS. *See also* PASSIVE DEVICE.

active immunity *See* IMMUNITY.

active site In biochemistry, the part of an enzyme molecule to which the SUBSTRATE binds. The active site is formed by the three-dimensional structure of the enzyme and the CHARGE distribution in the molecule. The substrate specificity of an enzyme is determined by the active site. An enzyme can have more than one active site. Enzyme inhibitors can also reversibly bind to the active site, thereby blocking the action of the enzyme. *See also* ENZYME, ENZYME INHIBITION.

active transport In biology, an energy-requiring process (usually involving ATP) where substances, usually molecules or ions, are moved across a membrane against a concentration gradient – that is, from a region of low concentration to one of higher concentration. The process involves 'pumps' of protein molecules in the membrane that carry specific ions across, such as the SODIUM PUMP. *Compare* DIFFUSION.

activity **1.** (α) A thermodynamic function used to calculate the EQUILIBRIUM CONSTANT for a REAL GAS. It is a correction factor that allows for the effect that the INTERMOLECULAR FORCES between gas molecules have on the equilibrium concentrations of reacting gases. In a reaction

$$A \Leftrightarrow B + C$$

the equilibrium constant (K) is given by

$$K = \alpha_B \alpha_C / \alpha_A$$

where α_A is the activity for A, etc.

The activity coefficient (γ) for a gas of pressure p is defined as

$$\gamma = \alpha / p$$

For a solution,

$$\gamma = \alpha X$$

where X is the MOLE FRACTION.

2. (A) The level of IONIZING RADIATION emitted by a radioactive material. Activity is usually measured in BECQUEREL, or becquerel per litre, although other units, for example the CURIE, are sometimes used. *See also* RADIOACTIVITY.

activity coefficient *See* ACTIVITY.

actomyosin A complex biochemical compound formed by the interaction of the proteins ACTIN and MYOSIN. This occurs in muscle in the presence of calcium ions and is the basis of MUSCULAR CONTRACTION. Actomyosin dissociates in the presence of ATP.

acute angle Any angle less than 90°.

acylating agent Any organic compound capable of adding an ACYL GROUP to a molecule. *See also* ACID CHLORIDE.

acylation The substitution of an ACYL GROUP (RCO–, where R is an ALKYL GROUP) into a molecule, usually in exchange for a hydrogen atom of a HYDROXYL GROUP. For example:

$$CH_3COCl + NH_3 \rightarrow CH_3CONH_2 + HCl$$

acyl chloride *See* ACID CHLORIDE.

acyl group The RCO– group of a CARBOXYLIC ACID (RCOOH) that remains when the –OH HYDROXYL GROUP has been removed. R is an ALKYL GROUP. Acyl groups are named after the carboxylic acid from which they are derived, for example ethanoyl, CH_3CO–, from ethanoic acid, CH_3COOH.

acyl halide A CARBOXYLIC ACID in which the –OH HYDROXYL GROUP has been replaced by a HALOGEN, leaving an ACYL GROUP (RCO–, where R is an ALKYL GROUP) attached to a halogen. An example is ethanoyl chloride, CH_3COCl. *See also* ACID CHLORIDES.

Adam's apple *See* LARYNX.

adaptive radiation A variation on the theory of evolution, which suggests that new species can evolve from a single ancestral type as a result of MIGRATION to new unoccupied ECOLOGICAL NICHES. New adaptations develop to accommodate the different ways of life, resulting eventually in new species. For example, DARWIN'S FINCHES on the Galapagos Islands probably descended from a single species on the South American mainland that adapted to suit different niches on the islands. *See also* SPECIATION.

Addison's disease *See* GLUCOCORTICOID.

addition polymerization *See* POLYMERIZATION.

addition reaction A chemical reaction in which a molecule is added to an UNSATURATED COMPOUND across a double or triple COVALENT BOND. An example is the addition of bromine to ethene to give 1,2-dibromoethane:

$$CH_2{=}CH_2 + Br_2 \rightarrow BrCH_2CH_2Br$$

See also ELECTROPHILIC ADDITION

additive In food, any substance added to improve the colour, flavour or nutrient value, or to prolong the shelf-life of the food. Additives can be natural or artificial. Their use is regulated, as some can cause side-effects, such as asthma, hyperactivity and cancer, in certain people. Additives approved by the European Community are called 'E-NUMBERS'.

Most additives used are synthetic flavourings or flavour enhancers (such as monosodium glutamate and sweeteners, which have no flavour of their own but strengthen the flavour of other substances) that do not have to be listed in detail on a product label. Texture enhancers, such as thickeners, emulsifiers (which bind fat and water) and stabilizers (which prevent separation of fat and water), are useful for producing low fat spreads. Added minerals and vitamins also classify as additives. *See also* FOOD PRESERVATION.

address A specific location in computer MEMORY, at which a single piece of data can be stored. Each address is identified by a number, which in PERSONAL COMPUTERS corresponds to one BYTE. The number of addresses determines the capacity of the computer memory.

adduct A compound formed by an ADDITION REACTION. In particular, an adduct is a compound formed between a LEWIS ACID and a LEWIS BASE.

adenine An organic base called a PURINE that occurs in NUCLEOTIDES. *See also* BASE PAIR, DNA, RNA.

adenoids A mass of lymphoid tissue at the back of the nose. *See* LYMPH NODE.

adenosine A PURINE NUCLEOSIDE, consisting of the organic base ADENINE and the sugar RIBOSE. In its phosphorylated forms (*see* PHOSPHORYLATION) adenosine is AMP, ADP and ATP.

adenosine diphosphate *See* ADP.

adenosine monophosphate *See* AMP.

adenosine triphosphate *See* ATP.

adenylate cyclase An enzyme that catalyses the formation of CYCLIC AMP from ATP.

ADH *See* ANTI-DIURETIC HORMONE.

adhesion The intermolecular force of attraction between molecules of one substance and those of another. This effect causes water to spread out and wet glass, for example. The adhesion of the water molecules to the glass molecules is greater than COHESION between the water molecules themselves. In the case of mercury on glass, cohesion between the mercury molecules is greater than the adhesion of mercury to glass. The mercury molecules pull inwards, away from the glass, and small drops of mercury on a glass plate are almost spherical. *See also* CAPILLARY EFFECT, SURFACE TENSION.

adiabatic (*adj.*) Describing a change in which there is no exchange of energy between a system and its surroundings.

adipose tissue, ***fatty tissue*** CONNECTIVE TISSUE consisting of cells containing large globules of fat that provide insulation, particularly around the kidneys, heart and in the inner layer of skin. Adipose tissue serves as an energy reserve.

ADP (adenosine diphosphate) The product formed by the PHOSPHORYLATION of AMP during energy-yielding biochemical reactions, and produced from the HYDROLYSIS of ATP. ADP can be phosphorylated to form ATP.

adrenal gland, ***suprarenal gland*** In vertebrates, either one of a pair of ENDOCRINE GLANDS found on top of the KIDNEY that produces hormones in response to stress situations.

Adrenal glands consist of two independent regions, the outer adrenal CORTEX and the inner adrenal MEDULLA. The adrenal cortex consists of the majority of the glands and is itself divided into three zones, secreting a number of STEROID hormones collectively called CORTICOSTEROIDS (or corticoids). The outermost layer of the cortex secretes ALDOSTERONE, which regulates water retention by the kidney. The middle zone of the adrenal cortex secretes GLUCOCORTICOIDS, including CORTISOL, in response to internal stress, such as low blood temperature or volume. The adrenal medulla produces the hormones ADRENALINE and NORADRENALINE, which prepare the body for action in response to the external stress, such as fear, anger and pain.

adrenaline, ***epinephrine*** A hormone, derived from the amino acid tyrosine, that is secreted by the ADRENAL GLAND in response to external stress, such as fear, anger and pain. It prepares the body for action by increasing blood flow to the heart and muscles, causing the heart rate to quicken, and dilating airways in the lungs to enable more oxygen to be delivered to cells of the body, while constricting blood vessels in the skin and gut. Adrenaline also increases the amount of sweat produced, causes hair to stand up, pupils to dilate and increases breakdown of GLYCOGEN to GLUCOSE in the liver. Both adrenaline and the similar NORADRENALINE are an important link between the ENDOCRINE SYSTEM and the NERVOUS SYSTEM because they are also NEUROTRANSMITTERS. Their production by the adrenal glands is regulated by the HYPOTHALAMUS via nervous connections. They mimic effects of the SYMPATHETIC NERVOUS SYSTEM.

adrenocorticotrophic hormone (ACTH), ***corticotrophin*** A POLYPEPTIDE secreted by the anterior PITUITARY GLAND which regulates growth of the CORTEX of the ADRENAL GLAND and stimulates production of its hormones. ACTH is commonly produced as a result of stress, which causes a substance, known as corticotrophin-releasing factor (CRF), to be released from the HYPOTHALAMUS. CRF initiates production of ACTH.

adsorption The taking up of a gas by the surface of a solid. Unlike ABSORPTION, the gas does not penetrate the solid material, but is held on the surface either by the formation of chemical bonds (called chemisorption) or by VAN DER WAALS' FORCES (called physisorption).

adsorption chromatography *See* CHROMATOGRAPHY.

advection The transport of energy by the horizontal bulk motion of a fluid.

advection fog Fog produced when warm moist air from above the sea is blown over cooler land. *See also* RADIATION FOG.

adventitious root A root of a plant that grows from a stem instead of other roots, or that grows in an unusual position. For example, adventitious roots of ivy grow sideways out of the stem to find support. *See also* FIBROUS ROOT, PROP ROOT.

aerial, *antenna* A device used to transmit or receive radio waves. Aerials have many shapes and sizes, but are generally most efficient when they have a size of the same order as the wavelength of the wave concerned. An aerial may be equipped with a parabolic dish that produces a parallel beam of radio waves from a source of radiation placed at its focus, or that collects and focuses a parallel beam of radio waves. *See also* YAGI.

aerobe Any organism that requires oxygen for RESPIRATION (breakdown of food to release energy). Most organisms are aerobic, with the exception of some bacteria. Some cells can function for short periods without oxygen but most die. Those organisms that can survive without oxygen are called ANAEROBES.

aerobic respiration *See* AEROBE, RESPIRATION.

aerodynamics The study of the flow of gases (such as air) over solid objects. Aerodynamics has particular applications in the design of vehicles, especially aircraft. It is also used in the study of the flight of birds and insects. The term aerodynamics is also used to refer to the aerodynamic properties of vehicles.

aerosol A COLLOID in which liquid particles are suspended in a gas. The term aerosol is also used to refer to the mechanism for producing an aerosol, in which a propellant gas forces liquid out of a tube through a fine nozzle. CHLOROFLUOROCARBONS (CFC's) have been used as the propellant, but alternatives such as BUTANE are now more common since concerns have emerged over damage to the OZONE LAYER from CFC's in the atmosphere.

afterbirth In mammals, the material shed from the UTERUS following the birth of their young. It includes the PLACENTA and EXTRAEMBRYONIC MEMBRANES.

afterburner *See* REHEAT SYSTEM.

agglutination The clumping or sticking together of cells such as bacteria and RED BLOOD CELLS. Agglutination is caused by ANTIBODIES reacting with their specific ANTIGEN. It can occur to help the body remove foreign cells such as bacteria, or it can occur due to a mismatch of antigens of the BLOOD GROUP SYSTEM, for example during blood transfusion.

agglutinin A cell-surface or PLASMA PROTEIN that acts as an ANTIBODY, causing clumping (AGGLUTINATION) of ANTIGENS on foreign cells such as bacteria. Agglutinins are often LECTINS.

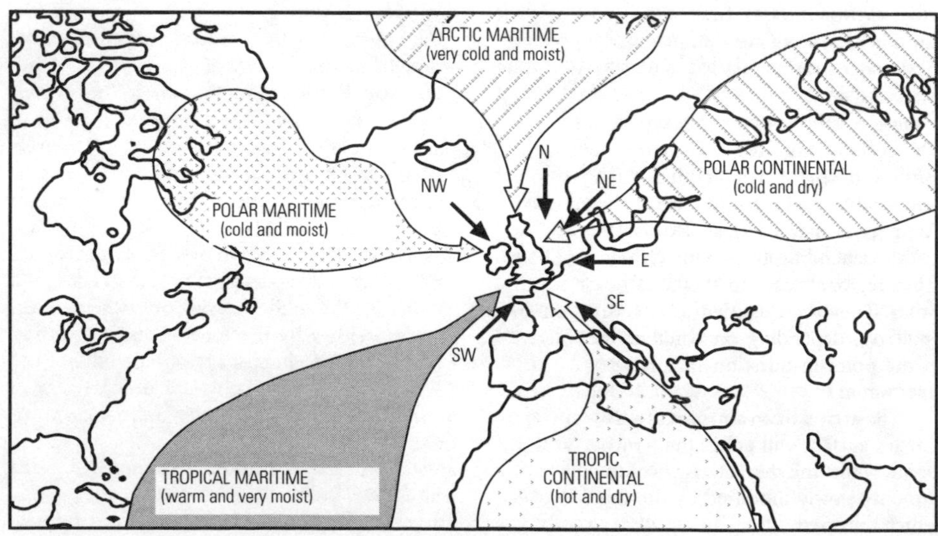

Air masses affecting the British Isles.

Agnatha The CLASS of vertebrates that consists of the jawless fish. *See* FISH.

agranulocyte A type of WHITE BLOOD CELL, which has a non-granular CYTOPLASM and compact nucleus, that can be phagocytic (*see* PHAGOCYTE). The main agranulocyte cell type in humans is the LYMPHOCYTE, which is involved in the body's defence mechanisms. *See also* GRANULOCYTE.

AIDS ***(acquired immune deficiency syndrome)*** A disease caused by the human immunodeficiency virus (HIV) in which the IMMUNE SYSTEM becomes deficient. Victims of AIDS usually die within 3 years of developing the disease, often from secondary infections such as pneumonia. At present only about 50 per cent of HIV-positive individuals develop fully blown AIDS within 10 years, but there is no evidence to suggest that HIV infection does not lead eventually to AIDS in 100 per cent of cases.

There is no cure for AIDS at present, but the drug AZT (zidovudine) interferes with the virus's ability to divide and can delay the onset and severity of AIDS. VACCINE development is difficult because there are many strains of HIV.

See also IMMUNOCOMPROMIZATION.

air The mixture of gases forming the Earth's atmosphere. Dry air contains 78 per cent nitrogen, 21 per cent oxygen, 0.9 per cent argon, 0.03 per cent carbon dioxide and traces of other NOBLE GASES. In addition, air usually contains a few per cent water vapour, though the concentration varies widely. *See also* ATMOSPHERE.

air mass A volume of air that typically extends for several hundred kilometres over the surface of the Earth. Within such regions temperature and HUMIDITY are fairly constant. Air masses may be classified as TROPICAL or POLAR depending on whether they originate from regions closer to the equator or to the poles. They are classified as CONTINENTAL or MARITIME depending on whether they arrive at the point in question mainly over land or over water.

The arrival of an air mass at a point on the Earth's surface will affect the weather at that point, whilst the characteristics of the air mass are only slowly modified by the regions over which they pass.

air speed indicator On an aircraft, an instrument used to display the speed of the aircraft through the air, typically by measuring the difference between HYDROSTATIC PRESSURE and DYNAMIC PRESSURE.

albedo The measure of the amount of light reflected on average by a planet or planetary satellite. An albedo of 1 indicates a perfectly reflecting surface, whilst a body with an albedo of 0 is perfectly black, absorbing all light falling on it.

albumin A group of PROTEINS produced by the LIVER that constitutes up to half of the content of human PLASMA PROTEIN. Albumins are coagulated by heat and are also found in egg white (albumen), milk and various animal and plant tissues.

alcohol Any one of a group of organic chemicals with the structure of an ALKANE but with one or more of the hydrogen atoms replaced by HYDROXYL GROUPS (–OH). An alcohol thus has the general formula $C_nH_{2n+1}OH$. The oxygen atom is added between the carbon–hydrogen bond. *Compare* ETHER.

Alcohols are classified by the position of their ALKYL GROUPS; primary alcohols contain CH_2OH, secondary alcohols contain CHOH and tertiary alcohols contain COH.

```
      H                H
      |                |
CH3 — C — OH     CH3 — C — OH
      |                |
      H               CH3

   Primary         Secondary

           CH3
            |
      CH3 — C — OH
            |
           CH3

         Tertiary
```

The nomenclature used for alcohols is based on that of the alkane forming the carbon skeleton, replacing the *-ane* suffix with *-anol,* for example propanol. The position of the hydroxyl group is indicated by a number placed before the *-ol,* for example propan-1-ol. Where there is more than one hydroxyl group this is also indicated in the name, for example ethane-1,2-diol has two hydroxyl groups and is called a dihydric alcohol. An alcohol with three hydroxyl groups is called trihydric and so on.

OXIDATION of primary alcohols yields ALDEHYDES, which can in turn be oxidized to CARBOXYLIC ACIDS.

$$\underset{\text{primary alcohol}}{RCH_2OH} \rightarrow \underset{\text{aldehyde}}{RC(=O)H} \rightarrow \underset{\text{carboxylic acid}}{RC(=O)OH}$$

Oxidation of secondary alcohols yields KETONES, which are not easily oxidized further. Alcohols react with acids to form ESTERS. HALOGENOALKANES can be manufactured by reacting alcohols with phosphorus HALIDES. The lower alcohols – methanol, ethanol, propanol, butanol – are liquids that mix with water. The higher members are oily liquids and the highest are waxy solids.

Ethanol is produced naturally during FERMENTATION and is used to manufacture alcoholic beverages. Other uses of alcohols are as solvents, in dye manufacture, in the cosmetics and medical industries, in detergents and in the manufacture of POLYESTER. Ethane-1,2-diol is used as an ANTIFREEZE. A common laboratory test for an alcohol is the evolution of hydrogen gas when sodium is added (*see* ALKOXIDE for the explanation of this).

alcoholic fermentation *See* FERMENTATION.

aldehyde, *alkanal* Any of a group of organic compounds containing the

$$-C(=O)H$$

group. The carbon of the CARBONYL GROUP (C=O) can be attached to another hydrogen atom as H-CO-H (written as HCHO), or to an ALKYL GROUP (R) as R-CO-H (written as RCHO). The general formula of an aldehyde is $C_nH_{2n}O$, the same as a KETONE, except that in the latter this comprises two alkyl groups. The nomenclature of aldehydes follows that of the ALKANE with the same carbon skeleton, with the ending *-ane* being replaced by *-anal*. The carbonyl carbon atom is given the positional number 1.

Aldehydes can be ALIPHATIC (which are usually colourless liquids) or AROMATIC (the higher members are solids). Examples of aldehydes include methanal, ethanal and benzenecarbaldehyde.

$$\underset{\text{methanal}}{H-C(=O)-H} \qquad \underset{\text{ethanal}}{CH_3-C(=O)-H}$$

$$\underset{\text{benzene-carbaldehyde}}{C_6H_5-CHO}$$

Aldehydes are formed by the oxidation of primary ALCOHOLS, hence their name, since the alcohol loses a hydrogen atom (*al*cohol *dehyd*rogenation). Aldehydes can be reduced back to primary alcohols and are themselves readily oxidized to the corresponding CARBOXYLIC ACID.

FEHLING'S TEST and TOLLEN'S REAGENT are used to test for aldehydes.

See also ALDOL CONDENSATION, CANNIZZARO REACTION, SCHIFF'S REAGENT.

aldol *See* HYDROXYALDEHYDE.

aldol condensation The reaction between two ALIPHATIC ALDEHYDES to yield a HYDROXYALDEHYDE (formerly known as aldol, hence the name). The hydroxyaldehydes easily lose water to yield unsaturated aldehydes. An example is the reaction of ethanal in dilute alkaline solution to give 3-hydroxybutanol, which is then dehydrated to give but-2-enal.

$$\underset{\text{ethanal}}{2\,CH_3-C(=O)H} \xrightarrow{\text{dilute alkaline solution}} \underset{\text{3-hydroxybutanal}}{CH_3-CH(OH)-CH_2-C(=O)H}$$

$$\xrightarrow{\text{easily dehydrated}} H_2O + \underset{\text{but-2-enal}}{CH_3-CH=CH-C(=O)H}$$

Only those aldehydes containing the $-CH_2CHO$ group will react in this way; other aldehydes undergo the CANNIZZARO REACTION. KETONES can also react in this way to produce unsaturated ketones but they do so less readily. These type of reactions where new carbon–carbon bonds are formed are useful in synthesizing large carbon skeletons from smaller molecules.

aldose, *aldo-sugar* A sugar containing an ALDEHYDE group (CHO). *See* MONOSACCHARIDE.

aldosterone A MINERALOCORTICOID hormone of the ADRENAL GLAND that regulates water retention in the kidney by controlling the distribution of sodium in the body tissue. It also affects centres in the brain, creating a sensation of thirst to stimulate the animal to seek water. *See also* ANGIOTENSIN.

aldo-sugar *See* ALDOSE.

aleuroplast A type of LEUCOPLAST that stores proteins in plants.

algae (*sing.* ***alga***) A collective term for a number of varied photosynthetic organisms (*see* PHOTOSYNTHESIS) normally living in aquatic or damp terrestrial conditions. They used to be classified with bacteria and fungi, but are now classified in the KINGDOM PROTOCTISTA. All algae are EUKARYOTES. There are several subdivisions according to their pigmentation: green algae (CHLOROPHYTA), which contain the same CHLOROPHYLL as higher plants; brown algae (PHAEOPHYTA), including the seaweeds; red algae (RHODOPHYTA); yellow-green algae (Xanthophyta); stoneworts (CHAROPHYTA); golden-brown algae (Chrysophyta); and the diatoms (BACILLARIOPHYTA). The CHLOROPLASTS of algae are varied, and most algae can reproduce sexually.

Algae are of considerable importance to humans because they cover the surface of the oceans and in total use more carbon dioxide from the atmosphere for photosynthesis than all land plants combined. This is important for carbon fixation at the first level of aquatic FOOD CHAINS, and to counteract GLOBAL WARMING (*see* GREENHOUSE EFFECT). Algae also produce more than half the oxygen released by plants and algae into the atmosphere.

Algae can be used as fertilizers, to assist in sewage breakdown, and as a direct food source, e.g. SINGLE CELL PROTEIN. Brown algae have a non-toxic acid in their cell walls that readily forms gels, and is used to thicken, for example, ice-cream, hand-cream, paint and confectionery.

CYANOBACTERIA were formerly known as blue-green algae and were grouped with algae under Prokaryotae. They are now classified as bacteria.

algal bloom An increased growth of ALGAE that may form in waters where NITRATES and PHOSPHATES are in excess from fertilizers and detergents. Algal blooms cause the water to smell and taste unpleasant, deplete oxygen and so cause the death of fish. *See also* EUTROPHICATION, SEWAGE DISPOSAL.

algebra The branch of mathematics that deals with the manipulation of symbols which represent variable numerical quantities. In science, these numerical quantities represent physical quantities, a number with an associated UNIT.

algorithm A predetermined set of mathematical instructions for solving a class of problems. More generally, an algorithm is a set of instructions for performing some task. Algorithms may include steps that are dependent on the truth of some logical statement, for example, 'if $x = 1$ then multiply y by 3, otherwise multiply y by 2'. In computing, an algorithm describes a series of steps to be completed by a PROGRAM. It may be represented visually by a FLOW CHART.

alicyclic (*adj.*) Describing any organic compound that is CYCLIC but does not possess an AROMATIC ring. These compounds are therefore cyclic with ALIPHATIC characteristics, for example CYCLOHEXANE. *See also* HETEROCYCLIC.

alimentary canal *See* DIGESTIVE SYSTEM.

aliphatic (*adj.*) Describing an organic chemical in which the carbon atoms are linked by COVALENT BONDS in straight chains, such as pentane, C_5H_{12}, or in branched chains, such as methylpropane, $CH_3CH(CH_3)CH_3$. This is in contrast to CYCLIC compounds.

alkali A base that dissolves in, or reacts with, water to produce HYDROXIDE ions, OH^-. Examples are sodium hydroxide, NaOH, which dissolves to form Na^+ and OH^- ions, and ammonia, NH_3, which produces NH_4^+ and OH^- ions.

alkali metal Any element, except hydrogen, from GROUP 1 (formerly Group IA) of the PERIODIC TABLE. The alkali metals are LITHIUM, SODIUM, POTASSIUM, RUBIDIUM, CAESIUM and FRANCIUM. They are classified by their electronic configuration, which is a NOBLE GAS electron structure plus a single electron in an outer S-ORBITAL. They are all highly reactive and readily lose an electron to form a positive ion with a single charge (M^+). Alkali metals form IONIC compounds and react with water to form HYDROXIDES plus hydrogen. Their reactivity increases with increasing ATOMIC NUMBER.

alkaline (*adj.*) Having the properties of an alkali.

alkaline earth Any element from GROUP 2 (formerly Group IIA) of the PERIODIC TABLE. The alkaline earth metals are beryllium, magnesium, calcium, strontium, barium and RADIUM. They are all chemically reactive, though less so than the ALKALI METALS. Reactivity increases with increasing ATOMIC NUMBER, though beryllium is more reactive than its position would suggest.

All compounds formed by alkaline earths are IONIC in character, with the metal atoms losing two electrons to form M^{2+} ions. This is due to their electron structure, which is that of a NOBLE GAS plus two electrons in an outer S-ORBITAL. Many, but not all, of the compounds are soluble in water, but the solubilities are generally lower than for compounds of the alkali metals.

alkaloid One of a group of organic substances found in plants, especially flowering plants, and which are usually poisonous. Many drugs used in medicine owe their properties to the presence of alkaloids. Examples include morphine, caffeine and nicotine. Alkaloids vary in their constitution but all are basic and combine with acids to form salts that are usually water-soluble.

alkanal *See* ALDEHYDE.

alkane, *paraffin* The general name for an ALIPHATIC HYDROCARBON having the general formula C_nH_{2n+2}. Alkanes have only single COVALENT BONDS and are therefore said to be SATURATED, for example:

```
                          H   H
                          |   |
CH3CH3 ethane   or    H - C - C - H
                          |   |
                          H   H
```

The first four alkanes of the group – methane, ethane, propane, butane – are gases, whilst those with higher RELATIVE MOLECULAR MASS are liquids or solids (the latter being those larger than $C_{16}H_{34}$). The liquids form the basis of petrol, kerosene and lubricating oil, whilst the solids form paraffin WAXES used in cosmetics and ointments. The names of all the alkanes end in *-ane* with the prefix referring to the number of carbon atoms present, for example the alkane with five carbon atoms is called pentane, the one with six is hexane.

Alkanes are found in PETROLEUM and NATURAL GAS. They are insoluble in water but soluble in benzene and chloroform. They are relatively chemically unreactive, compared to hydrocarbons with attached FUNCTIONAL GROUPS such as the HALOGENOALKANES. They are all flammable.

alkanone *See* KETONE.

alkene, *olefin* The general name for any ALIPHATIC HYDROCARBON that possesses one or more carbon-carbon double COVALENT BONDS and has the general formula C_nH_{2n}. Alkenes are therefore said to be UNSATURATED COMPOUNDS, for example:

```
                       H         H
                        \       /
CH2=CH2 ethene or        C  =  C
                        /       \
                       H         H
```

The names of all the alkenes end in *-ene* with the prefix referring to the number of carbon atoms present, for example the alkene with two carbon atoms is called ethene, the one with five is pentene.

The lower alkenes, such as ethene and propene, are gases obtained from PETROLEUM by CRACKING and provide the raw materials for most of the organic chemical industry. Products such as poly(ethene), polystyrene, polyvinyl chloride (PVC) and a vast range of detergents, paints and pharmaceuticals are all derived from the alkenes.

Alkenes can be reduced to ALKANES and they burn in air to give carbon dioxide and water. The alkenes are more reactive than the alkanes and react by ELECTROPHILIC ADDITION reactions, adding groups across the DOUBLE BOND. Their reactivity is due to the nature of the carbon-carbon double bond which consists of a SIGMA-BOND and a PI-BOND and is weaker than two single (sigma) carbon-carbon bonds. The double bond is therefore the site of most of the reactions of alkenes.

alkoxide Metal derivatives formed by the REDUCTION of ALCOHOLS by ALKALI METALS. For example, methanol is reduced by sodium to give sodium methoxide and hydrogen gas:

$$2CH_3OH + 2Na \rightarrow 2NaOCH_3 + H_2$$

This and similar reactions with other alcohols form the basis of the common laboratory test for an alcohol, which is the evolution of hydrogen gas from a liquid when sodium is added.

Alkoxides can also be formed as a result of the partial IONIZATION of alcohols in the presence of sodium hydroxide or potassium hydroxide:

$$ROH + HO^- \Leftrightarrow RO^- + H_2O$$

where RO^- indicates an alkoxide.

Alkoxides are used as catalysts and reagents in organic chemistry.

alkylation The introduction of an ALKYL GROUP into a HYDROCARBON chain or ring structure.

alkyl group (C_nH_{2n+1}) The group that remains after removing a hydrogen atom from an ALKANE. They are named after the alkane from which they are derived by replacing the *-ane* ending with *-yl*, for example methyl, ethyl. The symbol R is used to denote an unspecified alkyl group. The extent of chain branching in an alkyl group determines its classification as primary, secondary or tertiary.

```
    H            R′           R′
    |            |            |
  R—C—         R—C—         R—C—
    |            |            |
    H            H            R″

 primary     secondary     tertiary
```

alkyl halide *See* HALOGENOALKANE.

alkyne The general name for UNSATURATED HYDROCARBONS that possess one or more carbon-carbon triple COVALENT BONDS with the general formula C_nH_{2n-2}, for example:

$$CH{\equiv}CH \text{ ETHYNE or } H{-}C{\equiv}C{-}H$$

Lower alkynes such as ethyne are gases whilst those with higher RELATIVE MOLECULAR MASSES are liquids or solids.

allantois In birds and reptiles, an EXTRAEMBRYONIC MEMBRANE that acts as a bladder to store waste products. In humans it is less important and combines with the CHORION.

allele One of two (or sometimes more) variants of a GENE at a given position (LOCUS) on a CHROMOSOME. Variants are caused by a difference in the DNA of the gene. Each allele specifies a particular form of the characteristic coded for by the gene, for example, blue or brown eyes. DIPLOID organisms, such as humans, have two sets of chromosomes in the nucleus of each cell and two copies of each gene. If the two alleles occurring at a particular locus are the same they are called HOMOZYGOUS and if the two alleles are different they are called HETEROZYGOUS.

Some alleles are DOMINANT, which means that they hide the effects of other, RECESSIVE, alleles. For example, the allele for blue eyes is recessive and the allele for brown eyes is dominant. Therefore a heterozygous individual with one blue and one brown allele will have brown eyes. A recessive allele will only be expressed by an individual if both alleles are of the recessive type; that is the individual is homozygous recessive.

See also CODOMINANCE, HARDY–WEINBERG PRINCIPLE, LINKAGE, MENDEL'S LAWS, SEX LINKAGE.

allergic reaction An apparently pointless IMMUNE RESPONSE initiated by a non-threatening foreign protein. Such an ANTIGEN activates IgE-bearing (*see* IGE) B CELLS to secrete IgE antibody, which binds to MAST CELLS and BASOPHILS causing the release of allergic mediators, such as HISTAMINE. These mediators cause the typical symptoms of allergy, such as INFLAMMATION. Hayfever is a common allergic reaction.

allometric growth In biology, where the growth of a given feature progresses at a different rate to the growth of the entire organism. For example, the head of a human baby is relatively larger than the head of an adult. Some organs in an individual show allometric growth; for example, the organs of the REPRODUCTIVE SYSTEM grow very little early in life but then develop rapidly at the onset of puberty. The ratio of body surface area to total body volume is another example of allometric growth, as this ratio decreases as body size gets larger. *Compare* ISOMETRIC GROWTH. *See also* GROWTH.

allotrope Any one of two or more forms of an element or compound with different physical properties, but which exist in the same state of matter (solid, liquid or gas). *See* ALLOTROPY.

allotropy The existence of an element or compound in more than one form (called allotropes). Allotropes have different physical properties, but exist in the same state of matter (solid, liquid or gas). For example, the solid allotropes of carbon are diamond, graphite and AMORPHOUS carbon, such as coal, charcoal and soot. Both diamond and graphite are crystalline and all three allotropes are stable at all temperatures. Some substances have allotropes that are stable at different temperatures. An example is the allotropes of sulphur. Red sulphur is stable at higher temperatures, whilst yellow sulphur is stable at lower temperatures. A third, non-crystalline allotrope, amorphous sulphur also exists. This is unstable at all temperatures, but a high ACTIVATION ENERGY means that it is converted only extremely slowly to the other forms. Allotropy that involves only crystalline solids is called POLYMORPHISM.

Gaseous oxygen exists in two allotropes: O_2 ('normal' oxygen) and O_3 (OZONE). These differ in their molecular configurations.

See also ENANTIOTROPY, MONOTROPY.

alloy A material with metallic properties consisting of two or more metals or a metal with a nonmetal. An alloy may be a SOLID SOLUTION, a compound or a mixture of two or more crystalline solids. Alloys are very often used in engineering applications. They are often stronger, harder or more resistant to corrosion than their constituent metals.

The most common example of an alloy is steel, which consists of a few per cent of carbon in iron. Many steels also contain other elements, such as chromium and manganese. The carbon atoms are much smaller than the iron atoms, and become INTERSTITIAL ATOMS, occupying the gaps between the roughly spherical iron atoms. The carbon atoms have relatively little effect on the ELASTIC properties of the iron, which depend on the interatomic forces between the iron atoms. However, they do prevent the onset of PLASTIC behaviour, caused by imperfections or gaps in the LATTICE, called DISLOCATIONS. These dislocations can move through a metal: as one atom moves to fill a gap in the lattice it leaves a gap in its original location, so the dislocation effectively moves through the lattice in the opposite direction. This makes the metal far softer than it would have been without any dislocations. In small quantities, the carbon in steel 'pins' the dislocations, increasing the strength of the metal without destroying its ductility (*see* DUCTILE). Larger quantities of carbon in steel produce a very hard but brittle material called cast iron. Similar properties apply to other metal alloys: for example, aluminium-magnesium alloy has a high strength for its density. (This high strength with low weight is exploited in the manufacture of aircraft.)

Although they may have enhanced mechanical properties, alloys are generally poorer conductors of electricity and heat than pure metals. This is because the lattice structure is less highly ordered, making it harder for any free electrons to pass through the metal.

See also INTERMETALLIC COMPOUND.

alpha decay The spontaneous disintegration of an unstable atomic nucleus with the emission of an ALPHA PARTICLE. *See* ALPHA RADIATION.

alpha helix, *α-helix* A common type of structure of PROTEINS in which the POLYPEPTIDE chain is coiled into a corkscrew shape (helix). HYDROGEN BONDS form between successive turns of the helix, stabilizing the structure. *See also* BETA-PLEATED SHEET.

alphanumeric display A system containing a pattern of shapes that display various letters and numbers when the appropriate segments of the display are illuminated. Such displays may be based on LIGHT-EMITTING DIODES, GAS DISCHARGES or LIQUID CRYSTAL DISPLAYS.

alpha particle A helium–4 nucleus (a stable particle consisting of two protons and two neutrons), emitted during ALPHA RADIATION.

alpha radiation The emission of alpha particles, which are helium–4 nuclei (two protons and two neutrons bound together in a stable entity). Alpha radiation occurs when large nuclei, which are unstable due to the ELECTROSTATIC repulsion of the protons in the nucleus for one another, spontaneously disintegrate. Alpha particles are highly ionizing (*see* IONIZING RADIATION), and hence lose their energy very quickly. They have a range of only a few centimetres in air and can be stopped by a thin sheet of paper. When a nucleus emits an alpha-particle, it changes into a new nucleus with an ATOMIC NUMBER smaller by 2, and a MASS NUMBER smaller by 4. For example, the metal radium–226 decays to the gas radon–224 with the emission of an alpha particle.

alternating current (a.c.) Electric CURRENT flowing first in one direction, then the opposite one. Alternating current is used for most power supplies as it has the advantage that the voltage can be changed using TRANSFORMERS. *See also* ALTERNATOR, DIRECT CURRENT, OSCILLATOR.

alternation of generations A life cycle of some terrestrial plants, for example plants from the PHYLUM BRYOPHYTA, in which there are two distinct forms occurring alternately. The first form, the DIPLOID generation, produces HAPLOID SPORES by MEIOSIS and is called the SPOROPHYTE form. The second form, the haploid generation, produces GAMETES and is called the GAMETOPHYTE form. The gametes of the gametophyte fuse to form a diploid ZYGOTE, which develops into a new sporophyte. The sporophyte is dependent on the gametophyte for water and nutrients, so the generations alternate.

alternator An a.c. GENERATOR: a machine for transferring MECHANICAL ENERGY to ELECTRICAL ENERGY in the form of ALTERNATING CURRENT. It usually consists of a rotating magnet surrounded by coils in which an alternating current is produced. The magnet may be an ELECTROMAGNET, in which case it will be supplied with current by brushes rubbing against SLIP RINGS – rotating metal rings connected to the coils of the electromagnet. The electromagnet may be powered by the alternator itself, but in the case of larger machines, such as those used in power stations, a separate DYNAMO, called an EXCITER, is used. Cars are equipped with alternators to recharge their batteries. In such cases the alternating current generated is converted to DIRECT CURRENT (rectified) by a series of DIODES.

altitude 1. The height of an object, such as an aircraft above some specified surface, usually sea-level.

2. (*astronomy*) The angle of a star or similar above the horizon

3. (*mathematics*) The perpendicular distance between the point of a cone or pyramid and its base.

alum Any double SULPHATE containing a MONOVALENT metal A and a TRIVALENT metal B in the crystalline form $A_2SO_4.B_2(SO_4)_3.24H_2O$. In particular aluminium potassium sulphate, $Al_2(SO_4)_3.K_2SO_4.24H_2O$, commonly known as potash alum. Potash alum occurs naturally and is important as a MORDANT for dyes and in the processing of leather.

alumina *See* ALUMINIUM OXIDE.

aluminate Any salt containing the aluminate ion $[Al(OH)_4]^-$, formed by the reaction of ALUMINIUM HYDROXIDE with strong bases.

aluminium (Al) The element with ATOMIC NUMBER 13; RELATIVE ATOMIC MASS 26.98; melting point 660°C; boiling point 1,800°C; RELATIVE DENSITY 2.7. It is a chemically reactive TRIVALENT metal, forming compounds containing the Al^{3+} ion.

Aluminium is the most abundant metal in the Earth's crust. It is extracted from its main ore, BAUXITE, by the process of ELECTROLYSIS. In the main industrial method, the bauxite is first purified to obtain the OXIDE Al_2O_3, which is mixed with CRYOLITE, Na_3AlF_6, to lower its melting point. The molten mixture is then electrolysed with graphite electrodes. Molten aluminium is produced at the CATHODE (generally the lining of the CELL) and tapped off.

Aluminium reacts with atmospheric oxygen to form a hard surface layer of aluminium oxide, which prevents further chemical attack. Aluminium is thus far more useful as an engineering material than its reactivity would suggest and is highly corrosion resistant. In its pure form it is very soft, but when alloyed (*see* ALLOY) with other metals, such as magnesium, it forms light low-density alloys, which are used in the aerospace industry. Aluminium is also used for the manufacture of cooking utensils, overhead power cables and other applications where its lightness and/or high electrical CONDUCTIVITY may be exploited.

aluminium chloride ($AlCl_3$) A white solid that sublimes at 178°C (the ANHYDROUS salt); RELATIVE DENSITY 2.44. Aluminium chloride reacts violently with water and fumes in moist air. Aluminium chloride can be formed by passing hydrogen chloride over hot aluminium:

$$6HCl + 2Al \rightarrow 2AlCl_3 + 3H_2$$

Commercially, it is manufactured by passing chlorine over heated ALUMINIUM OXIDE and carbon,

$$6Cl_2 + 2Al_2O_3 + 3C \rightarrow 4AlCl_3 + 3CO_2$$

Aluminium chloride contains POLAR BONDS and acts as a LEWIS ACID. It is also a powerful drying agent. In industry, it is used as a catalyst, particularly in the CRACKING of oil.

aluminium hydroxide ($Al(OH)_3$) A white crystalline compound; RELATIVE DENSITY 2.4–2.5. Aluminium hydroxide is formed as a characteristic gelatinous white precipitate by the reaction of alkalis with aqueous aluminium salts:

$$Al^{3+} + 3OH^- \rightarrow Al(OH)_3$$

Aluminium hydroxide is AMPHOTERIC, and the precipitate will dissolve in an excess of alkali to form the ALUMINATE ion $[Al(OH)_4]^-$. Aluminium hydroxide decomposes on heating, progressively losing water to form aluminium oxide:

$$2Al(OH)_3 \rightarrow Al_2O_3 + 3H_2O$$

aluminium oxide, *alumina* (Al_2O_3) A white crystalline solid; melting point 2,020°C; boiling point 2,980°C. Aluminium oxide occurs naturally as the ore BAUXITE and in a more pure form in the gemstones ruby and sapphire. Aluminium oxide is insoluble and unreactive. It is a very hard material, widely used as an abrasive and in the manufacture of REFRACTORY materials.

aluminium potassium sulphates *See* ALUM.

aluminium sulphate ($Al_2(SO_4)_3$) A white crystalline compound, commonly occurring as the HYDRATE, $Al_2(SO_4)_3.18H_2O$. It loses water at 86°C and decomposes at 770°C. The hydrate is soluble in water and may be formed by the reaction of aluminium hydroxide with sulphuric acid:

$$Al_2(OH)_3 + 3H_2SO_4 \rightarrow Al_2(SO_4)_3 + 3H_2O$$

Aluminium sulphate is important in the treatment of sewage and drinking water and is also used as a fireproofing agent.

alveolus (*pl.* ***alveoli***) **1.** One of millions of air sacs in the lungs in which exchange of oxygen and carbon dioxide takes place between the air and the blood. *See also* RESPIRATORY SYSTEM.

2. A sac of secretory EPITHELIUM, for example in MAMMARY GLANDS.

amalgam An alloy of a metal with mercury. Amalgams are soft and have low melting points. Silver and gold amalgams were traditionally used in dentistry to fill decayed teeth, but fears about the toxicity of mercury have led to a decline in this use.

amber A yellow or brown fossil RESIN, derived from certain trees. Amber is used in jewellry and ornaments. It has the property of acquiring an electric charge when rubbed – the word 'electron' is derived from the Greek word for amber.

American Standard Code for Information Interchange *See* ASCII.

americium (Am) The element with ATOMIC NUMBER 95; melting point 994°C; boiling point 2,607°C; RELATIVE DENSITY 13.7. Americium occurs in minute quantities in uranium ores, and is made in commercial quantities in NUCLEAR REACTORS. It is used as an ALPHA PARTICLE source in smoke detectors and other devices that require a strong source of IONIZING RADIATION. Ten ISOTOPES with HALF-LIVES up to 7,700 years (americium–243) are known.

amide An organic compound derived from ammonia, NH_3, in which one or more of the hydrogen atoms has been replaced by an organic acid group.

A primary amide has one hydrogen of ammonia replaced ($RCONH_2$, where R is an ALKYL GROUP), a secondary amide has two hydrogens replaced ($(RCO)_2NH$) and a tertiary amide has all three replaced ($(RCO)_3N$). Primary amides are formed by the reaction of ammonia or AMINES with an ACID CHLORIDE, ANHYDRIDE or an ESTER. For example, ethanamide is formed in the reaction between ethanoyl chloride and ammonia:

$$CH_3COCl + NH_3 \rightarrow CH_3CONH_2 + HCl$$

or between ethyl ethanoate and ammonia:

$$CH_3COOC_2H_5 + NH_3 \rightarrow CH_3CONH_2 + C_2H_5OH.$$

Ethanamide (acetamide) is an example of a primary amide.

Amides are weakly basic and react with nitrous acid to form CARBOXYLIC ACIDS.

$$RCONH_2 + HNO_2 \rightarrow RCOOH + N_2 + H_2O$$

Secondary and tertiary amides are formed by treating primary amides or NITRILES with organic acids or their anhydrides.

See also HOFMANN DEGRADATION.

amine An organic compound derived from ammonia, NH_3, in which one or more of the hydrogen atoms has been replaced by an ALKYL GROUP. Amines are classified as primary, secondary or tertiary depending on whether one, two or three alkyl groups are present respectively. The general formula for an amine is $C_nH_{2n+3}N$. The names of amines are derived from the alkyl groups attached to the nitrogen followed by the ending *-amine*, such as methylamine, CH_3NH_2, ethylmethylamine, $C_2H_5NHCH_3$. Amines can be ALIPHATIC, AROMATIC or a mixture. They are weak bases and easily form complexes with LEWIS ACIDS (the complexing agent EDTA is an amine). Many amines have distinctive odours, for example ethylamine smells of rotting fish. Amines may be produced by the reduction of NITRILES,

$$RCN \rightarrow RCH_2NH_2$$

where R is an ALKYL GROUP. They may also be produced by the reduction of of AMIDES,

$$RCONH_2 \rightarrow RNH_2$$

The latter reaction is known as the HOFMANN DEGRADATION.

Primary and secondary amines react with ACID CHLORIDES and ANHYDRIDES to give AMIDES. Primary amines react with nitrous acid to yield molecular nitrogen, which effervesces, providing a useful test for primary amines.

$$RNH_2 + HNO_2 \rightarrow N_2 + ROH + H_2O + \text{other organic products}$$

Artificial sweeteners are derived from amines and the aromatic amines are used in the dyeing industry.

amino acid One of a group of water-soluble molecules mainly composed of carbon, oxygen, hydrogen and nitrogen containing a basic AMINO GROUP (NH_2) and an acidic CARBOXYL GROUP (COOH). There are 20 amino acids that make up all the different proteins known. Some of these 20 amino acids are called the ESSENTIAL AMINO ACIDS.

All amino acids have the same core structure (two carbon atoms, two oxygen atoms, a nitrogen and four hydrogen atoms) with a variable ALKYL GROUP (denoted R) attached to this. This can be as simple as another hydrogen atom, as in GLYCERINE, NH_2CH_2COOH, or more complex, as in TYROSINE, $C_6H_4OH.CH_2CH.(NH_2).COOH$. Some amino acids contain sulphur groups (for example METHIONINE). Many amino acids are neutral because they have one acidic and one basic group, but some have more basic (NH_2) groups, for example ARGININE, and some have more acidic (COOH) groups, for example ASPARTIC ACID.

All amino acids except glycerine form ISOMERS; all naturally occurring amino acids are of the L-form (*see* LAEVOROTATORY, OPTICAL ISOMERISM). Amino acids can join together to form a PEPTIDE or POLYPEPTIDE.

2-aminobenzoic acid, *anthranilic acid* ($C_7H_7NO_2$) An important DYESTUFFS intermediate that is manufactured by HOFMANN DEGRADATION. It is used in the synthesis of indigo dye and it can be diazotized (*see* DIAZOTIZATION) and used as a first component in AZO DYES.

amino group The NH_2 group. *See* AMINO ACIDS.

ammeter An instrument for the measurement of electric current. Many ammeters work by using the magnetic field produced when a current flows through a conductor. The most common of this type of ammeter is the MOVING-COIL GALVANOMETER.

Many modern ammeters are actually digital VOLTMETERS connected across a known small resistance; the voltage across the resistance is measured, from which the current can be found.

An ammeter is connected to a circuit by breaking the circuit and inserting the ammeter at the point where the current is to be measured. So as not to alter the size of this current, an ammeter should have as small a resistance as possible. To reduce further the resistance of the ammeter, and to enable it to measure larger currents, a RESISTOR, called a SHUNT, can be connected across the ammeter (in parallel with it).

ammonia, *nitrogen hydride* (NH_3) A colourless gas; melting point –74°C; boiling point –31°C. Ammonia is an irritant gas, with a characteristic smell and is highly soluble in water, where it forms ammonium ions, NH_4^+. The molecule has a TRIGONAL PYRAMIDAL shape: the nitrogen atom forms the apex of a triangle-based pyramid, with a hydrogen atoms at each of the three corners of the base. The nitrogen atom has a LONE PAIR of electrons, which has a profound effect on the properties of ammonia. HYDROGEN BONDS may be formed with other ammonia molecules and with water molecules.

Ammonia is manufactured commercially in the HABER PROCESS and is an important material in the manufacture of fertilizers, explosives and dyes. In nature, the formation of ammonia is vitally important in the NITROGEN CYCLE. Nitrogen-fixing bacteria (*see* NITROGEN FIXATION) convert atmospheric nitrogen into ammonia, which is used by nitrifying bacteria (*see* NITRIFICATION) to produce NITRITES and NITRATES.

ammonification The breakdown of proteins and amino acids by bacteria to produce ammonia. *See* NITROGEN CYCLE.

ammoniotelic (*adj.*) Describing an animal that excretes ammonia as its main nitrogenous waste product. *See* UREA.

ammonite An extinct aquatic mollusc, characterized by a coiled shell divided into chambers. Ammonites evolved quickly during the MESOZOIC ERA, and their fossils are used to date the rock strata in which they are found.

ammonium carbonate ($(NH_4)_2CO_3$) A white crystalline solid that decomposes slowly at room temperature and more rapidly on heating:

$$(NH_4)_2CO_3 \rightarrow 2NH_3 + CO_2 + H_2O$$

Ammonium carbonate can be formed by heating ammonium chloride and calcium carbonate:

$$2NH_4Cl + CaCO_3 \rightarrow (NH_4)_2CO_3 + CaCl_2$$

The ammonium carbonate is more volatile than the calcium chloride and SUBLIMES.

ammonium chloride (NH_4Cl) A white crystalline solid that SUBLIMES at 340°C; RELATIVE DENSITY 1.5. Ammonium chloride is soluble in water. It can be prepared by the reaction between ammonia and hydrogen chloride, either in aqueous solution or directly between the two gases:

$$NH_3 + HCl \rightarrow NH_4Cl$$

The gas phase reaction, in which the product appears as white fumes, is used commercially. Ammonium chloride is widely used as the ELECTROLYTE in ZINC-CARBON CELLS.

ammonium nitrate (NH_4NO_3) A white crystalline solid; melting point 197°C; boiling point 210°C; RELATIVE DENSITY 1.7. Ammonium nitrate is soluble in water and can be prepared by the reaction between ammonia and dilute nitric acid:

$$NH_3 + HNO_3 \rightarrow NH_4NO_3$$

Commercially, the same reaction is used to make large quantities of ammonium nitrate, with ammonia gas being bubbled through nitric acid. The compound is very important commercially, with large amounts used in the manufacture of fertilizers as it is a good source of nitrogen. It is also used in the manufacture of explosives.

ammonium sulphate ($(NH_4)_2SO_4$) A white crystalline solid, which decomposes at 235°C; RELATIVE DENSITY 1.7. Ammonium sulphate is manufactured by the reaction between ammonia and dilute sulphuric acid:

$$2NH_3 + H_2SO_4 \rightarrow (NH_4)_2SO_4$$

Ammonium sulphate is used commercially as a fertilizer.

amniocentesis A medical procedure in which a fine needle is inserted through the AMNION surrounding the foetus during pregnancy, and a sample of the AMNIOTIC FLUID removed. This fluid contains some foetal cells, which therefore allows the foetal chromosomes to be examined for abnormalities, such as DOWN'S SYNDROME. This is performed at about 16 weeks of pregnancy and there is some risk of miscarriage. *See also* CHORIONIC VILLUS SAMPLING.

amnion In most mammals, the innermost EXTRAEMBRYONIC MEMBRANE that usually expands to reach the CHORION and encloses the embryo. It contains AMNIOTIC FLUID.

amniotic fluid In mammals, the fluid surrounding the foetus during pregnancy, contained by the AMNION. The fluid provides a cushioning pad to protect the foetus from physical impact and also maintains a constant internal environment. The fluid is swallowed by the foetus and so is circulated, allowing some waste to be removed by the PLACENTA. *See also* AMNIOCENTESIS.

Amoeba (*pl.* ***amoebae***) One of the simplest organisms, a protozoan (*see* PROTOZOA) belonging to the KINGDOM PROTOCTISTA. It consists of a colourless PROTOPLASM from which extensions called pseudopodia (*see* PSEUDOPODIUM) form and engulf food. Amoebae reproduce by BINARY FISSION and possess a contractile VACUOLE for osmoregulation (*see* OSMOSIS). The pseudopodia are also used for movement. Movement similar to that of an amoeba by pseudopodia is called 'amoeboid'. An example of an amoeba is *Entamoeba histolytica*, which causes amoebic dysentery in humans.

amoebocyte Any animal cell with no fixed location that is free to move through body tissue. *See also* PORIFERA.

amorphous (*adj.*) Having no particular shape. The term especially refers to solids such as glass, where the molecules have no regular LATTICE arrangement. *See also* DISORDERED SOLID.

AMP (adenosine monophosphate) A NUCLEOTIDE component of DNA and RNA, and the product formed from the HYDROLYSIS of ATP and ADP. PHOSPHORYLATION of AMP yields ADP. AMP can be converted to CYCLIC AMP by the enzyme adenylate cyclase in response to the appropriate extracellular signals. Cyclic AMP is important in many biochemical pathways.

ampere (*abbrev.* amp; *symbol* A) The SI UNIT of electric current. Since a current produces a magnetic field and a current in a magnetic field experiences a force, two currents flowing close to one another will produce a force on one another. The size of this force is used to define the ampere: one ampere is equal to that current which, when flowing in two infinitely long parallel wires one metre apart in a vacuum, will produce a force of 2×10^{-7} N on each metre of their length. This force will be attractive if the currents are in the same direction, repulsive if they are in opposite directions. This definition of the ampere determines the strength of the magnetic field produced by a given current.

ampere-hour (Ah) A unit of charge, used to measure the storage capacity of a battery or electrochemical CELL. The capacity in ampere-hours is the current that can be provided by the battery multiplied by the number of hours for which this current can be supplied. One ampere-hour is equal to 3,600 COULOMBS.

ampere-turns In a SOLENOID or other magnetic system, the magnetizing current multiplied by the number of turns in the coil carrying that current.

Amphibia A CLASS of vertebrates, including toads, frogs, newts and salamanders. Amphibians live partly in water and partly on land. They have four legs (each with five digits), a moist smooth skin with scales and lay eggs that are not protected by a shell. Amphibians begin their life in water in the larval stages (tadpoles), then after METAMORPHOSIS live on land as adults, with lungs, and return to the water to breed. They are poikilothermic (cold-blooded; *see* POIKILOTHERMY) animals, so cannot maintain their own body temperature. They continue to grow throughout their life.

amphibian A member of the vertebrate CLASS AMPHIBIA.

amphoteric (*adj.*) Displaying the properties of both an ACID and a BASE. The OXIDES and HYDROXIDES of some TRANSITION METALS are amphoteric. For instance, zinc oxide is BASIC in that it will react with an acid to produce a salt plus water, for example,

$$ZnO + 2HCl\,(aq) \rightarrow ZnCl_2 + H_2O$$

but is ACIDIC in its reaction with alkalis, forming the complex ANION $Zn(OH)_4^{2-}$, for example,

$$ZnO + H_2O + 2NaOH \rightarrow Na_2\,Zn(OH)_4$$

amplifier An ANALOGUE electronic system that multiplies a voltage or current. The output is equal to the input signal multiplied by a constant factor called the GAIN. Amplifiers are used to boost small signals, such as the signal from a microphone, to higher levels, to feed to a loudspeaker, for example. As well as having a high gain, practical amplifiers should be free from DISTORTION and be able to operate over a range of frequencies. Simple amplifiers may be made from TRANSISTORS or VALVES, but more modern devices are complex INTEGRATED CIRCUITS. *See also* DIFFERENTIAL AMPLIFIER, INVERTING AMPLIFIER, OPERATIONAL AMPLIFIER, SUMMING AMPLIFIER.

amplitude The maximum distance from equilibrium reached by a wave or oscillating motion.

amplitude modulation A method of transmitting information in which the amplitude of a CARRIER WAVE is varied at the frequency of the signal. Amplitude modulation is often used in telecommunication systems, to convey information such as a speech signal or the brightness of a particular part of a television picture. Amplitude modulation systems are simple and can be designed to operate in narrower BANDWIDTHS than FREQUENCY MODULATION, but are more prone to INTERFERENCE. *See also* MODULATION, SINGLE SIDEBAND.

ampulla A swelling in the SEMI-CIRCULAR CANALS of the inner ear. It is concerned with balance.

amyl alcohol *See* PENTANOL.

amylase One of a group of enzymes that breaks down STARCH into its constituent sugars. It is found in humans in SALIVA and PANCREATIC JUICES.

amylopectin A POLYSACCHARIDE made up of GLUCOSE molecules in a branched structure. Amylopectin is a component of STARCH. *See also* AMYLOSE.

amyloplast A type of LEUCOPLAST that stores STARCH in plants.

amylose A straight-chained POLYSACCHARIDE made up of hundreds of GLUCOSE molecules. Amylose is a component of STARCH. *See also* AMYLOPECTIN.

anabolism The building up of body tissue. *See* METABOLISM.

anaemia A deficiency in the number of RED BLOOD CELLS in the blood or in their HAEMOGLOBIN content. Anaemia results in pallor, shortness of breath, lack of energy, dizziness and digestive disorders. *See also* PERNICIOUS ANAEMIA, SICKLE-CELL DISEASE.

anaerobe An organism that does not require oxygen for RESPIRATION. *Compare* AEROBE. *See also* GLYCOLYSIS.

anaerobic respiration *See* ANAEROBE, RESPIRATION.

anaesthetic A drug that is used to render a person insensitive to pain. Anaesthetics can be local by freezing or injection of the drug at the site to be treated, or they can be general to cause loss of consciousness, during operations. General anaesthesia has been induced by a number of different agents in the past, including ETHER, CHLOROFORM and DINITROGEN OXIDE. HALOTHANE, $CF_3CHBrCl$, is now commonly used since it has less side-effects.

anal fin *See* FIN.

analgesic A pain-relieving agent. OPIATE analgesics are the strongest of these drugs, whilst non-opiates such as ASPIRIN and PARACETAMOL are useful for less severe pain. They act by both preventing NERVE stimuli being sent to the brain and also by removing awareness of the pain.

analogue (*adj.*) In electronics, describing an electronic signal (voltage or current) where each value of the signal is used to represent a value of some continuously varying quantity. A microphone, for example, will produce a voltage that varies continuously in accordance with the changing pressure of the air. Analogue signals are extremely important in the interfacing of electronic systems to the outside world – that is, in converting information to or from electronic form. This interfacing is done using various TRANSDUCERS. *See also* DIGITAL.

analogue-to-digital converter A device used to convert ANALOGUE electronic signals to DIGITAL form for storage or processing. A series of COMPARATORS and POTENTIAL DIVIDERS can be used to construct an analogue-to-digital converter. Other devices are based on a counter circuit and a comparator: these count the number of pulses from an ASTABLE whilst a CAPACITOR charges until the voltage across it is equal to the voltage being converted, causing a change in the output of the comparator.

analysis 1. (*mathematics*) The branch of mathematics that deals with functions of continuously variable quantities, particularly the behaviour of such functions in the limit of vanishingly small changes, and differential CALCULUS.

2. (*chemistry*) The determination of the elements present in a compound or mixture. QUALITATIVE ANALYSIS establishes the components of a chemical sample, and the way they are combined in the case of a compound. QUANTITATIVE ANALYSIS measures the proportions of known components of a mixture.

A wide range of techniques are used for analysis from simple chemical tests that give characteristic results in the presence of certain materials to more sophisticated physical techniques, including MASS SPECTROSCOPY, CHROMATOGRAPHY and spectroscopic techniques, such as ATOMIC ABSORPTION SPECTROSCOPY. Chemical techniques of quantitative analysis may be classified as VOLUMETRIC ANALYSIS, such as TITRATION, or GRAVIMETRIC ANALYSIS.

See also ACTIVATION ANALYSIS, ELECTROPHORESIS, FLAME TEST, INFRARED SPECTROSCOPY, MASS SPECTROMETER, MICROWAVE SPECTROSCOPY, SPECTROSCOPY, THERMAL ANALYSIS, ULTRAVIOLET SPECTROSCOPY.

analytical (*adj.*) In mathematics, describing a procedure or result that follows the rules of ANALYSIS, and so is a precise solution in terms of algebraic quantities, as opposed to a numerical solution. *See also* GEOMETRY.

anaphase A stage of MITOSIS and MEIOSIS.

anatomy The study of the structure of the body and its component parts.

AND gate A LOGIC GATE that has a HIGH output only if all its inputs are high.

androecium The collective name for the STAMENS, the male part of a flower.

anechoic (*adj.*) Describing a room or a material that completely absorbs all sound falling on it. Anechoic chambers are used to test the properties of TRANSDUCERS such as microphones and loudspeakers.

anemometer A device for measuring wind speed. An anemometer normally consists of three cups mounted horizontally, which rotate at a rate proportional to the wind speed. This rotation drives a pointer across a scale calibrated in speed units.

anemophily POLLINATION of flowers by the wind. Flowers pollinated in this way are usually unscented and lack petals. The male and female flowers are often separate and are formed before the leaves to allow the pollen to be easily transported.

aneroid The pressure sensing element in an ANEROID BAROMETER.

aneroid barometer A type of BAROMETER consisting of a sealed metal vessel, called an aneroid, from which all the air is removed. The vessel has a thin metal lid, which is supported by a spring. Changes in the ATMOSPHERIC PRESSURE causes the lid to move against the spring by varying amounts. This movement is transmitted by the spring to a pointer on a calibrated scale. *See also* BAROGRAPH.

angiosperm A flowering plant, from the PHYLUM ANGIOSPERMOPHYTA, in which the SEEDS are contained within an OVARY.

Angiospermophyta A PHYLUM of the plant KINGDOM comprising all flowering plants. Flowering plants produce seeds protected within an ovary (*see* CARPEL). ANGIOSPERMS form most of the terrestrial vegetation found today.

There are two groups of angiosperms, MONOCOTYLEDONS and DICOTYLEDONS, which have one and two seed leaves respectively in the embryo. Most flowering plants, for example the buttercup, daisy and wallflower, are dicotyledons, which are broad-leaved, with flower parts arranged in fours or fives, and usually pollinated by insects.

Flowering plants are unique in undergoing DOUBLE FERTILIZATION, after which the OVULE develops into the seed and the ovary into a fruit. Angiosperms are found in a variety of habitats and more than 250,000 species exist.

angiotensin A PLASMA PROTEIN produced by the action of the enzyme RENIN. It is made by the kidney in response to low levels of sodium in the blood or reduced blood volume. Angiotensin stimulates production of ALDOSTERONE, which is involved in the regulation of water retention by the kidney.

angle A measure of rotation, or a measure of the space between two intersecting lines or planes. Angles are measured in DEGREES or RADIANS. One rotation through a full circle is equal to an angle of 360 degrees (360°) or 2π radians.

angle of depression The angle below the horizontal at which some point appears.

angle of elevation The angle above the horizontal at which some point appears.

angle of friction The angle between the NORMAL and the LINE OF ACTION of the overall force of contact between two surfaces that are sliding over one another. The TANGENT of the angle of friction is equal to the COEFFICIENT OF FRICTION. If θ is the angle of friction and μ is the coefficient of friction, then

$$\mu = \tan\theta$$

angle of incidence The angle at which a ray strikes a surface, measured from the NORMAL to the surface.

angle of reflection The angle at which a ray leaves a surface, having been reflected. The angle of reflection is measured from the NORMAL to the surface. *See* REFLECTION.

angle of refraction The angle, measured from the NORMAL, at which a ray leaves the boundary between two transparent materials, having been refracted (*see* REFRACTION).

angstrom (Å) A non-SI UNIT of length, equal to 10^{-10} m. It is still sometimes used to specify the wavelengths and intermolecular distances, but has largely been superseded by the nanometre (nm), which is 10^{-9} m.

angular acceleration The rate of change of ANGULAR VELOCITY with time. *See* ROTATIONAL DYNAMICS.

angular frequency In SIMPLE HARMONIC MOTION, 2π times the frequency measured in HERTZ.

angular magnification *See* MAGNIFICATION.

angular momentum (L) The ANGULAR VELOCITY of an object multiplied by its MOMENT OF INERTIA. For a point of mass m moving at a speed v at a distance r from the axis, with the motion at right angles to the line joining the object and the axis

$$L = mvr = m\omega r^2$$

where ω is the angular velocity and L the angular momentum. *See also* LAW OF CONSERVATION OF ANGULAR MOMENTUM, ROTATIONAL DYNAMICS.

angular velocity (ω) The rate of change of the angular position of an object with time. *See* ROTATIONAL DYNAMICS.

anhydride A compound that reacts with water to produce a new compound, rather than simply dissolving to form an aqueous solution.

For example, sulphur trioxide, SO_3, is the anhydride of sulphuric acid, H_2SO_4:

$$SO_3 + H_2O \rightarrow H_2SO_4$$

anhydrite A mineral form of ANHYDROUS calcium sulphate, $CaSO_4$. It is used in the chemical industry as a raw material for the manufacture of cement.

anhydrous (*adj.*) Containing no water, in particular no WATER OF CRYSTALLIZATION.

aniline The common name for PHENYLAMINE.

aniline dyes *See* AZO DYES.

animal A multicellular, eukaryotic (*see* EUKARYOTE) organism of the KINGDOM Animalia, lacking the rigid cell walls of plants and usually capable of movement for at least part of its life cycle. All animals are HETEROTROPHS. The oldest animal fossil on land was found in 1990 in Ludlow (Shropshire, UK); it was 440 million years old. There are 18 phyla (*see* PHYLUM) of animals, from the most primitive PORIFERA (SPONGES), to the largest phylum CHORDATA, which includes the VERTEBRATA. Vertebrates are the dominant animals of land, sea and air, not in numbers but in BIOMASS and other ecological terms. *See also* ANNELIDA, ARTHROPODA, CNIDARIA, ECHINODERMATA, MOLLUSCA, NEMATODA, PLATYHELMINTHES.

Animalia The KINGDOM consisting of ANIMALS.

anion A negatively charged ion. So called because it will be attracted to the ANODE in ELECTROLYSIS. *Compare* CATION.

anisotropic (*adj.*) Describing a medium, usually a crystalline solid, in which certain physical properties, such as electrical and THERMAL CONDUCTIVITY, are different in different directions. GRAPHITE (a crystalline ALLOTROPE of carbon) is an example: electric conduction can take place relatively easily along the planes of carbon atoms, but with much more difficulty across the planes.

annealing A process in which the metal is heated and then allowed to cool slowly. The result is that DISLOCATIONS are formed, under the influence of thermal vibrations. Some of these disappear as the material cools, but sufficient remain for the material to be soft and easily worked. If the material then undergoes sufficient PLASTIC deformations, it may become hard and brittle (WORK HARDENING), as the dislocations become tangled with one another or run up against the edges of the individual crystals in the POLYCRYSTALLINE metal. If the material is again annealed, the effects of work hardening are reversed and the material again becomes soft and easily worked. *See also* QUENCHING.

annelid A member of the PHYLUM ANNELIDA.

Annelida A PHYLUM consisting of invertebrate animals that have a soft, segmented body with an outer CUTICLE and possess bristles (CHAETAE). Examples include earthworms, lugworms and leeches. Some annelids have a distinct head, for example the lugworm. There are about 9,000 species living in water and soil.

The segmentation of the body is called 'metameric', in which there is a series of similar segments separated from one another by internal membranes. There is repetition of nerves, blood vessels and muscles in each segment, but some features, for example reproductive organs, are only repeated in a few segments. Annelidae possess longitudinal and circular muscles by which some move, aided by the bristles, and they feed on a variety of organisms and organic debris. Reproduction can be sexual or asexual (*see* BUDDING, FRAGMENTATION). Some are parasitic (*see* PARASITE) but they do not cause major problems for humans.

See also HIRUDINEA, OLIGOCHAETA, POLYCHAETA.

annihilation The complete destruction of matter, such as takes place when a particle and its antiparticle collide. The energy generated in such a collision is either carried away in the form of photons, usually GAMMA RADIATION, or MESONS.

annual plant A plant that completes its life cycle in one year and then dies. Annual plants, for example the common poppy, germinate from seeds, grow to maturity and produce seeds within one year or season. *Compare* BIENNIAL PLANT, PERENNIAL PLANT.

annulus A ring-shaped figure, the surface between two concentric circles. If the circles have radii r_1 and r_2, the area A of the annulus is

$$A = \pi(r_1^2 - r_2^2)$$

anode A positively charged ELECTRODE.

anodic oxidation The process that takes place at the ANODE of an ELECTROLYSIS, regarded as an OXIDATION. Since the anode is positively charged, electrons are removed from the ELECTROLYTE at this point. For example, in the electrolysis of copper sulphate with copper

ELECTRODES, copper metal is oxidized to form Cu^{2+} ions:

$$Cu \rightarrow Cu^{2+} + 2e^-$$

See also CATHODIC REDUCTION.

anodize (*vb.*) To apply a protective coating of ALUMINIUM OXIDE to a piece of aluminium, by making it the ANODE in an ELECTROLYSIS process. A solution of sulphuric acid or chromic acid is normally used as the ELECTROLYTE and the oxygen released at the anode reacts with the aluminium to produce a thin oxide layer.

anorexia nervosa An eating disorder, usually due to psychological problems or stress. It is characterized by an abnormal fear of obesity, leading to an excessive reduction in food intake and consequent wasting of the muscles.

antagonistic (*adj.*) Describing opposing actions or forces. Two muscles operating together to enable movement in opposite directions, for example at a joint, are antagonistic. In this case one muscle contracts while the other relaxes. Substances such as drugs or hormones are antagonistic if the action of one inhibits the action of the other. *Compare* SYNERGISTIC.

antenna (*pl.* **antennae**) **1.** (*biology*) Either one of a pair of appendages on the heads of insects, crustaceans, etc., that respond to touch and taste. Some antennae are specialized for swimming or attachment.

2. (*physics*) *See* AERIAL.

anterior (*adj.*) In biology, referring to the front of an organism. In lower animals this is the head end and in higher animals it is the front of the body. The term is also used to refer to the front of an organ or gland, for example anterior PITUITARY GLAND, or to a chamber, for example in the eye. In plants, anterior refers to leaves or flowers that are in front of and face away from the main stem. *Compare* POSTERIOR.

anther A structure in flowers that is responsible for the production of pollen grains. It is part of the STAMEN (the male reproductive structure) and consists of two lobes, containing pollen sacs, supported by a long stalk called the FILAMENT.

antheridium (*pl. **antheridia***) The male sex organ of fungi and plants without seeds, such as members of BRYOPHYTA. The male GAMETES are released from the antheridium, which is a sac, and swim down into the ARCHEGONIUM to fuse with the female gamete. This is part of the GAMETOPHYTE generation in plants showing ALTERNATION OF GENERATIONS.

Anthocerotae A CLASS of the PHYLUM BRYOPHYTA, consisting of the HORNWORTS. Hornworts are usually found in warm climates in damp conditions. Like the LIVERWORTS and MOSSES, to which they are related, they show ALTERNATION OF GENERATIONS.

Anthozoa *See* ACTINOZOA.

anthracene ($C_{14}H_{16}$) An AROMATIC HYDROCARBON; melting point 216°C; boiling point 351°C. Anthracene is separated from the high boiling point fractions of COAL TAR by FRACTIONAL DISTILLATION. It is a white crystalline substance with a slight blue FLUORESCENCE.

anthracite A type of hard, shiny coal consisting of over 90 per cent carbon with a low percentage of impurities. It is therefore a clean fuel that burns slowly without flame, smoke or smell. Anthracite gives out intense heat but is not suitable for open fires as it is hard to light and slow to burn.

anthranilic acid *See* 2-AMINOBENZOIC ACID.

antibiotic A chemical substance produced by a micro-organism that prevents the growth of other micro-organisms (but not viruses) and is used to combat many animal and human illnesses. The first antibiotic, PENICILLIN, was discovered in 1929 by Alexander Fleming (1881–1955). Penicillin now consists of a family of antibiotics obtained from moulds of the genus *Penicillium*. Some antibiotics, including penicillin, are called narrow-spectrum because they are only effective against a few PATHOGENS. Other antibiotics are broad-spectrum and can inhibit the growth of a wide variety of pathogens.

Many antibiotics have been discovered but only a few are medically useful and commercially viable. Their use may be restricted because of side-effects, for example toxicity and allergy. A pathogen may become resistant to a particular antibiotic (*see* ANTIBIOTIC RESISTANCE). Antibiotics are secondary METABOLITES, chemicals that are not essential for growth of the organism (as primary metabolites are) but often have a secondary role in, for example, the defence of the organism by being toxic, and their production is complex.

The action of antibiotics varies: streptomycin affects DNA, RNA and PROTEIN

SYNTHESIS; penicillin prevents formation of bacterial cell walls.

antibiotic resistance The inability of an ANTIBIOTIC to slow the growth of a PATHOGEN previously affected by it. This occurs when a micro-organism is repeatedly exposed to an antibiotic or is exposed to insufficient doses. Resistance can be inherited, so the micro-organism may not need to be exposed to the antibiotic to exhibit resistance. This has been a problem with PENICILLIN. There is therefore a continual need to find new types of antibiotics to overcome the problem of resistance. As well as searching for new natural antibiotics, more use is being made of GENETIC ENGINEERING to develop new strains or mutated strains of micro-organisms to be antibiotic producers.

antibody A protein secreted by a subclass of LYMPHOCYTES called B CELLS in response to the presence of a foreign substance (called an ANTIGEN), such as a viral or bacterial infection. This is only one of the ways the body can fight an infection. Antibodies are not restricted to the blood, but occur throughout the body. Different B cells produce antibodies with specificities for different antigens. This is called antibody diversity.

On its surface, a B cell possesses a specific antibody; when an appropriate antigen is presented, the B cell is stimulated to divide and secrete its antibody. Once produced, the antibody binds non-covalently to its specific antigen (some cross-reactivity can occur as antigens can have similarities), recognizing the overall three-dimensional shape of the antigen as well as its chemical make-up, and an antigen-antibody complex forms.

Antibodies can act in a number of ways to remove or destroy the foreign substance, for example by PRECIPITATION of the antigen-antibody complex, AGGLUTINATION of antigens, or NEUTRALIZATION of TOXINS produced by micro-organisms. Antibodies can remain in the blood after an infection and protect the body against future infection by the same organism.

See also IMMUNE RESPONSE, ANTITOXIN, VACCINATION.

antibonding orbital The higher energy of the two MOLECULAR ORBITALS formed when two atomic ORBITALS overlap. This orbital tends to push atoms apart, preventing closer bonding. The high energy of this orbital means that atoms only form COVALENT BONDS if this orbital is not filled. *See also* BONDING ORBITAL.

anti-bumping granules Small pieces of porous material that provide an irregular surface on which bubbles can form more easily, preventing the problems of bumping in chemistry experiments. *See* SUPERHEATED.

anticoagulant A substance that prevents the clotting of blood. *See also* BLOOD CLOTTING CASCADE.

anticodon A specific sequence of three NUCLEOTIDES carried by TRANSFER RNA that is complimentary to and can therefore form BASE PAIRS with a CODON sequence carried on MESSENGER RNA. *See also* PROTEIN SYNTHESIS.

anticyclone A region of high ATMOSPHERIC PRESSURE. Anticyclones are generally slow moving and correspond to a period of settled weather. Winds radiate from a calm centre in a clockwise direction in the northern hemisphere, and an anticlockwise direction in the southern hemisphere.

anti-diuretic hormone (ADH) A hormone produced by the HYPOTHALAMUS that is responsible for maintaining the correct salt/water balance in vertebrates. ADH is passed to the PITUITARY GLAND, where it is stored and secreted under the control of the hypothalamus. ADH reduces the amount of water lost from the kidney as urine and also raises blood pressure by constricting (tightening) ARTERIOLES. When water is in short supply, ADH secretion is increased allowing water to be conserved in the kidney. When water is plentiful, ADH secretion is reduced and the urine is more dilute so that more water can leave the body.

antifreeze An ALCOHOL added to water in the cooling system of INTERNAL COMBUSTION ENGINES, such as cars, to lower the point at which the water freezes. The most common antifreeze used is GLYCOL, which can be mixed with water in any concentration and lowers the point at which it freezes accordingly, thereby preventing freezing in cold weather.

antigen Any substance that induces the production of an ANTIBODY by the body's IMMUNE SYSTEM. Antigens are usually proteins or GLYCOPROTEINS, such as proteins on the surface of bacteria, viruses and pollen grains. An antigen that triggers the IMMUNE RESPONSE is said to be immunogenic. Not all antigens cause the initial induction of the immune response. Some initiate further production of antibodies in a response that has already been induced.

Antigens do not usually bind directly to antibodies but are instead presented to LYMPHOCYTES on the surface of ANTIGEN-PRESENTING CELLS. Body tissues and blood cells can also act as antigens and these have to be matched between donor and recipient for successful blood transfusions or organ transplants.

See also ANTIGENIC VARIATION, ALLERGIC RESPONSE, HISTAMINE, MAJOR HISTOCOMPATIBILITY COMPLEX.

antigen D *See* RHESUS FACTOR.

antigen-presenting cell (APC) A cell that presents, on its surface, fragments of ANTIGEN to T CELLS. Antigen-presenting cells are found in the SPLEEN and LYMPH NODES and trap antigens in the blood or LYMPH. The antigen degrades and fragments are presented in combination with molecules of the MAJOR HISTOCOMPATIBILITY COMPLEX, which may then activate T cells to divide. This may in turn activate B CELLS to produce a specific antibody. *See also* MACROPHAGE.

antigenic variation The ability of some PATHOGENS, for example the influenza virus, to change their surface ANTIGENS during an infection. This makes the fight against the disease and the search for a VACCINE more difficult.

anti-Markownikoff addition *See* MARKOWNIKOFF'S RULE.

antimatter Hypothetical matter made up of ANTIPARTICLES. Antimatter is not found in nature as contact with an equivalent amount of matter would result in the complete destruction of both. *See also* ANNIHILATION.

antimony (Sb) The element with ATOMIC NUMBER 51; RELATIVE ATOMIC MASS 121.8; melting point 630°C; boiling point 1,750°C; RELATIVE DENSITY 6.7. It is a METALLOID occurring mainly in the ore STIBNITE.

Antimony has a VALENCY of 5 and is used as a source of electrons in some N-TYPE SEMICONDUCTORS. The main use of the metal is as an alloying agent, while its compounds are used in fire-proofing, pigments and rubber technology.

antinode A region in a STANDING WAVE where the oscillations have a maximum amplitude. The distance from one antinode to the next is half the wavelength of the wave.

antiparticle A particle with the same mass but an opposite QUANTUM NUMBER as another particle. Every type of particle has an equivalent antiparticle with an opposite charge, or MAGNETIC MOMENT, or SPIN, etc. Thus antiprotons are negatively charged versions of the proton. The antiparticle of the electron is called the POSITRON. *See also* ELEMENTARY PARTICLE.

antipodal cell In plants, one of three cells within the EMBRYO SAC of a developing OVULE. The cells are at the opposite end to the MICROPYLE.

antiseptic A substance that prevents or inhibits the growth of micro-organisms. The first antiseptic used was CARBOLIC ACID and many substances used today are based on it.

antiserum A blood SERUM that contains ANTIBODIES to a specific ANTIGEN.

antitoxin An antibody that works by neutralizing TOXINS produced by micro-organisms so that they become inactive.

anus The opening of the RECTUM to the outside, at the end of the DIGESTIVE SYSTEM. Undigested food is removed (egested) from the body through the anus in the form of FAECES. This is controlled by a muscular SPHINCTER that in adults can be regulated voluntarily. Some simpler organisms have no anus and have only one opening to the digestive system. *See also* ANUS.

aorta The main ARTERY of vertebrates that carries oxygenated blood away from the heart. It branches to form smaller arteries that supply blood to the rest of the body except the lungs. Unlike most other arteries, the aorta has non-return valves to ensure the one-way flow of blood. *See also* CIRCULATORY SYSTEM.

APC *See* ANTIGEN-PRESENTING CELL.

aperture A gap in an otherwise solid or OPAQUE object; in particular one for the admission of light into a camera or other optical instrument. *See also* DIAPHRAGM.

aphelion For an object in orbit around the Sun, the point in the orbit at which it is furthest from the Sun.

Apicomplexa A PHYLUM from the KINGDOM PROTOCTISTA. The members (sporozoans) are mostly parasitic (*see* PARASITE) and have little movement. Reproduction is by multiple fission (*see* BINARY FISSION). One member of this phylum is *Plasmodium*, the parasite that causes MALARIA.

apocrine gland An EXOCRINE GLAND in which the apical part (the tip) of a cell breaks down during secretion. An example is the MAMMARY GLAND. *See also* HOLOCRINE GLAND, MEROCRINE GLAND.

apoenzyme An inactive enzyme that needs to associate with a COFACTOR in order to function.

apogee For an object in orbit around the Earth, the point in the orbit at which it is furthest from the Earth.

apoplast pathway *See* TRANSLOCATION.

apparent depth The depth that a transparent medium, such as water, appears to have when viewed from above. The apparent depth is less than the true depth by a factor of the REFRACTIVE INDEX of the material concerned. The effect is caused by light that has passed through the transparent material being refracted on entering the air between this material and the eye. *See also* REFRACTION.

apparent luminosity The amount of light energy received from a star per second per square metre. This depends on the distance of the star from the Earth and on the true LUMINOSITY of the star.

appendix In mammals, a small closed sac within the gut leading on from the CAECUM. In humans the appendix has no real use and frequently becomes inflamed and needs to be removed. However, it is important in HERBIVORES because it contains micro-organisms needed for digestion of CELLULOSE.

applications program A piece of computer SOFTWARE, such as a word-processing package, spreadsheet or accountancy program that does work directly for the end user. This is distinct from a SYSTEMS PROGRAM, which controls the computer or assists the programmer.

approximation A result that is only roughly correct, having been obtained either on the basis of approximate measurements, or on the basis of a calculation that gives only approximate results. Such calculations are often far simpler than completely correct solutions, which in any case are often not available.

aqua regia A mixture of concentrated hydrochloric and nitric acids in the ratio 1:3. It will dissolve all metals except silver, including the NOBLE METALS gold and platinum.

aqueous (*adj.*) Relating to water, particularly a solution in water.

aqueous humour In vertebrates, a watery fluid found in the eye in the space between the CORNEA and the lens. It is similar to CEREBROSPINAL FLUID in composition and is continuously renewed. It provides a link between the CIRCULATORY SYSTEM and the lens and cornea.

aquifer A rock formation that holds water.

Arachnida A CLASS of the PHYLUM ARTHROPODA, including SPIDERS, scorpions and ticks.

arachnoid membrane The middle of the three membranes that cover the brain and spinal cord. *See* MENINGES.

Araldite Trade name for an EPOXY RESIN used as a household adhesive and to mount specimens to be viewed in an ELECTRON MICROSCOPE.

arc A portion of a circle. If an arc subtends an angle of θ RADIANS, and has a radius of r, it will have a length of $r\theta$. An arc that is less than half a circle is sometimes described as a minor arc, whilst an arc that is more than half a circle is a major arc.

arc, electric *See* ELECTRIC ARC.

archegonium (*pl.* ***archegonia***) The female sex organ of MOSSES, LIVERWORTS, FERNS and most GYMNOSPERMS. It is a flask-shaped structure with the OVUM or egg cell at the base. The male GAMETE (formed in the ANTHERIDIUM) swims down the neck of the archegonium and fuses with the ovum. The archegonium and antheridium form part of the GAMETOPHYTE in plants showing ALTERNATION OF GENERATIONS. Once the male and female gametes have fused the SPOROPHYTE generation begins.

Archimedes' principle The UPTHRUST on any object immersed (partially or totally) in a fluid is equal to the weight of fluid displaced. *See also* BUOYANCY, FLOTATION.

area The measure of the size of a surface. For a rectangular surface, the area is defined as the product of the lengths of the two sides of the surface. Areas of other surfaces are described in terms of the total amount of rectangular area into which they can be divided.

area of outstanding natural beauty (AONB) An area of land worthy of CONSERVATION and protection due to its natural beauty.

arene A HYDROCARBON containing a BENZENE RING. The term is derived from *AR*OMATIC and ALK*ENE*.

areolar tissue *See* CONNECTIVE TISSUE.

Argand diagram A representation of COMPLEX NUMBERS as points on a two-dimensional surface. The distance of a point from the ORIGIN of an Argand diagram in the two dimensions of the plane is equal to the real and imaginary parts of the number. The real part is represented on the x-axis, and the imaginary part on the y-axis.

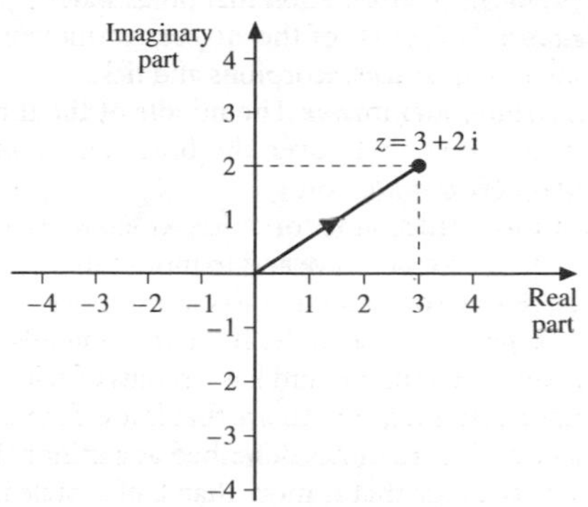

Argand diagram

arginine An amino acid essential for nutrition and the production of UREA.

argon (Ar) The element with ATOMIC NUMBER 18; RELATIVE ATOMIC MASS 39.95; melting point −189°C; boiling point −185°C. Argon is the most common inert gas in the atmosphere, making up about 1 per cent. It is separated from liquid air by FRACTIONAL DISTILLATION. It is used to provide an inert atmosphere in some welding processes and in light bulbs, as well as in some GAS DISCHARGE tubes.

argument Any number used as the starting point for the calculation of some function. For example, in the equation $\sin 30° = {}^1/_2$, 30° is the argument of the SINE function.

arithmetic The branch of mathematics that deals with numbers, as opposed to ALGEBRA, where the place of numbers is taken by abstract variable quantities. All the operations that can be performed on numbers are ultimately derived from the four fundamental processes of addition, subtraction, multiplication and division.

arithmetic logic unit (ALU) The part of a MICROPROCESSOR or computer that performs actual manipulations on numbers fed into it from some form of memory, with the output also being stored in memory.

arithmetic mean *See* MEAN.

arithmetic progression A SERIES of numbers in which each number is greater than the previous number in the series by a fixed amount. Thus if the *n*th member of the series is a_n, the $(n+1)$th member will be

$$a_{n+1} = a_n + k$$

where k is a constant called the common difference. For an arithmetic progression of N terms, the sum of the series will be

$$Na_1/2 + {}^1/_2 N(N-1)k$$

armature The moving part of a MOTOR, DYNAMO or other electromagnetic system.

aromatic (*adj.*) Describing any organic compound containing a BENZENE RING and some non-benzene compounds that are HETEROCYCLIC. Aromatic compounds undergo SUBSTITUTION REACTIONS. The ring usually contains nitrogen, sulphur or oxygen if it does not contain all carbon atoms. These compounds were originally named because of their fragrance, although not all of them are pleasant.

Aromatic compounds are more stable than other UNSATURATED COMPOUNDS (such as ALKENES) and thus less reactive, usually undergoing substitution rather than ADDITION REACTIONS. Examples include benzene, C_6H_6, and pyridine, C_6H_5N. Where an aromatic compound contains a FUNCTIONAL GROUP the reactivity of both the benzene ring and the functional group is altered. This is because the delocalized electrons of the ring and those of functional groups containing DOUBLE BONDS interact. This interaction controls the chemistry of aromatic compounds.

See also BENZENE, DELOCALIZED ORBITAL.

Arrhenius' equation An equation that describes the rate of any reaction involving an ACTIVATION ENERGY. Arrhenius' equation is

$$k = A\exp(-E/RT)$$

where k is the rate constant for the reaction, A is a constant for a given reaction, E is the activation energy per mole, R is the MOLAR GAS CONSTANT and T the ABSOLUTE TEMPERATURE. *See also* RATE OF REACTION.

arrow of time A concept arising from the idea of irreversibility in the SECOND LAW OF THERMODYNAMICS. Although the individual interactions between molecules could equally well happen backwards, the overall direction of the universe is from order to chaos. The fact that this feature is present only on the macroscopic scale (when large numbers of interacting particles are considered) and not on the microscopic scale is a key area of philosophical interest.

arsenic (As) The element with ATOMIC NUMBER 33; RELATIVE ATOMIC MASS 74.9; sublimes at 613°C; RELATIVE DENSITY 3.9. It is a powdery METALLOID that occurs quite widely in nature and is best known for its toxic properties. Arsenic is a cumulative poison, and although its use in certain pigments was once quite common, it is little used nowadays due to fears about its toxicity. It is produced as a by-product of the extraction of many metals from their ores.

arsenic hydride *See* ARSINE.

arsenic oxide Either of the two oxides arsenic(III) oxide, As_4O_6 (otherwise known as arsenic trioxide, arsenous oxide or white arsenic) and arsenic(V) oxide, As_2O_5.

Arsenic(V) oxide is a white solid that decomposes to arsenic(III) oxide at 315°C and has a RELATIVE DENSITY of 3.5. It is formed by the dehydration of arsenous acid,

$$2HAsO_3 \rightarrow As_2O_5 + H_2O$$

Arsenic(III) oxide is a white compound with three solid ALLOTROPES. It sublimes at 193°C and the OCTAHEDRAL form has a relative density of 3.9. It is formed by burning arsenic in air:

$$4As + 5O_2 \rightarrow 2As_2O_5$$

It is extremely toxic, and its former use as a poison for vermin has been vastly reduced on account of concerns over its accumulation in the FOOD CHAIN.

arsine, *arsenic hydride* (AsH_3) A colourless gas; melting point –116°C; boiling point –55°C. Arsine is manufactured by the reaction of strong acids with metal arsenides, for example,

$$3H_2SO_4 + 2K_3As \rightarrow 3K_2SO_4 + 2AsH_3$$

Arsine is used commercially in small quantities as a vehicle for depositing arsenic for DOPING in the manufacture of integrated circuits.

arsenic trioxide *See* ARSENIC OXIDE.

arsenide Any BINARY COMPOUND containing arsenic and a metal.

arsenous oxide *See* ARSENIC OXIDE.

arteriogram A RADIOGRAPH of the arteries, made with an injection of a CONTRAST ENHANCING MEDIUM.

arteriole A small branch of an artery.

artery In animals with a CIRCULATORY SYSTEM, a vessel that carries blood away from the heart to the rest of the body ('a' for artery, 'a' for away).

Arteries have thick muscular walls to withstand blood at high pressure, but also contain elastic fibres so they can expand to allow for the increase in blood pressure following contraction of the heart muscles. Arteries (except the PULMONARY arteries supplying the lungs) carry highly oxygenated blood to all the main organs of the body. Unlike veins, which all have valves to ensure the one-way flow of blood, the only arteries to have valves are the AORTA and the pulmonary arteries. *See also* ATHEROSCLEROSIS, PULSE.

arthropod A member of the PHYLUM ARTHROPODA.

Arthropoda A large PHYLUM of invertebrate animals with a hard EXOSKELETON, jointed legs and a segmented body. Among the classes of the Arthropoda phylum are CRUSTACEA, including Daphnia (the water flea), crabs, prawns and crayfish; INSECTA, including the locust, cockroach, housefly, butterfly and bee; ARACHNIDA, including the spider, scorpion and tick; CHILOPODA, including the centipede; and DIPLOPODA, including the millipede.

Arthropods make up three-quarters of all living animals and are well adapted to living in water or on land. They can be free-living or parasitic (*see* PARASITE). The exoskeleton is made mainly from CHITIN. Growth of arthropods occurs in stages after the moulting of their exoskeleton, called ECDYSIS.

articular (*adj.*) Relating to JOINTS or the structural components within joints.

articular cartilage *See* CARTILAGE.

artificial satellite *See* SATELLITE.

artificial selection The selected breeding by humans of plants or animals in order to develop particular characteristics, for example, disease-resistance in plants, improved milk production in cows, racehorse breeding, cat and dog breeding. Inbreeding (breeding of an animal with its close relatives) can result in harmful genes being expressed, which in an outbreeding population would be RECESSIVE. To avoid this, regular inbreeding is interspersed with occasional new genes by outbreeding.

Eugenics is the study of ways in which the human race can be improved by the selection or elimination of specific characters, for example to control the spread of genetic disorders.

aryl group A group obtained by removing a hydrogen atom from a HYDROCARBON of the BENZENE series.

ASCII Acronym for *A*merican *S*tandard *C*ode for *I*nformation *I*nterchange. ASCII is a standard for encoding letters, numbers and punctuation as a BINARY code that can be stored and manipulated by computers.

ascomycete A member of the ASCOMYCOTA PHYLUM of FUNGI.

Ascomycota A PHYLUM of the KINGDOM FUNGI. Ascomycetes are characterized by having septa (partitions) in their HYPHAE and SEXUAL REPRODUCTION is by the formation of ascospores in a structure called an ascus. Asexual reproduction is by non-motile SPORES called conidia (*see* CONIDIUM). The phylum includes yeast (*Saccharomyces*) and the moulds *Aspergillus* and *Penicillium*. Many members of the Ascomycota are a cause of food or crop spoilage.

ascorbic acid *See* VITAMIN C.

ascospore Any SPORE of an ASCOMYCETE, from the ASCOMYCOTA PHYLUM of FUNGI.

ascus In ASCOMYCETES, a cell within which ASCOSPORES are formed during SEXUAL REPRODUCTION.

asexual reproduction The production of offspring from a single parent, involving no fusion of GAMETES. The offspring are usually genetically identical to each other and to the parent. Since only a single parent is involved it can lead to a rapid increase in the population, which is a considerable advantage. The disadvantage is that there is no genetic VARIATION and therefore no opportunity to adapt to a changing environment. Many asexual organisms can also reproduce sexually, which increases the chance for genetic variation (*see* SEXUAL REPRODUCTION, FERTILIZATION). Asexual reproduction can be by BINARY FISSION, BUDDING, FRAGMENTATION, SPORULATION or VEGETATIVE REPRODUCTION. *See also* PARTHENOGENESIS.

aspartame An artificial sweetener (trade name Nutrasweet) consisting of two amino acids (ASPARTIC ACID and PHENYLALANINE) linked by a methylene ($-CH_2-$) group. It is 200 times sweeter than sugar and does not have the aftertaste that SACCHARIN has. It cannot be used for cooking as it breaks down on heating.

aspartic acid An amino acid that is a constituent of proteins and acts as a NEUROTRANSMITTER.

aspirin, *acetylsalicylic acid, acetylbenzoic acid* ($C_9H_8O_4$) A widely used pain-killing drug that acts by inhibiting PROSTAGLANDINS. It also reduces fever and INFLAMMATION, for example in arthritis. Side-effects of long-term usage are stomach bleeding and kidney damage, although more recently it has been suggested that an aspirin a day can reduce the risk of heart attacks and thrombosis.

assembler A computer PROGRAM that translates an ASSEMBLY LANGUAGE into MACHINE CODE. Generally, each instruction in assembly language translates into one instruction in machine code.

assembly language A low-level computer PROGRAMMING LANGUAGE closely related to a computer's internal code. It uses simple MNEMONIC abbreviations that are translated by an ASSEMBLER into MACHINE CODE. Most programs are now written in high-level languages, but assembly languages are sometimes used when a more efficient program is required.

associated production Any process in which a quark-antiquark pair of one of the heavier QUARKS (for example, the STRANGE quark) are produced, with the two members of the pair ending up in different HADRONS.

association In chemistry, the formation of a substance consisting of two different types of molecule held together by HYDROGEN BONDS in some new regular structure.

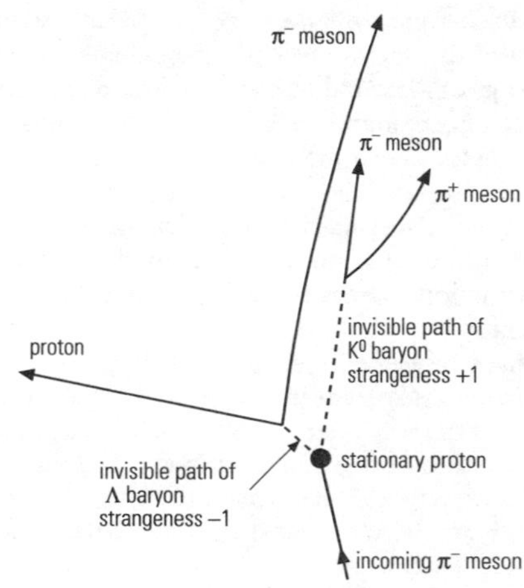

Associated production as seen in a bubble chamber.

association area A part of the brain that allows an individual to interpret information received by the visual, speech, auditory and other sensory areas of the brain in relation to previous experience. *See* CEREBRUM.

associative (*adj.*) Describing any mathematical operation with the property that the value of an expression is independent of the way in which the terms in the expression are grouped. Thus multiplication is associative, since

$$(a \times b) \times c = (c \times b) \times a = (a \times c) \times b$$

Addition is also associative; subtraction and division are not. *See also* COMMUTATIVE.

astable A digital electronic device that changes its output from HIGH to LOW and back at a rate determined either by the charging of a CAPACITOR or by a resonant device, such as a PIEZOELECTRIC quartz crystal.

astatine (At) The element with ATOMIC NUMBER 85; melting point 302°C; boiling point 377°C. Only unstable ISOTOPES are known, with the longest-lived (astatine–210) having a HALF-LIFE of 8 hours. Astatine is present in nature in minute amounts, being formed by the decay of longer lived radioactive elements.

asteroid belt The region between the orbits of Mars and Jupiter where most ASTEROIDS are found.

asteroid, *minor planet* Any one of a large number of small rocky bodies, less than 1,000 km in diameter, that orbit the Sun. Many of the smaller asteroids are irregularly shaped. Most asteroids orbit in a region of the Solar System called the asteroid belt, which lies between the orbits of Mars and Jupiter. However, a number have been found in other orbits, including some that cross the orbit of the Earth. It has been proposed that collisions of such asteroids with the Earth may have caused major climatic changes in the past, leading to the sudden extinction of many species, particularly the dinosaurs.

asthenosphere A layer of the Earth's MANTLE, that extends from about 70 km to 260 km below the surface. It is a soft, probably partially molten, layer on which TECTONIC PLATES move. SEISMIC WAVES slow down in this zone.

astigmatism An ABERRATION in mirrors and lenses (including the CORNEA of the eye) that arises when the surface is more strongly curved in one direction than another. Parallel rays of light are not brought to a focus at the same point, leading to DISTORTIONS in the image. Astigmatism in the eye can be corrected with spectacles that have a cylindrically curved surface.

astrocyte A star-shaped GLIAL CELL.

astrometric binary *See* BINARY STAR.

astrometry The branch of astronomy concerned with the measurement of the position of astronomical objects.

astronomical unit (AU) A unit of distance used in astronomy, the mean distance between the Earth and the Sun. One AU is equivalent to 1.50×10^8 km.

astronomy The study of celestial bodies, their positions, motions, nature and evolution. The main branches of astronomy are ASTROMETRY, CELESTIAL MECHANICS and ASTROPHYSICS. *See also* COSMOLOGY.

astrophysics The branch of astronomy concerned with the study of the physical and chemical properties of astronomical objects and phenomena. Astrophysics deals with the structure and evolution of stars, CLUSTERS and galaxies, and the properties of interstellar matter. Astrophysicists study the composition of celestial objects by analysing the electromagnetic radiation (such as light, radio waves and X-rays) they emit. *See also* ASTRONOMY, COSMOLOGY, SPECTROSCOPY.

asymptote The line towards which a curve approaches more and more closely, but never reaches. For example, the graph of the equation $y = 1/x$ is asymptotic to $y = 0$ as x tends to infinity.

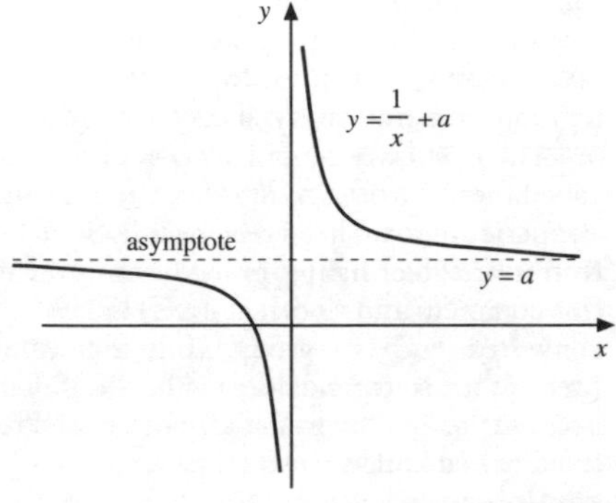

Asymptote.

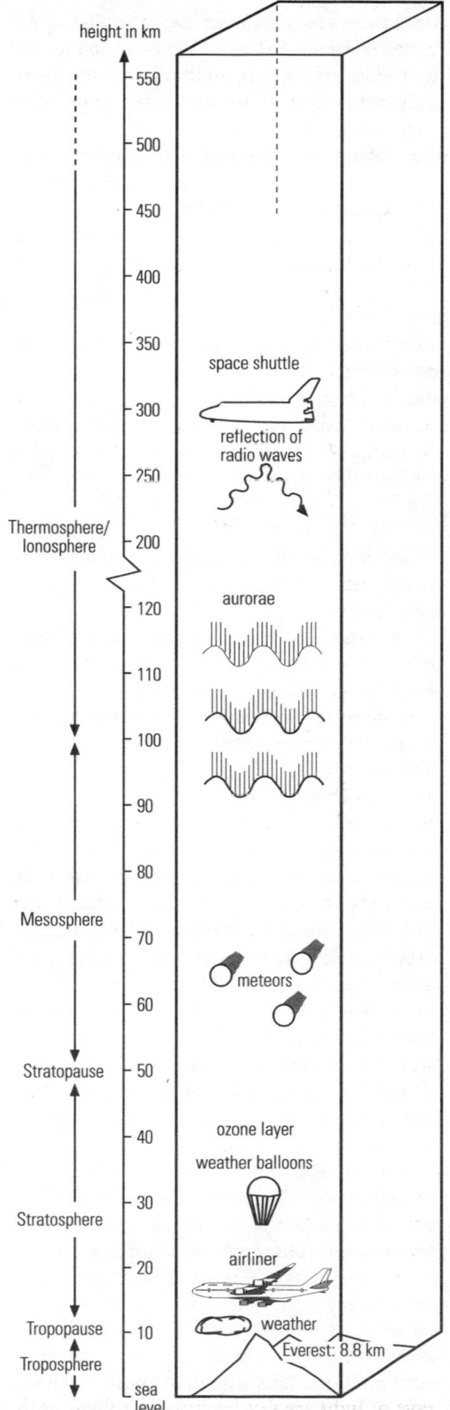

The structure of the Earth's atmosphere.

atactic polymer A POLYMER in which the substituted carbons have a random arrangement with respect to the carbon chain, if it is considered that the carbon atoms all lie in the same plane. This results in a sticky polymer unsuitable for manufacture. *Compare* ISOTACTIC POLYMER, SYNDIOTACTIC POLYMER.

atherosclerosis A disease in which FATTY ACIDS build up in the walls of arteries as a person ages. This reduces the flexibility and internal diameter of the artery, so increasing blood pressure and leading to heart disease. Diet plays an important role in this process. The disease is particularly prevalent in the Western world. *See also* CHOLESTEROL.

atmolysis The separation of different components in a mixture of gases, including different ISOTOPES of the same element, on the basis of their differing rates of DIFFUSION. *See also* GRAHAM'S LAW OF DIFFUSION.

atmosphere A layer of gas surrounding the surface of the Earth and some other planets. The Earth's atmosphere has a thickness of around 400 km but its density decreases with height.

The lowest layer of the atmosphere, about 10 km thick, is called the TROPOSPHERE. All clouds and weather systems occur within this region. All the movement of the atmosphere is driven by the energy absorbed from solar radiation. The Earth also radiates energy into space, and clouds in the troposphere play a major part in the energy balance, reflecting both incoming and outgoing radiation. By the top of the troposphere, called the TROPOPAUSE, the atmospheric pressure has fallen to about 1 per cent of its surface value and the temperature is typically –50°C.

The next layer, from 25 km to 100 km, is called the STRATOSPHERE. Within this layer the temperature rises, heated by energy absorbed from ultraviolet light emitted by the Sun. At this temperature 'normal' oxygen, O_2, is converted into OZONE, O_3, and the OZONE LAYER is thus formed. Recently, fears have been expressed over the depletion of this layer, the so-called OZONE HOLES, which have been observed over Antarctica and parts of Europe. Man-made chemicals, particularly CHLOROFLUOROCARBONS, which act as catalysts for the decomposition of ozone, are believed to be responsible. This has led to a reduction in the use of these materials in refrigerators and as the propellant in AEROSOL sprays. It is feared that reductions in the concentration of ozone over populated areas may allow ultraviolet light to reach the surface, resulting in an increase in the incidence of skin cancers.

The top of the stratosphere is called the STRATOPAUSE. Here the temperature is about 0°C and the pressure only about one thousandth of the surface pressure. Above the stratopause, from 50 km to 100 km is the MESOSPHERE, where temperature falls with increasing height to about –100°C and the pressure falls to 10^{-5} times the surface pressure. Above this lies the THERMOSPHERE, the lower region of which is called the IONOSPHERE as the gas here is ionized by the absorption of X-rays from the Sun. The energy absorbed causes temperatures to rise, but the density and pressure of the material in this layer is very low.

See also CLIMATE, WEATHER SYSTEMS, GREENHOUSE EFFECT, POLLUTION.

atmosphere, *standard atmosphere* (atm.) A unit of pressure equivalent to 101,325 Pa. This is equal to the pressure that will support a column of mercury 760 mm high, at 0°C, at sea level and latitude 45°.

atmospheric pressure The pressure produced by gravity acting on the atmosphere. At the surface of the Earth this produces a pressure of about 1×10^5 Pa (1 ATMOSPHERE), though the exact value varies from day to day. Changes in atmospheric pressure are responsible for WEATHER SYSTEMS. *See also* BAROMETER.

atom The smallest particle into which an element can be divided without losing its chemical identity. Atoms were originally believed to be indivisible objects, but it is now known that they comprise a small dense positive nucleus of protons and neutrons, surrounded by electrons.

Almost all the mass of an atom is contained in its nucleus: protons and neutrons have similar masses, while electrons are about 1/1,836 the mass of protons. In a neutral atom, the number of electrons is equal to the number of protons in the nucleus (called the ATOMIC NUMBER). The number and arrangement of the electrons in an atom is what gives each element its distinct chemical properties.

In a simple model of the atom, electrons are regarded as orbiting the nucleus in shells. However, a more sophisticated treatment recognizes the influence of quantum mechanics on the atom, and places the electrons in ORBITALS, each having a certain defined energy. The sequence in which these orbitals are filled, and the energies involved are crucial in explaining the chemical properties of each element.

Chemistry generally concerns itself with the reactions of neutral atoms or of ions that contain a few electrons more or less than the number needed to make the atom neutral.

See also ATOMIC THEORY, BOHR THEORY, CHEMICAL COMBINATION, DALTON'S ATOMIC THEORY, LIQUID DROP MODEL, MOLECULE, QUANTUM THEORY, RUTHERFORD–BOHR ATOM, RUTHERFORD SCATTERING EXPERIMENT, RYDBERG EQUATION, SHELL MODEL, WAVE NATURE OF PARTICLES.

atomic absorption spectroscopy A technique of chemical ANALYSIS based on the ABSORPTION SPECTRUM formed when white light is shone through a sample in the form of a vapour. The wavelengths of light absorbed are characteristic of the elements present.

atomic clock A clock that measures time by some periodic process occurring in atoms or molecules, such as atomic vibrations or the frequency of emission or absorption of electromagnetic radiation .

The caesium clock is based on the very precise frequency of radiation produced or absorbed by transitions between ENERGY LEVELS in caesium atoms. This frequency is used to define the second, and so caesium clocks are used in international timekeeping. Experiments with atomic clocks have been used to confirm the predictions of TIME DILATION in the SPECIAL THEORY OF RELATIVITY.

atomic emission spectroscopy A technique of chemical analysis in which a sample is IONIZED using an ELECTRIC ARC or flame and the wavelengths in the EMISSION SPECTRUM so produced are measured. These wavelengths are characteristic of the elements present.

atomic force microscope A microscope that produces an image using a diamond-tipped probe, which is moved over the surface of a sample and responds to the interatomic forces between the probe and the sample. The probe in effect 'feels' its way over the contours of the surface, and its up-and-down movements are transmitted to a computer that produces a profile of the sample. The atomic force microscope can resolve (*see* RESOLUTION) single molecules, and is useful for biological specimens as the sample does not have to be electrically conducting.

atomicity The number of atoms present in a single molecule of a given compound. For example, ethanol, C_2H_5OH, has an atomicity of 9.

atomic mass unit (amu) The unit in which nuclear masses are usually measured. The mass of one atom of carbon–12 (the isotope of carbon with MASS NUMBER 12) is defined to be 12 amu exactly. Nuclear masses can be measured using a MASS SPECTROMETER.

atomic number The number of protons in a nucleus.

atomic theory The idea that all materials are made up of small particles called atoms. The motion of these particles leads to KINETIC THEORY in physics, whilst the way in which they combine to form molecules is the foundation of DALTON'S ATOMIC THEORY in chemistry.

atomic volume The RELATIVE ATOMIC MASS of an element divided by its volume, usually expressed in $cm^3 mol^{-1}$.

atomic weapon Any weapon that derives its energy from NUCLEAR FISSION. *See also* NUCLEAR FALLOUT.

atomic weight *See* RELATIVE ATOMIC MASS.

ATP (adenosine triphosphate) The short-term energy storage and carrier molecule found in all living cells. It transfers energy from where there is plenty to where it is needed for cellular reactions.

Energy is released when one of the three phosphate groups of ATP is removed (catalysed by a number of enzymes) by a process called HYDROLYSIS. This yields ADP (adenosine diphosphate). Hydrolysis of ADP then yields AMP (adenosine monophosphate). The phosphate molecules can be added back to AMP to reconvert to ATP by a process called PHOSPHORYLATION.

atrium (*pl. atria*), ***auricle*** Either one of the two upper chambers in the heart. The walls of the atria are thin so that they can stretch to receive blood that returns from the body. The atria contract to force blood into the VENTRICLES. *See also* HEART.

attenuation The factor by which the intensity of a wave or the size of an electronic signal is reduced in passing through some device. In electronics, attenuation is often measured in DECIBELS.

attenuator A device designed to produce a certain ATTENUATION, particularly of an electronic signal.

auditory canal, *ear canal, external auditory meatus* In mammals and birds, a tube leading from the opening of the outer ear to the EARDRUM.

auditory nerve The nerve that transmits impulses from the ear to the brain.

auricle *See* ATRIUM.

aurora A luminous glow in the atmosphere caused by the IONIZATION of gases by high-energy charged particles in the SOLAR WIND. The MAGNETIC FIELD OF THE EARTH causes these particles to enter the atmosphere only in regions close to the magnetic poles. An aurora in the northern hemisphere is called aurora borealis, or northern lights. An aurora in the southern hemisphere is called aurora australis.

autocatalysis Any chemical reaction in which one of the products of the reaction is also a CATALYST for the reaction. Such a reaction will start slowly then accelerate as the products are produced and catalyse further reactions, finally slowing down as the reagents are used up.

autolysis The self-destruction of a cell or tissue, brought about by the action of enzymes released by the cell itself. *See also* LYSOSOME.

autonomic nervous system Part of the NERVOUS SYSTEM that is self-governing and controls the involuntary responses of smooth muscle (such as in the digestive tract and blood vessels), the heart and glands. It forms part of the EFFECTOR SYSTEM, which receives information from the CENTRAL NERVOUS SYSTEM and transmits it to EFFECTORS (muscles or glands) that stimulate the appropriate action.

The autonomic nervous system can be divided into the sympathetic system and the parasympathetic system. The sympathetic system responds to stress, for example by increasing the heart rate, increasing blood pressure, preparing the body for action. The parasympathetic system is important when the body is resting, for example by slowing the heart rate, decreasing blood pressure, stimulating the digestive tract. These two systems normally oppose one another.

Some control over the activities of the autonomic nervous system, for example control of bladder and anal SPHINCTERS, can be learned through training.

autophagosome *See* PHAGOSOME.

autosome A chromosome that is the same in males and females; that is, all chromosomes except the sex chromosomes. The sex chromosomes are called HETEROSOMES.

autotroph An organism that can manufacture organic compounds from inorganic molecules using light or chemical energy. Autotrophs can therefore exist independently of any external source of organic compounds, unlike HETEROTROPHS.

All green plants and some bacteria are photoautotrophs. Photoautotrophs obtain their energy from PHOTOSYNTHESIS, which uses light to convert carbon dioxide and water into sugars. Some bacteria use chemical energy, for example from sulphur-containing compounds, to synthesize organic compounds. These are chemoautotrophs (*see* CHEMOSYNTHESIS). Autotrophs are the primary producers in the FOOD CHAIN; they provide nourishment for all the other animals in the food chain, which are heterotrophic.

autotrophic nutrition The synthesis of organic compounds from inorganic molecules using light or CHEMICAL ENERGY. The organisms that are capable of this are called AUTOTROPHS and do not need any external source of organic compounds. Autotrophic nutrition is self-feeding. *See also* CHEMOSYNTHESIS, FOOD CHAIN, HETEROTROPH, PHOTOSYNTHESIS.

auxiliary circle A mathematical construction for the study of SIMPLE HARMONIC MOTION. In this construction, a point is considered to move around a circle of radius A at a constant ANGULAR VELOCITY ω. The mathematical projection of this point onto the x-axis then moves with simple harmonic motion,

$$x = A\cos \omega t$$

auxin Any one of a group of PLANT GROWTH SUBSTANCES. Auxins influence many aspects of plant growth, including TROPISMS, cell enlargement and growth of roots. They are the most common type of plant growth substance. The most common naturally occurring auxin is indoleacetic acid, which is made in the shoot and root tips and transported to other parts of the plant.

The short-distance cell-to-cell transport of auxin is by DIFFUSION, but long-distance transport is via the PHLOEM. The transport of auxin is polar (in one direction only), away from the tips. Auxins act by increasing the elasticity of the cell wall, so the cell expands when TURGOR pressure increases, and continues to enlarge until enough resistance is provided by the cell wall. Auxin also affects GENE EXPRESSION (of at least 10 genes). At higher concentrations, auxins can inhibit growth and cause death of a plant. PHOTOTROPISM can be explained in terms of auxin distribution.

Many synthetic auxins have been developed, for example, to help root development in cuttings, to prevent fruit drop in orchards, to achieve synchronous flowering (and therefore fruiting) in pineapple and as weedkillers where they can cause such rapid growth that the plant dies. Some synthetic auxins have different effects on different plants, which is useful in producing selective weedkillers to kill only the unwanted plants.

See also ABSCISSIC ACID, CYTOKININ.

avalanche breakdown The mechanism by which a gas conducts electricity. An avalanche breakdown results when a few stray ions of a gas in an electric field are accelerated so violently by the field that they collide with other molecules with enough energy to break them apart. This produces further IONIZATION and leads to a rapid increase in the number of ions present and a decrease in the resistance of the gas.

At low pressures, the molecules in a gas are more widely spaced, and a GAS DISCHARGE, as the phenomenon is then called, can be produced relatively easily. Each ion has space in which to accelerate before hitting another atom, so weaker electric fields are required. At higher pressures, the result is a spark once the electric field reaches the required value. In each case light is given off from the energy released as ions of opposite charges recombine and release energy.

average A term used in STATISTICS to indicate the typical member of a set of data. It usually refers to the arithmetic MEAN, which is obtained by adding all of a group of numbers together and dividing the total by the number of samples. A mean value is often expressed plus or minus the STANDARD DEVIATION. The term average is also used to refer to the

MEDIAN, which is the middle number in a set of numbers arranged in increasing or decreasing order, or the MODE, which is the most frequently occurring number in a group.

Aves A CLASS of vertebrates, consisting of the birds. Birds possess feathers on their skin, scales on their legs, a beak instead of teeth, wings, lungs and eggs with a large yolk and hard shell from which the young hatch. Birds form the largest group of land vertebrates and there are 8,500 species. They have two legs, each with three digits; the front legs are modified to form a wing. Most birds can fly but some cannot, for example the ostrich. Birds are HOMEOTHERMS, maintaining a body temperature of 41°C. Their hearing and eyesight are good but their sense of smell is poor. Males are often brightly coloured to attract females, and communication is by vision and sound. The eggs are hatched by the female in a nest and cared for over a period of time. The study of birds is called ornithology.

Avogadro constant, *Avogadro number* (L or N_A) The number of atoms in one MOLE of atoms, molecules in one mole of molecules, etc. It is equal to 6.022×10^{23}. The mass of this number of carbon–12 atoms is 12 g.

Avogadro number *See* AVOGADRO CONSTANT.

Avogadro's hypothesis A given number of molecules of any gas at a given temperature and pressure will occupy the same volume, regardless of the nature of the gas. In particular, one MOLE of any gas occupies a volume of 22.4 dm^3 under conditions of STANDARD TEMPERATURE AND PRESSURE (atmospheric pressure and a temperature of 0°C). This result was originally based on EMPIRICAL observations but is now seen to be a consequence of KINETIC THEORY.

axil In a plant, the angle between the stalk of a leaf or branch and the stem from which it grows.

axiom A mathematical statement from which other statements may be deduced, but which itself cannot be proved but rather is accepted as self-evident.

axis (*pl. **axes***) **1.** One of a set of lines from which CO-ORDINATES are measured in co-ordinate geometry. For example, in CARTESIAN CO-ORDINATES, the x-axis is the line along which the x co-ordinate increases.

2. The line about which an object rotates (an axis of rotation), or about which an object can be rotated (possibly only through certain angles) and appear unchanged (an axis of symmetry).

axon A NERVE FIBRE that is a long extension of a NEURONE, and conducts impulses away from the cell body to the SYNAPSE with another neurone or EFFECTOR organ, such as a muscle. Most neurones only have one axon (monopolar), some have two (bipolar) and others have several (multipolar). Some axons are insulated by a MYELIN SHEATH.

azeotropic Describing a liquid mixture, such as that of ethanol in water, that produces a minimum in the graph of boiling point against composition. Such a minimum will be the lowest temperature at which the liquid can boil and is lower than the boiling point of either material on its own. This mixture will be produced if an attempt is made to separate the mixture by FRACTIONAL DISTILLATION. For example, in the fractional distillation of ethanol from water, an azeotropic mixture is produced, with a composition of 96 per cent ethanol and a boiling point of 78.3°C, compared to a boiling point of 78.5°C for pure ethanol.

azo compound One of a group of organic compounds formed by reacting DIAZONIUM SALTS with AROMATIC AMINES or a PHENOL. The reaction is a diazo-coupling reaction in which a second BENZENE RING is added to the diazonium salt to give two benzene rings joined by the –N=N– group. The products formed are brightly coloured and form the basis of AZO DYES. The compounds are usually very stable. An example is METHYL ORANGE:

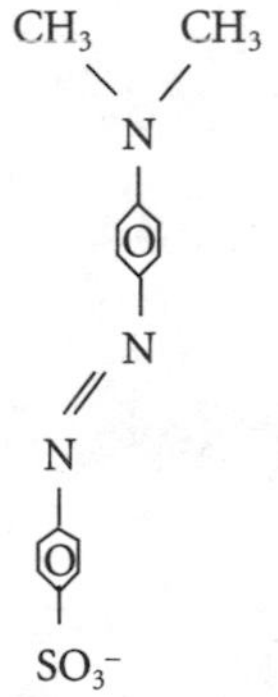

CH_3 CH_3
N
N
N
SO_3^-

azo dyes, *aniline dyes* A large group of DYES, constituting about half of all dyes made. Azo dyes are formed by the reaction of a

DIAZONIUM SALT with an AROMATIC AMINE or a PHENOL. They consist of two BENZENE RINGS joined by the –N=N– group. The simplest azo dyes derive from PHENYLAMINE (aniline) and they are therefore also known as aniline dyes. They are cheap to make and since they are stable it is easy to add groups in order to change their colour and fabric bonding properties. Azo dyes are usually red, brown or yellow and can be used on most types of fabrics. They are also used as pigments in the photographic industry.

AZT, ***zidovudine*** A drug used in the treatment of AIDS.

B

Bacillariophyta A PHYLUM of the KINGDOM PROTOCTISTA that consists of the diatoms. They are unicellular organisms that live in moist soil, freshwater or on the surface layers of the oceans, where they form a major component of PLANKTON. Their cell walls are composed chiefly of silica and are divided into two halves that fit one inside the other, like a box with its lid. The fossil remains of diatom shells are mined and used, for example, in abrasives, filters and fillers for paint and rubber.

bacillus (*pl.* ***bacilli***) Any rod-shaped bacterium (*see* BACTERIA). Bacilli are widespread in soil and air and cause diseases such as anthrax (*Bacillus anthracis*).

backbone *See* VERTEBRAL COLUMN.

backcross A mating between a parent and its offspring, used in genetics to determine CHROMOSOME MAPS.

back e.m.f. A voltage that opposes the current flowing in a circuit. The back e.m.f. of a coil is produced as a result of a change in the magnetic field in that coil, caused by a change in the current flowing. *See* SELF-INDUCTANCE. *See also* ELECTROMOTIVE FORCE, LENZ'S LAW.

background radiation The collective name for the many sources of IONIZING RADIATION. The most important of these are naturally occurring radioactive materials in rocks, soil and atmosphere. The other main source of background radiation is COSMIC RADIATION – high-energy charged particles, mostly protons, that enter the atmosphere from space.

Compared to these two sources of radiation, the radiation present from NUCLEAR REACTORS and NUCLEAR WEAPON tests represents only 1 or 2 per cent of the total exposure to ionizing radiation for the average human. In addition, individuals often experience significant doses of ionizing radiation from medical sources, mainly X-rays. The level of medical exposure can vary widely from one individual to another, though in the West it typically accounts for about 13 per cent of the lifetime dose. *See also* RADIOACTIVITY.

backing store A computer memory with the capability to hold a large amount of data, but with a relatively slow ACCESS TIME. Backing stores are used to archive programs and data. Typical backing stores are magnetic tapes and disks.

backup file *See* BACKUP SYSTEM.

backup system In computing, a duplicate system that can take over in the case of a fault in the main system. A backup file is a copy of a computer file taken in case the original is lost, destroyed or altered in any way.

bacteria (*sing.* ***bacterium***) Microscopic organisms with a single cell and lacking an organized nucleus. Bacteria are classified as in the KINGDOM PROKARYOTAE and are neither plant (most lack CHLOROPHYLL) nor animal.

Bacteria can be spherical (COCCUS) in shape, for example *Staphylococcus aureus;* rod-shaped (BACILLUS), for example *Escherichia coli;* spiral (spirilla), for example *Spirillum rubrum;* or comma-shaped (vibrios), for example *Vibrio cholerae.* Cells can be either Gram-positive or Gram-negative according to their ability to stain in Christian Gram's (1855–1938) stain of 1884, which is important in their classification (*see* GRAM'S STAIN). Their size varies between 0.5 and 2.0 μm, so they are visible with a light MICROSCOPE. Reproduction is by BINARY FISSION. Non-spherical bacteria can possess one or more FLAGELLA by which they move. Some have smaller filaments called pili, which make the cells sticky, and under certain conditions some secrete a capsule of non-living viscous material around the cell wall.

Bacteria are found in large numbers everywhere and their activities are very important. Most are SAPROTROPHS or PARASITES. In the soil they break down plant and animal tissues and their role is crucial to soil fertility and thus to all life. This ability is also important in SEWAGE DISPOSAL. Some bacteria are used in industry to bring about desirable chemical reactions, while others cause serious food and drink spoilage. They cause many serious diseases in animals and humans, but some are a source of ANTIBIOTICS. Some live harmlessly in sites within animals and humans, for example the skin and gut.

bacteriochlorophyll A type of CHLOROPHYLL found in some photosynthetic bacteria.

bacteriophage A virus that infects a bacterium (*see* BACTERIA). Bacteriophages can contain DNA or RNA as their genetic material in the centre of the phage surrounded by a protein coat. Phages can be very simple (RNA phages) or complex, consisting of head, collar and tail regions. Some phages are spherical in shape, while others are filamentous. The complex phages attach to bacterial cell walls by their tail, and inject their DNA into the bacteria cell leaving their protein 'ghost' outside. Virulent phages kill their host cell, whereas temperate phages, do not and instead integrate their DNA with the hosts and replicate with it for several generations. *See also* TRANSFECTION.

bacterium *See* BACTERIA.

Bakelite Trade name for certain PHENOL–FORMALDEHYDE RESINS. Bakelite was the first synthetic plastic to be made and is tough, hard, heat resistant and an electrical insulator. It has many uses, for example, telephone receivers, electric plugs and sockets.

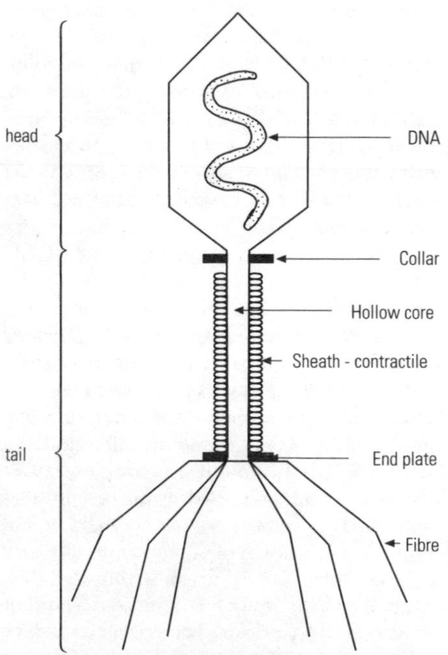

Structure of a complex bacteriophage.

balance A device for weighing. A beam balance consists of a lever balanced on a central pivot, with known and unknown masses being suspended on opposite sides of the pivot, usually at equal distances. The two masses are compared by using the pull of gravity on the masses to produce turning MOMENTS about the pivot: when the masses are equal, the beam is exactly horizontal. A top-pan balance does not compare one mass with another, but uses the pull of gravity on an unknown mass to deform a STRAIN GAUGE, with the mass being presented directly as a digital reading. *See also* CURRENT BALANCE.

ball bearing *See* BEARING.

ballistics The study of PROJECTILES. In particular ballistics is the study of bullets or shells fired from guns, but also of any object that is set moving by a sudden IMPULSE and then moves under the influence of forces, such as gravity and the force of resistance exerted by the air.

Balmer series A series of lines in the HYDROGEN SPECTRUM. The wavelengths are mostly in the visible part of the ELECTROMAGNETIC SPECTRUM, though the series extends slightly into the ultraviolet. Each line corresponds to a transition between the second ENERGY LEVEL and some higher level. *See also* BOHR THEORY.

band gap In the BAND THEORY of solids, the gap between one energy band and the next, particularly between a full VALENCE BAND and an empty CONDUCTION BAND.

band theory The branch of QUANTUM MECHANICS that explains the properties of solids in terms of ENERGY LEVELS. The electrons in a single atom exist in discrete, sharply defined energy levels. When the atoms come together to form a solid, these sharply defined energy levels become bands of allowed energies. These bands may be filled with electrons, or may be partially full or empty. Between the allowed bands are 'forbidden' bands.

The VALENCE ELECTRONS, those involved in chemical bonding, form the VALENCE BAND of a solid. In an ideal crystal, the valence band is fully occupied. The only electrons free to move through the solid, carrying heat and electricity, are those in partially full bands. The PAULI EXCLUSION PRINCIPLE forbids an electron from gaining energy unless there is an empty energy level for it to move to. If the valence band is full, the electrons must move to an unfilled band, called the CONDUCTION BAND. Conductors are those materials whose valence band and conduction band are unfilled, or whose conduction and valence bands overlap – in either case there are vacant energy levels.

In insulators, the valence band and conduction band are separated by a wide forbidden band. Such materials do not conduct because the electrons do not have enough energy to cross from one band to another. In semiconductors, the energy difference between the valence band and the next band is sufficiently small that thermal vibrations may give electrons enough energy to enter the empty conduction band. The energy required is called the BAND GAP. If the band gap has an energy corresponding to a PHOTON of visible light, light will be absorbed or emitted as electrons cross the band gap. This effect is exploited in PHOTODIODES and LIGHT EMITTING DIODES.

bandwidth The spread of frequencies needed to convey information in any TELECOMMUNICATIONS system, particularly those using radio waves. An important feature of any telecommunications system is the rate at which information is transmitted (for digital systems this is called the BAUD RATE). The faster the information is transmitted, the greater the bandwidth needed to transmit it and so the more widely spaced must be the chosen CARRIER WAVE frequencies.

Speech and music have a bandwidth of only a few kilohertz (*see* HERTZ), depending on how high a quality of reproduction is required. Television, on the other hand, requires a bandwidth of many megahertz if a complete picture is to be transmitted enough times per second for the impression of movement to be produced.

See also SIDEBAND.

bar A unit of pressure in the c.g.s. system (*see* C.G.S. UNITS). One bar is equal to 10^5 PASCAL, or approximately one ATMOSPHERE. The millibar (100 Pa) is used as the unit of pressure in METEOROLOGY.

bar chart A simple visual means of comparing numerical quantities, each quantity being represented by a bar the length of which is proportional to its value. In particular, bar charts are often used in statistics to show the number of times a particular event occurs, such as the number of times a certain total value is obtained when two dice are thrown. Such statistical bar charts are often referred to as HISTOGRAMS.

barium (Ba) The element with ATOMIC NUMBER 56; RELATIVE ATOMIC MASS 137.3; melting point 725°C, boiling point 1,640°C; RELATIVE DENSITY 3.5. Barium is an ALKALINE EARTH metal that occurs in the minerals barytes, $BaSO_4$, and witherite, $BaCO_3$. It forms an insoluble SULPHATE, which is used as BARIUM MEALS in RADIOGRAPHY. Barium is used in alloys, and its compounds are used in pigments, matches and fireworks.

barium carbonate ($BaCO_3$) A white powder; RELATIVE DENSITY 4.4; decomposes on heating. Barium carbonate occurs naturally, and being insoluble it can be formed as a precipitate by adding a carbonate to a dissolved barium salt, for example:

$$Na_2CO_3 + BaCl_2 \rightarrow BaCO_3 + 2HCl$$

Barium carbonate decomposes on heating to give BARIUM OXIDE:

$$BaCO_3 \rightarrow BaO + CO_2$$

barium hydroxide ($Ba(OH)_2$) A white solid; melting point 408°C; decomposes on further heating; RELATIVE DENSITY 4.5. Barium hydroxide occurs naturally and is slightly soluble in water. It can be produced as a precipitate in the reaction between sodium hydroxide and barium chloride:

$$2NaOH + BaCl_2 \rightarrow Ba(OH)_2 + 2HCl$$

barium meal A drink containing barium in the form of barium sulphate, given to a patient as a CONTRAST ENHANCING MEDIUM before taking a RADIOGRAPH of the stomach or intestines.

barium oxide (BaO) A white solid; melting point 1,920°C; boiling point 2,000°C; RELATIVE DENSITY 5.7. Barium oxide is insoluble in water and can be manufactured by heating BARIUM CARBONATE:

$$BaCO_3 \rightarrow BaO + CO_2$$

barium peroxide (BaO_2) A creamy white solid; melting point 450°C; decomposes on further heating; RELATIVE DENSITY 5.0. Barium peroxide can be manufactured by heating barium oxide in oxygen:

$$2BaO + O_2 \rightarrow 2BaO_2$$

On heating in air it decomposes to barium oxide:

$$2BaO_2 \rightarrow 2BaO + O_2$$

Barium peroxide reacts with acids to form hydrogen peroxide, for example:

$$BaO_2 + 2HCl \rightarrow H_2O_2 + 2HCl$$

barium sulphate ($BaSO_4$) A white powder; melting point 1,580°C; RELATIVE DENSITY 4.5. Barium sulphate occurs naturally and can be made as an insoluble precipitate in the reaction between sulphuric acid and barium chloride solution:

$$BaCl_2 + H_2SO_4 \rightarrow BaSO_4 + 2HCl$$

Barium sulphate is opaque to X-rays and is used in BARIUM MEALS to provide a CONTRAST ENHANCING MEDIUM in RADIOGRAPHY. The sulphate is very insoluble and is not absorbed into the body, so is non-toxic. Barium sulphate is also used as a white pigment in some paints.

bark The outer protective layer on the stems and roots of WOODY PLANTS. Bark is composed mainly of dead cells and varies in thickness. It is added to from the inside of the plant, and so the outer layer often cracks and is shed to allow for expansion of the stem. A number of chemicals are deposited by trees in their bark, many of which are used in medicine.

bar magnet A magnet in the shape of a bar or rod with a POLE at each end.

barn A unit of area used in particle physics to measure cross-sections of atomic nuclei in interactions. One barn is equivalent to 10^{-28} m^2.

barograph An instrument that gives a record of changes in ATMOSPHERIC PRESSURE over time. It is similar in construction to the ANEROID BAROMETER, but has an inked pointer that leaves a trace on a sheet of paper moving under it. Falling pressures indicate the arrival of a DEPRESSION, which usually means deteriorating weather, whilst rising pressures generally mean improving weather.

barometer Any instrument used to measure ATMOSPHERIC PRESSURE. *See* ANEROID BAROMETER, BAROGRAPH, FORTIN BAROMETER, MERCURY BAROMETER.

baryon Any HADRON with half integral SPIN. Baryons are composed of three QUARKS and carry a QUANTUM NUMBER called the BARYON NUMBER, which is believed to be conserved in all interactions. Baryons are either NUCLEONS (protons or neutrons) or short-lived particles known as HYPERONS that include nucleons in their decay products. *See also* LAMBDA PARTICLE, OMEGA-MINUS PARTICLE, SIGMA PARTICLE.

baryon number A QUANTUM NUMBER carried by all BARYONS and the QUARKS from which they are made. Each quark has baryon number $^1/_3$, whilst antiquarks all have baryon number $-^1/_3$. Thus a baryon (which comprises three quarks) has baryon number 1 and a MESON (which comprises a quark and an antiquark) has baryon number 0. It is thought that baryon number is exactly conserved in all interactions, but some GRAND UNIFIED THEORIES, predict a very slow rate of PROTON DECAY, leading to a violation of baryon number conservation.

barytes The chief ore of barium, the mineral form of BARIUM SULPHATE.

basalt A fine-grained basic IGNEOUS rock, the commonest type of LAVA.

base 1. (*chemistry*) (i) Any compound that will react with an ACID to produce a SALT plus water. An example is sodium hydroxide:

$$NaOH + HCl \rightarrow NaCl + H_2O$$

Most bases are the OXIDES or HYDROXIDES of metals – ammonia is an important exception. A base that is soluble in water is termed an alkali.

(ii) In the Lowry–Brønsted theory of acids and bases, a base is any compound that accepts a proton. For example, SODIUM CHLORIDE behaves as a base when it dissolves in ammonia:

$$NH_3 + NaCl \rightarrow NaNH_2 + HCl$$

(iii) In the Lewis theory of acids and bases, a base is defined as any substance that donates electrons. This encompasses not only the traditional base reactions, but also REDOX REACTIONS and the formation of CO-ORDINATE BONDS. Thus boron trichloride acts as a base when it forms a bond with ammonia:

$$H_3N{:} + BCl_3 \rightarrow H_3NBCl_3$$

See also ALKALI, BASICITY.

2. (*mathematics*) The flat surface at the bottom of a geometrical figure, such as the flat surface of a cylinder or cone, or the rectangular surface of a pyramid.

3. (*mathematics*) A description of a number system, the base of the system being the number of different values that can be represented by a single digit. Thus in the decimal (base–10) system there are 10 number symbols, the digits 0 to 9. A binary (base–2) system has only 2 digits, for which the symbols 0 and 1 are normally used. Hexadecimal (base–16) numbers are widely used in computing, the digits are given the symbols 0 to 9 and A to F, with A to F corresponding to 11 to 15 in the decimal system.

4. (*electronics*) The central ELECTRODE in a JUNCTION TRANSISTOR.

base exchange A chemical reaction in which one ANION changes places with another. An example is the precipitation of copper(II) carbonate in a solution of sodium sulphate from the reaction between solutions of copper sulphate and sodium carbonate:

$$CuSO_4 + Na_2CO_3 \rightarrow CuCO_3 + Na_2SO_4$$

basement membrane A thin layer of proteins secreted by animal EPITHELIAL cells, usually combined with COLLAGEN fibres.

base metal A common inexpensive metal, especially one used as a substitute for a NOBLE METAL, such as gold. Base metals are those which are chemically more reactive, so corrode rapidly.

base pair Two NUCLEOTIDE bases in DNA linked by HYDROGEN BONDS between the two strands of the double helix. The pairing is always between a PURINE and a PYRIMIDINE, so the bases ADENINE and THYMINE always link and CYTOSINE and GUANINE always link. Base-pairing also occurs during MESSENGER RNA TRANSCRIPTION and TRANSLATION. URACIL in RNA pairs with adenine.

base unit, *fundamental unit* Any unit the size of which is fixed by reference to some experimental measurement, as opposed to a DERIVED UNIT. In the SI system, the base units are the KILOGRAM (mass), METRE (length), SECOND (time), AMPERE (current), KELVIN (temperature) and MOLE (amount of substance).

basic (*adj.*) Having the properties of a base.

BASIC (Beginner's All-purpose Symbolic Instruction Code) An easy-to-learn HIGH-LEVEL computer programming language, popular with personal computer users.

basicity The number of hydrogen atoms in an acid that can be replaced by metal ions in the formation of a salt. Thus hydrochloric acid, HCl, has a basicity of 1 (monobasic), whilst sulphuric acid, H_2SO_4, has a basicity of 2 (dibasic).

basic lead carbonate *See* LEAD CARBONATE HYDROXIDE.

basic salt Any SALT formed by replacing some of the OXIDE or HYDROXIDE of a base with some other negative ion. The compound consists of a normal salt combined with a base in a simple molecular ratio. An example is basic lead carbonate, $2PbCO_3.Pb(OH)_2$, produced by the reaction of excess lead oxide with carbonic acid. *See also* ACIDIC SALT, NORMAL SALT.

basidiomycete A member of the PHYLUM BASIDIOMYCOTA.

Basidiomycota A PHYLUM of the KINGDOM FUNGI. Basidiomycetes are characterized by having septa (partitions) in their HYPHAE and they often form large fruiting structures. Examples are the field mushroom (*Agaricus campestris*) and the ink-cap toadstool (*Coprinus*). Reproduction is usually sexual by means of BASIDIOSPORES formed outside basidia (*see* BASIDIUM). Many members of the Basidiomycota are a cause of world-wide crop spoilage.

basidiospore A SPORE produced by the BASIDIOMYCOTA PHYLUM of FUNGI. It is formed by MEIOSIS within a BASIDIUM but is borne outside the basidium.

basidium (*pl.* ***basidia***) A specialized cell of the BASIDIOMYCOTA PHYLUM of FUNGI. It is concerned with the production of BASIDIOSPORES during sexual reproduction.

basophil A type of GRANULOCYTE (blood cell) with cytoplasmic granules that stain with basic dyes.

batch processing A computer program that once started runs uninterrupted and to completion. They work without the need for intervention to a rigorous schedule, such as every night. They are useful for repetitive tasks, such as the preparation of a payroll. *Compare* INTERACTIVE COMPUTING.

battery Several electrochemical CELLS connected together, usually in series to produce a larger voltage.

baud rate In a digital TELECOMMUNICATIONS system, the number of BITS sent per second. *See also* BANDWIDTH.

bauxite A aluminium ore containing mainly aluminium oxide, Al_2O_3. Bauxite is the most important source of aluminium in the Earth's crust.

B cell, *B lymphocyte* A type of LYMPHOCYTE that is formed in BONE MARROW and settles in the SPLEEN or a LYMPH NODE without passing through the THYMUS, as T CELLS do. B cells express a specific antibody on their surface and when this binds to a specific ANTIGEN it activates the cell to divide, producing a group of identical cells (*see* CLONE). These are then capable of producing the correct antibody needed to fight the invading antigen.

Following an infection some B cells do not produce much antibody but instead become 'memory cells' and circulate in the body ready to be reactivated to produce antibody quickly if a second exposure to the initial antigen occurs. Other B cells mature into 'PLASMA cells', which are specific for one antigen and are the main secretors of antibody if this antigen is encountered again.

See also MAJOR HISTOCOMPATIBILITY COMPLEX.

beam balance *See* BALANCE.

beam tube In a PARTICLE ACCELERATOR, the evacuated tube through which the particle beam passes.

bearing 1. A direction measured as an angle clockwise from some reference direction, usually north. Thus a bearing of 150° represents an angle of 150° around from north, sometimes written as S 30° E – that is, 30° east of south. Bearings are often expressed as three digit numbers, such as 090° for East.

2. (*physics*) A point at which a LOAD is supported, particularly if the load is able to rotate and steps have been taken to reduce friction by supporting the load on balls (a ball bearing) or rollers (a roller bearing).

beats An INTERFERENCE effect, often observed in sound, that occurs when two waves of slightly different frequency interfere. The two waves drift in and out of PHASE producing a periodic increase and decrease in loudness.

beauty In particle physics, *see* BOTTOM.

becquerel (Bq) The SI UNIT of radioactive ACTIVITY, one becquerel being an activity of one ionizing particle per second.

beehive shelf A cylindrical ceramic device with a hole in the top and the side, used to support a GAS JAR when collecting gas over water.

Beginner's All-purpose Symbolic Instruction Code *See* BASIC.

belt transect *See* TRANSECT.

Benedict's test A test for REDUCING SUGARS based on a modification of FEHLING'S TEST. Only one solution is used, containing sodium citrate, SODIUM CARBONATE and COPPER SULPHATE in water. This is added to the test sample and if a reducing sugar is present a rust-brown precipitate forms on boiling.

benign (*adj.*) Of a TUMOUR, not MALIGNANT; not threatening to life or health.

benzaldehyde *See* BENZENECARBALDEHYDE.

benzene (C_6H_6) A colourless, liquid HYDROCARBON derived from COAL TAR, with a characteristic smell; melting point 5°C; boiling point 80°C. Benzene is the simplest AROMATIC compound, consisting of six carbon atoms arranged so that they form a regular hexagon (with 120° angles and carbon–carbon bonds of equal length, intermediate between single and DOUBLE BOND lengths). The molecule is represented by a hexagon either with three double bonds or with a circle drawn inside the hexagon.

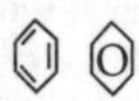

The PI ELECTRONS represented by the double bonds are not localized but are uniformly distributed around the ring. These are said to be delocalized and provide stability to the benzene ring. Reactions of benzene are usually ELECTROPHILIC SUBSTITUTION REACTIONS, which retain this delocalization of electrons, rather than ADDITION REACTIONS.

Benzene derivatives are named by prefixing 'benzene' with the name of the group attached, for example, methyl benzene, $C_6H_5CH_3$, 1,2-dimethylbenzene, $C_6H_4(CH_3)_2$. In the latter example the numbers are chosen to be as low as possible. In addition, the terms *ORTHO*, *META* and *PARA* are used to describe 1,2-, 1,3- and 1,4- disubstituted benzenes respectively, for example, *ortho*-dimethylbenzene (or *o*-dimethylbenzene). If there are different groups in the same ring they are named alphabetically, for example, methyl before nitro.

Benzene is used in the synthesis of many chemicals and as a solvent. Recently, it has become known that benzene vapours are dangerous, causing respiratory problems and cancer, and so derivatives such as METHYLBENZENE are more often used.

See also DELOCALIZED ORBITAL.

benzenecarbaldehyde, *benzaldehyde* (C_6H_5CHO) An ALDEHYDE that is a colourless liquid with a smell of almonds; boiling point 180°C. It occurs in nature in certain leaves and nuts and can be extracted from these or from TOLUENE. It is readily oxidized to BENZENECARBOXYLIC ACID. It is used in the manufacture of perfumes and dyes and as a solvent.

benzenecarboxylic acid, *benzoic acid* (C_6H_5COOH) A white solid occurring in some natural resins and oils; melting point 122°C; boiling point 249°C. It is a CARBOXYLIC ACID and is manufactured from the oxidation of TOLUENE or BENZENECARBALDEHYDE. It is used as a food preservative and an antiseptic.

1,4-benzenedicarboxylic acid, *terephthalic acid* ($C_8H_6O_4$) A DIBASIC ACID; melting point 300°C. It is used in the manufacture of terylene and is produced by the OXIDATION of dimethylbenzene.

benzene hexachloride *See* BHC.

benzene ring A term referring to the six carbon ring structure of benzene and found in many other AROMATIC COMPOUNDS.

benzoic acid The common name for BENZENECARBOXYLIC ACID.

beriberi A deficiency of vitamin B_1 (thiamine), which causes inflammation of the nerve endings. It particularly affects the ability to walk. *See also* VITAMIN B.

berkelium (Bk) The element with ATOMIC NUMBER 97. It does not occur in nature, and is named for Berkeley, California, where it was first synthesized. The longest-lived ISOTOPE (berkelium–247) has a HALF-LIFE of 1,400 years; all the others have half-lives of a few days or less.

Bernoulli effect The fall in pressure in a fluid as it accelerates. In an aircraft wing, for example, the curved upper surface forces air to travel more quickly than over the flat under surface. This acceleration results in a fall in pressure and the aircraft flies because the upward pressure on the underside of the wing is greater than the downward pressure on the upper surface. *See also* BERNOULLI'S THEOREM.

Bernoulli's theorem The total energy per unit volume in a fluid remains constant, so changes in pressure (POTENTIAL ENERGY) and speed (KINETIC ENERGY) compensate for one another. If the density of the fluid is *r* and it moves at speed *v*, with pressure *p*, then

$$^1/_2rv^2 + p = \text{constant}$$

See also BERNOULLI'S EFFECT.

berry A fleshy FRUIT with an outer skin (exocarp), often brightly coloured to attract birds, a thick fleshy wall (mesocarp) and an inner membrane surrounding many seeds. Examples include the tomato, grape and gooseberry.

beryllate Any compound containing the beryllate ion, BeO_2^{2-}, formed by the reaction of beryllium with strong alkalis, for example:

$$Be + 2NaOH \rightarrow Na_2BeO_2 + H_2$$

beryllium (Be) The element with ATOMIC NUMBER 4; RELATIVE ATOMIC MASS 9.0; melting point 1,285°C; boiling point 2,970°C; RELATIVE DENSITY 1.9. Beryllium is a hard, white ALKALINE EARTH metal with a low density. It is used to manufacture low density alloys and as a MODERATOR in some NUCLEAR FISSION systems. Beryllium is highly toxic.

beryllium oxide (BeO) A white solid; melting point 2,550°C; boiling point 4,120°C; RELATIVE DENSITY 3.0. Beryllium oxide is insoluble in water. It is AMPHOTERIC, forming salts with dilute acids, and reacting with strong alkalis to form BERYLLATES.

Bessemer converter A device, now largely obsolete, for the manufacture of steel from PIG IRON. The converter consists of a vessel with a REFRACTORY lining into which molten pig-iron is poured. Air is blown through the base of the furnace to OXIDIZE any impurities, in particular carbon. Additional materials, including some extra carbon, may then be added to control the properties of the steel. The converter is then tipped onto its side and the steel poured into moulds. The Bessemer converter has now largely been replaced by the OXYGEN FURNACE and the ELECTRIC ARC FURNACE. *See also* OPEN HEARTH FURNACE.

beta decay The emission of BETA PARTICLES, which may be either positive or negative, from an unstable atomic nucleus. Negative beta particles are fast moving electrons, produced by atomic nuclei with too many neutrons; a neutron turns into a proton, and an electron is emitted together with an antineutrino (*see* NEUTRINO). Positive beta particles are POSITRONS, produced when a nucleus with too few neutrons converts a proton into a neutron emitting a positron and a neutrino.

Beta radiation is less ionizing (*see* IONIZATION) than ALPHA RADIATION, but more ionizing than GAMMA RADIATION. Beta particles have a range of several metres in air but can be stopped by a layer of aluminium a few millimetres thick. In beta decay, the MASS NUMBER is unchanged, but the ATOMIC NUMBER is increased by 1 in negative beta decay and decreased by 1 in positive beta decay.

beta particle An electron or POSITRON produced in an atomic nucleus in BETA DECAY.

beta-pleated sheet, *β-sheet* A type of protein structure resulting from HYDROGEN BONDING. Hydrogen atoms from one side of the protein molecule link with the oxygen atoms of the side parallel to it, causing anti-parallel folding of the molecule. *See also* ALPHA HELIX.

beta radiation A stream of BETA PARTICLES, emitted from unstable nuclei by BETA DECAY.

beta-sheet *See* BETA-PLEATED SHEET.

BHC (benzene hexachloride), *1,2,3,4,5,6-hexachlorocyclohexane* ($C_6H_6Cl_6$) A powerful INSECTICIDE formed by the reaction of benzene with chlorine in sunlight or ultraviolet irradiation. BHC actually exists in several isomeric forms (*see* ISOMER), only one of which, gamma-BHC, is active against insects. The isomers have differing melting points, gamma-BHC is 113°C. Despite its success in the combat of a number of disease-carrying insects gamma-BHC now has restricted use due to its dangerous accumulation in the FOOD CHAIN.

bicuspid valve, *mitral valve* A flap of tissue in the heart between the left ATRIUM and VENTRICLE. It prevents blood flowing back into the atrium from the ventricle.

bidentate *See* LIGAND.

biennial plant A plant that completes its life cycle in 2 years and then dies. In the first season biennial plants store food in underground organs and in the second season they use this energy to produce flowers and seeds. Many root vegetables such as carrots are biennial. *Compare* ANNUAL PLANT, PERENNIAL PLANT.

bifocal (*adj.*) Describing a lens that is effectively two lenses with different powers made from a single piece of glass. *See also* PRESBYOPIA.

big bang The currently accepted model of the early Universe. According to the big bang model, the Universe was initially vastly hotter and denser than it is today. In the very early stages the temperature was high enough to create a 'soup' of QUARKS and antiquarks, electrons and POSITRONS and NEUTRINOS. As the universe expanded and cooled, the quarks began to fuse into protons and neutrons. Some of these then fused into simple nuclei: first into deuterium nuclei (1 proton, 1 neutron) and then into helium nuclei (2 protons, 2 neutrons). Helium now makes up about 25 per cent of the matter in the universe, the rest being mostly hydrogen. This matter was initially in the form of a PLASMA, but eventually the universe was cool enough for electrons and nuclei to combine into atoms faster than these atoms were ionized by collisions. At this point the universe became transparent. Much later, the clouds of gas began to collapse into regions that eventually became galaxies.

One problem with theories of the early universe is that little is known about how matter behaves under such extreme conditions. Theories also tend to have problems with predicting the correct degree of 'lumpiness'. On a small scale, stars are dense objects separated by large regions of space, and galaxies form CLUSTERS that appear to lie along string-like paths enclosing voids (empty regions of space). On a larger scale, the distribution of matter in the universe seems very uniform, and there are only very small variations in the intensity of the COSMIC MICROWAVE BACKGROUND. One solution to these problems is to suggest that the very early universe underwent a period of INFLATION – very rapid growth in which different regions of space became separated from one another.

See also BIG CRUNCH, COSMOLOGY, HUBBLE'S LAW.

big crunch The hypothetical end of the Universe if its density exceeds the CRITICAL DENSITY. *See* COSMOLOGY. *See also* BIG BANG.

bilateral symmetry A type of structure of an organ or organism in which there is only one plane through which the organ or organism can be cut to produce two halves that are mirror images of each other. *Compare* RADIAL SYMMETRY.

bile A brownish fluid made by HEPATOCYTES that assists in the breakdown and absorption of fats in the SMALL INTESTINE. Bile contains BILE SALTS, BILE PIGMENTS, CHOLESTEROL and LECITHIN. Bile is secreted from the liver into a BILE DUCT which, in humans, then feeds into the GALL BLADDER. It is stored here until it enters the small intestine under hormone regulation. The bile contents that are to be excreted (excess pigments, cholesterol and lecithin) are eliminated with the faeces. In addition to breaking down fat, bile has an ALKALINE PH, which is important for the functioning of some digesting enzymes.

bile duct A duct from the liver to the DUODENUM. *See* BILE, GALL BLADDER.

bile pigment A breakdown product of old RED BLOOD CELLS. Bile pigments form part of BILE. They are eliminated with the FAECES, which they colour.

bile salt A constituent of BILE that assists in the breakdown and absorption of fats. Examples are SODIUM CHLORIDE and SODIUM HYDROGENCARBONATE.

billion A thousand million (10^9). In the UK, one billion was originally equivalent to one million million, but the US definition is now used world-wide.

bimetallic strip A device made from strips of two metals of different expansivities securely fastened together. As the strip is heated or cooled, the two metals will expand or contract by different amounts, causing the strip to bend. This can be used as a temperature controlled switch, or THERMOSTAT, in which the bending of the strip makes or breaks an electric circuit.

binary (*adj.*) Describing a system that works in BASE–2; that is, all numbers contain only the digits 1 and 0. *See also* DECIMAL.

binary compound A compound containing just two elements.

binary fission A form of ASEXUAL REPRODUCTION occurring in single-celled organisms, such as bacteria and AMOEBAE. The cell divides into two daughter cells of equal size, each containing half the nuclear material (*see* NUCLEUS) and CYTOPLASM. The division can be transverse or longitudinal.

binary star A system in which two (or sometimes more) stars orbit around one another. Binary stars that are directly visible as two stars are called physical binaries to distinguish them from optical binaries, which are not linked but appear close together due to line-of-sight effects. Some binaries are observed as spectroscopic binaries, where the two components are inferred from the SPECTRUM received from what appears to be a single star. For astrometric binaries, the existence of a faint companion star is deduced from its influence on the motion of a brighter star. Eclipsing binaries are stars whose light output falls regularly as one member of the pair passes in front of (eclipses) its partner.

Binary stars are important in ASTROPHYSICS because their observation enables the masses of the stars in the system to be measured using Kepler's third law (*see* KEPLER'S LAW).

binding energy *See* NUCLEAR BINDING ENERGY.

binoculars Low-powered magnifying devices for viewing distant objects. They consist of two small telescopes mounted side by side (one for each eye). Inside each telescope is a pair of PRISMS to reverse the direction of travel of the light beam, so it passes along the length of the binoculars three times, producing an upright image and enabling the binoculars to be more compact.

binomial A mathematical expression containing two variables that are added or subtracted, in the form $a + b$ or $a - b$. *See also* BINOMIAL COEFFICIENT, BINOMIAL EXPANSION.

binomial coefficient Any one of the numbers that multiplies a term of the form $a^{n-r}b^r$ in the BINOMIAL EXPANSION $(a + b)^n$. Such a binomial coefficient is generally denoted by the symbol ${}^nC_r$, and is equal to

$${}^nC_r = n!/[r!(n-r)!]$$

binomial expansion A BINOMIAL raised to a power n, which need not be an integer, expanded according to the formula:

$$(a+b)^n = a^n + na^{n-1}b + n(n-1)a^{n-2}b^2/2! + n(n-1)(n-2)a^{n-3}b^3/3! + \dots$$

The series terminates after n+1 terms if n is a positive integer. If n is not a positive integer, the expansion is usually written for the binomial $(1+x)^n$, and converges to a LIMIT provided that $|x|<1$. *See also* BINOMIAL COEFFICIENT.

binomial nomenclature The system used for naming living organisms. It uses a two-part Latin name for every organism. The system was devised by Carolus Linnaeus (1707–78). The first part of the name denotes the GENUS and is written with a capital first letter, and the second part is the SPECIES. Both names are written in italics. The name for humans using this system is *Homo sapiens*. *See also* CLASSIFICATION.

biochemical evolution theory *See* ORIGIN OF LIFE.

biochemistry The study of the chemistry of living organisms. This includes the structure and functioning of, for example, enzymes, proteins and NUCLEIC ACIDS, and molecular biology. The study of biochemistry is fundamental to the understanding of life processes.

biodegradable (*adj.*) Describing any substance that can be broken down by natural biological processes, usually involving bacteria or fungi. Biodegradable substances, such as food and sewage, can therefore be recycled by the ECOSYSTEM. Many substances, such as plastics, are not biodegradable and present a problem of POLLUTION.

biological clock *See* BIORHYTHM.

biological control The control of pests by biological means instead of using chemicals. This can be achieved in a number of ways, for example by introducing a pest's predator or a disease-inducing organism, or by breeding resistant crops.

biology The study of life. Biology includes the study of any organism or system to which the term 'living' can be applied. It includes ANATOMY, BIOCHEMISTRY, BIOPHYSICS, BOTANY, CYTOLOGY, ECOLOGY, GENETICS, MICROBIOLOGY, PHYSIOLOGY, VIROLOGY and ZOOLOGY.

biomagnification The accumulation of harmful substances along the FOOD CHAIN. For example, the insecticide DDT, which has been used to combat mosquitoes, was found in low levels in PLANKTON, but higher levels in fish feeding on the plankton, and even higher levels in birds feeding on the fish. *See also* PESTICIDE.

biomass An energy production programme using crops grown for fuel or the waste products of other agricultural production. Crops grown for energy production have to compete for land with crops grown for food. Only in sparsely populated countries is there enough land to grow sufficient material to supply a significant fraction of the energy demand.

biome A broad category in a BIOSPHERE, consisting of all the plants and animals in a region subjected to and affected by a common set of climatic conditions, for example, tropical rain forest, desert, tundra (very cold, no trees and low plant growth) or DECIDUOUS forests. A localized area within a biome is called a HABITAT.

biophysics The study of the physics of biological processes. The laws of physics are applied to a biological system, for example the study of crystals of MACROMOLECULES or the use of MECHANICS to determine the strength of bones.

biorhythm Rhythmic changes in the behavioural or activity patterns of plants and animals, produced by hormones relating to environmental cycles. These patterns can be related to seasonal changes, such as HIBERNATION in winter. In insects and some other invertebrates a similar period of dormancy called DIAPAUSE can occur at any stage of development (most often the eggs or pupae) but usually only once in a lifetime. Seasonal flowering in plants and breeding in animals is controlled by changes in day length (*see* PHOTOPERIODISM). Bird MIGRATION is another example of a biorhythm.

Some biorhythm patterns are related to short cycles, for example the CIRCADIAN RHYTHM (24 hours) and CIRCALUNAR RHYTHM (28 days), and other patterns are annual. Biorhythms indicate the existence of a 'biological clock', which is probably a series of clocks running simultaneously but starting at different times. The mechanism of these 'clocks' is not understood. In humans, it is suggested that activity is controlled by intellectual, emotional and physical cycles with certain coincidence days being critical.

biosphere The narrow region over the Earth's surface that supports life, including the land (less than 20 km thick, 8 km above sea level) and the water and air around it. The biosphere is either aquatic or terrestrial and can be divided further into BIOMES and HABITATS.

biotechnology The use of living organisms in the large-scale industrial manufacture of food, drugs and other products. For example, yeast is used for FERMENTATION in the baking and brewing industries, and a range of bacteria and fungi are used in the dairy industry for converting milk into cheeses and yoghurts (*see* LACTIC ACID). Other food products that benefit from biotechnology are vinegar, sauerkraut and soy sauce. A number of enzymes made by micro-organisms are needed at various stages in food manufacture and are sometimes genetically engineered.

One of the major biotechnological advances in the medical world has been the manufacture of the hormone INSULIN to treat DIABETES. Using RECOMBINANT DNA technology, insulin can be made on a large scale. Previously, insulin had to be extracted from animal tissues. Human GROWTH HORMONE genetically engineered from micro-organisms has helped in the treatment of human dwarfism, and bovine somatotrophin (BST) is a genetically engineered hormone given to cows to increase milk yield.

See also GENETIC ENGINEERING.

biotic (*adj.*) Referring to the living elements in an ECOSYSTEM. Biotic elements include activities of animal populations, such as competition for food, water, shelter and mates. Predator–prey relationships affect the distribution of species, as do the activities of humans. Many plants depend on animals to disperse them or on insects to pollinate them. These are all biotic elements.

biotin *See* VITAMIN B.

bird A member of the vertebrate CLASS AVES.

birefringence, *double refraction* The property of some crystals, such as the mineral calcite, by which two refracted rays of light are formed from a single unpolarized ray (*see* POLARIZATION). The two refracted rays are polarized at right angles to each other: one (the ordinary ray) follows the normal laws of REFRACTION; the other (the extraordinary ray) follows different laws. *See also* PHOTOELASTICITY.

bismuth (Bi) The element with ATOMIC NUMBER 83; RELATIVE ATOMIC MASS 208.98; melting point 271°C; boiling point 1,560°C; RELATIVE DENSITY 9.8. It is the highest atomic number known to have a stable ISOTOPE. Bismuth is a white crystalline metal with a reddish tint. It is used in the manufacture of some THERMOPILES, and in certain alloys to lower the melting point.

bistable A LOGIC GATE with two outputs, one of which is HIGH and the other LOW. It also has two inputs, usually called set and reset. Making one of these inputs low puts the bistable into one of its two stable states, fixing one output as high and the other as low. When the inputs are both high, the bistable does not change states but remains in the state to which it was set. This is the basis of electronic memory. *See also* CLOCKED BISTABLE, SHIFT REGISTER.

bit A BINARY digit, a 1 or a 0, or a signal representing a binary digit.

bituminous coal *See* COAL.

bivalve A member of the CLASS PELECYPODA.

black body A hypothetical object that absorbs and emits all wavelengths of electromagnetic radiation perfectly. *See also* BLACK-BODY RADIATION.

black-body radiation The electromagnetic radiation emitted by a BLACK BODY at any temperature above ABSOLUTE ZERO. It extends over the entire range of wavelengths. A graph of the distribution of energies has a characteristic shape, with a maximum at a certain wavelength dependent on the temperature. Black body radiation is used as an approximation for the radiation given off by any hot object. *See also* STEFAN'S LAW, WIEN'S LAW.

black hole An astronomical object so dense that not even light can escape from its GRAVITATIONAL FIELD. Stars with cores of mass greater than about 2.5 SOLAR MASSES are believed to collapse without limit at the end of their lives, forming black holes. There is no firm evidence for the existence of black holes, but a number of intense X-ray sources have been discovered that can be explained in terms of the intense heating of matter as it falls into a black hole. *See also* HAWKING RADIATION.

bladder In bony fish, amphibians, mammals and some reptiles, a hollow, elastic-walled organ that stores urine before it is discharged. Urine enters the bladder via the URETER, one from each kidney, and in mammals leaves via the URETHRA. In other vertebrates (most reptiles, birds, amphibians and many fish), urine drains from the bladder into the CLOACA. *See also* URINARY SYSTEM.

blast furnace A device for extracting iron from its ores. Iron ore, COKE and LIMESTONE are fed into the top of the furnace and heated by the OXIDATION of the coke by oxygen blown upwards from the base of the furnace. The detailed chain of reactions is complex, but the carbon in the coke is oxidized to carbon dioxide whilst the iron ore is reduced to metallic iron. Molten iron, with dissolved carbon, and molten SLAG formed from the impurities fall to the bottom of the furnace, where they form separate layers. The hot gases leaving the top of the furnace are used to pre-heat the incoming air in HEAT EXCHANGERS called COWPER STOVES. The heated air is blown into the furnace through a ring of pipes called TUYERS. The molten metal, called PIG IRON, and the slag are regularly tapped off from the base of the furnace and fresh raw materials are added at the top, so the furnace can operate continuously.

blastocyst In mammals, the 64-cell stage of a fertilized OVUM. The blastocyst is a hollow cavity surrounded by a ball of cells. There is an inner cell mass that develops into the embryo, and an outer layer of cells called the TROPHOBLAST that develops into the EXTRA-EMBRYONIC MEMBRANES. It is at the blastocyst stage that IMPLANTATION into the wall of the UTERUS occurs. *See also* BLASTULA, EMBRYONIC DEVELOPMENT.

blastula In animals other than mammals, an early stage of development of a fertilized OVUM, when the ovum changes from being a solid mass of cells to a hollow ball of cells with a fluid-filled cavity. It is similar to the BLASTOCYST in mammals. *See also* EMBRYONIC DEVELOPMENT.

blende A mineral SULPHIDE of certain metals. In particular, zinc blende, which is ZINC SULPHIDE, ZnS, the chief ORE of ZINC.

blind spot The point in an eye where the OPTIC NERVE leaves the RETINA. At the blind spot, there are no RODS or CONES, so no visual images are transmitted.

blood The liquid circulating in the arteries, veins, capillaries or spaces of animals. It carries oxygen and nutrients to cells of the body and removes waste products such as carbon dioxide. It also has a role in the IMMUNE RESPONSE and in distributing heat in some animals.

An average human adult has 5.5 litres of blood. It consists of a watery colourless PLASMA that carries a number of blood cells serving different functions. The majority of the cells are RED BLOOD CELLS, which carry oxygen around the body. The main function of WHITE BLOOD CELLS is defence. PLATELETS are important in the BLOOD CLOTTING CASCADE. White blood cells can be divided into two further groups, GRANULOCYTES and AGRANULOCYTES.

See also BLOOD GROUP SYSTEM, BLOOD VESSEL, CIRCULATORY SYSTEM, HEART, SERUM.

blood clotting cascade A series of events that occurs following injury, in order to prevent excessive bleeding. Small cell fragments called PLATELETS aggregate and release an enzyme called thrombokinase when they come into contact with a damaged blood vessel. The vessel wall also releases this enzyme. In the presence of calcium ions and VITAMIN K, thrombokinase converts the inactive enzyme PROTHROMBIN into the active THROMBIN. This in turn converts the soluble PLASMA PROTEIN FIBRINOGEN into the insoluble FIBRIN, which forms a meshwork of protein fibres and blood cells over the wound. These dry to form a SCAB, which bacteria cannot enter and under which the wound can be repaired. In certain circumstances clotting has to be prevented and so anticoagulants, such as HEPARIN, which inhibits conversion of prothrombin to thrombin, may be administered. *See also* HAEMOPHILIA.

blood group system The classification of blood into groups according to molecules or ANTIGENS carried on the membrane of RED BLOOD CELLS. There are at least 14 blood group systems in humans, the main ones being the ABO and rhesus systems. In the ABO system there are two main antigens, A and B, which give rise to four possible blood groups: A (with antigen A only), B (with antigen B only), AB (with both antigens) and O (with neither antigen). In the rhesus system most individuals carry the RHESUS FACTOR and are called rhesus positive (Rh^+), while those who do not carry the factor are rhesus negative (Rh^-). Each of the ABO groups may or may not contain the rhesus factor.

It is important in blood transfusion to match the blood group, since an incompatible donor will cause an antibody response leading to AGGLUTINATION of the recipient's red blood cells, possibly resulting in death. Blood group O is considered to be the universal donor because there are no antigens to cause an antibody response in the recipient. Group AB is the universal recipient because it has both A and B antigens and therefore the recipient can tolerate blood from group A, B or O. The rhesus system can cause a problem during pregnancy if a rhesus negative mother carries a rhesus positive baby, resulting in RHESUS DISEASE. In Britain 46 per cent of the population are group O, 42 per cent group A, 9 per cent group B and 3 per cent group AB.

blood vessel A specialized tube for carrying blood around the body of multicellular organisms. There are three main types of blood vessels: arteries carry blood away from the heart ('a' for artery, 'a' for away); veins carry blood to the heart; and capillaries link arteries and veins. All arteries, except the PULMONARY arteries, carry oxygenated blood, and all veins, except the pulmonary vein, carry deoxygenated blood. The main artery leaving the heart is called the AORTA and further away from the heart small arteries are called ARTERIOLES. The main vein entering the heart is the

VENA CAVA and further from the heart small veins are called VENULES. Capillaries are grouped together in capillary beds (or networks) that link arteries and veins and serve as the main site of exchange between blood and body fluids (*see* LYMPH).

blue-green algae The former name of CYANOBACTERIA.

blue-green bacteria *See* CYANOBACTERIA.

blue-shift *See* DOPPLER EFFECT.

B lymphocyte *See* B CELL.

bob The mass in a SIMPLE PENDULUM.

body centred cubic A crystalline structure in which the UNIT CELL is a cube with an atom at the centre surrounded by one eighth of an atom at each corner. This structure has a CO-ORDINATION NUMBER of 8, leading to a relatively low PACKING FRACTION.

All the ALKALI METALS form crystals with this structure, which accounts for their low density compared to other metals of comparable atomic mass. The structure is also found in IONIC compounds where the two ions have equal charges and similar radii. If the ions differ much in radius, such as in sodium chloride, a different cubic structure is favoured with a co-ordination number of 6.

Bohr atom *See* BOHR THEORY.

Bohr effect In biology, the effect of carbon dioxide concentration on the association of oxygen with the RESPIRATORY PIGMENT HAEMOGLOBIN. Oxygen is more readily released from haemoglobin in the presence of carbon dioxide (in respiring tissues) and taken up more readily where the carbon dioxide concentration is low (in the lungs).

Bohr magneton A unit of MAGNETIC MOMENT equal to 9.27×10^{-24} Am^2. *See* MAGNETON.

Bohr theory, *Bohr atom* A simple model of ATOMS put forward in 1913 by Niels Bohr (1885–1962) to explain the LINE SPECTRUM of hydrogen. It has three basic postulates: (i) electrons revolve around the nucleus in fixed orbits without the emission of ELECTROMAGNETIC RADIATION; (ii) electrons can only occupy orbits in which the ANGULAR MOMENTUM of the electron is a whole number times $h/2\pi$, where h is PLANCK'S CONSTANT – in other words, within each orbit an electron has a fixed amount of energy; and (iii) when an electron jumps from one orbit to another, energy is emitted or absorbed as a PHOTON of electromagnetic radiation.

Whilst this model explains the broad features of the HYDROGEN SPECTRUM, it does not account for the fine detail and has been superseded by more complete quantum mechanical descriptions. It was however an important early step in the development of QUANTUM THEORY.

boil (*vb.*) To turn from a liquid to a vapour, with bubbles of gas being formed in the liquid.

boiling point The temperature at which a liquid boils, usually quoted at a pressure of one ATMOSPHERE. More technically, the one temperature for a given VAPOUR PRESSURE at which the liquid and its vapour can exist in equilibrium together. Boiling points increase with pressure, as it is harder for a bubble to form if the liquid is at high pressure. All materials expand on boiling and have boiling points that increase with increasing pressure. *See also* SATURATED VAPOUR PRESSURE.

boiling tube A large thick-walled glass tube, used to hold samples being strongly heated. *See also* TEST TUBE.

bolometer A device for the detection and measurement of infrared radiation. It consists of a blackened strip of platinum metal, the resistance of which increases as it heats up in response to absorbed radiation. *See also* THERMOPILE.

Boltzmann constant The constant k in the IDEAL GAS EQUATION

$$pV = NkT$$

where N is the number of molecules, p the pressure, V the volume and T the ABSOLUTE TEMPERATURE. The Boltzmann constant is equal to 1.38×10^{-23} JK^{-1}.

Boltzmann factor The factor $e^{-E/kT}$ in the MAXWELL–BOLTZMANN DISTRIBUTION. It describes the number of particles with an energy greater than E when the ABSOLUTE TEMPERATURE is T, where k is the BOLTZMANN CONSTANT.

bolus *See* DIGESTIVE SYSTEM.

bomb calorimeter A strong metal container used to measure the CALORIFIC VALUE of a fuel or food. A sample of known mass is burnt inside the container and the calorific value is calculated from the quantity of heat produced.

bond *See* CHEMICAL BOND.

bond energy The energy released when a chemical bond is formed, or, alternatively, the energy needed to break a chemical bond. Typical values are of the order of a few

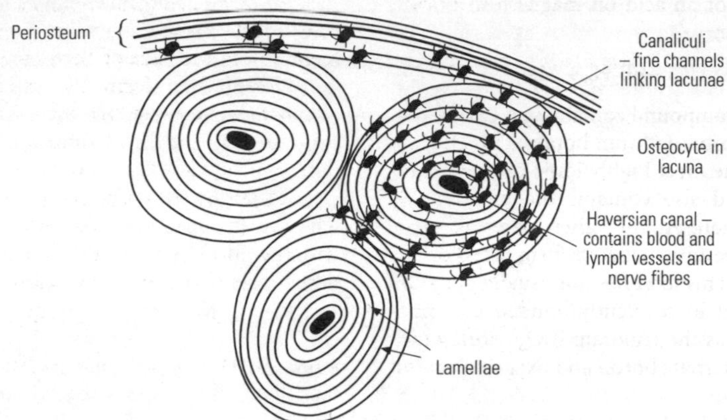

Compact bone.

hundred kilojoules per mole. Bond energies can be deduced from the HEAT OF FORMATION of the compound and the HEAT OF ATOMIZATION of the elements.

The energy of a bond may be slightly altered by its chemical environment, but bond energies generally give a fair indication of the strength of a particular bond, regardless of the molecule in which it occurs.

See also HESS'S LAW.

bonding orbital The lower energy of the two MOLECULAR ORBITALS formed when two atomic ORBITALS overlap. The bonding orbital is the structure that holds covalently bonded atoms together. For the bond to be stable, the bonding orbital must be full and the ANTIBONDING ORBITAL empty. Since each orbital can hold two electrons, this explains why COVALENT BONDS involve the sharing of pairs of electrons in an orbital. *See also* ANTIBONDING ORBITAL.

bone A hard CONNECTIVE TISSUE forming the skeleton of most vertebrates. Bone consists of COLLAGEN fibres, providing strength, impregnated with mineral ions (calcium phosphate and calcium carbonate), which provide the hardness.

Blood, LYMPH VESSELS and nerves are carried through the bone by small tubes called Haversian canals, and the other constituents of bone are arranged in concentric circles or lamellae around these canals. OSTEOCYTES are the bone cells and are found in spaces in the lamellae known as lacunae, which are linked by fine channels called canaliculi. Inside the long bones of limbs there is a hollow shaft called the diaphysis that contains BONE MARROW. Each end of a bone has an expanded head called the EPIPHYSIS. Bone can be a fine spongy network, as at the ends of long bones, designed to transmit the body's weight, or compact to resist bending, as in the shaft of long bones.

Bone can develop from CARTILAGE by the process of OSSIFICATION. Bone can also form directly from other connective tissues, and is then plate-like in shape, for example the skull. Bone shape can be remodelled during growth by OSTEOCLASTS.

bone marrow In vertebrates, a soft material in the centre of some bones that produces blood cells. Yellow marrow is found at the centre of the long bones and consists mainly of fat cells. It can make some WHITE BLOOD CELLS. Red marrow produces RED BLOOD CELLS and is located at the ends of the bone.

boot, *bootstrap* The initial stages of starting up a computer. The act of switching on the power automatically starts a small boot program that loads the OPERATING SYSTEM.

bootstrap *See* BOOT.

borane, *boron hydride* Any one of a family of BINARY COMPOUNDS of boron with hydrogen. All are readily oxidized by atmospheric oxygen. The simplest is diborane, B_2H_6, formed by the action of an acid on magnesium boride, for example:

$$MgB_2 + 6HCl \rightarrow B_2H_6 + MgCl_2 + Cl_2$$

borate Any compound containing a CATION that contains boron. Lithium borate, $LiB(OH)_4$, is amongst the most highly ionic of these compounds and also contains the most highly HYDRATED ion. At the other extreme, many borates have an ANHYDROUS form based on the BO_3^{2-} ion. This ion does not usually occur on its own but in covalently bonded combinations such as the triborate $B_3O_6^{3-}$ ion, which contains alternate boron and oxygen atoms in a ring.

borax Hydrated sodium borate, $Na_2B4O_7.10H_2O$. Borax is a white solid that loses water on heating; RELATIVE DENSITY 1.7. Borax occurs naturally and is an important source of boron. It is used directly in the manufacture of glass and ceramics and its ability to dissolve metal oxides makes it an important flux for soldering.

boric acid Any acid containing boron and oxygen. In particular, H_3BO_3, a white solid; melting point 169°C; RELATIVE DENSITY 1.4. It occurs naturally in small quantities, but can be manufactured by the reaction between SULPHURIC ACID and BORAX:

$$Na_2B_4O_7 + H_2SO_4 + 5H_2O \rightarrow 4H_3BO_3 + Na_2SO_4$$

On heating, H_3BO_3 loses water to form a polymeric form (*see* POLYMER) of boric acid, $(HBO_2)_n$.

boride Any BINARY COMPOUND containing a metal and boron, such as zinc boride, ZnB_2. Borides are unusual in that they combine the characteristic high melting points and hardness of ceramic materials with the good electrical and thermal conductivities of a metal.

Born–Haber cycle A cycle of reactions used to calculate the LATTICE ENERGY of an IONIC crystal LATTICE. The lattice energy (ΔH_L) of the compound NaCl, for example, is the ENTHALPY of formation of the solid lattice from its ions; that is:

$$Na^+ (g) + Cl^- (g) \rightarrow NaCl (s) \; \Delta H_L$$

The STANDARD ENTHALPY OF FORMATION (ΔH_f) of the solid is the enthalpy of the reaction

$$Na (s) + {}^1/_2Cl_2 (g) \rightarrow NaCl (s) \; \Delta H_f$$

which can be determined experimentally. The Born–Haber cycle involves equating this enthalpy to the sum of the enthalpies of the steps needed to form the solid from the elements. These are: (i) the atomization of sodium ΔH_1; (ii) the atomization of chlorine, ΔH_2; (iii) the IONIZATION of sodium (the first IONIZATION ENERGY) ΔH_3; (iv) the ionization of chlorine (the first ELECTRON AFFINITY) ΔH_4; and (v) the formation of the lattice ΔH_L. The lattice energy can therefore be calculated from

$$\Delta H_L = \Delta H_f - (\Delta H_1 + \Delta H_2 + \Delta H_3 + \Delta H_4)$$

boron (B) The element with ATOMIC NUMBER 5; RELATIVE ATOMIC MASS 10.8; melting point 2,079°C; boiling point 2,550°C; RELATIVE DENSITY 2.4. It is a non-metal, occurring as a brown powder or as clear crystals. In nature it occurs in the mineral BORAX. Boron is widely used in the ceramics and glass industries and its neutron-absorbing properties led to its use in CONTROL RODS in some NUCLEAR REACTORS.

boron carbide (B_4C) A hard black solid; melting point 2,350°C; boiling point 3,500°C (approx.); RELATIVE DENSITY 2.5. Boron carbide is made by the reduction of boron oxide with carbon:

$$2B_2O_3 + 4C \rightarrow B_4C + 3CO_2$$

Boron carbide is extremely hard and is used as an abrasive.

boron hydride *See* BORANE.

Bosch process An industrial process, now largely obsolete, for extracting hydrogen from WATER GAS, a mixture of carbon monoxide and hydrogen. Water gas and steam are passed over a catalyst at around 450°C. Iron, with chromium(III) oxide as a PROMOTER is often used as the catalyst. The carbon dioxide in the water gas is oxidized, increasing the hydrogen content at the expense of the poisonous carbon monoxide:

$$CO + H_2O \rightarrow CO_2 + H_2$$

The carbon dioxide is removed by POTASSIUM CARBONATE solution in a SCRUBBER. The Bosch process was formerly used to produce hydrogen for the HABER PROCESS in the manufacture of ammonia. However, the necessary hydrogen is now largely produced from natural gas.

Bose–Einstein condensation The accumulation of large numbers of BOSONS in the lowest available ENERGY LEVEL at low temperatures. The effect accounts for SUPERFLUIDITY in liquid helium and SUPERCONDUCTIVITY. At very low temperatures (close to ABSOLUTE ZERO), condensates form in which thousands of atoms accumulate to form what is effectively a single entity (a superatom). The effect only occurs with bosons, as FERMIONS are kept in higher energy levels even at low temperatures by the PAULI EXCLUSION PRINCIPLE.

boson Any particle with integral SPIN; that is, a spin which is an even number times $h/4\pi$, where h is PLANCK'S CONSTANT. *See also* BOSE–EINSTEIN CONDENSATION. *Compare* FERMION.

botany The branch of biology that deals with the study of plants. It includes many areas, such as the study of plant structure, function, geographical distribution and classification.

bottom, ***beauty*** A flavour of QUARK.

boundary layer A thin layer of fluid that moves with an object, such as an aircraft, as it passes through the fluid. In TURBULENT flow, the boundary layer becomes detached from the surface. By delaying the point at which the boundary layer becomes detached, it is possible to reduce the amount of turbulent flow and thus reduce DRAG.

Bourdon gauge A pressure measuring device in which a curved metal tube straightens out slightly under the influence of the applied pressure. A system of gears converts this small motion into a larger motion of a pointer across a scale.

bovine spongiform encephalopathy (BSE) A disease of cows (popularly called 'mad cow disease') similar to SCRAPIE in sheep and CREUTZFELDT-JAKOB DISEASE (CJD) in humans. All three diseases are characterized by the progressive degeneration of the CENTRAL NERVOUS SYSTEM and the spongy appearance of the brain upon examination after death. The spongy appearance is due to the presence of numerous holes in the tissue. It is thought to be caused not by a bacterium or a virus but by a 'prion protein', which is a self-replicating protein fragment, or an agent that can switch on a latent gene to cause production of the prion protein, which infects the brain and SPINAL CORD of its victims.

bovine somatotrophin A genetically engineered hormone given to cows to increase milk yield. *See also* BIOTECHNOLOGY.

Bowman's capsule A cup-shaped structure in the KIDNEY that surrounds the GLOMERULUS and forms part of the NEPHRON tubule. Certain smaller substances (of RELATIVE MOLECULAR MASS less than 68,000), such as GLUCOSE, vitamins, amino acids, hormones, UREA, URIC ACID and water, can pass through the wall of the glomerulus and into the Bowman's capsule as a filtrate. Blood cells and larger PLASMA PROTEINS remain in the blood. The cut-off size can be increased by hormonal or nervous signals increasing the blood pressure within the capsule. The filtrate then moves through the tubule, eventually forming urine. There are over a million Bowman's capsules in the outer region of the kidney. *See also* URINARY SYSTEM.

Boyle's law For a fixed mass of an IDEAL GAS at constant temperature, the pressure is inversely proportional to the volume, i.e. the pressure multiplied by the volume is a constant. *See also* CHARLES' LAW, GAS CONSTANT, IDEAL GAS EQUATION, PRESSURE LAW.

bract A small leaf in whose AXIL (between the bract and stem) a flower or branch develops.

Bragg's law In X-RAY DIFFRACTION, the relationship between the angles at which CONSTRUCTIVE INTERFERENCE is observed (measured from the surface of a crystal) and the separation of the planes of atoms in the crystal.

For a beam of X-rays of wavelength λ, striking a crystal at an angle θ to the planes of the crystal, which have a separation d between one plane and the next, constructive interference will occur if

$$2d \sin\theta = n\lambda$$

where n is a whole number.

brain In higher animals, a large mass of interconnected NEURONES, forming part of the CENTRAL NERVOUS SYSTEM and controlling the actions of the whole body. It is surrounded by the MENINGES and is protected further by the bones of the skull. Cavities called VENTRICLES are filled with CEREBROSPINAL FLUID, which supplies the brain with nutrients and respiratory gases and removes waste by exchanging these materials with the blood (there are numerous capillaries in the ventricles).

There are three main regions of the brain: the FOREBRAIN, the HINDBRAIN and the MIDBRAIN. *See also* BRAINSTEM, CEREBELLUM, CEREBRUM, GREY MATTER, HYPOTHALAMUS, MEDULLA OBLONGATA, PINEAL GLAND, SPINAL CORD, THALAMUS, WHITE MATTER.

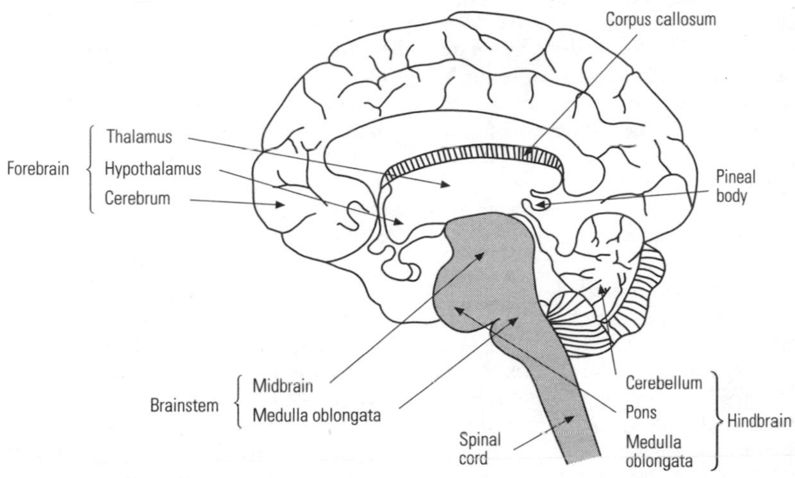

Cross-section through the human brain.

brainstem The region where the SPINAL CORD merges with the under surface of the brain. It consists of the MEDULLA OBLONGATA and the MIDBRAIN. It controls the body's vital functions, such as breathing, heart rate and blood pressure, and maintains wakefulness by a system of nerve cells which is spread diffusely throughout the brainstem. In many countries death of the brainstem is considered synonymous with death of the individual, although the heartbeat can be maintained by life-support equipment.

brake horsepower (bhp) The output of a machine, especially an INTERNAL COMBUSTION ENGINE, measured in HORSEPOWER using a DYNAMOMETER.

brass An alloy of copper with around 20 per cent zinc. Harder than copper but still soft enough to be worked easily, it is used for small metal parts but is too expensive and too weak to find widespread use.

breathing, *external respiration* In terrestrial animals, the process by which air is taken into the LUNGS to provide oxygen and then expelled to release gaseous wastes. It is sometimes called external respiration, to distinguish it from true RESPIRATION, which is internal and is the use of oxygen by cells to make energy.

Air passes through a series of tubes and organs (*see* RESPIRATORY SYSTEM) to reach the lungs, where oxygen diffuses from the air across the thin, moist EPITHELIAL layer of the alveoli (*see* ALVEOLUS), and from here across blood capillaries to combine with HAEMOGLOBIN in the blood for circulation all over the body. Likewise, carbon dioxide diffuses from the blood back across the alveoli for expulsion from the lungs. This process is also called gas exchange.

For air to enter the lungs (inspiration) the pressure inside them must be lower than the ATMOSPHERIC PRESSURE. This is brought about by the action of the INTERCOSTAL MUSCLES causing the ribs to move upwards and outwards. The DIAPHRAGM also moves and the volume of the THORAX increases to allow the lungs to fill with air and expand. Breathing out (expiration) occurs when the muscles relax, causing the ribs to move down and so forcing the air out of the lungs.

The rate of breathing is under involuntary control by the breathing centre in the HINDBRAIN and by CHEMORECEPTORS that detect the level of carbon dioxide in the blood. If carbon dioxide levels increase, for example during exercise, then NERVE IMPULSES stimulate the breathing centre in the brain to increase the breathing rate. There is also some voluntary control over breathing.

Human lungs have a total volume of about 5,000 cm^2 of air, but following a forced exhalation 3,500 cm^2 is the most that can be exchanged. This is called the vital capacity. A residual volume of 1,500 cm^2 therefore remains in the lungs, but this is continually circulated as it mixes with fresh air during normal breathing. The volume of air exchanged during normal breathing is called the tidal capacity, and is about 450 cm^2.

breeding The process of bearing offspring. *See also* ARTIFICIAL SELECTION, SEXUAL REPRODUCTION.

bremsstrahlung X-RAYS produced by rapidly decelerating electrons, as in an X-RAY TUBE, for example.

Brewster angle The ANGLE OF INCIDENCE at which light falls on a transparent material such that the reflected and REFRACTED RAY are at right angles to one another. At this angle, the reflected light will be completely POLARIZED, with its electric field parallel to the reflecting surface. *See also* BREWSTER'S LAW.

Brewster's law Light reflected from a transparent material is completely POLARIZED when the reflected and refracted rays are at right angles to one another. *See also* BREWSTER ANGLE.

bridge rectifier A RECTIFIER circuit containing four DIODES in an arrangement similar to the WHEATSTONE BRIDGE. It has the advantage that current flows from the ALTERNATING CURRENT supply to the load throughout the a.c. cycle. *See also* PN JUNCTION DIODE.

brine A concentrated solution of SODIUM CHLORIDE (common salt) in water. *See also* MERCURY-CATHODE CELL.

British thermal unit (Btu) An obsolete unit of energy, originally defined as the energy required to heat one pound of water by one degree FAHRENHEIT. One Btu is equivalent to 1,055 JOULES.

broadcasting The transmission of signals in all directions for general reception. *See also* RADIO.

bromic acid Name given to bromic(I) acid (hypobromous acid), HBrO, or, more usually, bromic(V) acid, $HBrO_3$.

Bromic(I) acid is a weak acid, but a strong OXIDIZING AGENT. It is formed by the action of hydrogen peroxide on bromine:

$$H_2O_2 + Br_2 \rightarrow 2HBrO$$

Bromic(V) acid, stable only in aqueous solution, is a reducing agent and a strong acid. It can be produced by the reaction between barium bromate(V) and sulphuric acid:

$$Ba(BrO_3)_2 + H_2SO_4 \rightarrow 2HBrO_3 + BaSO_4$$

bromide Any BINARY COMPOUND containing BROMINE.

bromine (Br) The element with ATOMIC NUMBER 35; RELATIVE ATOMIC MASS 79.91; melting point –7°C; boiling point 58°C. It is a HALOGEN and is a deep brown, volatile liquid, producing brown fumes of bromine vapour. It is extracted as sodium bromide from the salts present in sea water. Bromine is extracted from sodium bromide by ELECTROLYSIS. Silver bromide is an important component of many photographic materials. Large quantities of bromine are also used in the manufacture of halogenated HYDROCARBONS (*see* HALOGENATION).

bromine water A red-brown solution of bromine in water. It is used in organic chemistry to test for UNSATURATED COMPOUNDS that decolourize the water when they react with bromine, for example:

$$CH_2{=}CH_2 + Br_2 \rightarrow CH_2BrCH_2Br$$

2-bromo-2-chloro-1,1,1-trifluoroethane *See* HALOTHANE.

bromoethane (C_2H_5Br) A HALOGENOALKANE. Bromoethane undergoes SUBSTITUTION REACTIONS more readily than ELIMINATION reactions. Examples of such reactions are: the reaction with aqueous sodium hydroxide to give ethanol,

$$OH^- + C_2H_5Br \rightarrow C_2H_5OH + Br^-$$

the reaction with a CYANIDE to give propanenitrile,

$$NC^- + C_2H_5Br \rightarrow C_2H_5CN + Br^-$$

the reaction with sodium methoxide, $NaOCH_3$, in methanol, to give methoxyethane,

$$CH_3O^- + C_2H_5Br \rightarrow C_2H_5OCH_3 + Br^-$$

which is an example of WILLIAMSON ETHER SYNTHESIS.

bronchiole In vertebrates, one of many small tubes found in the lung that leads off from the larger BRONCHUS. It branches further and finally terminates in the alveoli (*see* ALVEOLUS).

bronchus (*pl.* ***bronchi***) In vertebrates, one of two tubes branching off from the TRACHEA, which carry the air into each of the lungs. To prevent their collapse during BREATHING, the bronchi possess cartilaginous rings for extra rigidity. Glands are present that secrete a sticky liquid called MUCUS to collect dust and other unwanted particles. These are then pushed out towards the mouth for swallowing, aided by CILIA on the walls of the bronchi. The bronchi divide further into small tubes called BRONCHIOLES and then into alveoli (*see* ALVEOLUS). *See also* RESPIRATORY SYSTEM.

bronze An alloy of copper with up to about 30 per cent tin. Bronze is far harder than tin and was important historically as the first alloy to be made by man, since it melts at a far lower temperature than iron.

Brownian motion The rapid, random motion of small particles suspended in a fluid, which is seen when such particles are viewed through a microscope. Brownian motion provides evidence for KINETIC THEORY, as it is caused by the bombardment of the particles by far smaller, but fast moving, molecules in the suspending fluid. Brownian motion was first observed in pollen grains suspended in water, but is now usually demonstrated with smoke particles in air.

brush border The surface of many eukaryotic (*see* EUKARYOTE) cells, particularly EPITHELIUM, as it appears under an ELECTRON MICROSCOPE due to the presence of millions of minute projections called MICROVILLI. These increase the surface area of, for example, the SMALL INTESTINE for absorption of food products.

Bryophyta A PHYLUM of the plant KINGDOM comprising three classes: HEPATICAE (liverworts), MUSCI (mosses) and ANTHOCEROTAE (hornworts). Bryophytes are small terrestrial plants growing in moist HABITATS, possessing no roots or vascular (conducting) system. Their life cycle shows ALTERNATION OF GENERATIONS. Plants of this phylum consist of simple stems and leaves, or a THALLUS, anchored by RHIZOIDS.

bryophyte A member of the PHYLUM BRYOPHYTA.

BSE *See* BOVINE SPONGIFORM ENCEPHALOPATHY.

bubble chamber An PARTICLE DETECTOR comprising a vessel containing liquid, often liquid hydrogen, held under pressure by a piston. When the piston is withdrawn, reducing the pressure, the liquid starts to boil, but it does so first on any CONDENSATION NUCLEI (irregularities in the liquid). The trail of IONIZATION left by any charged particle that has recently passed through the chamber acts as one source of condensation nuclei, giving rise to a trail of bubbles, which are photographed. The piston is then forced back into the chamber, causing the bubbles to collapse, ready for the next event.

bubble-jet printer A computer PRINTER that produces its text using a fine spray of ink droplets that are deflected by electric fields to form characters on the page.

bubble memory A form of magnetic computer MEMORY. Information is stored in microscopic regions (called bubbles) of reversible magnetic field in a magnetic material. Bubble memory is non-volatile – in other words, the information is not lost when the computer is switched off.

buccal cavity *See* MOUTH.

buckminsterfullerene A recently discovered ALLOTROPE of carbon. Each molecule contains 60 carbon atoms at the vertices of a roughly spherical structure, rather like a soccer ball. Several similar allotropes have since been discovered, known as collectively as fullerenes or as buckyballs if roughly spherical and buckytubes if elongated.

buckyball *See* BUCKMINSTERFULLERENE.

buckytube *See* BUCKMINSTERFULLERENE.

bud In plants, a small, undeveloped shoot containing immature leaves or petals. A bud may develop at the tip of a stem or branch, or in the AXILS of a leaf.

budding A method of ASEXUAL REPRODUCTION where a small outgrowth develops from the parent cell and eventually separates, sometimes after becoming more differentiated and forming buds of its own. Yeast and *Hydra* are examples of organisms that can reproduce by budding.

buffer 1. (*chemistry*) A solution designed to maintain a fairly uniform PH level despite other changes that may take place. Buffer solutions are made by mixing a WEAK ACID or a weak BASE with a salt of the same acid or base. For example, an acidic buffer solution may be made from ethanoic acid and sodium ethanoate. The sodium ethanoate is fully ionized in solution:

$$CH_3COONa \rightarrow CH_3COO^- + Na^+$$

while the ethanoic acid is only weakly ionized:

$$CH_3COOH + H_2O \Leftrightarrow CH_3COO^- + H_3O^-$$

The resulting mixture contains large numbers of ethanoate ions, mostly coming from the sodium ethanoate. The free ethanoate ions will tend to associate with any excess hydrogen ions thus reducing excess acidity, whilst any HYDROXIDE ions will be mopped up by the ethanoic acid. Many commercially made buffer solutions are available.

Natural buffers exist in many biological systems where enzymes are highly sensitive to changes in pH, and similar solutions are widely used in medicine. The most common natural buffers involve HYDROGENCARBONATE (HCO_3^-) and hydrogenphosphate (HPO_4^{2-}) ions.

2. (*computing*) A temporary computer memory in which data is stored while it is waiting to be used.

bug Any fault or error in a computer hardware or software.

bulb A small, underground stem with fleshy leaves, which are a food reserve, and roots growing from its base. Examples include the daffodil and onion. New plants arise from buds (small, undeveloped shoots) that grow between the leaves and then develop into new bulbs. *See also* VEGETATIVE REPRODUCTION.

bulimia nervosa An eating disorder, usually due to psychological problems or stress, characterized by episodes of overeating (binging) followed by self-induced vomiting, fasting, the use of laxatives or excessive exercise.

bulk modulus A measure of the COMPRESSIBILITY of a material when subjected to an increase in external pressure tending to decrease its volume. If a pressure change Δp causes a change in volume ΔV in a sample of volume V, then the bulk modulus is $V\Delta p/\Delta V$.

bumping The violent boiling of a SUPERHEATED liquid. *See also* ANTI-BUMPING GRANULES.

bundle of His The collective name for the PURKINJE FIBRES.

bunsen burner A simple gas burner widely used as a source of heat for chemistry experiments. Gas, usually METHANE, emerges from a small hole and is mixed with air, which enters via an adjustable hole. The gas and air mixture burn at the end of a short metal tube. A large amount of air produces a hot roaring blue flame with a paler cone of unburnt gas. Less air produces a quieter, cooler flame. If the air hole is closed, a highly luminous yellow flame is produced, which will leave a deposit of unburnt carbon on a cool surface.

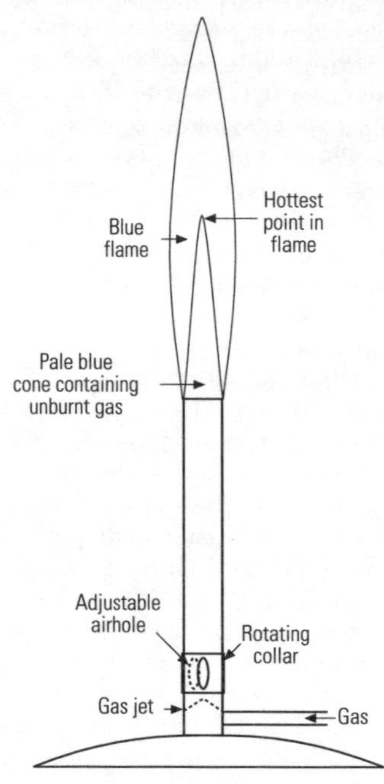

Bunsen burner.

buoyancy The UPTHRUST on a body immersed in a fluid, equal to the weight of the fluid displaced by the body. If the object is of a sufficiently low density, the upthrust will exceed the weight and the object will float. *See also* ARCHIMEDES' PRINCIPLE, FLOTATION.

burette A glass tube with a calibrated scale and a tap, used to measure the amount of material involved in certain reactions, such as TITRATIONS.

bus The central wiring that connects all the main components of a computers, such as the MICROPROCESSOR, memory chips and PORTS.

butadiene, ***buta-1,3-diene*** (CH_2=CH-CH=CH_2) A colourless gas; boiling point –5°C. It is made from PETROLEUM by CRACKING. It is used in the preparation of artificial RUBBER and RESINS. Butadiene is often combined with PHENYLETHENE (styrene) to give the COPOLYMER SBR (styrene-butadiene rubber) or with phenylethene and PROPENENITRILE (acrylonitrile) to make ABS (ACRYLONITRILE-BUTADIENE-STYRENE).

butane (C_4H_{10}) An ALKANE, the first in the series to exhibit isomerism (*see* ISOMER). The isomers are normal or *n*-butane ($CH_3CH_2CH_2CH_3$) and isobutane ($CH_3CH(CH_3)CH_3$). Both are colourless gases found in NATURAL GAS which may be separated by FRACTIONAL DISTILLATION. The boiling point of *n*-butane is –0.3°C and of isobutane –10.3°C.

Normal butane is obtained from natural gas and used as a domestic fuel for heating or portable cookers. Isobutane is a by-product of PETROLEUM manufacture.

butanol, *butyl alcohol* (C_4H_9OH) The fourth member in the series of ALCOHOLS, a colourless liquid important as a solvent for RESINS and lacquers. There are four structural ISOMERS of butanol.

butanone, *methyl ethyl ketone (M.E.K.)* (C_4H_8O, $CH_3COCH_2CH_3$) A colourless liquid with a pleasant odour; boiling point 80°C. It is one of the products of the DESTRUCTIVE DISTILLATION of wood. It is used as a solvent in the manufacture of RESINS and ACETATE film.

butene (C_4H_8) The fourth member of the HYDROCARBON series of ALKENES. It is a colourless gas with an unpleasant odour. Three ISOMERS are obtained from PETROLEUM by CRACKING, which have some industrial use as various POLYMERS.

butyl alcohol *See* BUTANOL.

byte, *word* A number made up of several (usually 4) BINARY digits. In computing, a byte is treated as a separate unit by the CENTRAL PROCESSING UNIT.

C

C The most popular computer programming language for PERSONAL COMPUTERS and WORKSTATIONS. Originally written for UNIX, it is used to write fast and efficient SYSTEM PROGRAMS for a variety of OPERATING SYSTEMS.

C_3 plant Any plant that produces a three-carbon intermediate (glycerate 3-phosphate) during the fixation of carbon dioxide in the light-independent stage of PHOTOSYNTHESIS. Most temperate plants and about 85 per cent of all plant species are C_3 plants. *Compare* C_4 PLANT. *See also* CALVIN CYCLE.

C_4 plant Any plant that can produce a four-carbon intermediate (oxaloacetic acid) during the fixation of carbon dioxide in the light-independent stage of PHOTOSYNTHESIS (*see* CALVIN CYCLE). Tropical plants, such as sugar cane, and many cereals are C_4 plants. In these plants, the C_4 system operates in addition to the C_3 PLANT system.

The advantage of the C_4 system is that carbon dioxide can be trapped at lower concentrations than that needed by C_3 plants. This is particularly useful for tropical plants, where carbon dioxide may be in short supply, and the leaves of C_4 plants are modified to suit this method of carbon dioxide fixation. The C_4 system is also more efficient at higher temperatures. PHOTORESPIRATION is avoided in the C_4 system.

cache memory A small area of very fast IMMEDIATE ACCESS MEMORY where frequently used instructions and data are stored. This reduces the ACCESS TIME of these instructions and allows the program to run more quickly.

CAD *See* COMPUTER-AIDED DESIGN.

cadmium (Cd) The element with ATOMIC NUMBER 48; RELATIVE ATOMIC MASS 112.4; melting point 321°C; boiling point 765°C; RELATIVE DENSITY 8.7. Cadmium is a soft silvery white metal. Chemically, it is similar to zinc and the two elements often occur together. Cadmium is widely used in batteries and for ELECTROPLATING. Its neutron-absorbing properties allow it to be used in CONTROL RODS in some NUCLEAR FISSION plants. Cadmium is a cumulative poison and its widespread use in industry has led to concerns about the pollution of water supplies by the dumping of cadmium compounds.

cadmium sulphide (CdS) A brown or yellow solid, which decomposes on heating; RELATIVE DENSITY 4.8. Cadmium sulphide occurs naturally as the mineral GREENOCKITE, which can be roasted to provide cadmium. In its AMORPHOUS form, it is PHOTOCONDUCTIVE and is used in the manufacture of LIGHT-DEPENDENT RESISTORS.

caecum In humans, a slight expansion between the small and large intestines, leading to the APPENDIX, that has no function. However, in HERBIVORES it is used (as is the appendix) to assist in the digestion of CELLULOSE in plant cell walls, which requires the presence of micro-organisms with the enzyme cellulase that live in the caecum.

caesium, *cesium* (Cs) The element with ATOMIC NUMBER 55; RELATIVE ATOMIC MASS 132.9; melting point 28°C; boiling point 678°C; RELATIVE DENSITY 1.9. Caesium is the most reactive of the ALKALI METALS, exploding violently on contact with water. The natural ISOTOPE, caesium–133, is stable and occurs in a number of minerals. There are 15 RADIOISOTOPES. Caesium–137 is a common product of NUCLEAR FISSION, and causes particular concern in the handling of nuclear waste. It has a HALF-LIFE of 20 years, and, being chemically similar to POTASSIUM, it is easily incorporated into the FOOD CHAIN. Caesium is used in PHOTOELECTRIC CELLS and as a catalyst.

calcite, *calcspar* The mineral calcium carbonate, $CaCO_3$. It is the main constituent of LIMESTONE and MARBLE.

calcitonin A HORMONE produced by the THYROID GLAND that reduces the levels of calcium ions in the blood. *See* PARATHYROID GLAND.

calcium (Ca) The element with ATOMIC NUMBER 20; RELATIVE ATOMIC MASS 40.1; melting point 840°C; boiling point 1,484°C; RELATIVE DENSITY

1.6. Calcium is a soft white ALKALINE EARTH metal that tarnishes rapidly in air. It is the fifth most abundant element in the Earth's CRUST, occurring mostly as the carbonate, $CaCO_3$, in limestone, coral, marble and CALCITE. It is extracted by reacting the carbonate with hydrochloric acid to form calcium chloride:

$$CaCO_3 + 2HCl \rightarrow CaCl_2 + CO_2 + H_2O$$

The fused chloride is then electrolysed.

Calcium is an essential element for living organisms. In animals it is an important constituent of bones and teeth.

calcium acetylide *See* CALCIUM DICARBIDE.

calcium carbide *See* CALCIUM DICARBIDE.

calcium carbonate ($CaCO_3$) A white solid; RELATIVE DENSITY 2.8; decomposes on heating. Calcium carbonate occurs naturally in limestone, chalk and marble, and as a purer crystalline form in the mineral CALCITE. The passage of rainwater through rocks containing calcium carbonate, which is very slightly soluble, is responsible for TEMPORARY HARDNESS in water. Industrially, calcium carbonate is roasted to produce calcium oxide (quicklime):

$$CaCO_3 \rightarrow CaO + CO_2$$

and is the main raw material in the SOLVAY PROCESS for the manufacture of sodium carbonate.

calcium chloride ($CaCl_2$) A white crystalline solid; melting point 772°C, boiling point 7,600°C; RELATIVE DENSITY 2.2 (ANHYDROUS). Calcium chloride is DELIQUESCENT and a number of HYDRATED crystalline forms exist. Anhydrous calcium chloride is used as a drying agent. Calcium chloride is produced as a by-product of the SOLVAY PROCESS. In the laboratory it can be prepared by the action of hydrochloric acid on calcium carbonate:

$$CaCO_3 + 2HCl \rightarrow CaCl_2 + H_2O$$

Fused calcium chloride is electrolysed as a source of metallic calcium.

calcium cyanamide ($CaCN_2$) A white solid that sublimes at 1,150°C. Calcium cyanamide can be prepared by heating calcium acetylide in nitrogen:

$$CaC_2 + N_2 \rightarrow CaCN_2 + C$$

Calcium cyanamide decomposes on contact with water to give calcium carbonate and ammonia:

$$CaCN_2 + 3H_2O \rightarrow CaCO_3 + 2NH_3$$

calcium dicarbide, *calcium acetylide, calcium carbide* (CaC_2) A white or grey solid; melting point 450°C; boiling point 2,300°C. Calcium dicarbide is an ionic salt that can be manufactured by reducing calcium oxide with coke at high temperatures:

$$CaO + 3C \rightarrow CaC_2 + CO$$

Calcium dicarbide reacts with water to produce ETHYNE:

$$CaC_2 + 2H_2O \rightarrow Ca(OH)_2 + C_2H_2$$

In addition to being an important source of ethyne for the synthesis of other organic reagents, this reaction is also used to produce a simple portable source of combustible gas in remote locations.

calcium fluoride (CaF_2) A white crystalline solid; melting point 1,360°C; boiling point 2,500°C; RELATIVE DENSITY 3.2. Calcium fluoride occurs naturally as the mineral fluorite and is an important source of fluorine, which is extracted by electrolysing the fused salt.

calcium hydroxide, *slaked lime* ($Ca(OH)_2$) A white solid that decomposes on heating; RELATIVE DENSITY 2.2. It is formed by the addition of water to calcium oxide:

$$CaO + H_2O \rightarrow Ca(OH)_2$$

Calcium hydroxide is slightly soluble in water, the solution being known as lime water. Calcium hydroxide is widely used in agriculture as an alkali to neutralize excess acidity in some soils.

calcium nitrate ($Ca(NO_3)_2$) A white solid; melting point 561°C; decomposes on further heating; RELATIVE DENSITY 2.5. Calcium nitrate is a DELIQUESCENT solid and several hydrated crystalline forms exist. It can be prepared by the reaction of nitric acid on calcium carbonate:

$$CaCO_3 + 2HNO_3 \rightarrow Ca(NO_3)_2 + CO_2 + H_2O$$

Calcium nitrate decomposes on strong heating:

$$2Ca(NO_3)_2 \rightarrow 2CaO + 4NO_2 + O_2$$

calcium oxide, *quicklime* (CaO) A white solid; melting point 2,600°C; boiling point 2,850°C; RELATIVE DENSITY 3.4. Calcium oxide can be prepared by burning calcium in air or industrially by heating calcium carbonate:

$$CaCO_3 \rightarrow CaO + CO_2$$

Calcium oxide is used in the manufacture of calcium hydroxide and in many metal extraction processes to form a SLAG with SILICA impurities in metal ores.

calcium phosphate ($Ca_3(PO_4)_2$) A white solid; decomposes on heating; RELATIVE DENSITY 3.1. Calcium phosphate is the major mineral constituent of animal bones, and is widely used as a fertilizer.

calcium sulphate ($CaSO_4$) A white solid; melting point 1,450°C; decomposes on further heating; RELATIVE DENSITY 3.0. Calcium sulphate occurs naturally as the mineral gypsum, which is comprised mostly of the DIHYDRATE $CaSO_4.2H_2O$. On heating gently, this loses water to form another HYDRATED form, $2CaSO_4.H_2O$, known as plaster of Paris. When water is added to plaster of Paris it forms a paste, which sets hard.

Calcium sulphate is an important raw material in the manufacture of sulphuric acid. It is only slightly soluble, and is a source of PERMANENT HARDNESS in water. In the laboratory, it can be prepared as a PRECIPITATE in the reaction between aqueous calcium chloride and sodium sulphate:

$$CaCl_2 + Na_2SO_4 \rightarrow CaSO_4 + 2NaCl$$

calcspar *See* CALCITE.

calculus The branch of mathematics that deals with functions that change in a continuous way, particularly the rate of change of a function with a change in an independent variable.

If y is a function of x, then the rate of change of y with respect to a change in x is defined as the DERIVATIVE of y, denoted by the symbol y' or dy/dx. The process of finding a derivative is called DIFFERENTIATION. The reverse process of finding a function with a particular derivative is called INTEGRATION.

Graphically, the process of differentiation may be regarded as equivalent to finding the gradient of the TANGENT to a curve at a given point, and integration as finding the area bounded by the curve and other specified lines.

See also ANALYSIS, INTEGRAL.

Calgon *See* WATER SOFTENING.

calibration 1. The process of using a measuring device to measure known quantities in order to check or adjust the instrument concerned.

2. A marking on the scale of an instrument representing a specified numerical value for the quantity being measured.

californium (Cf) The element with ATOMIC NUMBER 98. It does not occur in nature, but nine isotopes have been synthesized. The longest-lived ISOTOPE has a HALF-LIFE of 800 years. Some of its isotopes decay by SPONTANEOUS FISSION, which makes it useful as a neutron source.

calorie (cal) A unit of quantity of heat in C.G.S. UNITS. One calorie is the amount of heat needed to raise the temperature of one gram of water by 1°C. The calorie has been largely replaced by the JOULE; it is roughly equivalent to 4.2 J.

Calorie (kcal) A unit sometimes used to specify the energy value of foods. One Calorie (capital c) is equivalent to 1,000 CALORIES (small c).

calorific value The energy content of a fuel or food. It is the amount of heat generated by completely burning a given mass of fuel (which can be food) in a piece of apparatus called a bomb CALORIMETER. Thus it is equal to the HEAT OF COMBUSTION. Calorific value is measured in JOULES per kilogram.

calorimeter A container for performing experiments related to heat transfer and temperature changes, such as the measurement of SPECIFIC HEAT CAPACITY and LATENT HEAT. Calorimeters are generally made of a metal (a good conductor of heat, so the entire vessel reaches the same temperature), of known heat capacity. *See also* CALORIFIC VALUE.

Calvin cycle Another name for the light-independent reaction of PHOTOSYNTHESIS, after Melvin Calvin (1911–) who established the details of the reactions occurring. He did this by exposing plants to radioactively labelled carbon dioxide (carbon–14 dioxide), allowing them to photosynthesize and examining the products.

Carbon dioxide from the air diffuses into a leaf through the stomata (*see* STOMA) and then dissolves and diffuses through the CELL MEMBRANE and the CHLOROPLAST membrane into the STROMA of the chloroplast. The light-independent stage of photosynthesis occurs in the stroma. In most plants (C_3 PLANTS), carbon dioxide combines with a five-carbon compound called ribulose bisphosphate to

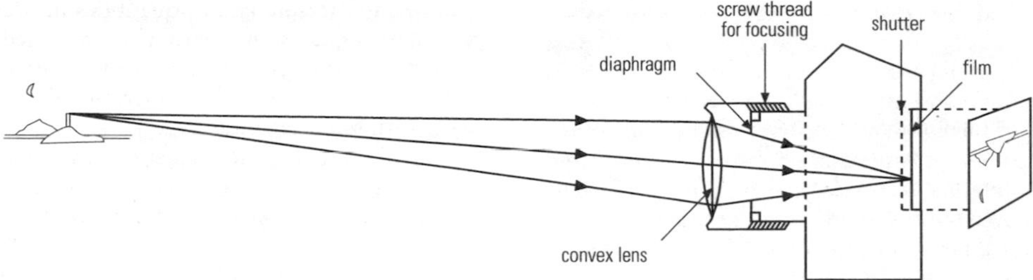

Principle of the camera.

form a 6-carbon intermediate that is unstable and breaks down into two molecules of the three-carbon compound glycerate 3-phosphate (GP). This is then converted, in the presence of ATP and reduced NADP, into TRIOSE phosphate, which combines to form HEXOSE sugars. These polymerize to yield STARCH for the plant to store. Some of the triose phosphate is used to regenerate ribulose bisphosphate.

In C_4 PLANTS, carbon dioxide combines with the substance phosphoenolpyruvate (PEP) instead of ribulose bisphosphate and produces the four-carbon oxaloacetic acid instead of GP.

calyx The collective term for the SEPALS of a flower.

cambium In plants, a layer of actively dividing cells in stems and roots that gives rise to the secondary growth occurring in woody PERENNIALS. It is a lateral MERISTEM responsible for increased girth. Vascular cambium is a layer of cells separating the XYLEM and PHLOEM within the VASCULAR BUNDLE, and gives rise to new or secondary xylem and phloem. In trees the secondary xylem laid down by the vascular cambium forms a new layer of wood annually on the outside of the old wood, and these form the annual rings seen when a tree is felled. Cork cambium gives rise to secondary layers of bark and cork.

Cambrian The PERIOD of geological time from 570 to 510 million years ago at the beginning of the PALAEOZOIC ERA. The first marine animals with hard shells evolved during this time.

camera A device used to record an optical image of an object. The object to be photographed is placed in front of a lens with PHOTOGRAPHIC FILM behind the lens. A shutter controls how long the film is exposed to the light, while an adjustable hole called an APERTURE, or DIAPHRAGM, controls the amount of light reaching the film.

The distance from the lens to the film can be varied to produce a sharp image on the film – a process called focusing. For a distant object, the lens to film distance is equal to the FOCAL LENGTH of the lens; for closer objects this distance will be greater. By changing the focal length of the lens the image size can be altered: the longer the focal length, the larger the image for a given object distance.

When the aperture size is decreased, the outermost rays of light are stopped from reaching the film. This has the effect of making those parts of the image that would not be perfectly sharp on the film appear sharper. The range of distances over which the image on the film is acceptably sharp is called the DEPTH OF FIELD.

cAMP *See* CYCLIC AMP.

canaliculus In zoology, a small channel, such as in bone.

cancer A group of diseases in which certain cells do not show the usual growth restraints and so continue to divide, resulting in unlimited growth of the tissue. Such growth may lead to a swelling or lump called a TUMOUR. If a tumour is cancerous it is called malignant, but tumours can also be benign (non-cancerous). Cancer is the biggest killer in the Western world: there are more than 100 types, named according to the tissue of their origin, for example CARCINOMAS (of EPITHELIAL origin), SARCOMAS (of CONNECTIVE TISSUE) and MYELOMAS (of BONE MARROW).

The causes of cancer are not fully known and are many and complex; there is not usually one direct cause. Some factors (CARCINOGENS) are known to be linked with an increased incidence of cancer; tumour-inducing ONCOGENES have also been identified in DNA. Some cancers are now treatable and potentially curable. Treatment has historically been by surgery, CHEMOTHERAPY or RADIOTHERAPY, but more recently MONOCLONAL ANTIBODIES attached to CYTOTOXIC drugs to target and kill specific cells have been used with some success. Prevention and early detection still remain the best options.

See also TRANSFORMATION.

candela (Cd) The SI UNIT of LUMINOUS INTENSITY. One candela is equal to the luminous intensity in a given direction of a source of MONOCHROMATIC radiation of frequency 5.4×10^{14} Hz and has a radiant intensity in that direction of 1/683 WATT per STERADIAN.

Cannizzaro reaction The interaction of two ALDEHYDES in the presence of dilute alkalis where one is reduced to the corresponding ALCOHOL and the other is oxidized to the corresponding acid. Many but not all aldehydes react in this way. *See also* ALDOL CONDENSATION. OXIDATION, REDUCTION.

cap *See* DIAPHRAGM.

capacitance The charge storage ability of a CAPACITOR; that is, the electric charge stored divided by the voltage between the plates. The unit of capacitance is the FARAD.

capacitation The final stage of maturation of a SPERM cell. It takes place in the female tract.

capacitor A device used to store electric charge. In its simplest form, a capacitor consists of two metal plates separated either by a vacuum or by some insulating material, called a DIELECTRIC. The amount of charge stored is directly proportional to the POTENTIAL DIFFERENCE (p.d.) between the two plates. The charge stored for each volt of potential difference is called the CAPACITANCE. The capacitance of a capacitor is directly proportional to the area of overlap of the plates and inversely proportional to their separation, thus the greatest capacitance is obtained by having large, closely-spaced plates. In practice these plates are made of metal foil separated by an insulator and folded or rolled into a many-layered construction.

If a capacitor stores a charge Q when there is a potential V between its plates, the capacitance is C, where

$$C = Q/V$$

For a capacitor with plates of area A separated by a distance d, with the space between the plates being filled by a dielectric of RELATIVE PERMITTIVITY ε_r, the capacitance C is

$$C = \varepsilon_0 \varepsilon_r A/d$$

where ε_0 is the PERMITTIVITY of free space (8.85×10^{-12} $\mathrm{Fm^{-1}}$).

Capacitors are often distinguished by the material used for their dielectric. Various types of plastic are often used, but for large capacitances in small packages, ELECTROLYTIC CAPACITORS are used.

When two or more capacitors are connected in SERIES, the charge that flows off one plate of one capacitor flows onto the first plate of the next capacitor and so on, thus each capacitor has the same charge. The total capacitance of this combination is the reciprocal of the sum of the reciprocals of the individual capacitances. When two or more capacitors are connected in PARALLEL, the total charge stored is just the sum of the charges that each capacitor alone would store if charged to the same voltage. Thus the total capacitance is simply the sum of the individual capacitances.

For capacitors C_1, C_2 and C_3 connected in series, the total capacitance C is

$$1/C = 1/C_1 + 1/C_2 + 1/C_3$$

For the same capacitors connected in parallel,

$$C = C_1 + C_2 + C_3$$

The rate of flow of charge on or off a capacitor is directly proportional to the difference between the potential difference across the capacitor and the voltage of any supply of charge to which it is connected, and is inversely proportional to the resistance in the circuit. The charge on a capacitor rises or falls exponentially, with a TIME CONSTANT equal to RC, where R is the resistance in the circuit and C the capacitance. For a capacitor of capacitance C, initially charged to a charge Q_o, discharging through a resistance R, the charge Q after time t is

$$Q = Q_o e^{-t/RC}$$

For a capacitor of capacitance *C*, initially uncharged, connected to a supply of voltage *V* by a resistance *R*, the charge *Q* after a time *t* is

$$Q = CV(1 - e^{-t/RC})$$

As well as storing charge, a capacitor stores energy, in that energy is needed to force the charges onto the plates. This energy will be released when the opposite charges on the two plates are provided with a conducting path from one plate to the other. The energy stored in a capacitor is equal to the average potential difference between the plates (which is half the final p.d.) multiplied by the charge on the plates.

Capacitors do not store much energy for their size, so are not an alternative to CELLS, but they can be used for short-term energy storage, for example in photographic flash guns, or in SMOOTHING in power supply circuits.

For a capacitor of capacitance *C* charged to a p.d. *V*, storing a charge *Q*, the energy stored *E* is

$$E = {}^1/_2QV = {}^1/_2CV^2 = {}^1/_2Q^2/C$$

See also VARIABLE CAPACITOR.

capillary In biology, the smallest blood vessel in animals (8–20 μm in diameter), with thin walls only one cell thick containing no muscle or elastic fibres. Capillaries are therefore permeable to nutrients, dissolved gases and waste products. They are grouped together in capillary beds (or networks) that link arteries and veins and serve as the main site of exchange between blood and body fluids (*see* LYMPH). Blood pressure in the arteries is reduced by capillaries and blood is deoxygenated in capillaries before entering the veins.

capillary effect The effect that causes most liquids, including water, to rise up a glass tube with a narrow bore (capillary tube). This arises due to the intermolecular attraction between the water and the glass (called ADHESION) being stronger than the forces between the water molecules, called COHESION in this context. The same effect causes water to spread out on a glass surface, provided it is free of oil or grease.

It is the capillary effect that causes water to form a curved surface, called a MENISCUS, in a tube, and which allows materials with fine cracks (POROUS materials) to draw up water. Many plants rely on this mechanism to draw water from their roots into their leaves, though the process of OSMOSIS is also important (*see* TRANSLOCATION). With some liquids, such as mercury, the cohesion is so strong that the liquid will form nearly spherical droplets on almost any surface. This results in the outward curving surface on mercury in a glass tube.

In a capillary tube, the height *h* by which the liquid inside the tube rises above that outside is

$$h = 2T/r\rho g$$

where *T* is the SURFACE TENSION, *r* the radius of the tube, ρ the DENSITY of the liquid and *g* the ACCELERATION DUE TO GRAVITY.

capillary tube A glass tube with a narrow bore. *See* CAPILLARY EFFECT.

carbaminohaemoglobin *See* HAEMOGLOBIN.

carbenium ion *See* CARBONIUM ION.

carbide A BINARY COMPOUND containing a metal and carbon. An example is silicon carbide (carborundum), SiC, which has a macromolecular structure (*see* MACROMOLECULE) with a high boiling point and is extremely hard. Silicon carbide is commonly used as an abrasive. The carbides of the ALKALI METALS and ALKALINE EARTHS are IONIC compounds, containing the C_2^{2-} ion. They react with water to produce ETHYNE (acetylene) and are known as DICARBIDES.

carbocation *See* CARBONIUM ION.

carbohydrate One of a large group of organic compounds with the general formula $C_x(H_2O)_y$. They are the main energy-providing components of the human diet. There are three main groups of carbohydrates: MONOSACCHARIDES, DISACCHARIDES and POLYSACCHARIDES. Monosaccharides, such as GLUCOSE and FRUCTOSE, are single sugars with the general formula $(CH_2O)_n$ that cannot be split into smaller carbohydrate units. Disaccharides, such as SUCROSE, MALTOSE and LACTOSE, are double sugars, where two monosaccharides are combined, which can be split into their single sugar components. Polysaccharides, such as STARCH and GLYCOGEN, consist of variable numbers of monosaccharides joined together in chains that can be branched or not and can fold for easy storage, and be broken down into their constituent disaccharides or monosaccharides for use. Some polysaccharides are structural, for example, CELLULOSE and CHITIN, and some are food reserves, for example, starch and glycogen.

Most carbohydrates can form ISOMERS (which have the same formula but a different arrangement of atoms) that gives them a different functional property. enzymes usually only react with one form. Naturally occurring carbohydrates have a D(+) form (*see* DEXTROROTATORY and OPTICAL ISOMERISM).

There are two tests that can be used to identify sugar in a solution, BENEDICT'S TEST and FEHLING'S TEST. Both rely on the ability of monosaccharides and some disaccharides to reduce copper(II) sulphate to copper oxide, causing a colour change on boiling. These sugars can also be classified as reducing or non-reducing. The test for starch is the addition of iodine in potassium iodide, which integrates into the starch polymer causing a colour change from yellow/orange to blue/black.

carbolic acid The common name for PHENOL.

carbon (C) The element with ATOMIC NUMBER 6; RELATIVE ATOMIC MASS 12.00; sublimes at 3,500°C; RELATIVE DENSITY 2.3 (graphite). Carbon occurs naturally in AMORPHOUS forms, such as coal, charcoal and soot and as the two crystalline allotropes graphite and diamond. A third allotrope, BUCKMINSTERFULLERENE, was discovered in 1995.

Carbon has two stable isotopes (carbon–12 and –13) and four radioactive ones. The radioactive isotope carbon–14 is used in RADIOCARBON DATING.

Carbon's most important chemical property is its ability to form long chain molecules. This is the starting point for all organic compounds, including the complex molecules responsible for all known living organisms.

See also CARBON CYCLE.

carbonaceous (*adj.*) Describing any mineral containing carbon other than in CARBONATES.

carbonate Any compound containing a metal CATION and the carbonate ANION, CO_3^{2-}. Carbonates are formed by the reaction of the many metals with carbonic acid, for example:

$$Mg + H_2CO_3 \rightarrow MgCO_3 + H_2$$

All carbonates except those of the ALKALI METALS decompose on heating (lithium carbonate also decomposes), for example:

$$MgCO_3 \rightarrow MgO + CO_2$$

They also react with acids to release carbon dioxide, this reaction being the classic test for the presence of the carbonate ion, for example:

$$MgCO_3 + 2HCl \rightarrow MgCl_2 + CO_2 + H_2O$$

Only the carbonates of the alkali metals (except lithium) are soluble. These form alkaline solutions, due to the formation of the HYDROGENCARBONATE (or bicarbonate) ion:

$$CO_3^{2-} + H_2O \rightarrow HCO_3^- + OH^-$$

carbon cycle The constant circulation of carbon between organic and inorganic sources in nature. This recycling maintains the balance between carbon dioxide in the atmosphere and carbon in organisms, and is vital for all forms of life.

Carbon dioxide in the atmosphere provides a major source of carbon that is used by photosynthetic plants (*see* PHOTOSYNTHESIS) and organisms to make CARBOHYDRATES. These photosynthetic plants and organisms then release oxygen back to the atmosphere. HETEROTROPHS (animals and fungi) eat the plants and other photosynthetic organisms, or other heterotrophs (*see* FOOD CHAIN), and through their RESPIRATION return carbon back to the atmosphere as carbon dioxide. Carbon dioxide is also released back to water by some organisms, for example algae, where an equilibrium with atmospheric carbon dioxide is maintained.

Part of any natural cycle includes not only the biological (BIOTIC or living) component but also a geological (ABIOTIC or non-living) component. These geological components include rocks and other deposits in the oceans and atmosphere, for example in the form of coal and oil under the ocean, peat in wetlands and limestone rocks, and they provide the largest reservoir of carbon.

The natural processes of photosynthesis and respiration balance each other out, but in recent years human intervention has disturbed this balance. The burning of FOSSIL FUELS and the destruction of large areas of tropical forest causes the atmospheric levels of carbon dioxide to rise, contributing to the GREENHOUSE EFFECT.

carbon dioxide (CO_2) A colourless gas; sublimes at –78°C. It is present in small quantities in the atmosphere, playing an important part in biological processes as it is produced by RESPIRATION and taken up in PHOTOSYNTHESIS.

In the laboratory carbon dioxide is prepared by the action of acids on carbonates, for example:

$$CaCO_3 + 2HCl \rightarrow CO_2 + CaCl_2 + H_2O$$

Industrially, carbon dioxide is produced as a by-product of the production of calcium oxide (quicklime) from calcium carbonate (in the form of limestone). It is stored as a liquid under pressure and used for a wide variety of purposes – as a fire-extinguishing gas, for instance, and in its solid form for refrigeration.

The increase of carbon dioxide in the atmosphere, from the burning of fossil fuels and the destruction of vast areas of forest, is thought to be greatest cause of GLOBAL WARMING.

See also CARBON CYCLE.

carbon disulphide (CS_2) A colourless liquid; melting point –110°C; boiling point 46°C; RELATIVE DENSITY 1.3. Carbon disulphide can be produced by the action of methane on sulphur:

$$CH_4 + 4S \rightarrow CS_2 + 2H_2S$$

It is used as a solvent for many organic materials, including rubber, and will also dissolve sulphur and phosphorus.

carbon fibre A COMPOSITE MATERIAL made in a similar way to GLASS-FIBRE REINFORCED PLASTICS, but the carbon fibres are even stronger. However, like glass fibre, the material does not lend itself to mass-production techniques, so is found only in relatively expensive high-technology products.

carbonic acid (H_2CO_3) A weak DIBASIC ACID formed by dissolving carbon dioxide in water. The acid exists only in equilibrium with dissolved carbon dioxide, and cannot be isolated from aqueous solution. It dissociates strongly into HYDROGENCARBONATE ions, HCO_3^-, which in turn dissociate weakly into CARBONATE ions, CO_3^{2-}. It therefore forms two series of salts: NORMAL SALTS (CARBONATES) and ACID SALTS (HYDROGENCARBONATES). Soda water is a solution of carbonic acid.

carboniferous (*adj.*) Describing a material containing NATIVE carbon, particularly coal.

Carboniferous A PERIOD of geological time during the PALAEOZOIC ERA. It began after the DEVONIAN period (about 345 million years ago) and extended until the PERMIAN era (about 280 million years ago). The period is significant for the widespread deposits of limestone and the deposits of coal from swamp vegetation. Amphibians became more abundant during this time and the first reptiles evolved.

carbonium ion, *carbenium ion, carbocation* Any positively charged ion containing a TRIVALENT carbon atom, R_3C^+, where R is an alkyl group. They are thought to be intermediates in many reactions, for example:

$$(CH_3)_3C\text{–}Br \underset{\text{in water and ethanol}}{\overset{\text{dissolved}}{\rightleftharpoons}} (CH_3)_2C^+\text{–}CH_3 + Br^-$$

2, bromo-2-methyl propane → dimethylethyl cation (carbonium ion)

$$\downarrow$$

$$(CH_3)_3C\text{–}OH$$

2-methylpropan–2–ol

Overall this reaction is a NUCLEOPHILIC SUBSTITUTION. Carbonium ions have a strong affinity for NUCLEOPHILES. The most stable is the tertiary carbonium ion where all three bonds lead to other carbon atoms. The stability decreases in secondary and more so in primary carbonium ions where only two and one bonds respectively lead to other carbon atoms.

carbon monoxide (CO) A colourless gas; melting point –199°C; boiling point –192°C. Carbon monoxide is produced by the combustion of carbon in limited supplies of oxygen:

$$2C + O_2 \rightarrow 2CO$$

It burns readily in air to give carbon dioxide:

$$2CO + O_2 \rightarrow 2CO_2$$

Carbon monoxide is toxic, as it replaces oxygen in HAEMOGLOBIN in the blood, and so prevents oxygen being transported to body tissues. The gas is very dangerous, as it is colourless and odourless and therefore difficult to detect. Its traditional use as a fuel gas has largely been replaced by methane, though it is

Some derivatives of carboxylic acids

General formula	Class name	Example	
R–CO–Cl	acid chloride or acyl chloride	CH_3COCl	ethanoyl chloride
R–CO–O–CO–R	acid anhydride	$(CH_3CO)_2O$	ethanoic anhydride
R–CO–OR′	ester	$CH_3COOC_2H_5$	ethyl ethanoate
R–CO–NH_2	amide	CH_3CONH_2	ethanamide
R–CN	nitrile	CH_3CN	ethanenitrile

still used as a REDUCING AGENT in some industrial processes.

carbon tetrachloride, ***tetrachloromethane*** (CCl_4) An organic chemical; boiling point 76°C; melting point –23°C. It is is manufactured from the CHLORINATION of methane or other HYDROCARBONS. It was used in the past as a dry cleaning agent but is now thought to be carcinogenic (*see* CARCINOGEN). It is used as a solvent and in fire extinguishers.

carbonyl chloride, ***phosgene*** ($COCl_2$) A colourless gas with a distinctive odour, produced by the reaction between carbon monoxide and chlorine:

$$CO + Cl_2 \rightarrow COCl_2$$

Carbonyl chloride is used as a chlorinating compound in the manufacture of some chlorinated hydrocarbons, and being highly toxic, was used as a chemical weapon in the first world war.

carbonyl group In organic compounds, a carbon atom attached by a DOUBLE BOND to an oxygen atom. The group is found in ALDEHYDES, KETONES and the related CARBOHYDRATES. *See also* HALOFORM REACTION, TRIIODOMETHANE TEST.

carboxyhaemoglobin *See* HAEMOGLOBIN.

carboxyl group The –COOH FUNCTIONAL GROUP found in CARBOXYLIC ACIDS and their derivatives.

carboxylic acid Any one of a group of organic compounds that contain the CARBOXYL GROUP (–COOH) attached to another group (R). The other group can be a hydrogen atom (HCOOH, methanoic acid) or a larger molecule, with up to 24 carbon atoms. When R is a straight chain ALKYL GROUP (such as CH_3, C_2H_5 or $CH_3(CH_2)_{16}COOH$) the compound is termed a FATTY ACID.

Although carboxylic acids contain the C=O CARBONYL GROUP and the OH HYDROXYL GROUP these groups behave differently when together as the –COOH carboxyl group. The hydroxyl group can be replaced by another FUNCTIONAL GROUP giving rise to compounds which are termed carboxylic acid derivatives. For example, RCOCl is an ACID CHLORIDE, RCOOR′ is an ESTER and $RCONH_2$ is an AMIDE. NITRILES contain a carbon-nitrogen TRIPLE BOND, RC≡N, and are also related to carboxylic acids.

Carboxylic acids are named after the ALKANE from which they are derived, replacing the ending *-ane* by *-anoic acid.* Examples include CH_3COOH, ethanoic acid and propanoic acid, CH_3CH_2. The carbonyl carbon atom is numbered one.

Carboxylic acids usually exist as hydrogen-bonded DIMERS (*see* HYDROGEN BOND), except in aqueous solution. They are weak acids, affected greatly by what is attached to the carboxyl group. They are able to neutralize alkalis and liberate carbon dioxide from CARBONATES. Carboxylic acids can be made by the OXIDATION of primary ALCOHOLS, RCH_2OH, or of ALDEHYDES, RCHO, using acidified DICHROMATE(VI):

$$RCH_2OH \rightarrow RCHO \rightarrow RCOOH$$

They can also be made by reacting GRIGNARD

REAGENTS with carbon dioxide.

$$CO_2 + C_2H_5MgBr \rightarrow C_2H_5COOH + HOMgBr$$

Many carboxylic acids occur naturally in plants and animals. The lower carboxylic acids have strong smells, for example, ethanoic (acetic) acid gives the smell to vinegar, while rancid butter smells of butanoic acid. Ethanoic acid is important in the manufacture of vinegar and VISCOSE. Animal and vegetable fats and oils (which are mostly ESTERS of GLYCEROL), along with fatty acids, are important in food products and diet.

carboxylic acid derivatives *See* CARBOXYLIC ACID.

carcinogen Any factor known to be linked with an increased incidence of cancer. Examples are chemicals (including smoking and asbestos dust), IONIZING RADIATION, viral infections and dietary or genetic factors. *See also* ONCOGENE, TRANSFORMATION.

carcinoma A MALIGNANT tumour of EPITHELIAL cells, for example of the skin or glandular tissue. *See also* CANCER.

cardiac muscle A specialized network of striated muscle fibres (*see* MUSCLE), found only in the heart, that is capable of rhythmic contraction and relaxation over a long period. The muscle is said to be MYOGENIC (the contraction is stimulated within the heart itself) and is involuntary. The fibres contain separate cells, each with one nucleus and irregular thickening of their surrounding membrane (sarcolemma), that form intercalated discs. A distinctive feature of this muscle is the branching and rejoining of fibres. *See also* HEART, PURKINJE FIBRES.

cardiac sphincter A ring of muscles at the entrance of the stomach that relax and contract to allow food to enter the stomach.

cardioid A heart-shaped closed curve. It represents the path of a point on the circumference of a circle that rolls on a fixed circle of equal radius. A cardioid can be expressed in POLAR CO-ORDINATES by the equation

$$r = a(1 + \cos\theta)$$

carnivore Any animal, for example a cat, dog, tiger or shark, that eats meat (mainly muscle) from other animals as the main part of its diet. Meat is more easily digested than the CELLULOSE in a herbivorous (*see* HERBIVORE) diet, and is richer in nutrients, but it is more difficult to obtain. The main adaptation for meat eating is in the jaw, which opens wider, and the teeth, which are very sharp. *See also* OMNIVORE.

Carnot cycle In physics, an ideal cycle of operations, leading to the greatest efficiency of conversion of heat energy into WORK attainable by any reversible HEAT ENGINE.

The Carnot cycle has four reversible stages, operating between two HEAT RESERVOIRS. An ideal gas starts at a high temperature and pressure, and expands isothermally (at a constant temperature, *see* ISOTHERMAL), converting energy from a high temperature reservoir into mechanical work. The gas then expands further adiabatically (with no exchange of heat energy to its surroundings, *see* ADIABATIC) cooling to a lower temperature. In the third stage, the gas is compressed isothermally, releasing any remaining heat it possess. Finally the gas is compressed adiabatically to its original state. Heat energy is taken from the high temperature reservoir and a smaller amount given up to the lower temperature. The difference between these two energies is the amount of mechanical work done.

Since every stage in the Carnot cycle is reversible, there is no overall change in ENTROPY and the Carnot engine must be the most efficient heat engine permitted by the SECOND LAW OF THERMODYNAMICS. Realistic heat engines are far less efficient, but are designed to approach the Carnot cycles as closely as possible.

See also INTERNAL COMBUSTION ENGINE.

Carnot engine A hypothetical HEAT ENGINE operating on the CARNOT CYCLE.

carotene A natural CAROTENOID pigment that is responsible for the orange, yellow and red colour of carrots, tomatoes and oranges.

carotenoid One of a number of coloured pigments found in many living organisms. Carotenoid pigments are lipids and can be yellow, orange, red or brown. They are frequently found in the CHLOROPLASTS of plants, where they can act as accessory pigments in PHOTOSYNTHESIS. In some ALGAE the carotenoid pigments are the main light-absorbing pigments used instead of CHLOROPHYLL. Carotenoids are also found in fruits, roots and petals, giving them their colour, and provide the autumn colours. The group includes CAROTENE and the XANTHOPHYLLS.

carotid artery Either one of two main arteries at each side of the neck that supplies the head with blood from the heart. *See also* CIRCULATORY SYSTEM.

carpel The essential female reproductive structure in a flowering plant (ANGIOSPERM). In the centre of the carpel is a slender stalk called the STYLE, which supports the STIGMA at the top. The stigma is specialized for receiving pollen; it often has hairs and produces a sticky secretion to trap pollen grains. At the base of the carpel is a hollow called the ovary, with thick walls protecting one or more OVULES, each enclosing the egg nucleus (the female gamete). The ovule is attached to the ovary wall by the funicle and the ovary is attached to the plant by the placenta. The ovary develops into the fruit wall after FERTILIZATION, and the ovule into the seed. There may be one or more carpels, which can be fused, as in the tulip, or not fused, as in the buttercup, and collectively the carpels are called the gynoecium. *See also* DOUBLE FERTILIZATION, POLLINATION.

carrier wave A radio wave used to carry some information, such as speech or a picture, or which may be switched between two similar frequencies to convey BINARY information. The process of imparting information on a carrier wave is called MODULATION. *See also* AMPLITUDE MODULATION, FREQUENCY MODULATION, PHASE MODULATION, PULSE CODE MODULATION, SIDEBAND.

carrying capacity In ecology, the maximum number of individuals of a given species that can be supported by a particular ecological area. When resources, such as food, are exhausted, the population will be reduced by death, emigration or reproductive failure.

Cartesian co-ordinates A mathematical system for locating a point in space by measuring its distance from three fixed lines, called axes. The axes are at right angles to one another and intersect at a fixed point, called the origin. These distances are called the co-ordinates of the point, and are usually written in the form (x,y,z). On a two dimensional surface only the x and y co-ordinates are needed.

cartilage A hard but flexible CONNECTIVE TISSUE that forms the embryonic skeleton and is replaced by bone except in areas of wear and tear, such as the intervertebral discs between the backbones and at bone endings (articular cartilage). In mammals cartilage is also found in the nose, ear and LARYNX, and in cartilaginous fish such as sharks it forms the skeleton. Cells called chondrocytes secrete the POLYSACCHARIDE-containing matrix of chondrin. COLLAGEN fibres are embedded within this, giving cartilage its strength. In adults cartilage contains no blood vessels.

cartilaginous fish *See* CHONDRICHTHYES.

casein The main protein constituent of milk. Caseinogen is the soluble form, which is precipitated by RENIN to casein. This is particularly important in young animals. Casein is also a major component of cheese.

caseinogen *See* CASEIN.

cast A metal object formed by pouring molten metal into a suitably shaped mould. *See* CAST IRON.

cast iron Iron with a high proportion of carbon (typically 3 per cent). It is made by melting PIG IRON, often with steel scrap to reduce the proportion of carbon. This material is then poured into moulds and allowed to cool. Cast iron is very hard, but is BRITTLE, so difficult to machine and prone to cracking.

catabolism The breaking down of living tissue into energy and waste products. *See* METABOLISM.

catalase An enzyme that catalyses (*see* CATALYST) the breakdown of hydrogen peroxide to water and oxygen. It is found in PEROXISOMES.

catalyst Any substance that changes the rate of a chemical reaction, without itself being permanently altered chemically by that reaction. For example, hydrogen peroxide, H_2O_2, decomposes to water and oxygen only very slowly at room temperature:

$$2H_2O_2 \rightarrow 2H_2O + O_2$$

but in the presence of manganese(IV) oxide, MnO_2, the reaction proceeds much more rapidly. The manganese(IV) oxide is itself unchanged by the reaction.

A catalyst is described as homogeneous if it exists in the same phase as the reagents, for example a liquid catalysing a reaction between liquids. A catalyst is heterogeneous if the reagents are in a different phase to itself, for example a solid catalysing reactions between gases. The term catalyst usually applies to substances that increase the rate of a reaction (positive catalysts). Those that slow a reaction down are termed negative catalysts.

The action of catalysts is complex, though in general they lower the ACTIVATION ENERGY of a chemical process. Many catalysts are TRANSITION

METALS, and platinum is particularly widely used (for instance in the CONTACT PROCESS for manufacturing sulphuric acid). The multiple OXIDATION STATES displayed by transition metals seem important in some catalytic processes. These allow a reaction to proceed via an intermediate step that involves the formation of compounds including the catalyst, which is regenerated at a later stage in the reaction.

Many catalytic processes use heterogeneous catalysts, for example, solid platinum is used as a catalyst in many gas phase reactions. It is important that the catalyst is finely divided to allow maximum contact with the gas. In such processes, VAN DER WAAL'S FORCES between the reagents and incompletely bound atoms in the surface of the solid catalyst appear to be important to the function of the catalyst.

Many biological reactions rely on complex catalysts, often specific to a particular reaction. These are called enzymes.

See also AUTOCATALYSIS.

catalytic converter A platinum-based CATALYST placed in the flow of the burnt gases leaving a PETROL ENGINE to ensure that they are fully oxidized (*see* OXIDATION). This reduces the levels of carbon monoxide and nitrogen oxides in the exhaust gases.

catalytic cracking *See* CRACKING.

catalytic reforming A process similar to catalytic cracking (*see* CRACKING) except that conditions are chosen to minimize the formation of straight-chain ALKANES and instead encourage the formation of branched-chain alkanes. These alkanes burn more easily in petrol engines with a lesser tendency to KNOCKING. Thus catalytic reforming is used to reduce the knocking and improve the OCTANE NUMBER of petrol.

cataphoresis *See* ELECTROPHORESIS.

catecholamine An amino acid derivative that can function as a NEUROTRANSMITTER or a HORMONE, for example, ADRENALINE and NORADRENALINE.

cathode A negatively charged ELECTRODE.

cathode ray A stream of electrons emitted by a CATHODE in a vacuum, such as in a CATHODE RAY TUBE. *See also* CATHODE RAY OSCILLOSCOPE.

cathode ray oscilloscope A device based on the CATHODE RAY TUBE and used to display how voltages change with time. The voltage to be measured is connected, via an AMPLIFIER, to the Y-PLATES (*see* CATHODE RAY TUBE), whilst a voltage that increases steadily with time, called the TIMEBASE, is connected to the X-PLATES. The timebase causes the spot on the screen to sweep at a steady rate from left to right, whilst the voltage being measured moves the spot up or down, thus effectively producing a graph of the voltage against time. When the spot reaches the right-hand end of the screen, the CONTROL GRID switches off the electron beam and the spot is returned to the left-hand edge of the screen, a process called flyback.

cathode ray tube The display device used in televisions, VISUAL DISPLAY UNITS and the CATHODE RAY OSCILLOSCOPE.

In the neck of a cathode ray tube there is an ELECTRON GUN comprising a heated filament together with two or more cylindrical ANODES. Electrons are released from the filament by THERMIONIC EMISSION. A beam of electrons is thus formed that can be focused to a narrow point on a PHOSPHOR-coated screen at the far end of the tube by varying the potential on the anodes. The electron gun also contains a CONTROL GRID, an ELECTRODE that can be made negative to reduce the number of electrons leaving the gun and thus control the brightness of the spot of light produced on the screen. The electron beam can be steered to any point on the screen, either electrostatically, by two pairs of plates called X-PLATES and Y-PLATES that steer the beam horizontally and vertically, or magnetically by coils placed around the neck of the tube.

Cathode ray tubes have the disadvantages of requiring high voltage power supplies and being rather large in terms of the amount of space required between the electron gun and the screen. They are gradually being replaced by LIQUID CRYSTAL DISPLAYS, but at present these are too expensive to replace large cathode ray tubes. *See also* CATHODE RAY OSCILLOSCOPE.

cathodic reduction The process that takes place at the CATHODE of an ELECTROLYSIS, regarded as a REDUCTION. Since the cathode is negatively charged, electrons are added to the ELECTROLYTE at this point and thus the electrolyte is reduced. For example, in the electrolysis of water, hydrogen ions are reduced to gaseous hydrogen:

$$2H^+ + 2e^- \rightarrow H_2$$

See also ANODIC OXIDATION.

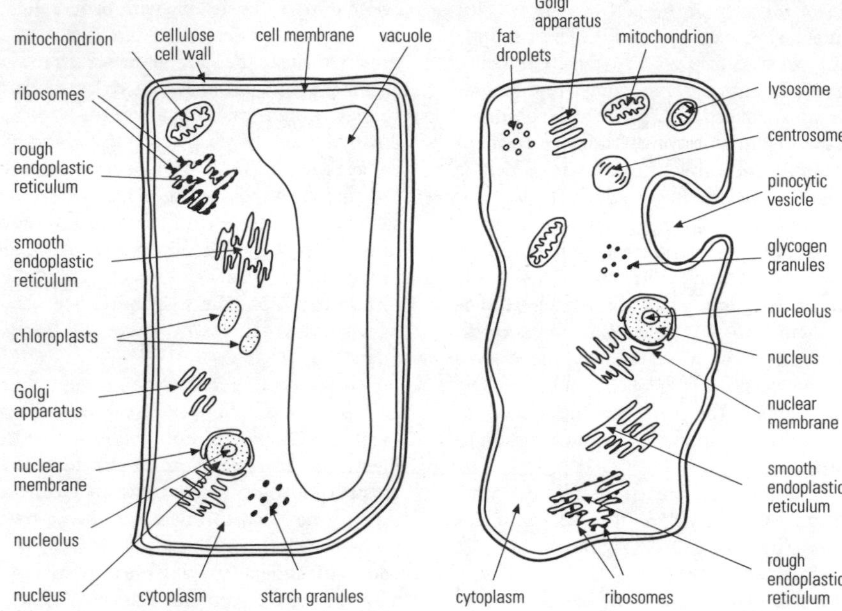

Generalized structure of the eukaryotic cell.
a) Plant cell b) Animal cell

cation A positively charged ion. So called because it will be attracted to the CATHODE in ELECTROLYSIS. *Compare* ANION.

cation pump *See* SODIUM PUMP.

caudal fin *See* FIN.

caustic 1. (*chemistry*) (*adj.*) Describing a highly reactive ALKALI, such as sodium hydroxide.

2. (*physics*) (*n.*) The curve that contains all the rays from a point source of light as they approach and depart from the focus of a lens or mirror that suffers ABERRATION.

caustic soda *See* SODIUM HYDROXIDE.

caustic potash *See* POTASSIUM HYDROXIDE.

Cavendish's experiment An experiment to determine the GRAVITATIONAL CONSTANT. Two small lead spheres are fastened to the ends of a rod, which is suspended from its midpoint by a fine fibre. The suspended spheres are allowed to rotate to-and-fro on this fibre and their period of oscillation is timed, enabling the stiffness of the fibre to be calculated. Two larger lead spheres are then placed near to the smaller ones in such a way that the gravitational attraction between two pairs of spheres twists the fibre. By measuring the deflection of the rod for different configurations of the four spheres and from the known stiffness of the fibre, the gravitational constant can be determined using NEWTON'S LAW OF GRAVITATION.

caveolae *See* POTOCYTOSIS.

CCD (charge coupled device) An array of small detectors based on the PHOTOELECTRIC EFFECT. CCD's are built as the top layer of an INTEGRATED CIRCUIT, which allows the build up of charge on each line of detectors to be 'read out' in turn. CCD's can be made very small and produced at a reasonable cost, allowing the production of small, cheap, television cameras. CCD's can also be used at very low light levels, which has made them very popular in astronomy.

cDNA *See* COPY DNA.

celestial mechanics The study of the motions of astronomical objects and the forces between them, based on NEWTON'S LAWS OF MOTION and NEWTON'S LAW OF GRAVITATION.

cell, *electrochemical cell* (*physics*) A device that connects chemical reagents to an electrical circuit. A cell consists of two ELECTRODES in

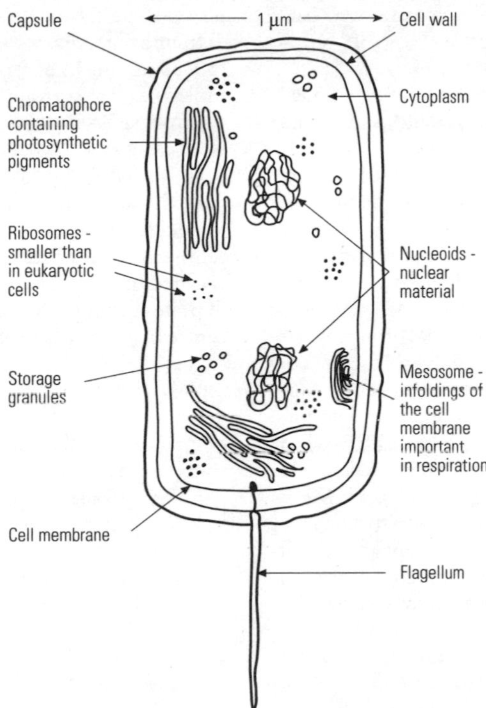

Structure of the prokaryotic cell – a generalized bacterial cell.

contact with an ELECTROLYTE – a conducting liquid or jelly. An electrolytic cell is one in which a chemical change takes place when a current is passed through the electrolyte (*see* ELECTROLYSIS). In a voltaic cell, a chemical reaction between the electrolyte and the electrodes produces an ELECTROMOTIVE FORCE, which drives a current around a circuit.

Some voltaic cells can be used once only: examples are the ZINC-CARBON CELL, which is cheap but stores relatively little energy, and the more expensive but longer lasting MANGANESE-ALKALINE CELL. Others are rechargeable by passing a current through them in the reverse direction, the chemical reactions can be reversed and electrical energy converted back to chemical energy. Examples of rechargeable cells are the LEAD-ACID CELL, used to provide high currents in cars for example, and the NICKEL-CADMIUM (or Nicad) cell, often used in electronic appliances. The amount of charge that a voltaic cell can drive around a circuit is measured in AMPERE-HOURS.

See also HALF-CELL.

cell (*biology*) The smallest mass of self-contained living matter of an animal or plant. Cells normally range in size from 10 to 30 µm. Some organisms consist of only one cell, for example, bacteria and PROTOZOA and many other micro-organisms, while others, for example humans, are made of billions of cells.

A cell consists of PROTOPLASM, which is the CYTOPLASM and nucleus, surrounded by a CELL MEMBRANE. In plant cells there is also a rigid CELL WALL made of CELLULOSE that gives the cell, and therefore the plant, support. The cell membrane regulates the entry of substances into and out of the cell.

In EUKARYOTES, the DNA is organized into chromosomes and contained within a clearly defined nucleus bounded by a membrane within the cell. Eukaryote cells also contain specialized structures called ORGANELLES (including the nucleus), such as ENDOPLASMIC RETICULUM, GOLGI APPARATUS, LYSOSOMES, MITOCHONDRIA and RIBOSOMES.

In PROKARYOTES, the DNA forms a coiled structure called a NUCLEOID and there is no nucleus and no major organelles.

See also CELL CYCLE, CELL DIVISION, EXTRACELLULAR MATRIX, INTERPHASE, MEIOSIS, MITOSIS, VACUOLE.

cell cycle In EUKARYOTES, the regular pattern of events occurring in a dividing CELL. There are two main parts to the cycle: INTERPHASE and MITOSIS (or sometimes MEIOSIS). During mitosis the nucleus divides. During interphase the cell ORGANELLES are duplicated and the DNA content is doubled. The cell cycle can be continuous for some cells (such as single-celled organisms) or it can cease for other cells after a period of time or at a stage of maturity. Much research has been focused on understanding the details of the cell cycle in the hope that it will lead to a greater understanding of cancer, in which the normal growth pattern is changed.

cell division The processes that result in the division of a living CELL into two daughter cells. Cell division is necessary for growth, repair and reproduction of an organism. In EUKARYOTES it is always preceded by division of the nucleus (by MITOSIS or MEIOSIS), followed by duplication of the cell ORGANELLES and the actual splitting of the CELL MEMBRANE. Plant cells divide by forming a cell plate, whereas animal cells divide by constriction. Nuclear division can occur without cell division, giving rise to multinucleate cells. In PROKARYOTES there is no nucleus and the one event of cell division separates the DNA as well as the cell membrane.

cell-mediated immunity (CMI) The response of an organism to invasion by a foreign object. The cells involved in CMI are PHAGOCYTES, such as MACROPHAGES, or T CELLS, such as CYTOTOXIC T cells. Antibodies play a more minor role in CMI. Foreign ANTIGENS are only recognized by T cells in association with cell antigens of the MAJOR HISTOCOMPATIBILITY COMPLEX. CMI contrasts with HUMORAL IMMUNITY, in which antibodies provide the main line of defence. The distinction between the two is not clear-cut, however, as cells are needed in the initiation of an antibody response and cell-mediated responses usually also involve the production of antibodies. *See also* IMMUNITY, IMMUNE RESPONSE.

cell membrane, *plasma membrane* A thin layer (10–80 nm thick) of PROTEIN and PHOSPHOLIPID molecules that surrounds CELLS and selectively controls substances passing through it. It is a SEMIPERMEABLE MEMBRANE: small molecules such, as water, GLUCOSE and AMINO ACIDS, can pass through the membrane, whereas large molecules, such as STARCH, cannot. The currently accepted structure of the cell membrane is called the FLUID MOSAIC MODEL.

cell plate A thin partition that forms across the centre of plant cells to effect the division of the CYTOPLASM. *See* CELL DIVISION.

cellulase *See* CELLULOSE.

cellulose A CARBOHYDRATE made of long chains of GLUCOSE that gives strength to plant cell walls. It is branched, with many chains running parallel to each other and cross-linked, which provides the strength needed to support plant cell walls. The cellulose in plant cell walls is important in the diet of many animals, although no vertebrate possesses the enzyme cellulase needed to break down cellulose. HERBIVORES (plant-eating animals) digest cellulose (the major part of their diet) by having specialized bacteria in their gut that can make cellulase and therefore digest cellulose. Humans cannot digest cellulose because they do not have the necessary gut micro-organisms or grinding teeth, but it provides a vital source of ROUGHAGE. The strength of cellulose has been utilized by humans, for example in the use of cotton, paper, plastic and cellophane.

cell wall The outer wall of a plant cell, made from CELLULOSE, that provides the plant with support and protection. When a cell is turgid (swollen with water, *see* PLASMOLYSIS) a HYDROSTATIC PRESSURE is exerted on the cell wall and this gives the plant the support and rigidity it needs. *See also* PLASMODESMA, TURGOR.

Celsius A TEMPERATURE SCALE in which the freezing point of water is defined as zero degrees Celsius (0°C), whilst the boiling point of water is 100°C, both temperatures being measured at ATMOSPHERIC PRESSURE. The degree Celsius is the same size as the KELVIN. *See also* ABSOLUTE TEMPERATURE.

Cenozoic The current period of geological time, which began about 65 million years ago. It is subdivided into the TERTIARY and QUATERNARY periods. It is sometimes called 'the age of mammals' as it is during this time that mammals emerged as the dominant group.

centi- Prefix used to denote one hundredth. For example, one centimetre is one hundredth of a metre (0.01 m).

central nervous system (**CNS**) The part of the NERVOUS SYSTEM that co-ordinates body functions by integrating the SENSORY SYSTEM and the EFFECTOR SYSTEM. The CNS receives information from the sensory NEURONES, interprets these and sends messages to the effector neurones to stimulate the appropriate action. In vertebrates, the CNS consists of a BRAIN and SPINAL CORD. The spinal cord is enclosed and protected by the spinal column surrounded by three membranes called the MENINGES. In many invertebrates the CNS consists mainly of ganglia (*see* GANGLION), and in some simple invertebrates there is no CNS but instead a simple network of nerve cells called a 'nerve net'. *See also* AUTONOMIC NERVOUS SYSTEM, CEREBROSPINAL FLUID.

central processing unit (**CPU**) That part of a computer that executes the programs and controls the other parts. It contains the ARITHMETIC LOGIC UNIT, which carries out all the calculations and logical operations, which are decoded and executed by a CONTROL UNIT. The CPU also contains an IMMEDIATE ACCESS MEMORY, which stores all the data the computer is currently working on.

centre A point that is equally distant from all points on a circle or on the surface of a sphere.

centre of gravity *See* CENTRE OF MASS.

centre of mass, ***centre of gravity*** The single point in an object at which the whole of the GRAVITATIONAL FORCE on that object can be considered to act. When an object is suspended, the centre of mass will always be directly below the point where the object is suspended. By suspending an object from several different points in turn, its centre of mass can be found.

centrifugal force An apparent (but not real) force used in some descriptions of CIRCULAR MOTION to balance the CENTRIPETAL FORCE as if the object moving in a circle were in equilibrium. This is sometimes a useful way of thinking about the motion of an object moving in a circle, but it must be remembered that there is no real equilibrium. Inside an object accelerating in a circular path, such an orbiting spacecraft, it is convenient to think of the centripetal force, which acts towards the centre of the circle, as being balanced by an equal and opposite centrifugal force. However, this will not provide a full description of the situation unless a second imaginary force called the CORIOLIS FORCE is also included.

centrifuge A machine for separating two different materials on the basis of their relative densities. A centrifuge may be used to separate the components in an EMULSION or a solid suspended in a liquid. The mixture is placed in a tube and rotated very rapidly in a horizontal circle. The CENTRIPETAL FORCE needed to make the mixture rotate forces the denser component outwards along the tube, displacing the less dense component, and collecting at the bottom of the tube. Centrifugation is used for separating different cell types or, for example, for separating blood PLASMA from the heavier RED BLOOD CELLS. *See also* ULTRACENTRIFUGE.

centriole In animal cells, an ORGANELLE that is a hollow cylinder. Centrioles are similar to CILIA. They arise in pairs at the CENTROSOME within the cell CYTOPLASM. During MITOSIS and MEIOSIS they separate and go to opposite poles of the cell, where they give rise to the spindle (*see* MITOSIS). In some higher plant cells a spindle forms even though there are no centrioles.

centripetal acceleration The acceleration of an object moving in a circle. *See* CIRCULAR MOTION.

centripetal force The force needed to make an object move in a circular path. Note that this is not a particular type of force, but a force doing a particular job: for a car going around a bend, the centripetal force is provided by the friction between the wheels and the road. For a satellite in orbit, the centripetal force is provided by gravity. *See* CIRCULAR MOTION. *See also* CENTRIFUGAL FORCE.

centroid The geometric centre of a body: the point from which the average distance of all other points is zero. This is the CENTRE OF MASS of a shape or body of uniform density.

centromere In EUKARYOTES, the region of a chromosome at which the two CHROMATIDS join and at which the spindle fibres attach during MITOSIS and MEIOSIS. Under the microscope it is visible as a constriction in the chromosome. There are no genes at the centromere.

centrosome In animal cells, a distinct region within the CYTOPLASM, situated close to the nucleus. The CENTRIOLES arise from here.

cephalopod A member of the CLASS CEPHALOPODA.

Cephalopoda A CLASS of the PHYLUM MOLLUSCA, including octopus, squid and cuttlefish. Cephalopods are mostly marine and possess a well-developed head surrounded by tentacles. Cephalopods are the most advanced of the molluscs. They have complex eyes, similar to those in vertebrates, and a highly developed NERVOUS SYSTEM. They possess a rasping tongue or radula for feeding.

cephalothorax The combined head and THORAX of many CRUSTACEANS and ARTHROPODS. *See also* SPIDER.

Cepheid variable A type of RED GIANT that varies in brightness over a few days. There is a definite link between the period of this variation and the LUMINOSITY of these stars, which enables their distance to be found. Once the distance to a Cepheid variable within a CLUSTER or galaxy is known, the distances of all the other members of that cluster or galaxy can be taken as being roughly equal.

cerebellum Part of the vertebrate HINDBRAIN, overlying the MEDULLA OBLONGATA, that controls the muscle movement needed for posture and locomotion. It is well developed in humans and birds to control the balance needed for walking and flight, respectively, but is smaller in lower animals.

cerebral hemisphere *See* CEREBRUM.

cerebrospinal fluid (CSF) A clear, colourless solution found in the VENTRICLES of the BRAIN and between the membranes of the MENINGES (surrounding and protecting the spinal cord) that acts as a shock-absorber for the CENTRAL NERVOUS SYSTEM. It contains glucose and mineral ions and a few WHITE BLOOD CELLS (but no protein), and so supplies nutrients to the central nervous system. It is secreted continuously by the choroid plexuses (projections of non-nervous EPITHELIUM in the ventricles of the brain) and reabsorbed by veins.

cerebrum Part of the vertebrate FOREBRAIN (the largest part of the human BRAIN) that co-ordinates the body's voluntary activities and some involuntary ones, including the senses and complex activities, such as reasoning, learning and memory. Its surface area is increased by being highly convoluted, allowing greater capacity for more complex activity. In lower animals the size of the cerebrum in relation to the body size decreases.

The cerebrum is divided into two halves, the cerebral hemispheres, which are joined by a strip of WHITE MATTER known as the corpus callosum. The outer layer of the cerebral hemisphere is called the cerebral CORTEX and is a particularly large area in humans, and is functionally the most important part of the brain. The cerebral cortex is made of GREY MATTER on the outside and white matter underneath this. Within this cortex different functional regions have been localized, for example, visual, speech and auditory areas. Linked to these areas are association areas, for example the visual association area, which allow an individual to interpret information received in relation to previous experience.

In humans the left cerebral hemisphere is associated with control of the right side of the body, and vice versa, due to the crossing over of nerve fibres as they enter the brain from the body.

See also THALAMUS.

cerium (Ce) The element with ATOMIC NUMBER 58; RELATIVE ATOMIC MASS 140.1; melting point 798°C; boiling point 3,433°C; RELATIVE DENSITY 6.8. Cerium is the most abundant of the LANTHANIDE metals and is used in some specialist alloys.

CERN (European Centre for Nuclear Research) A multinational institution, located in Geneva, Switzerland, home to many of the world's largest PARTICLE ACCELERATORS. *See also* LEP, SPS.

cervix The neck of the mammalian UTERUS that consists of a ring of muscle at the base of the uterus opening into the VAGINA. The cervix secretes MUCUS into the vagina.

cesium *See* CAESIUM.

Cestoda A CLASS of the PHYLUM PLATYHELMINTHES, consisting of flatworms. The Cestoda are all PARASITES and are of great economic importance. An example is the pork tapeworm, *Taenia solium,* which uses the human as its primary host and the pig as an intermediate, and causes anaemia, diarrhoea, weight loss and intestinal blockage and pain in humans.

CFC *See* CHLOROFLUOROCARBON.

c.g.s. units A system of physical units derived from the metric system and based on the centimetre, gram and second. It has been replaced by SI UNITS.

chaeta In invertebrates, bristles made of CHITIN used in locomotion. *See also* ANNELIDA.

chain reaction Any reaction in which the products of the reaction initiate further reactions of the same type, leading to an exponential growth in the rate of reaction.

In physics, a NUCLEAR FISSION process may lead to a chain reaction, where neutrons initiate fission, which produces further neutrons for more fission, and so on. Such reactions can be controlled using moderators to absorb excess neutrons, but if there is no control this can lead to a nuclear explosion. *See also* CRITICAL MASS, THERMONUCLEAR REACTION.

In chemistry, chain reactions usually involve RADICALS, atoms or molecules having unpaired electrons. There are three stages in a chain reaction – initiation, propagation and termination. During initiation the reactive species are formed, often radicals. These are then involved in subsequent reactions that yield similar or different reactive species, themselves able to repeat the process thus creating a chain of reactions. The chain is terminated when a reaction occurs in which the product is unable to react further.

An example of a chain reaction is the reaction between hydrogen and bromine to form hydrogen bromide (HBr). The reactive species here is the bromine radical (.Br) formed by the splitting of bromine (Br_2). This attacks hydrogen molecules

$$Br\cdot + H_2 \rightarrow HBr + .H$$

and then H· generated reacts with bromine

$$H\cdot + Br_2 \rightarrow HBr + .Br$$

The Br· can then react with another hydrogen so continuing the chain. The reaction is terminated if Br reacts with another .Br to form Br_2, thereby removing the reactive species. Chain reactions can be inhibited (the rate of production of the product reduced) by one of the products formed. Many organic reactions involving radicals (*see* HOMOLYTIC FISSION) proceed by a chain reaction, for instance the reaction of methane with chlorine.

The speed at which chain reactions occur is variable, some slowly and some accelerating with increasing number of reactive species to the point of an EXPLOSION.

change of state Any process in which substance changes from one of the STATES OF MATTER to another. In any change of state, LATENT HEAT is taken in or given out. *See* BOILING POINT, MELTING POINT, SUBLIMATION.

channel 1. A radio frequency allocated for a particular purpose. The spacing between channels is arranged to allow for the BANDWIDTH of the transmitted signal.

2. The semiconducting material through which CHARGE CARRIERS flow in a FIELD EFFECT TRANSISTOR.

chaos The state of a system whereby a small change in the initial state of the system leads to large changes in the final state. Since initial conditions can never be known precisely, it is impossible to predict the behaviour of chaotic systems over long periods of time, although they may obey essentially simple laws. An important example is in meteorology, where the pattern of air movement is so complex that it is difficult to confidently predict the weather more than a day or so ahead. It has been said that a butterfly flapping its wings in one part of the world may eventually lead to a hurricane somewhere else. Surprisingly, even quite simple systems can exhibit chaotic behaviour.

char (*vb.*) To reduce an organic material to carbon, either by the action of heating or by the action of a strong dehydrating agent (*see* DEHYDRATE), such as concentrated sulphuric acid.

charge A fundamental property of matter. Charges are of two types, positive and negative. Charged particles exert forces on one another: charges of the same type repel one another whilst opposite charges attract. The SI UNIT of charge is the COULOMB (C).

Electrons are negatively charged, whilst protons have an equal and opposite positive charge of 1.6×10^{-19} C. These are the basic units of electrically charged matter. If an object has equal numbers of electrons and protons it is electrically neutral, if it has an excess of electrons it has an overall negative charge, while if it has an excess of protons it has a positive charge. The flow of charged particles, in particular the flow of electrons, is what constitutes an electric current. The removal of a charge from an object is called discharging.

See also CAPACITOR, CHARGING BY FRICTION, CHARGING BY INDUCTION, COULOMB'S LAW, CURRENT, INDUCED CHARGE, MILIKAN'S OIL DROP EXPERIMENT, STATIC ELECTRICITY.

charge carrier A charged particle that moves through a substance to carry a current. In metals, the charge carriers are electrons. In SEMICONDUCTORS they are electrons or HOLES. In IONIC liquids and IONIZED gases the charge carriers are ions.

charge coupled device *See* CCD.

charge density The amount of charge per unit volume.

charged (*adj.*) The state of a CAPACITOR or electrochemical CELL when it is storing the greatest possible amount of charge. In the case of a capacitor, this is when the POTENTIAL DIFFERENCE across its plates is equal to the ELECTROMOTIVE FORCE of the charging supply. The opposite state, when the charge stored is zero, is described as discharged.

charging by friction The mechanism by which a pair of objects gain an equal and opposite charge when they are rubbed against one another. Electrons are transferred from one of the objects, leaving it with a positive charge, to the other, which gains a negative charge. *See also* LIGHTNING, VAN DE GRAAF GENERATOR.

charging by induction The process by which an object can be given a charge by placing it near an electrically charged object. If an uncharged conductor is placed near a positively charged rod, for example, electrons in the conductor are attracted towards the positive rod. The conductor remains electrically neutral overall, but the influence of the charged rod redistributes the charges in the conductor. The surface of the conductor closest to the rod will become negatively charged, leaving the opposite surface positively charged. If the conductor is then separated into two parts, they will have equal and opposite charges. Such charges are called induced charges.

Charles' law For a fixed mass of an ideal gas at constant pressure, the volume is proportional to the ABSOLUTE TEMPERATURE; that is, the volume divided by the absolute temperature is a constant. For a fixed mass of an ideal gas with an absolute temperature T in volume V, at constant pressure: V/T = constant. *See also* BOYLE'S LAW, GAS LAW, IDEAL GAS EQUATION, PRESSURE LAW.

charm A flavour of QUARK or the QUANTUM NUMBER associated with that quark. Charm is conserved in interactions involving the STRONG NUCLEAR FORCE but not in those involving the WEAK NUCLEAR FORCE. *See also* J/Ψ.

Charophyta A PHYLUM of the KINGDOM PROTOCTISTA that consists of the stoneworts. They grow submerged in HARD WATER and are usually covered in a thick, brittle crust of calcium carbonate.

chelate *See* CHELATING AGENT.

chelating agent Any compound that forms a complex consisting of charged ion(s) or metal atom(s) chemically bonded to and surrounded by chains of organic residues. This linkage of the ion or atom joins the organic chains to form complete rings. An example of a chelating agent is EDTA. The compound formed is called the chelate and the process is called chelation. EDTA complexes calcium ions by surrounding them with four oxygen atoms and two nitrogen atoms. In this configuration the metal ions are strongly attached to the organic molecules. Another example of chelation is the formation of complexes similar to copper tetrammine between copper ions and 1,2 diaminoethane, $[(H_2NCH_2CH_2NH_2)_2Cu]^{2+}$.

Chelating agents are used in chemical analysis and in determining the hardness of water. They are also used in the removal of toxic metal ions from the body, in lead poisoning (lead ions are chelated and since the complexes are soluble they can be safely excreted) and in THALASSAEMIA to remove excess iron (to prevent the build up of toxic levels). Chelates can act as carriers of essential metal ions, useful in the agricultural industry.

chelation *See* CHELATING AGENT.

chemical bond The force of attraction that holds atoms together in a molecule or LATTICE. Chemical bonds are of sufficient force that they can only be broken by a chemical reaction and not by thermal vibrations at the temperatures under consideration. Bonds are broadly categorized as IONIC BONDS and COVALENT BONDS. Ionic, or electrovalent bonds, arise from the ELECTROSTATIC forces of attraction between oppositely charged ions. In covalent bonding, pairs of atoms share electrons to form a bond that is directed in space.

Many molecules have chemical bonds intermediate between ionic and covalent. These are described as POLAR: the bond is regarded as covalent but the electron pair in the bond spends more time with one atom than the other. This polarizes the molecule, so it has a negative charge on one end, which acts as an ionic bond.

See also BOND ENERGY, CO-ORDINATE BOND, ELECTRONEGATIVITY, HYDROGEN BOND, METALLIC BOND, POLAR BOND, SHELL, VALENCY.

chemical combination The combination of elements to give compounds. There are three laws that govern chemical combination: the LAW OF CONSTANT PROPORTIONS, the LAW OF EQUIVALENT PROPORTIONS, and the LAW OF MULTIPLE PROPORTIONS. These laws were originally based on empirical evidence, but later recognized to be a consequence of the different masses of atoms and the simple ratios in which atoms combine to form compounds. *See also* DALTON'S ATOMIC THEORY.

chemical energy The energy that can be released in forming or breaking chemical bonds. As the forces involved in chemical reactions are ELECTROSTATIC in nature, this can be thought of as a form of ELECTRICAL POTENTIAL ENERGY.

chemical engineering The study of chemical processes applied to the commercial manufacture of materials, often on a large scale. It is also the design, construction and control of equipment for carrying out large-scale chemical processes.

chemical equation A shorthand for showing the reagents and products in a chemical reaction, and the number of molecules of each substance involved. For example, the reaction between hydrochloric acid and magnesium to give hydrogen and magnesium sulphate could be denoted by the equation:

$$2HCl + Mg \rightarrow MgCl_2 + H_2$$

The equation shows that two molecules of hydrogen chloride are needed for every atom of magnesium. It is also common to attach state symbols to the equation to indicate whether a substance is a solid (s), liquid (l), aqueous solution (aq) or a gas (g). With state symbols, the above equation would be written:

$$2HCl\,(aq) + Mg\,(s) \rightarrow MgCl_2\,(aq) + H_2\,(g)$$

chemical equilibrium The state of a system of containing two or more chemical substances, where the concentration of all substances is constant over time. Particularly applied to a chemical reaction that could proceed in either direction, such as the formation of ammonia:

$$N_2 + 3H_2 \Leftrightarrow 2NH_3$$

In this reaction, a chemical equilibrium exists when the concentrations of all three gases are constant. Chemical equilibrium is an example of a dynamic equilibrium: individual molecules continue to react, but the reactions in each direction take place at the same rate. *See also* CHEMICAL POTENTIAL, EQUILIBRIUM, EQUILIBRIUM CONSTANT, LE CHATELIER'S PRINCIPLE, REVERSIBLE REACTION.

chemical equivalent The mass of a specified substance that can displace or react with 1 g of hydrogen or 8 g of oxygen, either directly or indirectly. For example, the chemical equivalent of chlorine is 35.5 g, since that much chlorine reacts with hydrogen in the reaction

$$Cl_2 + H_2 \rightarrow 2HCl$$

For a compound where more than one reaction is possible, or an element with more than one VALENCY, the equivalent weight will not be unambiguously defined, thus carbon has an equivalent weight of 3 g in

$$C + 2H_2 \rightarrow CH_4$$

but 12 g in

$$2C + H_2 \rightarrow C_2H_2$$

chemical potential In any system with two or more chemical components, the chemical potential of each component is the rate of change of the Gibbs free energy (*see* FREE ENERGY) of the system with the amount of that component, all other factors being kept constant. Chemical potential is a useful quantity in studying the CHEMICAL EQUILIBRIUM of a system: equilibrium is achieved when all components have equal chemical potential. *See also* WATER POTENTIAL.

chemical reaction Any process in which one or more COMPOUNDS or ELEMENTS interact, producing different compounds or converting a compound into its constituent elements.

See also ADDITION REACTION, CHAIN REACTION, CHEMICAL COMBINATION, CHEMICAL EQUILIBRIUM, COMBUSTION, CONDENSATION REACTION, DISPLACEMENT REACTION, ELIMINATION REACTION, HEAT OF REACTION, HESS'S LAW, KINETICS, OXIDATION, RATE OF REACTION, REACTIVITY, REDUCTION, REDOX REACTION, REVERSIBLE REACTION, SUBSTITUTION REACTION.

chemiosmotic theory *See* ELECTRON TRANSPORT SYSTEM.

chemisorption The ADSORPTION of a gas by a solid in which the molecules of the adsorbed gas are held on the surface of the adsorbing solid by the formation of chemical bonds.

chemistry The science of the elements and the ways in which they interact with one another, particularly in the formation of chemical bonds. Chemistry is usually divided into physical chemistry, the study of the physical properties of substances and how they are affected by chemical change; organic chemistry, which deals with all molecules containing carbon (except oxides of carbon and metal carbides); and inorganic chemistry, which deals with all other compounds.

chemoautotroph *See* CHEMOSYNTHESIS.

chemonasty The NASTIC MOVEMENT of plants in response to chemicals.

chemoreceptor A RECEPTOR cell that responds to chemical changes in the internal or external environment. *See* SENSE ORGAN, TONGUE.

chemosynthesis A method of AUTOTROPHIC NUTRITION (self-feeding) in which chemical energy is used to synthesize organic compounds. It is similar to PHOTOSYNTHESIS, which uses light energy. Certain bacteria OXIDIZE a variety of inorganic chemicals to generate the energy needed for chemosynthesis and are vitally important in the recycling of minerals. Examples include the nitrifying bacteria of the NITROGEN CYCLE, which convert free nitrogen into a form that can be used by plants. *Nitrobacter* converts NITRITES (NO^{2-}) to NITRATES (NO^{3-}); *Thiobacillus* converts sulphur (S) to SULPHATE (SO_4^{2-}); and *Ferrobacillus* converts ferric (Fe^{2+}) to ferrous (Fe^{3+}). Such organisms are called chemoautotrophs. As they do not need light to function, they can survive where other organisms cannot.

chemotaxis *See* TAXIS.

chemotherapy The treatment of medical conditions, usually cancer, with chemicals.

chemotropism The directional growth of a plant (or part of it) in response to a chemical stimulus. *See* TROPISM.

chi-squared test A statistical test measuring the extent of deviation between an observed result and what was expected. From this it is possible to calculate whether the deviation is due to chance (and so is non-significant) or not (and so is significant). Chi is represented by the Greek letter χ^2, shown squared, and can be calculated by the following equation:

$$\chi^2 = \Sigma\,(O - E)^2/E$$

where Σ is 'the sum of', O is the observed result and E is the expected result.

Using a χ^2 table, and relating the χ^2 value to the number of degrees of freedom (which is the number of classes of results minus 1), the probability that the deviation is due to chance alone can be calculated. If the probability is less than 5 per cent then the observed deviation is considered significant. If the probability is more than 5 per cent then the deviation is not significant.

The chi-squared test is useful in genetics. For example, in *Drosophila* there are two types of wings: normal and vestigial. Normal wings are DOMINANT to vestigial wings and if two normal-winged individuals are crossed, the expected ratio of normal to vestigial wings would be 3:1 (*see* MENDEL'S LAWS). In practice the numbers of each wing-type may not correspond exactly to this ratio. Using the chi-squared test it can be calculated whether the actual numbers seen are significantly different from the expected ratio, or whether the difference can be explained by chance.

chiasma (*pl.* ***chiasmata***) A point where two CHROMATIDS from homologous pairs of chromosomes join as they wrap around one another during MEIOSIS. It is at chiasmata that chromatids may separate and rejoin during CROSSING-OVER and RECOMBINATION.

Chilopoda A CLASS of the PHYLUM ARTHROPODA, including the centipede.

chiral (Greek *cheir* = hand) A molecule that cannot be superimposed on its mirror image because it is asymmetrical (like the relationship of the right hand to the left hand). Any compound with four different atoms or groups attached to a single carbon is chiral. The central carbon atom is termed the chiral centre. The term 'chirality' describes the property of a molecule to exhibit non-identity with its mirror image ('handedness'). The term can be used to refer to any asymmetric object or molecule. A chiral molecule would exhibit OPTICAL ISOMERISM.

chirality *See* CHIRAL.

chitin A structural POLYSACCHARIDE that forms the hard, protective EXOSKELETON of insects and arthropods. It is a complex, long chain nitrogenous derivative of GLUCOSE. Chitin can also be soft and flexible, as in caterpillars. It is insoluble in water and protects the organism from many solvents, acids and alkalis. In

CRUSTACEANS, chitin combines with calcium carbonate to give extra strength.

chloral *See* TRICHLOROETHANAL.

chlorate Any salt containing an ANION that contains chlorine and oxygen. Most common is the chlorate(V) ion, ClO_3^-, but chlorate(I) (or hypochlorite) ClO^-, chlorate(III) (or chlorite) ClO_2^- and chlorate(VII) (or perchlorate) ClO_4^- ions also exist.

chlorenchyma PARENCHYMA tissue in plants that contains CHLOROPLASTS.

chloric acid Any acid comprising hydrogen, oxygen and chlorine. They are: chloric(I) acid (hypochlorous acid), HOCl; chloric(III) acid (chlorous acid), $HClO_2$; chloric(V) acid $HClO_3$; and chloric(VII) acid (perchlorous acid) $HClO_4$. Chloric(I) and chloric(III) acids are stable only in solution, chloric(V) and chloric(VII) acids can be isolated as ANHYDROUS liquids, but are unstable and explode on heating. All are OXIDIZING AGENTS with oxidizing strengths that increase with the oxygen content of the acid. Chloric(I) and(III) acids are weak acids, chloric(V) and(VII) acids are strong acids.

chloride Any BINARY COMPOUND containing chlorine. Non-metallic chlorides are covalently bonded (*see* COVALENT BOND). CARBON TETRACHLORIDE (tetrachloromethane), CCl_4, is an important solvent. The metallic chlorides are IONIC.

chlorination The incorporation of chlorine into an organic compound, by ADDITION REACTIONS or SUBSTITUTION REACTIONS.

chlorine (Cl) The element with ATOMIC NUMBER 17; RELATIVE ATOMIC MASS 35.5; melting point –101°C; boiling point –35°C. Chlorine is a greenish yellow gas, the most widely used of the HALOGENS. It is found in large quantities in seawater, where it occurs as dissolved sodium chloride, from which it is extracted by ELECTROLYSIS in a MERCURY-CATHODE CELL. It is a powerful OXIDIZING AGENT and is widely used for its bleaching properties. Industrially, it is used in the manufacture of many chlorinated organic compounds, such as polyvinyl chloride (PVC) which is an important PLASTIC.

chlorite Any salt containing the chlorate(III) (or chlorite) ion, ClO_3^-.

chlorobenzene (C_6H_5Cl) An AROMATIC compound; boiling point 132°C. It is used in the manufacture of DDT. Chlorobenzene consists of a BENZENE RING in which a hydrogen atom has been replaced by a chlorine atom, by ELECTROPHILIC SUBSTITUTION. It is prepared by the direct chorination of benzene in the presence of an iron catalyst.

chlorocruorin A green RESPIRATORY PIGMENT found in certain POLYCHAETES.

chloroethene *See* VINYL CHLORIDE.

chlorofluorocarbon (CFC) A synthetic chemical used in AEROSOLS as a propellant, in refrigerators and air conditioners as a coolant, and in foam packaging. Although CFC's are chemically inert, odourless, non-flammable and non-toxic, they can remain in the Earth's atmosphere for more than 100 years. This is destroying the OZONE LAYER. When CFC's are released into the atmosphere, ultraviolet radiation from the sun causes them to break down into chloride atoms that combine with OZONE. This decreases the ozone concentration, which allows harmful radiation from the Sun to reach the Earth. The more destructive CFC's have now been banned and replacements are being developed, and safer methods of disposal are being investigated for existing CFC's. *See also* GREENHOUSE EFFECT.

chloroform, ***trichloromethane*** ($CHCl_3$) A colourless liquid with a pleasant, sweet odour; boiling point 60–61°C. It is made by the CHLORINATION of methane. It was used as an ANAESTHETIC but has been replaced by less toxic chemicals. Chloroform is a good solvent for fats, waxes, rubber, for instance, and is used chiefly in the manufacture of CHLOROFLUOROCARBON refrigerants.

chlorophyll A green pigment present in most plants and ALGAE that is responsible for the capture of light during PHOTOSYNTHESIS, and for the coloration of plants. Chlorophyll is found within the CHLOROPLAST of higher plants and algae. In CYANOBACTERIA, where there is no chloroplast, chlorophyll is found on special photosynthetic membranes (THYLAKOIDS) lying free in the CYTOPLASM. Several chlorophylls exist (a, b, c, d and e); chlorophyll a is common to all plants (and the only one in cyanobacteria). In a few photosynthetic bacteria, another type of chlorophyll, bacteriochlorophyll, occurs. Chlorophyll is similar in structure to HAEMOGLOBIN but contains magnesium instead of iron.

Chlorophyta A PHYLUM from the KINGDOM PROTOCTISTA that consists of green ALGAE. Green algae can be unicellular or filamentous and contain CHLOROPHYLL a and b, like higher plants. Their CHLOROPLASTS are various shapes: in *Chlamydomonas* the chloroplasts are bowl-shaped and in *Spirogyra* they are spiral. Some green algae have flagella (*see* FLAGELLUM). Most are found in fresh water. Both ASEXUAL REPRODUCTION and SEXUAL REPRODUCTION can occur.

chloroplast In eukaryotic plant and algal cells, a PLASTID containing the green pigment CHLOROPHYLL that is the site of PHOTOSYNTHESIS. Chloroplasts are therefore found in most cells of plants and algae that are exposed to light. They also contain CAROTENOID pigments.

In higher plants chloroplasts are usually disc-shaped, but change their shape and position in relation to light intensity. There is an outer chloroplast envelope that is made up of two membranes. The outer membrane has a similar structure to the CELL MEMBRANE, but the inner membrane consists of a series of folds. Within the envelope lies the STROMA, which is a colourless, structureless matrix containing a stack of around 50 grana per chloroplast. Each granum is made of a number of closed, flattened sacs called THYLAKOIDS, and inside these are the photosynthetic pigments such as chlorophyll. Grana can be connected to one another by large intergranal thylakoids. Absorption of light energy occurs in the grana and the subsequent utilization of energy to form CARBOHYDRATES occurs in the stroma. The main food reserve is STARCH, which is stored as grains in the chloroplast. Chloroplasts also contain some DNA and RIBOSOMES.

In algae, chloroplasts can be various shapes and there may be one or more per cell, often associated with PYRENOIDS. Grana do not occur in algae, and thylakoids run across the stroma of the chloroplast as a whole. It is thought that chloroplasts were originally free-living CYANOBACTERIA that invaded larger non-photosynthetic cells and developed a symbiotic relationship (*see* SYMBIOSIS) with them.

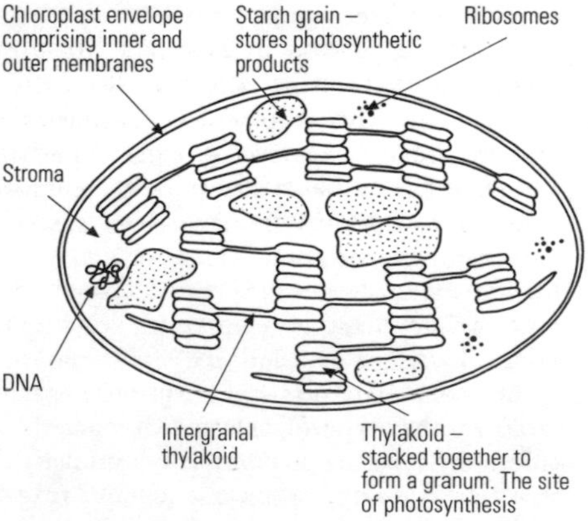

Structure of a chloroplast.

choanocyte In sponges, a cell bearing FLAGELLA, which circulate water thorough the body. *See* PORIFERA.

cholecalciserol *See* VITAMIN D.

cholecystokinin A HORMONE made by the SMALL INTESTINE in response to the presence of acid CHYME from the stomach. Cholecystokinin inhibits secretion of GASTRIC JUICE but stimulates production of PANCREATIC JUICE and BILE.

cholesterol A STEROL LIPID that is a component of all CELL MEMBRANES and a precursor of STEROID HORMONES. It is found throughout the body of animals, but is not present in higher plants or bacteria. In the diet cholesterol is obtained from dairy products and meat. A high level of cholesterol in the blood is thought to contribute to ATHEROSCLEROSIS. It is broken down by the liver into BILE SALTS and any excess is secreted in the BILE.

Chondrichthyes A CLASS of vertebrates, consisting of cartilaginous fish (which have a skeleton made of CARTILAGE), for example, the dogfish and ray. *Compare* OSTEICHTHYES.

chondrin A firm, elastic substance that forms the matrix of CARTILAGE. Embedded within this matrix are COLLAGEN fibres, which gives the cartilage its strength.

chondrocyte A CARTILAGE cell that secretes the matrix of CHONDRIN.

chord A straight line drawn between two points on a curve, especially a circle.

Chordata A PHYLUM of the animal KINGDOM consisting of animals that at some stage of their lives have a supporting rod of tissue (NOTOCHORD or VERTEBRAL COLUMN) running down their bodies. This phylum includes vertebrates, which are the dominant species of land, sea and air, in mass and ecological dominance rather than in numbers (*see* VERTEBRATA).

chordate A member of the PHYLUM CHORDATA.

chorion In most mammals, the outermost EXTRAEMBRYONIC MEMBRANE that is next to the uterine walls (*see* UTERUS). VILLI from part of the chorion (chorionic villi) invade the tissue of the mother, forming the TROPHOBLAST that later develops into the PLACENTA. In reptiles and birds the chorion and ALLANTOIS form a surface for gaseous exchange.

chorionic villus *See* CHORION.

chorionic villus sampling (CVS) In mammals, a procedure performed during early pregnancy to detect chromosomal abnormalities in the foetus. A small biopsy of chorionic villus tissue (*see* CHORION) is taken at 6–10 weeks, which is earlier than when the similar AMNIOCENTESIS analysis is carried out. There is some risk of miscarriage.

choroid A pigmented layer inside the white of the eye that is rich in blood vessels and supplies the RETINA.

choroid plexus A membrane that lines the VENTRICLES of the brain and secretes CEREBROSPINAL FLUID.

CHP *See* COMBINED HEAT AND POWER.

chromate(VI) Any salt containing the chromate(VI) ION, CrO_4^{2-}. The ion has a yellow colour. In solution, chromate(VI) salts are stable only under alkaline conditions. In the presence of acids, the chromate(VI) ion is converted to the DICHROMATE ion:

$$2CrO_4^{2-} + H^+ \rightarrow Cr_2O_7^{2-} + H_2O$$

chromatic aberration *See* ABERRATION.

chromatid In EUKARYOTES, one of two strands of CHROMATIN that make up chromosomes. Chromatids are joined together at a point called the CENTROMERE. In MITOSIS the chromatids are identical, but in MEIOSIS CROSSING-OVER can occur (exchange of chromosome fragments; *see* RECOMBINATION).

chromatin In EUKARYOTES, the material that chromosomes are made of. Chromatin consists of DNA and five different HISTONE proteins that are organized into two strands, called CHROMATIDS.

chromatogram The pattern of separated chemicals along a separating medium in CHROMATOGRAPHY.

chromatography An analytical technique used to separate the components of a mixture, which flow at different rates along some separating medium. The fixed material over which the mixture passes is called the stationary phase and is usually a solid or GEL. The moving fluid, often a liquid, but sometimes a gas, is called the moving phase. The mixture is dissolved in a solvent, called an ELUENT in this context, which then diffuses along the separating medium. In adsorption chromatography a column of aluminium oxide is often used as the solid phase. The different strengths of the interatomic force between the molecules being separated and the molecules of the stationary phase and the eluent are responsible for the separation of the mixture. The different components may then be identified, and in some cases may be determined quantitatively. *See also* CHROMATOGRAM, COLUMN CHROMATOGRAPHY, GAS CHROMATOGRAPHY, ION-EXCHANGE CHROMATOGRAPHY, PAPER CHROMATOGRAPHY, PARTITION CHROMATOGRAPHY, THIN-LAYER CHROMATOGRAPHY.

chromatophore **1.** In PROKARYOTES, a membrane-bounded vesicle (THYLAKOID) containing photosynthetic pigments. It serves a similar function to the CHLOROPLAST in EUKARYOTES.

2. *See* CHROMOPLAST.

chrome alum *See* POTASSIUM CHROMIUM SULPHATE.

chromic acid The hypothetical acid H_2CrO_4, containing the chromate(VI) ion, CrO_4^{2-}. Salts with this ion do exist, but the acid itself has not been isolated.

chromium (Cr) The element with ATOMIC NUMBER 24; RELATIVE ATOMIC MASS 52.0; melting point 1,900°C; boiling point 2,640°C; RELATIVE DENSITY 7.2. Chromium is a TRANSITION METAL and shows multiple VALENCY. It is fairly resistant to corrosion and is often used to ELECTROPLATE steel parts, giving them a hard shiny finish. It is also incorporated into many steels, particularly stainless steel. *See also* THERMIT PROCESS.

chromium oxide Any of the BINARY COMPOUNDS of chromium with oxygen. They are: chromium(II) oxide, CrO, a black powder; chromium(III) oxide, Cr_2O_3, a green crystalline solid; chromium(IV) oxide (chromium dioxide), CrO_2, a black powder; and chromium(VI) oxide (chromium trioxide), CrO_3, a red crystalline solid. Chromium(III) oxide is the most stable of the oxides. Chromium(VI) oxide is an extremely strong OXIDIZING AGENT and decomposes partially on heating:

$$4CrO_3 \rightarrow 2Cr_2O_3 + 3O_2$$

Chromium(VI) oxide is the only chromium oxide to be soluble in water, forming an acidic solution believed to contain a mixture of chromic acid, H_2CrO_4, and dichromic acid, $H_2Cr_2O_7$.

chromoplast, *chromatophore* A PLASTID that may arise from a CHLOROPLAST. It has coloured CAROTENOID pigments but no CHLOROPHYLL.

chromosome A structure in the cell nucleus that carries the genetic material DNA and RNA along with some proteins. It is only visible during CELL DIVISION. Higher organisms, such as humans, are DIPLOID, which means that there are two copies of each chromosome. Humans have 46 chromosomes, forming 23 pairs. The number of chromosomes differs between species. One set of chromosomes is derived from each parent and carried in the GAMETE. Some organisms have only one set of chromosomes and are called HAPLOID. More than two sets of chromosomes is called polyploidy (*see* POLYPLOID) and is rare in animals but arises spontaneously quite frequently in plants.

In EUKARYOTES, chromosomes consist of DNA and HISTONE that form CHROMATIN, which is organized into two strands (CHROMATIDS) joined together at a point called the CENTROMERE. Each chromosome has a characteristic banding pattern when stained with a routine coloured stain, and varies in size and shape within and between species. PROKARYOTES usually have only one circular chromosome per cell, and may have a small percentage of their DNA in another smaller loop called a PLASMID. In humans, sex is determined by the sex chromosomes, the X-chromosome and the Y-chromosome (*see* SEX DETERMINATION).

See also AUTOSOME, HETEROSOME, MUTATION, CHROMOSOME MAP, GENE, KARYOTYPE, LINKAGE, MITOSIS, MEIOSIS, POLYPLOID, RECOMBINATION, SEX LINKAGE.

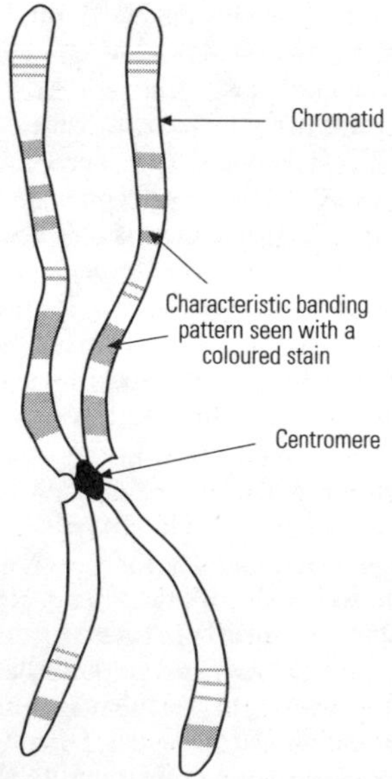

Structure of a chromosome.

chromosome map A linear map showing the arrangement of GENES on a CHROMOSOME. It is useful for identifying the relative positions of particular genes, and especially for studying abnormal genes. *See also* BACKCROSS.

chromosome mapping The techniques involved in determining a CHROMOSOME MAP. Chromosome mapping involves examining the PHENOTYPES of offspring obtained in appropriate breeding routines and studying LINKAGE. Mapping information can also be obtained by analysis of DNA sequences.

chrysalis *See* PUPA.

Chrysophyta A PHYLUM of the KINGDOM PROTOCTISTA that consists of the golden-brown ALGAE.

chyme The creamy and partly digested food that is released by the STOMACH into the SMALL INTESTINE.

cilia (*sing.* ***cilium***) Many thread-like structures on the surface of some cells that can contract to produce rhythmic co-ordinated waving movements. Cilia can be used for cell locomotion in single-celled organisms, such as *Paramecium*. In multicellular organisms they are present as ciliated EPITHELIUM, where they move the surrounding fluid or substances through ducts, such as FALLOPIAN TUBES or TRACHEA, often with the help of MUCUS secreted by single glandular cells or GOBLET CELLS.

Cilia have a characteristic arrangement of internal fibres: two central and nine peripheral pairs. The central fibres can be absent, in which case the structure is non-motile. In size cilia are about 0.2 μm in diameter and 10 μm long. They are similar to FLAGELLA, but the latter are longer, fewer and move in a different way.

ciliary muscles The ring of muscles that surrounds the EYE lens, which can contract to squeeze it into a fatter shape for viewing nearby objects.

ciliate A member of the PHYLUM CILIOPHORA.

Ciliophora A PHYLUM from the KINGDOM PROTOCTISTA, consisting of ciliate PROTOZOA. Ciliates possess CILIA, which they use to move and to trap food. Reproduction is by BINARY FISSION or CONJUGATION. They have two nuclei. Examples are *Paramecium* and *Vorticella.*

cinnabar A mineral form of mercury sulphide, HgS. It is an important ore of mercury.

circadian rhythm A BIORHYTHM found in most organisms that is related to the 24-hour day cycle, such as the sleeping/waking pattern. Other factors, such as hormone concentration and temperature, may also vary throughout the day, affecting behaviour and mood.

circalunar rhythm A BIORHYTHM that follows a 28-day cycle related to the phases of the moon.

circle A closed curve joining a set of points that are all equally distant from a fixed point, called the centre of the circle.

The boundary of the circle is called the circumference. The distance from any point on the circumference to the centre of the circle is the radius. Any portion of the circumference is an arc. A straight line joining two points on the circumference is a chord. A chord that passes through the centre of the circle is the diameter: the diameter is equal to twice the radius. The portion cut off by a chord is a segment; the portion cut off by two radii is a sector.

In CARTESIAN CO-ORDINATES, the equation of a circle with its centre at the origin is

$$x^2 + y^2 = r^2$$

where r is the radius.

circuit, electric Any closed conducting path around which an electric current can flow. *See also* CONTACT BREAKER, OPEN CIRCUIT, SHORT CIRCUIT.

circularly polarized (*adj.*) Describing a form of polarized light (*see* POLARIZATION) in which the vector describing the electric field rotates as the wave propagates, tracing out a spiral path. Circularly polarized light can be thought of as the superposition of two plane polarizations with a 90° phase difference. *See also* PLANE-POLARIZED.

circular motion An object moving along a circular path at a constant speed is accelerating due to the change in its direction. In other words, its velocity is changing though its speed may be constant. This acceleration is directed towards the centre of the circle and is called CENTRIPETAL ACCELERATION. To produce this acceleration, a force is needed; this force is called a CENTRIPETAL FORCE. For an object moving at a constant speed v around a circle of radius r, the centripetal acceleration is v^2/r. *See also* CENTRIFUGAL FORCE, CORIOLIS FORCE.

circulatory system In an animal, a network of vessels that distributes blood containing essential substances throughout the body. Circulatory system can also refer to other circulating body fluids, such as LYMPH.

In the simplest animals, such as NEMATODES, an open blood system exists where there are no vessels, and blood (called haemolymph) moves freely over the tissues through spaces collectively called the haemocoel. Insects have an open system with a heart providing some circulation into the haemocoel. In this open system blood is moved at low pressure and with little control.

Larger animals, including humans, have a closed blood system in which the blood is contained within the blood vessels and pumped around the body at high pressure by the heart. In humans, other mammals and birds, blood passes from the heart to the lungs and then back to the heart before it is ready for circulation around the body. This is called double circulation and is more efficient than the single circulation of fish, where blood circulates once around the body before returning to the heart and becomes sluggish as the pressure drops further from the heart. In a closed system, blood flows in one direction only, ensured by one-way valves in the heart, arteries and veins.

See also HEART, LYMPHATIC SYSTEM.

circumference The boundary line of a circle, and the length of this line. The circumference of a circle is equal to 2π times its radius.

circumpolar orbit An orbit that passes over the Earth's North and South Poles. For an orbit just above the atmosphere, the period is about 90 minutes, whilst the Earth rotates underneath the satellite every 24 hours. Thus the satellite has a view of every point on the Earth at least once every 12 hours. Such orbits are used by some weather satellites and also to view the

Earth for military and environmental monitoring. *See also* GEOSTATIONARY ORBIT, ORBIT.

cirrus High-level clouds, composed of ice crystals. They appear as feathery white wisps. *See also* WEATHER SYSTEMS.

CISC (**Complex Instruction Set Computing**) An approach to the design of microprocessing, in which the instruction set contains as many functions as the MICROPROCESSOR can handle. The complexity of all these instructions slows the microprocessor down, and CISC has now been superseded by RISC.

cisternum A cavity or vesicle in the cytoplasm of a living cell, formed by membranes of the ENDOPLASMIC RETICULUM or GOLGI APPARATUS.

cis/trans isomerism *See* GEOMETRIC ISOMERISM.

cistron A region of DNA containing structural genes responsible for the production of a complete POLYPEPTIDE. In modern terminology a cistron is considered to be equivalent to a gene. *See also* GENE EXPRESSION, GENE, OPERON.

citric acid An organic acid found in many plants. It exists in particularly high concentrations in citrus fruits, such as oranges and lemons. It has a sharp, sour taste. Citric acid is an intermediate in the KREBS CYCLE.

citric acid cycle *See* KREBS CYCLE.

CJD *See* CREUTZFELDT-JAKOB DISEASE.

class One of the subdivisions of a PHYLUM in the CLASSIFICATION of organisms. MAMMALIA is the class consisting of mammals, AVES is the class consisting of birds. Plant class names end in 'idae' and fungi class names end in 'mycetes'. Classes are further subdivided into ORDERS.

classification The organization of all organisms (living and extinct) into groups, for human convenience, based on similarities in their physiological, anatomical and biochemical characteristics. There are many ways of doing this, none of which are perfect.

In the 18th century, Carolus Linnaeus (1707–78) devised a natural classification scheme based on homologous characters (features with a similarity in origin, structure and position) rather than analogous characters (similar functions but with different origins). The divisions (or ranks) in classification used today are based on those of Linnaeus. The first division is a KINGDOM, of which there are usually five: ANIMALIA, PLANTAE, FUNGI, PROTOCTISTA (these four are all EUKARYOTES) and PROKARYOTAE (PROKARYOTES). The divisions below a kingdom are PHYLUM, CLASS, ORDER, FAMILY, GENUS, SPECIES.

The written name given to an organism consists of two parts (BINOMIAL NOMENCLATURE): the genus (with a capital letter) and the species (with a small initial letter) both written in italics, for example, *Homo sapiens* is the name for humans. The species is the lowest level in the classification scheme, but within a species different populations may exist (for example, different breeds of dog).

An artificial classification scheme is based on the differences between organisms and is used to identify species. It uses keys that are a series of divisions based on a single character, for example, wings present/wings absent, from which a species can eventually be identified. In this method of classification, groups may contain unrelated forms.

clathrate A solid mixture in which individual molecules of one compound are trapped within the crystal LATTICE of another.

Clausius statement of the second law of thermodynamics The SECOND LAW OF THERMODYNAMICS forbids any system in which heat energy only flows from a region of low temperature to one of higher temperature. For example, a refrigerator cannot work without a supply of extra energy, usually electrical, which is also deposited in the warm part of the system.

clavicle In vertebrates, the collar bone. *See* PECTORAL GIRDLE.

clay A fine-grained deposit, consisting chiefly of silicates of aluminium and/or magnesium.

cleavage *See* EMBRYONIC DEVELOPMENT.

cleavage plane In a non-metallic crystal, a surface along which the crystal will break cleanly and relatively easily. If an attempt is made to break the crystal in other directions, it may shatter into smaller fragments rather than breaking cleanly. A knowledge of the cleavage planes of diamond is important for breaking naturally occurring diamonds, which usually have no particular shape, into gemstones for jewellry. *See also* SLIP PLANE.

climate The general long-term pattern of weather as it varies from place to place on the planet or from year to year. The temperature of the Earth's surface is greater near the equator than near the poles, where the Sun hits the Earth at an angle (so the intensity of radiation reaching the surface is less). Since the axis about which the Earth rotates is not at right angles to the Earth's orbit around the

Sun, the northern and southern hemispheres each receive more heat in one half of the year (called summer) than in the other (winter).

Heat energy absorbed from the sun sets up major CONVECTION CURRENTS within the atmosphere, which form closed loops of convection called HADLEY CELLS. The convection currents set up by solar heating produce areas of high and low pressure. Air does not flow directly from high to low pressure, but tends to rotate. In the northern hemisphere the rotation is clockwise around areas of high pressure and anticlockwise around areas of low pressure (which are called DEPRESSIONS). This rotation (which is in the opposite direction in the southern hemisphere) is caused by the CORIOLIS FORCE: the Earth is spherical so the surface rotates faster near the equator than at the poles. Air moving from the equator towards the poles thus appears to be moving quickly relative to its surroundings, as if there was a force on it.

In addition to these global patterns of air flow, the climate of a region is determined by the distribution of land and sea (affecting humidity), by altitude (affecting temperature and rainfall) and ocean currents (affecting temperature).

See also ATMOSPHERE, WEATHER SYSTEM.

climax community *See* SUCCESSION.

clinker Non-VOLATILE waste products, particularly in a furnace or boiler. In particular the solid residue from burning coal or coke.

cloaca In birds, reptiles, amphibians, many fish and some marsupials, a chamber containing all excretory products (digestive and urinary) and into which the reproductive tracts enter. Products can be stored in the cloaca before being discharged from the body. Placental mammals do not have a cloaca. *See also* BLADDER, URINARY SYSTEM.

clocked bistable, *flip flop* In electronics, a BISTABLE with an additional 'clock' input. Pulses on the clock input change the output from one state to another. By connecting the output of one bistable to the clock input of another a counter can be made as the second bistable will receive clock pulses at half the rate of the first. This is the basis of digital counting and timing circuits. *See also* SHIFT REGISTER.

clock rate In computing, the frequency of a computer's internal clock. The clock synchronizes processes by generating pulses at a constant rate. The speed at which a computer executes an instruction depends on its clock rate.

clone A group of genetically identical offspring produced by ASEXUAL REPRODUCTION, involving the development of an entire organism from a single cell. This is useful in reproducing certain plants and is theoretically possible in humans (although unlikely to be done). In 1997 a sheep was cloned from one cell.

close packed (*adj.*) Describing a pattern in a crystal that maximizes the numbers of atoms that will fit into a given volume. The CO-ORDINATION NUMBER of a close packed structure is 12. *See also* CUBIC CLOSE PACKED, HEXAGONAL CLOSE PACKED.

cloud A mass of minute water droplets or ice crystals suspended in the atmosphere. The droplets or crystals are formed when water vapour condenses on tiny dust or salt particles.

cloud chamber A device that shows the path taken by IONIZING RADIATION, particularly ALPHA PARTICLES. The chamber contains a vapour (usually alcohol), which is cooled to the point where it is about to condense. Droplets of liquid will form first on any CONDENSATION NUCLEI present, such as trails of ionized particles left by an alpha or BETA PARTICLE (beta particles leave only very faint tracks).

club moss A member of the PHYLUM LYCOPODIOPHYTA.

cluster In astronomy, a group of stars (star cluster) or galaxies (galaxy cluster). *See also* GLOBULAR CLUSTER.

CMI *See* CELL-MEDIATED IMMUNITY.

Cnidaria A PHYLUM of mainly marine invertebrates, commonly known as COELENTERATES. Cnidaria includes *Hydra*, the Portuguese man of war (CLASS HYDROZOA), jellyfish (class SCYPHOZOA), sea anemones and marine coral (class ACTINOZOA). Coelenterates have two structural forms, an attached 'polyp' form with an opening at the opposite end to the attachment, and a free-swimming 'medusa' form that is umbrella-shaped with an opening in the middle of the underside (*see* CTENOPHORA). One or both of these forms may occur in the life cycle of a coelenterate. *Hydra* (which is a freshwater coelenterate) exists mainly in the polyp phase but is unusual because it moves by extending its body in a series of somersaults. Jellyfish are mostly free-swimming medusa. They feed by trapping or stinging organisms with their tentacles.

Coelenterates contain specialized stinging cells called cnidoblasts that allow them to adhere to or penetrate their prey. The stings of these animals can be painful and sometimes fatal to humans. Polyps reproduce asexually by BUDDING, but *Hydra* can reproduce sexually in the winter. Other polyp forms produce medusa that can reproduce sexually.

cnidoblast A specialized stinging cell found only in CNIDARIA.

CNS *See* CENTRAL NERVOUS SYSTEM.

coagulation The process of particles coming together, particularly of particles in a LYOPHOBIC sol, which tend to clump into larger particles.

coal A soft black SEDIMENTARY rock formed by the compaction of dead plant material growing over 200 million years ago. Coal burns slowly producing heat and is therefore a useful domestic fuel and is also used in the chemical industry. It contains carbon in varying proportions giving rise to a number of different types of coal, for example anthracite contains more than 90 per cent carbon, bituminous coal contains 80 per cent and peat about 50 per cent.

Many important products are derived from coal, including COAL GAS, COAL TAR, COKE and many by-products that form the basis of organic chemistry. The burning of coal contributes to atmospheric pollution and ACID RAIN.

coalesce (*vb.*) To stick together. The term is used in MECHANICS to describe objects that combine and move with a common velocity after an INELASTIC COLLISION.

coal gas An inflammable gas produced when coal is heated in the absence of air. It contains many compounds, but chiefly hydrogen, methane and carbon monoxide. It is sometimes used as a fuel and also for gas lighting. Coal gas has been mostly replaced by NATURAL GAS as a domestic energy source.

coal tar A black, oily substance produced by the DESTRUCTIVE DISTILLATION of bituminous coal (*see* COAL). Further DISTILLATION yields various fractions of oil with the residue forming PITCH. Products such as ANTHRACENE and tar are derived from coal tar. A large number of other products for use in the medical and dyeing industries can be obtained by yet further distillation.

cobalt (Co) The element with ATOMIC NUMBER 27; RELATIVE ATOMIC MASS 58.9; melting point 1,495°C; boiling point 2,870°C; RELATIVE DENSITY 8.9. Cobalt is a light-grey TRANSITION METAL. Its ions are pink when HYDRATED, but blue in the ANHYDROUS state – this feature is sometimes used as a test for small quantities of water.

Cobalt is a FERROMAGNETIC metal and its alloys are widely used in the manufacture of PERMANENT MAGNETS. The radioactive ISOTOPE cobalt–60 is widely manufactured and used as a source of GAMMA RADIATION, particularly for RADIOTHERAPY (the BETA PARTICLES also produced are stopped by an absorbing layer).

COBOL (**Common Business Oriented Language**) A high-level computer programming language, designed for the business community and one of the most widely used language. It is an easy language to learn and understand as it was created to read like English.

coccus (*pl.* ***cocci***) A spherical-shaped bacterium (*see* BACTERIA). Examples include *Streptococcus,* which associates in straight chains and is a common cause of sore throats in humans, and *Staphylococcus,* which associates in clusters and is found on the skin and MUCOUS MEMBRANE in humans where it can cause abscesses.

cochlea A spiral-shaped structure in the inner ear concerned with hearing, both sound detection and PITCH analysis. The cochlea comprises two narrow canals filled with the fluids perilymph and endolymph, and is coiled to save space. Within the cochlea is a specialized structure called the organ of Corti that consists of fine sensory hairs.

Movement of a membrane between the middle and inner ear (the 'oval window'), caused by vibrations of the bone OSSICLES, causes the perilymph behind this to move, which in turn displaces another membrane (called the 'round window') that is also between the inner and middle ear. These pressure waves cause movement of the sensory hairs, and this sets up an ACTION POTENTIAL that is transmitted to the brain along the auditory nerve. The pitch is determined by which part of the cochlea is stimulated, and the loudness by how many sensory hairs are stimulated.

codominance In genetics, the expression of a mixture of two ALLELES when neither one of the pair is truly DOMINANT. For example, when HETEROZYGOUS red- and white-flowered snapdragons are crossed, the next generation of snapdragons has pink flowers.

codon In genetics, a triplet of NUCLEOTIDES in DNA or RNA that determines the order in which amino acids are placed during PROTEIN SYNTHESIS. MESSENGER RNA is the template for protein synthesis and the codons are complementary to, and therefore form BASE PAIRS with, ANTICODON triplets in TRANSFER RNA, which then leads to the production of a protein with a specific chain of amino acids.

Some codons do not code for an amino acid and are called STOP CODONS. Stop codons, for example UAG, UAA and UGA, terminate protein synthesis. The codon AUG, a START CODON, initiates protein synthesis. There are 64 possible codons, which is more than enough to code for the 20 amino acids, and in fact some amino acids are coded for by several codons.

coefficient Any number that multiplies some other number or variable in an equation.

coefficient of drag A mathematical representation of the amount of DRAG produced by a given shape. It is lower for more STREAMLINED shapes. The coefficient of drag does not depend on the size or speed of motion of the object. For TURBULENT flow of an object of cross-sectional area *A* through a fluid of density *r* at a speed *v*, the force *F* is

$$F = {}^1/_2 C_D \rho v^2$$

where C_D is the coefficient of drag.

coefficient of expansion *See* EXPANSIVITY.

coefficient of friction Usually given the symbol μ in equations, the ratio of the DYNAMIC FRICTION to the NORMAL REACTION. It is a measure of the roughness of a surface. Values for the coefficient of friction generally vary between about 0.01 for smooth surfaces to about 0.8 for rough surfaces. Much larger coefficients of friction can occur in situations where there is a strong force of ADHESION – such as very smooth metal or glass surfaces. For a normal reaction of *N*, the frictional force *F* is given by

$$F = \mu N$$

See also FRICTION.

coefficient of restitution A quantity used to describe the degree of elasticity in partially ELASTIC COLLISIONS. The coefficient of restitution is the RELATIVE VELOCITY of the colliding objects immediately after impact divided by their relative velocity before the collision. For a perfectly elastic collision, the coefficient of restitution is 1, whilst in a totally INELASTIC COLLISION (where as much KINETIC ENERGY is lost as is allowed by the LAW OF CONSERVATION OF MOMENTUM), the coefficient of restitution is 0. *See also* HYPERELASTIC COLLISION.

Coelenterate Any member of the PHYLUM CNIDARIA. Coelenterates are aquatic, diploblastic (the body wall is composed of two layers with a jelly-like substance between them), radially symmetrical and have one opening to the outside surrounded by stinging cells. Coelenterates include *Hydra,* jellyfish, sea anemones and marine coral and the sea-gooseberry (*see* CTENOPHORA).

coelom The main body cavity that separates the DIGESTIVE SYSTEM and associated organs from the body wall. It is a fluid-filled cavity lined with mesoderm (*see* GERM LAYER) and is only absent in the simplest animals.

coenzyme An organic molecule, often a vitamin derivative, that acts as a COFACTOR in an enzyme reaction. It acts without binding, or only temporarily binding, to the enzyme, unlike a PROSTHETIC GROUP. Examples include COENZYME A, coenzyme Q (*see* ELECTRON TRANSPORT SYSTEM), FAD and NAD. Coenzymes are frequently essential for the removal of end-products of enzyme reactions that would otherwise cause inhibition of the enzyme.

coenzyme A A derivative of PANTOTHENIC ACID that is a COENZYME acting as a carrier of ACYL GROUPS in the OXIDATION of FATTY ACIDS (*see* KREBS CYCLE).

coenzyme Q *See* ELECTRON TRANSPORT SYSTEM.

coercive force The magnetic field that must be applied in a reverse direction to remove the magnetism of a PERMANENT MAGNET.

cofactor A non-protein substance that is needed for some enzymes to function efficiently. Cofactors can be COENZYMES, PROSTHETIC GROUPS or activators. Activators are substances other than coenzymes and prosthetic groups that are needed to activate an enzyme, for example, calcium ions are needed to activate THROMBOKINASE (which converts PROTHROMBIN to THROMBIN in blood clotting; *see* BLOOD CLOTTING CASCADE). An enzyme–cofactor complex is called a haloenzyme and the inactive enzyme on its own is called an apoenzyme.

coherent (*adj.*) Describing a wave that has a steadily varying PHASE, or two or more waves that have a constant phase difference. Light from a laser has a steadily varying phase, but light from other sources has jumps in phase, due to the light being emitted by one atom after another with no link between the phases produced by each atom. *See also* DIVISION OF AMPLITUDE, DIVISION OF WAVEFRONT.

cohesion The attractive force between molecules of the same type. Cohesion is responsible for SURFACE TENSION. *See also* ADHESION, CAPILLARY EFFECT.

cohesion tension theory *See* TRANSPIRATION STREAM.

coil In contraception, *see* INTRAUTERINE DEVICE.

coke The residue remaining after coal has been strongly heated in an airtight oven. Coke consists of 90 per cent carbon mixed with inorganic material and is a widely used domestic and industrial fuel.

colchicine An ALKALOID that prevents SPINDLE formation during MEIOSIS so that the chromosomes cannot separate, resulting in multiple sets of chromosomes. *See* POLYPLOID.

cold front A region where cold air attempts to force its way under warmer air, producing rain showers and possibly thunderstorms. *See* WEATHER SYSTEMS. *See also* WARM FRONT.

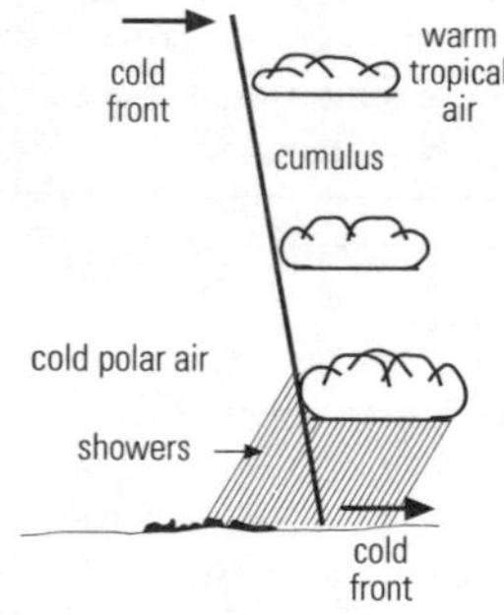

Formation of a cold front.

cold fusion NUCLEAR FUSION at room temperatures. A process by which DEUTERIUM nuclei undergo fusion by absorption into palladium atoms was claimed to have been observed in 1989. Subsequent work failed to confirm this discovery and the process is now generally believed not to take place.

collagen In vertebrates, a major structural protein found in CONNECTIVE TISSUE, where it is made by FIBROBLASTS. Collagen forms fibres that provide strength but little elasticity, and it is the main constituent of LIGAMENTS and TENDONS. It is also found in CARTILAGE and bone.

collector In a JUNCTION TRANSISTOR, the ELECTRODE at which CHARGE CARRIERS arrive having passed through from the EMITTER and the BASE.

collenchyma Simple (one cell-type only) plant tissue consisting of elongated cells with cell walls thickened by additional CELLULOSE at the corners, providing extra strength and support. Collenchyma is particularly important in growing stems because the cells can stretch.

colliding beam experiment A particle physics experiment in which two beams collide head-on, as opposed to a beam colliding with a fixed target. It has the advantage that the total MOMENTUM is zero, so all the energy of the beams can be harnessed for the production of new particles, rather than being used to give those particles the KINETIC ENERGY they need to conserve momentum.

colligative properties Those properties of a solution that are dependent on the concentration of particles in the solution, but not on the nature of those particles. Examples are the LOWERING OF VAPOUR PRESSURE, the DEPRESSION OF FREEZING POINT, the ELEVATION OF BOILING POINT and OSMOSIS.

collimator A tube with an adjustable slit at one end and a CONVERGING LENS at the other, with adjustments to allow the slit to be placed in the FOCAL PLANE of the lens. A collimator is used to produce a parallel beam from the light that DIFFRACTS from the slit, which provides a COHERENT source of light for SPECTROSCOPY.

collision cross-section The cross-sectional area of the volume swept out by a molecule or some other particle as it moves through space. If another particle enters this volume, there may be a collision. Collision cross-sections are used in the calculation of collision rates in particle physics, and for the calculation of the MEAN FREE PATH of a molecule in a gas.

colloid A mixture containing small particles of one material suspended in another, often of a different phase. The particles in a colloid have sizes between one and one hundred nanometres, so are larger than the individual molecules that occur in a solution and smaller than the particles which are found in precipitates and can be removed by a filter.

Examples include AEROSOLS and FOAMS (gas and liquid mixtures), EMULSIONS (liquid mixtures such as milk and paint), and solvents and GELS (solids dispersed in a liquid). A colloid containing solid particles suspended in a liquid is more accurately called a SOL.

Particles in a colloid can be separated by passing them through a porous material. The material that forms the separate individual particles is sometimes called the dispersed phase, to distinguish it from the continuous phase which forms a single connected body of material.

See also LYOPHILIC, LYOPHOBIC.

colon Part of the LARGE INTESTINE between the CAECUM and the RECTUM. It is in the colon that any water and minerals in digestive secretions not absorbed by the ILEUM are reabsorbed and the remainder formed into FAECES. The colon also absorbs vitamins, some of which, for example biotin (*see* VITAMIN B) and VITAMIN K, are produced by bacteria, such as *Escherichia coli*, that live in the colon.

colorimeter A device for comparing the colour, usually of a solution, against a calibrated filter. Colorimeters are used to measure the concentration of a coloured component in a solution.

colostrum In mammals, the milk produced by the MAMMARY GLANDS during the first few days following the birth of their offspring. It consists of a clear fluid containing water, proteins, vitamins and antibodies, but is low in fat and sugar. Colostrum plays an important role in PASSIVE IMMUNITY. *See also* LACTATION.

colour 1. (*optics*) The name given to the physiological sensation caused by different wavelengths of visible light. The eye can distinguish three wavelength bands which produce the sensations of the three primary colours: red (the longest wavelengths), green (intermediate wavelengths) and blue (short wavelengths). When the eye receives wavelengths between these bands, or a mixture of two of these colours, the secondary colours are seen. These are yellow (red plus green), cyan (green plus blue) and magenta (blue plus red). The mixture of all three primary colours produces the sensation of white, whilst the sensation of black is produced by an absence of any visible light.

Objects appear coloured because they reflect the different colours of light falling on them by differing amounts. A red object is red because when viewed under white light only red light is reflected. Viewed under green light, for example, it will appear dark as there is no red light falling on it to be reflected. Colour filters work in a similar way, modifying the light by absorbing certain colours. A yellow filter, for example, absorbs blue light whilst transmitting red and green. For this reason yellow and blue are known as complementary colours – added together they will produce white light. The other pairs of complementary colours are green/magenta and red/cyan.

See also COLORIMETER, RETINA.

2. (*particle physics*) A quantity possessed by QUARKS and GLUONS in QUANTUM CHROMODYNAMICS, analogous to the electrical charge in ELECTROMAGNETISM.

colour filter A device that absorbs some wavelengths of visible light whilst allowing others to pass. A yellow filter, for example, is transparent to the red and green parts of the spectrum, but absorbs blue. *See also* COLOUR.

colour television A television system in which three ELECTRON GUNS are used. electrons are sent through a SHADOWMASK – a metal screen with holes just behind the PHOSPHOR coating. This ensures that electrons from each gun can only reach certain parts of the phosphor screen. These regions are coated with different phosphors, which glow red, green and blue when struck by electrons. In this way, each electron gun can be used to build up a picture in each of the PRIMARY COLOURS.

column chromatography CHROMATOGRAPHY in which the ELUENT runs down a glass column containing the STATIONARY PHASE, such as tightly packed powdered alumina.

coma 1. (*physics*) An ABERRATION in images formed by lenses and CURVED MIRRORS when the rays forming the image are at a large angle to the PRINCIPAL AXIS. Comas are so called because the image of a point is shaped like a comet.

2. (*astronomy*) The cloud of luminous gas around a comet when it is close to the sun.

3. (*medicine*) A deep, prolonged unconsciousness in which the person is unable to respond to external stimulus, usually a result of injury or disease.

combination In mathematics, a selection of some objects from a larger group, where no account is taken of the order in which the objects are selected. Thus in selecting three

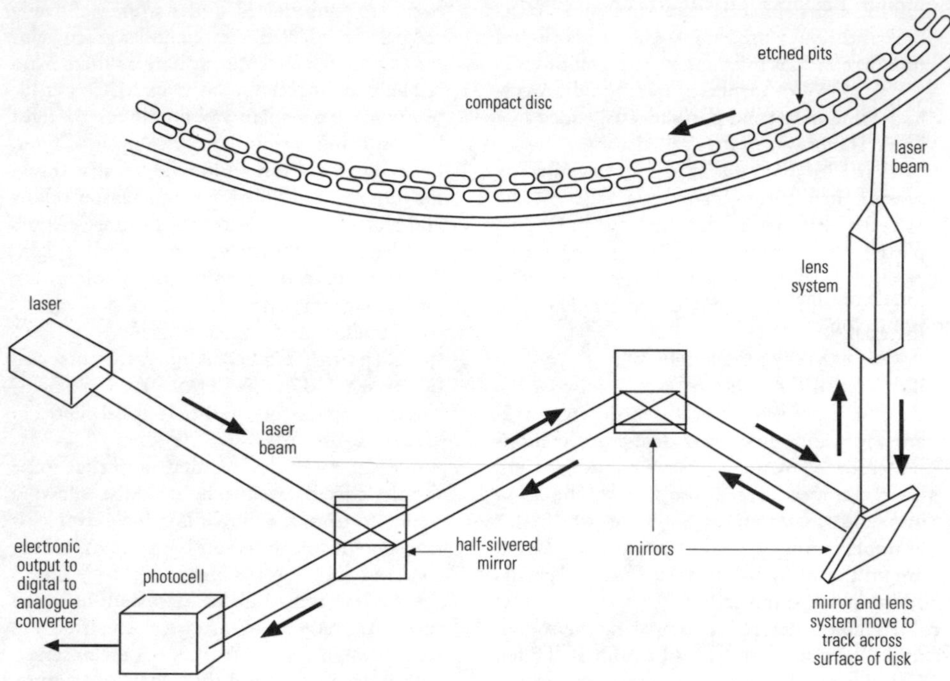

Compact disc player.

letters of the alphabet, abc and acb are counted as the same combination.

The number of combinations of *r* objects that can be made from a set of *n* objects is denoted by ^nC_r and is equal to

$$^nC_r = n!/r!(n-r)!$$

Compare PERMUTATION.

combined heat and power (CHP) The combined production of heat and electricity in a power station. The steam produced to drive turbines in the power station is afterwards used to heat nearby buildings. The high costs of CHP plants have meant that few such systems have been built, despite the greater EFFICIENCY in energy usage.

combined pill *See* PILL.

combustion Any chemical process that is self-supporting and releases heat. In particular, the rapid combination of any compound with oxygen from the air with the production of heat and light. *See also* EXPLOSION, FLAME, HEAT OF COMBUSTION, SPONTANEOUS COMBUSTION.

comet A small body of rock and ice that orbits the Sun. Most comets are believed to spend most of their time in very distant orbits that make up a region called the OORT CLOUD. Occasionally, a gravitational disturbance will cause a comet to leave this cloud and fall into a highly elliptical orbit, bringing it very close to the Sun. As it heats up, some of the comet's material is vaporized and then IONIZED by solar radiation, producing a glowing tail, which always points away from the Sun. As the comet moves away from the Sun it cools and fades from view. Comets in short orbits lose a lot of their material on each pass and rapidly become fainter. Some comets make only one pass of the Sun and are never seen again.

commensalism In biology, an association between two species in which only one partner (the commensal) benefits but the other (the host) is not harmed. Commensalism is a variation of SYMBIOSIS. *See also* MUTUALISM, PARASITISM.

Common Business Oriented Language *See* COBOL.

common denominator A number by which two or more terms in a mathematical equation are all divided. Rewriting an equation so that all terms have a common denominator is an important step in solving some kinds of equation. For example, $^1/_4 + ^1/_6$ can be rewritten as $^3/_{12} + ^2/_{12}$, where 12 is a common denominator. Since 12 is the lowest number that could be used in this case, it is called the lowest common denominator (l.c.d.).

common ion effect A salt will be less soluble if the solution already contains ions of a type produced when the salt is dissolved. *See also* SOLUBILITY PRODUCT.

common salt *See* SODIUM CHLORIDE.

community All of the POPULATIONS of plant or animal species that live together within an ecological area, or HABITAT. A community is often identified by a dominant feature, either a physical feature, such as a swamp, or a dominant species, such as an oak woodland. *See also* SUCCESSION.

commutative (*adj.*) Describing a mathematical operation performed on two numbers where the numbers can always be exchanged without any change in the result. Thus addition is commutative since

$$a + b = b + a$$

Multiplication is also commutative; division and subtraction are not. *See also* ASSOCIATIVE.

commutator A device for reversing electrical connections. An example is the SPLIT-RING COMMUTATOR used in many ELECTRIC MOTORS and DYNAMOS.

compact disc A system for the digital storage of information. A compact disc contains music, visual images or computer data encoded as a series of BINARY 1's and 0's recorded as small hollows or pits in a thin layer of aluminium sandwiched between transparent plastic layers. The disc is read by the reflection of a low power laser beam.

The digital nature of the data and the error correction systems built into the way the data is encoded mean that the data can be read perfectly unless the disc is badly damaged. There is also no wear as there is in a conventional vinyl record, as there is no mechanical contact between the compact disc and the reading device.

companion cell *See* PHLOEM.

comparator A circuit that compares two voltages and gives a digital output depending on whichever voltage is the larger. Comparator circuits are often based on OPERATIONAL AMPLIFIERS. Provided there is no FEEDBACK, the output of the operational amplifier will be high if the NON-INVERTING INPUT is at a higher potential than the INVERTING INPUT and low (or negative) if the non-inverting input is at the lower potential.

compass *See* MAGNETIC COMPASS.

competition In biology, the fight between organisms or species for resources that are in short supply. In animals competition might be for food, water or shelter. In plants, competition might be for light or minerals. Competition results in the survival of or adaptation of some individuals or species instead of others. It is the basis of DARWINISM.

compiler A computer program that translates a HIGH-LEVEL LANGUAGE into MACHINE CODE. Each instruction in the high-level language translates into several instructions in machine code. Different high-level languages use different compilers.

To run a program using a compiler, the compiler first translates the original SOURCE CODE program into a separate file in machine code. The computer then runs the program by executing the instructions in the machine code file. The machine code file can be run as often as necessary, without the original source code being present. This is faster than using an INTERPRETER, which translates and executes the source code itself in a single step. However, compiled programs have the disadvantage that whenever a change is made to the source program it must be recompiled.

complement 1. The complement of an angle is 90° minus the angle concerned. In other words, an angle and its complement add up to 90°.

2. In SET theory, all the elements that are not part of a specified set.

complementary angles Any pair of angles that add up to 90°; the two other angles in a right-angled triangle.

complementary colour The colour that must be added to a specified colour to make white light. Thus blue and yellow, for example, are complementary colours.

complementary DNA *See* COPY DNA.

complementary pair A pair of JUNCTION TRANSISTORS, one NPN and the other PNP, used together.

complex conjugate *See* CONJUGATE.

Complex Instruction Set Computing *See* CISC.

complex ion An ion, usually a metal ion, that has a number of molecules or ions bound to it by CO-ORDINATE BONDS. HYDRATED ions of TRANSITION METALS are an example. Some of these ions will form similar complexes with ammonia molecules in place of water, for example the copper tetrammine ion, $[Cu(NH_3)]^{2+}$. Another example is the hexacyanoferrate(III) ion, $Fe(CN)_6^{3-}$, which consists of an Fe^{3+} ion surrounded by six CN^- ions. *See also* LIGAND.

complex number Any number formed by adding a REAL NUMBER and an IMAGINARY NUMBER. A complex number can be expressed as $x + iy$, where x is the real part and iy is the imaginary part. Both x and y are real numbers and i is the square root of -1. Complex numbers are often represented as points on a two-dimensional plane called an ARGAND DIAGRAM, with the real and imaginary parts representing the x and y CO-ORDINATES of the point.

Complex numbers are widely used in physics. This may be a matter of convenience, when dealing with ALTERNATING CURRENT, for example, but equations of QUANTUM PHYSICS, such as SCHRÖDINGER'S EQUATION, have complex COEFFICIENTS and require that the WAVEFUNCTION be represented by complex numbers.

See also CONJUGATE.

component 1. (*mathematics*) One of a set of VECTORS that add to give some other vector. In particular, one of three vectors parallel to the axes of a CARTESIAN CO-ORDINATE system, described as the components of the vector formed by adding them together. The process of finding the components of a vector in specified directions is called resolving (*see* RESOLVE).

2. (*physics*) That part of a FORCE that acts in a particular direction. The component of a force F in a direction at an angle θ to the direction of the force is $F\cos\theta$.

composite material Any engineering material made from two or more different materials, designed to exploit the advantages of each without suffering their weaknesses. *See also* CARBON FIBRE, GLASS-FIBRE REINFORCED PLASTIC, REINFORCED CONCRETE.

compound A substance made up of two or more elements that cannot be separated by physical means. In a compound, the quantity of the elements present is fixed and is a simple ratio, though more than one compound may exist containing the same elements, for example, water (H_2O) and hydrogen peroxide (H_2O_2). *See also* CHEMICAL COMBINATION.

compound lens An optical lens made from several pieces of glass.

compound microscope A MICROSCOPE that uses two lenses. The first lens, called the OBJECTIVE, is used to form an enlarged REAL IMAGE of the object. This image is then viewed through a second lens, called the EYEPIECE, which acts as a magnifying glass, further enlarging the object. The MAGNIFICATION of the microscope can be altered by adjusting the power of the two lenses, but the useful magnification is limited by the fact that DIFFRACTION effects cause the light to spread out as it enters the objective. It is therefore not possible to see details smaller than the wavelength of the light used to view them. *See also* SIMPLE MICROSCOPE.

compressibility The tendency of a material to change its density with pressure. The term is particularly applied to the effects arising from the significant changes in the density of air around an aircraft as it approaches the SPEED OF SOUND. The compressibility of a material is measured by its BULK MODULUS.

compression A pushing force, tending to make an object smaller.

compression ratio In an INTERNAL COMBUSTION ENGINE, the factor by which the volume of the gas in the cylinder of the engine is reduced during the COMPRESSION STROKE. Higher compression ratios lead to greater EFFICIENCY but also increase problems of PRE-IGNITION.

compression stroke The motion of a PISTON in an INTERNAL COMBUSTION ENGINE to compress and heat the air or fuel/air mixture prior to the POWER STROKE. *See also* COMPRESSION RATIO.

computer A digital electronic system containing a CENTRAL PROCESSING UNIT and memory. Much of the memory is RAM into which program instructions are read from a storage unit such as a DISK DRIVE. The physical machinery of a computer is described as hardware, whilst the programs it uses are called software. Modern computer systems allow data to be entered by a keyboard or by using a hand-held control called a MOUSE, which moves a pointer on the screen, called a VISUAL DISPLAY UNIT (VDU). *See also* MAINFRAME, MINICOMPUTER, PERSONAL COMPUTER, SUPERCOMPUTER.

computer-aided design (CAD) The use of computers to create drawings and graphics. CAD is extensively used to design three-dimensional models in architecture and engineering, which may be used to test these designs by computer simulation. *See also* FINITE ELEMENT ANALYSIS.

concave (*adj.*) Describing an inwardly curving surface, particularly in a lens or mirror.

concentrated (*adj.*) Describing a solution having a concentration at or close to the maximum that can be achieved at a particular temperature; that is, a solution that is SATURATED or nearly so.

concentration The amount of SOLUTE dissolved in a specified amount of solution. Concentration is normally expressed in terms of the number of MOLES of solute per decimetre cubed of solution. A solution with a concentration of 1 mol dm^{-3} is described as molar. *See also* HENRY'S LAW, MOLALITY.

concentration gradient The difference in concentration of a substance between two regions. *See also* DIFFUSION.

concentric circles Two or more circles with a common centre.

condensation 1. The process by which a vapour turns into a liquid.

2. Droplets, usually of water, formed by this process.

condensation nucleus A small imperfection or piece of dirt that may act as a centre for a change of state, such as the condensation of a SUPERSATURATED VAPOUR or the boiling of a SUPERHEATED LIQUID.

condensation polymerization *See* POLYMERIZATION.

condensation reaction In organic chemistry, a reaction in which two compounds are joined together to make a larger molecule with the loss of a small molecule, such as water. Such reactions can be considered to be addition-elimination reactions. Many POLYMERS are made by condensation POLYMERIZATION. *See also* ADDITION REACTION, ELIMINATION REACTION.

condense (*vb.*) To turn from a gas or vapour into a liquid.

condensed matter Solids and liquids, those states of matter where the spacing between the molecules is of the same order as their size.

condenser 1. *condensing lens* (*optics*) A lens of short FOCAL LENGTH designed to concentrate as much light as possible onto an object, to illuminate it as brightly as possible.

2. (*chemistry*) A device for the condensing of a vapour, often in the form of a coiled glass tube surrounded by an outer glass jacket through which cold water is passed. *See also* LIEBIG CONDENSER.

3. (*electronics*) An obsolete word for CAPACITOR.

condensing lens *See* CONDENSER.

conditioning A form of LEARNING that involves the association of two stimuli. An example is the conditioned reflex, which involves two stimuli presented together, and is temporary, involuntary and reinforced by repetition.

The conditioned reflex was demonstrated by the experiments of the Russian scientist Ivan Pavlov (1849–1936). Dogs learned to associate the ringing of a bell with the arrival of food, and responded to the bell by involuntary salivation even if no food arrived. Another form of conditioning is trial and error learning, or operant conditioning, as described by Edward Thorndike (1874–1949) and B. Skinner (1903–1990). Here an animal learns a pattern of behaviour based on what occurs after the action. The frequency of a voluntary response is increased by giving a reward for the correct response. *See also* HABITUATION, IMPRINTING.

condom A rubber sheath placed over the erect PENIS preventing entry of SPERM into the VAGINA during sexual intercourse. It is used as a method of CONTRACEPTION. It is 97 per cent effective if used with a spermicide (jelly or cream preparations that kill sperm) but only 85 per cent effective if used on its own. The condom also provides protection against sexually transmitted diseases such as AIDS.

conductance In electricity, the reciprocal of resistance. A material that has a large resistance has a low conductance and vice versa. This means it is sometimes more convenient to think in terms of conductance, or CONDUCTIVITY, the reciprocal of RESISTIVITY. The unit of conductance is the SIEMENS.

conduction The flow of electricity or heat through a material. *See* CONDUCTOR, THERMAL CONDUCTION.

conduction band In the BAND THEORY of solids, an energy band that is not completely occupied by electrons, so an electron in that band can gain a small amount of energy and move through the material, carrying heat or electricity. *See also* VALENCE BAND.

conductivity The reciprocal of RESISTIVITY. The unit of conductivity is the SIEMENS per metre. *See also* CONDUCTANCE, THERMAL CONDUCTIVITY.

conductor, electrical Any material through which current can flow. All metals are conductors, as their structure contains VALENCE ELECTRONS that are free to move through a LATTICE of positive IONS. Metals conduct less well as their temperature increases, since thermal vibrations make the lattice less regular, making collisions between the electrons and the lattice more likely. For similar reasons, pure metals, which have highly regular lattice structures, usually conduct electricity (and heat – *see* THERMAL CONDUCTION) better than alloys. Graphite is one of the few nonmetallic solids that conducts electricity.

Molten IONIC materials and solutions of ionic salts all conduct electricity. In all these cases there are both positive and negative ions free to move through the liquid and carry the charge. Once these ions reach an ELECTRODE, the ions gain or lose electrons and chemical reactions may take place (*see* ELECTROLYSIS). In ionic solids, these ions are locked into a rigid lattice and so are not free to carry charge. Molten metals also conduct electricity.

Gases do not normally conduct electricity, but may do so if they are IONIZED – that is, if they contain a number of ions as well as uncharged atoms or molecules. Sources of IONIZATION include IONIZING RADIATION from RADIOACTIVE materials, flames or very high voltages, which may produce an AVALANCHE BREAKDOWN.

See also BAND THEORY, INSULATOR, SEMICONDUCTOR.

cone 1. (*mathematics*) A solid shape produced by the set of all lines that pass through a fixed point, called the vertex of the cone, and a circle (called the base). Unless otherwise specified, it is usually assumed that the line joining the vertex to the centre of the circle is at right angles to the plane of the circle – this is called a right cone.

For a cone of vertical height h and radius of base r, the volume V is given by:

$$V = {}^{1}/_{3}\,\pi r^2 h$$

For a cone of slant height l, the area A of the curved surface is:

$$A = \pi r l$$

2. (*botany*) The reproductive structure of CONIFERS. A cone, for example the cone of a pine tree, consists of numerous scales called SPOROPHYLLS that overlap around a central axis. There are usually separate male and female cones. *See also* CONIFEROPHYTA.

3. (*zoology*) Light-sensitive cells in the RETINA of the human eye. Cones are sensitive to colour and are used mostly for day vision. *See* RETINA.

congenital disease A disease present at birth, which may be a result of genetic factors, injury, infection or environmental factors.

congruent (*adj.*) Describing two or more geometric figures of the same size and shape. Such figures can be placed on top of one another by the mathematical operations of ROTATION, REFLECTION and TRANSLATION.

conical pendulum A variation on the SIMPLE PENDULUM in which the BOB moves in a circle rather than from side-to-side.

conic section Any of the curves that can be formed by the intersection of a plane and a cone. The four main types of conic sections are the circle (if the plane is perpendicular to the axis of a right circular cone); the PARABOLA (if the plane is parallel to the edge of the cone); the ELLIPSE (if the intersection follows a closed curve that is not a circle); and the HYPERBOLA (if an open curve is produced that is not a parabola).

conidium (*pl.* ***conidia***) A non-motile SPORE of some fungi formed during ASEXUAL REPRODUCTION.

conifer A member of the PHYLUM CONIFEROPHYTA.

Coniferophyta A large PHYLUM of plants that have exposed seeds. There are many species, including giant redwoods, firs and pines. Conifers were once the dominant form of vegetation but now the ANGIOSPERMS are. There are fossil remains of conifers that are 350 million years old.

The reproductive structures of conifers are CONES (not flowers as in angiosperms). Female and male cones are usually separate and pollen grains from the male cone are carried by the wind to the female cone. The seeds develop (sometimes over a period of years) on the female cone, and are only released when the scales of the cone open in dry conditions that favour DISPERSAL. Conifers bear no fruit.

Conifers are adapted to dry conditions in many ways, and are often found where water is inaccessible. Conifers are fast growing trees and so many species are an important source of timber and wood pulp for paper. Conifers are often planted to prevent land erosion.

conjugate 1. ***complex conjugate*** One of a pair of COMPLEX NUMBERS with the same real part but opposite imaginary parts: that is, $a + ib$ and $a - ib$.

2. ***conjugate angle*** One of a pair of angles that add up to 360°.

conjugate angle *See* CONJUGATE.

conjugated protein *See* PROTEIN.

conjugate solutions The pair of liquids formed when two liquids that are only partially MISCIBLE are mixed together. If the liquids are A and B, with A being denser, the conjugate solutions will form with a SATURATED solution of A in B floating above a saturated solution of B in A.

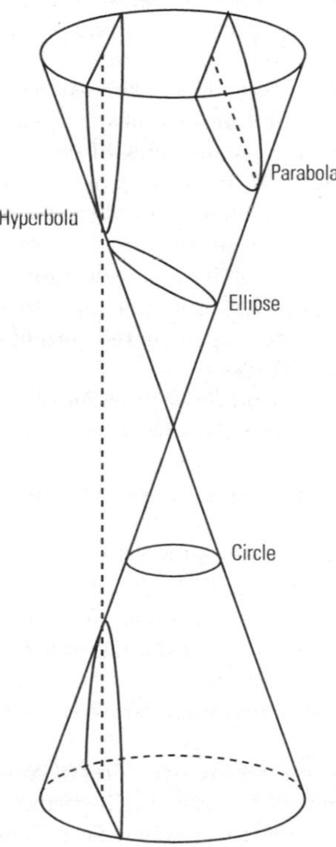

Conic section.

conjugation In bacteria, the equivalent of SEXUAL REPRODUCTION. The DNA is exchanged or donated between bacteria by passing through a tube called the pilus.

conjunctiva In vertebrates, an outer protective membrane of the eye that is an extension of the eyelid EPITHELIUM and lies over the CORNEA. The lachrymal glands produce antiseptic tears to nourish and lubricate the cornea and conjunctiva, because they have no blood supply. Conjunctivitis is inflammation of the conjunctiva due to infection.

conjunctivitis Inflammation of the CONJUNCTIVA.

connective tissue Animal tissue developed from the embryonic MESODERM (*see* EMBRYONIC DEVELOPMENT) which is made up of a variety of cells and connective tissue fibres (usually COLLAGEN) embedded in the EXTRACELLULAR MATRIX. Connective tissue provides support in CARTILAGE and bone and in the transport system of blood. The strength and elasticity of connective tissue is provided by collagen and ELASTIN fibres which are produced by FIBROBLASTS. ADIPOSE connective tissue provides insulation. Loose connective tissue (areolar tissue) binds many other tissues together (such as MENINGES of the central nervous system and bone PERIOSTEUM). TENDONS and LIGAMENTS are also connective tissue. As well as providing support, connective tissue provides some defence due to the presence of tissue MACROPHAGES and MAST CELLS.

conservation The protection of the natural world from the effects of human activities. This includes the protection of endangered species and valuable natural environments, such as the rainforests, protection from pollution, reduction of the GREENHOUSE EFFECT and the recycling of glass, paper, plastics and some metals. Many organizations exist to promote conservation.

National parks and nature parks have been established to protect particular environments. The first national park was Yellowstone National Park, USA, set up in 1872. The first in the UK was set up in 1949. National parks vary in size and are sites of natural beauty as well as historical or scientific interest. Some national parks are wilderness areas with no traffic or buildings, but others (in the UK) are areas of restricted development. On a smaller scale SITES OF SPECIAL SCIENTIFIC INTEREST (SSSIs), AREAS OF OUTSTANDING NATURAL BEAUTY (AONBs) and country parks are areas of land worthy of particular protection.

Some endangered species are protected by law, although this is difficult to enforce. Other ways of protecting species are commercial farming, for example of mink and deer, to provide sought-after goods without the need to kill wild animals, and zoos provide a safe environment for breeding endangered species for later reintroduction into the wild. In the same way botanical gardens provide a place where a diverse number of plants, unlikely to be encountered naturally, can be viewed by members of the public. Botanical gardens and other organizations keep seed banks to conserve plants that are rare in the wild.

constant A quantity that does not change or depend on other quantities.

constant flow method A way of measuring the HEAT CAPACITY of a liquid. The power of an electric heater is calculated from the current through it and the voltage across it whilst the fluid passes the heater at a measured rate. The heat capacity can then be found from the temperature difference in the fluid produced by the heater.

constant volume gas thermometer A thermometer that uses the changes in pressure of a fixed amount of gas held in a constant volume to measure temperature. *See also* PRESSURE LAW.

constitutive (*adj.*) Describing an enzyme that is synthesized all the time, regardless of the availability of SUBSTRATE. *Compare* INDUCIBLE.

constructive interference The effect produced when two or more waves arrive IN PHASE, in which the waves add to produce a wave with an amplitude equal to the sum of the amplitudes of the original waves. *Compare* DESTRUCTIVE INTERFERENCE. *See also* INTERFERENCE.

contact breaker A switch, often operated by the rotation of some machinery, such as an INTERNAL COMBUSTION ENGINE, designed to break the flow of current in a device such as an INDUCTION COIL.

contact force Any force that results from two solid objects touching each other. The contact force is usually a combination of the NORMAL REACTION, which acts at right angles to the surface and prevents one object from entering the other, and friction, which acts along the surfaces and prevents them from sliding along one another.

contact inhibition A phenomenon seen in TISSUE CULTURES when cells growing in a monolayer stop moving and dividing when they come into contact with other cells. This means that they only fill the available space. Cancer cells lose this regulatory ability. *See also* DENSITY-DEPENDENCE.

contact lens A small MENISCUS lens designed to have the same effect as spectacles but resting directly on the CORNEA (the curved front surface of the eye).

contact process An industrial process for the manufacture of sulphuric acid. A mixture of sulphur dioxide, SO_2, and air are passed over a hot vanadium(V) oxide catalyst. The sulphur dioxide is oxidized to sulphur trioxide, SO_3:

$$2SO_2 + O_2 \rightarrow 2SO_3$$

The sulphur trioxide is then dissolved in sulphuric acid producing OLEUM:

$$H_2SO_4 + SO_3 \rightarrow H_2S_2O_7$$

The oleum is then diluted with water to give concentrated sulphuric acid.

continent A large mass of land rising above the ocean floor. The boundary of a continent lies at the edge of the continental shelf, so offshore islands are considered part of the nearby continent. There are seven continents: Africa, Antarctica, Asia, Australia, Europe, North America and South America. *See also* CONTINENTAL DRIFT, PANGAEA.

continental (*adj.*) Describing an AIR MASS that has travelled mostly over land.

continental drift The slow motion of the continents from one part of the globe to another. It is thought that the Earth's continents once formed a single large landmass, given the name Pangaea, and then moved relative to each other. The phenomenon of continental drift may be explained by the motion of TECTONIC PLATES.

continuous ambulatory peritoneal dialysis *See* DIALYSIS.

continuous data In statistics, measured numbers that can take on any numerical value so cannot be represented fully by INTEGERS. Most physical data is continuous. *Compare* DISCRETE DATA.

continuous phase *See* COLLOID.

continuous spectrum *See* SPECTRUM.

contraception The deliberate prevention of pregnancy while maintaining a sexual relationship. Contraception can be a simple barrier method, where the GAMETES are prevented from meeting. Barrier methods include the CONDOM and the DIAPHRAGM.

Some methods of contraception rely on hormonal interference of the female's MENSTRUAL CYCLE, the main one being the oral PILL which uses synthetic hormones to mimic pregnancy. Other similar preparations include the mini pill, and the morning after pill taken after unprotected intercourse (*see* PILL). Hormone implants can be surgically placed under the skin to release hormones over a few months (depo-provera-progestin) or years (norplants release female hormones over 5 years). Another method is the INTRAUTERINE DEVICE (IUD), which is a plastic or copper coil inserted into the UTERUS. Non-reversible methods of contraception are male vasectomy and female tubal ligation (*see* STERILIZATION).

contractile root A thickened root on the base of CORMS or bulbs that contracts to pull the plant deeper into the ground. Some contractile roots store CARBOHYDRATE for use by the plant.

contrast enhancing medium An X-ray absorbing material introduced into a patient to show up more clearly the differences between various organs or structures. *See also* ARTERIOGRAM, BARIUM MEAL, RADIOGRAPH.

control grid In a CATHODE RAY TUBE, or any other device that produces a beam of electrons by the process of THERMIONIC EMISSION, an ELECTRODE used to control the flow of electrons. If the control grid is more negative than the heated CATHODE, it limits the number of electrons able to reach the ANODE or form a beam.

control rod In a NUCLEAR REACTOR, a rod made of a neutron-absorbing material, usually boron or cadmium, which can be inserted into the reactor to control the CHAIN REACTION by absorbing excess neutrons.

control unit In a computer or microprocessor, the part of the CENTRAL PROCESSING UNIT that decodes, synchronizes and executes instructions.

convection The motion of a fluid (usually air or water) in a GRAVITATIONAL FIELD driven by changes in density resulting from temperature changes. Fluids expand and become less dense as they are heated, so a source of heat near the base of a volume of fluid will set up convection currents carrying energy throughout the fluid. If the heat is applied near the top of the fluid there will be no convection. An important exception to the rule that fluids expand on heating is water in the temperature range 0°C to 4°C. For this reason, lakes and ponds freeze from the surface down rather than freezing solid as soon as the air temperature falls below freezing point.

convection current The movement of a volume of fluid caused by CONVECTION. *See also* WEATHER SYSTEMS.

conventional current A current imagined to flow in the direction that would be taken by positive charges – that is, from the positive terminal of the battery around the circuit to the negative terminal.

converging lens An optical lens that is thicker at the centre than at the edges and brings parallel light rays together at a point. The distance between the lens and the point at which parallel rays of light are brought together is called the FOCAL LENGTH.

An object placed more than two focal lengths from a converging lens will form an inverted (upside down), diminished REAL IMAGE located at a distance between one and two times the focal length of the lens. This is the arrangement in a camera. If an object is placed in front of a converging lens at a distance between one and two focal lengths, a real image will be formed which is inverted and enlarged (is larger than the original object). This is the system used in a PROJECTOR.

If an object is placed in front of a converging lens at a distance of less than one focal length, the lens will behave as a magnifying glass, producing an enlarged, upright, VIRTUAL IMAGE, which can be seen by looking at the object through the lens.

Compare DIVERGING LENS.

convex (*adj.*) Describing an outward curving surface, particularly in a lens or mirror.

cooling by evaporation When a liquid evaporates, the fastest molecules escape from the surface, resulting in a reduction of the average energy of the remaining molecules and so producing a lower temperature in the liquid.

cooling correction A calculation designed to compensate for any heat losses, such as in an experiment to measure HEAT CAPACITY.

Cooper pair *See* SUPERCONDUCTIVITY.

co-ordinate bond, *dative bond* A COVALENT BOND in which both the bonding electrons are donated to the bond by one of the atoms. An example is the bond between boron trichloride and ammonia:

$$H_3N: + BCl_3 \rightarrow H_3NBCl_3$$

The nitrogen atom donates a LONE PAIR of electrons (shown as :), to form a bond with the boron atom.

co-ordinate geometry *See* GEOMETRY.

co-ordinates A set of numbers that represent the position of a point measured from some fixed point called the ORIGIN. To describe the position of a point in an *n*-dimensional space, *n* co-ordinates are needed. CARTESIAN CO-ORDINATES or POLAR CO-ORDINATES are generally used, though other systems are possible.

co-ordination compound Any compound containing CO-ORDINATE BONDS, particularly a salt containing an inorganic COMPLEX ION, such as potassium hexacyanoferrate(III), $K_3[Fe(CN)_6]$ in which the cyano groups (CN^-) form co-ordinate bonds with the iron atom.

co-ordination number In a crystalline structure, the number of nearest neighbours surrounding any atom or ion. In the sodium chloride structure, the co-ordination number is 6, each sodium ion has six nearby chlorine ions and vice versa. In a BODY CENTRED CUBIC structure, the co-ordination number is 8, whilst CLOSE PACKED structures have a co-ordination number of 12.

copolymerization The POLYMERIZATION of two or more different MONOMERS resulting in a compound with properties distinct from the original monomers.

copper (Cu) The element with ATOMIC NUMBER 29; RELATIVE ATOMIC MASS 63.5; melting point 1,083°C; boiling point 2,582°C; RELATIVE DENSITY 8.9. Copper is a TRANSITION METAL with a characteristic reddish brown colour. It is extracted from its ores by ELECTROLYSIS. Many copper compounds have a characteristic blue or green colour, though copper oxide is black.

Copper is a good conductor of heat and electricity and is widely used as the conductor in electrical cables. It is also used to manufacture domestic water pipes. Brass, a hard alloy of copper and zinc, is used for the manufacture of small metal objects, but the high cost of copper limits its use for larger structures. *See also* BRONZE.

copper chloride Either of the salts copper(I) chloride (cuprous chloride), CuCl, or copper(II) chloride (cupric chloride), $CuCl_2$.

Copper(I) chloride is the less stable of the two compounds. It is a white solid; melting point 430°C; boiling point 1,490°C; RELATIVE DENSITY 3.4. Copper(I) chloride can be produced in a reaction of copper(II) chloride with excess copper in a concentrated solution of hydrochloric acid, to form the complex ion $CuCl_2^-$:

$$CuCl_2 + Cu + 2HCl \rightarrow 2H^+ + 2CuCl_2^-$$

On diluting the acid, this solution produces a PRECIPITATE of the insoluble chloride:

$$CuCl_2^- \rightarrow CuCl + Cl^-$$

Copper(II) chloride is a brownish yellow powder; melting point 620°C; decomposes on further heating; relative density 3.4. It also occurs as a blue-green HYDRATED form, $CuCl_2.2H_2O$. Copper(II) chloride can be formed by burning copper in chlorine:

$$Cu + Cl_2 \rightarrow CuCl_2$$

Copper(II) chloride dissolves in water to give a solution that is blue when dilute but changes through green to brown at higher concentrations, due to the replacement of the blue hydrated copper ion $[Cu(H_2O)6]^{2+}$, by brown chlorinated complexes, such as $CuCl_4^{2-}$.

copper losses In electromagnetic systems, particularly TRANSFORMERS, energy losses in the electrical rather than the magnetic parts of the system, such as the coils of a transformer, which are normally made of copper. *Compare* IRON LOSSES.

copper oxide Either of the two oxides of copper, copper(I) oxide, Cu_2O, or copper(II) oxide, CuO.

Copper(I) oxide is a red powder; melting point 1,235°C; decomposes on further heating; RELATIVE DENSITY 3.4. It can be formed by the partial thermal decomposition of copper(II) oxide:

$$4CuO \rightarrow 2Cu_2O + O_2$$

Copper(I) oxide disproportionates (*see* DISPROPORTIONATION) in dilute acids to form a solution of copper(II) ions plus a precipitate of metallic copper, for example:

$$Cu_2O + H_2SO_4 \rightarrow Cu SO_4 + Cu + H_2O$$

Copper(II) oxide is a black powder; decomposes on heating; relative density 6.3. It can be formed by heating copper carbonate, which occurs naturally:

$$CuCO_3 \rightarrow CuO + CO_2$$

Copper(II) oxide is insoluble in water, but reacts with acids to give blue solutions of copper(II) salts, for example:

$$CuO + 2HNO_3 \rightarrow Cu(NO_3)_2 + H_2O$$

It can be reduced to metallic copper by heating in a stream of hydrogen:

$$CuO + H_2 \rightarrow Cu + H_2O$$

copper(II) sulphate ($CuSO_4$) A white powder in its ANHYDROUS form, but more familiar as HYDRATED blue crystals, $CuSO_4.5H_2O$. It decomposes on heating; RELATIVE DENSITY 3.6 (anhydrous), 2.3 (hydrated). Copper(II) sulphate can be prepared as an aqueous solution by mixing copper carbonate with dilute sulphuric acid:

$$CuCO_3 + H_2SO_4 \rightarrow CuSO_4 + CO_2$$

Copper(II) sulphate is widely used as a fungicide and wood preservative.

copy DNA, *complementary DNA (cDNA)* DNA formed from an RNA template and therefore complementary to it. It is produced by the action of the enzyme REVERSE TRANSCRIPTASE. It is initially single-stranded but can be converted to double-stranded DNA by the action of the enzyme DNA POLYMERASE.

core 1. (*geology*) The central part of a star or planet. The Earth's core is believed to be composed of nickel and iron, and to be partly liquid with a temperature in excess of 6,000°C.

2. (*physics*) The iron centre of a SOLENOID or other electromagnetic device.

Coriolis force An imaginary force that makes a moving object appear to follow a curved path when viewed against a rotating background treated as stationary. The Coriolis force is used to simplify calculations involving rotating systems, which can then be treated as INERTIAL REFERENCE FRAMES. It is the Coriolis force that creates the rotating flow of air in WEATHER SYSTEMS as air flows from high to low pressure, and the rotation of water flowing down a plug hole – in both cases the motion is described with reference to the Earth as being at rest, though it is in fact rotating. *See also* CENTRIFUGAL FORCE, CENTRIPETAL FORCE, CIRCULAR MOTION.

cork Part of the bark of the stems and roots of most trees and shrubs. Cork that is used commercially comes from the cork oak tree, which has very thick layers of bark.

corm In plants, a short, thick, round underground stem surrounded by protective scale-like leaves that acts as a food store. Gladioli and crocuses, for example, have corms. New buds form in the scale leaves and use the food supply to develop into leafy, flowering shoots. The old corm then withers and a new corm forms at the base of the shoot, on top of the old one (*see* VEGETATIVE REPRODUCTION).

cornea The outer layer of the human eye, lying underneath the protective CONJUNCTIVA. The cornea is curved and acts as a fixed lens carrying out most of the REFRACTION (or bending) of the light entering the eye, before it reaches the lens. The cornea receives its nourishment and lubrication from antiseptic tears produced by the lachrymal (tear) glands, because it has no blood vessels. Corneal grafts can be performed in humans to replace diseased or opaque parts that are impairing eyesight.

corolla The collective term for the petals of a flower.

corona The region of high temperature, low density PLASMA around the Sun. The corona is too faint to be seen normally, but is visible during a total ECLIPSE of the Sun. It is distorted by the Sun's magnetic field.

corona discharge An area of IONIZATION around a conducting object in an electric field that allows charge to leak away but is not strong enough to produce a spark.

corpus callosum In the brain, the band of WHITE MATTER that joins the two halves of the CEREBRUM.

corpus luteum, *yellow body* In mammals, a temporary ENDOCRINE GLAND in the OVARY that is formed by the action of LUTEINIZING HORMONE on the ruptured GRAAFIAN FOLLICLE after OVULATION (egg release). It secretes the steroid hormone PROGESTERONE that prepares the uterine wall for pregnancy. If pregnancy does not occur, the corpus luteum breaks down. If pregnancy does occur, the life of the corpus luteum is prolonged by the hormone HUMAN

CHORIONIC GONADOTROPHIN, until the PLACENTA is fully developed at about 3 months of pregnancy. *See also* MENSTRUAL CYCLE.

correlation A measure of the extent to which two quantities are related. For example, in a population of people there will be a strong correlation between their height and their weight.

The degree of correlation can be observed by plotting the two quantities on a SCATTER DIAGRAM. The greater the correlation, the closer to a single straight line the points will lie. If the line has a positive gradient the correlation is described as positive, with a perfect correlation being given the value of +1. For a negative gradient the correlation is negative.

Correlation can be measured numerically by a correlation COEFFICIENT. This can be calculated in a number of ways depending on the nature of the data.

correlation coefficient *See* CORRELATION.

corrode (*vb.*) Of a metal, to form compounds by reaction with surrounding materials, particularly with oxygen, moisture and acidic gases in the atmosphere. This reduces the strength and electrical CONDUCTIVITY of metallic objects and limits their life when exposed to the atmosphere and atmospheric pollutants, such as sulphur dioxide.

Corrosion can be reduced either by coating the metal with an unreactive layer (such as paint, oil or plastic – *see* POWDER COATING) or by SACRIFICIAL CORROSION. Aluminium is remarkable in that the aluminium oxide, Al_2O_3, formed by the initial corrosion of an exposed aluminium surface is hard enough to prevent further reaction.

corrosion 1. Powdery material produced when a metal corrodes.

2. The process by which a metal corrodes.

cortex 1. In animals, the outer layer of some organs, such as the ADRENAL GLAND, brain and kidney.

2. In plants, the region beneath the outermost EPIDERMIS. *See also* MEDULLA.

corticoid *See* CORTICOSTEROID.

corticosteroid, ***corticoid*** A collective term for a group of STEROID HORMONES secreted by the CORTEX of the ADRENAL GLAND. Corticosteroids are either GLUCOCORTICOIDS, concerned with GLUCOSE METABOLISM (such as CORTISOL) or MINERALOCORTICOIDS, concerned with mineral metabolism (such as ALDOSTERONE). They are produced in response to stress situations under the control of the pituitary ADRENOCORTICOTROPHIC HORMONE.

corticotrophin *See* ADRENOCORTICOTROPHIC HORMONE.

corticotrophin-releasing factor *See* ADRENOCORTICOTROPHIC HORMONE.

cortisol, ***hydrocortisone*** A major GLUCOCORTICOID hormone of humans and other mammals secreted by the cortex of the ADRENAL GLAND in response to internal stress (such as low blood temperature or volume). It raises blood pressure and promotes GLUCONEOGENESIS (the conversion of protein and fat into glucose) and is under the control of the pituitary ADRENOCORTICOTROPHIC HORMONE.

cortisone A GLUCOCORTICOID.

cosecant A function of angle. In a right-angled triangle, the cosecant of an angle is defined as the length of the HYPOTENUSE divided by the length of the side opposite to the angle. The cosecant of an angle is the reciprocal of its SINE.

cosine A function of angle. In a right-angled triangle, the cosine of an angle is defined as the length of the side adjacent to the angle divided by the length of the HYPOTENUSE. The cosine of an angle is equal to the SINE of the COMPLEMENTARY ANGLE.

cosmic microwave background ELECTROMAGNETIC RADIATION received from all directions in space with a SPECTRUM characteristic of a BLACK BODY at a temperature of 2.7 K. This is interpreted as electromagnetic radiation from the hot early universe, Doppler shifted (*see* DOPPLER SHIFT) to longer wavelengths as the universe has expanded. The cosmic microwave background provides strong evidence for the BIG BANG theory of the early universe. *See also* COSMOLOGY.

cosmic radiation High-energy charged particles arriving at the Earth from space. Cosmic radiation consists mostly of protons, but also includes electrons and other atomic nuclei. When these so-called 'primary rays' enter the Earth's atmosphere, collisions with oxygen and nitrogen nuclei generate other ELEMENTARY PARTICLES and GAMMA RADIATION. The origin of cosmic radiation is uncertain, though SUPERNOVAE or other violent processes are probably involved, given the very high energies of some of the particles.

cosmological principle The idea that there is no privileged central position for observing the UNIVERSE and that the universe would appear basically the same viewed from any other point. HUBBLE'S LAW satisfies this principle, as an observer standing in some distant galaxy would also believe it to be true. *See also* COSMOLOGY.

cosmology The study of the Universe as a whole, particularly the early stages in the evolution of the universe, which was believed to begin with a large explosion called the BIG BANG.

Cosmology also concerns itself with the end of the universe. If the density of the matter observed is greater than the CRITICAL DENSITY, the expansion will eventually be overcome by gravity and the universe will end in a BIG CRUNCH, like the big bang in reverse. If the density is less than critical it will expand forever. Theories that predict a period of INFLATION suggest that the universe is just poised between these two possibilities, with the positive KINETIC ENERGY of the expansion exactly balancing the negative GRAVITATIONAL POTENTIAL ENERGY.

The observed matter in the universe accounts for a few per cent of the mass needed to reach the critical density, and astronomers are currently engaged in searches for the 'MISSING MASS'. Some believe that this may be in the form of WIMP'S (weakly interacting massive particles) not yet discovered by particle physics. Others believe that the matter is in the form of MACHO'S (Massive Compact Halo Objects), possibly burnt-out stars orbiting in the regions where GLOBULAR CLUSTERS are found (GALACTIC HALOS).

See also COSMIC MICROWAVE BACKGROUND, COSMOLOGICAL PRINCIPLE, HUBBLE'S LAW.

cosmozoan theory *See* ORIGIN OF LIFE.

cotangent A function of angle. In a right-angled triangle, the cotangent of an angle is defined as the length of the side adjacent to the angle divided by the length of the side opposite to the angle. The cotangent of an angle is the reciprocal of its TANGENT.

cotyledon, *seed leaf* A structure in the embryo of a SEED PLANT that may provide food for the growing embryo. The number of cotyledons in an embryo is an important means of classification in flowering plants (ANGIOSPERMS). Most plants have two cotyledons and are called DICOTYLEDONS, but some have a single one and are called MONOCOTYLEDONS. In GYMNOSPERMS there may be up to a dozen cotyledons in each seed.

The cotyledons may remain below ground (hypogeal) following germination, which most monocotyledons do, or may spread out above the soil (epigeal) following germination, forming the first green leaves, as most dicotyledons do. Where an ENDOSPERM (food source) is present in the embryo seed the cotyledons are thin, but in some plants, for example peas and beans, the cotyledons provide the main food supply and are therefore large. Cotyledons often store LIPIDS that form a high percentage of the dry weight of a seed, for example in walnuts, coconuts and sunflowers. Less frequently, the cotyledons store protein, for example in LEGUMES and nuts.

coulomb The SI UNIT of electric charge. One coulomb is the amount of charge carried in one second by a current of one AMPERE. *See also* FARADAY.

Coulomb's law For point charges, the size of the attractive or repulsive force between two charged particles is proportional to the size of each charge and inversely proportional to the square of the distance between them. Algebraically, the force F between two point charges, q_1 and q_2 is

$$F = q_1 q_2 / 4\pi\varepsilon_0 r^2$$

where ε_0 is the PERMITTIVITY of free space (equal to 8.85×10^{-12} $\mathrm{Fm^{-1}}$), and r is the separation of the two charges.

counter In physics, any device for detecting and counting IONIZING RADIATION, PHOTONS, etc. They generally work by electronically counting the current or voltage pulse created when a charged particle or photon causes IONIZATION.

countercurrent system Any physical or biological system in which two fluids flow in opposite directions along vessels close to one another, so that exchange, for example of heat or contents, can occur. The level of substance or heat drops in one fluid and rises in the other.

Some animals living in cold climates reduce heat loss from their feet by countercurrent HEAT EXCHANGERS. A capillary network exists in the extremity of the limb and warm blood entering the limb from the body passes alongside vessels containing cold blood returning to the body from the limb. The heat from the warm blood enters the cold blood as it returns to the body and the limb is kept at a lower temperature than the body, so reducing heat loss from the body.

Other examples are the LOOP OF HENLE in the kidney and the exchange of respiratory gases in the gills of bony fish.

couple A pair of parallel forces that are equal in size and opposite in direction, but do not act along the same line. Whilst a couple will not cause the CENTRE OF MASS of an object to accelerate, it will cause an ANGULAR ACCELERATION (that is, cause the object to turn). The MOMENT of a couple is equal to the size of the forces multiplied by the distance between their lines of action.

covalency The number of COVALENT BONDS an atom can form. Generally this is equal to the number of unpaired electrons in the outer orbitals. Thus carbon, with four electrons in the 2s and 2p ORBITALS, forms four covalent bonds, whilst nitrogen, with an extra electron, has a covalency of 3.

covalent bond A chemical bond in which electrons are shared between two atoms, giving each one a share in the other's electrons.

Molecules with covalent bonds tend to be relatively small (unless they are giant structures) and forces between them tend to be weak. Therefore the compounds tend to have low melting and boiling points, and are likely to be gases, volatile liquids or solids with low melting points. Covalent compounds do not conduct electricity, and are generally insoluble in water (although some are hydrolysed by water) but soluble in organic solvents.

In a covalent bond, electron ORBITALS in two neighbouring atoms overlap to form a MOLECULAR ORBITAL of lower energy. Since the orbitals involved in covalent bonding are P-ORBITALS or sp HYBRID ORBITALS, these bonds are aligned in space in a way that is related to the directions of these orbitals. This means that covalent bonds between two atoms are at specified angles to one another. For example, each covalent bond in a carbon atom that forms four such bonds has a bond angle of 109.5°. If the STEREOCHEMISTRY of a complex molecule demands a different bond angle, the bond will be substantially weaker.

Where two atoms share one pair of electrons a single covalent bond is formed, represented by a single line drawn between the atoms, such as in the structure of the ALKANES. Where the two atoms share two pairs of electrons a DOUBLE BOND forms, such as exists in the ALKENES, represented by two lines, and a TRIPLE BOND forms if three pairs of electrons are shared, such as in the ALKYNES.

Compare IONIC BOND. *See also* COVALENT CRYSTAL, FAJAN'S RULES, POLAR BOND, RESONANCE HYBRID, SHELL.

covalent crystal A crystalline MACROMOLECULE with all molecules being attached to their neighbours by a regular pattern of COVALENT BONDS. Diamond and silicon dioxide are important examples. Covalent crystals are hard and have high melting points. *See also* IONIC SOLID.

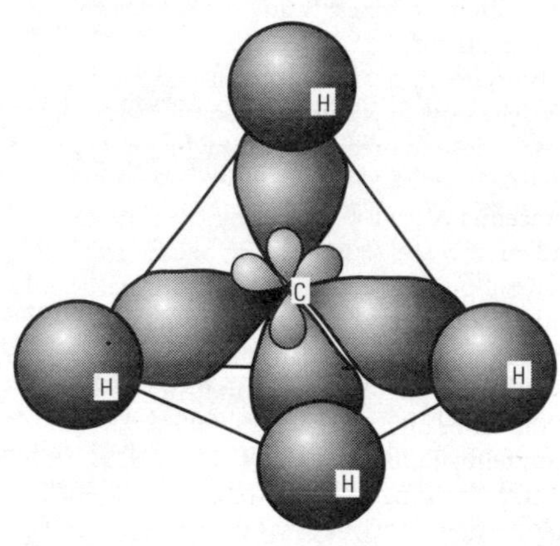

Covalent bond.

covalent radius An effective size associated with an atom involved in a COVALENT BOND. The covalent radius is half the separation between the nuclei of two similar atoms bonded together covalently. The bond length in a bond between two different atoms is then found by adding their covalent radii. This calculation works well for non-POLAR BONDS, but must be used with caution. The bond length in the POLAR MOLECULE HCl, for example, is not equal to the sum of the covalent radii of hydrogen and chlorine as measured by studying molecules of H_2 and Cl_2.

Cowper's gland In male mammals, one of a pair of glands below the PROSTATE GLAND. Cowper's gland secrete a sticky fluid into the URETHRA that contributes to SEMEN. *See also* SEXUAL REPRODUCTION.

Cowper stove A large HEAT EXCHANGER used in the steel industry to pre-heat air entering a BLAST FURNACE.

CPU *See* CENTRAL PROCESSING UNIT.

cracking A process used in the PETROCHEMICAL INDUSTRY for breaking larger ALKANES obtained from the FRACTIONAL DISTILLATION of CRUDE OIL into smaller alkanes and ALKENES. This is achieved by heating the alkanes to high temperatures of around 800°C, which breaks the carbon-carbon bonds. Often a CATALYST is used that allows lower temperatures (around 500°C) to be employed. This is termed catalytic (cat) cracking. Cracking is the method by which alkenes and also petrol are manufactured.

creation theory *See* ORIGIN OF LIFE.

creep The slow flow shown by some plastics and metals close to breaking point when subject to a LOAD.

Cretaceous A PERIOD of geological time at the end of the MESOZOIC ERA. It began after the JURASSIC period (about 140 million years ago) and extended until the PALAEOCENE ERA (about 65 million years ago). This period saw the first ANGIOSPERM plants on land and the extinction of the dinosaurs.

Creutzfeldt–Jakob disease (CJD) A disease of humans causing progressive degeneration of the CENTRAL NERVOUS SYSTEM. It appears to be caused by a 'prion protein', which is found in the brains of affected people after they die. The brains of these people have numerous holes in the tissue, which gives a spongy appearance similar to that seen in SCRAPIE of sheep. The disease usually affects older people because it has a very long incubation period.

During 1996 several cases of younger people dying from a disease with similar symptoms to CJD, but with a different brain appearance, led to a possible link between this new human disease and BOVINE SPONGIFORM ENCEPHALOPATHY, as a result of consumption of contaminated beef products.

crista *See* MITOCHONDRIA.

critical angle The ANGLE OF INCIDENCE at a boundary at which TOTAL INTERNAL REFLECTION first takes place. When light passes from a more dense to a less dense medium, some of the light does not pass the boundary, but is internally reflected. As the angle of incidence increases, the intensity of the reflected beam increases, until, at the critical angle, none of the light passes through the boundary and the whole beam is internally reflected.

critical density The density above which there is enough matter in the Universe for it eventually to collapse on itself as a result of gravity (the 'BIG CRUNCH'). *See* COSMOLOGY.

critical mass The mass of a piece of FISSILE material above which a CHAIN REACTION is no longer prevented by the escape of neutrons from the surface. Many NUCLEAR WEAPONS are triggered by a chemical explosive forcing together two smaller masses to form a lump of greater than critical mass.

critical point *See* CRITICAL TEMPERATURE.

critical solution temperature The temperature above which (for an upper critical solution temperature) or below which (for a lower critical solution temperature) two liquids are completely MISCIBLE and can be mixed together in any proportions.

critical temperature, *critical point* The temperature at which there is no distinction between the liquid and gas states of a substance. Above the critical temperature a gas cannot be liquefied by pressure alone. *See also* GAS.

crop In birds, an expanded part of the digestive tract between the OESOPHAGUS and the stomach. It is used to store food, especially seeds. Digestion begins in the crop, with the moistening of the food.

crossing-over A RECOMBINATION process occurring during MEIOSIS that results in pairs of chromosomes twisting around one another and exchanging segments (*see* CHIASMA, CHROMATID). The new combinations are called recombinants. Crossing-over provides the genetic variation that forms the basis of evolution. *See also* CROSS-OVER VALUE.

cross-over value, *recombination frequency* The proportion of recombinants formed in a group of offspring as a result of CROSSING-OVER This value can be used to determine the distance between genes on a chromosome. The closer together two genes are, the less likely it is that they would be separated by crossing-over, and therefore there would be fewer recombinants.

cross-pollination The transfer of pollen from the male part of one plant to the female part of another plant. *See* POLLINATION.

cross-section The surface formed by cutting through an object, at right angles to its axis of symmetry. The area of this surface is also called the cross section. *See also* COLLISION CROSS-SECTION.

crucible An open vessel made of a REFRACTORY material used to heat materials to high temperatures.

crude oil Unrefined PETROLEUM.

crust The solid surface layer of the Earth. *See* GEOLOGY.

Crustacea A CLASS of the PHYLUM ARTHROPODA comprising 26,000 species, including crabs, lobsters, prawns, shrimps, woodlice, crayfish, barnacles and the water flea *Daphnia*. All crustaceans possess an EXOSKELETON made of CHITIN impregnated with calcium carbonate, making it hard and impervious to water, acids, alkalis and solvents. The body is segmented, each part bearing appendages that serve a number of purposes. For example, crabs have eight pairs of thoracic appendages for walking and feeding, abdominal appendages for swimming and two pairs of antennae. Some crustaceans, for example *Daphnia*, are filter feeders, filtering particles from the water with hair-like bristles; others use modified appendages as mouth parts to feed on any organic matter; a few are parasitic. Most crustaceans have separate sexes and lay eggs.

crustacean A member of the CLASS CRUSTACEA.

cryogenics The study of materials and processes at temperatures close to ABSOLUTE ZERO. *See* SUPERCONDUCTIVITY, SUPERFLUIDITY.

cryolite A mineral form of sodium aluminoflouride, Na_3AlF_6. It is used in the production of aluminium from its ore BAUXITE.

cryoscopic constant *See* DEPRESSION OF FREEZING POINT.

cryostat A vessel for storing a material at low temperatures, usually immersed in a liquefied gas such as liquid helium or liquid nitrogen. Cryostats are generally many-walled vessels with a vacuum between the walls to prevent heat entering by conduction or CONVECTION, and also usually incorporate reflective coatings to reflect THERMAL RADIATION. *See also* VACUUM FLASK.

crypt of Lieberkühn One of several intestinal glands between the VILLI of the ILEUM. Crypts of Lieberkühn contain specialized cells (Paneth cells) that release enzymes involved in digestion.

crystal A piece of solid material throughout which the atoms are arranged in a single regular arrangement called a LATTICE. This arrangement is apparent in many naturally occurring crystals, which have symmetrical shapes reflecting the long-range ordering of their atoms.

Crystals are grouped into seven crystal systems, each one based on a different geometric shape. The classes are: cubic, HEXAGONAL, MONOCLINIC, ORTHORHOMBIC, TETRAGONAL, TRICLINIC and TRIGONAL. A given mineral will always crystallize in the same system, although the crystals may not always be the same shape.

Compare DISORDERED SOLID. *See also* BODY-CENTRED CUBIC, CLEAVAGE PLANE, CLOSE PACKED, COVALENT CRYSTAL, CRYSTALLIZATION, CRYSTALLOGRAPHY, CUBIC CLOSE PACKED, HEXAGONAL CLOSE PACKED, IONIC SOLID, ISOMORPHIC, SLIP PLANE, WATER OF CRYSTALLIZATION.

crystalline (*adj.*) Describing a material having the structure of a crystal, though the material itself may be POLYCRYSTALLINE, i.e. made of many small crystals with irregular shapes.

crystallization The process of forming crystals, particularly when used to purify a material or extract it from a solution. Since the SOLUBILITY of most materials increases with temperature, a typical crystallization process may begin with a warm solution that is cooled until it becomes SATURATED, at which point crystals will usually start to form. Some solutions become SUPERSATURATED, in which case it may be necessary to add a SEED CRYSTAL to start the process. *See also* FRACTIONAL CRYSTALLIZATION.

crystallography The study of crystals. Early crystallography was aimed at discovering the arrangement of atoms or molecules within a crystal, but more recently similar techniques have been used to map the density of electrons within the molecules forming organic crystals. X-RAY DIFFRACTION and NEUTRON DIFFRACTION are important techniques in this study. *See also* X-RAY CRYSTALLOGRAPHY.

crystal oscillator An OSCILLATOR that has its frequency determined by the RESONANCE of a PIEZOELECTRIC crystal, usually of quartz. Such oscillators are very stable, particularly if the temperature of the crystal is THERMOSTATICALLY controlled. They are commonly used in clocks and watches.

CSF *See* CEREBROSPINAL FLUID.

Ctenophora A CLASS of the PHYLUM CNIDARIA. The body form is neither polyp nor medusa and movement is by means of CILIA fused in rows (combs). Ctenophoras possess specialized sticky cells called lasso cells that are used for capturing prey. They do not possess stinging cells nor do they penetrate their prey as other COELENTERATES do. Examples include comb-jellies and sea-gooseberries.

cube 1. A three-dimensional figure having 12 sides all of equal length and at right angles to one another. The volume of a cube is equal to the third power of its length.

2. The third power of any number, thus the cube of *a* (written a^3) is $a \times a \times a$.

cubic close packed, *face-centred cubic* (*adj.*) Describing a crystal structure in which each layer of atoms is CLOSE PACKED, with each atom surrounded by six others in that layer. The next layer is placed so that it lies above gaps in the previous layer with the third layer again lying in gaps in the second layer, but also above the gaps in the first layer. The atoms in the fourth layer lie directly above those in the first layer, so labelling the layers of atoms as A, B and C, the packing can be described as ABCABC...

Seen from an angle, this structure can be seen as being cubic, with the UNIT CELL having an eighth of an atom at each corner and half an atom in the middle of each face. This accounts for the alternative name for this structure: face centred cubic. Many metals occur with this structure, including calcium and copper, though the HEXAGONAL CLOSE PACKED structure is also common.

cubic equation An equation containing POWERS of some quantity up to and including the third. For example:

$$ax^3 + bx^2 + cx + d = 0$$

cuboid A solid shape in which all the sides are at right angles to one another, but of different lengths. If the sides have lengths *a*, *b* and *c*, the volume of the cuboid will be *abc*.

cumene process An industrial method used for the manufacture of PHENOL and PROPANONE. It is more economical to make both products together than separately (from BENZENE and PROPENE respectively). Cumene itself, which is (1-methylethyl)benzene, $PhCH(CH_3)_2$, is obtained by the reaction of benzene with propene in the presence of a catalyst, such as hydrogen fluoride. The cumene is then oxidized and roughly equal amounts of both phenol and propanone result.

cumulative frequency In STATISTICS, a measure of the total number of objects in the sample for which some quantity is less than or equal to a certain value. *See also* OGIVE.

cumulonimbus A dense, heavy cloud that appears as a mountain or a huge vertical tower. They are usually associated with heavy rain and sometimes thunderstorms. *See also* WEATHER SYSTEMS.

cumulus 1. (*meteorology*) A heaped, fluffy cloud, produced by CONVECTION or a COLD FRONT. *See also* WEATHER SYSTEMS.

2. (*biology*) In mammals, the mass of cells that surround the developed OVUM.

cupric chloride *See* COPPER CHLORIDE.

cuprous chloride *See* COPPER CHLORIDE.

cupula A flat gelatinous plate found within the AMPULLA of the inner ear that is concerned with balance (*see* EAR).

curie (Ci) A unit of ACTIVITY, now superseded by the BECQUEREL. One curie is equivalent to 3.7×10^{10} Bq.

Curie point The temperature above which a material loses its FERROMAGNETIC properties. *See* PERMANENT MAGNET.

curium (Cm) The element with ATOMIC NUMBER 96; melting point 1,340°C (approx.). Curium does not occur naturally, but has been synthesized in fairly large amounts. The longest lived ISOTOPE (curium–247) has a HALF-LIFE of 17 million years. Curium is used as a heat source to generate electricity in some artificial satellites.

current The flow of electric charge through a CONDUCTOR. The size of the current through a certain cross-section of conductor is equal to the rate of flow of charge. The unit of current is the AMPERE.

CHARGE CARRIERS may be ELECTRONS, HOLES, or IONS. The amount of charge is measured in a unit called the COULOMB (C) and the size of the charge on one electron or PROTON is 1.6×10^{-19} C. The current is the number of coulombs passing a given point in one second. When a current flows through a conductor, the material becomes hot; the charge carriers collide with one another or with the surrounding LATTICE, producing random thermal vibrations.

See also AMMETER, CONDUCTION, CONVENTIONAL CURRENT, CURRENT DENSITY, KIRCHHOFF'S LAWS.

current balance A weighing device in which two parallel wires carry currents, with one of the wires pivoted so it is free to move. The force from the interaction between the currents is then balanced against known weights. *See also* BALANCE.

current density The amount of electric current flowing per unit area. This is a useful concept, as a conductor of a certain material can generally handle a fixed current density without overheating, regardless of the thickness of the conductor in question.

curvature of space The effect of gravity in the GENERAL THEORY OF RELATIVITY, which accounts for change in the motion of light and of massive objects under the influence of a GRAVITATIONAL FIELD. The path followed by light in curved space is called a GEODESIC LINE.

curve A line on a graph that is not straight; that is, where a given change in the *x* CO-ORDINATE does not always produce the same change in the *y* co-ordinate.

curved mirror Any mirror with a surface that is not flat, although in many applications the amount of curvature is very slight. Curved mirrors are usually of one of two types, parabolic or spherical.

Rays of light leaving a point in front of a curved mirror called the PRINCIPAL FOCUS of the mirror will be reflected to form a parallel beam. This effect is used in car headlights – a bulb is placed at the principal focus and all the light rays that hit the mirror leave it travelling in a single direction. The same effect is used in reverse in a REFLECTING TELESCOPE – parallel rays of light are brought together at the focus if they are travelling parallel to the PRINCIPAL AXIS of the mirror (the line joining the focus to the centre of the mirror). If the parallel rays of light are at a small angle to the principal focus they will be brought together (or focused) at a point in the FOCAL PLANE. Thus a REAL IMAGE is formed in the focal plane.

See also PARABOLIC MIRROR, SPHERICAL MIRROR.

Cushing's syndrome *See* GLUCOCORTICOID.

cuticle A tough, non-cellular, outer layer (the waxy layer) of plants that prevents loss of water and provides protection. In higher plants the cuticle is continuous except for the stomata (*see* STOMA) and LENTICELS. It is secreted by cells of the EPIDERMIS.

In invertebrates, such as insects and other arthropods, the cuticle acts as a protective EXOSKELETON. The insect cuticle often contains additional compounds and is of greater complexity than plant cuticle.

CVS *See* CHORIONIC VILLUS SAMPLING.

cyan A blue-green colour.

cyanamide 1. Any salt containing the cyanamide ion CN_2^{2-}.

2. The parent acid of cyanamide salts, H_2NCN. Cyanamide is a colourless crystalline solid; melting point 42°C; boiling point 140°C; RELATIVE DENSITY 1.3. It is formed by the reaction of carbon dioxide on sodium amide:

$$2NaNH_2 + CO_2 \rightarrow H_2NCN + 2NaOH$$

It is a weak acid, and in acidic solutions it hydrolyses (*see* HYDROLYSIS) to form UREA:

$$H_2NCN + H_2O \rightarrow H_2NCONH_2$$

cyanide Any salt or COMPLEX ION containing the cyanide ion, CN^-, such as potassium cyanide, KCN, and potassium hexacyanoferrate(III), $[Fe(CN)_6]^{3-}$. Cyanide salts can be formed by the action of bases on hydrogen cyanide, for example:

$$KOH + HCN \rightarrow KCN + H_2O$$

Cyanides are highly toxic because the ion has the ability to form stable complexes with the iron in haemoglobin, preventing the uptake of oxygen.

cyanobacteria, *blue-green bacteria* A group of single-celled prokaryotic organisms of the KINGDOM PROKARYOTAE, which were formerly called blue-green algae are found in damp surfaces of rocks and trees, aquatic habitats and in the soil. They are thought to be the oldest form of life. Many species can photosynthesize (*see* PHOTOSYNTHESIS) but their pigments are not contained in CHLOROPLASTS but THYLAKOID membranes within the cell. Some associate symbiotically with fungi to form LICHENS, and some are important in the NITROGEN CYCLE due to their ability to fix atmospheric nitrogen.

Cyanobacteria can be a problem in waters polluted with NITRATES and PHOSPHATES from fertilizers and detergents, because they multiply rapidly forming ALGAL BLOOMS that use all the available oxygen. When the bacteria die after a bloom, they can release toxins poisonous to fish and other animals (*see* POLLUTION).

cyanocobalamin *See* VITAMIN B.

cyanogen ($(CN)_2$) A gas with a strong characteristic odour; melting point –28°C; boiling point –21°C. Highly toxic, it can be produced by the OXIDATION of hydrogen cyanide in the presence of a silver catalyst:

$$4HCN + O_2 \rightarrow (CN)_2 + H_2O$$

It is an important FEEDSTOCK in the manufacture of some fertilizers.

cyanuric acid ($(HNCO)_3$) A white crystalline material, unstable on heating. Its structure contains alternate carbon and nitrogen atoms in a six-membered ring.

cycads *See* GYMNOSPERM.

cyclic (*adj.*) Describing any organic compound in which any of the carbon atoms are linked in a ring structure. Cyclic compounds can be ALICYCLIC, AROMATIC or HETEROCYCLIC.

cyclic AMP (cAMP) A cyclic NUCLEOTIDE produced from ATP by the action of the enzyme ADENYLATE CYCLASE. It is an important second messenger, and is produced in response to other signals, for example hormones. It determines the rate of many biochemical pathways.

cycloalkane Any SATURATED HYDROCARBON that contains a ring of carbon atoms. The general formula is C_nH_{2n}. An example of a cycloalkane is CYCLOHEXANE.

cyclohexane (C_6H_{12}) A colourless, inflammable liquid; boiling point 81°C; melting point 6.5°C.

H_2
C
H_2C CH_2
H_2C CH_2
C
H_2

It is produced by the REDUCTION of BENZENE with hydrogen in the presence of a nickel catalyst. It is an intermediate in the preparation of nylon and it is used as a solvent and paint remover.

cyclone Alternative name for a DEPRESSION, a region of low ATMOSPHERIC PRESSURE. Severe cyclones in the tropics (tropical cyclones) are accompanied by strong winds and can develop into hurricanes or typhoons.

cyclotron A PARTICLE ACCELERATOR in which charged particles are accelerated in a spiral path inside two hollow D-shaped electrodes, called dees. A magnetic field is applied at right angles to the plane of the dees, which keeps the particles moving in a circular path. The particles are accelerated by an electric field in the evacuated gap between the dees, which then reverses direction whilst the particles are within the dees. As the particles move faster the radius of their path increases, so the time taken to complete one orbit remains constant; thus there is no need to alter the frequency of the electric field or the strength of the magnetic field. After several thousands of revolutions, the particles reach the perimeter of the dees, where they are deflected onto the target. The energies that can be achieved by the cyclotron are limited by the RELATIVISTIC increase in mass, which upsets the balance between increasing speed and path radius. *See also* SYNCHROCYCLOTRON, SYNCHROTRON.

cylinder 1. (*mathematics*) A shape made by a line moving in a circle around a second line. The two lines are usually taken to be parallel, in which case the cylinder is more fully described as a right cylinder. For a cylinder of radius r and height h, the volume is $2\pi rh$.

2. (*engineering*) A space, straight sided and circular in cross-section, in which a PISTON can move.

cylindrical co-ordinates *See* POLAR CO-ORDINATES.

cytidine In biochemistry, a PYRIMIDINE NUCLEOSIDE consisting of the organic base CYTOSINE and the sugar RIBOSE.

cytochrome A protein forming part of the ELECTRON TRANSPORT SYSTEM. Electrons are transferred to the next cytochrome in a series of electron carriers, resulting in the REDUCTION of oxygen (O_2) to oxygen ions (O^{2-}), which combine with hydrogen ions to form water (H_2O) during aerobic RESPIRATION. The passage of electrons along the carrier chain results in the release of energy that is used to make ATP. Cytochromes are located in inner mitochondrial membranes, ENDOPLASMIC RETICULUM and THYLAKOIDS of CHLOROPLASTS.

cytochrome oxidase *See* ELECTRON TRANSPORT SYSTEM.

cytokinin A PLANT GROWTH SUBSTANCE that promotes CELL DIVISION in the presence of AUXINS. Cytokinins are derivatives of ADENINE (*see* NUCLEOTIDES) and are found in actively dividing tissues, such as fruits and seeds. They can delay ageing of leaves. They are thought to operate by increasing the METABOLISM of NUCLEIC ACIDS and PROTEIN SYNTHESIS.

cytology The study of cells and their functions, particularly through MICROSCOPY.

cytoplasm All the PROTOPLASM of a living cell excluding the nucleus. It is a transparent, slightly viscous fluid composed of a soluble jelly-like part (the cytosol) and the ORGANELLES embedded within this.

In many cells the cytoplasm is made up of two layers: the ectoplasm, a dense gelatinous outer layer containing few granules and associated with cell movement; and the endoplasm (or plasmasol), forming an inner layer more fluid in nature and containing many granules and most of the organelles. In plants the ectoplasm is equivalent to the CELL MEMBRANE.

cytoplasmic (*adj.*) Relating to CYTOPLASM.

cytoplasmic streaming The directional movement of the CYTOPLASM of certain cells, which allows substances to move through the cell. *See also* HYPHA.

cytosine An organic base called a PYRIMIDINE occurring in NUCLEOTIDES. *See also* DNA, RNA.

cytoskeleton A network of protein filaments (thread-like structures) and MICROTUBULES found within the CYTOSOL of eukaryotic cells (*see* EUKARYOTE), giving the cell its shape and internal organization. The cytoskeleton enables movement of materials throughout the cell and is also concerned with cell locomotion. ACTIN is an important constituent of the cytoskeleton.

cytosol, *hyaloplasm* The liquid part of the CYTOPLASM.

cytotoxic (*adj.*) Capable of killing cells. The term cytotoxic is used particularly to describe drugs that destroy cells, for example in CHEMOTHERAPY, or in reference to a subset of T CELLS.

D

daisywheel A wheel with letters in raised type around the rim, which is used in some types of computer PRINTERS. The quality is high but the number of characters available is limited and the time taken for the wheel to be rotated to the correct position for each letter makes printing slow.

Dalton's atomic theory The earliest contribution to an understanding of chemistry in terms of ATOMS. In 1805, John Dalton (1766–1844) postulated that: (i) elements are all composed of indivisible small particles (atoms); (ii) the atoms of an element are all identical to one another, but differ from the atoms of other elements; (iii) these atoms can neither be created nor destroyed; and (iv) compounds are formed by the combination of atoms of different elements in fixed ratios. *See also* LAW OF CONSTANT PROPORTIONS.

Dalton's law of partial pressures In any mixture of gases, the total PRESSURE exerted by the mixture is equal to the sum of the PARTIAL PRESSURE of each gas on its own (that is, the pressure that each gas would exert if it were present alone).

damping Any force that always opposes motion and increases with the speed of the motion, particularly in the context of oscillating systems. If no other forces acted other than those which caused a system to return to its EQUILIBRIUM position, oscillation would continue for ever. In practice, however, this is never the case. Provided the damping forces are small (that is, the system is UNDERDAMPED) they will produce an exponential decrease in the amplitude and a small reduction in the frequency. Larger damping forces – systems which are OVERDAMPED – will result in no oscillation at all, but the system slowly returns to equilibrium after it has been displaced. An example of overdamping is that of a pendulum immersed in treacle or thick oil – it will no longer swing to and fro, but simply return to equilibrium. The borderline state between underdamping and overdamping is called critical damping, and gives the most rapid approach to equilibrium. *See also* Q-FACTOR.

Darlington pair In electronics, a pair of TRANSISTORS connected together in such a way that the EMITTER current of the first transistor in the pair provides the BASE current of the second, producing a much larger GAIN.

Darwinism The theory of EVOLUTION proposed by Charles Darwin (1809–1882). It suggests that new species can arise and old species can become extinct by a process called NATURAL SELECTION. Darwin's work still forms the basis of modern day theories on evolution. *See also* COMPETITION.

Darwin's finches A group of about 14 species of finches, unique to the Galapagos Islands and studied by Charles Darwin (1809–1882). Each species had adapted to fill a different ECOLOGICAL NICHE. *See also* ADAPTIVE RADIATION.

database A large collection of information that is structured and stored in a computer independent of any particular application. Information can be retrieved by means of category headings.

data compression In computing, techniques for reducing the amount of data needed for the same amount of information, in order to reduce storage space.

dative bond *See* CO-ORDINATE BOND.

day The time taken for the Earth to revolve once on its axis. The solar day is the time taken for the Earth to revolve once relative to the Sun. The mean solar day is the average value of the solar day for one year. It is divided into 24 hours and is the basis of our civil day.

The sidereal day is the time taken for the Earth to revolve once relative to the stars. It is shorter that the mean solar day by four minutes, due to the Earth's orbit around the Sun.

d-block element A TRANSITION METAL; that is, an element with the outer electrons in D-ORBITALS. *See also* PERIODIC TABLE.

DBS *See* DIRECT BROADCAST SATELLITE.

d.c. motor An electric motor designed to operate from a DIRECT CURRENT supply. Most d.c.

motors comprise a set of rotating coils, called an ARMATURE, wound on an iron CORE. The ends of the coils are connected to a series of metal segments arranged in a ring – an assembly known as a SPLIT-RING COMMUTATOR. Brushes, fixed contacts, usually made of graphite, rub against the segments of the commutator and carry current in and out of the coils in the armature. The armature assembly rotates in a magnetic field provided either by PERMANENT MAGNETS or by ELECTROMAGNETS. The current in the magnetic field produces a force, which causes the coils to rotate (this is sometimes called the MOTOR EFFECT). The commutator reverses the current direction in each coil every 180° so the force produced always makes the coil rotate in a single direction.

If the coils are electromagnets, the motor can also be made to run off an ALTERNATING CURRENT supply – the reversal of the magnetic field caused by a change in polarity of the supply is compensated by a simultaneous reversal in the direction of the current in the armature.

D.C. motors with electromagnets may have the FIELD COILS (coils that provide the magnetic field in which the armature rotates) connected either in series (series-wound motors) or in parallel with the armature (shunt-wound motors). As the motor speed increases, the movement of the armature through the magnetic field produces a BACK E.M.F. in the armature, reducing the current through it. In a series-wound motor, this also reduces the magnetic field and thus the TORQUE produced by the motor. Series-wound motors thus produce very high torque at low speeds, so are used in electric vehicles. Shunt-wound motors are used where it is important that the speed does not change too much with the mechanical load on the motor.

DDT, *dichlorodiphenyltrichlorethane* (systematic name ***1,1,1-trichloro-2,2-di(4-chlorophenyl)ethane***) A synthetic PESTICIDE discovered in 1939 that has been used worldwide to kill organisms such as lice, fleas and mosquitoes (to combat MALARIA). It has caused many problems because it is persistent and accumulates along FOOD CHAINS. The use of DDT is banned in most countries, although it is still in use in developing countries where insect-borne diseases are a problem. Despite being banned, the persistence of DDT means that it is still found in many organisms. Many insects have developed DDT resistance. *See also* BIOMAGNIFICATION.

deamination Removal of the AMINO GROUP, $-NH_2$, from an AMINO ACID to form ammonia, urea or uric acid, depending on the animal, for excretion in the URINE. In vertebrates this process occurs in the liver, to remove unwanted amino acids.

de Broglie wavelength The wavelength that a particle appears to have when it is exhibiting wave-like properties (*see* WAVE NATURE OF PARTICLES). The de Broglie wavelength λ is given by

$$\lambda = h/p$$

where h is PLANCK'S CONSTANT and p the MOMENTUM of the particle.

debye A measurement of DIPOLE MOMENT used to state the degree to which a molecule is POLAR. One debye is equal to 3.34×10^{-30} Cm. Highly POLAR MOLECULES, such as caesium chloride, have dipole moments of about 10 debye, whilst a more typical polar molecule, such as hydrogen chloride, has a dipole moment of about 1 debye.

decahydrate A HYDRATE containing 10 parts water to one part of the compound. For example, sodium carbonate forms the decahydrate $NaCO_3.10H_2O$.

decant (*vb.*) To pour a liquid carefully from one vessel to another, so only the required part of the liquid is transferred, leaving behind any solid material, or the denser of two IMMISCIBLE liquids.

decay constant In radioactivity, the probability per second of an atomic nucleus decaying. In a sample of radioactive nuclei, which originally contains N_0 nuclei with decay constant λ, after a time t, the number of nuclei which remain in their undecayed state will be N, where

$$N = N_0 e^{-\lambda t}.$$

decay series *See* RADIOACTIVE SERIES.

deci- Prefix denoting one tenth. For example, a decimetre is one tenth of a metre (0.1 m).

decibel (dB, dBA) A unit for measuring the ratio of two signal levels on a logarithmic scale, commonly used in ACOUSTICS and electronics. A difference of signal strength of 10 dB represents a factor of 10, 20 dB represents a factor of 100, etc, while 3 dB represents a factor of (roughly) 2. Sound levels are often measured in decibels (dBA), with 0 dBA being the quietest sound that can be heard, 10 dBA being 10 times louder, 20 dBA 100 times louder and so on. The 0 dBA level is called the THRESHOLD OF HEARING. *See also* PHON.

decidua *See* ENDOMETRIUM.

deciduous (*adj.*) In botany, describing plants that lose their leaves seasonally, for example in the autumn. This is in contrast to evergreen plants, where the leaves are retained all year.

deciduous teeth, *milk teeth* In mammals, the first set of the two sets of teeth that mammals have. *See* TOOTH.

decimal 1. A number, or part of a number, smaller than 1, represented after a decimal point, such as 0.67.

2. (*adj.*) Describing a number written with a BASE of 10.

decomposer A general term for any organism in the FOOD CHAIN that feeds on dead material or excrement, breaking down complex organic compounds into simple organic or inorganic ones, enabling the recycling of nutrients to the soil or atmosphere. Decomposers are mostly saprophytic (*see* SAPROTROPH) bacteria and fungi and play a vital role in the NITROGEN CYCLE, CARBON CYCLE and PHOSPHOROUS CYCLE.

decomposition A chemical reaction in which a compound is broken down into its elements or into simpler compounds, usually under the action of heat.

decomposition voltage The voltage that must be applied to make an electric current flow in ELECTROLYSIS.

dee The hollow D-shaped ELECTRODE in a CYCLOTRON.

deep mining The extraction of a mineral ore by drilling a vertical shaft to reach the mineral, which is then lifted to the surface. *See also* DRIFT MINING, OPEN-CAST MINING.

defecation Elimination of FAECES from the body.

deflagration A sudden and spontaneous bursting into flames, such as observed with phosphorous in oxygen.

deforestation The destruction of forests by humans without replanting new trees or allowing for a cycle of regeneration. Deforestation can be to provide trees to use as timber or fuel, or to clear land for agricultural purposes or mining. Deforestation of both tropical rainforests and temperate forests is a great ecological problem. More than half of the world's rainforests have been destroyed.

Some of the consequences of deforestation are soil erosion, flooding, drought and GLOBAL WARMING. Also, the salt level in the ground may rise to the surface, making the ground unsuitable for farming. It can lead to DESERTIFICATION, which, along with other factors such as overgrazing and intensive cultivation, then leads to soil infertility.

degassing The process of removing dissolved gases from a liquid or gases absorbed into a solid, usually by heating. In particular, degassing is the removal of gas from the walls of a vacuum vessel.

degaussing A term for the removal of any permanent magnetism. *See* PERMANENT MAGNET.

degree of freedom 1. Any one of a set of independent variables needed to define the state of a physical system (such as pressure, temperature, etc.)

2. A way in which a molecule can possess energy independent of any other degree of freedom. Thus a MONATOMIC gas – one composed of molecules each containing only one atom – has three degrees of freedom, motion in each of three independent directions, and a MOLAR HEAT CAPACITY $3R/2$, where R is the MOLAR GAS CONSTANT. A molecule containing two or more atoms in a line has an extra two degrees of freedom, due to rotational motion about two axes at right angles to one another and to the line of the molecule. Molecules with three or more atoms not arranged in a straight line have a further degree of freedom. The molar heat capacity of an IDEAL GAS at constant pressure is $R/2$ times the number of degrees of freedom active in the material.

In solids, there are potentially six degrees of freedom: three due to KINETIC ENERGY from motion in three dimensions and three due to POTENTIAL ENERGY from departure from EQUILIBRIUM in three dimensions. QUANTUM effects mean that these six degrees of freedom are not fully active in most solids at ordinary temperatures – the molecules can have only certain quantized energies. This makes the molar heat capacity less than the expected value of $3R$.

See also PHASE RULE.

dehiscent (*adj.*) Describing a fruit that opens to shed its seeds, for example the poppy and pea.

dehydrate (*vb.*) To remove water from a substance. This may be from a solid containing WATER OF CRYSTALLIZATION, or from any

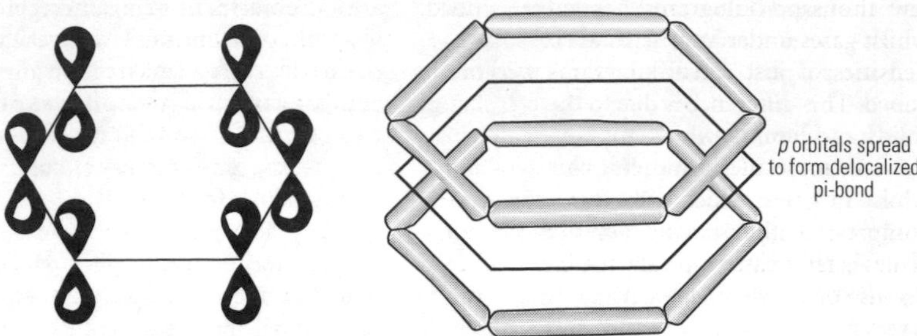

Delocalized orbital.

substance where the removal of water requires a chemical reaction, for example the dehydration of ethanol by concentrated sulphuric acid:

$$C_2H_5OH \rightarrow C_2H_4 + H_2O$$

See also HYDRATION.

deletion A chromosomal MUTATION in which a portion of a chromosome is lost.

deliquescent (*adj.*) Describing a solid that absorbs water vapour from the atmosphere and dissolves in this water to form a concentrated solution. Sodium hydroxide is an example of a deliquescent material.

delivery tube A glass tube, bent at one end, used with a GAS JAR to collect gas produced in a chemical reaction.

delocalized orbital One of a number of MOLECULAR ORBITALS that overlap to effectively produce a single large orbital that can hold as many electrons as the original orbitals. This overlap increases the stability of the molecule.

In metals, the VALENCE ELECTRON orbitals overlap to form a CONDUCTION BAND. In BENZENE, and other AROMATIC compounds, the P-ORBITALS at right angles to the plane of the aromatic ring overlap from one atom to the next, forming a ring around which electrons can move freely.

delta An area of land at the mouth of a river, formed by silt deposited by the river that the sea currents cannot remove. As the river weaves through the land, it splits and enters the sea through a number of mouths.

demodulation The reverse process to MODULATION, extracting information from a CARRIER WAVE.

dendrite One of many short projections from the cell body of a NEURONE that conducts NERVE IMPULSES from other neurones towards its own cell body. Impulses are passed to the dendrite of one neurone from the tip of an AXON of another neurone during a SYNAPSE.

dendrochronology The dating of fallen and fossilized trees by comparison of their growth rings. Dendrochronology is used to provide a check on radiocarbon dates, enabling those to be recalibrated to take account of the changes in carbon–14 concentration in the atmosphere. *See also* RADIOCARBON DATING.

dendron An extension from the body of a NEURONE that branches into DENDRITES and conducts NERVE IMPULSES from other neurones towards its own cell body.

denitrification The process by which NITRATES in the soil are converted back to atmospheric nitrogen. Anaerobic bacteria, such as *Pseudomonas denitrificans* and *Thiobacillus denitrificans*, bring about this conversion, particularly in waterlogged soil. Because most organisms cannot utilize atmospheric nitrogen (*see* NITROGEN CYCLE) denitrification reduces soil fertility. To avoid this, soil is ploughed and dug to improve drainage and aeration. *See also* NITRIFICATION.

denominator A number or quantity in a mathematical equation by which some other number is divided, such as y in the expression x/y.

dense (*adj.*) Having a high DENSITY.

density The mass of a substance contained in a given volume. This depends both on the mass of the molecules or atoms from which the material is made and on their separation.

Most solids or liquids have densities of a few thousand kilograms per metre cubed, whilst gases under ATMOSPHERIC PRESSURE have densities of just a few kilograms per metre cubed. This difference is due to the fact that in solids and liquids, which are not easily compressed, molecules are pretty closely packed, whilst in gases, which are much more easily compressed, the spacing between the molecules is far greater, typically ten times as great. *See also* DENSITY CAN, RELATIVE DENSITY, SPECIFIC VOLUME.

density can, *eureka vessel* A container with an overflow used to measure the volume of an object of known mass in order to find its DENSITY. The density can is filled with water to the level of the overflow and the object is lowered into it. The volume of water that then overflows is equal to the volume of the object.

density-dependence A method by which cells or organisms naturally regulate the size of their population. One or more factor(s) either speeds up the increase in the size of a population when its density is low, or decreases the expansion when the population density is high. The effect of the factor must be proportional to the population density for there to be true density-dependence. Cells that show CONTACT INHIBITION are exhibiting density-dependent inhibition.

dentine *See* TOOTH.

dentition The type and number of teeth in a species. *See* TOOTH.

deoxyribonuclease *See* DNASE.

deoxyribonucleic acid *See* DNA.

deoxyribose A PENTOSE sugar ($C_5H_{10}O_4$) that is a component of DNA.

dependent variable *See* VARIABLE.

depleted uranium Uranium from which the uranium–235 ISOTOPE has been removed. Depleted uranium is a dense and fairly hard material, and has been used in the manufacture of armour for military vehicles and for the manufacture of bullets and shells. *See also* ENRICHED URANIUM.

depletion layer An area close to a junction between P-TYPE SEMICONDUCTORS and N-TYPE SEMICONDUCTORS, in which there are few free CHARGE CARRIERS as FREE ELECTRONS, and HOLES diffuse across the junction to cancel one another out. *See also* PN JUNCTION DIODE.

depression, *cyclone* In meteorology, a region of lower than average ATMOSPHERIC PRESSURE. As air enters a depression it rotates anticlockwise in the northern hemisphere, clockwise in the southern hemisphere, as a result of the CORIOLIS FORCE. Depressions typically bring about a mixing of POLAR and TROPICAL AIR MASSES, leading to the arrival of a WARM FRONT, bringing lowering cloud then rain, followed by a COLD FRONT bringing showers. *See also* ANTICYCLONE, WEATHER SYSTEMS.

depression of freezing point The amount by which the freezing point of a solvent is reduced by the presence of dissolved molecules or ions. It is a COLLIGATIVE PROPERTY, independent of the nature of the dissolved particles, but proportional to their molar concentration, with a constant of proportionality known as the cryoscopic constant. Measurements of freezing point depression for measured masses of a material dissolved in a solvent can be used to estimate RELATIVE MOLECULAR MASSES.

depro-provera-progestin A long-lasting contraceptive preparation which is administered as an injection every three months.

depth of field The range of distances away from a camera at which an object can be placed and still produce an acceptably sharp image on the film.

derivative 1. (*mathematics*) The function that measures the rate of change of some other function. It is found by differentiation (*see* CALCULUS). If $y = x^n$, the derivative of y with respect to x, usually written as dy/dx, is nx^{n-1}.

2. (*chemistry*) A chemical compound derived from some other compound by a straightforward reaction, which usually retains the structure and some of the chemical properties of the original compound.

derived unit Any UNIT that is not a BASE UNIT. For example, the SI UNIT of speed, the metre per second, is derived from the metre and second. Some derived units are given special names – the SI unit of charge, for example, is called the COULOMB, but the definition of the coulomb is derived from those for the AMPERE and the second.

dermatology The study of SKIN and its disorders.

dermis The inner layer of the SKIN. The dermis contains blood vessels, hair follicles, nerves (RECEPTORS for touch, pressure, pain and temperature) and SEBACEOUS GLANDS and SWEAT GLANDS embedded in CONNECTIVE TISSUE. *See also* SENSE ORGAN.

desert An arid region capable of supporting only very few life forms.

desertification The creation of deserts as a result of climatic changes or human intervention. Desertification can be caused by DEFORESTATION, overgrazing or intensive cultivation that leads to soil infertility. It can be reversed by replanting trees or grasses and improving soil fertility and water retention.

desiccant A substance, usually a solid, that can be used to desiccate (remove all water from) a substance. Silica gel and ANHYDROUS calcium chloride are common desiccants.

desiccate (*vb.*) To completely remove all water from a substance.

desiccator A vessel with a sealed lid used to remove water from a material by exposing it to a DESICCANT.

destructive distillation A type of DISTILLATION process in which the substance being heated partly or fully decomposes, leaving a solid or viscous liquid. Volatile liquids can be collected as usual after condensation. Destructive distillation is used, for example, in the heating of coal to produce COAL TAR.

destructive interference The situation in which two waves arrive at the same point exactly out of PHASE. The waves tend to cancel one another out and the resulting wave has an amplitude equal to the difference in amplitude between the two original waves, or zero if the two waves had equal amplitudes. In order to produce destructive interference, TRANSVERSE waves must either be POLARIZED in the same direction, or as is often the case with light, unpolarized (really a mixture of all polarizations). The waves must also be COHERENT if destructive interference is to occur constantly at a given point. *See also* CONSTRUCTIVE INTERFERENCE, INTERFERENCE.

detergent A cleansing agent that is capable of wetting a variety of surfaces and removing dirt, usually found in oily or greasy deposits. Detergents are said to be surface active. They consist of molecules with long HYDROCARBON chains ('tails'), which are soluble in oil, attached to a salt group (the 'head'), such as SULPHATE, which is water-soluble. The detergent molecules surround the oil or grease drops and the hydrocarbon chain is able to penetrate the droplet whilst the salt group remains in the water. The salt groups become negatively charged and so then do the oil droplets, which thus repel one another and remain in suspension until they are removed by rinsing.

SOAP is a detergent but has the disadvantage of forming insoluble salts with the magnesium and calcium in hard water, leaving a scum. A range of synthetic detergents derived from petroleum have soluble magnesium and calcium salts to avoid this problem. Another problem with detergents is their disposal. Early detergents could not be degraded by the bacteria in sewage works and therefore caused foaming in rivers. New detergents have now been developed which are more easily broken down by bacteria, although there is still a problem if they escape the normal sewage process.

In addition many detergents contain added PHOSPHATES, bleaches or fluorescent substances as whiteners. These can cause pollution and EUTROPHICATION. More recently detergents are being used which contain no phosphates or bleaches and are thus more environmentally friendly.

See also SURFACTANT.

determinant A mathematical function of a square array of numbers. For a 2 x 2 array,

$$\begin{pmatrix} ab \\ cd \end{pmatrix}$$

the determinant is $ad - bc$. For a 3 x 3 array, the determinant can be found in terms of 2 x 2 determinants. For example, for the array

$$\begin{pmatrix} a_1 b_1 c_1 \\ a_2 b_2 c_2 \\ a_3 b_3 c_3 \end{pmatrix}$$

the determinant is given by

$$a_1(b_2c_3 - b_3c_2) - a_2(b_1c_3 - b_3c_1) \\ + a_3(b_1c_2 - b_2c_1)$$

The determinant of a 4 x 4 array can be found in terms of a 3 x 3 determinant in the same way. *See also* MATRIX.

detritivore An organism in the FOOD CHAIN that feeds on the organic debris (detritus) from decomposing plants and animals and excrement. Detritivores are usually larger than DECOMPOSERS and digest their food internally. Examples are earthworms, maggots and woodlice. *Compare* SAPROTROPHS, which feed on the excrement or dead bodies of others and digest their food externally.

detritus Organic debris from decomposing plants and animals. *See also* DETRITIVORE.

deuterium The naturally occurring heavy ISOTOPE of hydrogen, hydrogen–2 (one proton, one neutron).

Devonian A PERIOD of geological time during the PALAEOZOIC ERA. It extended from the end of the SILURIAN period (about 395 million years ago) to the CARBONIFEROUS era (about 345 million years ago). The period is characterized by extensive river and desert deposits and saw the first land plants and land insects.

dew Water vapour from the atmosphere condensed on the ground as the air cools at night. Dew tends to be thicker when the sky is clear, since THERMAL RADIATION is not reflected back to the ground. *See also* WEATHER SYSTEMS.

dew point The temperature at which air becomes SATURATED with water vapour. *See also* HUMIDITY.

dextrin A POLYSACCHARIDE formed during the HYDROLYSIS of STARCH.

dextrorotatory (*adj.*) Describing the ENANTIOMER of a compound exhibiting OPTICAL ISOMERISM that rotates PLANE-POLARIZED light to the right (clockwise). This used to be denoted by the prefix *d*- but (+) is now used. This is not to be confused with the prefix D- used to indicate the configuration of CARBOHYDRATES and AMINO ACIDS.

dextrose *See* GLUCOSE.

diabetes A name commonly used to refer to the disease *Diabetes mellitus,* in which there is a failure in the production of the hormone INSULIN by the PANCREAS. Insulin usually regulates blood sugar levels and excess glucose is stored in the liver. In diabetes, excess glucose is not stored so the blood sugar level rises. The kidney cannot absorb all the glucose passing through it, so the excess is secreted in the urine. Diabetes can potentially cause kidney failure, blindness and death in severe cases. Other symptoms include weight loss, thirst and coma.

Treatment of diabetes is by administration of insulin, either orally or by injection. The insulin can be obtained from pigs or calves, or it can be produced synthetically, since its structure is well known. Human insulin can be produced from bacteria by GENETIC ENGINEERING. Mild cases of diabetes can be controlled by diet.

Approximately 4 per cent of the world population have diabetes. The disease can start early in life, which is usually insulin-dependent diabetes caused by the autoimmune (self) destruction of certain cells in the pancreas, or can begin later in life, which is usually insulin-independent. Onset can also occur in pregnancy, called *Gestational diabetes,* leading to the birth of large babies, but this condition can be temporary.

Another form of diabetes is *Diabetes insipidus,* in which there is a failure to secrete ANTI-DIURETIC HORMONE, resulting in an excess of urine production.

diageotropic *See* GEOTROPISM.

dialysate *See* DIALYSIS.

dialysis The method by which small and large molecules in a mixed solution are separated. The mixed solution is placed inside a semipermeable bag (one that allows molecules of a certain size to pass through its pores) and surrounded by water. Small molecules diffuse out of the bag into the water, which is repeatedly changed, leaving the larger molecules inside the bag. In kidney failure, this principle is used in RENAL dialysis to remove toxic substances from the bloodstream.

In haemodialysis, the patient's blood is passed through a pump where it is separated by a semipermeable membrane from the dialysis fluid (dialysate). Toxic substances, for example urea, are filtered out, but RED BLOOD CELLS and WHITE BLOOD CELLS remain in the blood. Loss of useful substances, such as GLUCOSE and salt is minimized by adding them to the dialysate so that an equilibrium exists between the blood and dialysate.

Another method called continuous ambulatory peritoneal dialysis (CAPD) uses the membrane enclosing the PERITONEAL CAVITY as the semipermeable membrane. The dialysis fluid is pumped into the cavity and slowly out again, during which time toxic substances have diffused from the blood to the peritoneal fluid. The patient on CAPD can move around during dialysis.

Both methods of dialysis are expensive and are not always suitable. Kidney transplants are more effective and desirable in cases of chronic kidney failure.

diamagnetism The property of all substances that have a RELATIVE PERMEABILITY slightly less than 1. An applied magnetic field is slightly weaker inside the material than it would be if the material were not present. A bar of diamagnetic material will align itself at right angles to a magnetic field. Diamagnetism is caused by the interaction between the electrons in the atom and the external magnetic field. The phenomenon occurs in all substances, although in many cases it is masked by the much greater effects of PARAMAGNETISM or ferromagnetism (*see* FERROMAGNETIC).

diameter The distance between two opposite points on a circle, twice the RADIUS.

diamond A crystalline ALLOTROPE of CARBON. Diamond is the hardest known mineral and is widely used in cutting and drilling tools, and as a gem. Diamond has a MACROMOLECULAR structure with each carbon atom making four COVALENT BONDS with its neighbours. The rigidity of this structure is responsible for the hardness of diamond and its high THERMAL CONDUCTIVITY.

diapause In insects and some other invertebrates, a period of reduced metabolic activity (*see* METABOLISM) that can occur at any stage of development (usually eggs or pupae) but usually only once in a lifetime. It often coincides with the winter months and increases the organism's chances of survival. *See also* BIORHYTHM.

diaphragm 1. In general, a thin sheet separating off a region of space, but able to move.

2. In a CAMERA, an adjustable circular APERTURE made from overlapping sheets of metal.

3. (*biology*) In mammals, a sheet that is made of muscle and TENDON and separates the THORAX from the ABDOMEN. It is arched when resting but flattens during INSPIRATION, so reducing the pressure within the thorax to allow air to be drawn into the lungs.

4. ***cap, Dutch cap*** A method of CONTRACEPTION that consists of a rubber dome placed over the female CERVIX during sexual intercourse, preventing entry of sperm into the UTERUS. The diaphragm is 97 per cent effective if fitted correctly, used with a spermicide (cream or jelly preparations that kill sperm) and left in place for 6–8 hours after intercourse.

diaphysis The hollow shaft in the centre of a limb bone that contains BONE MARROW.

diatom Any member of the PHYLUM BACILLARIOPHYTA.

diatomic (*adj.*) Describing a molecule comprising just two atoms, such as chlorine, Cl_2, or hydrogen chloride, HCl.

diazo compounds *See* DIAZONIUM COMPOUNDS.

diazonium compounds, *diazo compounds* A class of compounds containing the RN=NX group, where R is an ALKYL or ARYL GROUP, and X is some other ion or group. The most important diazo-compounds are DIAZONIUM SALTS, which form the basis of the manufacture of AZO DYES.

diazonium salts DIAZONIUM COMPOUNDS derived from the base RN=NOH, where R is an ARYL GROUP. They are made from the reaction of AROMATIC AMINES with nitrous acid (HNO_2) in a process called DIAZOTIZATION.

The $-N^+{\equiv}N$ group formed in the process of diazotization can be easily replaced. For example, if a diazonium salt in water is heated the $-N^+{\equiv}N$ group is replaced by an OH (hydroxyl) group, resulting in the formation of a PHENOL.

$$C_6H_5N^+{\equiv}N\,HOSO_3^- + H_2O \rightarrow C_6H_5OH + N_2 + (HO)_2SO_2$$

The $-N^+{\equiv}N$ group can also be replaced by chlorine or bromine by treatment with copper(I) chloride or bromide in what is termed the Sandmeyer reaction. Similarly, treating with potassium iodide substitutes iodine.

Diazonium salts undergo coupling reactions with phenols or amines in which the $-N^+{\equiv}N$ group of the salt is not replaced but is instead added to by a second BENZENE RING. The products thus formed are brightly coloured and form the basis of the manufacture of AZO DYES.

$$C_6H_5N^+{\equiv}N + C_6H_5OH \rightarrow C_6H_5N{=}NC_6H_4OH$$

diazotization The process by which DIAZONIUM SALTS are made. A reaction mixture of dilute sulphuric and aqueous sodium nitrate is used to dissolve an AMINE. This results in the production of nitrous acid, HNO_2, which reacts with the amine. AROMATIC amines react at low temperatures to produce diazonium salts. The diazonium salts produced by ALIPHATIC amines are unstable and decompose rapidly.

$$C_6H_5NH_2 + HONO + (HO)_2SO_2 \rightarrow C_6H_5N^+(\equiv N)HOSO_3^- + 2H_2O$$

dibasic acid Any acid having a BASICITY of 2, such as sulphuric acid (H_2SO_4), carbonic acid (H_2CO_3). The acid can form a NORMAL SALT (such as a SULPHATE, SO_4^{2-}) or an ACID SALT (such as a hydrogensulphate HSO_4^-).

dicarbide, *acetylide* A BINARY COMPOUND formed by ALKALI METALS and ALKALINE EARTHS in combination with the dicarbide ion, C_2^{2-}. They react with water to produce ETHYNE (acetylene). For example, calcium dicarbide, CaC_2:

$$CaC_2 + 2H_2O \rightarrow C_2H_2 + Ca(OH)_2$$

dichlorodiphenyltrichloroethane *See* DDT.

dichromate(VI) Any salt containing the dichromate(VI) ion, $Cr_2O_7^{2-}$. The ion has an orange colour. In solution, the dichromate(VI) ion is stable only under acid conditions. In the presence of alkalis, they are converted to the CHROMATE(VI) ion:

$$Cr_2O_7^{2-} + 2OH^- \rightarrow 2CrO_4^{2-} + H_2O$$

dicotyledon A flowering plant that possesses two COTYLEDONS, or seed leaves, in the embryo. Most flowers, vegetables, fruit trees and forest trees (except CONIFERS) are dicotyledons. The cotyledons in dicotyledonous plants are usually surrounded by an ENDOSPERM that provides the food for the embryo, sometimes by passing it on to the cotyledon to be used by the embryo.

Dicotyledons usually have broad leaves, with net-like veins, and can be small or large plants. The flower parts are arranged in groups of five and the VASCULAR BUNDLE forms a regular ring arrangement in the centre of the stem. POLLINATION is usually by insects. After GERMINATION, most dicotyledons spread out above the soil forming the first green leaves, and are called 'epigeal'. *See also* MONOCOTYLEDON.

dictyosome *See* GOLGI APPARATUS.

dielectric An electrically insulating material, in particular one that causes a displacement of charge – but not a flow of charge – in an applied electric field. The inclusion of a dielectric between the plates of a CAPACITOR increases the CAPACITANCE over that in a vacuum by a factor known as the RELATIVE PERMITTIVITY. The molecules in a dielectric are POLARIZED by the electric field. This produces surface charges on the dielectric that are of the opposite sign to the charges on the adjacent capacitor plates. The result of this is to weaken the electric field in the dielectric, so that the same charge on the capacitor plates now leads to a lower POTENTIAL DIFFERENCE between them, i.e. to a greater capacitance. *See also* PERMITTIVITY.

dielectric constant *See* RELATIVE PERMITTIVITY.

diesel engine A type of INTERNAL COMBUSTION ENGINE that burns a lightweight fuel oil. It is similar in principle to the four-stroke PETROL ENGINE, but only air is drawn into the engine during the INDUCTION STROKE. The air is then compressed to a greater extent than in a petrol engine, and at the top of the COMPRESSION STROKE fuel is injected directly into the engine. Here, the compressed air is at a high enough temperature for the fuel to ignite without the need for a SPARK PLUG.

Diesel engines tend to be used on larger vehicles such as trucks and trains. The greater compression of the air means they are more efficient, however the need for a stronger construction makes them heavier and slower to accelerate.

diet The range of foods eaten by an animal. Diet varies greatly between species. HERBIVORES eat green plants, CARNIVORES eat meat from other animals, and OMNIVORES eat both plants and animal meat. In a mammalian diet, the main constituents are CARBOHYDRATES and FATS for energy, PROTEINS for growth and repair, VITAMINS and MINERALS for specific functions, water (which generally makes up 70 per cent of the total body weight of the mammal) and ROUGHAGE for efficient digestion.

Humans need nine ESSENTIAL AMINO ACIDS that we cannot make within our bodies and therefore need to take in the form of proteins from meat or vegetables. Minerals needed for various functions are numerous, but include calcium from dairy foods for bones and teeth; chlorine and sodium in common salt for maintenance of ANION/CATION balance; SULPHATES from dairy foods and meat for proteins; potassium from meat, fruit and vegetables for nerves and muscle action.

The amount of energy from food required by an individual varies according to gender, age and activity. Excess energy intake leads to obesity (overweight) and too little can lead to a condition called maramus in young children, where the child is irritable, does not grow and becomes thin due to lack of energy from carbohydrates and fat. A lack of protein in the diet of young children results in a condition called Kwashiorkor, where the body is swollen, the hair is soft and changes colour and growth is retarded. A high intake of fat (especially saturated fat; *see* FATTY ACIDS) has been related to heart disease, although other factors, for example, lack of exercise, smoking and stress, play a part.

Anorexia nervosa and bulimia nervosa are self-inflicted eating disorders, seen mostly in teenage girls and young women, that are usually due to psychological problems or stress. Anorexia nervosa is characterized by lack of eating and consequent wasting of the muscles. Bulimia nervosa results in a similar wasting away but the victim overeats (binges) and then causes self-induced vomiting.

dietary fibre *See* ROUGHAGE.

diethylamine *See* ETHYLAMINE.

differential amplifier An AMPLIFIER whose output depends on the difference between two inputs, called the INVERTING INPUT and NON-INVERTING INPUT. If the voltages at these inputs are V_- and V_+, the output voltage will be $A(V_+ - V_-)$, where A is the GAIN. *See also* OPERATIONAL AMPLIFIER.

differentiation The mathematical process of finding the DERIVATIVE of a function. *See* CALCULUS.

diffract (*vb.*) Of a wave, to spread out by DIFFRACTION.

diffraction The spreading out of any wave motion as it passes through an APERTURE, or when obstructed by a narrow barrier. The waves do not pass through with a sharp edge, like a shadow, but spread out, or diffract. The amount of diffraction is greatest for small apertures and for long wavelengths and depends on the ratio of wavelength to aperture size. For this reason diffraction is not normally noticed on an everyday basis with light – the wavelength is too short.

One consequence of diffraction is that no system, such as a telescope for example, can ever form a perfect image of an object – the light entering the telescope has WAVEFRONTS that are limited by the lens or the mirror of the telescope. The larger the aperture compared to the wavelength, the less the image will be affected by diffraction (the RESOLUTION is higher). RADIO TELESCOPES, which operate at longer wavelengths than optical telescopes, have relatively poor resolution, so are less able to distinguish as separate two objects that appear close to one another in the sky.

On a simple level, diffraction can be explained by recognizing the fact that it is not possible to have a wavefront with a perfectly sharp edge. A more sophisticated explanation relies on HUYGENS' CONSTRUCTION, which provides a detailed description of the distribution of the amplitude of a wave that has passed through an aperture. In the case of FRAUNHOFER DIFFRACTION of light from a single slit, there is a broad central peak in the AMPLITUDE before it falls to zero, but further out it increases again, reaching about one third of its maximum value (a subsidiary maximum). There are therefore a succession of points where the amplitude reaches zero, with smaller and smaller peaks in amplitude between each minimum. The zero points are evenly spaced. As the slit is made narrower, the pattern becomes wider and less bright. The first minimum is explained by Huygens' construction in terms of light from each point in one half of the slit interfering destructively (*see* DESTRUCTIVE INTERFERENCE) with light from the corresponding point in the bottom half of the slit.

For Fraunhofer diffraction from a single slit, of width α, provided that the wavelength λ is much less than the width of the slit, the points of zero amplitude will be at angles θ to the straight ahead direction, where

$$\theta = n\lambda/\alpha$$

where n is a whole number.

See also DIFFRACTION GRATING, FRESNEL DIFFRACTION, INTERFERENCE, YOUNG'S SLIT EXPERIMENT.

diffraction envelope A gradual overall variation in intensity of some patterns of light, such as YOUNG'S FRINGES, caused by the DIFFRACTION of the light from the individual slits through which light passes to make the pattern.

diffraction grating A mirror or transparent piece of material on which many regularly spaced lines are ruled to form a series of slits. Light diffracts from each slit, but only interferes constructively (*see* CONSTRUCTIVE INTERFERENCE) in those directions in which the PATH DIFFERENCE between light from one slit and the next is a whole number of wavelengths.

For a grating illuminated by light striking the grating at 90°, the light will leave the grating at angles θ such that

$$d\sin\theta = n\lambda$$

where λ is the wavelength of the light, *d* is the grating spacing (the distance between adjacent slits) and *n* is a whole number called the order of the diffraction maximum. Thus different wavelengths leave the grating at different angles. If the light does not hit the grating square on, but at an angle ϕ, this introduces a PHASE difference between light leaving one slit and the next, modifying the formula to

$$d(\sin\phi + \sin\theta) = \lambda$$

The RESOLUTION of a grating is its ability to distinguish between two similar wavelengths and is proportional to the number of slits in the grating. If there are *N* slits, the smallest wavelength separation detectable will be δλ, where

$$\delta\lambda = \lambda/N$$

diffuse (*vb.*) To spread out by DIFFUSION.

diffusion The spontaneous and random movement of molecules or particles in a fluid (gas or liquid) from a region where they are at a high concentration to one where they are at a low concentration, until a uniform concentration or dynamic EQUILIBRIUM is achieved. Once at a uniform concentration, the molecules will continue to move in random motion, but there is no net diffusion. The concentration gradient is the difference in concentration of a substance between two regions. The rate at which one material diffuses into another is determined by the average speed of its molecules and by the MEAN FREE PATH.

In biological systems, diffusion often occurs across epithelial layers (*see* EPITHELIUM) or CELL MEMBRANES, and since a greater surface area leads to faster diffusion, areas specialized for this purpose often have VILLI to increase their surface area. Sometimes channels occur within a cell membrane or carrier molecules exist to speed up diffusion of specific substances. Diffusion is important in the transport of nutrients, respiratory gases and NEUROTRANSMITTERS within and between cells.

Compare ACTIVE TRANSPORT. *See also* GRAHAM'S LAW OF DIFFUSION.

digestion In animals, the break down of food both physically and chemically (by enzymes) into its basic constituents ready for absorption to convert to energy. Digestion usually occurs in the STOMACH and INTESTINES. *See also* DIGESTIVE SYSTEM.

digestive system, *alimentary canal, gut* The system of cavities, tubes, organs and glands associated with DIGESTION (*see following page*).

In humans the system begins at the MOUTH, where food is mixed with saliva and the first stage of digestion occurs. Food is chewed into a round ball called a 'bolus' and this is pushed by the TONGUE to the back of the mouth and to the PHARYNX. In this region the OESOPHAGUS and TRACHEA meet and the bolus passes down the oesophagus aided by a number of reflexes that prevent it entering the trachea. The food is lubricated by secretions from glands lining the oesophagus, and then passes into the STOMACH. It is here that storage and further digestion is carried out. The resulting fluid (CHYME) passes to the DUODENUM and further breakdown, and most absorption of nutrients, occurs in the SMALL INTESTINE. What remains passes to the LARGE INTESTINE for water absorption and storage of FAECES before their excretion.

The digestive system has a good blood supply through which nutrients from digested food are carried to the LIVER for use by the body cells.

In simpler organisms, the digestive system can be a cavity with a single opening, for example, jellyfish and *Hydra*, where some of the contents taken in are absorbed and the rest expelled through the same opening.

See also GALL BLADDER, GIZZARD, ILEUM.

digit A single numeral or symbol used to represent a number. DECIMAL numbers are all represented using the digits 0 to 9.

digital (*adj.*) Describing a system in which only whole numbers are handled, normally only two values, corresponding to BINARY 1 and 0 are used. *Compare* ANALOGUE.

digital signal An electrical signal represented by two values only: high and low (or 1 and 0).

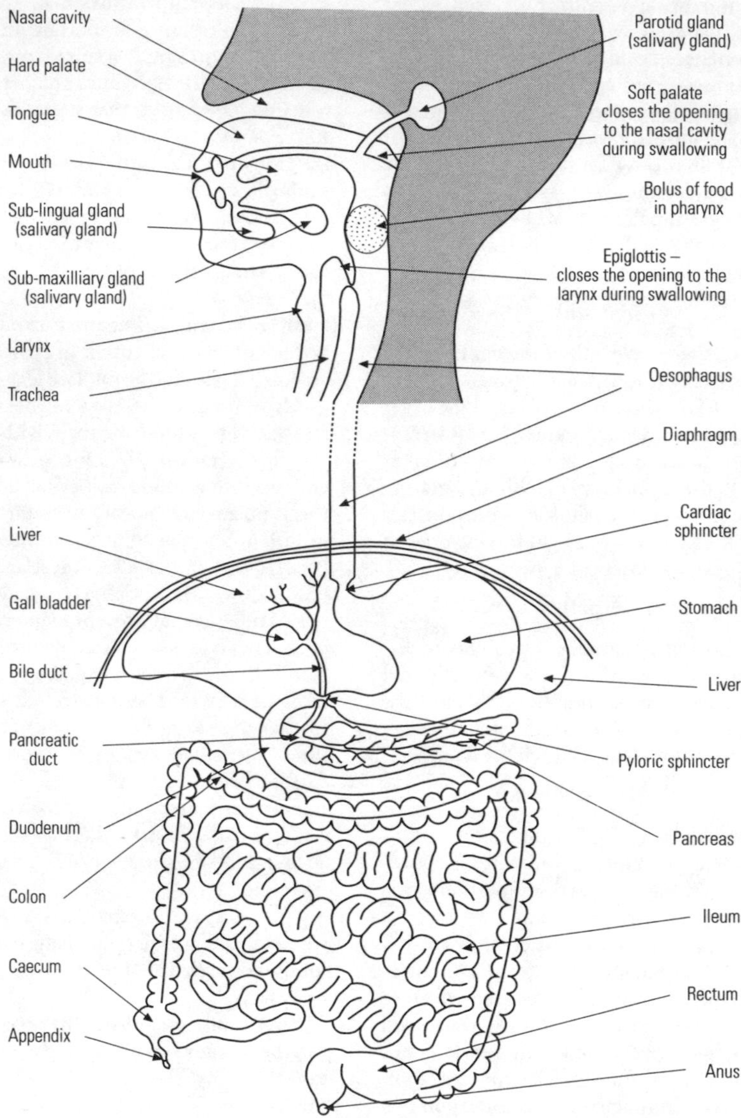

Human digestive system.

High is a voltage close to the supply voltage, while low is a voltage close to zero. A digital system has the advantage that any small change in the signal, the addition of electrical noise for example, will not prevent the signal from being recognized as a one or a zero. Digital systems are therefore much less prone to interference than systems based on ANALOGUE signals.

The disadvantage of this technique is that most signals that we wish to handle are analogue in nature, and must be converted into a binary number several digits long to give a reasonably continuous range of values. These binary digits are then dealt with by repeated circuits (PARALLEL PROCESSING) or one after another (SERIAL LOGIC). Modern digital electronic circuits can provide complex processing power at high speed in a small space with low cost and low power consumption. This, combined with the other advantages of digital systems, means that they are increasingly used.

digital-to-analogue converter A device that converts DIGITAL signals into ANALOGUE signals. With a suitable choice of resistors and each input connected to either a HIGH or LOW voltage, a SUMMING AMPLIFIER, which adds these signals together in appropriate proportions, can be used as the basis of a digital-to-analogue converter.

dihybrid cross In genetics, a cross between two pure-bred individuals of the same species that differ in two characteristics. The characteristics could, for example, be seed colour and shape. The F_1 GENERATION yields offspring identical for those characteristics (e.g. green, round seeds) because the DOMINANT ALLELES are expressed. When the F_1 offspring are crossed with one another, the offspring would show different characteristics (*see* PHENOTYPES) in the ratio of 9:3:3:1, according to MENDEL'S LAWS. In practice, not all dihybrid crosses give this ratio. *See also* MONOHYBRID CROSS.

dihydrate A HYDRATE containing two parts water to one part of the compound. For example, calcium sulphate forms the dihydrate $CaSO_4.2H_2O$.

dihydric (*adj.*) Describing a compound with two HYDROXYL GROUPS. For example, GLYCOL is a dihydric ALCOHOL.

dilead(II) lead(IV) oxide, *red lead* (Pb_3O_4) A red powder (black when hot); decomposes on heating; RELATIVE DENSITY 9.1. Dilead(II) lead(IV) oxide can be made by heating lead(II) oxide to about 400°C in air:

$$6PbO + O_2 \rightarrow Pb_3O_4$$

The oxide decomposes back to lead(II) oxide on heating. The structure is complex and largely covalent (*see* COVALENT BOND), often containing less oxygen than the formula would suggest. The material was once widely used as a pigment in paints, but this has been discontinued because of concerns about the toxicity of lead.

dilute 1. (*adj.*) Having a low concentration.

2. (*vb.*) To add more solvent to a solution, reducing the concentration. *See also* HEAT OF DILUTION, OSTWALD'S DILUTION LAW.

dilution The factor by which the volume of a dilute solution is greater than the original concentrated solution, or liquid solute. Thus a dilution of 5 indicates a solution in which 5 litres of dilute solution has been produced from 1 litre of the original material.

dimensions The fundamental physical quantities that describe any system, and the powers to which these quantities are raised to obtain a derived physical quantity. In a mechanical system, the basic physical quantities are usually taken to be mass (M), length (L) and time (T). Using these dimensions, the derived physical quantity of volume will have the dimension L^3, velocity is L/T (or LT^{-2}), and MOMENTUM, which is a product of mass and velocity, will have the dimensions MLT^{-2}.

dimer 1. A structure in which two identical molecules are held together, either by COVALENT BONDS or more loosely by HYDROGEN BONDS. Dinitrogen tetroxide, N_2O_4, is a dimer of nitrogen dioxide, NO_2, for example.

2. Two MONOMERS joined together by POLYMERIZATION.

4-dimethylamino-4′-azobenzene sodium sulphonate *See* METHYL ORANGE.

dimethylbenzene, *xylene* (C_8H_{10}) A colourless liquid with a characteristic smell. It is a mixture of three ISOMERS and is widely used as a solvent.

diminished (*adj.*) Describing an IMAGE that is smaller than the original object.

dinitrogen oxide, *nitrous oxide* (N_2O) An odourless gas; melting point –91°C; boiling point –89°C. It can be prepared by heating ammonium nitrate:

$$NH_4NO_3 \rightarrow N_2O + 2H_2O$$

Dinitrogen oxide will support the combustion of many compounds, for example:

$$Mg + N_2O \rightarrow MgO + N_2$$

It decomposes on heating:

$$2N_2O \rightarrow 2N_2 + O_2$$

Inhalation of dinitrogen oxide causes a sense of well-being, followed by unconsciousness, hence its alternative name of laughing gas, and its use as an anaesthetic.

dinitrogen tetroxide (N_2O_4) A pale brown liquid; melting point 11°C; boiling point 21°C; RELATIVE DENSITY 1.5. Dinitrogen tetroxide exists in equilibrium with nitrogen dioxide:

$$N_2O_4 \Leftrightarrow 2NO_2$$

In the liquid phase the mixture consists almost entirely of the DIMER, whereas in the gas the MONOMER predominates. Dinitrogen tetroxide can be produced by condensing nitrogen dioxide produced in the reaction between concentrated nitric acid and copper:

$$Cu + 4HNO_3 \rightarrow Cu(NO_3)_2 + 2NO_2 + 2H_2O$$

diode An electronic device that allows current to pass in one direction only. *See* PN JUNCTION DIODE, THERMIONIC DIODE, VARICAP DIODE. *See also* BRIDGE RECTIFIER, RECTIFIER.

dioecious The presence of male and female reproductive organs on different individuals. This term is often used to refer to plants that have male and female flowers on separate plants of the same species. This arrangement favours cross-fertilization (*see* FERTILIZATION). *Compare* MONOECIOUS.

dioptre (D) A measure of the refractive power of an optical lens, equal to the reciprocal of the FOCAL LENGTH in metres. Thus, for example, a 10 D lens has a focal length of 1/10 m or 0.1 m. A negative power represents a DIVERGING lens.

diploblastic (*adj.*) Of an animal, having a body that develops from just two of the three GERM LAYERS – the ECTODERM and ENDODERM. COELENTERATES are diploblastic. *Compare* TRIPLOBLASTIC.

diploid (*adj.*) The presence of two sets of chromosomes in the nucleus of a cell. Most animal cells are diploid, except the GAMETES in sexually reproducing species, which are HAPLOID. *See also* POLYPLOID.

Diplopoda A CLASS of the PHYLUM ARTHROPODA, including the millipede.

dipole A magnetic system containing two POLES, such as a BAR MAGNET, or an ELECTROSTATIC system containing two equal but opposite charges.

At a distance from the dipole that is large compared to the separation of the charges or poles, the field produced decreases with increasing distance by a factor equal to the inverse third power of the distance from the dipole. Thus if the distance is doubled, the field falls to one eighth of its original strength. *See* MAGNETIC DIPOLE.

dipole moment The strength of a dipole. For an electric dipole, the dipole moment is equal to the size of the charges multiplied by the distance between them. *See also* DEBYE.

direct broadcast satellite (DBS) A television system in which individual homes are equipped with a receiving AERIAL and parabolic reflector (satellite dish) for the reception of television signals from a satellite in GEOSTATIONARY ORBIT.

direct current (d.c.) A steady electric current flowing always in the same direction. *Compare* ALTERNATING CURRENT.

directory A list of computer FILE names, together with the information that enables them to be retrieved from the BACKING STORE by the OPERATING SYSTEM.

disaccharide A double sugar formed by the combination of two MONOSACCHARIDES, with the loss of a water molecule (a CONDENSATION REACTION). A disaccharide can be split into its single sugars by the addition of water (a HYDROLYSIS reaction). Disaccharides are sweet, soluble and crystalline. Examples include SUCROSE (cane sugar or table sugar, $C_{12}H_{22}O_{11}$; GLUCOSE plus FRUCTOSE), maltose (glucose plus glucose) and LACTOSE (glucose plus GALACTOSE). *See also* CARBOHYDRATE, POLYSACCHARIDE.

discharge (*vb.*) To remove the electric charge from an object. This is most easily done by connecting it to the EARTH (a process called earthing or grounding). Electrons flow either to or from the earth until the earthed object has no overall charge.

discharged (*adj.*) The state of a CAPACITOR or cell when it has no stored charge. *See also* CHARGED.

discrete data In statistics, quantities that can only take on particular values, with values between these having no meaning, for example the number of children in a family can only ever be a whole number. *Compare* CONTINUOUS DATA.

disease An impairment of the normal functioning of an organism. Causes of disease can be congenital (inborn) or acquired through infection or injury. Many diseases have a known causative agent (a PATHOGEN), such as a virus, bacteria, fungus or PROTOZOA, that results in a characteristic set of symptoms, for example a rash in the case of chickenpox. Some organisms cause their effects by releasing toxins or venoms. For some diseases such as cancer, arthritis and multiple sclerosis, the cause is complex or even unknown.

If many people suffer from the same disease at the same time it is called an epidemic. Some diseases only occur in certain parts of the world (endemic), other diseases occur all over the world (pandemic). Sporadic refers to irregular outbreaks of diseases in particular places.

disintegration A term sometimes applied to the decay of a radioactive nucleus. *See* RADIOACTIVITY.

disk drive In a computer, a device that holds a disc of a magnetic material on which PROGRAM data or FILES are stored as a series of magnetized regions. A HARD DISK has a fixed disc of magnetic material, whilst a FLOPPY DISK is removable, so programs can be purchased separate from the machine on which they will be used and data from one computer can be transferred to another.

Disk Operating System *See* DOS.

dislocation An imperfection in the LATTICE structure of a solid, particularly a metal. A dislocation arises as a result of a misalignment of atomic planes as the crystals of the material are forming, with a plane of atoms coming to a halt in the middle of the atom or forming a 'spiral staircase' structure. *See also* ALLOY.

disordered solid A solid in which the atoms are not arranged in a regular crystalline LATTICE. Some materials, such as glass, are naturally disordered; others form disordered solids only if they are cooled so rapidly that crystals do not have time to form.

dispersal The movement of seeds, eggs and offspring away from their parents in order to prevent unfavourable competition or overcrowding. There are various methods of dispersal, for example by insects, wind, animals (that eat or carry seeds) and locomotion (of animal offspring themselves).

dispersed phase *See* COLLOID.

dispersion The variation of the REFRACTIVE INDEX of a medium with the wavelength of the light passing through it. Thus the amount of REFRACTION will be slightly different for different colours of light. This effect is used in a PRISM to split white light into the colours from which it is made up.

dispersive (*adj.*) Describing a material that shows DISPERSION.

displace (*vb.*) To push out of the way, as in a DISPLACEMENT REACTION.

displacement The volume, mass or weight of fluid pushed out of the way by an immersed object. In particular, the water displaced by a ship, where the displacement is equal to the weight of water contained in the volume of the hull below the water level. By ARCHIMEDES' PRINCIPLE, this will be equal to the weight of the ship.

displacement reaction A chemical reaction in which one element, particularly a metal, is replaced by another, more reactive, element. For example, iron filings will displace copper in copper sulphate solution, with the formation of a PRECIPITATE of metallic copper:

$$CuSO_4 + Fe \rightarrow FeSO_4 + Cu$$

This reaction shows that iron is more reactive than copper. *See also* ELECTROCHEMICAL SERIES.

disproportionation A process in which the same chemical compound is both oxidized and reduced (*see* OXIDATION, REDUCTION). For example, copper(I) compounds in aqueous solution will form copper(II) compounds (an oxidation) and a precipitate of metallic copper (a reduction), for instance:

$$Cu_2SO_4 \rightarrow Cu + CuSO_4$$

dissipate (*vb.*) In physics, to turn energy, in particular electrical energy, into heat.

dissipative force A force, such as friction, that always opposes the direction of motion, and converts KINETIC ENERGY to heat. Dissipative forces produce DAMPING in SIMPLE HARMONIC MOTION.

dissociate (*vb.*) To break up, for example into ions or into the elements from which a substance is made.

dissociation The breaking up of a compound into simple compounds or the breaking up of an IONIC compound into separate ions when it is dissolved or melted. For example, copper(II) sulphate, $CuSO_4$, dissociates into copper ions, Cu^{2+}, and sulphate ions, SO_4^{2-}. *See also* DISSOCIATION CONSTANT, HEAT OF DILUTION, OSTWALD'S DILUTION LAW.

dissociation constant The EQUILIBRIUM CONSTANT for the dissociation of a molecule into ions, $AB \Leftrightarrow A^+ + B^-$. The dissociation constant is $[A^+][B^-]/[AB]$, where the square brackets, [ ], denote CONCENTRATION. *See also* PK.

dissolve (*vb.*) The breaking up of a solid into individual ions or molecules when placed in a solvent, thus producing a liquid solution. The term also refers to the distribution of the molecules of a gas through the liquid, again producing a liquid solution. *See also* HEAT OF SOLUTION, SOLUBILITY, SOLUBLE.

distil (*vb.*) To perform a DISTILLATION process.

distillate *See* DISTILLATION.

distillation A process used to purify liquids or to separate mixtures of liquids by virtue of their differing boiling points. The liquid is heated to the required temperature and the vapour is collected into a separate vessel (the condenser), where it is cooled and condensed back to a liquid. The liquid collected (the distillate) is pure, leaving solid impurities in the distillation vessel. For separating mixtures of liquids with similar boiling points, such as CRUDE OIL, a process called FRACTIONAL DISTILLATION is used. Distillation is also used in the manufacture of alcoholic beverages. *See also* DESTRUCTIVE DISTILLATION, MOLECULAR DISTILLATION.

distortion An unwanted change introduced in an optical image or an electronic signal. Particularly, a measure of the extent to which the output from an AMPLIFIER is not simply a constant times the input. *See also* ABERRATION.

distribution, *frequency distribution* In statistics, a set of measurements that show how many times a quantity takes a particular value, or a value that lies in a particular range.

disulphuric(VI) acid ($H_2S_2O_7$) A colourless crystalline solid; melting point 35°C; decomposes on further heating; RELATIVE DENSITY 1.9. Disulphuric(VI) acid is strongly HYGROSCOPIC. Commonly found in association with sulphuric acid, it is formed by dissolving sulphur trioxide in sulphuric acid:

$$H_2SO_4 + SO_3 \rightarrow H_2S_2O_7$$

The solution of sulphuric and disulphuric acids is called OLEUM and is an important material in the CONTACT PROCESS. Oleum is also used in the SULPHONATION of organic molecules.

dithionate Any salt containing the dithionate ion, $S_2O_6^{2-}$. Dithionates can be formed from the OXIDATION of sulphites by manganese(IV) oxide, for example:

$$4Na_2SO_3 + MnO_2 + 2H_2O \rightarrow 2Na_2S_2O_6 + 4NaOH + Mn$$

divalent (*adj.*) Having a VALENCY of 2.

diverging lens A lens that is thinner in the middle than at the edges and causes parallel rays of light to spread out as if they had come from a point behind the lens called the VIRTUAL FOCUS. The FOCAL LENGTH of a diverging lens is usually taken as being negative and is the distance of the virtual focus from the lens. An object viewed through a diverging lens always appears diminished (smaller in size than the original object) and is a VIRTUAL IMAGE. *Compare* CONVERGING LENS.

division In the CLASSIFICATION of plants, a major grouping made up of several CLASSES; equivalent to a PHYLUM.

division of amplitude A technique for obtaining two or more COHERENT light beams from a non-coherent source by using partial REFLECTION to split a single beam into two. The beams are then brought together again to interfere after travelling along different paths. *See also* INTERFERENCE.

division of wavefront A technique for obtaining two or more COHERENT light beams from a non-coherent source by using DIFFRACTION to spread a small section of a wavefront over a large angle. These sections are then brought together again to interfere after travelling along different paths. *See also* INTERFERENCE.

DNA, *deoxyribonucleic acid* A complex, very large molecule that contains all the information for building and controlling a living organism. It is a double-stranded NUCLEIC ACID made of NUCLEOTIDES with the bases ADENINE, GUANINE, CYTOSINE and THYMINE (never URACIL, which is found only in RNA) and a PENTOSE sugar that is always DEOXYRIBOSE. Except in bacteria, DNA is found in the nuclei of cells, arranged in CHROMOSOMES. The molecular structure of DNA was first proposed by James Watson (1928–) and Francis Crick (1916–) in 1953, for which they were awarded the Nobel prize.

The molecule consists of two polynucleotide strands (each of millions of nucleotides) linked to each other by base pairing (*see* BASE PAIR) and HYDROGEN BONDS. The bases adenine and thymine always link and cytosine and guanine always link. The linking of one PURINE and one PYRIMIDINE in this way allows the same spacing between the strands throughout the length of the molecule. So the DNA is like a ladder, with the base pairs

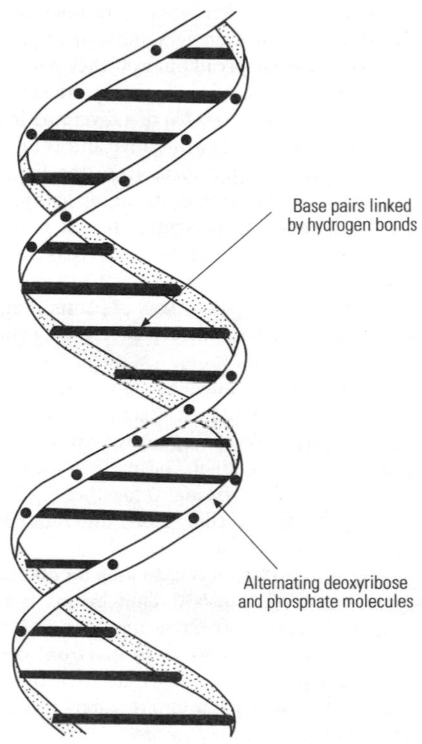

The DNA structure is like a ladder with alternating deoxyribose and phosphate molecules forming the uprights and base pairs forming the rungs. The ladder is twisted to form a double helix. The uprights run in opposite directions to each other. Base pairings are always adenine-thymine and cytosine-guanine.

forming the rungs and the deoxyribose and PHOSPHATE groups forming the uprights. In addition, the two chains forming the upright run in opposite directions and are called anti-parallel. The ladder is then twisted into a double helix.

The hereditary information is stored as a specific sequence of bases. Individual AMINO ACIDS are coded for by a set of three bases called a CODON. The precise sequence of bases therefore determines the amino acids that are made and therefore the PROTEINS that are produced by the cell. This sequence is called the GENETIC CODE and, because of the importance of the proteins it codes for, it controls the whole organism. In order for the genetic information to be passed on from cell to cell and generation to generation, DNA must be able to replicate. It does this by SEMI-CONSERVATIVE REPLICATION.

See also COPY DNA, GENE, GENETIC ENGINEERING, PROTEIN SYNTHESIS, RECOMBINANT DNA, X-RAY CRYSTALLOGRAPHY.

DNA fingerprinting *See* GENETIC FINGERPRINTING.

DNA ligase *See* LIGASE.

DNA polymerase *See* POLYMERASE.

DNAse, *deoxyribonuclease* One of many enzymes that hydrolyse (*see* HYDROLYSIS) DNA by breaking down the sugar–phosphate bonds. *See* RESTRICTION ENDONUCLEASE.

dodecahedron A POLYHEDRON with 12 plane faces.

dodecahydrate A HYDRATE that contains 12 parts water to one part of the compound. For example, trisodium phosphate forms the dodecahydrate $Na_3PO_4.12H_2O$.

dolomite A common METAMORPHIC rock containing calcium and magnesium carbonates.

domain 1. (*physics*) A region within a FERROMAGNETIC material in which all the atomic magnets are aligned in the same direction. *See also* PERMANENT MAGNET.

2. (*mathematics*) The set of values that a quantity may take, particularly the independent variable in a FUNCTION. For example, the domain of the inverse sine function is –1 to +1, and the function is meaningless for values outside this range.

domain wall In a FERROMAGNETIC material, the boundary between one DOMAIN and the next, which is usually magnetized in the opposite direction, with the direction of magnetization varying gradually through the thickness of the wall.

dominant In genetics, an ALLELE that is expressed in preference to another in the HETEROZYGOUS form. The allele that is masked, and therefore not expressed, is RECESSIVE. *See also* CODOMINANCE.

donor atom 1. An atom that provides a conduction electron in the DOPING of a SEMICONDUCTOR to form an N-TYPE SEMICONDUCTOR.

2. An atom that contributes a pair of electrons to form a CO-ORDINATE BOND, for example the nitrogen atom in the ammonium RADICAL, NH_4^+.

donor impurity An element added to a SEMICONDUCTOR in a DOPING process that releases FREE ELECTRONS, producing an N-TYPE SEMICONDUCTOR.

dopamine ($C_8H_{11}NO_2$) In biochemistry, a NEUROTRANSMITTER that is an intermediate in the synthesis of ADRENALINE. There are special areas in the brain that particularly use dopamine to transmit nerve impulses. Patients with the tremors of Parkinson's disease show degeneration of such areas.

doping The addition of small quantities of impurity to a SEMICONDUCTOR to alter its electric properties. By doping different parts of a silicon wafer in different ways, complex INTEGRATED CIRCUITS can be built up. *See also* N-TYPE SEMICONDUCTOR, P-TYPE SEMICONDUCTOR.

Doppler effect The apparent change in observed frequency (or wavelength) of a wave due to relative motion between the source of the wave and the observer. An example is a police car siren that increases in PITCH (frequency) as it moves towards a stationary observer, and decreases as it moves away. As the car approaches, the wavefronts emitted in the direction of the observer are 'bunched' more closely together, since the source is chasing the wavefronts in that direction. As a result, the wavelength is reduced and the frequency increased. On moving away, the wavefronts are further apart, leading to a longer wavelength and lower frequency.

With visible light, and other electromagnetic WAVES, the same effects are observed, though the details are modified by the SPECIAL THEORY OF RELATIVITY. If a light source moves towards the observer, the light appears more blue, blue being of a shorter wavelength than the other colours. This phenomenon is called a blue-shift. If the source and the observer are moving away from one another, the light appears more red (a RED-SHIFT).

For a wave source moving towards the observer with a speed u and the observer moving towards the source with a speed v, then if the speed of the waves is c and the frequency of the source is f, the observer will hear a frequency f', where

$$f' = f(c + v)/(c - u)$$

For light, if the relative motion of the source towards the observer is v,

$$f' = f\{(1 + v/c)/(1 - v/c)\}^{-1}$$

See also HUBBLE'S LAW.

d-orbital The third lowest energy ORBITAL for a given PRINCIPAL QUANTUM NUMBER. They exist only for principal quantum number of 3 or greater. There are five d-orbitals for each principal quantum number. Four of these consist of four lobes in an X-shape, each lying in a single plane, whilst the fifth d-orbital consists of two lobes surrounded by a TORUS.

dormancy In plants, a phase of reduced activity shown by seeds, spores and some buds that often aids survival of the plant during unfavourable conditions. Some seeds will remain dormant for many years, even during apparently favourable conditions. Particular requirements, such as a period of cold (to prevent GERMINATION until after winter), a period of light, or a period of 'after-ripening' (for internal changes to take place), are needed for dormancy to be broken. Dormancy can often be broken artificially. *See also* BIORHYTHM.

dorsal (*adj.*) **1.** Of an animal, relating to the back or spine.

2. Of a plant, relating to the back of an organ or part.

dorsal fin The FIN on the back of an aquatic vertebrate.

dorsal root *See* SPINAL CORD.

DOS (**Disk Operating System**) An operating system for desktop computers. It was designed specifically for use with disk storage.

dose The amount of IONIZING RADIATION absorbed by a living organism, usually human, over a specified period of time. It is usually specified in terms of the amount of energy deposited by the radiation, measured in GRAY. Some forms of ionizing radiation are more biologically harmful than others, and to take account of this, the DOSE EQUIVALENT is often considered.

dose equivalent The DOSE of IONIZING RADIATION absorbed by a human, multiplied by a factor that takes account of the relative damage done by different types of ionizing radiation. This factor is 20 for ALPHA PARTICLES and 1 for BETA PARTICLES. For GAMMA RADIATION, the factor varies between 0.60 and 0.28, depending on the energy of the gamma rays concerned. The SI UNIT of dose equivalent is the SIEVERT.

dot-matrix A computer PRINTER in which characters are formed from a series of dots produced by the action of a number of pins which press onto an inked ribbon under electronic control. The characters can be formed quite quickly, but the quality of print is not as high as can be achieved by other methods.

double bond A COVALENT BOND formed when two atoms share two pairs of electrons. Two sets of ORBITALS overlap to form a bond stronger than the usual, single, covalent bond. This involves the formation of a PI-BOND and a SIGMA-BOND.

double circulation *See* CIRCULATORY SYSTEM.

double decomposition A reaction between two soluble salts in aqueous solution, in which an exchange of ions forms an insoluble salt, which is produced as a PRECIPITATE, for example:

$$Na_2CO_3\ (aq) + CuSO_4\ (aq) \rightarrow CuCO_3\ (s) + Na_2SO_4\ (aq)$$

double fertilization An event unique to flowering plants, in which two male nuclei fuse with two female nuclei. When a POLLEN grain lands on a STIGMA of a flowering plant, it absorbs water and an outgrowth called a POLLEN TUBE grows inwards down the STYLE towards the OVULE (*see* CARPEL). The tube transports two male nuclei to the ovule, and when it enters the EMBRYO SAC (usually through a small hole called the MICROPYLE) it disintegrates and the nuclei are released. One nucleus fuses with the female egg nucleus to give rise to a DIPLOID ZYGOTE, and the other fuses with two polar nuclei (*see* POLAR NUCLEUS) to form a TRIPLOID PRIMARY ENDOSPERM NUCLEUS. The latter provides nourishment for the zygote and the process seems to be a way of ensuring that nourishment is only provided if there is a zygote to use it. *See also* FERTILIZATION, GERMINATION, POLLINATION.

double glazing The fitting of a window with two layers of glass. This reduces heat loss from a room by providing extra glass-air boundaries and so increasing the THERMAL CONTACT RESISTANCE.

double refraction *See* BIREFRINGENCE.

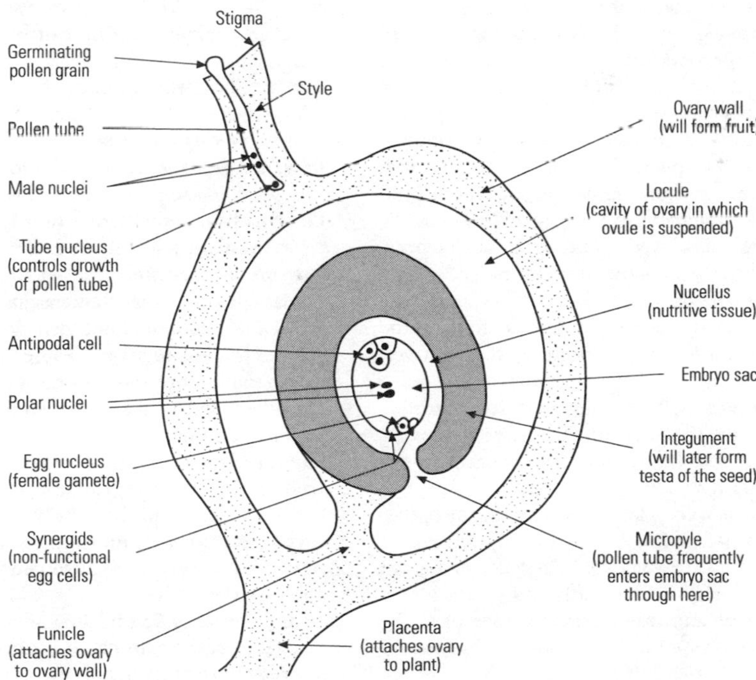

Mature carpel during double fertilization.

double salt A crystalline structure containing two NORMAL SALTS, such as potassium aluminium sulphate, $KAl(SO_4)_2.12H_2O$.

down A flavour of QUARK, with a charge of $-^2/_3$ in units of the electron charge, found in protons and neutrons.

Down's syndrome A genetic disorder characterized by mental retardation, altered facial features and reduced life expectancy. It is caused by the presence of three copies of chromosome 21, which happens when the pair of chromosomes does not separate at MEIOSIS, resulting in one GAMETE with 24 chromosomes and one with 22, instead of 23 each. This is called non-disjunction. Down's syndrome occurs in 1:700 births, and the incidence increases with maternal age.

drag A DISSIPATIVE FORCE caused by a solid object moving through a fluid, such as air or water. Provided the speed of the motion is slow, and the solid has a smoothly curved or STREAMLINED shape, there will be a smooth flow of the fluid around the solid. This type of flow is called LAMINAR FLOW – there is a gradual change in the speed of the fluid around the moving object. In this case the drag will be proportional to the speed of motion through the fluid and to the VISCOSITY of the fluid.

At higher speeds, or for less streamlined objects, the flow will become TURBULENT, with vortices – irregular spiralling patterns of fluid in the wake of the moving object. Usually there is still a layer of laminar flow, called the BOUNDARY LAYER, close to the surface of at least part of the object. Turbulent flow consumes much more energy than laminar flow, so the force increases much more rapidly with velocity, being roughly proportional to the square of the velocity. The size of the force is proportional to the area of the object moving through the fluid, and also depends on the shape of the object.

See also COEFFICIENT OF DRAG, STOKES' LAW.

drain In electronics, the ELECTRODE at one end of the CHANNEL in a FIELD EFFECT TRANSISTOR, from which CHARGE CARRIERS leave the channel.

DRAM (**dynamic random access memory**) The memory chips used by most computers to provide IMMEDIATE ACCESS MEMORY. BITS are stored as electric charges on a grid of tiny electric circuits. By specifying the row and column of the circuit on which charge is held, each bit's charge, or lack of charge can be accessed independently. The charge gradually dwindles, so DRAM must be read and rewritten constantly. This type of RAM is slower but cheaper than SRAM.

drift chamber A development of the SPARK CHAMBER that enables the trails of ionizing particles to be measured more precisely. The wires are placed further apart and low voltages are used, so that the IONS created by the passage of the charged particle move more slowly and do not cause the further ionization that leads to a spark. The voltage and the spacing of the wires are chosen so that the electric field just compensates for the energy lost by the ions in collisions with gas molecules. In this way, the time of arrival of the ions gives a measure of the position at which they were produced, enabling the path of the particles to be reconstructed.

drift mining The extraction of a mineral ore by drilling a horizontal shaft into the side of a hill to reach a mineral deposit. *See also* DEEP MINING, OPEN-CAST MINING.

drift tube A hollow metal ELECTRODE through which a charged particle can move without influence from external electric fields, such as in a LINEAR ACCELERATOR.

drift velocity The imbalance in speed that causes the CHARGE CARRIERS to produce a net flow of charge though a conductor.

Drift velocities are very small compared to the speed of random motion of charge carriers. In a typical metal, for example, electrons have random speeds of around 10^5 ms^{-1}, whilst the largest current that a wire made of such a metal could carry without melting requires the charges to move with an average speed of only a few millimetres per second. For a wire of cross-sectional area A made of a material with n charge carriers per metre cubed, carrying a current I, the drift velocity v is

$$v = I/nAe,$$

where e is the charge on an electron.

driving force The external force that drives a FORCED OSCILLATION.

driving frequency The frequency at which a system is forced to oscillate by some external force. *See* FORCED OSCILLATION.

dross The coating of solid impurities that forms on the surface of a metal when it is melted. The metal can be refined by skimming off the dross before being allowed to re-solidify, usually in moulds.

drug Any substance administered to animals or humans that alters a biological function. Most drugs are used in medicine, for example antibiotics, immunosuppressives, sedatives and pain-relievers. Other drugs are used socially, for example nicotine in tobacco, alcohol and heroin. All drugs are potentially harmful.

drupe, *stone fruit* A fleshy fruit similar to a BERRY, with an outer skin (exocarp) and a fleshy middle layer (mesocarp) but containing a single seed (berries have many seeds) that is surrounded by a hard woody layer (endocarp). Examples include the cherry, plum, coconut (the outer wall is dry when mature) and blackberry (which comprises many drupes together).

dry cell *See* ZINC-CARBON CELL.

D-T reaction A NUCLEAR FUSION process involving DEUTERIUM (hydrogen–2, which is naturally occurring) and TRITIUM (hydrogen–3, which is radioactive with a HALF-LIFE of about 12 years, so must be manufactured artificially). At high enough temperatures (many million kelvin), these nuclei collide violently enough to fuse, producing a helium–4 nucleus and a neutron. The neutron released can be captured by lithium to produce further tritium.

ductile (*adj.*) Able to be drawn out into a long strand, such as a wire. Ductility is an important property of many metals.

Dulong and Petit's law An experimental rule that states that the MOLAR HEAT CAPACITY of a solid tends to $3R$ at high temperatures, where R is the MOLAR GAS CONSTANT. The theoretical basis for this lies in EQUIPARTITION OF ENERGY between six DEGREES OF FREEDOM, ELASTIC and KINETIC ENERGY in each of three dimensions.

Dumas' method A technique for finding the RELATIVE MOLECULAR MASS of a volatile substance. A sample of the substance is placed in the bottom of a glass vessel and then vaporized by heating to a known temperature in a water or oil bath, driving out all the air. The vessel is then sealed and weighed. By comparing this with the mass of the vessel when full of air, the density of the substance in its gaseous form at a known temperature can be found, and from this its molecular mass.

dummy leads In electrical engineering, a pair of wires that do not carry a signal, but which follow the same path as the wires that do carry a signal. In this way, any effect on the signal caused by changes in the environment can be measured. In a RESISTANCE THERMOMETER, for example, dummy leads are used to compensate for any changes in resistance caused by changes in temperature of the wires connecting the temperature sensing element to the rest of the instrument.

duodenum In vertebrates, a short length of SMALL INTESTINE immediately after the stomach, where most digestion and some absorption of food occurs. The inner lining of the walls of the duodenum, like the rest of the small intestine, consists of projections called VILLI, which are folds to increase its surface area. Food enters the duodenum through a ring of muscle at the base of the stomach called the pyloric SPHINCTER, and mixes with BILE juice from the liver and PANCREATIC JUICES, both of which neutralize the acid CHYME entering from the stomach and aid the digestion of fat.

duplex (*adj.*) Describing a TELECOMMUNICATIONS path along which messages can be sent in both directions simultaneously. *See* SIMPLEX.

duplication In biology, a chromosomal MUTATION in which a portion of, or an entire CHROMOSOME is copied (*see* MUTATION).

dura mater The outermost of the three membranes that cover the brain and spinal cord. *See* MENINGES.

Dutch cap *See* DIAPHRAGM.

dyes, *dyestuffs* Coloured compounds that can be applied to fibres, paper, hair, etc. to give colour that is resistant to washing. They absorb PHOTONS in the visible region of the ELECTROMAGNETIC SPECTRUM. There are numerous dyes, some natural, such as indigo and cochineal, and many synthetic. Dyes vary in their chemical composition and the method in which they are applied. Most are organic compounds but can be applied with inorganic substances. Direct dyes (such as Congo red) chemically bond to the fabric forming a coloured compound. Indirect dyes require the fabric to be first treated with another compound. Vat dyes are insoluble coloured compounds formed from colourless soluble compounds on exposure to air in the dyeing vat. These then become enmeshed in the fibres of the fabric. In acid dyes (such as naphthol green) the acidic group of the dye attaches to the basic AMINO GROUP of protein fibres, such as in wool or silk. AZO DYES form the most important group of industrial dyes.

dyestuffs *See* DYES.

dynamic equilibrium *See* EQUILIBRIUM.

dynamic friction The FRICTION between two objects moving relative to one another. *See also* LIMITING FRICTION, STATIC FRICTION.

dynamic pressure The pressure an object experiences due to its motion through a fluid, caused by there being more molecular collisions per second on the side that is moving forwards than on the other side. This extra pressure can be used to measure the speed of an object through the fluid – an example of this is the PITOT-STATIC system used to measure the speed of an aircraft through the air by comparing the pressure on a forward facing pipe (the pitot head) with that on a sideways opening (the static vent) where there is no dynamic pressure. The difference in pressures is displayed as a speed on an instrument called an air speed indicator. *See also* HYDROSTATIC PRESSURE.

dynamic random access memory *See* DRAM.

dynamics The branch of MECHANICS concerned with moving and accelerating objects under the action of forces. *Compare* STATICS.

dynamo A device that converts mechanical energy into electrical energy. It is based on the principle that if a conductor moves in a magnetic field, a current will flow in the conductor. The simplest dynamo comprises a coil of conducting wire (called an ARMATURE) rotated between the poles of an ELECTROMAGNET. The mechanical energy of rotation is converted into electrical energy in the form of a current flowing in the armature. The term dynamo is usually taken to mean a device that generates a DIRECT CURRENT. An ALTERNATING CURRENT generator is usually called an ALTERNATOR.

dynamometer A machine for measuring the power output of an engine, particularly of an INTERNAL COMBUSTION ENGINE. The engine is made to do WORK, by driving a paddle wheel through a fluid for example, and the power delivered is measured, usually from the force applied by the engine.

dyne A unit of force in the c.g.s. system. The force needed to make a MASS of 1 g accelerate at 1 cm^{-2}. One dyne is equivalent to 10^{-5} newtons.

dysprosium (Dy) The element with ATOMIC NUMBER 66; RELATIVE ATOMIC MASS 162.5; melting point 1,412°C; boiling point 2,567°C; RELATIVE DENSITY 8.6. Dysprosium is a member of the LANTHANIDE series. The element is highly magnetic, but it is too scarce and too difficult to purify for this property to be used commercially.

E

e, *exponential* In mathematics, the BASE of NATURAL LOGARITHMS, approximately 2.713. e is an IRRATIONAL number, but it can be represented by the SERIES

1 + 1/2! + 1/3! + 1/4! +….

E102 *See* TARTRAZINE.

ear In animals, an organ of hearing and balance. The ear consists of three parts. The outer ear is a flap of CARTILAGE (the pinna) that collects, amplifies and focuses the air vibrations that constitute sound, and directs them along the AUDITORY CANAL to a membrane called the EARDRUM. Sound causes the eardrum to vibrate. The middle ear is an air-filled cavity containing small bones called EAR OSSICLES. Vibration of the eardrum causes movement of the ossicles, which in turn causes a second membrane called the 'oval window' to vibrate. The pressure inside the middle ear cavity is controlled by the EUSTACHIAN TUBE. The inner ear is a fluid-filled region consisting of a COCHLEA, which is concerned with hearing, and the SEMI-CIRCULAR CANALS, which are concerned with balance. Amphibians and some reptiles have no outer ear (the eardrum is in the skin) and fish have no ear. *See also* SENSE ORGAN.

ear canal *See* AUDITORY CANAL.

eardrum, *tympanic membrane* A thin membrane separating the outer EAR from the

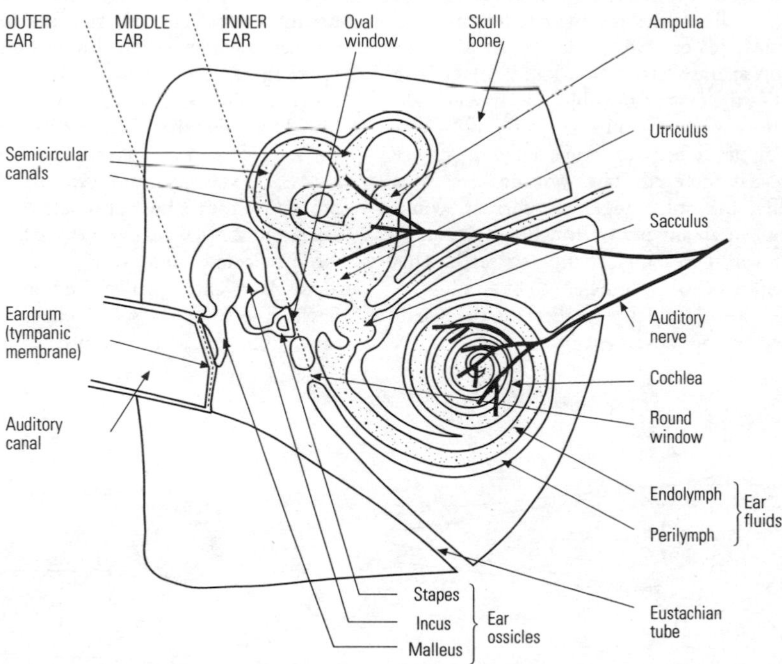

The human ear.

middle ear. Sound causes the eardrum to vibrate and this movement is detected by the bone EAR OSSICLES of the middle ear.

ear ossicle One of three small bones in the middle EAR. The three ear ossicles are called the malleus (hammer), incus (anvil) and stapes (stirrup), connected to one another and held in place by muscles. Vibrations of the EARDRUM are detected by the ossicles, which in turn vibrate and cause movement of a membrane called the 'oval window', which separates the middle and inner ear. Sound vibrations are amplified here more than 20 times. Birds, reptiles and many amphibians have only one ear ossicle.

Earth The third planet from the Sun. It has a mean diameter of 13,000 km, a mass of 6.0 x 10^{24} kg, and mean density of about 5,500 kg m^{-3}. The Earth's orbit lies between those of Venus and Mars, and it has a mean orbital radius of 1 AU (149,500,000 km). It has an orbital period of one year and a rotational period of 24 hours.

See also ATMOSPHERE, GEOLOGY.

earth A connection to the Earth to DISCHARGE an object; the act of making this connection.

As well as shielding the equipment from electric fields, earthing an object provides protection against ELECTRIC SHOCK. If a LIVE wire touches the case of the object, a large current will flow to Earth and this should blow any FUSE or circuit breaker connected to the power supply. If the earth wire were not present a smaller, but still dangerous, current would flow through any person touching the equipment whilst in contact with the earth. This current may not be sufficient to blow a fuse.

earthquake A sudden vibration of the Earth caused by the movement of TECTONIC PLATES, particularly along SUBDUCTION ZONES. *See also* RICHTER SCALE, SEISMIC WAVE, SEISMOLOGY.

earth science Any branch of science concerned with the study of the Earth. The earth sciences include GEOCHEMISTRY, GEOGRAPHY, GEOLOGY, GEOPHYSICS, METEOROLOGY, OCEANOGRAPHY and PALAEONTOLOGY.

ebullioscopic constant *See* ELEVATION OF BOILING POINT.

eccentricity A number describing the shape of any CONIC SECTION, in particular a measure of the amount by which an ELLIPSE differs from a circle (which has an eccentricity of 1). The equation for an ellipse can be written in POLAR CO-ORDINATES as

$$r = r_0(1 + e\cos\theta)$$

where e is the eccentricity.

ecdysis The periodic shedding of the EXOSKELETON of arthropods or the outermost layer of skin of reptiles to allow new growth. *See also* ARTHROPODA.

Echinoderm A member of the PHYLUM ECHINODERMATA.

Echinodermata A PHYLUM of spiny-skinned marine invertebrates that includes starfish, sea urchins and sea cucumbers. ECHINODERMS have a water vascular (conducting) system with an opening for transporting substances around the body. They have small water-filled sacs called tube-feet that may have suckers on their ends, by which they move. Most feed on MOLLUSCS, CRUSTACEANS and ANNELIDS. Some can reproduce asexually, but others have separate sexes and FERTILIZATION is external.

echo A reflected sound wave.

eclipse An event in which the Sun, Moon and Earth line up. In a solar eclipse, the Moon blocks out the Sun as seen from part of the

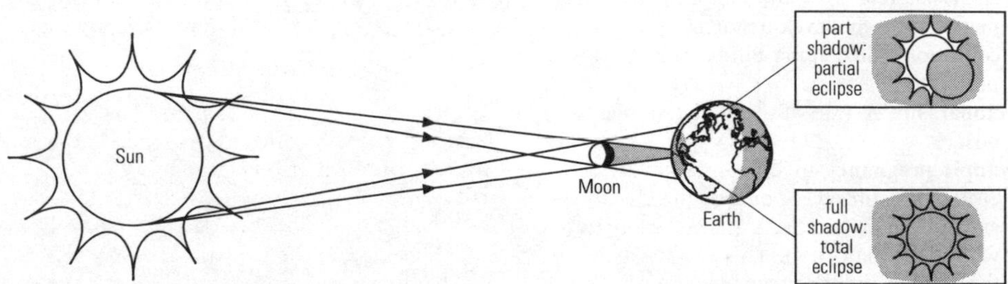

Solar eclipse.

Earth's surface. In a lunar eclipse the Moon moves into the Earth's shadow and appears much darker than normal.

eclipsing binary *See* BINARY STAR.

ECM *See* EXTRACELLULAR MATRIX.

ecological niche The position any one species occupies within an ecological HABITAT, including not just the physical environment but all interactions between living, non-living and behavioural components.

ecological pyramid A diagrammatic representation of the levels in a FOOD CHAIN. *See* PYRAMID OF BIOMASS, PYRAMID OF ENERGY, PYRAMID OF NUMBERS.

ecology (Greek *oikos* = house) The study of the relationship between organisms and the environment in which they live (the study of ECOSYSTEMS). Ecology includes the study of living (BIOTIC) elements and non-living (ABIOTIC) elements and is concerned with factors such as feeding habits, behaviour, competition between individuals or between whole COMMUNITIES, and can include CONSERVATION and management of POLLUTION.

ecosystem The interactions between the living (BIOTIC) and non-living (ABIOTIC) elements within a biological system, for example a lake or a forest, forming an ecological unit. The living elements constitute a FOOD CHAIN. The balance in an ecosystem is delicate and removal of one species can destroy the balance. Because energy is lost at each stage of the food chain, the flow of energy through an ecosystem is in one direction only (unlike minerals or elements, such as carbon, nitrogen and phosphorous, which are recycled).

ectoderm In animals, the outermost GERM LAYER of a developing embryo. The ectoderm develops into a neural tube, which is the precursor of the CENTRAL NERVOUS SYSTEM, and also forms EPIDERMIS (e.g. skin and hair). Failure of the neural tube to develop normally leads to a condition called spina bifida. *See also* ENDODERM, MESODERM.

ectoparasite A PARASITE that lives outside its host.

ectopic pregnancy In humans, the IMPLANTATION of a fertilized egg outside the UTERUS (for example, in the FALLOPIAN TUBE or, more rarely, within the ABDOMEN). This condition is life-threatening and the pregnancy must be terminated.

ectoplasm *See* CYTOPLASM.

ectotherm *See* POIKILOTHERM.

ectothermy *See* POIKILOTHERMY.

edaphic factor Any factor relating to the physical or chemical composition of SOIL in a particular area, especially in relation to the ECOSYSTEM in that area.

eddy current An electric current induced in a conductor moving through a magnetic field, or which is in an area of changing magnetic field, which circulates in that conductor without doing any useful work. Eddy currents are a potential source of energy loss in many electromagnetic machines. To minimize the effects of eddy currents, iron cores are often made in a LAMINATED construction, from thin sheets of iron coated with an insulating varnish. In this way the eddy currents are broken up into smaller loops without much reduction in the total volume of magnetic material.

EDTA *See* ETHYLENEDIAMINETETRAACETIC ACID.

effector A cell or organ by which an animal responds to internal or external stimuli received via the NERVOUS SYSTEM. An effector can be a MUSCLE or a GLAND.

effector system, *motor system* The part of the NERVOUS SYSTEM that carries NERVE IMPULSES away from the CENTRAL NERVOUS SYSTEM to a body muscle or gland, called an EFFECTOR, which then acts upon the impulse. The effector system has two parts: the AUTONOMIC NERVOUS SYSTEM (self-governing), which activates involuntary responses; and the SOMATIC nervous system, which activates voluntary responses.

effervescence The production of a gas in a liquid by a chemical reaction, with the formation of many small bubbles and a characteristic fizzing sound.

efficiency A measurement of the performance of a machine. Efficiency is the fraction of the energy applied to a machine that appears in the desired form at the output of the machine.

$$\text{Efficiency} = \text{useful work done}/\text{total energy input}$$

Efficiency is sometimes multiplied by 100 to give a percentage efficiency.

efflorescence The loss of WATER OF CRYSTALLIZATION to the atmosphere from a crystalline solid, leading to the formation of a powdery solid, for example, sodium carbonate:

$$Na_2CO_3.10H_2O \rightarrow Na_2CO_3.H_2O + 9H_2O$$

effort The FORCE applied to a MACHINE to move a LOAD.

effusion The flow of a gas through a small opening, comparable in size to the spacing between the molecules in the gas, so that the gas cannot be considered to flow through the opening as a fluid, but rather passes through one molecule at a time. The rate of effusion is proportional to the ROOT MEAN SQUARE speed of the molecules, and thus at a given temperature is inversely proportional to the square root of the MOLECULAR MASS.

egestion, *elimination* The removal of undigested material from the DIGESTIVE SYSTEM in the form of FAECES. This is different from EXCRETION because the material has never been inside the body cells. In most animals faeces are removed through the ANUS.

egg The female GAMETE. After FERTILIZATION by a male gamete, the egg divides to form an EMBRYO. In animals, the eggs can develop inside the female (VIVIPARY or OVOVIVIPARY) or can be deposited by her (OVIPARY). In birds and reptiles, the eggs are protected by a shell and the embryo is nourished by a YOLK. *See* OVULE, OVUM.

egg membrane *See* EXTRAEMBRYONIC MEMBRANE.

einsteinium (Es) The element with ATOMIC NUMBER 99. It does not occur naturally but has been synthesized by the bombardment of lighter nuclei with charged particles from PARTICLE ACCELERATORS. All known ISOTOPES have HALF-LIVES of less than one year. Einsteinium has not found uses to justify its manufacture in commercial quantities.

elaioplast A type of LEUCOPLAST that stores oils in plants.

elastic (*adj.*) Describing a substance or a process by which the substance is deformed by an applied force, but returns to its original shape once the force is removed. *See also* ELASTIC, HOOKE'S LAW, PLASTIC.

elastic collision A collision in which the total KINETIC ENERGY of the colliding bodies before and after the collision are the same. Collisions involving objects that bounce off each other are partially elastic, whilst collisions between atoms and between sub-nuclear particles are totally elastic unless some energy is used to excite the atom or create a new particle of higher mass. *Compare* HYPERELASTIC COLLISION, INELASTIC COLLISION. *See also* COEFFICIENT OF RESTITUTION.

elastic energy, *strain energy* The energy of an object that has been elastically deformed in some way, such as a stretched spring. The elastic energy E of a spring stretched through a distance x by a force F is

$$E = {}^1/_2 Fx$$

elastic limit The maximum TENSILE STRAIN that can be applied to an ELASTIC material beyond which the material does not return to its original length when the stretching force is removed. *See also* FATIGUE, HOOKE'S LAW.

elastic modulus Any one of the three measures of the stiffness of an ELASTIC material: the YOUNG MODULUS, MODULUS OF RIGIDITY or BULK MODULUS. *See also* SHEAR MODULUS.

elastin A protein of animal CONNECTIVE TISSUE that is very ELASTIC. It is found in LIGAMENTS, arteries and lungs.

electret 1. A plastic material containing POLAR molecules that are aligned in such a way that the material is electrically POLARIZED and carries charges on its surface, even in the absence of an electric field.

2. A microphone based on such a material. A charged plastic DIAPHRAGM is moved by a sound wave and induces charges on a fixed metal ELECTRODE nearby.

electrical conductor *See* CONDUCTOR, ELECTRICAL.

electrical energy Energy carried by the flow of charge around an electric circuit. For a total charge Q moving through a POTENTIAL DIFFERENCE V the energy gained or lost is QV. On a domestic or industrial scale, electrical energy is also sometimes measured by multiplying the power in kilowatts by the time in hours over which that power is used. *See also* KILOWATT-HOUR.

electrical potential energy The energy of a charge in an electric field. The change in electrical potential energy E of a charge q moving through an electrical POTENTIAL DIFFERENCE V is

$$E = qV$$

See also ELECTRIC POTENTIAL.

electric arc A highly luminous spark between two ELECTRODES, which themselves become vaporized by the heat of the arc. Arcs may be used as a source of heat for welding, or as a highly intense source of light. Carbon electrodes are often used, driven forward mechanically as the surface is burnt away by the arc. Arc lights have generally been replaced by more reliable light sources.

electric arc furnace A device used to melt iron as part of a steel making process. The furnace is a cylindrical vessel with a lid containing graphite ELECTRODES. An ELECTRIC ARC between the electrodes and the surface melts the material in the furnace, and the molten material is then removed by tipping the furnace onto one side.

The process does not introduce any oxygen and so can be closely controlled chemically. Steel scrap is often added to the furnace, the chemical composition of which is determined from a sample of the molten steel and then adjusted as required.

See also BESSEMER CONVERTER, OPEN HEARTH FURNACE, OXYGEN FURNACE.

electric charge *See* CHARGE.

electric field The region of influence around any electric CHARGE that causes any other charge to experience a force proportional to its charge. Numerically, the strength of an electric field at a point is the force per unit charge experienced by a vanishingly small charge placed at that point. The SI UNIT of electric field is NC^{-1} (NEWTON/COULOMB) or Vm^{-1} (VOLT/METRE).

The electric field produced around a point charge, or outside a spherically symmetric distribution of charge is proportional to the size of the charge and inversely proportional to the distance from the centre of charge. The force F on a charge q in an electric field E is

$$F = Eq$$

The field E at a distance r from a point charge q is

$$E = q/4\pi\varepsilon_0 r^2$$

Between two parallel metal plates with opposite charges (as in a CAPACITOR for example) the electric field will be constant. The electric field between two parallel plates between which there is a distance d and a POTENTIAL DIFFERENCE V is

$$E = V/d$$

See also ELECTRIC POTENTIAL.

electric field lines Lines drawn on a diagram that show the direction of the force on a positively charged particle placed at that point in an ELECTRIC FIELD. These field lines are drawn from positive charges to negative charges. The closer the lines the greater the electric field strength (*see* ELECTRIC POTENTIAL). Electric field lines always leave a conducting surface at right angles to that surface.

electric field strength *See* ELECTRIC POTENTIAL.

electric flux A measure of the total amount of ELECTRIC FIELD passing through an area. If the field is E, and it passes through area A at an angle θ, then the electric flux Ψ is

$$\Psi = EA\cos\theta$$

electricity The general term describing all effects caused by electric CHARGE, whether at rest (electrostatic; *see* STATIC ELECTRICITY) or in motion (electric CURRENT).

electric motor Any device for converting electrical energy into rotational motion. Generally, motors use the interaction between electric currents and magnetic fields to produce a force (*see* MOTOR EFFECT). *See also* D.C. MOTOR.

electric potential A measure of the WORK needed to move an electric charge to a particular point, or the energy that would be produced if the charge moved away from that point. The electric potential at a given point is defined as the work done per unit charge in bringing a vanishingly small positive charge to that point from infinity. The SI UNIT of electric potential is the VOLT. The electric potential V at a distance r from a point charge q is

$$V = q/4\pi\varepsilon_0 r$$

where ε_0 is the PERMITTIVITY of free space.

The electric field strength at any point is equal to minus the rate of change of electric potential with distance at that point. Between a pair of parallel plates, the electric field strength is equal to the difference in electric potential between the plates divided by the separation between them. The relationship between an electric field E and electric potential V is

$$E = -\mathrm{d}V/\mathrm{d}x$$

See also ELECTRICAL POTENTIAL ENERGY.

electric power The rate at which energy is carried by an electric circuit. The power is equal to the current multiplied by the voltage.

electric shock The reaction of the human body to an electric current flowing through it. The reaction to electric shock depends on the size of current flowing through the body and the duration of current flow. Much of the body's resistance comes from the skin and this is much lower when the skin is wet, which accounts for the particular dangers of handling electrical apparatus with wet hands. Currents of a few milliamperes or so mimic nerve signals and cause tingling sensations and involuntary muscle contraction. Above about 50 mA, currents through any part of the body become painful, and those through the chest can interfere with breathing and heart operation. Large currents may also cause burning particularly at the points where the current enters and leaves the body. *See also* EARTH, ELECTROCUTION.

electrochemical cell *See* CELL.

electrochemical equivalent The mass of a specified material released or dissolved when one COULOMB of charge flows in an ELECTROLYSIS experiment. The electrochemical equivalent is the mass of the atom in grams divided by the charge of the ion in coulombs (gC^{-1}). *See also* ELECTROLYSIS.

electrochemical series, *electromotive series* The list of metallic elements, plus hydrogen, in order of increasing (more positive) STANDARD ELECTRODE POTENTIAL. Elements above hydrogen are described as electropositive, whilst those below hydrogen are said to be electronegative. For the more common metals, the electrochemical series is: potassium, calcium, sodium, magnesium, aluminium, zinc, iron, lead, hydrogen, copper, silver, platinum.

The standard electrode potentials effectively compare the energy needed to form one coulomb of ions in aqueous solution from the sold metal. Since many other processes involve a similar chemical change, the electrochemical series has applications well beyond the considerations of electrochemical CELLS, and this series is effectively the REACTIVITY SERIES for metals. It also gives a far better indication of the relative energies in chemical processes than the IONIZATION ENERGIES of the metals, since ionization energies relate to gaseous atoms being turned into gaseous ions, whilst the electrochemical series relates to the more common process of solid metals being converted to aqueous ions.

Any metal that lies higher than another one in the electrochemical series (that is, has a more negative standard electrode potential) will displace the lower metal in a DISPLACEMENT REACTION. Those metals that lie below hydrogen can be deposited on a CATHODE in an ELECTROLYSIS of an aqueous solution. The metals above hydrogen will release hydrogen in an electrolysis of an aqueous solution, and electrolysis can only be used to extract such metals if a molten salt is used rather than an aqueous solution.

electrochemistry The part of chemistry that deals with the interaction between electricity and chemical materials, particularly electrochemical CELLS. *See also* ELECTROLYSIS.

electrocution An ELECTRIC SHOCK causing death, usually by the current interfering with the nerves of the heart. Lower currents, if sustained for long enough, may cause death by interfering with the motion of the chest, preventing breathing.

electrode 1. A conductor through which an electric current passes in or out of an ELECTROLYTE (as in an electrolytic cell; *see* CELL), a gas (as in a discharge tube) or a vacuum (as in a THERMIONIC VALVE).

2. In a SEMICONDUCTOR, an EMITTER or COLLECTOR of electrons or HOLES.

electrode potential The POTENTIAL DIFFERENCE between an IONIC solution and a metal ELECTRODE. The difference between the electrode potential of the two electrodes in a CELL is what gives rise to voltage of that cell. *See also* HALF-CELL, REDOX POTENTIAL, STANDARD ELECTRODE POTENTIAL.

electrodynamics The study of the ELECTRIC FIELDS and MAGNETIC FIELDS produced around moving charges.

electrolyse (*vb.*) To bring about chemical decomposition by ELECTROLYSIS.

electrolysis The chemical change effected by the passage of an electric current through an IONIC liquid (*see* ELECTROLYTE). In electrolysis, ions are attracted to the oppositely charged ELECTRODE (*see* ANODE, CATHODE, ANION, CATION). When the ions reach the electrode they may gain electrons at the cathode or lose them at the anode and form neutral atoms or molecules. Alternatively, they may react with the electrode, which is then IONIZED and dissolves in the liquid. This latter reaction generally takes place at the anode, as the electrodes tend to be metals that readily give up electrons to form positive ions. Which of the possible reactions takes place depends on the relative reactivities of the substances involved. The more reactive substance will generally form ions or remain in its ionic form. When water acidified with sulphuric acid (to ionize the water so it will conduct) is electrolysed using platinum electrodes, hydrogen gas is formed at the cathode and oxygen gas and water are liberated at the anode, since hydrogen is more reactive than platinum and the HYDROXIDE ion (OH^-) gives up electrons more readily than the SULPHATE ion (SO_4^{2-}).

Electrolysis is used to extract some metals from their ores – for example the extraction of aluminium from bauxite. Electrolysis is also used in ELECTROPLATING, where the object to be coated is the negative electrode in a solution of a salt of the coating metal.

See also ANODIC OXIDATION, CATHODIC REDUCTION, ELECTROCHEMICAL EQUIVALENT, FARADAY'S LAWS OF ELECTROLYSIS, VOLTAMMETER.

electrolyte Any conducting liquid, through which electric charge flows by movement of IONS. Electrolytes are molten IONIC compounds or solutions of ionic salts or of compounds that ionize in solution. *See also* MOLAR CONDUCTIVITY.

electrolytic capacitor A CAPACITOR in which the insulating layer is formed on one of the plates by ELECTROLYSIS. The result is an insulating layer that is very thin, giving large CAPACITANCES. A disadvantage is that the layer may not be perfectly insulating, so small LEAKAGE CURRENTS may flow. It is also important to use electrolytic capacitors with the voltage in the correct direction. If it is reversed, the leakage current may result in electrolytic processes destroying the insulating layer.

electrolytic cell *See* CELL.

electromagnet A MAGNET that relies on INDUCED MAGNETISM in the iron core of a SOLENOID.

electromagnetic force The force responsible for interactions between electrically charged particles. The electromagnetic force is explained in terms of the exchange of PHOTONS. The electromagnetic force is responsible for all the non-gravitational interactions between atoms (rather than atomic nuclei) and thus causes friction, contact forces between solid objects, and all forms of chemical bonding. The full quantum mechanical theory of electromagnetism is called QUANTUM ELECTRODYNAMICS (Q.E.D.) and is an important example of the successful links that have been made between QUANTUM MECHANICS and the SPECIAL THEORY OF RELATIVITY in the second half of the 20th century.

electromagnetic induction The process by which an ELECTROMOTIVE FORCE is produced in a circuit when the MAGNETIC FLUX LINKAGE in a circuit changes. The flux linkage in a circuit may change in one of two ways: either part of the circuit moves, as in a DYNAMO, effectively producing a change in the area of the circuit; or the field may change, if the magnet producing it is moved or the current in an ELECTROMAGNET changes, as in a TRANSFORMER for example.

For a wire of length l moving through a magnetic field B at a speed v, with the wire, its motion, and the magnetic field direction all at right angles, the induced e.m.f. will be

$$E = Blv.$$

See also FARADAY'S LAW OF ELECTROMAGNETIC INDUCTION, LENZ'S LAW, MUTUAL INDUCTANCE, SELF INDUCTANCE.

electromagnetic radiation Energy resulting from the acceleration of electric charge, that propagates through space in the form of ELECTROMAGNETIC WAVES. Alternatively, electromagnetic radiation can be thought of as a stream of PHOTONS travelling at the speed of light (c) each with energy hc/λ, where h is PLANCK'S CONSTANT and λ is the wavelength of the associated wave. *See also* BLACK BODY RADIATION, ELECTROMAGNETIC SPECTRUM, RADIATION.

electromagnetic spectrum The spread of different frequencies and wavelengths of ELECTROMAGNETIC WAVES with particular reference to the similarities and differences between different parts of this range. In order of increasing frequency, it comprises RADIO WAVES, INFRARED radiation, VISIBLE LIGHT, ULTRAVIOLET radiation, X-RAYS and GAMMA RADIATION.

electromagnetic wave WAVES of energy composed of oscillating electric and magnetic fields, in PHASE but at right angles to one another, propagating through space as a TRANSVERSE WAVE. Since the associated fields are capable of existing in empty space, electromagnetic waves can travel through a vacuum. The detailed properties of the wave depend upon its frequency and wavelength: wavelengths from several kilometres down to 10^{-15} m have been observed. This range is called the ELECTROMAGNETIC SPECTRUM.

Electromagnetic waves travel through a vacuum at a constant speed (the SPEED OF LIGHT) of 3 x 10^8 ms^{-1}. In other materials they travel more slowly than this, depending on the RELATIVE PERMITTIVITY and RELATIVE PERMEABILITY of the medium.

electrometer An instrument for measuring the ELECTRIC POTENTIAL of electrostatic charges (*see* STATIC ELECTRICITY), or any VOLTMETER with a very high resistance. Historically, such instruments measured the electrostatic force between a charged electrode and one at EARTH potential, and were not very sensitive. Modern instruments use OPERATIONAL AMPLIFIERS with FIELD EFFECT TRANSISTORS in the input stages to combine high sensitivity with very high resistance.

electromotive force (e.m.f.) The amount of electrical energy given to each COULOMB of charge that is driven around a circuit by the source of e.m.f. The SI UNIT of electromotive force is the VOLT. An e.m.f. may be provided by an electrolytic CELL (which converts chemical energy into electrical energy) or an electrical GENERATOR, which relies on the relative motion of a wire though a magnetic field (*see* ELECTROMAGNETIC INDUCTION).

electromotive series *See* ELECTROCHEMICAL SERIES.

electron The lightest charged LEPTON. Electrons are negatively charged ELEMENTARY PARTICLES that occur in all atoms. Atoms consist of a central nucleus surrounded by orbiting electrons. The number and distribution of the electrons in an atom (the electron structure) is responsible for the chemical properties of the atom. An electron that has become detached from an atom is known as a free electron.

The mass of an electron is 9.1 x 10^{-31} kg and it has a charge of 1.6 x 10^{-19} C. This charge is the fundamental unit of negative charge, and the electron is the basic particle of electricity. In metals, an electric current consists of a movement of free electrons.

See also CHEMICAL BOND, MILLIKAN'S OIL DROP EXPERIMENT, NEUTRON, NUCLEUS, POSITRON, ORBITAL, PROTON.

electron affinity 1. Qualitatively, the tendency of an atom or molecule to gain electrons and form a negative ion.

2. Quantitatively, the energy required for the formation of negative ions from one mole of an element. Energy is sometimes released in the formation of such an ion, so for the more reactive nonmetals, the electron affinities are negative. If the electron affinity is positive, the reaction can only proceed if the energy required is released at some other stage of the reaction, such as the formation of an IONIC SOLID.

electron degeneracy pressure The pressure arising from the influence of the PAULI EXCLUSION PRINCIPLE on electrons. This means that extra ENERGY is required to compress a material containing FREE ELECTRONS, such as the PLASMA in a star, regardless of temperature. *See* FERMI DISTRIBUTION. *See also* WHITE DWARF.

electron diffraction The spreading of a beam of electrons as they pass through a narrow aperture or crystal LATTICE, analogous to the DIFFRACTION of visible light. This effect can only be explained in terms of the WAVE NATURE OF PARTICLES. Electron diffraction is used to measure bond lengths and angles, and in the study of solid surfaces. *See also* NEUTRON DIFFRACTION, X-RAY DIFFRACTION.

electronegativity A measure of the tendency of an atom in a COVALENT BOND to attract the electrons in the bond towards itself. Thus if the two elements in the bond have equal electronegativities the bond will not be POLAR, otherwise it will be polar with the more electronegative atom located at the negatively charged end.

Electronegativities can be calculated in a number of ways, but the most commonly used is the PAULING ELECTRONEGATIVITY based on the energies of covalent bonds.

See also ELECTROPOSITIVITY.

electron gun A device for generating a beam of electrons, as in a CATHODE RAY TUBE. An electron gun consists of an electrically heated metal CATHODE together with a number of cylindrical ANODES in an evacuated tube. Electrons are released from the cathode in a process known as THERMIONIC EMISSION, and are accelerated and focused by the anodes. A CONTROL GRID may also be incorporated to alter the intensity of the beam.

electron-hole pair An electron and a HOLE produced at a point in a SEMICONDUCTOR, by the action of light, thermal vibrations or IONIZING RADIATION. The electron and the hole then move in opposite directions under the influence of any electric field. However, an electron may meet another hole and recombine, thus there will be a dynamic EQUILIBRIUM controlling the number of CHARGE CARRIERS.

electronic mail, *e-mail* A message that can be sent electronically from one computer to another via a computer NETWORK.

electronics The science of circuits that contain ACTIVE DEVICES; that is, devices that behave in a more complex way than simple RESISTORS, INDUCTORS and CAPACITORS.

electron lens A set of ELECTRODES, or an ELECTROMAGNET, designed to focus a beam of electrons in the same way that a conventional lens focuses light. Electron lenses are used in ELECTRON MICROSCOPES, in which the beam of electrons are focused by an arrangement of coils that produce a variable magnetic field, and in CATHODE RAY TUBES.

electron microscope An instrument, developed in 1933, that magnifies objects using a beam of electrons. The electrons are accelerated by a high voltage through the object (held in a vacuum) and focused by powerful ELECTROMAGNETS (*see* ELECTRON LENS), instead of optical lenses, onto a fluorescent screen for viewing. A camera may be built in to record what is seen on the screen. A beam of electrons of a DE BROGLIE WAVELENGTH of typically 0.005 nm is used, compared to the wavelength of about 500 nm of the light rays used in a light microscope. Therefore the RESOLVING POWER of the electron microscope is vastly improved to about 1 nm, and objects can be magnified more than 500,000 times.

The sample to be viewed is fixed and embedded in ARALDITE (an EPOXY RESIN), which enables ultra-thin sections to be cut. Living specimens cannot be viewed. There are several types of electron microscope.

See also ELECTRON-PROBE MICROANALYSER, SCANNING ELECTRON MICROSCOPE, SCANNING TRANSMISSION ELECTRON MICROSCOPE, SCANNING TUNNELLING MICROSCOPE, TRANSMISSION ELECTRON MICROSCOPE, ULTRAMICROTOME.

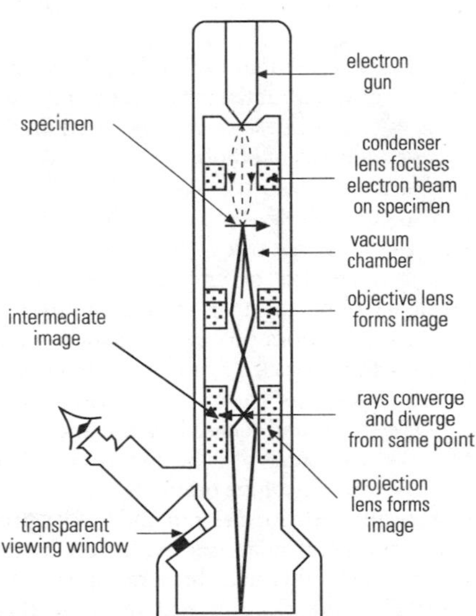

The electron microscope.

electron number A QUANTUM NUMBER believed to be conserved in all interactions. The electron and its NEUTRINO are assigned an electron number of one, their ANTIPARTICLES –1, while all other particles have an electron number of 0.

electron-probe microanalyser A modified ELECTRON MICROSCOPE in which the object emits X-rays when it is hit by electrons, the different intensities of which indicate the presence of different chemicals. The specimens can be examined without being destroyed.

electron transport system, *respiratory chain* The third stage of cellular RESPIRATION in which hydrogen atoms from the KREBS CYCLE are converted to ATP, so providing energy. The hydrogen atoms are carried from the Krebs cycle by the electron carriers NAD or FAD, and transferred to a series of other carriers at lower energy levels. At each transfer the energy released is used to produce ATP. The other carriers are COENZYME Q and a series of CYTOCHROMES. The hydrogen atoms are split into their protons and electrons and the electrons are passed along the carrier chain and recombine with hydrogen at the end, which links with oxygen to form water. The whole process of forming ATP through OXIDATION of hydrogen atoms is called OXIDATIVE PHOSPHORYLATION and occurs in the MITOCHONDRIA.

It is thought that electron carriers and their enzymes attach, in a precise sequence, to particles within the inner mitochondrial membrane. In addition, within this membrane there is a mechanism for actively transporting protons from the cell matrix to the space between the mitochondrial membranes, creating a gradient of hydrogen ions across the

Summary of the electron transport system.

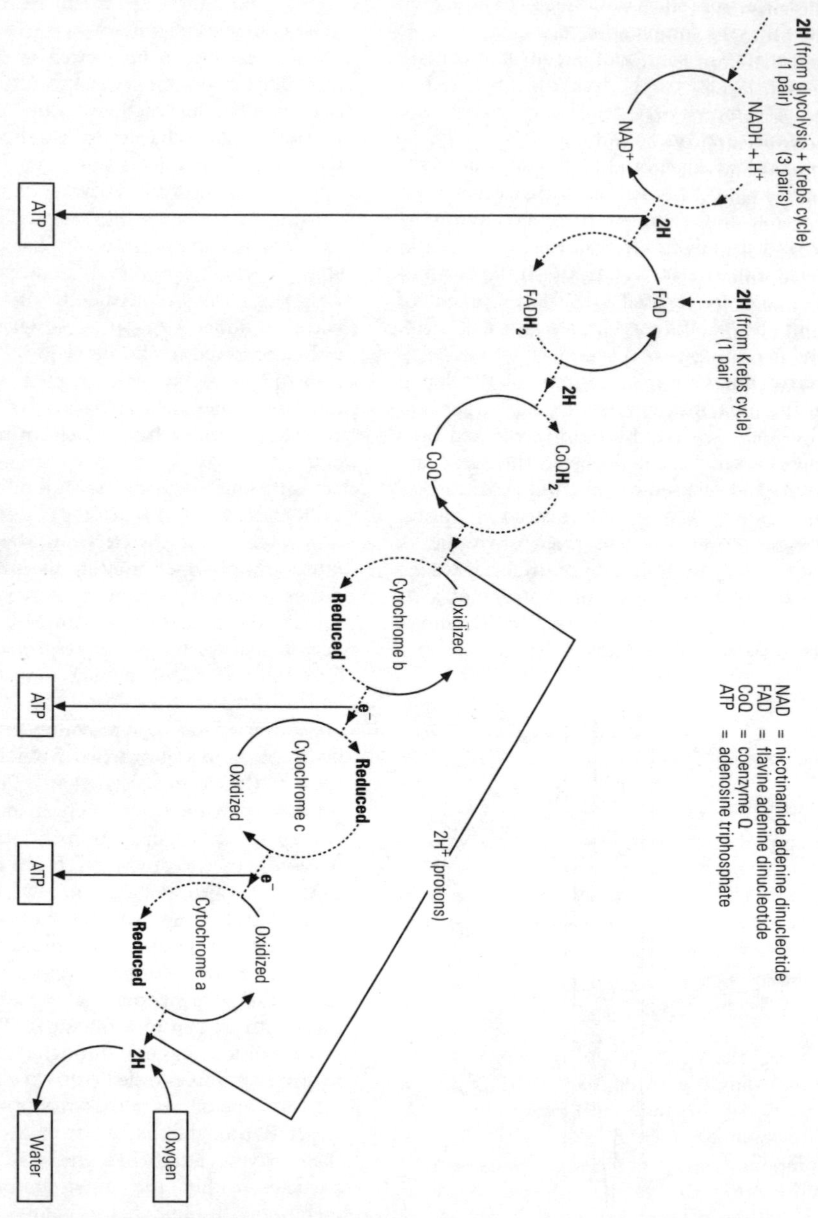

membrane that provides the energy to synthesize ATP. This is called the CHEMIOSMOTIC THEORY. The respiratory inhibitor cyanide inhibits the enzyme (cytochrome oxidase) that links the hydrogen atoms and oxygen to form water, and therefore hydrogen atoms build up and aerobic respiration ceases.

See also GLYCOLYSIS.

electron-volt (eV) A unit of energy used in atomic and nuclear physics. One eV is the energy gained by an electron when it moves through a POTENTIAL DIFFERENCE of one volt. It is equivalent to 1.6×10^{-19} JOULES.

electrophile (Greek = electron loving) Any ion or molecule lacking in electrons and so will tend to gain electrons in a chemical reaction, acting as a LEWIS ACID, or a REDUCING AGENT. In organic chemistry, electrophiles often attach themselves to negatively charged parts of a molecule. *See also* ELECTROPHILIC ADDITION, ELECTROPHILIC SUBSTITUTION.

electrophilic addition An ADDITION REACTION that is triggered by attack from an ELECTROPHILE. This type of reaction is typical of the ALKENES, for example the addition of hydrobromic acid (HBr) or bromine to ethene, giving bromoethane and 1,2-dibromoethane respectively:

$$CH_2{=}CH_2 + HBr \rightarrow CH_3CH_2Br$$

$$CH_2{=}CH_2 + Br_2 \rightarrow BrCH_2CH_2Br$$

See also MARKOWNIKOFF'S RULE.

electrophilic substitution A SUBSTITUTION REACTION in which an atom or group in a molecule is replaced by an ELECTROPHILE (which accepts electrons to form a new COVALENT BOND). The general reaction can be considered to be as follows:

$$RX + E^+ \rightarrow RE + X^+$$

where R is a HYDROCARBON, metal or METALLOID group and the electrophile (E) and X can be a variety of organic or inorganic CATIONS. Electrophilic substitution involving AROMATIC compounds is termed electrophilic aromatic substitution. Examples of this include NITRATION, SULPHONATION and FRIEDEL-CRAFTS REACTION of aromatic compounds.

electrophoresis, *cataphoresis* A technique used for separating molecules on the basis of their charge. The mixture to be separated is placed on a gel or paper in a BUFFER solution at a given pH, and an electric current is applied through the buffer. Substances of different sizes can be separated because they diffuse at different rates, and can be compared with known standards. Electrophoresis is widely used in biology as an analytical technique, to separate different length strands of DNA for example. The process is also sometimes used as a way of attracting paint particles to metallic objects.

electroplating An ELECTROLYSIS technique in which metal is deposited on the CATHODE. For example, when copper sulphate is electrolysed with a copper ANODE, the anode dissolves and copper is deposited on the cathode. Metal objects are often electroplated with silver.

electropositivity The tendency of an atom to lose electrons and form positive ions, or, in a COVALENT BOND, to lose hold of the electrons in the bond, leading to that atom finding itself at the positive end of a POLAR BOND. The opposite of ELECTRONEGATIVITY.

electroreceptor A RECEPTOR cell that detects electric fields. *See* SENSE ORGAN.

electroscope Any device designed to measure ELECTRIC POTENTIAL. *See also* GOLD-LEAF ELECTROSCOPE.

electrostatic (*adj.*) Related to, producing or caused by STATIC ELECTRICITY.

electrostatic precipitator A device fitted to chimneys, for instance in power stations, to remove small particles from smoke. The smoke is passed between two oppositely charged grids; the smoke particles become charged and are attracted to the oppositely charged grid. These grids are then brushed or shaken to collect the dust.

electrostatics (*adj.*) The study of electric charges at rest. *See* STATIC ELECTRICITY.

electrovalency The number of electrons that may be lost or gained by the atoms of a given element in forming an ion. For S-BLOCK and P-BLOCK elements, the electrovalency is determined by the number of electrons that must be lost or gained to achieve a NOBLE GAS configuration. Thus the electrovalency of sodium is 1 (it loses an electron) and for chlorine is also 1 (though in this case an electron is gained). The TRANSITION METALS often form more than one type of ion, for example iron exists with electrovalency of 2, Fe^{2+}, and 3, Fe^{3+}.

electroweak force The single interaction now known to account for the effects of the WEAK NUCLEAR FORCE and the ELECTROMAGNETIC FORCE. This example of unification of forces has been an important part of the progress made in theoretical physics since the 1960s. The electroweak unification was suggested by Glashow, Weinberg and Salam in 1967, and the W BOSON and Z BOSON predicted were discovered at CERN in 1983. *See also* GRAND UNIFICATION THEORY, STANDARD MODEL.

element A substance that cannot be broken down into more fundamental constituents by normal chemical means, being made of ATOMS all of the same type (that is, having the same ATOMIC NUMBER). About 110 elements are known, though there are claims for a few more. Of these, 91 are found in nature.

elementary particle, *fundamental particle* Any one of the indivisible particles from which all MATTER is believed to be made up, plus the GAUGE BOSONS that are responsible for the forces between the particles of matter. For every particle of matter there is a similar ANTIPARTICLE, with the same mass, and HALF-LIFE (if unstable), but opposite values of all QUANTUM NUMBERS, such as CHARGE, which remain the same in given types of interaction. Elementary particles are divided into two categories: LEPTONS, which do not feel the STRONG NUCLEAR FORCE, and QUARKS, which do. *See also* BARYON, FOUR FORCES OF NATURE, HADRON, HYPERON, SUPERSTRING THEORY, TACHYON, THEORY OF EVERYTHING.

elevation of boiling point The increase in the boiling point of a solvent caused by the presence of dissolved ions or molecules. This is a COLLIGATIVE PROPERTY, independent of the nature of the particles present, and proportional to their molar concentration, with a constant of proportionality called the ebullioscopic constant. Measurement of the elevation of boiling point caused by dissolving a known mass of substance in a solvent can be used to estimate its RELATIVE MOLECULAR MASS.

elimination 1. (*chemistry*) A reaction in which an atom or group in a molecule is eliminated. HALOGENOALKANES can undergo elimination reactions in the presence of a NUCLEOPHILE, for example the formation of an ALKENE from 2-bromo-2-methylpropane in the presence of a hydroxide ion by the elimination of H^+ and Br^- from the latter:

$$\mathrm{H{-}\overset{\displaystyle H}{\underset{\displaystyle H}{C}}{-}\overset{\displaystyle CH_3}{\underset{\displaystyle CH_3}{C}}{-}Br + HO^- \rightarrow H_2O + Br^- + (H)_2C{=}C(CH_3)_2}$$

Tertiary, and to a lesser extent secondary, halogenoalkanes undergo elimination reactions.

See also NUCLEOPHILIC SUBSTITUTION REACTIONS.

2. (*biology*) *See* EGESTION.

ELISA *See* ENZYME-LINKED IMMUNOSORBANT ASSAY.

ellipse A closed curve that can be formed as a CONIC SECTION, or as the set of points such that the total distance of every point from two fixed points, called the foci, is constant. An ellipse has the appearance of a flattened circle, with the degree of flattening measured by the ECCENTRICITY of the ellipse. The orbits of satellites are ellipses or circles. Most planets have orbits that are almost circular, but comets travel in highly elliptical orbits.

eluent A SOLVENT used in CHROMATOGRAPHY.

elution The process of using a SOLVENT to remove an adsorbed substance (*see* ADSORPTION) from an adsorbing medium, particularly in CHROMATOGRAPHY. In chromatography the rate at which material is dissolved and redeposited is the key to separation of different substances.

e-mail *See* ELECTRONIC MAIL.

embryo In mammals, the name given to a developing fertilized egg before it becomes a FOETUS at 7 weeks of pregnancy. In plants, the young developing plant after FERTILIZATION of an egg cell (*see* OVULE, SEED). *See also* BLASTOCYST, EMBRYONIC DEVELOPMENT, SEXUAL REPRODUCTION.

embryology The study of EMBRYONIC DEVELOPMENT.

embryonic development In an animal, the growth and development of a fertilized egg to become a fully formed FOETUS ready for birth. There are three stages of development: cleavage, gastrulation and organogenesis. Cleavage is the mitotic division (*see* MITOSIS) of the fertilized egg to form a ball of identical cells. Gastrulation is where the ball of cells arranges into definite layers called GERM LAYERS, of which there are three in most animals. These are the outer ECTODERM, middle MESODERM and inner ENDODERM. In organogenesis, the cells of the germ layers differentiate.

Some cell types are derived from more than one lineage. External factors can have an effect on embryo development, for example the rubella (German measles) virus causes brain damage and deafness, while smoking, alcohol and nutrition have less defined effects.

See also PREGNANCY.

embryo sac A large cell within the OVULE of a SEED PLANT within which DOUBLE FERTILIZATION occurs and the embryo subsequently develops. The embryo sac is formed from one of the cells of the NUCELLUS and its nucleus divides by MITOSIS until there are typically eight nuclei. One of these nuclei develops into the egg cell, which is the female GAMETE that fuses with the male gamete at fertilization. Three other nuclei become ANTIPODAL CELLS and two are POLAR NUCLEI that fuse to form the PRIMARY ENDOSPERM NUCLEUS. Two others degenerate.

embryo transfer *See IN VITRO* FERTILIZATION.

e.m.f. *See* ELECTROMOTIVE FORCE.

emission spectrum A SPECTRUM produced as excited atoms or molecules return to lower ENERGY LEVELS, giving off electromagnetic radiation of specific wavelengths, with all other wavelengths being absent.

emissivity The ability of a surface to give off THERMAL RADIATION when heated. A BLACK BODY, which radiates perfectly, has an emissivity of 1. A non-emitting surface has an emissivity of 0.

emitter The ELECTRODE in a JUNCTION TRANSISTOR from which charge carriers are injected into the BASE before arriving at the COLLECTOR.

empirical (*adj.*) Describing a result that is known purely from experiment, without any theoretical understanding or explanation.

empirical formula The simplest type of chemical formula. It gives the relative proportions of the elements present in a compound, but gives no indication of the RELATIVE MOLECULAR MASS or the structure. For example, the empirical structure of ethane is CH_3, giving no indication of the presence of two carbon atoms and six hydrogen atoms in each molecule. The full MOLECULAR FORMULA of ethane is C_2H_6. The empirical formula is obtained by QUANTITATIVE ANALYSIS of the proportion of each element present. *See also* STRUCTURAL FORMULA.

emulator A piece of computer hardware or software that allows one device to do the functions of another. Emulators are used to allow programs written for one computer to be run on another.

emulsion A mixture of two liquids, with one liquid forming small droplets suspended in the other. *See also* COLLOID, IMMISCIBLE.

emulsion test *See* LIPID.

enantiomer (Greek *enantios* = face-to-face) One of a pair of optical ISOMERS. *See* OPTICAL ISOMERISM.

enantiotropy A form of ALLOTROPY in which one of the allotropes is stable at high temperature and the other is stable at lower temperatures. Transitions between one state and the other can be brought about by heating or cooling the material, though such transitions may happen only slowly. The allotropy of SULPHUR is an example of this, with the rhombic form being stable at low temperatures and the monoclinic form at high temperatures. *See also* MONOTROPY.

encephalin One of a variety of PEPTIDES produced by nerve cells in the brain that acts as a natural pain-killer in a similar way to OPIATE drugs. However, encephalin is not addictive as it is quickly degraded by the body. *See also* ENDORPHINS.

endemic *See* DISEASE.

endocarp The innermost layer of the wall of a fruit. The endocarp may be hard and woody, as in a DRUPE.

endocrine gland A ductless GLAND that secretes its product (such as a hormone or enzyme) directly into the blood by DIFFUSION. Such glands include the PANCREAS, PITUITARY GLAND, ADRENAL GLAND, THYROID GLAND, OVARY and TESTIS. *Compare* EXOCRINE GLAND.

endocrine system One of the two major co-ordinating systems of animals (the other being the NERVOUS SYSTEM), consisting of a series of ENDOCRINE GLANDS that secrete products such as hormones and enzymes directly into the bloodstream. Compared to the speed of nerve impulses, the endocrine system is a slower means of co-ordination. It controls growth, breeding and maintenance of body fluid composition, for which a slower rate is appropriate. The study of the endocrine system is called endocrinology. In some instances, there is an overlap between the nervous and endocrine systems shown by neurosecretory (neuroendocrine) cells, which are cells of the nervous system (for example in the brain and HYPOTHALAMUS) that can both conduct nerve impulses and secrete hormones (which travel along the AXON and then in the blood to target cells). ADRENALINE is another example of overlap because it can act as a hormone and a NEUROTRANSMITTER.

endocrinology The study of the ENDOCRINE SYSTEM.

endocytosis A term used to refer to bulk intake of material into a cell, often used as a collective term for PHAGOCYTOSIS and PINOCYTOSIS.

endoderm The innermost GERM LAYER of the developing animal embryo. It gives rise to the digestive system, bladder, lungs, liver, pancreas and thyroid gland. *See also* ECTODERM, MESODERM.

endolymph Fluid found in the inner ear. *See* COCHLEA, SEMI-CIRCULAR CANAL.

endometrium The inner lining of the UTERUS. The endometrium is a glandular tissue that changes structure in response to hormone stimulation during the MENSTRUAL CYCLE and pregnancy. During the menstrual cycle, the endometrium thickens and its blood supply increases in preparation for pregnancy. If FERTILIZATION of the OVUM (egg) does not occur, the endometrium is shed and rebuilds again during the next cycle. If fertilization does occur, the ZYGOTE (fertilized ovum) implants into the endometrium, which then thickens further and becomes known as the decidua. In early pregnancy, the decidua produces many hormones and proteins thought to be important in early embryo survival. Part of the decidua becomes the PLACENTA that supports the foetus during pregnancy.

endoparasite A PARASITE that lives inside its host.

endoplasm *See* CYTOPLASM.

endoplasmic reticulum (ER) In biology, a complex system of membranous stacks (cisternae) and tubes within the CYTOPLASM of eukaryotic cells. It is an extension from and is continuous with the nuclear membrane, and is only visible under the ELECTRON MICROSCOPE. Some membranes are called rough ER (rER). These are most abundant in rapidly growing cells, are lined with RIBOSOMES and are the site of PROTEIN SYNTHESIS. Smooth ER (sER) has no associated ribosomes and is concerned with LIPID synthesis and is present in cells producing lipid-related secretions. The membranes of the ER are thought to regulate exchange of materials passing through them, like the CELL MEMBRANE. The ER stores proteins needed by the cell (in the LUMEN) and organizes them into transport vesicles that bud from the main membrane to be carried elsewhere in the cell. A specialized stack of membranous sacs similar in structure to smooth ER but more compact is the GOLGI APPARATUS.

endorphin A PEPTIDE produced by the PITUITARY GLAND and HYPOTHALAMUS. Endorphins include some amino acid sequences found in hormones. They are considered to be natural morphine-like pain-killers and operate in ways similar to both hormones and NEUROTRANSMITTERS (e.g. relief of pain, reduction of breathing and heart rate, water conservation). They reduce the perception of pain by reducing the transmission of nerve impulses. They also affect the release of sex hormones from the pituitary gland. Their effects are short-lived because they are rapidly degraded by the body.

endoscope A device for viewing the inside of otherwise inaccessible structures, such as the inside of the human body, by sending light along a bundle of FIBRE OPTICS.

endoskeleton An internal skeleton of vertebrates consisting of bone or cartilage. It provides support, protection and a system of levers to which muscles are attached to enable movement (*see* SKELETON). *Compare* EXOSKELETON.

endosperm The food supply in the seeds of most flowering plants, containing starch, fat and protein. The endosperm is a mass of triploid (three sets of chromosomes) cells resulting from mitotic divisions (*see* MITOSIS) of the PRIMARY ENDOSPERM NUCLEUS. Some seeds have no endosperm because the food has been used up by the embryo before GERMINATION.

endothelial (*adj.*) Relating to ENDOTHELIUM.

endothelium (*pl.* ***endothelia***) A single layer of flattened cells derived from embryonic MESODERM. Endothelia are found lining blood vessels, lymph vessels and the heart. Similar to EPITHELIUM.

endotherm *See* HOMEOTHERM.

endothermic (*adj.*) Describing a chemical reaction in which heat energy is taken in; that is, one in which the HEAT OF REACTION is positive.

The formation of hydrogen iodide is an endothermic reaction:

$$\frac{1}{2}H_2 + \frac{1}{2}I_2 \rightarrow HI, \Delta H = 26 \text{ kJ mol}^{-1}$$

endothermy *See* HOMEOTHERMY.

end-point In a TITRATION, the point in a reaction at which an INDICATOR shows that enough of a second REAGENT has been added to have used up all of a fixed quantity of the first. For example, the indicator may show that enough acid has been added to completely neutralize a given quantity of alkali. In practice, a slight excess of the second reagent will be needed to bring about the change in the indicator. The reaction is actually complete a little sooner, at a point called the EQUIVALENCE POINT.

energy A measure of the ability of an object or a system to do WORK. Energy can be broadly divided into two classes: POTENTIAL ENERGY, which arises from an objects state or position (in a gravitational field, for example); and KINETIC ENERGY, which is a function of motion. The amount of work done, or the energy that gives the capability to do work, is measured in a unit called the JOULE.

The importance of the idea of energy comes from the LAW OF CONSERVATION OF ENERGY, which states that the total energy of any closed system is conserved.

As a result of the increasing demands for energy there has been a constant search for new sources. In the developed world, the chief energy source is the burning of FOSSIL FUELS. In the light of the problems with fossil fuels, increasing attention is being given to RENEWABLE energy sources. NUCLEAR FISSION is a major energy source, but there are fears about the safety of the radioactive materials involved, whilst NUCLEAR FUSION appears to be a long way from producing energy on a commercial scale.

See also CHEMICAL ENERGY, HEAT, INTERNAL ENERGY, MECHANICAL ENERGY.

energy band A range of energies that electrons can have in a solid. *See* BAND THEORY.

energy barrier The amount of energy that may need to be provided before a physical or chemical process can take place, even though the process as a whole may release energy. *See* ACTIVATION PROCESS.

energy level Any one of the permitted energy states in which an atom or molecule may exist. Under the rules of QUANTUM MECHANICS, a system can only have certain fixed energies, and each different atom or molecule has a series of possible energy levels. When the system moves from one energy level to another, the energy is absorbed or emitted, often as a PHOTON of light or other electromagnetic radiation. The distinct energies of the allowed levels mean that only photons of certain discrete energies are involved. This gives rise to absorption and emission spectra characteristic of the atoms or molecules involved. *See also* ATOMIC ABSORPTION SPECTROSCOPY, ATOMIC EMISSION SPECTROSCOPY, BAND THEORY, SPECTROSCOPY, WAVE NATURE OF PARTICLES.

engine Any machine that converts chemical energy, from the burning of a fuel, to mechanical work via heat energy. Engines are very widely used, as a great deal of energy is available from fuels. However, the SECOND LAW OF THERMODYNAMICS means that a large amount of energy must be carried away in the hot gases produced by burning this fuel. The result is low efficiency, typically 20 to 25 per cent. *See also* CARNOT CYCLE, DIESEL ENGINE, HEAT ENGINE, INTERNAL COMBUSTION ENGINE, PETROL ENGINE.

enlarged (*adj.*) Increased in size, particularly used in the context of images formed by CONVERGING LENSES and mirrors.

enlargement A mathematical TRANSFORMATION applied to an object, in which the distance of every point on the object from a fixed point, called the centre of enlargement, is multiplied by some factor. This factor is called the scale factor or simply the enlargement. The new object is similar to the original object.

enriched uranium Uranium in which the concentration of the FISSILE ISOTOPE, uranium–235, has been increased. Natural uranium is a mixture of isotopes but mostly it contains uranium–238, which is not fissile. To produce a fuel for a reactor the uranium must be enriched. This is done by using DIFFUSION or CENTRIFUGE processes, which separate out the two different masses of isotope. Since the mass difference is small, the enrichment process must be repeated many times to produce a useful fuel. *See also* DEPLETED URANIUM, FISSION POWER.

enterokinase An enzyme in the SMALL INTESTINE that activates TRYPSIN from its inactive form trypsinogen.

enthalpy A measure of energy, commonly used in the study of chemical reactions. It is the

INTERNAL ENERGY of the molecules of a material plus the pressure at which the material is held multiplied by the volume occupied. It is a useful quantity because it is constant in any chemical reaction that takes place at a constant pressure. For a substance where the molecules have energy U and occupy a volume V at a pressure p, the enthalpy is H, where

$$H = U + pV$$

The internal energy of the molecules alone is not constant if there is any change in volume during the reaction. If a gas is produced, for example, work will have to be done at the expense of the internal energy of the molecules in pushing back the surrounding atmosphere to make room for this gas. Enthalpy is also constant during the FREE EXPANSION of a gas.

enthalpy of combustion *See* HEAT OF COMBUSTION.

enthalpy of formation *See* HEAT OF FORMATION.

enthalpy of reaction *See* HEAT OF REACTION.

entomology The study of INSECTS.

entomophily POLLINATION by insects.

entropy A measure of the degree of disorder in a thermodynamic system. Typically, this concept is applied to a system containing many molecules. The most probable state in which those molecules are found is the state that can be achieved in the largest number of ways. Entropy can be defined by the equation

$$S = k \ln W$$

where S is the entropy, k is the BOLTZMANN CONSTANT and W is the number of ways in which the molecules in the system can be arranged to produce the specified state.

For any system containing a large number of particles, the most probable state becomes overwhelmingly likely to be the state in which the system is found. This leads to the statement that any irreversible change (not able to happen in reverse) is one which produces an increase in the total entropy of the system. Since such systems tend to move from ordered (low entropy) to disordered (high entropy) states, the entropy of a system can be used as a measure of the extent to which the energy in a system is available for conversion to WORK.

See also FREE ENERGY, SECOND LAW OF THERMODYNAMICS.

E number One of a group of ADDITIVES (not including flavourings) approved by the European community. E numbers do not have to be listed with the ingredients of a product. Some E numbers are more often referred to by their name, for example E102 is TARTRAZINE. Some can cause side-effects in some people, for example tartrazine, used to provide an orange colour, is known to cause hyperactivity and worsen asthma.

environment The sum of conditions in which an organism lives. This includes all the BIOTIC (living) and ABIOTIC (non-living) factors. The term 'internal environment' is sometimes used to refer to an organism's internal conditions. Environment is also used to mean the total global environment, rather than that relating to an individual.

enzyme A biological molecule that alters the rate of a reaction (usually speeds it up) without undergoing a chemical change itself (and can therefore be used repeatedly). An enzyme is a natural CATALYST that does not alter the final EQUILIBRIUM of a reaction, only the speed at which it is achieved. Enzyme reactions are always reversible.

Most enzymes are globular proteins with a three-dimensional structure that provides a specific ACTIVE SITE where the SUBSTRATE molecule it acts upon fits, like a LOCK-AND-KEY MECHANISM. Modern interpretation of this theory suggests that the three-dimensional shape of the active site changes as the substrate binds. This is called 'induced fit'. An enzyme–substrate complex forms until the substrate is altered or split, and then its shape changes and it no longer fits into the active site, so the enzyme falls away. All enzymes act on specific substrates but some will bind a variety of similar substrates while others are very specific and bind only one.

Enzymes need different and precise conditions in which to function. Deviations from the optimum pH and temperature will cause the shape of the enzyme to change, which will eventually make it non-functional. The concentration of both enzyme and substrate has an effect on enzyme activity. A very low concentration of enzyme is needed for a reaction, and if there is a excess of substrate the RATE OF REACTION is proportional to the enzyme concentration.

The affinity of the enzyme for the substrate is variable. Some enzymes are CONSTITUTIVE while others are INDUCIBLE, dependent on whether or not the CISTRON(S) encoding the enzyme is expressed (*see* GENE EXPRESSION). Some enzymes need a COFACTOR for them to function. There are six recognized categories of enzymes based on their functions: OXIDOREDUCTASES, TRANSFERASES, HYDROLASES, LYASES, ISOMERASES and LIGASES. The nomenclature often gives information regarding an enzyme's activity, for example peptidases break down PEPTIDES.

Enzymes have many uses outside the body, in medicine and industry and as research tools in molecular biology, and they can be extracted from bacteria and even modified by GENETIC ENGINEERING for a particular purpose.

See also COENZYME, COFACTOR, PROSTHETIC GROUP.

enzyme inhibition The reduction of the RATE OF REACTION of an ENZYME by an INHIBITOR, which can be reversible or non-reversible. Heavy metal ions, for example mercury and silver, cause non-reversible inhibition and therefore permanent damage to enzymes by altering their shape when they break the SULPHIDE bonds. Reversible inhibition is temporary and the enzyme function returns once the inhibitor is removed.

Competitive inhibitors function by having a structure similar to the SUBSTRATE, and they bind and remain in the ACTIVE SITE. Non-competitive inhibitors attach elsewhere on the enzyme molecule and change the enzyme's shape so that the substrate can no longer bind. Cyanide is a non-competitive inhibitor that attaches to the copper PROSTHETIC GROUP of the enzyme cytochrome oxidase, inhibiting RESPIRATION. In many metabolic reactions the end-product of a pathway may inhibit the enzyme at the start of the pathway; this is an example of NEGATIVE FEEDBACK. *See also* HOMEOSTASIS.

enzyme-linked immunosorbant assay (ELISA) A laboratory technique using monoclonal antibodies (*see* MONOCLONAL ANTIBODY) that determines the amount of ANTIGEN in a given sample. The test has many uses, such as detecting drugs in athletes' urine, pregnancy testing kits and the AIDS test. The technique immobilizes a monoclonal antibody on a laboratory dish and passes a test solution over it. If the appropriate antigen is present it will bind. Then a second antibody with an enzyme attached is added that will bind only to the original antibody bound to the antigen. A SUBSTRATE is added that changes colour when reacted with the enzyme, which can be detected and measured.

Eocene The second EPOCH of the TERTIARY PERIOD. It began after the PALAEOCENE epoch (about 54 million years ago) and extended until the OLIGOCENE epoch (about 38 million years ago). The Eocene epoch saw the appearance of whales.

eon The longest division of geological time, made up of a number of ERAS.

eosinophil A type of GRANULOCYTE (blood cell) with cytoplasmic granules that stain with acid dyes. Eosinophils increase in certain parasitic infections and allergies.

epicentre The point on the surface of the Earth directly above the focus of an EARTHQUAKE. It is the point at which the most damage occurs.

epidemic *See* DISEASE.

epidermal growth factor *See* GROWTH FACTOR.

epidermis The outermost layer of cells covering an organism's body, providing protection and prevention of water loss. Plants and many invertebrates have an epidermis that is a single layer of cells, often with a non-cellular, tough outer CUTICLE that prevents desiccation (*see* PARENCHYMA). In vertebrates, the epidermis consists of a surface cornified layer (stratum corneum), which is a tough waterproof layer of dead cells impregnated with KERATIN, that provides protection from bacteria and prevention of loss of water. The cornified layer is constantly worn away and replaced by the living cells of the granular layer (stratum granulosum) below. As cells move through the granular layer they incorporate keratin and eventually die. The deepest layer of the epidermis is the MALPIGHIAN LAYER, from which actively dividing cells move to the granular layer. There are no blood vessels in the epidermis. *See also* SKIN.

epididymis A 6-metre long coiled tube in the TESTIS of male vertebrates (one for each testis). SPERM pass into the epididymis from the SEMINIFEROUS TUBULES, where they are made, to be stored for 18 hours, during which time they gain their motility. If the sperm are not used they are reabsorbed by the epididymis (after 4 weeks). Sperm are passed from the epididymis during sexual intercourse into a muscular tube called the VAS DEFERENS. *See also* SEXUAL REPRODUCTION.

epigeal (*adj.*) Describing seed GERMINATION where the COTYLEDONS emerge from the soil following germination and form the first green leaves.

epiglottis A small flap of CARTILAGE found at the back of the mouth in the throat that moves during swallowing to prevent food entering the TRACHEA (windpipe), which would cause choking. The epiglottis is necessary because the trachea and OESOPHAGUS (the tube carrying food to the stomach) both meet at the PHARYNX. The movement of the epiglottis is complex.

epinephrine *See* ADRENALINE.

epiphysis 1. *See* PINEAL GLAND.

2. The end of a long bone. The epiphysis is initially separated from the shaft of the bone by a section of CARTILAGE that eventually ossifies (see OSSIFICATION), so that the two portions merge together.

epithelial (*adj.*) Relating to EPITHELIUM.

epithelium (*pl.* ***epithelia***) Animal tissue composed of cells firmly held together in single or compound sheets or tubes by a minimal amount of intercellular substance. Epithelia line cavities, tubes and exposed surfaces of the body, and provide a protective function. Most epithelial tissue is derived from embryonic ECTODERM. One surface of epithelia is always attached to a BASEMENT MEMBRANE, while the other is free.

Epithelia are classified as columnar, cubicle and squamous (flattened), according to the height of the cell relative to its breadth, and according to whether the sheet is one cell thick (simple) or many cells thick (stratified). Most epithelia are specialized for absorption or secretion. Columnar epithelial cells often have minute finger-like projections on their free surface called MICROVILLI, which increase their surface area for absorption. Further fine cytoplasmic projections called CILIA can be found on these cells, which aid movement of fluid or substances through ducts. Stratified epithelium consists of a germinative layer attached to the basement membrane that undergoes MITOSIS to create new cells. The new cells push old ones to the surface, where they become flattened and are shed. Stratified epithelium may be thickened with KERATIN, as in skin, to provide protection. A specialized type of stratified epithelium called transitional epithelium is found in structures that need to stretch, such as the bladder. Transitional epithelium consists of three to four layers, flattened at the surface, that are not shed. *See also* MUCOUS MEMBRANE.

epoch A length of geological time; a subdivision of a PERIOD.

epoxy resin A synthetic RESIN which is a THERMOSETTING PLASTIC. It is used as a tough adhesive and in paints. Araldite is a trade name for a commonly used household adhesive which is also used to mount specimens to be viewed in an ELECTRON MICROSCOPE.

EPROM (erasable programmable read-only memory) ROM that can be erased (usually by exposure to ultraviolet light) so new information can be stored.

equation 1. (*mathematics*) Any mathematical statement containing an equals sign (=). An equation represents the equality of two expressions of variables and/or constants. They are generally used to determine one of the quantities in the equation in terms of the others. Equations are often used in this way in physics. *See also* INEQUALITY, LINEAR EQUATION, POLYNOMIAL, QUADRATIC EQUATION, SIMULTANEOUS EQUATION.

2. (*chemistry*) *See* CHEMICAL EQUATION.

equation of motion Any equation that enables the position of an object to be calculated at some future time by knowing its initial position and velocity and how the forces on that object vary with position and velocity. An important class of motion is that where the acceleration is constant. For an object with a constant acceleration a, starting speed u and final speed v, covering distance s in time t:

$$v = u + at$$
$$s = ut + at^2/2$$
$$v^2 = u^2 + 2as$$

equation of state In physics, any equation describing the behaviour of a material under different conditions. In particular, the change in volume under differing temperatures and pressure.

equator On the Earth, or any other planet, a line drawn around the surface equally distant from the two POLES.

equilateral (*adj.*) Describing a shape, especially a triangle, where all the sides are the same length.

equilibrium The state of an object or system in which all effects – forces, interactions, reactions, etc. – are balanced so that there is no net change.

In mechanics, an object or system is said to be in a static equilibrium when the total force acting is zero, and when the MOMENTS of the forces, taken about any axis, add to zero. In this case, the system will not accelerate (*see* ACCELERATION) – in particular, if the system starts at rest, it will remain at rest (*see* NEWTON'S LAWS OF MOTION). These conditions for equilibrium are particularly important in the branch of mechanics called STATICS, the study of structures or systems at rest.

If all the forces act in a single plane, the conditions for equilibrium are met if the COMPONENTS of the forces taken in any two non-parallel directions add up to zero, and the moments of the forces about any point add up to zero. If only three non-parallel forces act, the condition that the moments add to zero can be stated as a requirement that the lines along which the three forces act must cross at a single point.

A system is in a state of THERMAL EQUILIBRIUM if there is no net exchange of heat within it or between it and its surroundings. A system is in CHEMICAL EQUILIBRIUM if the rate of a forward reaction is exactly equal to the rate of the reverse reaction. The latter is an example of a dynamic equilibrium, where two opposing processes balance each other out, keeping a system unchanged.

See also STABLE EQUILIBRIUM, THERMODYNAMIC EQUILIBRIUM, UNSTABLE EQUILIBRIUM.

equilibrium constant In a REVERSIBLE REACTION,

$$xA + yB \Leftrightarrow mC + nD$$

the reaction will proceed in one direction until EQUILIBRIUM is reached. The reaction then proceeds at the same rate in the forward and backward directions. At this point, the CONCENTRATIONS of the REAGENTS, indicated by square brackets [ ], will obey the equation

$$[A]^x[B]^y/[C]^m[D]^n = K$$

where K is the equilibrium constant.

equinox One of two dates in the year when night and day are the same length. They occur in spring and autumn, when the Sun is directly above the equator.

equipartition of energy The principle that states that thermal energy will be distributed equally between the available DEGREES OF FREEDOM.

equipotential A line joining points that are at the same ELECTRIC POTENTIAL. Equipotentials are always at right angles to ELECTRIC FIELD LINES. A conductor will always be at a single potential throughout its volume.

equivalence of mass and energy The concept that when an object gains energy, in whatever form, it gains mass. This is summed up in the famous equation

$$E = mc^2$$

where E is the energy, m the mass and c the SPEED OF LIGHT. The mass changes predicted are normally too small to be detected, but can be measured in atomic nuclei (*see* NUCLEAR BINDING ENERGY). *See also* SPECIAL THEORY OF RELATIVITY.

equivalence point In a TITRATION, the point at which sufficient REAGENT has been added to have reacted completely with the fixed quantity of the second reagent. For example, in adding an acid to an alkali, the equivalence point is reached when the alkali has been completely neutralized. *See also* END-POINT.

ER *See* ENDOPLASMIC RETICULUM.

era A major division of geological time, consisting of a number of PERIODS but shorter than an EON.

erasable programmable read-only memory *See* EPROM.

erbium (Er) The element with ATOMIC NUMBER 68; RELATIVE ATOMIC MASS 167.3; melting point 1,529°C; boiling point 2,863°C; RELATIVE DENSITY 9.0. It is a member of the LANTHANIDE series. Difficulties in purifying it from the other chemically similar lanthanides has meant that it has found little commercial use.

erg A unit of energy in the C.G.S. SYSTEM, the work done when a force of one DYNE moves through a distance of one centimetre. One erg is equivalent to 10^{-7} JOULES.

erosion The wearing down of rock by the action of wind or water.

error In computing, a mistake or fault that makes a computer produce an unexpected result or stop running altogether. Errors may be introduced by the user or may be a result of a BUG in the software.

erythrocyte Another name for a RED BLOOD CELL.

escape velocity The speed an object must have to be launched into space and completely escape the gravitational field of the planet from which it was launched. The escape velocity is such that the object has sufficient KINETIC ENERGY to lose to compensate for the increase in GRAVITATIONAL POTENTIAL ENERGY in moving away from the surface of the planet. The escape velocity is independent of the mass being launched. For the Earth it is about 11 kms^{-2}. The escape velocity from the surface of a body of mass M and radius R is $(2GM/R)^{-1/2}$, where G is the GRAVITATIONAL CONSTANT.

essential oil A natural oil with a pleasant odour obtained from plants and used in perfumes and flavourings.

essential amino acid An AMINO ACID that is needed by humans but cannot be made by them and must therefore be included in their diet. The nine essential amino acids are LYSINE, PHENYLALANINE, LEUCINE, THREONINE, METHIONINE, ISOLEUCINE, TRYPTOPHAN, HISTIDINE and VALINE.

ester An organic compound formed when an acid and ALCOHOL react together with the elimination of water. For example, ethanoic acid, CH_3COOH, reacts with ethanol, C_2H_5OH, to give the ester ethyl ethanoate:

$$CH_3COOH + C_2H_5OH \rightarrow CH_3COOC_2H_5 + H_2O$$

The general formula of an ester is RCOOR′, where R and R′ are HYDROCARBONS. Esters occur naturally in fruit and many have a characteristic fruity odour. Esters are used as solvents, food flavourings and to provide the scent in perfumes. *See also* ESTERIFICATION, SAPONIFICATION.

esterification A term used to describe the formation of an ESTER from an ALCOHOL and an acid. If equal amounts of acid and alcohol are used an equilibrium is set up such that equal amounts of the four species (acid, alcohol, ester and water) exist. The equilibrium can be altered to give good yields of ester and also the ester can be hydrolysed (*see* HYDROLYSIS) back to a mixture of acid and alcohol (such as in SAPONIFICATION).

estuary A wide river mouth entering the sea where fresh water and salt water mix.

ethanal, *acetaldehyde* (CH_3CHO) A colourless, inflammable liquid ALDEHYDE; boiling point 20.8°C. It is made from the oxidation of ethanol or ethene. Ethanal is miscible with water, ALCOHOL and ETHER. It can be oxidized to give ethanoic acid and reduced to give ethanol. It is used in the manufacture of many organic chemicals.

ethane (CH_3CH_3) A gas with no colour or odour; boiling point –89°C. It is the second member of the ALKANE series of HYDROCARBONS. Ethane can be made by the reduction of ethene or ethyne by hydrogen under pressure, in the presence of a nickel catalyst. It is found in natural gas and forms an explosive mixture with air.

ethane-1,2-diol, *ethylene glycol, glycol* ($(CH_2OH)_2$) A thick, colourless, odourless, sweetish, liquid ALCOHOL; boiling point 197°C. It is used as an ANTIFREEZE and coolant for engines because it mixes with water and lowers the freezing point below 0°C. It is also used in the preparation of ETHERS and ESTERS, including POLYESTER fibres.

ethanedioic acid, *oxalic acid* ($C_2H_2O_4$, HOOC–COOH) A poisonous substance found in rhubarb and as oxalate salts in wood sorrel. Ethanedioic acid causes paralysis of the NERVOUS SYSTEM. It is used as a bleaching agent in the textile and leather industries and domestically for removing stains such as rust and blood. It is also used for metal cleaning.

ethanoate, *acetate* Salts or ESTERS of ETHANOIC ACID. They are used in the textile industry to make acetate RAYON, a synthetic fabric made from CELLULOSE treated with ethanoic acid, and in the photographic industry as acetate film, a clear sheet made from cellulose ethanoate.

ethanoic acid, *acetic acid* (CH_3COOH) A colourless liquid CARBOXYLIC ACID with a pungent odour (characteristic of vinegar); boiling point 119°C; melting point 17°C. It is made by oxidation of ethanal or ethanol. Vinegar contains 5 per cent or more ethanoic acid and is produced by FERMENTATION. Ethanoic acid forms large ice-like crystals when solid, and when in this form is sometimes called 'glacial acetic acid'.

ethanol, *ethyl alcohol* (C_2H_5OH) The ALCOHOL found in alcoholic drinks such as beer, wines and spirits. Ethanol is a colourless liquid with a pleasant odour; boiling point 78°C. It is miscible with water and burns in air with a pale blue flame.

Ethanol is produced rarely in nature, by the FERMENTATION of CARBOHYDRATES by yeast cells or bacteria. In the past pure ethanol was manufactured by the fermentation of sugar or

STARCH by yeast enzymes, but now it comes mainly from ethene derived from petroleum. In Brazil, where petroleum is expensive, it is economically advantageous to produce ethanol by fermentation of sugar cane, for use as a motor fuel.

Ethanol can be oxidized to ethanal or ethanoic acid. It reacts with acids to give ESTERS and with chlorine to give trichloroethanal (chloral). ETHER is produced by the action of sulphuric acid on ethanol. In industry, ethanol is used as a solvent and in the manufacture of those chemicals mentioned (particularly ethanal). It is also used in foodstuffs.

ethanoylation, ***acetylation*** A process for incorporating the ethanoyl (acetyl) group, CH_3CO–, into organic compounds that contain the groups –OH, –NH_2 or –SH.

ethene, ***ethylene*** (C_2H_4) A gaseous HYDROCARBON; melting point –169°C; boiling point –105°C. It is the first of the series of ALKENES, obtained from petroleum by CRACKING. As with the other alkenes, ethene reacts by ELECTROPHILIC ADDITION reactions, adding groups across the carbon-carbon DOUBLE BOND. It can be reduced by hydrogen to give ethane.

Ethene is widely used in industry, for example in the manufacture of plastics, detergents, paints and pharmaceuticals. POLY(ETHENE) and POLYVINYL CHLORIDE are examples of plastic products derived from ethene. Ethene is also a by-product of plant METABOLISM. It acts as a PLANT GROWTH SUBSTANCE, stimulating the ripening of fruit. It is therefore useful commercially as a spray to ripen fruit, such as tomatoes and grapes. Ethene also stimulates ABSCISSION (drop) of leaves, fruit and flowers and is used commercially to promote fruit loosening.

ether 1. (*chemistry*) One of a series of organic chemicals in which an oxygen atom is inserted between the carbon atoms of two HYDROCARBON groups. An ether thus has the general formula ROR′ (where R is a hydrocarbon group) or $C_nH_{2n+2}O$, which is the same as the formula for ALCOHOLS except for the position of the oxygen atom. Ethers are less reactive than alcohols. The nomenclature for the ethers is based on the ALKANE forming the longest carbon chain in the molecule but with the prefix *alkoxy-* indicating the RO– group attached, for example $CH_3OC_3H_7$ is 1-methoxypropane, $C_2H_5OC_2H_5$ is ethoxyethane. The latter was called diethyl ether, commonly termed ether, and is used as an anaesthetic, antiseptic and a solvent. *See also* WILLIAMSON ETHER SYNTHESIS.

2. (*physics*) The hypothetical medium that was once thought to fill all space and support the propagation of ELECTROMAGNETIC RADIATION through space. *See also* MICHELSON–MORLEY EXPERIMENT.

ethernet A form of LOCAL AREA NETWORK. One computer wishing to communicate with another sends a signal onto the network and the link is established when the signal is acknowledged. Computers can send signals any time they do not hear another computer sending. If two computers try to send at the same time, the data becomes garbled and the collision is thus detected. The two computers wait a random period of time before resending. *Compare* TOKEN RING NETWORK.

ethoxyethane *See* ETHER.

ethyl acetate *See* ETHYL ETHANOATE.

ethyl alcohol The common name for ETHANOL.

ethylamine, ***monoethylamine*** ($CH_3CH_2NH_2$) In organic chemistry, any compound derived from a primary AMINE in which one of the hydrogen atoms of ammonia has been replaced by an ETHYL GROUP. Two hydrogen atoms of ammonia can be replaced by ethyl groups and the resulting compound is termed diethylamine $(CH_3CH_2)_2NH$. If three are replaced the compound is called triethylamine $(CH_3CH_2)_3N$. These ethylamines are all colourless with a strong smell of ammonia and show the typical properties of ALIPHATIC amines.

ethylenediaminetetraacetic acid (EDTA) A CHELATING AGENT used in chemical analysis.

ethylene glycol, ***glycol*** The common name for ETHANE-1,2-DIOL.

ethyl ethanoate, ***ethyl acetate, acetic ester, acetic ether*** ($CH_3COOC_2H_5$) A colourless liquid ESTER with a fruity odour; boiling point 77°C. It is formed by the reaction of ethanol and ethanoic acid. It is an important solvent and is also used in cosmetics and as an artificial essence.

ethyl group In organic chemistry, the group C_2H_5, sometimes written as Et–.

ethyne, ***acetylene*** (C_2H_2) A colourless, inflammable gas; boiling point –84°C; melting point –82°C. The first member of the ALKYNE series of HYDROCARBONS, it is produced on mixing water with calcium carbide. Ethyne was first used in early gas lamps. It gives off an intense heat on burning and is therefore used in some welding torches. It is used today as the starting point for the manufacture of a number of organic chemicals such as PROPENENITRILE, VINYL CHLORIDE and ETHANOIC ACID.

Euclidean space A flat space in which the normal laws of geometry apply. In particular, the angles of a triangle add up to 180°. This will not be true for a triangle drawn on a curved surface. Non-Euclidean spaces in four dimensions, three of space and one of time, are used to describe physical space in the GENERAL THEORY OF RELATIVITY.

eudiometer A vessel that measures changes in the volumes of gases during chemical reactions. From this, changes in the number of molecules present may be deduced. A simple eudiometer comprises a graduated glass tube closed at one end and sealed at the other by being immersed in mercury. A pair of ELECTRODES are provided, between which a spark can be formed to initiate a reaction.

eugenics The study of the ways in which the human race can be improved, especially by ARTIFICIAL SELECTION.

Euglenophyta A PHYLUM from the KINGDOM PROTOCTISTA. Members of this phylum are characterized by their possession of CHLOROPHYLL a and b (which enables PHOTOSYNTHESIS) and FLAGELLA. The products of photosynthesis are stored as a CARBOHYDRATE called paramylon, which is not found in other organisms. Reproduction is by BINARY FISSION.

Euglena is the best example of an organism in this phylum. *Euglena* can undergo a flowing movement because of strips of protein called the pellicle inside the cell membrane (there is no CELL WALL) that allow a change in shape.

eukaryote (*Eu* = true, *karyo* = nucleus) An organism possessing a clearly defined membrane-bound NUCLEUS in its cells, and other cell ORGANELLES, such as MITOCHONDRIA and CHLOROPLASTS, which are lacking in the simple cells of PROKARYOTES. All organisms are eukaryotes except BACTERIA and CYANOBACTERIA. The genetic material within the nucleus is arranged in CHROMOSOMES, which are not seen in prokaryotes. The RIBOSOMES of eukaryotic cells are larger and denser than prokaryotes, and certain proteins exist (such as HISTONES, ACTIN, MYOSIN) that are not found in prokaryotes.

eureka vessel *See* DENSITY CAN.

europium (Eu) The element with ATOMIC NUMBER 63; RELATIVE ATOMIC MASS 152.0; melting point 852°C; boiling point 1,529°C; RELATIVE DENSITY 5.2. Europium is a member of the LANTHANIDE series and is used to make PHOSPHORS for CATHODE RAY TUBES in colour television systems.

Eustachian tube A narrow tube connecting the PHARYNX to the middle ear. On swallowing, air can enter or leave the middle ear via this tube. This ensures that the pressure inside the middle ear cavity remains the same as that of the atmosphere, which is necessary to avoid stretching of the EARDRUM and sound impairment.

eutectic mixture A SOLID SOLUTION of two or more substances, with the lowest freezing point for any possible mixture of the components. The freezing point of a eutectic mixture is called the eutectic point. On a graph of temperature against composition for a mixture of two substances, which may be either solid or liquid, there are generally four regions. Below the eutectic point, the mixture is entirely solid. This temperature may be lower than the melting point of either substance alone. At higher temperatures, one substance may occur as a solid – generally the one present in higher proportions – together with a liquid that may be regarded as a saturated solution of the substance present as a solid. At higher temperatures, both substances will be liquid.

The FREEZING MIXTURE formed by adding sodium chloride to water is an example of a eutectic mixture. A lower liquid temperature can be achieved than would be achieved from either material alone. Not all mixtures form eutectics.

eutectic point The freezing point of of a EUTECTIC MIXTURE.

eutrophication Excessive enrichment of lakes and rivers by NITRATES and PHOSPHATES. Eutrophication can occur naturally or as a result of human activities. The natural accumulation of salts into a body of water is slow and is usually counter-balanced by loss of salts through natural drainage.

The more serious cause of eutrophication is artificial. Artificial eutrophication can be caused by the addition of nitrates and phosphates from fertilizers (washed from the

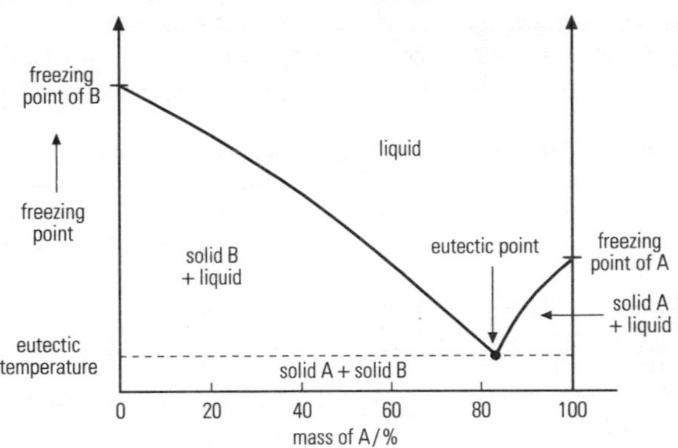

Eutectic mixture.

soil by rain), sewage and detergents. This enrichment causes the growth of ALGAL BLOOMS, which prevent light reaching deeper regions of the water and so aquatic plants die because they are unable to photosynthesize. The dead plants and algae are decomposed by saprophytic bacteria (*see* SAPROPHYTE), which use all the available oxygen in the water, leading to the death of other species such as fish.

See also POLLUTION.

evaporation The process by which a liquid turns into a gas at a temperature below the boiling point as some of the faster moving molecules escape from the surface of the liquid. Unlike boiling, evaporation takes place only on the liquid surface. The rate of evaporation can be increased by heating, by blowing air over the liquid surface so molecules cannot re-enter the liquid, or by increasing the surface area. A liquid that evaporates easily is called volatile. *See also* COOLING BY EVAPORATION.

event In PROBABILITY theory, one of the possible outcomes of some trial or procedure, such as drawing an ace from a pack of cards.

evergreen In botany, plants that retain their leaves all year round. *Compare* DECIDUOUS.

evolution The slow, continuous process by which changes in life forms take place. Such changes can be in the appearance of organisms (microevolution) or in the origin and extinction of species (macroevolution). The emergence of new species is called SPECIATION. There are many theories regarding the ORIGIN OF LIFE, but the process of evolution is the one most widely accepted by scientists today.

Two main theories of evolution were put forward in the 19th century, the first by the French naturalist Jean-Baptiste Lamarck (1744–1829). He suggested that useful characteristics acquired by an organism during its lifetime would be inherited by its offspring, while disuse of other characteristics would result in their eventual disappearance from the species. This theory influenced the more widely believed theory of the English naturalists Charles Darwin (1809–1882) and Alfred Wallace (1823–1913), who independently suggested that evolution occurs through NATURAL SELECTION. A combination of Darwin's theory and Mendel's theories (*see* MENDEL'S LAWS) on genetics gives us today's theory of evolution, which is called neo-Darwinism.

Factors other than natural selection may play a role in evolution, such as sexual selection (the choosing of sexual partners by characteristics considered to be attractive, therefore increasing their frequency of inheritance) and GENETIC DRIFT.

The process of evolution does not seem to be constant but occurs in periods of rapid change and relative stability. Much of the evidence to support the theory of evolution comes from the examination of fossils, which appear to show a gradual and progressive change from simple to complex forms, although some intermediate forms are yet to be found. Comparative molecular biology (e.g. amino acid sequences) and biochemical studies also support the theory of evolution. Other evidence for evolution is found in comparative anatomy and embryology (some different adult species have similarities in their embryos, suggesting common ancestry).

See also ADAPTIVE RADIATION, GENETICS, LAMARCKISM, VARIATION.

exchange In a telephone system, the switching centres that channel the signals to their intended recipients.

excited state A QUANTUM state, in an atom for example, with an energy higher than the lowest allowed energy. *See also* ENERGY LEVEL, GROUND STATE, WAVE NATURE OF PARTICLES.

exciter A DYNAMO used to provide the power supply for the rotating ELECTROMAGNET of a large ALTERNATOR.

excluded volume In a gas, the volume taken up by the molecules themselves. *See* IDEAL GAS.

excretion The removal of waste metabolic products from an organism. This is different from EGESTION, which is the removal of waste that has never been inside body cells. Excretion involves the removal of, for example, carbon dioxide from the lungs and nitrogenous compounds and water in the form of urine from the kidney. In plants and simple animals, excretion is by DIFFUSION rather than by specialized organs. *Compare* SECRETION.

exhaust stroke The stage in the operation of a PETROL ENGINE at which burnt gases are forced out of the engine.

exitance The amount of ELECTROMAGNETIC RADIATION per unit area emitted by a hot surface. *See also* BLACK-BODY RADIATION.

exocarp The outer skin of a fruit.

exocrine gland A gland of EPITHELIAL origin that transports its secretion(s) to an epithelial surface either directly or, more usually, via ducts. Exocrine glands can be classified according to their type of duct, which may be a simple tube or compound with many branches. Exocrine glands can also be classified according to whether or not their cells break down after secreting their contents. Merocrine glands (such as the SALIVARY GLANDS) remain intact after secreting their product, but holocrine gland (for example sebaceous or SWEAT GLANDS) cells are destroyed with the discharge of their secretion, and in apocrine glands (such as the MAMMARY GLANDS) the apical part of the cell breaks down during secretion. *Compare* ENDOCRINE GLAND.

exoskeleton A hard, external skeleton of ARTHROPODS, such as crabs and insects. *See also* CHITIN, CRUSTACEAN.

exosphere The outermost layer of the ATMOSPHERE, beginning at about 400 km.

exothermic (*adj.*) Describing a chemical reaction in which heat is given out; that is, one in which the HEAT OF REACTION is negative. COMBUSTION processes are exothermic, for example:

$$H_2 + {}^1/_2O_2 \rightarrow H_2O, \Delta H = -289 \text{ kJ mol}^{-1}$$

expansion 1. (*physics*) An increase in size, particularly as a result of an increase in temperature. *See also* THERMAL EXPANSION, FREE EXPANSION.

2. (*mathematics*) The rewriting of a mathematical expression without brackets, or as a sum of a larger number of simpler expressions. For example, $a^2 + 2ab + b^2$ is the expansion of $(a + b)^2$.

expansivity, *coefficient of expansion* A measure of the increase in length (linear expansivity), area, or volume (bulk expansivity) of a material when it is heated. The linear expansivity of a material is the increase in length for a 1°C temperature rise, divided by the original length. *See also* THERMAL EXPANSION.

experiment A set of measurements or observations, often performed on equipment designed specifically for this purpose, and designed to suggest or to test a theory. If the theory satisfies a sufficient range of experimental tests, it may then be used to predict what will happen in similar situations. If an experiment provides results that genuinely contradict the theory, the theory must be revised.

In the design of any experiment, it is important to ensure that only those factors being studied can change, and that all other factors remain constant throughout the experiment.

expiration Breathing out. *See* BREATHING.

explosimeter A device designed to measure the concentrations of explosive or inflammable gas. Current flows through a thin wire, which becomes hot. In the presence of inflammable gases, even in low concentrations, the OXIDATION of the gas heats the wire still further and increases its resistance. These changes in resistance are detected using a WHEATSTONE BRIDGE circuit.

explosion A violent and rapid COMBUSTION process in which the reaction accelerates to a rate that is limited only by the speed of movement of the molecules. In a gas this is equivalent to the SPEED OF SOUND in the gas. Such rapid combustion produces a rise in temperature leading to an increase in pressure as the reaction products do not have time to escape the immediate vicinity of the explosion. This high pressure is responsible for the destructive power of explosive materials.

exponent, *index* A number that expresses how many times another number must be multiplied by itself. For example, in the expression x^n, n is the exponent of x.

exponential 1. *See* E.

2. (*adj.*) In mathematics, frequently used to describe any rapid growth or decay, but more correctly describing the function e^x where e = 2.713 (the BASE of NATURAL LOGARITHMS). Exponential functions have an important common ratio property – each time the independent variable increases by a certain amount, the function changes by a fixed factor.

The exponential function $y = e^x$ can be defined in various ways, including as the solution to the equation $dy/dx = x$ which has $y = 1$ when $x = 0$.

In an exponential growth of population, the population may double after 10 years, then double again in the following 10 years and so on. In radioactivity, the level of radioactivity falls by a constant factor in equal time intervals. *See also* HALF-LIFE, HALF-THICKNESS, TIME-CONSTANT.

external auditory meatus *See* AUDITORY CANAL.

external respiration *See* BREATHING.

extracellular matrix (ECM) A network of proteins and POLYSACCHARIDES in which animal and plant cells are embedded and form TISSUES. ECM forms part of CONNECTIVE TISSUE, plant and bacterial CELL WALLS and the EXOSKELETON of ARTHROPODS. Structural proteins such as COLLAGEN and ELASTIN and cell adhesion molecules are components of ECM. ECM also contains many signalling molecules that play an important role in the regulation of cell functions, for example growth, differentiation, division and cell death.

extraction 1. (*mineralogy*) The separation of a metal from its ore.

2. (*chemistry*) The separation of one substance from a MIXTURE. This may be carried out on the basis that one of the materials in the mixture will dissolve in a solvent that will not dissolve the others – this is called SOLVENT EXTRACTION.

extraembryonic membrane In mammals, one of the membranes derived from and surrounding the EMBRYO during pregnancy. The extraembryonic membranes are the CHORION, AMNION, ALLANTOIS and YOLK SAC. *See also* EMBRYONIC DEVELOPMENT, TROPHOBLAST.

extraordinary ray *See* BIREFRINGENCE.

eye The SENSE ORGAN responding to light. The human eye is roughly spherical and protected by bony sockets in the skull (the orbits) and attached to the skull by the rectus muscles, which allow the eyeball to rotate. The whole eyeball (except the CORNEA) is covered by the sclerotic coat (sclera, the white of the eye), which is made of COLLAGEN fibres and so protects and maintains the shape of the eyeball.

Light passes through the external layers of the eye, the CONJUNCTIVA and the CORNEA, and then enters the eye through an aperture called the PUPIL at the centre of the coloured diaphragm, the IRIS. Most of the REFRACTION (bending) of the light entering the eye is carried out by the cornea, which is curved and acts as a fixed lens. Behind the pupil is the lens, which carries out the final focusing of the light entering the eye. The lens can alter its shape due to the action of CILIARY MUSCLES (radial and circular muscles) that surround it. These alter the tension of suspensory ligaments supporting the lens and so allow it to change shape (fatter for near objects, thinner for distant ones). This allows objects at different distances to be focused, an ability called 'accommodation'. In front of the lens is an ANTERIOR chamber containing a clear liquid called AQUEOUS HUMOUR, and behind the lens is a larger posterior chamber containing clear, jelly-like VITREOUS HUMOUR, which helps to maintain the shape of the eyeball.

The lens focuses the light entering the eye onto a layer called the RETINA. Light-sensitive cells here (RODS and CONES) convert the light they receive into NERVE IMPULSES that pass along the OPTIC NERVE to the brain. Internal reflection of light is prevented by a pigmented layer inside the sclera called the choroid, which is rich in blood vessels and supplies the retina. In some nocturnal animals, light is reflected by a layer called the TAPETUM.

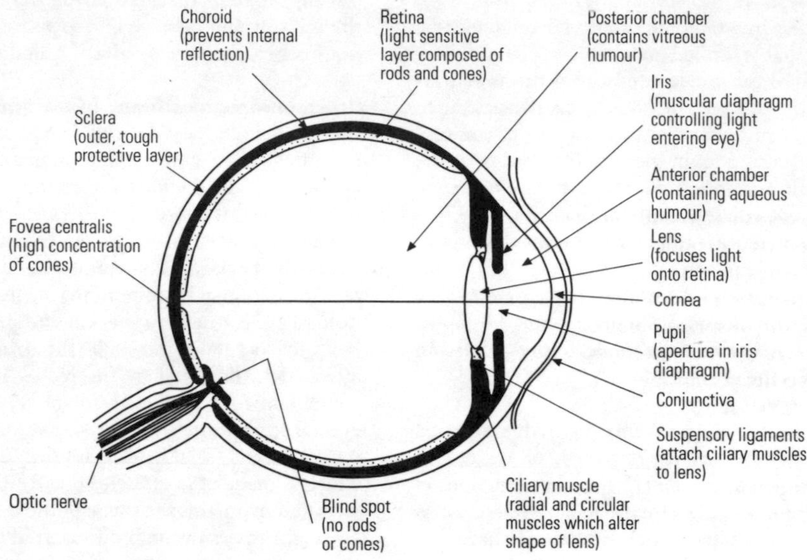

The human eye.

Most invertebrates have simple eyes with no lens. Insects have compound eyes made of many OMMATIDIA.

eyepiece In any optical instrument, such as a microscope or telescope, the lens placed closest to the eye. Generally, this acts as a magnifying glass, producing an enlarged image of the image formed in the instrument.

F

F_1 generation, *first filial generation* In genetics, the first generation of a cross between pure breeding parents.

F_2 generation, *second filial generation* In genetics, the offspring obtained by crossing members of the F_1 GENERATION.

face-centred cubic *See* CUBIC CLOSE PACKED.

facsimile *See* FAX.

F-actin *See* ACTIN.

factor In mathematics, a number or algebraic expression by which some other number or algebraic expression can be divided, leaving no remainder. For example, 10 is a factor of 60.

factorial A function of a positive INTEGER found by multiplying together all the integers up to and including the given integer. The factorial is denoted by the symbol !, for example:

$$4! = 1 \times 2 \times 3 \times 4$$

0! is given the value 1.

factorize (*vb.*) To find the FACTORS of a number or an expression. For example the expression $x^2 - x - 6$ can be factorized to $(x - 3)(x + 2)$.

facultative parasite A PARASITE that is able to survive without its host.

FAD (flavine adenine dinucleotide) A derivative of the vitamin riboflavin (*see* VITAMIN B) that acts as an electron carrier in the ELECTRON TRANSPORT SYSTEM. It is also a PROSTHETIC GROUP in some enzymes.

faeces The remains of food and other waste material from the DIGESTIVE SYSTEM of animals. Faeces consist mostly of residual material from digestive juices, some undigested ROUGHAGE, cells from the intestinal walls, bacteria and water. They are passed to the outside via the ANUS.

Fahrenheit A TEMPERATURE SCALE obsolete in science but still in everyday use in the UK and US. The fixed points are taken as the freezing point of water, 32°F, and the boiling point of water, 212°F, both at a pressure of one ATMOSPHERE.

Fajan and Soddy's Group Displacement Law An EMPIRICAL law summarizing the changes in ATOMIC NUMBER that take place during ALPHA DECAY and BETA DECAY. In an alpha decay, the new element produced is two places to the left of the decaying element in the PERIODIC TABLE. In a beta decay, the element produced is one place to the right of the original element.

Fajan's rules A set of three rules indicating the circumstances in which an IONIC compound may show significant COVALENT characteristics. Bonding is likely to become covalent if: (i) the charge on the ions is high, tending to allow electrons to be transferred from the negative ion to the positive ion; (ii) the positive ion is small or the negative ion large, again favouring electron transfer; and (iii) the positive ion has an electron configuration that is not that of a noble gas, and so is not particularly stable.

Fallopian tube, *oviduct* In female mammals, one of two tubes leading from the OVARY to the UTERUS, down which the ova (*see* OVUM) travel, aided by the muscular movements of the tube and CILIA lining it. The ovum is often fertilized in the Fallopian tube and sometimes implants there instead of the uterus, causing an ECTOPIC PREGNANCY. *See also* FERTILIZATION.

family One of the subdivisions of ORDERS in the CLASSIFICATION of organisms. Family names end in '-idae' for animals and '-aceae' for plants and fungi. Families are groups of related genera (*see* GENUS).

farad (F) The SI UNIT of CAPACITANCE. One farad is the capacitance of a CAPACITOR that stores one COULOMB of charge for each VOLT applied. In practice this is a very large capacitance, and most capacitors have a capacitance in the order of microfarads (μF).

faraday A unit of CHARGE particularly used in the study of ELECTROLYSIS. One faraday is the charge on one MOLE of electrons, i.e. about 96,487 COULOMB.

Faraday screen A metal enclosure used to protect sensitive electrical circuits from the influence of outside electric fields. INDUCED CHARGES within the screen arrange themselves to cancel out the electric field produced by any external charges.

Faraday's constant (F) The amount of CHARGE in one FARADAY: 96,487 COULOMBS.

Faraday's law of electromagnetic induction The induced ELECTROMAGNETIC FORCE in a circuit is equal to the rate of change of MAGNETIC FLUX LINKAGE in that circuit. If the rate of change of flux linkage through a circuit is $d\phi_n/dt$, then the induced e.m.f. in the circuit will be

$$E = -d\phi_n/dt$$

Faraday's laws of electrolysis Two laws governing the mass of substance dissolved or liberated from a solution in an ELECTROLYSIS. (i) The mass of the substance is proportional to the charge flowing. (ii) This mass is equal to the charge in FARADAYS multiplied by the number of electron charges (positive or negative) carried by the ion concerned multiplied by the relative atomic mass of the ion.

fast breeder reactor A type of NUCLEAR REACTOR in which a blanket of uranium–238 is placed around the reactor CORE. Uranium–238 absorbs NEUTRONS that would otherwise escape and after two BETA DECAYS it becomes plutonium–239, a FISSILE material. The plutonium may be extracted by chemical processing and then used to refuel the reactor.

fat A LIPID mixture consisting of GLYCEROL and FATTY ACIDS, mostly triglycerides (*see* LIPID), which are solid at room temperature – unlike OILS, which are liquid fats. In many animals, fats are stored to provide an energy reserve and also to give some protection and insulation to the animal and its internal organs. Fats are an essential constituent of an animal diet, but too much fat has been linked to heart disease in humans.

fatigue The failure or weakening of a material under STRAINS that are applied and removed repeatedly. Even though such strains may not reach the ELASTIC LIMIT of the material, small cracks will form and grow in some materials under these circumstances.

fatty acid An organic compound (a CARBOXYLIC ACID) made with a straight HYDROCARBON chain of up to 24 carbon atoms with a CARBOXYL GROUP (–COOH) at one end. Carbon atoms can be joined by double or single bonds. Where there is a DOUBLE BOND, only one hydrogen instead of two is carried by the carbon. If a fatty acid chain has a double bond, it is said to be unsaturated (polyunsaturates), for example, oleic and linoleic acids. Saturated fatty acids, for example, palmitic and stearic acids, have a single bond between the carbon atoms. The carbon atoms therefore carry all the hydrogen atoms possible. The more double bonds the fatty acid chains contain, the lower the melting point of the fat, for example, oil has many double bonds and lard has none. Polyunsaturates (such as in margarine) are thought to be less likely to contribute to cardiovascular disease than saturated fats (butter has both). Fatty acids usually combine with GLYCEROL to form LIPIDS. *See also* PROSTAGLANDIN, SOAP.

fatty tissue *See* ADIPOSE TISSUE.

fault A sudden discontinuity between layers of rock. Major faults are formed where TECTONIC PLATES meet and relative movement of rocks on each side of the fault leads to EARTHQUAKES.

fax (abbreviation for ***facsimile***) A TELECOMMUNICATIONS system for the transmission of written or printed text. Documents are scanned and converted into an electric signal, which is sent via the telephone network to a receiving machine where an exact replica of the document is created.

f-block element A LANTHANIDE or ACTINIDE; that is, an element with outer electrons in an F-ORBITAL.

feedback Taking the output of an electronic system (usually an AMPLIFIER) and feeding part of that output back to the input. *See* NEGATIVE FEEDBACK, POSITIVE FEEDBACK.

feedstock Any compound used as a starting point in an industrial chemical process.

Fehling's test A test used on organic substances to determine which are REDUCING AGENTS. It is usually used to detect REDUCING SUGARS and ALDEHYDES. The test involves heating the sample with a fresh solution of copper(II) sulphate, sodium hydroxide and sodium potassium tartrate. The presence of a reducing sugar is indicated by the production of a red precipitate. *See also* BENEDICT'S TEST.

Fermat's principle, *principle of least time* Any ray of light will take the path through an optical system that takes the least time. Also, all the rays leaving one point and arriving at another (in forming an image, for example) will have taken equal times to pass through the system.

The laws of REFLECTION and REFRACTION can be derived from this principle, which is also used to calculate the path of rays through complex optical systems.

fermentation The process by which sugars are broken down by bacteria or yeasts, in the absence of oxygen. This process is considered to be analogous to ANAEROBIC RESPIRATION. In nature the role of fermentation is to remove the hydrogen ions formed at the end of GLYCOLYSIS to allow the process to continue, since in the absence of oxygen the hydrogen ions cannot be used further to yield energy. The products of fermentation can be alcohol and carbon dioxide (alcoholic fermentation) or LACTIC ACID (lactate fermentation).

The process of fermentation has been utilized by humans in the baking and brewing industries, in cheese and yoghurt manufacture and also in the production of some antibiotics. In brewing, the alcohol ethanol is the important product, and in baking the carbon dioxide is more important, the bubbles causing bread to rise.

In animals, lactate fermentation is more common since lactic acid can be used to release energy if oxygen becomes available again, which allows them to withstand short periods without oxygen. A common situation where lactate fermentation occurs is in the muscles during strenuous exercise, where the build up of lactic acid causes cramps and prevents the muscle operating (*see* OXYGEN DEBT).

fermi (fm) A unit of length, used in nuclear physics. One fermi is equal to 10^{-15} m.

Fermi distribution A mathematical description of the energies of FERMIONS, such as electrons in metals or in certain stars. Although the forces between electrons are often small enough for them to be thought of as forming a gas, the effects of QUANTUM MECHANICS, particularly the PAULI EXCLUSION PRINCIPLE, mean that many of the particles have higher energies than would otherwise be the case. Unless the temperature is very high, it is a good approximation to say that all the ENERGY LEVELS from zero up to a fixed maximum, called the FERMI ENERGY, are filled whilst the rest are empty. This distribution of energies is often very little affected by temperature, as the energy of any thermal vibrations is small compared to the Fermi energy, thus only a few particles very close to the Fermi energy can be promoted into higher energy levels. In stars, the Fermi energy increases as a star becomes smaller and this can provide a pressure, called the ELECTRON DEGENERACY PRESSURE, which prevents the star from collapsing completely under its own gravitation.

Fermi energy The energy up to which ENERGY LEVELS are filled in the FERMI DISTRIBUTION. The Fermi energy is the energy at which the probability of an energy level being occupied is $^1/_2$.

fermion Any particle, such as an electron, with half-integer SPIN; that is spin that is an odd number times $h/4\pi$, where h is PLANCK'S CONSTANT. *See* SPIN. *See also* FERMI DISTRIBUTION, PAULI EXCLUSION PRINCIPLE.

fermium (Fm) The element with ATOMIC NUMBER 100. It does not occur naturally, but has been synthesized in PARTICLE ACCELERATOR experiments. The longest lived ISOTOPE (fermium–257) has a HALF-LIFE of only 10 days – too short for the element to be commercially useful.

fern A plant of the PHYLUM FILICINOPHYTA.

ferricyanide The old name for HEXACYANOFERRATE(III).

ferrite Any of a number of mixed oxides of the form $MO.Fe_2O_3$, where M is typically a FERROMAGNETIC element such as iron, cobalt or nickel. Ferrites have ceramic structures and are electrical insulators, but they are also ferromagnetic. This means they can be used as cores in SOLENOIDS at high frequencies without causing EDDY CURRENT losses.

ferrocyanide The old name for HEXACYANOFERRATE(II).

ferroelectric (*adj.*) Describing a DIELECTRIC that retains its electrical POLARIZATION when the polarizing electric field is removed. Many ferroelectric materials are also PIEZOELECTRIC.

ferromagnetic (*adj.*) Describing a strongly magnetic material, such as iron. In such materials the imbalance in ANGULAR MOMENTUM of the electrons in each atom causes the atom to behave like a tiny magnet. Neighbouring atoms line up to form DOMAINS, regions where all the atoms are pointing in the same direction. *See also* CURIE POINT, HARD, PERMANENT MAGNET.

fertilization The fusion of two GAMETES in sexual reproduction to form a single cell (or ZYGOTE) combining the genetic material of both gametes.

Some organisms are DIOECIOUS, the OVUM (egg) coming from the female and SPERM coming from the male individual. In mammals, the ovum is released from the OVARY and begins to move down the FALLOPIAN TUBE and the sperm swim towards it through the UTERUS. A small proportion of the sperm released by the male reaches the tubes and only one fertilizes the egg. This is achieved by the ACROSOME REACTION, in which the outer membrane of the egg is softened so that the sperm can penetrate it. The outer membrane of the egg then thickens to form a fertilization membrane, which prevents entry of a second sperm. The head and middle piece of the sperm enter the ovum (the tail separates), the two nuclei fuse and cell division begins immediately. If the ovum is not fertilized within 24 hours it dies and is shed in the menstrual flow (*see* MENSTRUAL CYCLE). In some species the female stores the sperm as a thick MUCUS plug and fertilization is delayed until a time when survival of offspring is optimum.

Some organisms are MONOECIOUS (or HERMAPHRODITE), where both sex organs are carried by one individual. In this case either self-fertilization (the gametes come from the same individual) or cross-fertilization (the gametes come from two individuals) can occur. Self-fertilization occurs in some plants but rarely in animals. Fertilization in some plant species is preceded by POLLINATION.

External fertilization is common in aquatic vertebrates, where both sexes release their gametes into the water.

See also DOUBLE FERTILIZATION, IMPLANTATION, PARTHENOGENESIS, SEXUAL REPRODUCTION.

fetch-execute cycle The cycle used by a computer's CENTRAL PROCESSING UNIT to process the instructions in a program. In the first step of the cycle, the next instruction is retrieved from the IMMEDIATE ACCESS MEMORY and placed in the instruction register. The instruction is then decoded and executed, and the next instruction retrieved.

fibre In biology, a thin elongated cell, such as nerve or muscle fibres. Fibre also refers to the structure of molecules such as COLLAGEN and ELASTIN. *See also* ROUGHAGE, SCLERENCHYMA.

fibreglass *See* GLASS-FIBRE REINFORCED PLASTIC.

fibre optic, *optical fibre* A solid glass or plastic fibre, typically thinner than the thickness of a human hair, that can transmit light. The light does not leave the fibre when it hits the edge, but passes along the fibre by the process of TOTAL INTERNAL REFLECTION. Light passing along such a fibre can be modulated to convey information (*see* MODULATION).

Many TELECOMMUNICATIONS systems use optical fibres in place of conventional electrical cables, particularly along trunk routes linking one major city with another. The relatively high frequency of light means that a higher BANDWIDTH is available than with an electrical signal. Higher rates of data transmission are therefore possible in fibres, which also take up less space and are cheaper than electric cables. Bundles of fibre optics can be used to convey an image from one place to another along a flexible cable, such as in an ENDOSCOPE.

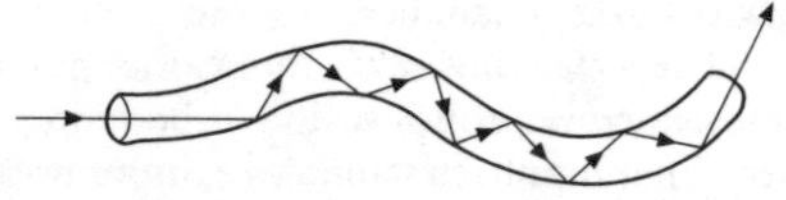

Fibre optic.

fibrin An insoluble blood protein, formed following injury, that prevents excessive bleeding. It is formed as part of the BLOOD CLOTTING CASCADE from the soluble plasma protein FIBRINOGEN by the action of the enzyme THROMBIN. Fibrin forms a meshwork of protein fibres and blood cells over the wound that dry to form a scab under which the wound can be repaired and bacteria cannot enter.

fibrinogen A soluble PLASMA PROTEIN made by the vertebrate liver, that is involved in the BLOOD CLOTTING CASCADE. *See also* FIBRIN.

fibroblast A spindle-shaped cell characteristic of CONNECTIVE TISSUE. It produces fibres of the proteins COLLAGEN and ELASTIN to provide strength and elasticity.

fibrous protein *See* PROTEIN.

fibrous root A type of root system in plants formed from ADVENTITIOUS ROOTS that consist of many branching roots (no main root) growing from the base of a stem. MONOCOTYLEDONS have this type of root.

field A region of influence around a mass, charge or current (including internal currents in magnetic atoms), where another mass, charge or current will experience a force. *See*

ELECTRIC FIELD, GRAVITATIONAL FIELD, MAGNETIC FIELD.

field coil In a motor or other electromagnetic machine, coils that provide a magnetic field.

field effect transistor (FET) An electronic device similar in its uses to a JUNCTION TRANSISTOR, but constructed in a different way. A field effect transistor has a CHANNEL of semiconducting material (*see* SEMICONDUCTOR) along which a current flows between terminals called the SOURCE and the DRAIN. A third ELECTRODE, called the GATE, is formed either on an insulating layer (in an IGFET – insulated gate FET) or from a REVERSE BIASED PN JUNCTION DIODE (JFET – junction FET). As the voltage on the gate is increased (made more negative in the case of an n-channel FET), its electric field narrows the region in which the charge carriers can move along the channel, thus reducing the current in the channel.

The chief advantage of the field effect transistor over the junction transistor is its very high input resistance (the input voltage divided by the input current). At high powers, FET's are also less prone to the problems of THERMAL RUNAWAY.

filament 1. In flowering plants, the stalk of the STAMEN.

2. In fungi, the thread-like structures that form HYPHAE.

3. A general term used to describe long thread-like structures or molecules, for example, ACTIN and MYOSIN.

Filicinophyta A PHYLUM of the plant KINGDOM consisting of the ferns. Fern leaves are called 'fronds', and ferns can be various shapes and sizes. More than 7,000 species exist. There are many small ferns that provide ground cover in moist areas, but there are some very tall ferns, such as tropical tree ferns.

The life cycle of ferns shows ALTERNATION OF GENERATIONS. For most of its life, a fern is a short stem (RHIZOME) with roots and leaves growing from it. This is the SPOROPHYTE generation. For the short-lived GAMETOPHYTE generation, a fern is a small heart-shaped plant. SPORES are carried on the underside of the leaves in sacs that split to release their contents. The spores develop into the prothallus, and during this gametophyte stage GAMETES fuse to produce a fertilized egg that develops into a new frond and root. Ferns are VASCULAR PLANTS.

file In computing, a collection of related data or a program stored together in computer memory under a single name.

film badge A device that records the total level of IONIZING RADIATION to which it has been exposed. It contains a piece of photographic film covered by a number of 'filters' of different materials and thicknesses. Ionizing radiation has a similar effect on a photographic film to light. When processed, the badge gives information about the amount and type of radiation to which it has been exposed. Film badges are often worn by those who work with ionizing radiation to monitor the level of radiation they are receiving.

filter 1. A fine POROUS material through which a liquid can pass, but solid particles cannot. Filters are widely used to remove solid particles from liquids.

2. *light filter* A piece of coloured glass or plastic that is transparent to certain wavelengths of light but absorbs others. Thus a red filter, for example, allows long–wavelength (red) visible light to pass, but absorbs other wavelengths.

filter funnel A cone-shaped funnel used to hold a piece of FILTER PAPER.

filter paper POROUS paper used as a FILTER in chemistry.

filtration The act of passing a material through a FILTER.

fin In fish, a projection from the body that provides stability and locomotion. The term usually refers to fins of bony and cartilaginous fish (*see* FISH). There are several types of fin: pectoral and pelvic (for steering and breaking), dorsal, anal and caudal (for rolling). The caudal, or tail fin, is also used for propulsion.

finite (*adj.*) Describing a quantity or number that is neither large without limit, or small without limit. *Compare* INFINITE, INFINITESIMAL.

finite element analysis An engineering technique for predicting how a complex structure will behave under the influence of loads or other forces. The object is modelled on a computer, broken down into a finite number of more simple shapes which interact with one another in simple ways. The computer than calculates the effects which these interacting elements will have on one another. The success of the method depends on having sufficient computing power to make the elements small enough to enable a realistic model of the system under test to be constructed. *See also* COMPUTER-AIDED DESIGN.

first filial generation *See* F_1 GENERATION.

first law of thermodynamics When heat is supplied to an isolated system, the amount of heat energy, ΔQ, equals the increase in INTERNAL ENERGY, ΔU, plus the mechanical WORK, ΔW, done by the system:

$$\Delta Q = \Delta U + \Delta W$$

For a gas at constant pressure p, the work is done by a change in the volume of the gas, ΔV, so

$$\Delta Q = \Delta U + p\Delta V$$

This statement is equivalent to the LAW OF CONSERVATION OF ENERGY.

fish A general term for an aquatic vertebrate of fresh or sea water that obtains oxygen through GILLS and uses FINS as a means of locomotion. There are three groups: bony fish (OSTEICHTHYES), for example cod and tuna; cartilaginous fish (CHONDRICHTHYES), for example sharks and rays; jawless fish (AGNATHA), for example lampreys and hagfish.

Bony fish constitute the majority of fish (20,000 species), and have a skeleton of bone and a body covered in scales. They have a number of mobile fins that control their movements (dorsal, pectoral, pelvic and caudal). In many bony fish their buoyancy is adjusted by a swim bladder. Most bony fish lay eggs; some retain the eggs inside their body and give birth to live young.

Cartilaginous fish have a CARTILAGE skeleton, a large sensitive nose and a series of open gill slits along the neck region. There are about 600 species. Jawless fish have a NOTOCHORD instead of a backbone and resemble primitive vertebrates before the true fishes with jaws evolved.

fissile (*adj.*) Describing an ISOTOPE, such as uranium–235 or plutonium–239, that exhibits INDUCED FISSION.

fission power The use of NUCLEAR FISSION to produce energy on a commercial scale. *See* NUCLEAR REACTOR.

fixed oil A natural LIPID found in animals and plants. *See also* OIL.

fixed point A temperature that can easily be maintained, so it can be used as the basis of a TEMPERATURE SCALE. Fixed points usually involve two or more phases of a particular material being in THERMAL EQUILIBRIUM, often at a specified pressure.

flagellate A member of the PHYLUM ZOOMASTIGINA. Euglenoid flagellates are members of the phylum EUGLENOPHYTA.

flagella (*sing.* ***flagellum***) A long thread-like structure used for locomotion of unicellular organisms or individual cells, such as the SPERM of multicellular organisms. Flagella are very similar to CILIA but are longer (100 μm), occur in fewer numbers and their action of movement is different. Unlike cilia, flagella can function independently, usually moving in one plane or occasionally in a spiral motion. Often a single flagellum is found at the rear of a cell that is propelled forwards by it, although sometimes (such as in bacteria) flagella are found in groups called tufts. The external structure of flagella in eukaryotic cells (*see* EUKARYOTE) is as found in cilia – two central fibres and nine peripheral pairs – but this arrangement is not found in prokaryotic cells (*see* PROKARYOTE).

flame A luminous region of hot gas around a COMBUSTION PROCESS. The material in the flame is IONIZED and the colour of the flame is characteristic of the ions present.

flame test A simple test for certain CATIONS. The material to be tested is dipped in concentrated hydrochloric acid, to form small quantities of the ionic chloride, which is usually the most VOLATILE salt. This is then vaporized in a BUNSEN BURNER flame. The electron arrangements of certain metals cause certain colours to be produced in the flame when they are ionized. Thus sodium produces an orange flame, whilst potassium is purple and copper blue-green.

flame-front The edge of a FLAME. In the case where the substances reacting to produce the flame are both gases, the flame front marks the boundary of the region where the reaction is taking place. If the flame front reaches the speed of sound in the gas, the COMBUSTION process becomes an EXPLOSION.

flammable (*adj.*) Able to burn.

flash memory In computing, a type of read-only memory (*see* ROM) that can be erased and reused without removal from the computer.

flash point For a VOLATILE flammable liquid, the temperature above which the VAPOUR PRESSURE of the liquid is high enough for the vapour to burn in air if it is ignited by a spark.

flask A round or cone-shape vessel with a narrow neck, used to hold liquid REAGENTS.

flatworm A member of the PHYLUM PLATYHELMINTHES.

flavine adenine dinucleotide *See* FAD.

flavour In particle physics, the property that distinguishes one type of QUARK from another. The six flavours of quark, which make up the STANDARD MODEL of particle physics, are UP, DOWN, STRANGE, CHARM, TOP and BOTTOM.

Fleming's left-hand rule A rule to find the direction of the force on a current in a magnetic field. If the fingers of the left hand are held at right angles to one another with the first finger pointing in the direction of the magnetic field and the second finger pointing in the direction of the CONVENTIONAL CURRENT, the thumb will point in the direction of the force.

flint Hard, brittle mineral found in chalk and limestone. It consists of fine-grained quartz.

flip flop *See* CLOCKED BISTABLE.

flocculent (*adj.*) Describing a PRECIPITATE with the appearance of wool floating in a solution.

floodplain Area of land in a river valley that is periodically flooded. The river sediment deposited during times of flooding make floodplains extremely fertile.

floppy disk In computing, a medium-capacity magnetic storage device, designed to be easily removed from a computer, to transfer data from one machine to another, or to allow a large amount of data to be stored using several disks.

floret A small flower, usually making up part of a composite flower head.

flotation The tendency of an object to float to the surface of a fluid in which it is immersed. This is caused by a force called UPTHRUST, which is due to the pressure of the lower parts of an immersed object being greater than those on the upper parts. The upthrust is equal to the weight of the fluid displaced (pushed out of the way) by the immersed object. If the average density of the immersed object is greater than the density of the fluid, the upthrust is greater than the weight, and the object will float.

Ships float even if they are made of steel, which is much denser than water. This is because the part of the ship below the water – much of which is air inside the ship – weighs the same as the water it displaces. When a ship is fully loaded, it will float lower in the water because of the need to displace a greater amount of water to balance the greater weight of the ship.

The fact that an object appears lighter when immersed in a fluid is also used as a way of measuring density. The difference between the weight an object appears to have in air and when immersed in water is equal to the weight of water displaced by the object. From this, its volume, and thus its density can be found (*see* ARCHIMEDES' PRINCIPLE).

flow chart A diagrammatic representation of an ALGORITHM, especially as used in computer programming. Steps in the algorithm that require a decision to be made on the basis of the validity of some logical statement are usually drawn in diamond-shaped boxes, whilst other steps are indicated by rectangular boxes.

flower The reproductive structure in flowering plants (ANGIOSPERMS). A flower consists of four sets of modified leaves, SEPALS, PETALS, STAMENS and CARPELS, attached to a RECEPTACLE, which is

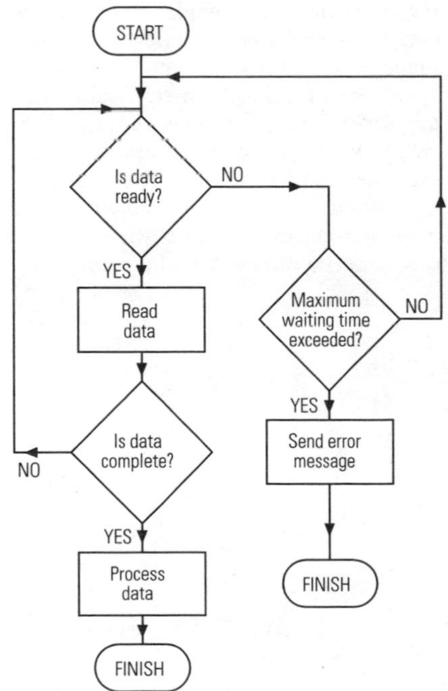

An example of a simple flow chart.

the modified end of the stem (PEDUNCLE). Flowers differ in their colour, size, number and arrangements of their parts depending on the method of POLLINATION. Insect-pollinated flowers usually have brightly coloured petals, whereas wind-pollinated flowers may have small or no petals and long STIGMAS to collect POLLEN. Flowers can contain male and female organs together (*see* HERMAPHRODITE), or the male and female organs occur in separate flowers (*see* MONOECIOUS) or on separate plants (*see* DIOECIOUS). The sepals and petals together form the PERIANTH, which protects the reproductive organs and attracts pollinators.

The structure of a flower can be represented symbolically in a floral formula using letters for parts of the flower, for example *K* to represent the CALYX and *P* perianth, and numbers to indicate how many of a part are present. A floral diagram can be used with this formula to show the arrangement of the parts. A flower with RADIAL SYMMETRY is called actinomorphic, for example the buttercup (the petals and sepals are of similar size), and a flower with BILATERAL SYMMETRY is called zygomorphic, for example the white dead nettle (unequal sepals and petals of different shapes).

A floret is a small flower, usually making up part of a composite flower head, and there are often two types of florets in one flower, for example the daisy has yellow and white florets.

Although the flower is thought of as the main reproductive structure, leaves, stems and roots can also carry out this function.

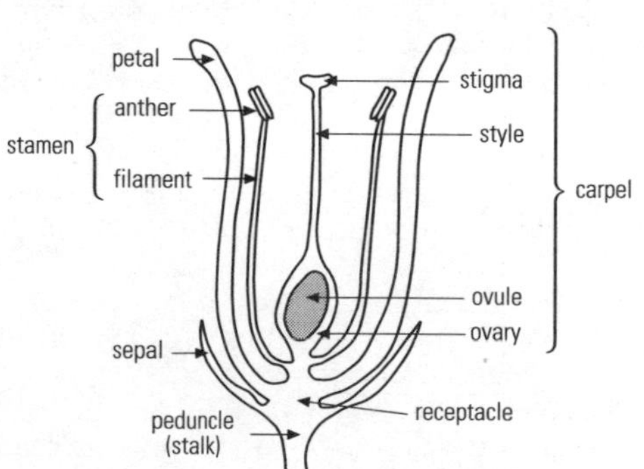

Cross-section of a generalized flower.

See also DOUBLE FERTILIZATION, INFLORESCENCE.

flue A chimney or other opening through which the products of combustion, particularly from burning gases in air, can escape into the atmosphere.

fluid Any substance that can flow: a LIQUID or a GAS.

fluidization The process of blowing a gas, often air, through a powdered solid, such as coal, so it behaves like a fluid. Fluidized beds are layers of fluidized material, such as solid catalysts for some gas phase reactions and coal burnt in large furnaces. Fluidized coal may be delivered to the furnace along pipes from a nearby crushing mill.

fluid mechanics The science of FLUIDS, their motion and interaction with solid matter. The BERNOULLI EFFECT and the onset of TURBULENCE are both of critical importance in studying fluid flow. *See also* BERNOULLI'S THEOREM.

fluid mosaic model In biology, the currently accepted structure of a CELL MEMBRANE. A cell membrane has two main components, PROTEIN and PHOSPHOLIPID. A phospholipid molecule is composed of a HYDROPHOBIC (water-repelling) tail end and a HYDROPHILIC (water-loving) head end. In the fluid mosaic model the phospholipids form a bimolecular layer in which the hydrophobic tail ends associate together at the centre of the membrane and the hydrophilic heads extend towards the surface. The proteins are arranged in a mosaic fashion dotted throughout the membrane: some on the surface, some extending into the phospholipid layer and some extending across it. Since the phospholipid layer seems to be capable of movement the term 'fluid' was added to the model. In this model it is suggested that the proteins provide structural support and also give specificity to the cell (for recognition by antibodies, hormones, etc.). The proteins also assist ACTIVE TRANSPORT across the cell membrane.

fluorescence An effect in which ultraviolet light is absorbed and then re-emitted immediately as visible light. This is used in whitening agents that are added to paper and some washing powders to produce a brighter appearance. *Compare* PHOSPHORESCENCE, where the effect takes place over a longer period of time.

fluorescence microscopy A modification of the use of a light MICROSCOPE in which the tissue to

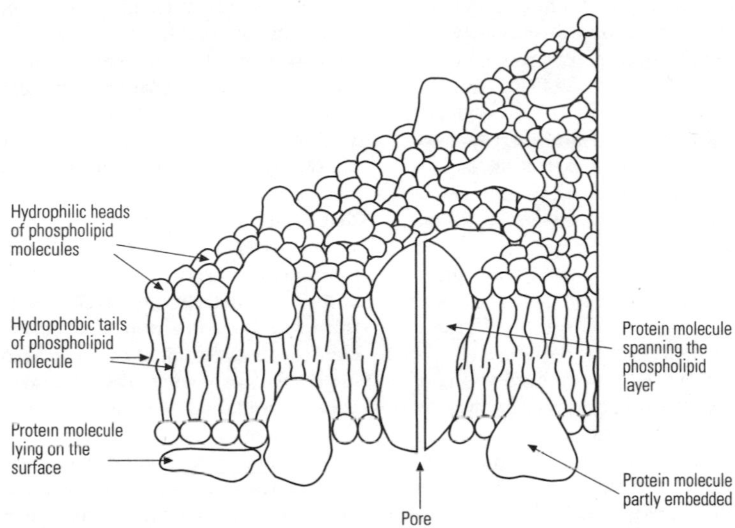

The fluid mosaic model of the cell membrane.

be viewed is stained with a fluorescent dye that is used as the light source. This can be added to a marker such as an antibody to highlight specific cells or proteins.

fluoride Any BINARY COMPOUND containing fluorine. All elements except nitrogen and the NOBLE GASES readily form fluorides, due to the highly reactive nature of fluorine. Non-metallic fluorides are covalently bonded (*see* COVALENT BOND), but highly POLAR, due to the extreme ELECTRONEGATIVITY of fluorine. The metallic fluorides contain the F^- ion. In many countries, small quantities of the fluoride ion are added to drinking water to combat tooth decay.

fluorine (F) The element with ATOMIC NUMBER 9; RELATIVE ATOMIC MASS 19.0; melting point –220°C; boiling point –188°C. Fluorine can be extracted from its main ore, fluorite, by electrolysis of the molten ore, which contains mostly calcium fluoride. The lightest and most reactive of the HALOGENS and the most ELECTRONEGATIVE (*see* ELECTRONEGATIVITY) of all elements, fluorine is too reactive to find much use as an element. However, fluorine compounds are widely used, for example PTFE (polytetrafluoroethylene) is an important PLASTIC.

fluorite A mineral form of calcium fluoride, CaF_2.

fluorocarbons Compounds containing carbon and fluorine, with the formula C_nF_{n+2}. They may be produced by replacing the hydrogen with fluorine in HYDROCARBONS. An example is fluoroethane, C_2F_6, formed by the fluorination of ethane:

$$C_2H_6 + 6F_2 \rightarrow C_2F_6 + 6HF$$

The stability of fluorocarbons makes them useful as fire-extinguishing gases and those with higher RELATIVE MOLECULAR MASS form good lubricants. *See also* CHLOROFLUOROCARBONS.

flyback The period during a TIMEBASE or RASTER when the electron beam is being returned to its starting position. *See also* CATHODE RAY OSCILLOSCOPE.

flywheel A heavy wheel used to store KINETIC ENERGY, between the POWER STROKES of a PETROL ENGINE for example.

FM *See* FREQUENCY MODULATION.

foam A COLLOID consisting of a gas suspended in a liquid.

focal length The distance between a lens or curved mirror and the plane in which parallel rays of light are brought to a FOCUS. In the case of a DIVERGING LENS or CONVEX mirror, the rays appear to emerge from a VIRTUAL FOCUS, and the focal length is often quoted as a negative number.

focal plane A plane associated with a lens or curved mirror in which rays of light that are parallel to one another before striking the lens or mirror are brought to a FOCUS.

focus 1. (*n.*) A point at which rays of light come together.

2. (*vb.*) To adjust the arrangement of an optical system so that rays of light come together at a chosen point – on the film in a camera for example.

foetus The name given to the unborn young of mammals after the first bone cells appear in the CARTILAGE. In humans this is at about 7 weeks of pregnancy, by which time the foetus is almost fully formed (internal organs formed, limbs, fingers, toes, eyelids). *See also* EMBRYO.

fog Water droplets in the air at ground level, reducing the visibility below 1,000 m. *See also* ADVECTION FOG, MIST, RADIATION FOG, WEATHER SYSTEMS.

fold A bend in the layers of rock.

folic acid *See* VITAMIN B.

follicle Any cluster of cells that protects and nourishes an enclosed cell or structure. For example a hair follicle surrounds the root of a hair. *See also* GRAAFIAN FOLLICLE.

follicle-stimulating hormone (FSH) A gonadotrophic GLYCOPROTEIN produced by the PITUITARY GLAND that in females stimulates development of the GRAAFIAN FOLLICLE within the OVARY. Mature ova, or egg cells, are released from the Graafian follicle under the further control of another hormone called LUTEINIZING HORMONE. In males FSH controls the production of SPERM by the TESTIS. FSH also stimulates OESTROGEN production. It is inhibited by PROGESTERONE.

food chain A sequence of organisms showing the order in which energy and nutrients pass through an ECOSYSTEM. Each organism in the chain feeds on, and is dependent on, the one preceding it, and is then itself eaten by the succeeding organism. There are usually three or four, and possibly up to six links, in a food chain.

The primary producers are the photosynthetic organisms (*see* PHOTOSYNTHESIS): green plants and some bacteria that are autotrophic (*see* AUTOTROPH) and manufacture sugars from raw materials using SOLAR ENERGY. All other organisms in the chain are consumers (HETEROTROPHS), which obtain their energy by consuming other organisms. Primary consumers are those that feed on the primary producers, for example HERBIVORES and some plant parasites. Secondary consumers are CARNIVORES that feed on herbivores, and tertiary consumers feed on other carnivores. Each of these levels in the food chain is called a trophic level and only a small proportion of energy passes from one level to the next because a lot of energy is lost during RESPIRATION. When producers and consumers die, organisms called DECOMPOSERS and DETRITIVORES break down the dead material, enabling nutrients to be recycled to the soil or atmosphere.

The food chain is an over simplification because many organisms have more varied diets than it suggests (*see* OMNIVORE). All the food chains in a COMMUNITY may interact (*see* FOOD WEB).

See also BIOMAGNIFICATION, PRODUCTIVITY, PYRAMID OF BIOMASS, PYRAMID OF ENERGY, PYRAMID OF NUMBERS.

food preservation Various methods used to prevent the spoilage of food by micro-organisms (such as bacteria, mould or yeast), oxidation of fats (making them rancid) or the action of enzymes. Food preservation can be achieved by the use of modern chemical additives, but traditional methods are still used. Methods include canning, which is the sealing of food after destroying micro-organisms and enzymes with high temperatures; pickling, which uses vinegar (ethanoic acid) to stop growth of mould; curing or salting (soaking in salt) of meats; preserving in sugar; drying or freeze-drying (the latter is less damaging to the food); heat treatment (*see* PASTEURIZATION); refrigeration (below 3°C) for cooked foods, which slows down spoilage; deep freezing (–18°C), which stops most spoilage. *See also* ADDITIVE.

food web The interaction of all of the FOOD CHAINS in a COMMUNITY. One organism may occur in several chains at different TROPHIC LEVELS, creating a complex feeding system.

forbidden band *See* BAND THEORY.

force Any agency that tends to change the state of rest or motion of a body; that is, one that tends to cause a body to accelerate (*see*

ACCELERATION). Force is defined as being proportional to the rate of change of MOMENTUM of a body. It is a VECTOR quantity, and the SI unit of force is the NEWTON.

For a body of mass *m*, travelling at velocity *v*, the momentum is *mv*. The force is given by $d(mv)/dt$. If mass remains constant, then the force is equal to $m dv/dt = ma$, where *a* is acceleration.

See also CONTACT FORCE, FOUR FORCES OF NATURE, NEWTON'S LAWS OF MOTION, STRESS, STRAIN.

forced oscillation The motion of a system that would exhibit SIMPLE HARMONIC MOTION were it displaced from its EQUILIBRIUM position and released, but which is driven by some external force (called the driving force), which is itself oscillating. The frequency at which the system would oscillate were it simply displaced and left to oscillate is called the NATURAL FREQUENCY of the system and the frequency at which the driving force oscillates is called the driving frequency.

If the driving frequency is large compared to the natural frequency, then the driven system will oscillate very little and almost exactly out of PHASE with the driving force. If the driving force is well below the natural frequency, the response will again be small, but now in phase with the driving force. When the driving frequency is close to the natural frequency, there will be a much greater response, reaching a maximum when the two frequencies are equal.

See also RESONANCE.

forebrain The largest of the three parts of the human BRAIN. The forebrain contains the CEREBRUM, which co-ordinates all of the body's senses and higher mental activities such as reasoning, learning and memory, and the THALAMUS, which is a relay centre for other regions of the brain and is associated with pain and pleasure. *See also* HINDBRAIN, MIDBRAIN.

foreskin *See* PENIS.

formaldehyde *See* METHANAL.

formalin A solution of METHANAL in water, used as a preservative for biological specimens.

formic acid See METHANOIC ACID.

formula 1. (*chemistry*) A representation of the composition of a compound using symbols for the elements present. Subscripts are used to indicate the number of atoms. The way in which these symbols are used vary according to how much detail is conveyed. Thus ethane may be shown as an EMPIRICAL FORMULA, CH_3, which represents simply the ratio of carbon atoms to hydrogen; as a MOLECULAR FORMULA, C_2H_6, which shows how many atoms of each element are present in a molecule; or as a STRUCTURAL FORMULA, such as CH_3CH_3, which gives some indication of how the atoms are arranged.

2. (*mathematics*) An algebraic equation that expresses the rule for finding one quantity in terms of some others. For example, the formula for the area of a circle is $A = \pi r^2$, where *A* is the area and *r* the radius.

Fortin barometer A version of the MERCURY BAROMETER equipped with an adjustment to allow the level of the open mercury surface to be set to a reference mark and a VERNIER scale for precise measurement of the height of the mercury column.

forward biased (*adj.*) Describing a junction between a P-TYPE SEMICONDUCTOR and a N-TYPE SEMICONDUCTOR to which a voltage is applied enabling CHARGE CARRIERS to carry a current across the junction – that is, with the p-type material positive and the n-type negative. *See also* PN JUNCTION DIODE, REVERSED BIASED.

fossil The remains of an animal or plant preserved in some way, usually in rocks but also in ice, peat and tar. Fossils are generally found in the layers of rock that form from the slow deposition of mud and silt; the oldest fossils are found in the lower layers. The scientific study of fossils is called PALAEONTOLOGY.

fossil fuels Petroleum oil, coal and natural gas. Petroleum and natural gas are produced in SEDIMENTARY rocks from the decay of marine life. Coal forms as a result of similar geological processes compressing decayed forests. There is increasing pressure to reduce the consumption of fossil fuels as they are NON-RENEWABLE. The burning of such fuels also releases carbon dioxide into the atmosphere, which is believed to be responsible for the GREENHOUSE EFFECT.

four forces of nature The four fundamental forces by which all matter is believed to interact. They are the gravitational force (*see* GRAVITY), the ELECTROMAGNETIC FORCE, the STRONG NUCLEAR FORCE and the WEAK NUCLEAR FORCE. Other forces, such as FRICTION are just large-scale effects of these four forces acting on a microscopic level.

At a fundamental level, all matter is believed to be made up of a limited number of ELEMENTARY PARTICLES, which cannot be divided into smaller objects. These particles interact with one another by the exchange of other particles called GAUGE BOSONS. Different elementary particles are each subject to the interactions carried by one or more of the gauge bosons, so 'feel' one or more of the four forces.

See also ELECTROWEAK FORCE, GRAND UNIFIED THEORY, STANDARD MODEL, THEORY OF EVERYTHING.

four-stroke cycle The sequence of events in most PETROL ENGINES, which requires the PISTON to make two movements in each direction to complete the cycle. *See* PETROL ENGINE. *See also* TWO-STROKE CYCLE.

fovea centralis The region of the RETINA of the eye with the greatest concentration of CONES. *See* RETINA.

fraction A part of a whole. Fractions are expressed in numerical terms as the QUOTIENT of one two numbers, e.g. $^3/_4$. The number at the top of the fraction is called the numerator and the number below is called the denominator. A proper fraction is one in which the numerator is less than the denominator. If the numerator is greater than the denominator, the fraction is described as improper. A decimal fraction is one where the denominator is some power of 10, e.g. 234/1000, usually written as 0.234.

fractional crystallization A process for separating two substances with different solubilities in a solvent. A solution containing the two substances is cooled until crystals form, and this process is repeated several times. At each stage, the proportion of the more soluble material in the liquid phase (called the MOTHER LIQUOR) will increase whilst the crystals contain an increasing proportion of the less soluble material.

fractional distillation A process of DISTILLATION used to separate a mixture of liquids having similar boiling points, by repeated boiling and cooling. The liquid mixture is heated to boiling and the vapours rise into a fractionating column and condense to form a liquid again. The most volatile liquid rises first but then on cooling falls down the column and is reheated when it meets the vapours from the second most volatile liquid. This process occurs repeatedly and eventually a gradient forms in the fractionating column with the most volatile fractions at the top and the least at the bottom. The required fractions can be tapped off the column at the appropriate position. Fractional distillation is an important part in the commercial processing of CRUDE OIL.

fractionating column *See* FRACTIONAL DISTILLATION.

fragmentation A method of ASEXUAL REPRODUCTION in some relatively undifferentiated organisms, such as algae, sponges and worms, that involves dividing into sections and separating to form new individuals. It is a form of regeneration in some organisms, where as a result of injury the separated fragments will regenerate parts.

frame of reference A set of directions in space and time together with a reference point called an origin. The position of any other point is described by how far away it is from the origin in each of the directions. The comparison between how events are seen by observers who use frames of reference in motion relative to one another is at the heart of the theory of RELATIVITY. *See also* INERTIAL REFERENCE FRAME.

frame shift mutation *See* MUTATION.

francium (Fr) The element with ATOMIC NUMBER 87; melting point 27°C; boiling point 677°C. Only RADIOACTIVE ISOTOPES are known, the longest lived of which (francium–223) has a HALF-LIFE of only 21 minutes. Minute amounts of francium occur in nature as a result of the decay of heavier elements.

Fraunhofer diffraction A form of DIFFRACTION in which the light rays that interfere are parallel on reaching and on leaving the APERTURE. They must then either have been FOCUSED by a lens or else the light source and the screen on which the diffraction is observed must be sufficiently far away for the rays to be effectively parallel. *Compare* FRESNEL DIFFRACTION.

free electron An electron in a METAL or SEMICONDUCTOR that is not bound to any single atom but is free to move, carrying its charge and energy through the material. The large numbers of free electrons in metals make them good conductors of electricity (charge flow) and heat (energy flow).

free energy A measure of the energy released or absorbed during a reversible process. The Gibbs free energy change, ΔG, in a reaction under constant temperature and pressure is defined as:

$$\Delta G = \Delta H - T\Delta S$$

where ΔH is the ENTHALPY change, ΔS is the change in ENTROPY and *T* is the ABSOLUTE TEMPERATURE. This quantity is useful since a chemical reaction will only proceed spontaneously if ΔG is negative. In other words, the heat released by the reaction, ΔH, must increase the entropy of the surroundings by a sufficient amount to compensate for any entropy decrease within the REAGENTS, or if the reaction is ENDOTHERMIC, the entropy of the reagents must increase sufficiently to compensate for the entropy decrease of the surroundings. If ΔG is positive, energy must be supplied to the reaction. In practice, reactions in which ΔG is small, either positive or negative, are neither impossible nor do they go to completion, but are regarded as REVERSIBLE REACTIONS.

The Helmholtz free energy change, ΔF, is defined by:

$$\Delta F = \Delta U - T\Delta S$$

where ΔU is the change in INTERNAL ENERGY. The Helmholtz free energy is a measure of the maximum work that may be done by a reversible process at constant temperature.

See also CHEMICAL POTENTIAL.

free expansion Any process by which a gas expands without doing any WORK against the surroundings, such as is the case where a gas expands freely into a vacuum. Most gases cool in these circumstances, as work is done against attractive INTERMOLECULAR FORCES. Thus cooling in a free expansion is a method by which gases can be cooled sufficiently to turn them into liquids.

At higher pressures and temperatures, the repulsive part of the intermolecular forces is more effective and gases become warmer in a free expansion. With hydrogen and helium this happens at room temperature and pressure. Before liquification, such gases must be pre-cooled to a temperature called the INVERSION TEMPERATURE, below which the gas cools in a free expansion.

free fall The motion of an object on which gravity is the only force acting. Because the gravitational force acting on an object increases with its mass, whilst the acceleration produced by that force decreases with mass, the acceleration produced on an object by a gravitational field does not depend on the mass of the falling object. On Earth, for example, all objects in free fall have an acceleration of roughly 9.8 ms^{-2}. A feather will fall more slowly than a coin, but this is due to air resistance – place them both in a vacuum, so they are genuinely falling freely, and they will accelerate together. This concept was first demonstrated in an experiment performed in 1604 by Gallileo (1564–1642), when he dropped two balls of different mass from the Leaning Tower of Pisa.

The motion of an object in free fall is not affected by any horizontal motion – a ball launched horizontally will hit the ground at the same time as one that is simply dropped, provided the ground is flat. If the horizontal speed is large enough, the curvature of the Earth will have to be taken into account, thus an object will fly further than expected before reaching the ground. If the horizontal speed is very large, it is possible for an object to be constantly falling toward the centre of the Earth without ever hitting it. In this case the object is in orbit, but it is also still in free fall.

An astronaut inside an orbiting spacecraft feels weightless for just the same reason as the occupant of a lift with a broken cable. In each case the container and its occupant are falling freely with the same acceleration, so there is no CONTACT FORCE between them. The only difference is that the lift will eventually hit the ground, whilst the spacecraft will remain in orbit until something is done to bring it back to Earth.

free radical An atom or a group of covalently bonded atoms (*see* COVALENT BOND), with an unpaired VALENCE ELECTRON. Free radicals are highly reactive and most are capable of only a limited period of independent existence, sometimes only microseconds, before they recombine with other atoms. They are produced by HOMOLYTIC FISSION, in which a covalent bond is broken by the action of heat or light.

freezing mixture A mixture of two or more components designed to produce a solution with a temperature below 0°C. The most common example is a mixture of common salt (sodium chloride), ice and water. The energy needed to dissolve the salt and that needed to melt the ice both come from the INTERNAL ENERGY of the molecules and the material cools. The ice continues to melt as a result of the EUTECTIC MIXTURE formed between the ice and salt, which is initially above its EUTECTIC POINT. The mixture will continue to cool until all the ice has melted, all the salt dissolved or else the eutectic temperature is reached.

frequency **1.** (*mathematics*) The number of times a particular event occurs or the number of times a variable is found to have a particular value or to lie within a particular range.

2. (*physics*) The number of oscillations completed in one second. The frequency of waves or of an oscillating system is measured in HERTZ (Hz).

frequency distribution *See* DISTRIBUTION.

frequency modulation (FM) A type of MODULATION system in which the frequency of a CARRIER WAVE is varied, to convey information such as a speech or music signal. Frequency modulation requires a wider BANDWIDTH than AMPLITUDE MODULATION, but is less prone to interference so is increasingly used for broadcast radio systems. *See also* FREQUENCY SHIFT KEYING. PHASE MODULATION, PULSE CODE MODULATION.

frequency shift keying (FSK) A BINARY form of FREQUENCY MODULATION, widely used in the transmission of DIGITAL data along radio links and telephone lines.

Fresnel diffraction A type of DIFFRACTION in which the waves that interfere to produce a diffraction pattern are not parallel when they come together. *Compare* FRAUNHOFER DIFFRACTION.

Fresnel's biprism A variation on YOUNG'S DOUBLE SLIT EXPERIMENT in which the double slit is replaced with a flat glass PRISM (called a biprism), which refracts light (*see* REFRACTION) to form two images of a single slit. Light from these two images then interferes (*see* INTERFERENCE).

friction The force produced when one object slides over another (DYNAMIC FRICTION) or which prevents one object sliding over another (STATIC FRICTION). Friction always acts to prevent or reduce the speed of the motion, and so tends to reduce the KINETIC ENERGY of the system, converting it into heat.

Friction can be explained in terms of the forces between the atoms of the two surfaces. Whilst these surfaces may appear smooth they are often quite rough on an atomic scale. Very smooth surfaces, such as a pair of very flat glass sheets, often produce surprisingly large frictional forces. This is contrary to the usual idea that friction is less for smooth surfaces, which follows from lumps on one rough surface becoming jammed in hollows in the other.

The details of friction are very difficult to model mathematically; a simple and surprisingly effective model for dynamic friction takes the frictional force as being proportional to the NORMAL REACTION, and as independent of the area of the sliding surfaces and of the speed at which they slide over one another.

See also COEFFICIENT OF FRICTION, LIMITING FRICTION.

Friedel-Crafts reaction In organic chemistry, a reaction used industrially in the manufacture of HYDROCARBONS and KETONES. The reaction involves the introduction of an ACYL GROUP or ALKYL GROUP to AROMATIC compounds by ELECTROPHILIC SUBSTITUTION. An example is the reaction between ACID CHLORIDES, RCOCl (where R is an alkyl group) and benzene which, in the presence of a LEWIS ACID (usually aluminium chloride, $AlCl_3$) produces ketones (phenylethanone in this example).

$$C_6H_5H + CH_3COCl \xrightarrow{AlCl_3} C_6H_5COCH_3 + HCl$$

This is Friedel-Crafts ACYLATION.

Friedel-Crafts ALKYLATION occurs with HALOGENOALKANES. For example the reaction between benzene and chloroethane in the presence of a Lewis acid produces ethylbenzene:

$$C_6H_5H + CH_3CH_2Cl \xrightarrow{AlCl_3} C_6H_5CH_2CH_3 + HCl$$

The Lewis acid is a catalyst and generates the ELECTROPHILE needed for the reaction, either an alkyl (R^+) or acyl (RCO^+) CATION.

front In meteorology, the boundary between two AIR MASSES. *See also* COLD FRONT, WARM FRONT.

front-silvered (*adj.*) Describing a mirror with the reflecting coating on the front of a glass support.

frost Ice formed as air becomes SATURATED when it cools at night, with the temperature of the air being below the freezing point of water. *See also* WEATHER SYSTEMS.

froth flotation A process for the separation of some ORES from waste soil by forming a froth of the mixture together with oil and water. Air is blown through the mixture and the less dense material is trapped in the bubbles of the froth whilst denser material sinks to the bottom.

fructose ($C_6H_{12}O_6$) A MONOSACCHARIDE that combines with GLUCOSE to form SUCROSE. Fructose contains a KETONE group and is therefore called a ketose sugar. It is a component of fruit and NECTAR, which plants use to attract animals to assist in seed dispersal.

fruit The ripened OVARY in flowering plants containing and protecting one or more SEEDS. Fruits are often edible, which aids DISPERSAL of seeds because the seeds pass through the guts of animals undigested. Simple fruits, for example the peach, are formed from one ovary, whereas multiple fruits, for example the blackberry, are formed from the ovaries of several flowers. A fruit consists of a wall or PERICARP that is divided into layers. Fruits may open to shed their seeds (DEHISCENT) or remain unopened (INDEHISCENT).

Fruits are classified as fleshy or dry. Examples of fleshy fruits are the BERRY or DRUPE. There are many examples of dry fruits, including LEGUMES, NUTS and ACHENES. A soft fruit, or PSEUDOCARP, can be formed from parts other than the ovary, for example the strawberry develops from the RECEPTACLE and its true fruits are the pips on the outer surface.

See also BERRY.

FSH *See* FOLLICLE-STIMULATING HORMONE.

fuel Any material that burns to provide a source of heat or energy.

fulcrum A fixed PIVOT about which a LEVER rotates.

fullerene *See* BUCKMINSTERFULLERENE.

fuming sulphuric acid *See* OLEUM.

function If a change in x produces a change in y, then x is said to be a function of y, expressed as

$$y = f(x)$$

In this case, y is called the dependent variable and x the independent variable. The independent variable, can take on any value within a specified range, called the domain. For example, the expression $y = x + 3$ is a function that gives a value which is always greater than the independent variable by 3.

functional group A small group of atoms linked together in an arrangement that determines the chemical properties of the group. In organic compounds the functional group is attached to a carbon skeleton but it is the functional group that determines the REACTIVITY of the compound. Examples of functional groups include the CARBOXYL GROUP COOH, the AMINE group NH_2 and the HYDROXYL GROUP OH (ALCOHOLS).

fundamental The lowest FREQUENCY of STANDING WAVE that can be supported by a system.

fundamental constant Any quantity believed to have the same value throughout all space and time and which does not depend on the value of any other such quantity. Examples of fundamental constants include the SPEED OF LIGHT in a vacuum and the charge and mass of an electron. *See also* BOLTZMANN CONSTANT, GRAVITATIONAL CONSTANT, PERMEABILITY, PERMITTIVITY, PLANCK'S CONSTANT, RYDBERG CONSTANT.

fundamental particle *See* ELEMENTARY PARTICLE.

fundamental unit *See* BASE UNIT.

Fungi A large KINGDOM of eukaryotic organisms that have no CHLOROPHYLL, and many reproduce by SPORES. The kingdom Fungi includes the phyla (*see* PHYLUM) ZYGOMYCOTA, ASCOMYCOTA and BASIDIOMYCOTA. Fungi (*sing.* fungus) were once classified with plants, but they have no leaves or roots and have the POLYSACCHARIDE CHITIN in their cell walls, which is never found in plants.

Fungi usually have no distinct cell boundaries and the main body of a fungus is called a MYCELIUM, which consists of a mass of thread-like HYPHAE. The spores by which a fungus reproduces are contained within a structure called a SPORANGIUM that grows at the ends of the hyphae. ASEXUAL REPRODUCTION or SEXUAL REPRODUCTION can take place. The lack of chlorophyll means fungi cannot photosynthesize (*see* PHOTOSYNTHESIS) and so obtain their nutrients as SAPROTROPHS or PARASITES.

Fungi are important to humans, for example in the production of ANTIBIOTICS (such as PENICILLIN from the mould *Pencillium*), in the production of alcohol and bread (YEAST), and in the decomposition of sewage and soil materials. However, they are also a cause of economic loss, causing decay of stored foods (for example moulds on bread and fruit) and deterioration of materials such as leather and wood.

Fungi also cause a number of plant (and to a lesser extent animal) diseases, for example, Dutch elm disease and powdery mildew, which attacks cereal crops.

See also MYCORRHIZA.

fungicide A PESTICIDE that kills unwanted fungi.

fungus (*pl. **fungi***) Any member of the KINGDOM FUNGI.

funicle The stalk that attaches the OVULE to the ovary wall in the CARPEL of a flowering plant. *See also* DOUBLE FERTILIZATION.

fuse A safety device that uses the heating effect of an electric current. A fuse is made from a thin piece of wire designed to get hot and melt once a certain current is exceeded. This protects the rest of a circuit from the damage that might result from overheating elsewhere caused by excessively large currents, produced by a SHORT-CIRCUIT for example.

fusion power The proposed extraction of energy on a commercial scale from NUCLEAR FUSION. Most schemes are based on the D-T REACTION, which is the fusion of deuterium and tritium nuclei.

fuzzy logic A type of logic used in artificial intelligence. Fuzzy logic allows for various degrees of imprecision, in contrast to the true-or-false stance of classical logic. This enables a computer to respond to more complex and inputs. Fuzzy logic has been used in numerous small appliances that need to be able to adjust to a range of situations.

G

G-actin *See* ACTIN.

gadolinium (Gd) The element with ATOMIC NUMBER 64; RELATIVE ATOMIC MASS 157.3; melting point 1,312°C; boiling point 3,273°C; RELATIVE DENSITY 7.9. Gadolinium is a LANTHANIDE metal and occurs only in small quantities in the Earth's CRUST. It has some uses in the manufacture of semiconductors and of high temperature and FERROMAGNETIC alloys.

gain The amount by which a signal leaving an AMPLIFIER is greater than the input signal. Sometimes measured in DECIBELS.

galactic halo The roughly spherical region around a galaxy in which gas clouds and GLOBULAR CLUSTER are found.

galactose ($C_6H_{12}O_6$) A MONOSACCHARIDE that combines with GLUCOSE to form LACTOSE. Galactose contains an ALDEHYDE group and is therefore called an ALDOSE sugar.

galaxy A large volume of space containing millions of STARS held together by gravity. Galaxies occur in various shapes, but the most common is a flat disc with stars arranged in spiral arms. The MILKY WAY, a starry band across the sky, is our own galaxy, seen edge on. Galaxies are surrounded by smaller, roughly spherical groupings of stars called GLOBULAR CLUSTERS. *See also* QUASAR.

gall bladder A small muscular sac, forming part of the DIGESTIVE SYSTEM. Not all vertebrates have a gall bladder. In humans, the gall bladder is located under the liver and is connected to the SMALL INTESTINE by the BILE DUCT. It receives and stores BILE from the liver until its release, under intestinal hormone control, into the small intestine. GALLSTONES can occur if the CHOLESTEROL content of bile is too great; gall stones block the bile duct and often need surgical removal.

Galilean relativity The RELATIVITY PRINCIPLE as applied to NEWTON'S LAWS OF MOTION, in which the motion of an object in one INERTIAL REFERENCE FRAME could be computed from a knowledge of its motion in another inertial frame by the addition of the RELATIVE VELOCITY of the two frames. *See also* GALILEAN TRANSFORMATION.

Galilean satellite Any of the four largest planetary satellites of JUPITER, discovered by Galileo shortly after the invention of the telescope. They are IO, Europa, Ganymede and Callisto.

Galilean transformation Addition of the RELATIVE VELOCITY of two INERTIAL REFERENCE FRAMES to convert the VELOCITY of an object in one frame to its velocity in the other. *See also* GALILEAN RELATIVITY.

gallium (Ga) The element with ATOMIC NUMBER 31; RELATIVE ATOMIC MASS 69.7; melting point 30°C; boiling point 2,403°C; RELATIVE DENSITY 5.9. Gallium is a soft metal, widely used in the electronics industry in the manufacture of gallium arsenide, used in high speed semiconductors for its high electron mobility.

gallstone A small, hard concretion, mainly of CHOLESTEROL and mineral salts, that sometimes forms in the GALL BLADDER or one of its ducts.

galvanizing A way of protecting steel articles from corrosion by dipping them in molten zinc, which then acts as a SACRIFICIAL CATHODE. Galvanized materials are easily recognized by the characteristic patterns formed by the crystals of zinc produced as the coating solidifies.

galvanometer A sensitive instrument for detecting and measuring small electric currents. *See* MOVING-COIL GALVANOMETER. *See also* AMMETER.

gamete A cell of male or female origin that is involved in SEXUAL REPRODUCTION. A gamete of one sex combines with the gamete of the opposite sex to produce the beginnings of a new organism. Most gametes are HAPLOID (have one set of chromosomes), so when two gametes fuse a full set (two sets) of chromosomes is restored in the DIPLOID offspring. In higher organisms, such as humans, the female gamete is called an OVUM (or egg) and is produced in the OVARY. The male gamete is called a SPERM and is produced in the TESTIS. *See also* FERTILIZATION.

gametophyte One form of a plant that shows ALTERNATION OF GENERATIONS. The gametophyte is the HAPLOID generation that produces GAMETES by MITOSIS. *See also* SPOROPHYTE.

gamma-BHC *See* BHC.

gamma camera An array of GAMMA RADIATION detectors used to detect the motion of a gamma ray source around a patient's body. A typical source may be a gamma ray-emitting gas, inhaled to check for blockages within the lungs, for example.

gamma radiation, *gamma ray* High energy, short wavelength ELECTROMAGNETIC RADIATION, emitted as an atomic nucleus rearranges itself into a lower energy state after an ALPHA DECAY or BETA DECAY. Most alpha decays and many beta decays produce gamma rays.

Gamma radiation is only weakly IONIZING, so has a very long range in air, effectively falling off in an INVERSE SQUARE LAW as the radiation spreads out. A dense material such as lead will provide some reduction in the intensity of gamma rays, provided a thickness of several centimetres is used. The intensity of gamma radiation falls off exponentially (*see* EXPONENTIAL) in an absorbing material, at a rate that depends on the nature of the absorbing material and on the distance from the gamma ray source. The thickness of a given material which is needed to reduce the level of radiation by a factor of one half is called the HALF-THICKNESS.

gamma ray *See* GAMMA RADIATION.

ganglion (*pl.* ***ganglia***) A solid cluster of nervous tissue made up of NERVE CELL bodies and SYNAPSES enclosed in a sheath. In many invertebrates, for example insects, crustaceans and worms, the CENTRAL NERVOUS SYSTEM consists of these ganglia including some in the head that are well-developed and are analogous to the vertebrate brain. Ganglia are also found in vertebrates, but mainly outside the central nervous system.

gas The state of matter in which a substance will expand to fill its container. In gases, the molecules are much more widely spaced than they are in solids, so the forces between them are much weaker; thus the density of a gas is much less than for a solid.

At high temperatures and pressures, the distinction between liquid and gas can disappear. The temperature above which this happens is called the CRITICAL TEMPERATURE. Oxygen is an example of a substance that is above its critical point at room temperature and cannot be turned into a liquid simply by compressing it – such gases are called PERMANENT GASES.

See also BOYLE'S LAW, CHARLES' LAW, DALTON'S LAW OF PARTIAL PRESSURE, IDEAL GAS, IDEAL GAS EQUATION, KINETIC THEORY, PRESSURE LAW, TRANSPORT COEFFICIENT.

gas chromatography CHROMATOGRAPHY in which the moving phase is a gas. Often this is a vaporized volatile sample, carried along by hydrogen, passed over an oily HYDROCARBON on a solid support. This technique is known as gas liquid chromatography (GLC). The rate at which the materials in the sample pass along the solid phase varies, and their time of arrival at the far end of the absorbing medium is measured, typically by a hot wire that detects changes in the THERMAL CONDUCTIVITY of the gas.

gas constant *See* MOLAR GAS CONSTANT.

gas discharge The flow of electric current through a gas, often at reduced pressure. The electric field must be strong enough to accelerate ions rapidly enough for them to create further ionization when they collide with gas molecules; thus gas discharges generally occur only at relatively high voltages. As the ions recombine with electrons, light is given out, with a colour characteristic of the gas used. *See also* GLOW DISCHARGE.

gas giants Large planets, composed chiefly of hydrogen and other gases. In order of distance from the Sun, they are JUPITER (the largest), SATURN, URANUS and NEPTUNE. The gas giants show active weather systems and cloud patterns, including Jupiter's GREAT RED SPOT, which is believed to be a long-lived storm system. All the gas giants are known to have RING SYSTEMS.

gas jar A thick walled cylindrical glass container open at one end, used for the collection of gases produced in reactions. The open end of a gas jar is ground flat, so a disc of glass coated with grease can be used to cover the jar and prevent the gas escaping.

gas laws The three laws, BOYLE'S LAW, CHARLES' LAW and the PRESSURE LAW, that between them describe the properties of IDEAL GASES. They contain the same information as is contained in the IDEAL GAS EQUATION, but relate only to a fixed mass of gas, i.e. a fixed number of molecules.

gas liquid chromatography *See* GAS CHROMATOGRAPHY.

gasoline The term for PETROL in the USA.

gastric juice An acidic product of secretory cells that are found in gastric pits lining the wall of the STOMACH. Gastric juice is made up of mostly water mixed with hydrochloric acid and it provides the acid environment needed for the digestive enzymes to function and to kill bacteria brought in with the food. *See also* SECRETIN, STOMACH.

gastric pit *See* STOMACH.

gastrin In mammals, a hormone secreted by the STOMACH and DUODENUM that stimulates the production of GASTRIC JUICE.

Gastropod A member of the CLASS GASTROPODA.

Gastropoda A CLASS of the PHYLUM MOLLUSCA, including snails, slugs and winkles. Gastropods often have a single shell (univalve) that is usually coiled. The head is distinct with tentacles and eyes. They possess a rasping tongue (radula) for feeding. *Compare* PELECYPODA.

gastrulation *See* EMBRYONIC DEVELOPMENT.

gas turbine A JET ENGINE in which some of the energy of the hot gases leaving the engine is used to drive a TURBINE attached to a shaft that drives a compressor forcing fresh air into the engine. The shaft may also be used to extract energy from the engine for other purposes – to drive a propeller on a ship or aircraft (where such engines are called TURBOPROP engines), or to generate electricity.

Modern aircraft often use gas turbine engines, and modern airliners are fitted with HIGH-BYPASS, or turbofan, engines. In these engines, much of the air from the compressor leaves the engine without having fuel burnt in it. This provides increased fuel economy, whilst the relatively cool air surrounds the air heated by combustion and reduces the noise level of the engine. In military aircraft, fuel consumption and noise are of less concern than power and additional fuel may be burnt in the engine after the main turbine in a system known as an afterburner or REHEAT SYSTEM.

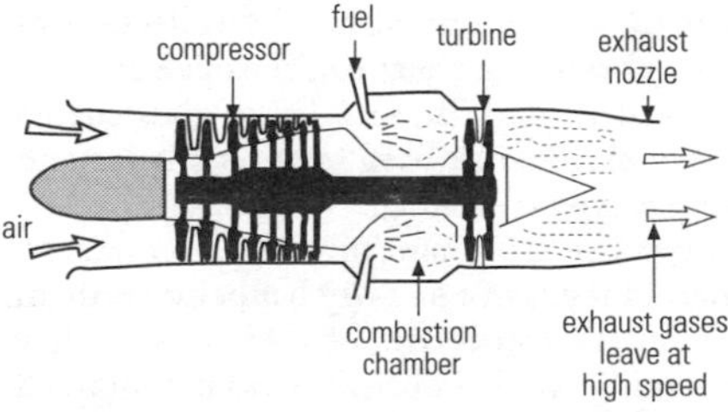

Gas turbine.

gate 1. In electronics, the ELECTRODE in a FIELD EFFECT TRANSISTOR at which an applied ELECTRIC FIELD controls the current flowing in the CHANNEL from the SOURCE to the DRAIN. The electrode with a similar function in a THYRISTOR or TRIAC is also called the gate.

2. *See* LOGIC GATE.

gauge boson A particle that is exchanged between two particles, carrying energy and MOMENTUM. This is the quantum mechanical (*see* QUANTUM MECHANICS) interpretation of a force acting between the two particles. The ELECTROMAGNETIC FORCE acts by the exchange of PHOTONS, the WEAK NUCLEAR FORCE by the exchange of W BOSONS and Z BOSONS and the STRONG NUCLEAR FORCE by the exchange of GLUONS. The name GRAVITON is used for the hypothetical exchange particle in a quantum theory of gravitation (*see* GRAVITY), but as yet there is no completely satisfactory theory of quantum gravity, nor any experimental evidence for the graviton.

gauss Unit of MAGNETIC FIELD strength in the C.G.S. SYSTEM. One gauss is equal to 10^{-4} TESLA.

Gaussian curve *See* NORMAL DISTRIBUTION CURVE.

Gauss' law The total ELECTRIC FLUX through any closed surface is equal to the total charge enclosed by the surface divided by the PERMITTIVITY of free space.

Gay-Lussac's law In any reaction where the REAGENTS and the products are all gases, the ratios of the volumes of the reagents and the products will be simple whole numbers, provided the temperature and pressure remain constant. For example, in the formation of ammonia:

$$N_2 + 3H_2 \rightarrow 2NH_3$$

1 volume of nitrogen combines with 3 of hydrogen to form 2 volumes of ammonia. Gay-Lussac's law was originally an EMPIRICAL law, but can now be understood in terms of AVOGADRO'S HYPOTHESIS and the molecular nature of chemical reactions.

gears A pair of rotating wheels, forming a MACHINE based on the LEVER principle. The wheels are provided with teeth to make them mesh (rotate together). One wheel, the pinion, is smaller than the other and rotates more quickly. The teeth act as a pivot between the two rotating wheels. The rotational speeds of the wheels are inversely proportional to the number of teeth on the wheel.

$$\frac{\text{rotation rate of wheel A}}{\text{rotation rate of wheel B}} = \frac{\text{number of teeth on B}}{\text{number of teeth on A}}$$

gedankenexperiment (German = thought experiment) An imaginary experiment, carried out with idealized apparatus, designed to illustrate a particular concept. Many complex concepts in QUANTUM MECHANICS are explained by *gedankenexperiments. See also* HEISENBERG'S MICROSCOPE, SCHRÖDINGER'S CAT.

Geiger counter A RADIATION DETECTOR containing a GEIGER-MÜLLER TUBE and its associated electronics.

Geiger-Müller tube A device for detecting IONIZING RADIATION using the flow of current in a sample of low pressure gas. *See* RADIATION DETECTOR.

gel (*n.*), ***gelatinous*** (*adj.*) A substance having some of the stiffness of a solid when subject to small deforming forces, but able to flow like a liquid under the influence of larger forces. Gels are formed as an intermediate stage in the coagulation of some COLLOIDS, with the particles in the colloids joining up to form long strands. *See also* THIXTROPIC.

gelatinous *See* GEL.

gene The basic unit of inheritance encoded by a specific length of DNA controlling one particular function or characteristic, for example eye colour. It was thought that one gene encoded one ENZYME, but it is now known that this is not always the case because proteins are made up of several POLYPEPTIDES each encoded by a separate gene. Today, a gene is considered to be a length of DNA encoding any molecular cell product. In higher organisms genes are located on CHROMOSOMES.

The position of a gene within a DNA molecule is called the LOCUS and a gene may have two (or sometimes more) variants called ALLELES, each specifying a particular form of the characteristic, for example blue or brown eyes. Genes can undergo MUTATION and RECOMBINATION. The full set of genes carried by a cell, an individual or the range of genes carried by a particular species is called the GENOME.

See also CISTRON, GENETIC CODE, GENETIC ENGINEERING, GENE REPLACEMENT THERAPY, LINKAGE, SEX LINKAGE, POLYGENE.

gene amplification The repeated duplication of a specific region of DNA, present in low amounts, until there is sufficient to use for conventional genetic analysis.

The procedure in gene amplification used is called the polymerase chain reaction (PCR). A small sample of DNA is heated and cooled in the presence of POLYMERASE enzymes (which allow it to replicate) and an excess of NUCLEOTIDES. The DNA strands separate, new paired strands are formed, and the DNA is reassembled. The process is repeated until enough DNA is present to analyse.

PCR typically takes about 20 cycles of 30 minutes. The technique allows very small amounts of DNA to be examined, for example to test for genetic defects in a single cell taken from an embryo, which would otherwise be impossible to analyse.

See also GENETIC ENGINEERING.

gene bank A stored collection of genetic material, for example frozen SPERM, frozen ova (*see* OVUM), seeds and bacterial cultures, collected for future use in, for example, medicine, breeding, agriculture and re-stocking endangered species.

gene expression, *gene regulation* The mechanisms controlling whether or not a particular GENE is transcribed (used to make a POLYPEPTIDE; *see* TRANSCRIPTION). It is now accepted that one gene codes for one polypeptide, but the genes need to be switched on and off as necessary. The Jacob–Monod theory suggests the concept of an OPERON, a group of adjacent genes that act together. This theory explains the control of some enzymes that are only produced when needed, and enzymes where the production is switched off when necessary. The theory does not fully explain all eukaryotic gene expression, which still has to be elucidated.

gene pool All the different ALLELES possessed by all the members of a given species or population at any one time.

general theory of relativity A theory proposed by Albert Einstein (1879–1955) in 1915, which extended the ideas of the SPECIAL THEORY OF RELATIVELY to cover observers in NON-INERTIAL FRAMES OF REFERENCE; that is, observers who would see one another as accelerating.

One of the predictions of the general theory is the deflection of light by gravity. This was shown to be true during a SOLAR ECLIPSE in 1919, when it was possible to observe the apparent shift in the position of stars that appeared in the same part of the sky as the sun. The effect of gravity on light also led to the prediction of BLACK HOLES.

Another result that pointed to the truth of the theory was the explanation of the PRECESSION of the PERIHELION of Mercury. The orbit of the planet Mercury is a flattened circle, or ELLIPSE. This ellipse does not remain fixed in space as predicted by NEWTONIAN MECHANICS, but instead rotates very gradually. Much of this rotation could be explained in part by the gravitational influence of other planets, but the general theory of relativity explained the rest of the motion.

The mathematics behind the general theory of relativity is complex, but is based on the idea that the gravity associated with a massive object is not a force in the conventional sense, but rather is a distortion of the space through which the object moves. In the general theory of relativity, the universe is described in four dimensions of space and time. This SPACE-TIME continuum obeys non-Euclidean geometry (*see* EUCLIDEAN SPACE).

See CURVATURE OF SPACE, GRAVITATIONAL WAVE, PRINCIPLE OF EQUIVALENCE.

generator A machine for converting mechanical energy into electrical energy. The term is also used to refer to the combination of a generator with a small INTERNAL COMBUSTION ENGINE to provide a source of electricity away from a mains supply. A generator may be a DYNAMO or an ALTERNATOR, but modern machines are usually alternators.

gene regulation *See* GENE EXPRESSION.

gene replacement therapy The treatment of diseases caused by GENE defects by manipulating the gene outside the body and replacing the functional (repaired) gene into the individual. This technique has been carried out successfully in animals and some trials have begun in humans as a treatment for cancer, cystic fibrosis and IMMUNODEFICIENCY caused by an enzyme defect in LYMPHOCYTES. Blood disorders such as THALASSAEMIA, in which the cause is a gene defect, are most likely to be curable by this technique because the cells are easily obtainable and have been grown successfully in the laboratory. There are stringent requirements that have to be met before a trial can be carried out, one of which is to ensure that the replaced genes are not transmitted to the GERM LINE (which would then interfere with evolution) *See also* GENETIC ENGINEERING.

genetic code The way in which the information contained within an organism's genetic material (DNA) is stored. The code is in the sequence of bases on the NUCLEOTIDE strands of DNA. Each AMINO ACID is coded for by a set of three bases called a CODON. The same codons code for the same amino acids in almost all organisms. The genetic code is said to be degenerate because several codons can code for the same amino acid. Three codons, UAA, UAG and UGA (the abbreviations stand for the names of the bases, for example U is URACIL), are called nonsense or STOP CODONS because they do not code for an amino acid and cause PROTEIN SYNTHESIS to come to an end. Another codon, AUG, is called the START CODON because this is where protein synthesis begins. The code is non-overlapping, so that a sequence GAUCAGUGA would be read as GAU-CAG-UGA and not GAU-AUC-UCA, etc. *See also* ALLELE.

genetic drift A change in the GENE frequencies in a population resulting from factors other than NATURAL SELECTION, emigration or immigration. For example, the accidental death of a disproportionate number of speckled moths would reduce the frequency of that ALLELE controlling colour within the population as a whole. Genetic drift applies to small populations because the change in allele frequency would be insignificant in a larger population. In some cases an allele could be eliminated by genetic drift, for example if only one speckled moth survived in a population of ten moths and it did not lay eggs. *See also* EVOLUTION.

genetic engineering The deliberate biochemical manipulation of GENES, or DNA itself, so that the genes are spliced (divided), altered, added or removed or transferred from one organism to another. Genetic engineering has enabled large-scale production of useful chemicals such as ANTIBIOTICS, INTERFERON and INSULIN.

The manipulation of DNA involves cutting and rejoining it with special enzymes called RESTRICTION ENDONUCLEASES and DNA LIGASE respectively. Any combined DNA from two different organisms is called RECOMBINANT

DNA. Using this technology, fragments of human DNA can be inserted into bacterial PLASMID DNA. By growing cultures of the bacteria containing the plasmid with its foreign DNA fragment, many copies of the foreign DNA can be obtained. In this way, for example, large amounts of human INSULIN are made; the DNA fragment joined to the plasmid contains the gene for insulin production, and replication within the bacterial plasmid allows large-scale production of the gene. If bacteria are used that secrete insulin, then large-scale production of a pure protein is possible. There is no danger of contamination, for example with HIV, as is possible when extracts from other human donors are used, when insulin is produced in this way, nor does it contain foreign material likely to cause immune rejection (*see* IMMUNE SYSTEM). If the required portion of DNA is not available, then COPY DNA can be made from MESSENGER RNA, taken from cells known to synthesize the protein in question, by using the enzyme REVERSE TRANSCRIPTASE.

Plants and bacteria can be modified by genetic engineering to achieve practical ends, for example production of drugs by bacteria, better growth of plants or disease resistance. In the future it is likely that some genetic diseases will be curable by genetic engineering, for example by GENE REPLACEMENT THERAPY. In addition to the benefits of genetic engineering, however, there are potential hazards and ethical issues. Organisms with altered genes could be dangerous in places outside their intended location. Gene manipulation could affect the natural balance of evolution and could be dangerous in the wrong hands.

See also GENE AMPLIFICATION, BIOTECHNOLOGY, TRANSLATION.

genetic fingerprinting, *DNA fingerprinting* A method for revealing the pattern of parts of the genetic material, DNA, that are unique to an individual (except identical twins, triplets, etc.) that can then be used, like fingerprinting, to identify individuals. The technique was discovered by Alex Jeffreys (1950–) of Leicester University, England, and has been of great importance in identifying criminals, including rapists, and establishing paternity. The material needed for testing can just be a spot of blood, a sample of skin, a few SPERM or a hair.

The method involves separating DNA from the sample, breaking it into smaller fragments with enzymes, and comparing the banding patterns of the component chemicals between individuals. Much of the DNA is similar between individuals but some regions are unique and it is these that are compared. The patterns obtained can be expressed in digital code, which provides greater accuracy, and the information is accepted as a legal means of identification. Other uses of the technique include detecting some inherited diseases and monitoring BONE MARROW transplants.

genetics The study of inheritance, GENES and their effects. It includes the manipulation of genes in GENETIC ENGINEERING. The work of Gregor Mendel (1822–1884) forms the basis of genetics. *See* MENDEL'S LAWS.

genitalia In mammals, the external sexual reproductive organs. In males the genitalia are the PENIS and testes (*see* TESTIS); in females they are the VULVA and CLITORIS.

genome The full set of GENES carried by a cell or an individual, or the range of genes carried by a particular species. In humans, the genome is about 3 x 10^9 BASE PAIRS and an international effort, called the Human Genome Project, to map all the genes is expected to be completed by 2005.

genotype The genetic constitution of an organism, or the set of ALLELES inherited by it. *Compare* PHENOTYPE. *See also* VARIATION.

genus One of the subdivisions of families (*see* FAMILY) in the CLASSIFICATION of organisms. The subdivision genus comes between family and SPECIES. Members of a genus are thought to have a common ancestral origin and share many characteristics. *See also* BINOMIAL NOMENCLATURE.

geochemistry The study of the chemistry of the Earth. It includes the study of the abundance and distribution of the naturally occurring elements and their ISOTOPES.

geodesic line The path of shortest time between two points, and the non-Euclidean equivalent of a straight line. *See* CURVATURE OF SPACE, EUCLIDEAN SPACE.

geography The study of the Earth's surface – its features, climate and physical conditions and the way humans interact with these. Geography is subdivided into physical geography and human geography.

geology The study of the solid part of the Earth. The Earth contains a central CORE of molten iron and nickel, surrounded by a layer of molten rock called MAGMA. This forms a region called the MANTLE. On top of this lies the Earth's crust, or LITHOSPHERE, only a few hundred kilometres thick compared to the Earth's radius of 6,400 km. The crust is made up of a number of distinct regions called TECTONIC PLATES.

In addition to the IGNEOUS rock that has solidified from the core, much of the rock found on the Earth's surface is SEDIMENTARY. Some sedimentary rocks are deposited by the processes of EROSION, whereby wind or water breaks down rock in one place and deposits it elsewhere to form shale or sandstone (depending on the size of the particles involved).

Other sedimentary rocks, such as LIMESTONE, are formed from the shells of dead marine life. Pressure then converts such material into rock, though it is generally softer than igneous rocks such as granite. Folding of such rocks by plate tectonic activity can lead to them experiencing greater heat and pressure, forming harder rocks, which are described as METAMORPHIC.

geometric isomerism, *cis/trans isomerism* In chemistry, ISOMERS that contain the same groups joined together in the same order (unlike structural isomers) and differ only in the geometry of their bonds. This occurs in compounds with a DOUBLE BOND or with certain ring structures, where there is restricted rotation of the atoms joined by the bonds, thereby giving rise to more than one possible geometric arrangement. Geometric isomers differ in their chemical reactions.

Geometric isomerism often occurs where there is a >C=C< double bond or a carbon ring. In these cases the terms *cis* and *trans* are used to distinguish the isomers. These derive from the Latin meaning 'on this side of' and 'across' respectively, so that the prefix *cis* is applied when the groups at the end of a double bond are on the same side of the plane of the bond. The term *trans* is used when similar groups are on opposite sides of the plane of the double bond:

Geometric isomerism also occurs where there are >C=N– or –N=N– double bonds and then the prefixes *syn-* and *anti-* are used.

```
CH3   H          CH3   H
  \  /             \  /
   C                C
   ||               ||
   C                C
  /  \             /  \
CH3   H           H    CH3
```

cis-but-2-ene *trans*-but-2-ene

geometric mean *See* MEAN.

geometric progression A SERIES of numbers in which each number is a constant multiple of the previous number in the series. Thus if the nth member of the series is a_n, the $(n+1)$th member will be

$$a_{n+1} = ka_n$$

where k is a constant called the common ratio. For a geometric progression of N terms, the sum S_N of the series will be

$$S_N = (1 - a_n)/(1 - a)$$

geometry The study of the properties of points, lines and planes, and of curves, shapes and solids. The study of two-dimensional shapes on flat surfaces is called plane geometry, and this, along with solid geometry (in three dimensions) make up pure geometry. In analytical, or co-ordinate geometry, problems are solved using algebraic methods. *See also* EUCLIDEAN SPACE.

geophysics The branch of science that applies the principles of physics and mathematics to the study of the Earth, its interior and climate.

geostationary orbit A circular ORBIT over the equator, moving in the same direction as the Earth's rotation and with the same period, i.e. 24 hours. As seen from the rotating Earth, an artificial satellite in such an orbit always appears in the same place in the sky. An altitude of 36,000 km corresponds to such an orbit.

Geostationary orbits are very important for communications satellites and also for DIRECT BROADCAST SATELLITE systems, which broadcast television signals straight into the homes of the viewers, as the receiving AERIALS do not then need to track the motion of the satellite. Geostationary orbits are also used by some weather satellites, as they provide a constant view of one half of the Earth.

See also CIRCUMPOLAR ORBIT.

geothermal energy Heat extracted from the Earth's CORE by pumping water through hot rocks. The hot water can either be used directly, for heating, or to generate electricity. In some countries, such as Iceland, this is an important source of energy. Experimental schemes have been tried elsewhere, but the technical problems inherent in pumping water in and out of deep rocks are too great at present for such schemes to be viable in most locations.

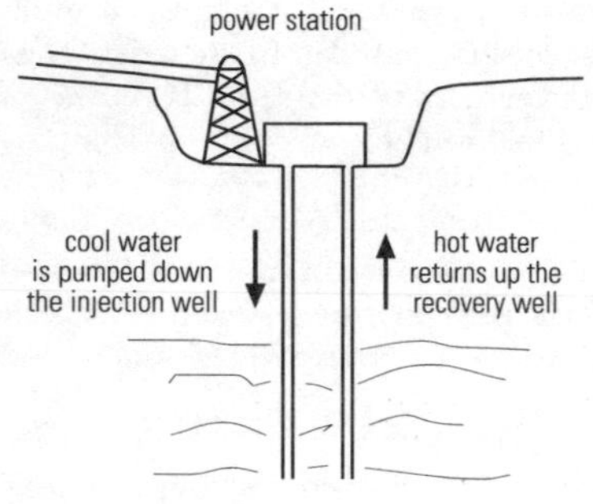

Geothermal energy.

geotropism The response of a plant to gravity. Shoots are negatively geotropic (they grow upwards) and roots are positively geotropic (they grow downwards). Leaves grow at right angles and are said to be diageotropic. *See also* TROPISM.

germanium (Ge) The element with ATOMIC NUMBER 32; RELATIVE ATOMIC MASS 72.6; melting point 937°C; boiling point 2,830°C; RELATIVE DENSITY 5.4. Its existence was predicted from a gap in Mendeleyev's PERIODIC TABLE, and the element was discovered in 1886.

Germanium is a semiconductor and was much used in early semiconductor devices. It has a relatively small BAND GAP (*see* BAND THEORY), which means that a fairly large number of CHARGE CARRIERS are produced by thermal vibrations unless it is kept at a low temperature. This has lead to its replacement by silicon in many electronic applications, though it is still used for small signal DIODES and some more specialized applications.

germination The initial stages of development and growth in a SEED, SPORE or POLLEN grain. Germination begins when the right conditions of water, oxygen, temperature, sometimes light and other factors needed to break DORMANCY are met. The uptake of water by the seed (through a small hole called the MICROPYLE) initiates germination. The embryo plant differentiates into the RADICLE (young root), PLUMULE (young shoot) and COTYLEDON (seed leaves). Food is provided for the growing embryo by the ENDOSPERM or cotyledon. The radicle appears first; its apical MERISTEM pushes through the soil and may develop into the entire root system (or ADVENTITIOUS ROOTS may develop).

The cotyledons may remain below ground (hypogeal) or spread out above the soil (epigeal) (*see* COTYLEDON). In epigeal germination, the plumule develops after the cotyledons have grown above ground, and forms the shoot bearing the first true leaves. In hypogeal germination, for example in the broad bean, the leaves develop on the plumule in the seed and the cotyledons remain below ground when the plumule grows up through the soil. Germination is considered to have ended when the first true leaves appear.

germinative layer *See* MALPIGHIAN LAYER.

germ layer In animals, one of the main recognizable layers to appear in a developing embryo after gastrulation (*see* EMBRYONIC DEVELOPMENT). There are three germ layers in most animals: the ECTODERM (outermost layer), the MESODERM (middle layer; not present in more primitive animals) and the ENDODERM (inner layer).

germ line The cell line in the development of many animals that goes on to form the GAMETES.

gestation In most mammals, the period between conception and birth. The gestation period varies between species: in humans it is 9 months, in elephants it is 18–22 months and in cats it is 60 days. In humans the gestation period is commonly called PREGNANCY.

getter A device for removing small amounts of gas from vacuum vessels in order to improve the vacuum. A getter consists of a wire made of a reactive metal, such as magnesium. The vessel is evacuated and sealed and the getter vaporized by passing a current through it. The metal atoms react with any air molecules remaining in the vessel to form a non-volatile OXIDE or NITRIDE.

geyser A hot underground spring that periodically erupts through the surface as a result of a build up of pressure, sending a fountain of steam and water into the air.

GFRP *See* GLASS-FIBRE REINFORCED PLASTIC.

gibberellin A PLANT GROWTH SUBSTANCE that promotes cell elongation and therefore growth. Gibberellin was originally isolated from a fungus and over 50 types are now known. Unlike AUXINS, gibberellins can stimulate growth in dwarf varieties and restore them to normal size, largely by causing elongation of the stem. They can also break the DORMANCY in certain buds and seeds and induce flowering. They have no effect on TROPISMS (the directional growth of a plant in response to an external stimulus). Gibberellin and auxin have some SYNERGISTIC effects but also some ANTAGONISTIC ones.

Gibbs free energy *See* FREE ENERGY.

giga- (G) Prefix used before a unit to indicate that the size of the unit is multiplied by 10^9. For instance, one gigawatt (GW), is one billion watts.

gill In biology, the main respiratory organ of fish. Water passes over the gills, providing oxygen that diffuses across a gill membrane to the circulation, and carbon dioxide passes out into the water.

ginkos *See* GYMNOSPERM.

gizzard In some animals, for example birds and reptiles, a muscular grinding organ that forms part of the DIGESTIVE SYSTEM. Food is ground in the gizzard before passing to the stomach for the main digestion. The wall is very tough and hardened with a layer of the protein KERATIN to prevent damage during the grinding process. Grit or stones may be swallowed to aid grinding in the gizzard.

glacier A mass of ice formed by the accumulation and compaction of snow. Glaciers move slowly downwards under their own weight and are constantly replenished at source.

gland An organ (or sometimes cells within an organ) specialized for producing and secreting HORMONES, ENZYMES or other chemicals. Glands can vary in size, for example tear glands are very small and the PANCREAS is large. In animals there are two types of glands: EXOCRINE GLANDS and ENDOCRINE GLANDS. Together the glands form the ENDOCRINE SYSTEM, which is one of the two major co-ordinating systems of animals (the other being the NERVOUS SYSTEM). Glands exist in plants, where they are always small, sometimes comprising only a single cell.

glans penis *See* PENIS.

glass Any of a number of transparent brittle materials. Common glass is made by melting together sand (mostly silicon dioxide) with lime (calcium oxide) and soda (mostly sodium carbonate). AMORPHOUS materials are sometimes referred to as glasses, as glass is an example of an amorphous material. In some ways glasses are more like liquids than solids, they have no melting point, but simply flow more readily as they are heated. *See also* DISORDERED SOLID.

glass-fibre reinforced plastic (GFRP) A COMPOSITE MATERIAL consisting of glass fibres embedded in a plastic RESIN. Glass is very strong for its density, but cannot be used alone as a structural material since it is brittle – cracks travel rapidly through the glass resulting in a catastrophic failure. The plastic resin is tough and prevents the spread of cracks from one fibre to the next.

glial cell A non-conducting nerve cell that provides support and protection for NEURONES. SCHWANN CELLS, astrocytes (which attach neurones to blood vessels), microglial cells (PHAGOCYTES) and OLIGODENDROCYTES are all examples of glial cells.

global warming An increase in the overall temperature of the Earth, believed to be caused by the increasing level of GREENHOUSE GASES, particularly carbon dioxide, in the atmosphere. *See* GREENHOUSE EFFECT.

globular cluster A roughly spherical grouping of a few hundred stars found in orbit around a GALAXY. Globular clusters are of particular interest to astronomers as they provide a group of stars all at much the same distance from the Earth. *See also* GALACTIC HALO.

globular protein *See* PROTEIN.

globulin Any one of a group of globular PROTEINS characterized by a spherical shape, solubility in weak salt solutions and a metabolic (*see* METABOLISM) rather than structural function. ENZYMES, some HORMONES and HAEMOGLOBIN are all globular proteins. In animals globulins are found in blood PLASMA. They are also found in plant seeds.

glomerulus A tight knot of blood capillaries forming part of a NEPHRON in the KIDNEY. Blood enters the glomerulus at high pressure, from ARTERIOLES of the RENAL artery, and certain substances including water, amino acids and sugar, are forced out through the capillary wall to form a filtrate that enters the BOWMAN'S CAPSULE. The filtrate then moves through the tubule of the nephron. Much of the water and sugars are reabsorbed by the LOOP OF HENLE and the remaining waste materials remain in the fluid that becomes URINE.

glow discharge A GAS DISCHARGE in a low pressure gas, producing a steady luminous glow.

glucagon A POLYPEPTIDE hormone (29 amino acids) produced by the PANCREAS which is involved in the regulation of blood sugar levels. Glucagon is produced by the α-cells of the ISLETS OF LANGERHANS and it converts GLYCOGEN, stored in the liver, to GLUCOSE in response to a drop in the concentration of glucose in the blood. This process is called GLYCOGENOLYSIS. Glucagon acts antagonistically with INSULIN.

glucocorticoid Any one of a group of CORTICOSTEROID hormones that are concerned with GLUCOSE metabolism. They are secreted by the CORTEX of the ADRENAL GLAND, and examples include CORTISOL and CORTISONE. Glucocorticoids raise blood pressure and blood sugar levels. They raise blood sugar levels by increasing formation of glucose from fat and proteins, by inhibiting INSULIN and also by increasing the rate of glycogen formation in the liver.

Too little glucocorticoid results in Addison's disease, characterized by low blood sugar levels, low blood pressure and fatigue. Over-production of glucocorticoid results in Cushing's syndrome, characterized by high blood sugar levels due to excessive break down of protein (and therefore tissue and muscle wasting) and high blood pressure.

gluconeogenesis The conversion of proteins and fats into GLUCOSE by the LIVER. *See also* GLYCOGENESIS, GLYCOGENOLYSIS, PANCREAS.

glucose, *dextrose* ($C_6H_{12}O_6$) A MONOSACCHARIDE. In humans, glucose is found in the blood and is a source of energy for the body because it is used to generate ATP. Glucose is made by the HYDROLYSIS of STARCH or SUCROSE. Levels of glucose in the blood are regulated by the LIVER and the pancreatic hormones GLUCAGON and INSULIN. *See also* GLUCONEOGENESIS, GLYCOGEN, GLYCOGENESIS, GLUCONEOGENESIS, PANCREAS.

gluon The GAUGE BOSON that is exchanged in the STRONG NUCLEAR FORCE, named as the glue that holds QUARKS together in a HADRON. *See also* QUANTUM CHROMODYNAMICS.

glycerate 3-phosphate A substance produced in C_3 PLANTS as an intermediate in the light-independent stage of PHOTOSYNTHESIS. *See* CALVIN CYCLE.

glyceride *See* LIPID.

glycerine *See* GLYCEROL.

glycerol, *glycerine, propan-1,2,3-triol* ($HOCH_2CH(OH)CH_2OH$) An ALCOHOL that is a sweet, colourless liquid extracted from animal and vegetable oils. Glycerol reacts with FATTY ACIDS to form LIPIDS. It is used in ANTIFREEZE solutions, explosives and cosmetics.

glycogen A POLYSACCHARIDE (made of branched chains) of the sugar GLUCOSE that is stored in the liver as a carbohydrate source until needed. Glycogen is the animal equivalent of STARCH. It is converted back to glucose by the pancreatic hormone GLUCAGON (a process known as GLYCOGENOLYSIS), and glucose is converted to glycogen by INSULIN (GLYCOGENESIS). These two hormones therefore act antagonistically to regulate blood sugar levels.

glycogenesis The conversion of GLUCOSE in the blood to GLYCOGEN by the LIVER to be stored until needed, when blood glucose levels fall. *See also* GLUCONEOGENESIS, GLYCOGENOLYSIS, PANCREAS.

glycogenolysis The break down of GLYCOGEN in the liver to GLUCOSE by GLUCAGON. Glucagon is a HORMONE produced by the PANCREAS in response to a fall in blood sugar levels and is secreted directly into the blood, and so to the LIVER. *See also* GLUCONEOGENESIS, GLYCOGENESIS.

glycol *See* ETHANE-1,2-DIOL.

glycolipid A LIPID that is covalently linked to a CARBOHYDRATE. Glycolipids are found in CELL MEMBRANES. There is a wide variation in the composition and complexity of glycolipids.

glycolysis (*Glycol* = sugar, *Lysol* = breakdown) The series of reactions resulting in the break down of the six-carbon sugar GLUCOSE to two molecules of the three-carbon compound pyruvate. The process requires no oxygen and is the only form of RESPIRATION and ATP synthesis in anaerobic organisms (*see* ANAEROBE). In AEROBES, it is the first stage of cellular respiration and the pyruvate formed then enters the KREBS CYCLE, which requires oxygen. Glycolysis occurs in the CYTOPLASM of cells, usually in the MITOCHONDRIA. There is an overall gain of two molecules of ATP for each glucose molecule broken down, and two pairs of hydrogen atoms either go into the ELECTRON TRANSPORT

Summary of glycolysis.

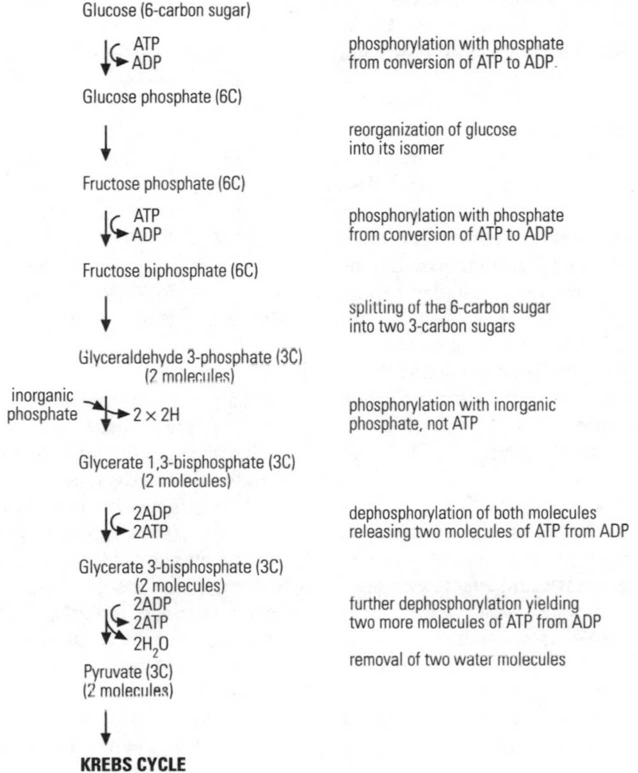

SYSTEM to generate a further six molecules of ATP if oxygen is present, or are removed by the FERMENTATION process if no oxygen is present.

glycoprotein A protein covalently linked to a CARBOHYDRATE. The protein can form the bulk of the molecule, as in cell-surface glycoproteins, or the carbohydrate can represent the major part. The addition of sugar residues to proteins is called glycosylation and occurs in the GOLGI APPARATUS or ENDOPLASMIC RETICULUM of a cell.

glycosylation The addition of sugar residues to protein. *See* GLYCOPROTEIN.

goblet cell A cell that is specialized for the production of MUCUS. Goblet cells are found in some epithelia (*see* EPITHELIUM), for example in the intestines.

goitre A swelling of the THYROID GLAND.

gold (Au) The element with ATOMIC NUMBER 79; RELATIVE ATOMIC MASS 197.0; melting point 1,064°C; boiling point 2,807°C; RELATIVE DENSITY 19.3. Gold is a TRANSITION METAL with a characteristic yellow colour.

The low reactivity of gold means that it does not tarnish, so it is widely used in jewellry and for electrical contacts. For jewellry it is alloyed with other metals such as silver and copper, the purity of the gold being measured in carats on a scale from 0 (no gold) to 24 (pure gold).

gold-leaf electroscope An ELECTROSCOPE in which the electrical repulsion between a conductor made of a metal plate and a thin sheet of gold, causes the gold to move away from the plate against the force of gravity.

Golgi apparatus, *Golgi body, dictyosome* A stack of membranous sacs found in the CYTOPLASM of a eukaryotic cell (*see* EUKARYOTE), similar in structure to smooth ENDOPLASMIC RETICULUM (ER). The Golgi apparatus is named after its discoverer, Camillo Golgi (1843–1926).

The Golgi apparatus is well developed in secretory cells and NEURONES, and is small in muscle cells. It is thought to play some role in the production of secretory material. Many molecules travel through the Golgi apparatus and are modified or assembled inside the sacs on their way to other organelles or ER, for example carbohydrate may be added to protein to form GLYCOPROTEINS such as MUCIN. With an ELECTRON MICROSCOPE vesicles can be seen budding from the Golgi apparatus.

In cells of vertebrates, there is usually only one Golgi apparatus, but in invertebrates and plants there may be several, called dictyosomes.

Golgi body *See* GOLGI APPARATUS.

gonad In animals, the sex organ producing the GAMETES required for SEXUAL REPRODUCTION. A male gonad is the TESTIS and a female gonad is the OVARY.

gonadotrophic (*adj.*) Describing a HORMONE that stimulates reproductive activity of the GONADS (the TESTIS or the OVARY).

gout A disease characterized by painful inflammation of the joints, caused by crystalline deposits of URIC ACID.

Graafian follicle In mammals, a fluid-filled structure in the mammalian OVARY, that may be as big as 1 cm, and within which the OVUM (egg cell) develops until it matures. The development of the ovum is under the control of the FOLLICLE-STIMULATING HORMONE (FSH). The mature ovum is released (OVULATION) when the follicle ruptures under the control of the LUTEINIZING HORMONE (LH). The Graafian follicle is then transformed, again under LH control, into the CORPUS LUTEUM. FSH is inhibited by PROGESTERONE, secreted by the corpus luteum, which therefore prevents the maturation of further Graafian follicles.

gradient A measure of the slope of a straight line on a GRAPH. The gradient is found by choosing two points on the line and is calculated as the change in y CO-ORDINATE between those points, divided by the change in x coordinate. *See also* STATIONARY POINT, TANGENT.

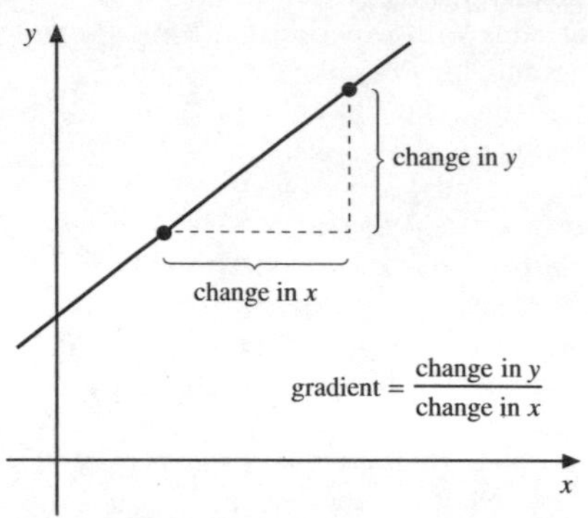

Gradient.

grafting The TRANSPLANTATION of a small portion of living tissue onto another tissue on either the same or a different organism.

Graham's law of diffusion The rate of DIFFUSION of a gas is inversely proportional to the MOLECULAR MASS. Thus light gases, such as hydrogen, diffuse more quickly than heavier ones, such as carbon dioxide.

grain boundary In a POLYCRYSTALLINE material, the boundary between one orientation of the crystal LATTICE and another.

gram A unit of mass, nowadays defined as one thousandth of a KILOGRAM.

Gram's stain A method of staining, devised by Christian Gram (1855–1938) in 1884, that differentiates BACTERIA into those that take up the stain and those that do not. This is of great value in the differentiation of otherwise similar bacteria. The procedure involves staining a smear of bacteria with a crystal violet solution, rinsing this off with Gram's iodine solution and applying 95 per cent ethanol until most of the dye has been removed. Gram-positive bacteria retain the dye while Gram-negative bacteria do not. A counterstain is applied that only stains the Gram-negative bacteria. The staining reflects differences in the bacterial cell membrane, but there are other differences between the two groups.

grand unified theory (GUT) A hypothetical unification of the ELECTROWEAK FORCE and STRONG NUCLEAR FORCE, and possibly GRAVITY too. Such theories make few predictions that can be tested at energies available from current or conceivable PARTICLE ACCELERATORS. The predictions they do make are also sometimes at odds with experimental observation, suggesting, for example, that the proton may be unstable and decay into lighter particles. Despite intensive searches for PROTON DECAY, no evidence has been found. *See also* STANDARD MODEL, THEORY OF EVERYTHING.

granite A hard, IGNEOUS rock.

granulocyte, *polymorphonuclear leucocyte* One of the two types of WHITE BLOOD CELLS in humans. The other type of white blood cell is the AGRANULOCYTE. Granulocytes are characterized by their granular CYTOPLASM and lobed, darkly staining NUCLEUS. They are about 8 μm in diameter and act as PHAGOCYTES. They form about 60–70 per cent of human white blood cells. In humans, the cytoplasmic granules of most granulocytes do not take up acid or basic dyes and are called NEUTROPHILS. Others stain strongly with basic dyes (BASOPHILS) or acid dyes (EOSINOPHILS). *See also* AGRANULOCYTES.

granum (*pl.* ***grana***) *See* CHLOROPLAST.

graph A way of showing how one variable behaves as a function of another. Points are plotted on a two-dimensional surface according to the values of the two variables. The variables are usually referred to as *x*, measured along a horizontal AXIS, and *y*, measured up a vertical axis. The independent variable is generally plotted along the x-axis, with the dependent variable up the y-axis. If statistical or experimental data is plotted, a series of measurements will build up a SCATTER DIAGRAM. A mathematical function may be represented by a line or curve on the graph.

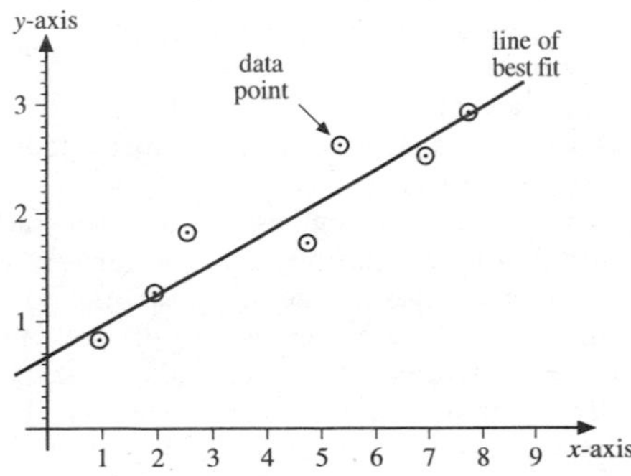

Graph.

graphite An ALLOTROPE of carbon in which carbon atoms are arranged in a hexagonal pattern held together by COVALENT BONDS, similar to those found in AROMATIC compounds. The individual layers are held together by much weaker VAN DER WAALS' bonds, which accounts for the flaky nature of graphite. The delocalized nature of the covalent bonding means that graphite conducts electricity along the layers of atoms, but not from one layer to the next. Graphite is also a good conductor of heat.

Graphite is used as the 'lead' in pencils, in electrical contacts, as a solid lubricant and as the MODERATOR in some nuclear power stations.

graticule A small measuring device, like a ruler, placed in the eyepiece lens of a light microscope, that can be used to determine the exact size of a feature being viewed.

gravimetric analysis Any technique of QUANTITATIVE ANALYSIS that involves measuring the mass of substance in a reaction. For example, the amount of barium in a sample could be measured by dissolving the sample in hydrochloric acid, and then adding sulphuric acid to precipitate the barium as barium sulphate, which is dried and weighed.

gravitation *See* GRAVITY.

gravitational anomaly A small change in the GRAVITATIONAL FIELD STRENGTH on the surface of the Earth caused by the uneven distribution of mass in the locality.

gravitational constant (*G*) The constant that measures the overall strength of the gravitational force, the force between two 1 kg masses 1 m apart. It is equal to 6.67×10^{-11} Nm^2kg^{-2}. *See also* CAVENDISH'S EXPERIMENT, GRAVITY, NEWTON'S LAW OF GRAVITATION.

gravitational field The region of influence around any mass that will cause any other mass within that region to experience a force proportional to its mass. *See also* GRAVITATIONAL POTENTIAL.

gravitational field strength The force per unit mass experienced by an object placed at the point where the GRAVITATIONAL FIELD strength is being measured. The SI UNIT of gravitational field strength is the Nkg^{-1} (newton per kilogram) or ms^{-2} (metre per second squared).

Since GRAVITATIONAL MASS and INERTIAL MASS appear to be the same for all objects, an object released in a gravitational field (on the surface of the Earth for example) will always fall with the same acceleration (the ACCELERATION DUE TO GRAVITY) regardless of its mass, provided no other forces act. This acceleration will be equal to the gravitational field strength. The gravitational field strength on the surface of the Earth is about 9.8 Nkg^{-1}, but it varies slightly from place to place due partly to the non-spherical shape of the Earth and partly to variations in the density of nearby rocks. These local variations, called gravitational anomalies, have been used as a way of studying the geology of the underlying rocks.

At a distance *r* from the centre of a body of mass *M* and radius *R*, if *r* is greater than *R*, the gravitational field is

$$g = GM/r^2$$

gravitational mass MASS measured as an object's response to a gravitational field, as opposed to INERTIAL MASS. *See also* PRINCIPLE OF EQUIVALENCE.

gravitational potential The gravitational potential at a point is the WORK needed to bring a 1 kg mass from infinity to that point. Because gravity is always an attractive force, systems tend to collapse under gravity rather than expand, thus gravitational potentials are always negative. The difference in gravitational potential between two points is a measure of the energy needed per unit mass to move an object from one point to the other.

At a distance *r* from the centre of a body of mass *M* and radius *R*, if *r* is greater than *R*, the gravitational potential is

$$V_g = -GM/r$$

and the relationship between GRAVITATIONAL FIELD *g* and gravitational potential V_g is

$$g = -\mathrm{d}V_g/\mathrm{d}x$$

gravitational potential energy The energy of a mass in a GRAVITATIONAL FIELD. For an object of mass *m* moving through a height *h* in a gravitational field *g*, the change in gravitational potential energy is *mgh*.

gravitational wave A wave of distortion of SPACE-TIME, carrying energy at the speed of light and predicted by the GENERAL THEORY OF RELATIVITY. Evidence for the existence of gravitational waves has recently been obtained form the changes in motion of a BINARY STAR, one element of which is a PULSAR. As the double star system loses energy by gravitational waves, its period should change. This can be detected in such a system by observing changes in the pattern of pulses from the pulsar, which are not quite regular, but influenced by the DOPPLER EFFECT as a result of the orbital motion.

graviton The hypothetical GAUGE BOSON responsible for GRAVITY in a QUANTUM THEORY of gravity. No such theory has yet been satisfactorily developed.

gravity, *gravitation* The force of attraction between all objects dependent on their MASS.

Gravity is a weak force, so it is normally only noticeable when at least one of the masses is very large, such as in the case of a planet or star. Whilst the GENERAL THEORY OF RELATIVITY provides a good explanation of gravity at large scales, it is widely believed that at small scales and very high energy there must be a theory that links gravity and QUANTUM MECHANICS and explains gravity in terms of the exchange of a particle called the GRAVITON.

At the scales currently studied by particle physics, and for laboratory-sized objects, gravity is generally too weak a force to be important. It only becomes important at large scales because it is a long range force with no cancellation between the effect of positive and negative charges as there is with the ELECTROMAGNETIC FORCE. The gravitational force acts on every particle in an object but can be considered as a single force acting at a point called the CENTRE OF MASS or centre of gravity.

See also FREE FALL, GRAND UNIFIED THEORY, GRAVITATIONAL FIELD, GRAVITATIONAL FIELD STRENGTH, GRAVITATIONAL POTENTIAL, GRAVITATIONAL WAVE, NEWTON'S LAW OF GRAVITATION.

gray (Gy) The SI UNIT of absorbed IONIZING RADIATION. One gray is equal to an energy of one JOULE absorbed from the radiation. *See also* DOSE.

Great Red Spot A long-lived cloud feature in the atmosphere of JUPITER, believed to be some form of storm system.

greenhouse effect A supposed increase in the average surface temperature of the Earth as a result of changes in the composition of its atmosphere. Since the temperature of the Earth is much lower than that of the Sun, whilst the electromagnetic radiation absorbed from the Sun mostly lies in or near the visible part of the

The greenhouse effect.

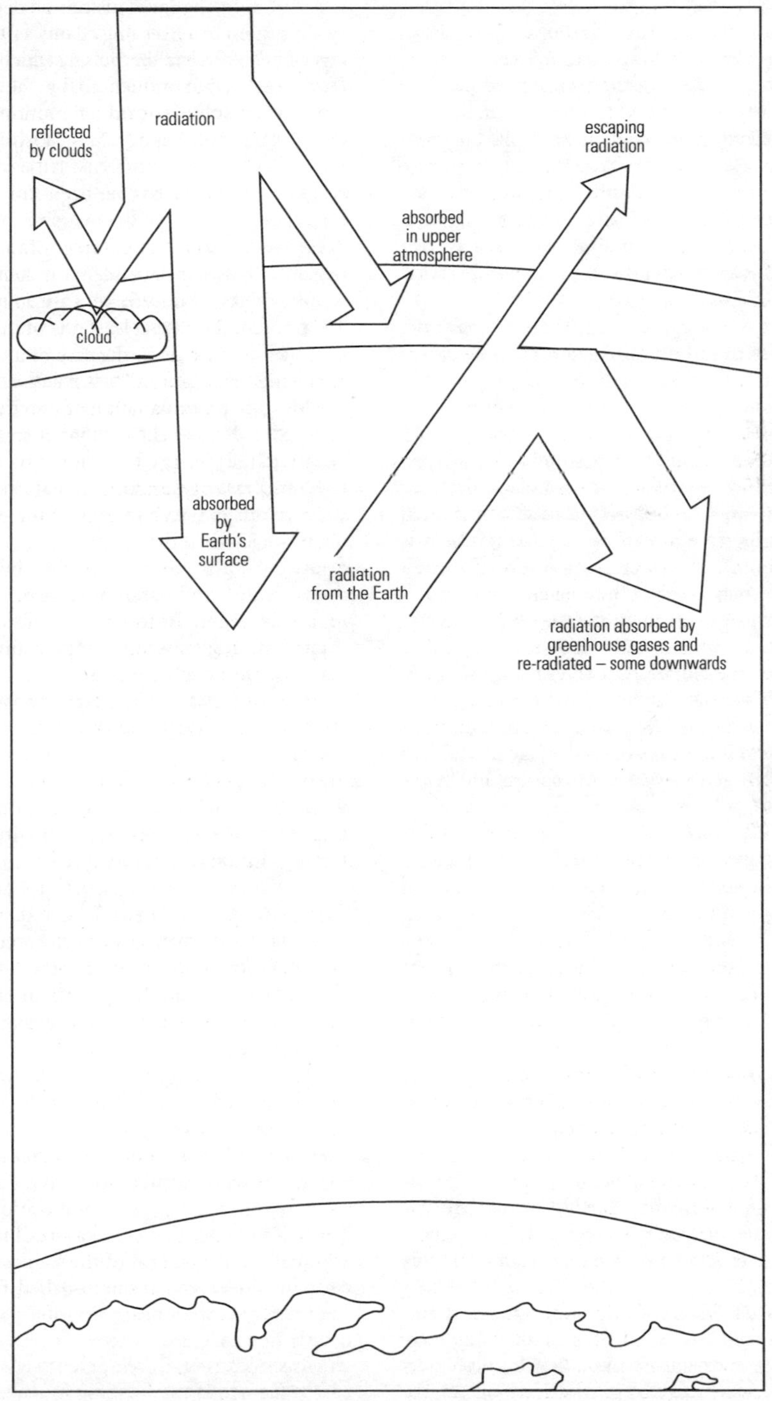
radiation
reflected by clouds
escaping radiation
absorbed in upper atmosphere
cloud
absorbed by Earth's surface
radiation from the Earth
radiation absorbed by greenhouse gases and re-radiated – some downwards

ELECTROMAGNETIC SPECTRUM, the radiation given off by the Earth is in the infrared part of the electromagnetic spectrum. Some of this is absorbed by gases in the atmosphere, particularly carbon dioxide and water vapour, thus the concentration of such gases (called GREENHOUSE GASES) has a marked effect on the surface temperature of the Earth. This effect is called the greenhouse effect since the Earth's atmosphere behaves in a way similar to the glass in a greenhouse, which also allows in visible radiation but absorbs infrared.

Fears have been expressed over the increase in carbon dioxide in the atmosphere caused by the burning of increasing quantities of FOSSIL FUELS over the last century, and attempts are being made to reduce carbon dioxide emission by increasing use of RENEWABLE energy sources. The systematic destruction of vast areas of forest has also contributed to the increase of carbon dioxide levels. It is believed that the increasing levels of carbon dioxide may increase the average temperature of the planet, a phenomenon called GLOBAL WARMING, which may also lead to other changes in climate. Other greenhouse gases include methane (a by-product from agriculture), water vapour (as a by-product from industry) and CHLOROFLUOROCARBONS (CFCs; from refrigerators, AEROSOL sprays and POLYSTYRENE).

The problem with continued global warming is that it will cause the polar ice caps to melt, resulting in a rise in sea levels and consequent flooding of low-lying land, which could include whole countries and many world capital cities. A change in the climate would also affect crop growth. It is not clear exactly what or how rapid the consequences of the greenhouse effect will be, because many effects will interact with one another, but the issue raises great concerns about the future of humankind.

greenhouse gas Any gas, such as carbon dioxide or methane, which contributes to the GREENHOUSE EFFECT in the Earth's atmosphere by absorbing infrared radiation.

greenockite A mineral form of cadmium sulphide, CdS.

grey matter Tissue of the vertebrate BRAIN and SPINAL CORD, so called because of its appearance. It consists of many cell bodies, DENDRITES, SYNAPSES, GLIAL CELLS and blood vessels. It forms an inner layer in the spinal cord and some regions of the brain, but is the outer layer of the cerebral cortex of higher primates (*see* CEREBRUM). *See also* WHITE MATTER.

Grignard reagents A group of ORGANOMETALLIC COMPOUNDS of the type RMgHal where R is an ALKYL GROUP and Hal a HALIDE. Grignard reagents are formed by the REDUCTION of HALOGENOALKANES with magnesium in the presence of dry ETHER. They are used in organic synthesis to form carbon-carbon bonds, thereby building up carbon skeletons. They add to the CARBONYL GROUP of ALDEHYDES and KETONES to give secondary and tertiary ALCOHOLS respectively. Their reaction with carbon dioxide yields CARBOXYLIC ACID and AMIDES and NITRILES give ketones. An example of a Grignard reagent is ethylmagnesium bromide, C_2H_5MgBr. Grignard reagents are named for the French chemist Victor Grignard (1871–1935).

ground (*US*) *See* EARTH.

ground state The lowest energy state of a system, such as an atom, from which it can be excited to higher energy states. *See* ENERGY LEVEL. *See also* EXCITED STATE.

Group A column within the PERIODIC TABLE, containing elements with similar chemical properties.

group A set of elements, usually represented by **A**, **B** etc., combined in pairs by some operation, represented by ⊗ and having the following properties: (i) ASSOCIATIVE:

$$\mathbf{A} \otimes \mathbf{B} = \mathbf{B} \otimes \mathbf{A}$$

(ii) closed: the operation performed on any two members of the group gives a result that is also a member of the group; (iii) identity: there is a single member of the group, **I**, which leaves all others unchanged:

$$\mathbf{I} \otimes \mathbf{A} = \mathbf{A}.$$

(iv) inverse: for every member of the group there is another member such that the two together lead to the identity

$$\mathbf{A} \otimes \mathbf{A}^{-1} = \mathbf{I}$$

where $\mathbf{A}^{-1}$ is the inverse of **A**.

group displacement law *See* FAJAN AND SODDY'S GROUP DISPLACEMENT LAW.

growth The increase in size and weight of an organism during development. More accurately, growth is the increase in dry mass and not fresh mass (which includes water), but in practice the former is difficult to measure because the organism would have to be destroyed. If the growth of most populations, organisms or organs is plotted against time on a graph then a S-shaped growth curve is produced. This represents initial slow growth (lag phase), then fast growth (EXPONENTIAL growth) and then non-existent growth (and even negative growth) towards death.

The growth of a population occurs in several stages: the lag phase is where the growth of a small number of individuals is slow; the exponential phase is where enough individuals are available for reproduction and growth is rapid; the stationary phase is where factors within a given area become limiting and growth of the population stabilizes; and the death phase is where the numbers of individuals dying exceeds the numbers of growing individuals.

The rapid growth of an organism is usually limited by genetic and environmental factors, although some plants show unlimited growth during their lives. Factors that may limit growth include the supply of food, water, oxygen or light, disease, predation or the accumulation of toxic waste. In some populations growth is density-dependent, so that at a certain population density all the available resources are used up. Growth may be a modification of the S-curve, for example in humans the rapid growth phases are in the early years and at adolescence.

Some populations show a J-shaped growth curve, where the initial density increases very rapidly in an exponential manner and then stops suddenly due to environmental resistance or other limiting factors. Some organisms, for example ARTHROPODS, show intermittent growth because their EXOSKELETON cannot expand; they have to moult periodically during growth.

See also ALLOMETRIC GROWTH, ISOMETRIC GROWTH.

growth factor A general term referring to a group of substances that affect the growth of cells. Growth factors can signal a variety of effects, such as CELL DIVISION, differentiation, locomotion or survival. All growth factors bind RECEPTORS on the cell surface and are linked to intracellular EFFECTOR molecules, such as enzymes, which set off a cascade of signals. Examples include T-cell growth factor (TCGF), which is required for the proliferation of T CELLS, and epidermal growth factor (EGF), which stimulates many cells to divide.

growth hormone, *somatotrophin* A hormone produced by the anterior PITUITARY GLAND that controls body METABOLISM and growth generally. The production of growth hormone is itself regulated by the HYPOTHALAMUS. Low levels in humans results in dwarfism and high levels in gigantism.

guanine An organic base called a PURINE that occurs in NUCLEOTIDES. *See also* DNA, RNA.

guanosine A PURINE NUCLEOSIDE, consisting of the organic base GUANINE and the sugar RIBOSE.

guard cell A specialized cell in plants that controls gas exchange and water loss. Guard cells contain CHLOROPLASTS and are found in pairs at intervals along the underside of leaves. They are crescent-shaped and surround pores called stomata (*see* STOMA) that are the main route of water loss in a plant and the site of carbon dioxide and oxygen exchange. The pores can be opened or closed by the guard cells changing shape (by changes in turgidity). The inner wall of the guard cell is thicker and less flexible than the outer wall, so when water is taken up by OSMOSIS the cells become turgid and kidney-shaped, causing the stomata to open. When they are less turgid the stomata close. In this way water loss can be adjusted, for example they can be closed during warm weather, to prevent water loss through evaporation, and at night when PHOTOSYNTHESIS cannot take place.

gut *See* DIGESTIVE SYSTEM.

gymnosperm A plant of the PHYLUM CONIFEROPHYTA, whose seeds are not contained within an OVARY and instead lie exposed. This is in contrast to ANGIOSPERMS (flowering plants), where the seeds are contained in an ovary. The reproductive structures of gymnosperms are CONES, not flowers. Although still used, the term gymnosperm is no longer formally part of a classification scheme. Gymnosperms include conifers, cycads and ginkgos. The cycads were mostly large, palm-like plants found in the tropics and subtropics that are now extinct. The maidenhair tree is the sole survivor of the ginkgos.

gynoecium The female part of a flower, the collective name for the CARPELS.

gypsum A mineral form of calcium sulphate, consisting mostly of the DIHYDRATE $CaSO_4.2H_2O$.

gyre A large, permanent circulation of ocean currents.

gyroscope A rapidly rotating metal wheel, that will tend to retain the same direction in space regardless of any motion of its support. Such devices are used to measure and control the direction of aircraft and spacecraft. *See also* LAW OF CONSERVATION OF ANGULAR MOMENTUM, PRECESSION.

H

Haber-Bosch process *See* HABER PROCESS.

Haber process, *Haber-Bosch process* An industrial process for the manufacture of AMMONIA from nitrogen and hydrogen:

$$3H_2 + N_2 \Leftrightarrow 2NH_3$$

The reaction is highly reversible and under the conditions generally used the yield is only a few per cent.

A balance has to be struck between the higher yield at low temperatures and the unacceptably long time taken to approach this yield even in the presence of a CATALYST. In practice, a temperature of about 450°C is used, at a pressure of 250 atmospheres in the presence of an iron catalyst. The ammonia is extracted from the unreacted gases by cooling it below its boiling point and the unreacted nitrogen and hydrogen are recycled.

habitat A localized area within a BIOME, for example a freshwater pond, a cave or a woodland. Habitats are varied and continually changing and may contain many species or only one. A microhabitat is one that has very specific living conditions, for example the area under a stone. The habitat largely provides for the needs of its members, which are POPULATIONS of individuals together forming a COMMUNITY.

habituation A form of learned behaviour in which an animal learns to ignore a repetitive stimulus that is neither harmful nor beneficial. *See also* CONDITIONING, IMPRINTING, LEARNING.

Hadley cell A closed loop of CONVECTION within the Earth's atmosphere. *See* CLIMATE.

hadron Any of a class of subatomic particles that are influenced by the STRONG NUCLEAR FORCE. Hadrons can be subdivided into BARYONS (which include PROTONS and NEUTRONS) and MESONS. It is now known that all hadrons are composed of particles called QUARKS held together by the strong nuclear force. Hadrons other than protons and neutrons may be produced in experiments using a high energy beam from a PARTICLE ACCELERATOR.

haem A complex organic molecule containing iron. Haem combines with the protein GLOBULIN to form HAEMOGLOBIN.

haematite An iron ore containing iron(III) oxide, Fe_2O_3.

haemocoel The body cavity of some invertebrates, such as ARTHROPODS and MOLLUSCS, through which blood flows. *See* CIRCULATORY SYSTEM.

haemocyanin A blue RESPIRATORY PIGMENT found in MOLLUSCS and CRUSTACEANS.

haemodialysis *See* DIALYSIS.

haemoerythrin A red/brown RESPIRATORY PIGMENT found in ANNELIDS.

haemoglobin In vertebrates and some invertebrates a RESPIRATORY PIGMENT found in the RED BLOOD CELLS that is used to transport oxygen from the lungs to the body tissues. It is made of a HAEM group, which contains iron, and GLOBULIN (a protein). In humans it has a RELATIVE MOLECULAR MASS of 68,000. In other species, the relative molecular mass ranges from 16,000 to 3 million. The haem group is always the same, but the globulin varies from species to species. In humans there are four haem groups (this also varies between species), each able to carry one oxygen molecule, so the amount of oxygen that can be carried in the blood is increased compared to species with fewer haem groups.

Haemoglobin combines easily with oxygen in regions of high oxygen concentration (e.g. in the lungs or GILLS) to form oxyhaemoglobin (which is bright red compared with the red colour of haemoglobin alone) and is transported in the blood to the body tissues, and is easily released where oxygen is at low concentration.

Foetal haemoglobin differs slightly from that of an adult and combines with oxygen even more easily, which allows it to obtain oxygen from the mother's haemoglobin in the placenta.

The carbon dioxide concentration affects the release and uptake of oxygen by haemoglobin (*see* BOHR EFFECT). Some carbon dioxide is carried by haemoglobin as carbaminohaemoglobin. Haemoglobin can also combine with carbon monoxide to form carboxyhaemoglobin, but this reaction is irreversible and results in death if too much carbon monoxide is inhaled.

See also MYOGLOBIN, THALASSAEMIA.

haemolymph The watery fluid that forms the blood of certain invertebrates. *See* CIRCULATORY SYSTEM.

haemolytic disease of the newborn *See* RHESUS DISEASE.

haemophilia An inherited disease in which the normal BLOOD CLOTTING CASCADE does not function. It is a sex-linked condition carried by females but only expressed in males. Sufferers lack factor VIII, one of the substances needed to form THROMBOKINASE, and therefore the insoluble blood protein FIBRIN does not form and bleeding does not stop. Treatment is to administer factor VIII, which can now be made by RECOMBINANT DNA technology. *See also* SEX LINKAGE.

hafnium (Hf) The element with ATOMIC NUMBER 72; RELATIVE ATOMIC MASS 178.5; melting point 2,230°C; boiling point 4,602°C; RELATIVE DENSITY 13.3. Hafnium is a TRANSITION METAL. It has found some use in the nuclear industry and in some high temperature alloys.

hahnium The name proposed for the element with ATOMIC NUMBER 105. Only very short-lived ISOTOPES exist and very little is know about the properties of this element.

hail Frozen water droplets that fall from CUMULONIMBUS clouds, often in association with a THUNDERSTORM. The CONVECTION CURRENTS in the clouds cause raindrops to move up and down the cloud, freezing, collecting a layer of water, freezing again and so on, until they become too massive to be supported by the rising air in the cloud. *See also* WEATHER SYSTEMS.

hair A fine, long outgrowth from mammalian skin that consists of dead cells impregnated with KERATIN and MELANIN, which give the hair colour. Hair grows from follicles within the DERMIS, which are produced by inpushings of the MALPIGHIAN LAYER. Cells at the base of the follicle divide to produce hair. Muscles at the base of the hair can cause it to become erect, trapping air and providing insulation for some mammals. Hair can also have a sensory role.

half-cell A single ELECTRODE immersed in an ELECTROLYTE. Two half-cells, connected together by a salt bridge can be used for the comparison of ELECTRODE POTENTIALS. *See also* REDOX HALF-CELL.

half-life The time taken for one half of the radioactive nuclei originally present in a sample to decay (*see* RADIOACTIVITY). The half-life τ is related to the DECAY CONSTANT λ by the formula

$$\tau = \ln 2/\lambda$$

half-thickness The thickness of a given material needed to reduce the intensity of GAMMA RADIATION by one half. The half-thickness of a material depends on the energy SPECTRUM of the gamma radiation concerned.

halide Any BINARY COMPOUND containing a HALOGEN; that is, a FLUORIDE, CHLORIDE, BROMIDE or IODIDE.

Hall effect The production of a POTENTIAL DIFFERENCE between the edges of a conductor or semiconductor carrying a current in a magnetic field. The MAGNETIC FORCE on the CHARGE CARRIERS produces a build-up of charge at one side of the conductor. The potential difference between the two edges of a conductor in a magnetic field, measured at right angles both to the field and to the direction of current flow, is called the HALL VOLTAGE. The electric field resulting from this build-up of charge limits the size of the Hall voltage. In metals it is very small, even for strong fields. Semiconductors have a far lower number of free charge carriers per metre cubed and so produce larger Hall voltages.

Halley's comet The brightest and best known of the periodic COMETS. It has a period of 76 years and last reached PERIHELION in 1986.

Hall probe A device used to measure magnetic fields, using the HALL EFFECT in a small piece of semiconductor.

Hall voltage Voltage produced by the HALL EFFECT. The Hall voltage V_H produced between the edges of a conductor of thickness t carrying a current I in a magnetic field B and having n free charge carriers per metre cubed, each of charge e is given by

$$V_H = BI/net$$

haloenzyme *See* COFACTOR.

haloform reaction The reaction between the CH_3CO– group, found in carbonyl compounds (*see* CARBONYL GROUP), and a HALOGEN in alkaline solution, to yield $CHHal_3$ (where Hal stands for halogen). If the halogen is iodine the reaction is termed the TRI-IODOMETHANE TEST.

halogen Any of the elements from group 17 (formerly group VII) of the PERIODIC TABLE: FLUORINE, CHLORINE, BROMINE, IODINE and ASTATINE. They all exhibit similar chemical properties, forming salts containing an ion with a single negative charge. They combine with hydrogen to form POLAR MOLECULES, which dissociate in water to form acids.

The reactivity of the halogen decreases going down the series: fluorine is the most electronegative element (*see* ELECTRONEGATIVITY) and is highly reactive, while iodine is relatively unreactive and forms mostly IONIC compounds. Melting and boiling points increase with increasing atomic mass: fluorine and chlorine are gases at room temperature, whilst bromine is a volatile liquid, and iodine and astatine are solids.

halogenation The incorporation of a HALOGEN into an organic compound, by ADDITION or SUBSTITUTION REACTIONS. *See also* MARKOWNIKOFF'S RULE.

halogenoalkane, ***alkyl halide*** An ALKANE in which a hydrogen atom has been replaced by a HALOGEN atom. A halogenoalkane has the general formula $C_nH_{2n+1}Hal$, where Hal is a halogen. The terms primary, secondary and tertiary used to describe ALKYL GROUPS are applied to halogenoalkanes in the same way, although these terms do not strictly refer to the halogenoalkane itself. The chemistry of the halogenoalkane is determined by the halogen FUNCTIONAL GROUP rather than the alkane skeleton to which it is attached. The melting and boiling points of halogenoalkanes are higher than those of their corresponding alkane.

Halogenoalkanes can be manufactured by reacting ALCOHOLS with phosphorus halides or sulphur dichloride oxide. For example, for chloroalkanes:

$$ROH + PCl_5 \rightarrow RCl + HCl + POCl_3$$

$$ROH + SCl_2O \rightarrow RCl + HCl + SO_2$$

where R is an alkyl group.

Bromoalkanes and iodoalkanes can be manufactured by reacting alcohols with concentrated aqueous hydrogen bromide or hydrogen iodide respectively. For example:

$$C_2H_5OH + HBr \rightarrow C_2H_5Br + H_2O$$

Halogenoalkanes undergo ELIMINATION and SUBSTITUTION REACTIONS (*see* BROMOETHANE).

Useful halogenoalkanes include dichloromethane as a solvent, chlorofluoromethane as a refrigerant and aerosol propellant, bromochlorodifluoromethane as a fire extinguisher and 2-bromo-2-chloro-1,1,1-trifluoroethane (HALOTHANE) as an anaesthetic. CHLOROFORM (trichloromethane) was used as an anaesthetic and CARBON TETRACHLORIDE (tetrachloromethane) as a dry cleaning agent although these latter two are no longer in use since they are thought to induce cancer.

halosere A series of plant SUCCESSIONS originating on land emerging from the sea, such as a salt marsh.

halothane, ***2-bromo-2-chloro-1,1,1-trifluoroethane*** ($CF_3CHBrCl$) A general ANAESTHETIC, now widely used since it is relatively nontoxic. It is a colourless liquid with an odour similar to CHLOROFORM.

handshake In computing, an exchange of signals to co-ordinate the activities of two computers.

haploid (*adj.*) Describing the presence of one set of CHROMOSOMES in the nucleus of a cell. GAMETES of higher organisms are haploid (as a result of MEIOSIS) while the other body cells are DIPLOID. Some plants are haploid, for example mosses and liverworts. *See also* POLYPLOID.

haptonasty *See* THIGMONASTY.

haptotropism *See* THIGMOTROPISM.

hard (*adj.*) **1.** Describing any material that is not easily scratched.

2. Describing FERROMAGNETIC materials in which the DOMAIN WALLS (boundaries between one DOMAIN and the next) are held in place (by carbon atoms in steel for example), so the material retains its magnetism when the magnetizing field is removed. Magnetically hard alloys are used in the manufacture of PERMANENT MAGNETS and for materials such as magnetic tapes and computer disks, which use magnetic fields to store information.

hard copy Computer jargon for material printed on paper rather than displayed on a screen.

hard disk A high capacity magnetic storage device, normally incorporated into a computer and not designed to be changed.

hardness 1. Resistance to scratching.

2. The presence of dissolved calcium and magnesium salts in water. *See* HARD WATER.

hardware The actual mechanical, electrical and electronic components of a computer system. *Compare* SOFTWARE.

hard water Water containing dissolved salts of calcium and magnesium. These react with soap to form insoluble compounds, which produce a scum on the water surface and prevent the soap from forming a lather. The main cause of hardness is the presence of carbonate salts, which dissolve in the water when it passes through porous rocks such as limestone. These are precipitated (*see* PRECIPITATION) when the water is boiled (in kettles for example), forming limescale. These deposits are poor conductors of heat and deposits on heating elements lead to reduced efficiency of heating and failure of heaters, due to the high temperatures reached. *See also* ION EXCHANGE, PERMANENT HARDNESS, SOFT WATER, TEMPORARY HARDNESS, WATER SOFTENING.

Hardy–Weinberg Equilibrium *See* HARDY–WEINBERG PRINCIPLE.

Hardy–Weinberg Principle, *Hardy–Weinberg Equilibrium* A mathematical representation of the theoretical relative frequencies of ALLELES within a given POPULATION. The principle assumes an isolated population with no NATURAL SELECTION, MUTATION or GENETIC DRIFT (so there is no gene flow into or out of the population).

These conditions are never met in a real population, but the principle is used in the study of gene frequencies. For example, there may be a DOMINANT allele 'A' and a RECESSIVE allele 'a'. In diploid individuals the combinations 'AA', 'Aa', 'aA' and 'aa' are possible. It might be assumed that the HOMOZYGOUS recessive 'aa' would in time be eliminated by natural selection, but the proportion of dominant to recessive alleles remains the same because the HETEROZYGOUS state 'Aa' or 'aA' acts as a reservoir for the recessive allele 'a'.

If p and q are the frequencies of the dominant 'A' allele and recessive 'a' allele' respectively', then the Hardy–Weinberg principle is expressed as:

$$p^2 + 2pq + q^2 = 1.0$$

This formula is useful in determining the frequency of any allele in a population.

harmonic A whole number multiple of a given FREQUENCY, such as the FUNDAMENTAL frequency, in a system that supports STANDING WAVES. The frequency that is twice the fundamental is called the second harmonic, that which is three times the fundamental is called the third harmonic etc.

Haversian canal Any of the tubes in the BONE that contain the blood and LYMPH VESSELS and NERVES.

Hawking radiation ELECTROMAGNETIC RADIATION arising from the creation of ELECTRON-POSITRON pairs in the intense GRAVITATIONAL FIELD of a BLACK HOLE with only one member of the pair falling into the black hole.

H-bomb *See* HYDROGEN BOMB.

HCG *See* HUMAN CHORIONIC GONADOTROPHIN.

heart A muscular organ that, by rhythmic contractions, pumps blood around the body of an animal with a CIRCULATORY SYSTEM.

In mammals the heart consists of four chambers: two atria (*see* ATRIUM) and two VENTRICLES. Oxygenated blood enters the left atrium from the lungs (via the PULMONARY vein) and passes then through the BICUSPID VALVE into the left ventricle, where it is pumped out through the main ARTERY, the AORTA, to the general circulation. The right atrium receives deoxygenated blood from the circulation from the main vein, the VENA CAVA, passes through the tricuspid valve into the right ventricle, and then to the lungs via the pulmonary artery. The aorta and pulmonary artery also have non-return valves called semi-lunar valves. Oxygenated and deoxygenated blood never mix because the right and left sides of the heart are completely separate.

The muscle of the mammalian heart is a specialized CARDIAC MUSCLE that generates the HEARTBEAT from within (myogenic) rather than via an external NERVE IMPULSE (neurogenic, as in insects).

A blood clot in the coronary arteries (which supply blood to the heart muscle itself) can lead to a heart attack, where the heart is deprived of blood. The severity of a heart attack depends on the position of the clot, whether it is in the main artery or a branch. Susceptibility to heart attacks is affected by factors such as smoking, diet (animal fats and

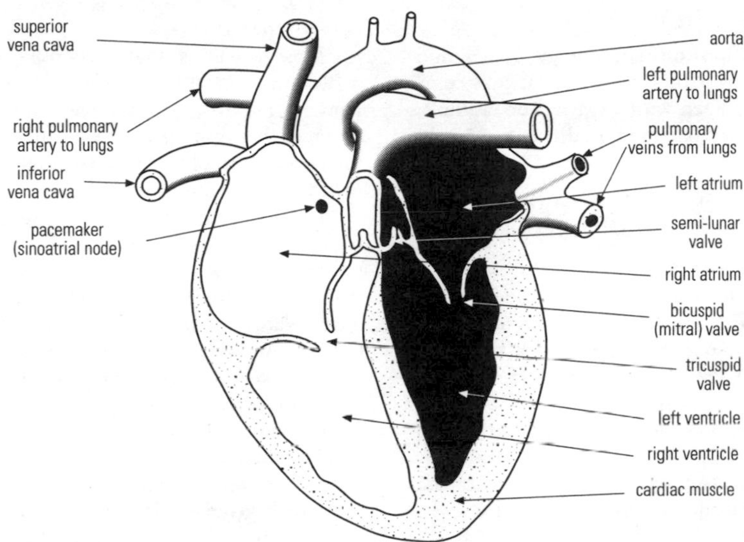

Structure of the human heart.

high salt cause ATHEROSCLEROSIS), stress and lack of exercise.

Fish have a single circulation (*see* CIRCULATORY SYSTEM) with only two chambers in the heart, an atrium and a ventricle. Amphibians and reptiles have a double circulation, with two atria and one ventricle, so blood enters separately from the lungs and body but oxygenated and deoxygenated blood then mixes in a single ventricle.

See also PACEMAKER, PURKINJE FIBRES.

heart attack *See* HEART.

heartbeat The regular contractions and relaxation of the HEART and the accompanying sounds made by the opening and closing of the valves. *See also* PACEMAKER, PULSE.

heat The energy that is transferred from one body or system to another as a result of a difference in temperature. In SI UNITS, heat, like all other forms of energy, is measured in JOULES. *See also* CONVECTION, HEAT CAPACITY, LATENT HEAT, LAW OF CONSERVATION OF ENERGY, THERMAL CONDUCTION, THERMAL RADIATION, THERMODYNAMICS.

heat capacity The amount of heat needed to change the temperature of an object by one degree CELSIUS (or one KELVIN, which is the same size). If the temperature is increasing, this much heat energy will have been taken in by the object; if it is decreasing, the energy will have been given out. *See also* CONSTANT FLOW METHOD, METHOD OF MIXTURES, MOLAR HEAT CAPACITY, RATIO OF SPECIFIC HEATS, SPECIFIC HEAT CAPACITY.

heat engine Any machine for converting HEAT energy to mechanical WORK. *See* CARNOT CYCLE, KELVIN STATEMENT OF THE SECOND LAW OF THERMODYNAMICS.

heat exchanger A device for transferring heat energy from one fluid to another, without contact between the two fluids. In many applications, a hot liquid, often water, needs to be cooled, giving up its heat to the surrounding air. The liquid flows through pipes to which thin metal plates are attached. Air is then forced over the plates carrying away heat. *See also* COUNTERCURRENT SYSTEM.

heat of atomization The energy needed to convert one mole of a given substance from its normal form at the temperature specified to atoms infinitely far apart from one another. For example, the heat of atomization for hydrogen is 433 kJ mol^{-1}:

$$H_2 \rightarrow 2H, \Delta H = 433 \text{ kJ mol}^{-1}$$

heat of combustion The energy of reaction when one MOLE of a given substance is completely combined with oxygen. For example, the heat of combustion for ethane is –1,556 kJ mol^{-1}:

$$C_2H_6 + 3^1/_2O_2 \rightarrow 2CO_2 + 3H_2O \quad \Delta H = -1{,}566 \text{ kJ mol}^{-1}$$

See also CALORIFIC VALUE.

heat of dilution The energy change when one mole of a given substance is diluted (*see* DILUTE) from one specified CONCENTRATION to another. At low concentrations, the ions or molecules in a solution no longer interact with one another and the heat of dilution falls to zero.

heat of formation The energy released or the ENTHALPY change when one MOLE of a given compound is formed from its elements in their usual states at the temperature specified. For example, the heat of formation of ammonia, NH_3, is –46 kJ mol^{-1}:

$$^1/_2N_2 + 1^1/_2H_2 \rightarrow NH_3, \Delta H = -46 \text{ kJ mol}^{-1}$$

The minus sign indicates that energy is given off in this process. *See also* STANDARD HEAT OF FORMATION.

heat of ionization The energy needed to completely ionize (*see* IONIZATION) one mole of a given compound with the ions being formed in an infinitely dilute aqueous solution.

heat of neutralization The energy of reaction when one MOLE of hydrogen ions are completely reacted with a BASE. In the case of a neutralization of a strong acid by a strong base, the heat of reaction is essentially the heat of reaction for the formation of water from aqueous ions:

$$H + (aq) + OH^- (aq) \rightarrow H_2O\ (l), \quad \Delta H = 57.3 \text{ kJ mol}^{-1}$$

For weak acids or weak bases, the heat of neutralization is different by an amount needed to ionize the weak acid or base, which depends on the HEAT OF IONIZATION of these compounds.

heat of reaction The amount of energy given out, or taken in, in a given chemical reaction. It is usually quoted in terms of the number of joules of energy per MOLE of reacting substances. The symbol used for a heat of reaction is ΔH, and this is taken as negative if heat is given out (that is, lost from the reaction).

Most reactions that take place spontaneously are EXOTHERMIC, with ΔH negative. However, ENDOTHERMIC reactions may take place provided they result in an increase in ENTROPY – the breaking up of a crystal for example, or an increase in the number of molecules.

Heats of reaction are sometimes referred to as enthalpies of reaction. The difference between the two terms is only significant if gases are involved, as the ENTHALPY includes a term relating to the energy needed to push back the surrounding atmosphere if a gas is given off, keeping the reacting vessel at a constant pressure, usually specified as ATMOSPHERIC PRESSURE.

See also HESS'S LAW

heat of solution The energy change when one mole of a given substance is dissolved in a large volume of a specified SOLVENT, usually water. The volume of solvent must be such that the dissolved substance is at such a low concentration that the addition of more solvent produces no further change.

heat pump A machine for removing heat from one system and depositing it in a second, hotter system. Energy must be supplied to make the heat flow in this direction (*see* SECOND LAW OF THERMODYNAMICS).

Refrigerators and air conditioning systems are examples of heat pumps. Heat pumps have also been proposed as a more energy efficient means of heating buildings in cold weather, but, because of the increased complexity compared to direct heating, they are not widely used.

heat radiation *See* THERMAL RADIATION.

heat reservoir A hypothetical object of infinite HEAT CAPACITY, which heat can enter or leave without producing a change in temperature.

heat treatment The heating and cooling of a material, usually a metal or alloy, under controlled conditions to produce changes in the mechanical properties of the material which remain after it has returned to room temperature. *See* ANNEALING, QUENCHING.

heavy metal The collective term for metals of high atomic mass, particularly those TRANSITION METALS that are toxic and cannot be processed by living organisms, such as lead, mercury, and cadmium.

heavy water Water in which both hydrogen atoms have been replaced with the DEUTERIUM ISOTOPE. Heavy water is used as a MODERATOR in some forms of nuclear reactor.

Heisenberg's microscope A thought experiment designed to illustrate the effect OF HEISENBERG'S UNCERTAINTY PRINCIPLE on experimental observations. A particle is viewed through a microscope but the image is affected by DIFFRACTION at the lens of the microscope, producing an uncertainty in position. To reduce this, shorter wavelength light could be used, but this will result in a greater uncertainty in the MOMENTUM transferred to the particle when this light is scattered into the microscope, since the only way the exact path of the PHOTON could be known is by using a microscope with a very small opening, leading to greater diffraction so again a larger uncertainty in position.

Heisenberg's uncertainty principle A consequence of the WAVE NATURE OF PARTICLES. Loosely stated, this means that the more accurately we know the position of a particle the less sure we can be of the associated wavelength and hence the MOMENTUM, and vice versa. More precisely, if the position is known with an uncertainty Δx and the momentum with an uncertainty Δp then

$$\Delta x \Delta p \geq h/\pi$$

where h is PLANCK'S CONSTANT. *See also* HEISENBERG'S MICROSCOPE, VIRTUAL PARTICLE.

helium (He) The element with ATOMIC NUMBER 2; RELATIVE ATOMIC MASS 4.0; boiling point –269°C; does not solidify at ATMOSPHERIC PRESSURE. Helium is the second most abundant element in the universe, making up almost 25 per cent of the atoms. It is scarce on Earth as it occurs only as a gas and its low mass means that it easily escapes from the Earth's gravitational pull. Helium was first discovered from its ABSORPTION SPECTRUM in sunlight.

Helium is found in some porous rocks in association with the production of ALPHA PARTICLES from radioactive ores. It is used as a filling in airships and high-altitude balloons, where it has replaced inflammable hydrogen. The gas is also used to dilute oxygen when fed to deep-sea divers under high pressure. Helium has a very low boiling point, about 4 K (–269°C), which has also led to the widespread use of liquid helium in low temperature refrigeration systems, particularly for superconducting magnets (*see* SUPERCONDUCTIVITY).

See also SUPERFLUIDITY.

helix A spiral in three dimensions. A helix can be described by the PARAMETRIC EQUATION:

$$x = r\cos\omega t,\ y = r\sin\omega t,\ z = kt.$$

The pitch of a helix is the distance moved along the axis of the helix in the course of one complete rotation. For the above equation the pitch is equal to $2\pi k/\omega$.

Helmholtz coils A pair of flat parallel coils separated by a distance equal to their radius. Helmholtz coils give a magnetic field that changes very little in the space between the coils.

Helmholtz free energy *See* FREE ENERGY.

hemisphere One half of a sphere, formed by the intersection between a sphere and a plane passing through its centre. The volume of a hemisphere is $2\pi r^2$, where r is the radius of the hemisphere, the distance between the centre of the flat surface and any point on the curved surface.

henry (H) The SI UNIT of SELF-INDUCTANCE and MUTUAL INDUCTANCE. An inductance of one henry will result in an induced ELECTROMOTIVE FORCE (e.m.f.) of one VOLT when the current producing the induced e.m.f. is changing at the rate of one AMPERE per second.

Henry's law The ratio of the CONCENTRATIONS of a given molecule in two phases is always constant at a given temperature. This accounts for the constant nature of the PARTITION CONSTANT, where the two phases are the two SOLVENTS. This also accounts for the fact that the SOLUBILITY of a gas is proportional to its pressure – here the two phases are the gas itself and the gas molecules in solution. This law does not apply to any case where the molecules change their nature in the two phases, for example in the case of hydrogen chloride or ammonia, which both form ions in aqueous solution.

heparin A natural ANTICOAGULANT in the body that inhibits the action of THROMBIN.

hepatic (*adj.*) Relating to the LIVER.

Hepaticae A CLASS of the PHYLUM BRYOPHYTA, consisting of the liverworts. Liverworts are small flat or leafy plants similar to MOSSES and are found in damp conditions. They show ALTERNATION OF GENERATIONS.

hepatic portal vein The vein that transports blood rich in soluble digested food from the INTESTINE to the LIVER.

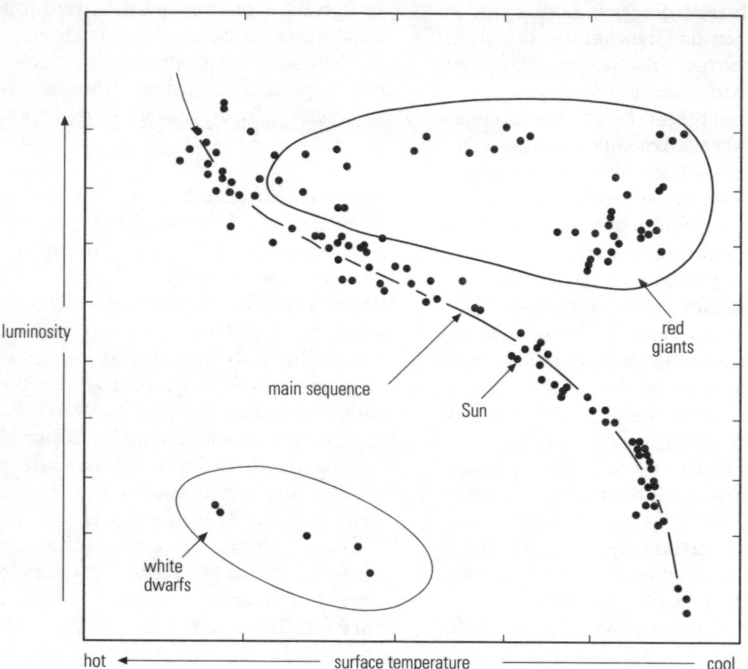

Hertzsprung–Russell diagram.

hepatocyte One of many cuboidal cells in the LIVER, with a large nucleus and containing many MITOCHONDRIA and GLYCOGEN granules in its CYTOPLASM. Hepatocytes also have MICROVILLI on their surface to facilitate exchange of materials between the blood and liver.

heptane (C_7H_{16}) The seventh member of the ALKANE series. There are nine possible ISOMERS. *n*-Heptane ($CH_3[CH_2]_5CH_3$) is a colourless, inflammable liquid with a boiling point of 98°C. Heptane is obtained from PETROLEUM and is used in defining the OCTANE NUMBER of PETROL.

herbaceous perennial *See* PERENNIAL PLANT.

herbicide A PESTICIDE that kills weeds. Herbicides make up 40 per cent of all pesticides used. They act by stimulating AUXIN production and disrupting plant growth (*see* PLANT GROWTH SUBSTANCES).

herbivore An animal whose diet consists of green plants, seeds or fruit. The largest group of herbivores is the ZOOPLANKTON on the surface of the ocean, which feed on small ALGAE. RUMINANTS such as cows, sheep and deer are herbivorous mammals and digest CELLULOSE as the major part of their diet. They can digest cellulose because they have specialized bacteria in their digestive system that are able to make the enzyme cellulase, which breaks down the cellulose in plant material. The teeth of ruminants are adapted to break down plant food more thoroughly, and their stomachs consist of four chambers to allow the bacteria breaking down cellulose to operate at the correct PH, isolated from the mammal's own GASTRIC JUICES. *See also* CARNIVORE, OMNIVORE.

hermaphrodite An organism that contains both female and male sex organs. In plants this means that STAMENS and CARPELS are in the same flower. In animals, for example, earthworms, an individual can produce both SPERM and ova (*see* OVUM). Despite this, hermaphroditic species usually cross-fertilize.

hertz (Hz) The SI UNIT of FREQUENCY. One hertz is a frequency of one oscillation per second.

Hertzsprung–Russell diagram A plot of the LUMINOSITY of various stars against their SPECTRAL CLASS (which is effectively a measure of temperature). Most stars lie in a diagonal band from top left (hot bright stars) to bottom right (cool dim stars) called the main sequence. *See also* MAIN SEQUENCE STAR.

hesperidium A type of berry with a thick leathery outer layer and fluid-containing segments inside. Examples include the citrus fruits.

Hess's Law An application of the LAW OF CONSERVATION OF ENERGY. This law states that the overall energy change involved in any chemical process is independent of the route by which that process takes place. Thus if the reaction A→B can proceed via two possible intermediates, X and Y, the HEATS OF REACTION for A→X and X→B will add to the same total as those for A→Y and Y→B, this total being the heat of reaction for A→B. The law is useful since it allows heats of reaction to be calculated for reactions that have not been observed directly.

heterocyclic (*adj.*) Describing any organic compound that is CYCLIC but which contains different atoms within a closed ring structure. *See also* ALICYCLIC, AROMATIC.

heterogametic *See* SEX DETERMINATION.

heterogeneous (*adj.*) Describing a substance whose properties vary from one place to another. *Compare* HOMOGENEOUS.

heterogeneous catalyst A CATALYST that is in a different phase from the reaction it catalyses. Usually a solid catalyst, such as finely divided platinum, is used to catalyse gaseous reactions.

heterolytic fission The breaking of a COVALENT BOND to give two oppositely charged fragments. The electrons forming the bond are thus shared unequally between the fragments. This is particularly common in compounds with polar bonding (*see* POLAR BOND). An example is the dissociation of hydrogen chloride in an aqueous solution:

$$HCl \rightarrow H^+ + :Cl^-$$

In contrast to HOMOLYTIC FISSION the electrons in a heterolytic reaction remain paired. Such reactions involve IONS rather than RADICALS (as in homolytic fission). Examples of heterolytic reactions are those of HALOGENOALKANES.

heterophagosome *See* PHAGOSOME.

heterosome A CHROMOSOME that is not the same in males and females; in other words, the sex chromosomes. *Compare* AUTOSOME *See also* SEX DETERMINATION.

heterosporous (*adj.*) Describing plants that have two types of SPORES. Microspores give rise to male GAMETOPHYTE generations, and megaspores give rise to female gametophytes (*see* ALTERNATION OF GENERATIONS). All SEED PLANTS are heterosporous and some CLUB MOSSES and FERNS. *Compare* HOMOSPOROUS.

heterotroph An organism that feeds on other animals or plants to obtain energy. All animals and fungi are heterotrophs. Heterotrophs can be HERBIVORES, CARNIVORES, OMNIVORES, SAPROTROPHS, DETRITIVORES and PARASITES. *Compare* AUTOTROPH.

heterozygous (*adj.*) In genetics, describing the presence of two different ALLELES at a particular LOCUS on a CHROMOSOME in a DIPLOID cell or organism. This is in contrast to the HOMOZYGOUS condition, where the alleles at a given locus are identical. RECESSIVE alleles are not expressed in the heterozygous state. Individuals in an outbreeding population will be heterozygous for some traits and homozygous for others.

heuristics The use of trial and error to solve a problem, as opposed to using a formal, mathematically rigorous ALGORITHM.

1,2,3,4,5,6-hexachlorocyclohexane *See* BHC.

hexacyanoferrate(II) Any salt containing the complex hexacyanoferrate(II) ion $Fe(CN)_6^{4-}$. The salts are generally yellow in colour.

hexacyanoferrate(III) Any salt containing the complex hexacyanoferrate(III) ion $Fe(CN)_6^{3-}$. These salts are usually blood red in colour, though an important blue pigment, Prussian blue, is the DOUBLE SALT $KFe[Fe(CN)_6]$.

hexadecimal (*adj.*) Describing a number system that works in base–16.

hexagonal close packed (*adj.*) Describing a crystalline structure in which each layer of atoms is CLOSE PACKED, with each atom surrounded by 6 others in that layer. The second layer of atoms lies above the gaps in the first, whilst the atoms in the third layer lie above gaps in the second and directly above the atoms in the first. If the layers of atoms are labelled *A* and *B*, the structure can be described as *ABAB*... Many metals occur as hexagonal close packed structures, including magnesium and zinc, though the CUBIC CLOSE PACKED structure is also common.

hexane (C_6H_{14}) The sixth member of the ALKANE series. There are five ISOMERS of hexane. n-Hexane, $CH_3[CH_2]_4CH_3$, is a colourless liquid with a boiling point of 69°C. It is used as a solvent.

hexose A MONOSACCHARIDE containing six carbon atoms in the molecule.

hibernation A state characterized by greatly reduced metabolic processes (reduced breathing and heart rate, fallen body temperature) that some animals enter to survive the winter. *See also* BIORHYTHM.

Higgs boson A hypothetical particle in the STANDARD MODEL of particle physics. The Higgs boson is responsible for giving mass to all the charged LEPTONS and QUARKS and is regarded as one of the more unsatisfactory aspects of the standard model. Its mass is about 300 times that of the proton, too heavy to have been found in the present generation of particle physics experiments, but within the reach of those now being planned.

high In DIGITAL electronics, a voltage level or other signal corresponding to a BINARY 1.

high-bypass engine, *turbofan* A GAS TURBINE in which much of the air leaves the back of the engine without being used to burn fuel.

highest common factor The largest INTEGER that is a FACTOR for all of the integers specified. For example, the highest common factor of 15, 30 and 60 is 5.

high-level language A computer PROGRAMMING LANGUAGE designed to be relatively simple to understand. They typically use words and grammar that resemble the language of the user. They are independent of the MACHINE CODE of any particular computer and must be translated by a COMPILER or INTERPRETER before being executed by a computer.

high-temperature superconductor A material that remains a SUPERCONDUCTOR at temperatures that can be reached using liquid nitrogen (77 K/–196°C) rather than liquid helium (4 K/–269°C). The ultimate goal is to produce a material which is superconducting at room temperature. The high-temperature superconductors discovered so far are ceramic materials and are chemically rather unstable. Their poor mechanical properties have made them less commercially valuable than was originally hoped.

hindbrain One of three regions of the human brain. The hindbrain contains the MEDULLA OBLONGATA that controls activities such as RESPIRATION, HEARTBEAT and blood pressure. Overlying this is the CEREBELLUM that controls the muscle movement needed for posture and locomotion. *See also* FOREBRAIN, MIDBRAIN.

hip girdle *See* PELVIC GIRDLE.

Hirudinea A CLASS of the PHYLUM ANNELIDA, consisting of leeches (e.g. *Hirudo*). Leeches possess suckers by which they attach themselves to a host to feed (they are therefore temporary PARASITES). They have no chaetae (bristles) and no distinct head and they are HERMAPHRODITES. Movement is by means of the suckers. Leeches are important medically. *Compare* POLYCHAETA, OLIGOCHAETA. *See also* ANNELIDA.

histamine A derivative of the amino acid histidine, released by MAST CELLS in response to the appropriate ANTIGEN during an inflammatory or allergic response. Histamine causes dilation of local blood vessels and an increase in their permeability to allow through, for example, ANTIBODIES, FIBRINOGEN and NEUTROPHILS, which are needed for repair or recovery of the site. During an allergic response, such as hayfever (allergy to pollen), the release of histamine causes some inflammation and this is responsible for the characteristic itching and sneezing. *See also* ALLERGIC REACTION, INFLAMMATION.

histidine An AMINO ACID, the precursor of HISTAMINE.

histiocyte A MACROPHAGE found within TISSUE.

histochemistry The study of the distribution of molecules within the cells and matrices of TISSUES using staining techniques and MICROSCOPY.

histogram A BAR CHART that displays the frequency with which particular values or ranges of values are found in a set of data. The areas of the bars are proportional to the frequency, and successive ranges are placed side by side.

histology The study of TISSUE structure mostly by means of staining and MICROSCOPY.

histone One of a group of proteins that is important in the packaging of eukaryotic DNA (*see* EUKARYOTE). Histones, together with DNA, form CHROMATIN, which is the major component of CHROMOSOMES.

HIV (human immunodeficiency virus) The virus that causes AIDS. It is a RETROVIRUS that originated in Africa and was first identified in 1983. The virus infects T-helper cells (*see* T

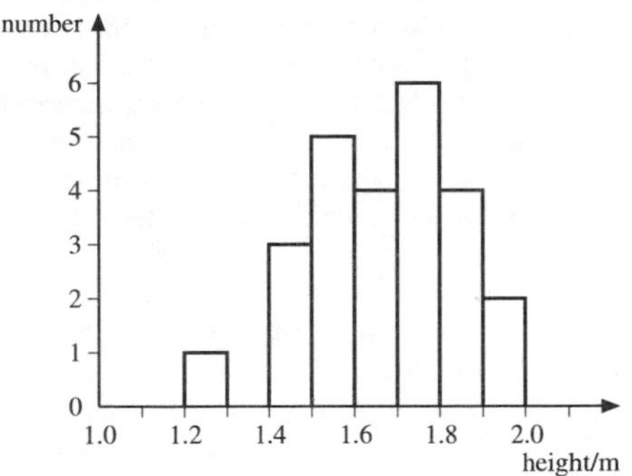

Histogram.

CELL), which are needed for the IMMUNE SYSTEM to function. An infected (HIV-positive) person is unable to fight the HIV and is also susceptible to other diseases because the immune system is deficient.

The virus is transmitted by blood or sexual secretions and therefore intravenous drug users, haemophiliacs, surgical patients receiving blood and those taking part in unprotected sex are all at high risk. The virus can remain dormant for about 8 years but the HIV-positive person is then a carrier of the virus, through blood and sexual secretions only. Pregnant women can pass the virus onto their children through the placenta or breast-milk. Transmission is prevented by the use of CONDOMS during sexual intercourse, and by screening donated blood and blood products for the virus.

Hoffman clip A metal clip used to squash a rubber tube to stop or control the flow of material through the tube.

Hofmann degradation The transformation of an AMIDE to a primary AMINE containing one carbon less than the original amide. This occurs when amides react with chlorine or bromine solutions in the presence of excess alkali. For example:

$$RCONH_2 \xrightarrow[NaOH]{Br_2} RNH_2 + CO_2 + HBr + NaBr$$

In this reaction the HALOGEN first replaces a hydrogen atom of the NH_2^- group to form a chloro- or bromo-amide. This reacts with alkali to give an isocyanate (compound with an –N=C=O group), which then decomposes to give an amine and carbon dioxide. The Hofmann degradation is used in the preparation of 2-aminobenzoic acid which is important in the DYESTUFFS industry.

Hofmann voltammeter A VOLTAMMETER designed to collect gases given off in ELECTROLYSIS.

hole A vacant space in the electron structure of a SEMICONDUCTOR. A neighbouring electron may move to fill this hole, which then appears to move through the semiconductor in the opposite direction to the electron flow. Holes can be thought of as behaving like positive CHARGE CARRIERS.

holmium (Ho) The element with ATOMIC NUMBER 67; RELATIVE ATOMIC MASS 164.9; melting point 1,472°C; boiling point 2,700°C; RELATIVE DENSITY 8.8. Holmium is a LANTHANIDE metal, known to be magnetic, but too difficult to separate from other elements to be commercially useful.

Holocene The current EPOCH of geological time, which began about 10,000 years ago, following the last period of glaciation in the PLEISTOCENE epoch. During this time the climate became warmer and humans developed significantly.

holocrine gland An EXOCRINE GLAND in which cells are destroyed with the discharge of their secretion, for example sebaceous and SWEAT GLANDS. *See also* APOCRINE GLAND, MEROCRINE GLAND.

hologram A three-dimensional image recorded on a flat piece of photographic film. In the simplest method of producing a hologram, a beam of COHERENT light, usually from a laser, is split into two by a semi-reflecting mirror. One beam (the signal beam) is diffracted by the object being recorded onto a piece of photographic film or plate. The other beam (the reference beam) falls directly onto the film, where it interferes with the signal beam. The INTERFERENCE pattern thus produced forms the hologram.

Once the film has been processed, the hologram can be viewed by illuminating it with a coherent beam of light (usually of the same wavelength as the original beam). As the viewer changes his angle of view, he sees the object from a different perspective. Each small part of the hologram effectively contains a whole image of the object, but only from a single viewpoint.

As with any other interference effect, distances comparable to the wavelength of light are critical. This has lead to interference holograms, used to study vibrations of solid objects. Because holograms are hard to make, they are also used as security labels in devices such as credit cards.

holography The process of making HOLOGRAMS.

homeostasis (*homeo* = same, *stasis* = standing) The maintenance of a constant internal environment within an organism. Homeostasis is important for efficient functioning of ENZYMES, which affects the entire organism.

There are many control systems to maintain homeostasis. Control of blood sugar levels is by the liver and involves interconversion of GLUCOSE and GLYCOGEN. This is regulated by the hormones INSULIN and glycogen from the PANCREAS. Respiratory gases (carbon dioxide and oxygen) are controlled by respiratory centres in the brain. Blood pressure is controlled by the rate of the HEARTBEAT. The HEART pumps blood and ensures the distribution of blood with its supply of essential materials to cells. OSMOREGULATION controls SOLUTE concentration and total body volume in an organism. Salt and PH levels must also be maintained.

Control of body temperature is called HOMEOTHERMY. Mammals and birds maintain a constant body temperature regardless of the external temperature and are called HOMEOTHERMS. Invertebrates, fish, amphibians and reptiles have a fluctuating body temperature, and obtain most of their heat from outside the body, and they are called POIKILOTHERMS.

homeotherm, *endotherm* An animal exhibiting HOMEOTHERMY. *Compare* POIKILOTHERM.

homeothermy (*homeo* = same, *thermo* = heat) The maintenance of a constant body temperature regardless of the external temperature and usually higher than it. Homeothermy is characteristic of mammals and birds, which are called homeotherms or endotherms (warm-blooded). Most homeotherms have an insulating layer of fat or fur that helps prevent heat loss and heat gain. The HYPOTHALAMUS controls other ways of maintaining a constant body temperature (in response to messages received from skin sensors), such as reducing the metabolic rate, sweating, panting and vasodilation (widening of blood vessels) to increase heat loss, and increasing the metabolic rate, shivering and vasoconstriction (narrowing of blood vessels) to reduce heat loss. Some means of homeothermy are behavioural, such as HIBERNATION or being nocturnal to avoid daytime heat. Homeothermy enables a greater efficiency of metabolic functions in contrast to POIKILOTHERMY, in which the body temperature fluctuates.

homogametic *See* SEX DETERMINATION.

homogeneous (*adj.*) Describing a substance that is the same throughout, such as a material made from a single compound. *Compare* HETEROGENEOUS.

homogeneous catalyst A CATALYST that is in the same phase as the reaction it catalyses.

homoiotherm *See* HOMEOTHERM.

homologous (*adj.*) Describing a similarity in some aspect, such as structure, position or functional properties. Chromosomes that are similar in appearance are called homologous and associate in pairs during MEIOSIS. In CLASSIFICATION, structures similar in appearance or origin are said to be homologous. In chemistry, organic chemicals form HOMOLOGOUS SERIES.

homologous series A group of organic compounds with similar chemical properties, forming a series in which the members differ from each other by a constant RELATIVE MOLECULAR MASS. ALKANES form such a series where methane is the first member and the series progresses by the addition of $-CH_2-$ through to ethane, propane, butane, pentane. ALKENES, ALKYNES, ALCOHOLS, ALDEHYDES, KETONES, and CARBOXYLIC ACIDS form similar homologous series. The physical properties of members of such a series show a steady change along the series.

homolytic fission The breaking of a COVALENT BOND to give two neutral atoms or FREE RADICALS (fragments with unpaired electrons). An example is the splitting of chlorine into free radicals by the action of ultraviolet light:

$$Cl_2 \rightarrow 2Cl^-$$

Many of the reactions of SATURATED HYDROCARBONS involve homolytic fission since the ELECTRONEGATIVITIES of carbon and hydrogen are similar and carbon-carbon and carbon-hydrogen bonds exert an equal attraction on the bonding electrons. Such reactions are usually initiated by other radicals, for example fluorine atoms released by the splitting of the weak bond in molecular fluorine.

The term homolytic reaction is also used to refer to the reaction in which radicals react with each other to form a covalent bond. Examples of homolytic reactions are those between ALKANES and chlorine or bromine, which are important in the manufacture of solvents, anaesthetics and refrigerants.

Compare HETEROLYTIC FISSION.

homosporous (*adj.*) Describing those plants that have one type of SPORE that gives rise to GAMETOPHYTES bearing both male and female reproductive organs. Examples include ferns and horsetails. *Compare* HETEROSPOROUS.

homozygous (*adj.*) In genetics, referring to the presence of two identical ALLELES at a particular LOCUS on a CHROMOSOME in a DIPLOID cell or organism. An individual homozygous for a particular trait will always pass this trait onto their offspring. RECESSIVE alleles are only expressed in the homozygous state. *See also* HETEROZYGOUS.

Hooke's Law For certain materials, up to a point called the ELASTIC LIMIT, the amount by which the length of the material increases in response to an applied force is proportional to the size of that force. If the force is removed, the material will return to its original size and shape.

For example, if a spring is stretched by 2 cm by a force of 1 N, it will stretch by 4 cm with a force of 2 N. For a force F, the extension x of a spring will be such that

$$F = kx$$

where k is a constant, called the spring constant.

hormone Any molecule (usually of small molecular mass) secreted directly into the blood by ENDOCRINE GLANDS and transported to a target cell or organ, causing a specific response. A hormone is sometimes referred to as a chemical messenger and, unlike PROSTAGLANDINS, its effects are not just local. The PITUITARY GLAND controls the co-ordination of hormone action, and the HYPOTHALAMUS controls the pituitary gland.

Hormones regulate a number of body functions, for example general metabolism and growth (hormones from the THYROID GLAND), responses to stress or danger (hormones from the ADRENAL GLAND), blood sugar levels (hormones from the PANCREAS) and reproductive functions (hormones from the TESTIS, OVARY and PLACENTA).

Some hormones are amines, for example ADRENALINE, NORADRENALINE and THYROXINE, and may consist of only a few AMINO ACIDS, others are POLYPEPTIDES of less than 100 amino acids, for example OXYTOCIN, ANTI-DIURETIC HORMONE, INSULIN and GLUCAGON, and others are larger (more than 300 amino acids) and constitute PROTEINS, for example PROLACTIN, FOLLICLE-STIMULATING HORMONE, LUTEINIZING HORMONE, THYROID-STIMULATING HORMONE, ADRENOCORTICOTROPHIC HORMONE and GROWTH HORMONE.

Some hormones are derived from LIPIDS and are called STEROID HORMONES, for example OESTROGEN, PROGESTERONE, TESTOSTERONE, CORTISONE and ALDOSTERONE. These are lipid-soluble and can pass through the cell membrane, so that when the hormone reaches its target cell it forms a complex with a RECEPTOR molecule, passes through the cell membrane and influences some activity within the cell.

Other hormones are proteins that are water-soluble and cannot pass through the cell membrane. They act by binding to a receptor molecule on the target cell membrane, which then activates a second messenger (a molecular signal) that initiates a specific chemical change within the cell (but is not exported by it). The water-soluble hormones usually mediate short-term effects, while lipid-soluble steroids mediate longer-term effects.

Hormones can act in a number of other ways (*see* individual entries). Hormones do not differ much from one species to another but their effects may differ. The use of the term hormone in the botanical context has now been replaced by the term PLANT GROWTH SUBSTANCE. *See also* ENCEPHALINS, ENDORPHINS, PHEROMONES,.

hormone-replacement therapy (HRT) The oral administration of the hormones OESTROGEN and PROGESTERONE to women to help them overcome the effects of the reduction of these hormones during MENOPAUSE. Symptoms of the menopause, such as anxiety, hot flushes, irregular bleeding and osteoporosis (thinning of bone leading to increased fractures), may be controlled by HRT but the value of HRT is controversial and there are some side-effects.

hornwort A member of the PHYLUM ANTHOCEROTAE.

horsepower (hp) A unit of power, obsolete in science but still widely used for measuring the power output of INTERNAL COMBUSTION ENGINES. One hp is equal to 746 WATTS. *See also* BRAKE HORSEPOWER.

horseshoe magnet A MAGNET in the form of a horseshoe, a bar with its ends bent round so the ends are next to one another, with a POLE at each end of the bar.

horsetail A member of the PHYLUM SPHENOPHYTA.

hot-air balloon A cloth container filled with hot air from which a basket is suspended to carry passengers and fuel to heat the air. A balloon rises because the density of the warm air in the balloon is less than that of the surrounding cooler air. The balloon stops rising when the average density of the warm air plus the passengers and their basket is equal to the density of the surrounding air. *See also* FLOTATION.

HRT *See* HORMONE-REPLACEMENT THERAPY.

Hubble constant The ratio between the speed of a galaxy and its distance from the Earth (*see* HUBBLE'S LAW). Experimental values vary due to the difficulty in estimating the distance of a galaxy, but generally range from 50 to 150 kms^{-1} per megaparsec (*see* PARSEC).

Hubble Space Telescope A large astronomical TELESCOPE placed in orbit around the Earth in 1990. It operates in the visible, ultraviolet and near-infrared regions of the spectrum.

Hubble's law Distant galaxies appear to be receding with a speed proportional to their distance from the Earth. Evidence for this comes from a measurement of the RED-SHIFT in the SPECTRAL LINES of stars in distant galaxies. Difficulties in measuring the distances of these distant galaxies lead to considerable uncertainties in the measurement of the HUBBLE CONSTANT, the constant of proportionality linking a galaxy's distance with the speed of its motion. Hubble's law provides important evidence for the BIG BANG theory. *See also* COSMOLOGY, DOPPLER EFFECT.

human chorionic gonadotrophin (hCG) A PEPTIDE hormone secreted by the TROPHOBLAST and that is detected in the urine of pregnant women and forms the basis of pregnancy tests. Its role is to ensure the continued production of OESTROGEN and PROGESTERONE by the CORPUS LUTEUM until the PLACENTA fully develops.

Human Genome Project *See* GENOME.

human immunodeficiency virus *See* HIV.

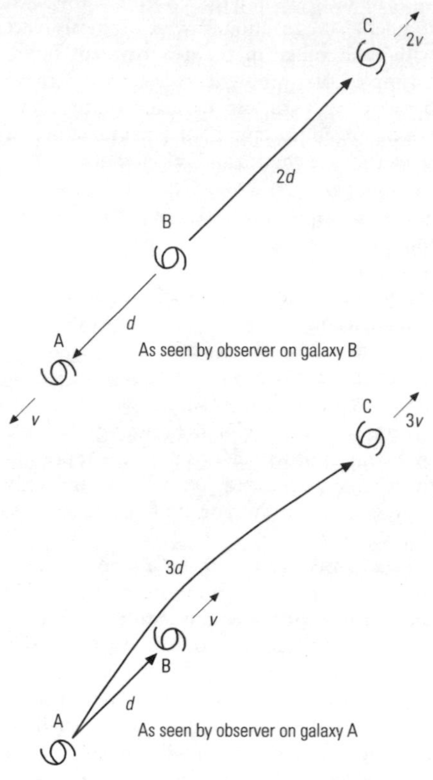

Hubble's law.

humidity The amount of water contained by an AIR MASS. This is often expressed as a RELATIVE HUMIDITY – the proportion of water vapour in the air compared to the maximum amount of water vapour the air can hold as a gas (at which point the air is said to be saturated). Humidity may also be expressed as a DEW POINT – the temperature to which the air would have to be cooled to become saturated. The humidity of the air depends on the amount of water it has passed over, and its temperature, since water evaporates more readily at high temperatures. Instruments for measuring humidity are called HYGROMETERS. *See also* WET AND DRY BULB HYGROMETER.

humoral immunity The response of an organism to invasion by a foreign object in which an ANTIBODY is produced. *Compare* CELL-MEDIATED IMMUNITY. *See also* IMMUNE RESPONSE, IMMUNITY.

humus An organic mixture in SOIL, consisting of dead plant and animal material and animal waste products that are broken down by DECOMPOSERS and DETRITIVORES. Humus has a high carbon content, is rich in minerals and is often acidic. It is important in helping water retention in sandy soil, and allows better aeration and drainage in clay soils. In water-logged areas humus cannot be broken down fully by the aerobic decomposers and accumulates as peat.

Hund's rule of maximum multiplicity A consequence of the repulsion between electrons in a given ORBITAL. The rule states that the electrons in p-, d- or f-orbitals tend to arrange themselves with one electron in each available orbital before a second electron enters any of the orbitals. Thus iron, for example, which has six 3d electrons, will have one D-ORBITAL with two electrons and four with a single electron.

hurricane A tropical CYCLONE characterized by strong surface winds in excess of 120 kmh^{-1}. Winds spiral around a low pressure centre in a clockwise direction in the northern hemisphere and anticlockwise in the southern hemisphere. Hurricanes are called typhoons in the North Pacific.

Huygens' construction A system for predicting the position of a WAVEFRONT from that of the previous wavefront. Each point on a wavefront is imagined to be a source of circular waves (called SECONDARY WAVELETS), and the combined effect of these wavelets gives the position of the next wavefront. In DIFFRACTION, the constructive and destructive interference (*see* INTERFERENCE) between different parts of the wavefront restricted by the aperture account for the observed effects.

hyaloplasm *See* CYTOSOL.

hybrid The offspring resulting from a cross between individuals of different SPECIES. In animals, hybrids are usually infertile and therefore cannot reproduce. An example that occurs naturally is the mule, which is a cross between a male horse and a female donkey. In plants, hybrids are often produced to combine desirable characteristics.

hybridization 1. (*chemistry*) The process of forming a HYBRID ORBITAL.

2. (*biology*) The production of a HYBRID individual or of hybrid cells by cell fusion. *See also* MONOCLONAL ANTIBODY.

3. (*biology*) In GENETIC ENGINEERING, the production of a DNA hybrid. *See* RECOMBINANT DNA.

hybridoma *See* MONOCLONAL ANTIBODY.

hybrid orbital A superposition of the WAVE-FUNCTIONS of a number of ORBITALS with the same energy to form a composite orbital. This is the way orbitals sometimes behave in the formation of COVALENT BONDS. Thus carbon, containing two electrons in a 2s orbital and two in 2p orbitals, forms four sp^3 hybrid orbitals, each containing one electron. These orbitals have the shape of a lobe directed towards the corners of a TETRAHEDRON centred on the carbon atom, and it is these orbitals that produce the well-known TETRAHEDRAL structure of carbon's covalent bonds.

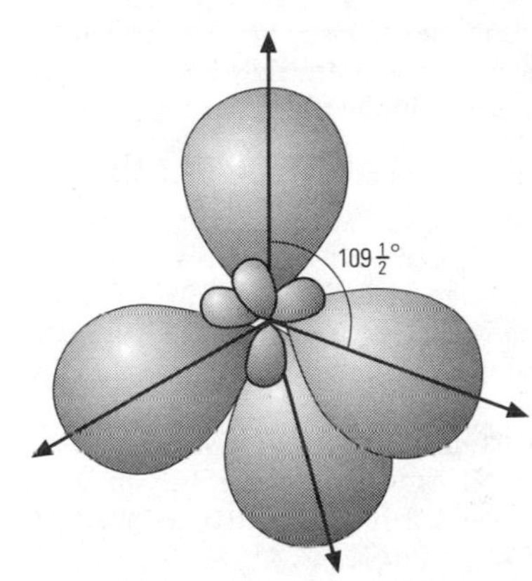

Hybrid orbital.

hydrate A substance containing a particular compound in association with water, particularly WATER OF CRYSTALLIZATION.

hydrated (*adj.*) Describing a compound that contains water, particularly WATER OF CRYSTALLIZATION, or an ion that has water molecules bound to it by hydrogen bonding (*see* HYDROGEN BOND).

hydration The combination of a substance with water to produce a single product. It is the opposite of dehydration. *See also* HYDRATE.

hydraulic Any device in which a liquid, such as oil, is used to transmit FORCES from one place to another. By making the PISTONS that produce and receive the pressure different sizes, the pressure can be used to exert forces of different sizes, producing a MACHINE. Such systems are used in the brake systems of cars and to control many large machines, from aircraft to excavators.

hydraulics The study of fluids in motion or at rest. It is the practical application of HYDROSTATICS and HYDRODYNAMICS.

hydrazine (N_2H_4) A colourless liquid; melting point 1.4°C; boiling point 114°C; RELATIVE DENSITY 1.0. Hydrazine is prepared by the reaction between ammonia and sodium(I) chlorate:

$$2NH_3 + NaClO \rightarrow N_2H_4 + H_2O + NaCl$$

Hydrazine is a powerful REDUCING AGENT and is used as such in some rocket fuels.

hydride Any BINARY COMPOUND containing a metal and hydrogen. They are unusual in containing the H^- ion. They are formed only by the more reactive metals and decompose on contact with water, for example:

$$KH + H_2O \rightarrow KOH + H_2$$

hydrocarbon A chemical compound consisting of hydrogen and carbon only. The carbon atoms can be linked by COVALENT BONDS in a long chain (ALIPHATIC) or in a ring (CYCLIC). Examples of hydrocarbons are the ALKANES, ALKENES and ALKYNES.

hydrochloric acid (HCl) A solution of HYDROGEN CHLORIDE gas in water. It is an important MINERAL ACID, widely used in industry.

hydrocortisone *See* CORTISOL.

hydrodynamics The study of the motion of fluids, particularly liquids.

hydroelectric power The generation of electricity from falling water, normally obtained by damming a river and allowing the water from the resulting lake to fall through a TURBINE. Such schemes are very effective in the production of electricity in mountainous regions, such as Scotland and Switzerland, but in developing countries there have been concerns about the environmental impact of flooding large areas of land.

hydrogen (H) The element with ATOMIC NUMBER 1; RELATIVE ATOMIC MASS 1.0; melting point –259°C; boiling point –253°C. By far the most abundant element in the universe, making up about 75 per cent of the atoms. Hydrogen is also widespread on Earth, forming many compounds, including water, from which it can be extracted by ELECTROLYSIS. In its elemental form, hydrogen normally occurs as a gas. However, it is believed to have a metallic phase at extreme temperatures and pressures, such as at the core of the GAS GIANTS.

Hydrogen is present in all ACIDS and these will react with the more reactive metals to release hydrogen gas. The presence of hydrogen in the laboratory is confirmed by lighting a sample of the gas: it burns with a characteristic squeaky pop, caused by the high speed of sound in the gas. The heavy ISOTOPES of hydrogen, DEUTERIUM and TRITIUM, are important in the production of energy by NUCLEAR FUSION. *See also* BOSCH PROCESS.

hydrogenation The addition of hydrogen to a substance, usually in the presence of a catalyst and at high pressures. Hydrogenation of UNSATURATED COMPOUNDS is of great industrial use, for example, in the manufacture of margarine and in the petroleum industry.

hydrogen azide (HN_3) A colourless liquid; melting point –80°C; boiling point 37°C; RELATIVE DENSITY 1.1. Hydrogen azide can be prepared from the reaction between sodium amide and sodium nitrate:

$$2NaNH_2 + NaNO_3 \rightarrow HN_3 + 3NaOH$$

Hydrogen azide is an extremely strong REDUCING AGENT.

hydrogen bomb, *H-bomb* A weapon that derives its energy from NUCLEAR FUSION of hydrogen nuclei.

hydrogen bond A weak bond formed between molecules that have a hydrogen atom in a POLAR COVALENT BOND. The positive charge on the hydrogen atom produces an attraction for the electronegative atom (*see* ELECTRONEGATIVITY) in a neighbouring atom.

Hydrogen bonding is particularly noticeable in hydrogen fluoride, HF, water, H_2O, and ammonia, NH_3. All three compounds have melting and boiling points much higher than methane, CH_4, which does not contain a POLAR BOND. Hydrogen bonding is also responsible for the unusual behaviour of water, which expands on cooling below 4°C, and further expands on freezing. Individual water molecules are believed to form groups such as $(H_2O)_3$.

Hydrogen bonds are much weaker than covalent bonds, having energies of typically a few tens of kilojoules per MOLE. They are easily broken by heating to temperatures much above room temperature, but are extremely important in holding together the structure of many biological molecules.

hydrogen bromide (HBr) A colourless gas; melting point –86°C; boiling point –66°C. Hydrogen bromide can be made by heating bromine in hydrogen in the presence of a platinum catalyst:

$$H_2 + Br_2 \rightarrow 2HBr$$

Hydrogen bromide is highly soluble in water, dissociating to form a strong acid.

hydrogencarbonate Any salt containing the hydrogencarbonate ion, HCO_3^-. Hydrogencarbonates are ACIDIC SALTS, the best known being sodium hydrogencarbonate (sodium bicarbonate), used in indigestion remedies.

hydrogen chloride (HCl) Colourless fuming gas; melting point –115°C; boiling point –85°C. Hydrogen chloride gas can be made by burning hydrogen (obtained from ELECTROLYSIS of water) with chlorine (from the electrolysis of BRINE) or by heating sodium chloride with concentrated sulphuric acid:

$$NaCl + H_2SO_4 \rightarrow Na_2SO_4 + HCl$$

Hydrogen chloride is a covalent gas (*see* COVALENT BOND), but is highly soluble in water, fully dissociating to give a strong acid known as hydrochloric acid.

hydrogen cyanide (HCN) A colourless gas; melting point –14°C, boiling point 26°C. It is a weak acid, formed by the action of strong acids on metal cyanides, for example:

$$NaCN + HCl \rightarrow HCN + NaCl$$

Hydrogen cyanide is extremely toxic. It is used in the manufacture of some plastics and is also produced when they are burnt in a limited oxygen supply, as often happens in fires involving soft furnishings.

hydrogen electrode A platinum ELECTRODE, coated with finely divided platinum (platinum black) over which hydrogen is passed at a pressure of one ATMOSPHERE, in contact with hydrogen ions at a concentration of one mole per decimetre cubed. *See also* STANDARD ELECTRODE POTENTIAL.

hydrogen fluoride (HF) A colourless liquid; melting point –83°C; boiling point 20°C; RELATIVE DENSITY 1.0. Hydrogen fluoride can be made by the action of sulphuric acid on calcium fluoride:

$$H_2SO_4 + CaF_2 \rightarrow 2HF + CaSO_4$$

HF is unusual in being a liquid and only forming a weak acid when dissolved in water (hydrofluoric acid). Both these features are a consequence of the highly reactive nature of fluorine, which forms a strong polar bond.

hydrogen iodide (HI) A colourless gas, melting point –51°C, boiling point –36°C. Hydrogen iodide can be formed by heating iodine vapour with hydrogen in the presence of a platinum catalyst:

$$H_2 + I_2 \rightarrow 2HI$$

Hydrogen iodide dissolves in water to form a moderately strong acid.

hydrogen peroxide (H_2O_2) A colourless liquid; melting point –1°C; boiling point 150°C; RELATIVE DENSITY 1.4. Hydrogen peroxide can be manufactured by adding acid to barium peroxide, for example:

$$H_2SO_4 + BaO_2 \rightarrow H_2O_2 + BaSO_4$$

Hydrogen peroxide decomposes readily to give oxygen and water, a reaction accelerated by the presence of light or metal ions, and which can be explosive in the case of pure hydrogen peroxide:

$$2H_2O_2 \rightarrow 2H_2O + O_2$$

Hydrogen peroxide is often supplied as an aqueous solution with its strength measured in 'volumes'. Thus one litre of '50 volume' hydrogen peroxide can be oxidized to give 50 litres of oxygen at STANDARD TEMPERATURE AND PRESSURE.

Hydrogen peroxide is an important oxidizing and bleaching agent and has been used as an oxidizer in some rocket fuels, usually with hydrogen:

$$H_2O_2 + H_2 \rightarrow 2H_2O$$

hydrogen spectrum The EMISSION SPECTRUM of HYDROGEN. As hydrogen has only a single electron its spectrum is fairly simple. It comprises a number of series of spectral lines – the Lyman series in the ultraviolet; the BALMER SERIES in visible light; and the Paschen, Ritz and Brackett-Pfund series in infrared.

Each series corresponds to the electron moving to a particular ENERGY LEVEL from one of the higher levels. In the Lyman series, the electron transitions are all between the

GROUND STATE and higher levels, in the Balmer series they all connect with the energy level above the ground state, in the Paschen series with the next higher energy level and so on. This pattern was first described empirically in the RYDBERG EQUATION, and the successful explanation of this pattern was an early triumph for QUANTUM MECHANICS.

See also BOHR THEORY, RUTHERFORD-BOHR ATOM.

hydrogen sulphide (H_2S) A colourless gas with a distinctive smell of rotten eggs; melting point –86°C; boiling point –61°C. Hydrogen sulphide is generally produced in the laboratory in KIPPS' APPARATUS, usually by the reaction between an acid and iron(II) sulphide:

$$FeS + 2HCl \rightarrow H_2S + FeCl_2$$

Hydrogen sulphide is highly toxic, though its characteristic smell makes it more detectable and therefore less dangerous than some toxic gases. It is a mild REDUCING AGENT and forms a black PRECIPITATE in the presence of many metal ions that form insoluble sulphides, for example:

$$H_2S + PbSO_4 \rightarrow PbS + H_2SO_4$$

hydrogensulphite Any compound containing the hydrogensulphite ion, HSO_3^-. Hydrogensulphites are ACIDIC SALTS and mild REDUCING AGENTS.

hydrolase One of a group of ENZYMES that is involved in HYDROLYSIS reactions, such as LIPASES.

hydrolysis A chemical DECOMPOSITION brought about by the presence of water. For example, iron(III) chloride is hydrolysed to form iron(III) hydroxide and hydrochloric acid:

$$FeCl_3 + 3H_2O \rightarrow Fe(OH)_3 + 3HCl$$

Other examples include ATP breakdown and digestion.

hydrometer A device for measuring the DENSITY of a fluid by means of a weighted float with a long stem. The level to which the float sinks in the liquid is a measure of the liquid's density. Hydrometers are often used to determine the concentrations of certain solutions, such as sugar and ethanol in water when brewing beer, or sulphuric acid in a LEAD-ACID CELL.

hydrophilic (*adj.*) (Greek = 'water-loving') Describing a molecule that exerts a strong INTERMOLECULAR FORCE towards water molecules (and other POLAR solvents), usually as the result of ionic or highly polar bonding (*see* IONIC BOND, POLAR BOND). Such materials are generally soluble in water or form LYOPHILIC SOLS. DETERGENT molecules function by having a polar end, which has a strong affinity for water molecules, and a non-polar end, which has a strong affinity for non-polar materials such as grease. *Compare* HYDROPHOBIC.

hydrophobic (*adj.*) (Greek = water-hating) Describing a molecule that does not exert a strong INTERMOLECULAR FORCE on water molecules. Such materials are generally IMMISCIBLE with water or form LYOPHOBIC SOLS. HYDROCARBONS are an example. *Compare* HYDROPHILIC.

hydrophyte A plant that lives wholly or partially under water, or in waterlogged soil. Examples include swamp, marsh and aquatic plants such as rushes, reeds and water lilies. Since there is a shortage of oxygen in such conditions, hydrophytes possess aeration tissue that contains large spaces between the cells of the stems and leaves. This gives the plants buoyancy, causing the leaves to rise to the surface to get maximum light, and also oxygen can be stored in the spaces from where it can diffuse to the roots for use in RESPIRATION. Stomata (*see* STOMA) are less or absent in submerged leaves (as water loss is not a problem), XYLEM is poorly developed (as a water conducting vessel is not needed) and supporting SCLERENCHYMA tissue is reduced (the water provides support and the plant needs to be flexible to move with the current). *See also* XEROPHYTE.

hydrosere A series of plant SUCCESSIONS originating in fresh water.

hydrosol A SOL in which solid particles are suspended in water.

hydrosphere The water on the surface of the Earth, incorporating the oceans, seas, rivers, etc.

hydrostatic pressure PRESSURE produced by FLUIDS. Hydrostatic pressure acts in all directions through a fluid. A fluid will transmit any pressure applied at one point to all other points in the fluid. One important source of hydrostatic pressure is GRAVITY, which pulls down on a fluid producing a pressure due to the weight of the fluid. An example of this is ATMOSPHERIC PRESSURE, produced by gravity acting on the Earth's atmosphere. For hydrostatic pressure produced by gravity:

pressure = depth of fluid × density × strength of gravity

hydrostatics The study of the effects of fluids at rest, particularly the effects of HYDROSTATIC PRESSURE on an immersed object.

hydrostatic skeleton A type of SKELETON found in certain invertebrates, such as earthworms, where the body cavity is full of fluid. Because liquid cannot be compressed, the structure is resilient and contraction of circular and longitudinal muscles alternately in one region that then passes along the body causes peristaltic movement (*see* PERISTALSIS).

hydrotropism The directional growth of a plant (or part of it) in response to water. Roots are positively hydrotropic, growing towards water, while stems and leaves show no response. *See also* TROPISM.

hydroxide Any compound containing a metal and the hydroxide ion, OH^-. Hydroxides are formed by the reaction between a metal and water, though this only happens at high temperatures with the less reactive metals, for example:

$$Ca + 2H_2O \rightarrow Ca(OH)_2 + H_2$$

The hydroxides of the ALKALI METALS are all soluble and form alkali solutions. The ALKALINE EARTHS form hydroxides that are basic but only slightly soluble. TRANSITION and P-BLOCK metal hydroxides are generally insoluble and are AMPHOTERIC, reacting with strong bases to form negative ions. For example, aluminium hydroxide will react with a base to form the aluminate ion:

$$Al(OH)_3 + OH^- \rightarrow Al\ (OH)_4^-$$

hydroxonium ion The ion H_3O^+ formed in the dissociation of water and present in acidic solutions. They are formed by the association of a proton (hydrogen ion) with a water molecule.

hydroxyaldehyde, *aldol* The product obtained from a reaction between two ALIPHATIC ALDEHYDES. The term aldol is derived from *ald-* for –CHO and *-ol* for –OH. The simplest hydroxyaldehyde is 3-hydroxybutanal ($CH_3CHOHCH_2CHO$), which is formed from the reaction of ethanal in dilute alkali solution. The name aldol is often used to refer to 3-hydroxybutanal. *See* ALDOL CONDENSATION.

2-hydroxybenzoic acid *See* SALICYLIC ACID.

2-hydroxypropanoic acid *See* LACTIC ACID.

hydroxyl group (–OH) An atom of hydrogen and an atom of oxygen bonded together in a covalently bonded compound (*see* COVALENT BOND). Hydroxyl groups are found in ALCOHOLS and PHENOLS, where they react as a single entity. The presence of a single hydroxyl group is termed monohydric, two is dihydric, three trihydric and more is polyhydric. GLYCOL is a dihydric alcohol and GLYCEROL is a trihydric alcohol.

Hydrozoa A CLASS of the PHYLUM CNIDARIA, including *Hydra* and the Portuguese man-of-war.

hygrometer An instrument for measuring HUMIDITY. One common form is the WET AND DRY BULB HYGROMETER.

hygroscopic (*adj.*) Describing a compound that can absorb water vapour from the atmosphere.

hyper- (*prefix*) Indicates above, higher, over, excessive. For example, hyperactive is extremely active, HYPERGLYCAEMIA is high levels of glucose. *Compare* HYPO-.

hyperbola A CONIC SECTION in which the plane intersects the CONE at an angle to the axis of cone that is smaller than that made by the cone itself. This results in an open curve, that is asymptotic (*see* ASYMPTOTE) to two straight lines. If these lines are at right angles, the curve is called a rectangular hyperbola, and can also be represented by the function $y = 1/x$.

hyperelastic collision Not a collision in the normal sense, but a process in which the KINETIC ENERGY after the event is greater than before. This means that some energy must have been converted to kinetic energy from some other form, for example from chemical energy in an explosion. *See also* COEFFICIENT OF RESTITUTION.

hyperglycaemia An abnormally high level of sugar in the blood.

hyperon Any BARYON more massive than the NEUTRON. All hyperons have very short HALF-LIVES and decay into NUCLEONS.

hyperthyroidism *See* THYROID GLAND.

hypertonic solution A SOLUTION with a higher OSMOTIC PRESSURE compared to some other solution. In a pair of solutions, the hypertonic solution is the one with the greater osmotic pressure. *Compare* HYPOTONIC SOLUTION, ISOTONE.

hypha (*pl.* ***hyphae***) One of many thread-like filaments that together form the main body (or MYCELIUM) of a FUNGUS. Hyphae grow in length and by having side-branches, which form a network of tubes containing CYTOPLASM and nuclei. Substances, including food, pass along the hyphae by movement of the cytoplasm in a process called cytoplasmic streaming. In some higher fungi (ASCOMYCETES and BASIDIOMYCETES) the hyphae are divided at intervals by septa (cross-walls) but there is still a central pore through which cytoplasm and nuclei can pass. In lower fungi (e.g. bread mould) no such divisions exist. At the ends of hyphae a structure called a SPORANGIUM grows that contains the SPORES by which the fungus reproduces.

hypo *See* SODIUM THIOSULPHATE.

hypo- (*prefix*) Indicates below, beneath, lower, incomplete. For example, HYPOGLYCAEMIA is low levels of GLUCOSE, hypothermia is low body temperature. *Compare* HYPER-.

hypobromous acid *See* BROMIC ACID.

hypochlorite Any salt containing the chlorate(I) ion, ClO^-.

hypochlorous acid *See* CHLORIC ACID.

hypogeal (*adj.*) Describing seed GERMINATION where the COTYLEDONS remain below the ground.

hypoglycaemia An abnormally low level of sugar in the blood. This occurs in DIABETES.

hypophysis *See* PITUITARY GLAND.

hypotenuse The longest side in a right-angled triangle.

hypothalamus A small region of the brain that plays a vital role in linking the NERVOUS SYSTEM and the ENDOCRINE SYSTEM and controls the activities of the latter. In the nervous system the hypothalamus regulates the autonomic (involuntary) responses and controls behavioural patterns (such as feeding, thirst, sleeping and aggression) and physiological stability (such as water balance and temperature). It has a rich supply of blood vessels and monitors blood composition so it can adjust it when necessary.

In the endocrine system, the hypothalamus regulates the action of the PITUITARY GLAND, which in turn regulates many other ENDOCRINE GLANDS. It produces ANTI-DIURETIC HORMONE and OXYTOCIN, which are passed to the posterior pituitary for storage and secretion. Another example of hypothalamus control within the endocrine system is in its production of THYROTROPHIN-releasing factor (TRF), which passes to the anterior pituitary where it regulates production of THYROXINE.

See also HOMEOTHERMY.

hypothesis A statement that has not been proved but which is used as a starting point for a logical or mathematical argument. If this argument leads to an obviously false conclusion, the hypothesis or the argument must be incorrect, a procedure called *reductio ad absurdum*, literally 'reduction to the absurd'.

hypothyroidism *See* THYROID GLAND.

hypotonic solution A SOLUTION with a lower OSMOTIC PRESSURE compared to some other solution. In a pair of solutions the hypotonic solution is the one with the lower osmotic pressure. *Compare* HYPERTONIC SOLUTION, ISOTONE.

hysteresis loop The shape of a graph of the level of INDUCED MAGNETISM in a FERROMAGNETIC material against the applied magnetic field over successive cycles of magnetization first in one direction then the other. The area of the loop is proportional to the energy needed to move the material through one magnetic cycle.

I

ice age Any period of glaciation (*see* GLACIER) in the history of the Earth, in particular the PLEISTOCENE EPOCH.

iceberg A floating mass of ice, of which only about one-fifth is above the water.

icon A small picture on a computer screen, designed to make a file or program instantly recognizable.

icosahedron A POLYHEDRON with 20 plane faces.

ICSH *See* INTERSTITIAL CELL-STIMULATING HORMONE.

ideal gas A hypothetical gas that obeys the GAS LAWS perfectly at all temperatures and pressures. For this to happen, the forces between the molecules of the gas must be negligible except during collisions, and the volume of the gas molecules must be negligible compared to the volume occupied by the gas itself. REAL GASES behave in approximately this way provided the temperature is not too low, so the KINETIC ENERGY of the molecules is large compared to the POTENTIAL ENERGY of the INTERMOLECULAR FORCES, and provided that the density is not too high so that the volume of the molecules (the EXCLUDED VOLUME) is not too large a fraction of the total volume of the gas. *See also* BOYLE'S LAW, CHARLES' LAW, IDEAL GAS EQUATION, JOULE'S LAW, KINETIC THEORY, PRESSURE LAW.

ideal gas equation, *universal gas equation* The EQUATION OF STATE for an IDEAL GAS. One particular consequence of the ideal gas equation is that the volume taken up by a gas at a given temperature and pressure depends only on the number of molecules present and not on the nature of those molecules. In particular, at STANDARD TEMPERATURE AND PRESSURE (0°C and one ATMOSPHERE), one MOLE of molecules of any gas will take up a volume of 22.4 dm^{-3}.

For a gas containing n moles of molecules, at a pressure p in a volume V:

$$pV = nRT$$

where T is the ABSOLUTE TEMPERATURE and R is the MOLAR GAS CONSTANT, $R = 8.31$ $JK^{-1}mol^{-1}$, or if the gas contains N molecules:

$$pV = NkT$$

where k is the BOLTZMANN CONSTANT, $k = 1.38 \times 10^{-23}$ $Jmol^{-1}$.

ideal gas temperature scale A TEMPERATURE SCALE that defines ABSOLUTE TEMPERATURE as being proportional to the pressure exerted by a fixed mass of IDEAL GAS held in a constant volume. The size of the temperature unit, the KELVIN, is fixed by defining the TRIPLE POINT of water to have a temperature of 273.15 K. The ideal gas scale is used in the CONSTANT VOLUME GAS THERMOMETER.

identity 1. Any relationship between mathematical functions that is always true regardless of the value taken by any of the independent variables. For example

$$\sin^2\theta + \cos^2\theta = 1$$

is true for any value of θ.

2. In a mathematical GROUP, the element that leaves all other elements unchanged, so that

$$\mathbf{I} \otimes \mathbf{A} = \mathbf{A}$$

for any **A**, where **I** is the identity for the operation $\otimes$.

IgA A class of IMMUNOGLOBULIN that often has two subunits and is found mostly in MUCUS secretions, for example milk and saliva.

IgD A class of IMMUNOGLOBULIN, the function of which is uncertain.

IgE A class of IMMUNOGLOBULIN that is found on BASOPHILS or MAST CELL surfaces and is associated with ALLERGIC REACTIONS, for example hayfever, asthma and parasitic infections.

IGFET (insulated gate field effect transistor) *See* FIELD EFFECT TRANSISTOR.

IgG A class of IMMUNOGLOBULIN that consists of one subunit, represents 70 per cent of human immunoglobulin and is the main immunoglobulin of the secondary IMMUNE RESPONSE. IgG is the only immunoglobulin to cross the PLACENTA.

IgM A class of IMMUNOGLOBULIN that consists of five subunits and is produced early in response to an infection.

igneous (*adj.*) Describing a rock that has solidified from LAVA or MAGMA.

ignition temperature The minimum temperature at which a substance will burn in air.

ignition tube A disposable glass tube, used in the laboratory for melting or boiling REAGENTS.

ileo-caecal valve The valve that controls the passage of food from the SMALL INTESTINE to the COLON. *See also* LARGE INTESTINE.

ileum The region of the vertebrate SMALL INTESTINE between the DUODENUM and COLON that is mostly concerned with the absorption of food. The wall of the ileum is specialized for this by being extensively folded and having numerous projections facing inwards called VILLI, which themselves have minute projections called MICROVILLI that together form a BRUSH BORDER. This provides the large surface area over which absorption can take place. Most absorption is by ACTIVE TRANSPORT rather than DIFFUSION, which would also allow molecules to leave the body. The ileum has a good blood supply that carries water-soluble materials, such as sugars, amino acids and minerals, to the LIVER, where their release to other parts of the body is regulated. Fats are absorbed by the lacteal (a small vessel in the villi) instead of blood capillaries and are transported in LYMPH VESSELS, entering the blood nearer to the heart.

Most of the water that is drunk is absorbed by the stomach and most of the water in digestive juices is absorbed by the ileum. The rest is reabsorbed by the COLON.

See also DIGESTIVE SYSTEM.

illuminance The amount of light per unit area falling on a surface. The SI UNIT of illuminance is the LUX.

image 1. (*physics*) A pattern of light rays coming from an object and passing through an optical system such that the light rays from each point on the object meet at, or appear to have come from, a single point on the image. *See also* REAL IMAGE, VIRTUAL IMAGE.

2. (*mathematics*) The point or set of points to which some other point or set of points is mapped by a REFLECTION. The image of any point will be at the same distance from the line of the reflection with the line between object and image being perpendicular to the line of reflection.

imaginary number A quantity used to represent the square ROOT of a negative number. All such numbers can be represented as a REAL NUMBER multiplied by the square root of –1, which is given the symbol *i*. A number that is made up of the sum of a real and an imaginary number is called a COMPLEX NUMBER. Imaginary numbers obey all the usual rules of algebra, with the additional rule

$$i^2 = -1$$

immediate access memory Computer memory in the CENTRAL PROCESSING UNIT that stores data and programs in current use. *See also* CACHE MEMORY, DRAM, SRAM.

immiscible (*adj.*) Describing two liquids that will not mix together, such as hexane and water. When stirred, an EMULSION is formed, which will separate if left to stand, with the less dense material floating to the surface. *See also* MISCIBLE.

immune response The body's reaction to invasion by a foreign object, for example an organism, pollen grains, transplanted tissues or cancerous cells. It involves the recognition of ANTIGENS on the surface of the foreign material and the production of ANTIBODIES to combat the invasion. This antibody response is called HUMORAL IMMUNITY, which involves B CELLS. The immune response also involves CELL-MEDIATED IMMUNITY, in which T CELLS play a major role.

The first exposure to an antigen results in a primary immune response, in which the antigen is processed by a phagocytic cell (*see* PHAGOCYTE) such as a MACROPHAGE, so that it can be recognized by T helper cells, which then activate B cells. These then differentiate into either PLASMA cells, which secrete antibody, or into memory cells, which are needed for future invasion by the same antigen. The primary immune response is characterized by the production of antibodies, usually of the IGM class (*see* IMMUNOGLOBULIN), but the response is slow and low compared to the secondary immune response, when the memory B cells are stimulated by a second exposure to an antigen and rapidly secrete more antibody (usually IGG). The antibodies bind to the antigen, forming complexes that are destroyed by macrophages or neutralized in a variety of ways. *See also* IMMUNITY.

immune system The components involved in the IMMUNE RESPONSE.

immunity The ability of an organism to resist an infection by harmful micro-organisms, and also to fight against cancerous cells. There is some innate immunity in animals that occurs naturally, for example, provided by the skin barrier and stomach acids, and acquired immunity that occurs as a result of exposure to ANTIGENS throughout life. Acquired immunity involves the production of antibodies (*see* ANTIBODY) in HUMORAL IMMUNITY and also CELL-MEDIATED IMMUNITY (CMI), involving cells other than antibodies. The two responses are not exclusive, as other cells are involved in the initiation of the antibody response and antibodies are usually present in CMI responses.

There are four main categories of acquired immunity. (i) Active natural immunity, in which the body's memory B CELLS (*see* LYMPHOCYTES) respond to repeated exposure of an antigen by multiplying and releasing specific antibodies. (ii) Active induced immunity, in which an antigen is given by VACCINATION to induce an IMMUNE RESPONSE. (iii) Passive natural immunity, where antibodies are acquired by a foetus from its mother across the PLACENTA or in the COLOSTRUM. (iv) Passive induced immunity, in which specific antibodies are given by intravenous inoculation to a person being treated for a particular disease, for example tetanus and diphtheria in humans. Passive immunity is only temporary.

Some diseases, such as AIDS, attack the immune system and are therefore difficult to fight. The study of immunity is called immunology.

immunization *See* VACCINATION.

immunocompromization The lack of a fully functional IMMUNE SYSTEM due to some impairment acquired later in life. This is in contrast to IMMUNODEFICIENCY, which is inborn. A person could be immunocompromised as a result of an infection such as AIDS, or as a result of pregnancy, DIABETES or old age. People on immunosuppressive drugs (*see* TRANSPLANTATION) would be immunocompromised.

immunodeficiency The impairment of one or more aspects of the IMMUNE SYSTEM, for example as a result of severe combined immune deficiency (SCID), which people are born with. With SCID, a BONE MARROW transplant to replace the defective immune cells is the only chance of survival. Patients with AIDS are often referred to as immunodeficient but because AIDS is an acquired condition they are more correctly described as immunocompromised (*see* IMMUNOCOMPROMIZATION).

immunogenic (*adj.*) Describing a substance that triggers an IMMUNE RESPONSE. *See also* ANTIGEN.

immunoglobulin A human GLYCOPROTEIN that is the ANTIBODY in an immune reaction to combat a foreign substance. There are five classes of immunoglobulin (IGA, IGD, IGG, IGE, IGM), which differ in their structure, the degree to which they polymerize and when they are produced.

The basic immunoglobulin unit consists of two heavy and two light POLYPEPTIDE chains, which form two 'Fab' regions containing the ANTIGEN-binding site, and an 'Fc' region that determines the biological properties of the immunoglobulin, for example where the antibody binds. Specificity of the antibody is determined by a sequence of AMINO ACIDS at one end of the Fab region.

Normal immunoglobulin containing most of the body's antibodies can be obtained from pooled blood PLASMA and given to patients at high risk of contracting a disease for a short period of time. A specific immunoglobulin can also be recovered from convalescing patients for administration to people at risk from a fatal disease, for example rabies.

immunology The study of IMMUNITY.

immunosuppressive drug Any drug designed to suppress an IMMUNE RESPONSE. *See also* TRANSPLANTATION.

impedance The total RESISTANCE of a circuit to the passage of electric current. In a circuit carrying ALTERNATING CURRENT, it is equal to the peak value of the voltage divided by the peak current, or equivalently the ROOT MEAN SQUARE voltage divided by the r.m.s. current. The SI UNIT of impedance is the OHM. If the circuit has a total resistance R in series with a REACTANCE X, then the impedance Z is given by

$$Z^2 = R^2 + X^2$$

impermeable (*adj.*) Unable to be penetrated by water. The term is used to describe some kinds of rock, such as granite, compared to limestone, for example, through which water can penetrate.

implantation In mammals, the process by which a developing EMBRYO attaches to the wall of the UTERUS and stimulates the development of the PLACENTA. Following FERTILIZATION, the fertilized OVUM (ZYGOTE) divides by MEIOSIS until a BLASTOCYST is produced (the 64-cell stage). This takes about 7 days to travel down the FALLOPIAN TUBE, while dividing, before it implants in the ENDOMETRIUM lining the uterus. The outer layer of cells of the blastocyst develop into the TROPHOBLAST and later into the EXTRAEMBRYONIC MEMBRANES. It is from one of these membranes, the CHORION, that VILLI develop that invade the surrounding maternal tissue and eventually form the placenta. Delayed implantation can occur in some species, for example, badgers and polar bears, where the blastocyst forms but may not implant for many months until, for example, food supplies or the weather have improved.

imprinting A form of LEARNING that is an involuntary response to a specific stimulus at a sensitive time in an animal's development. The main example is the specific recognition by a young animal of its mother and of its own species. This imprinting can occur very soon after birth. Goslings learn to recognize the first moving object they see, so they can easily become imprinted on individuals (even of another species) other than their mother. Imprinted behaviour is fixed and not easily changed. *See also* CONDITIONING, HABITUATION.

impulse 1. (*physics*) The effect that causes a change in MOMENTUM: the force applied to an object multiplied by the time for which that force acts.

2. (*biology*) *See* NERVE IMPULSE.

inactive electrode An ELECTRODE used in ELECTROLYSIS made from a material such as platinum or graphite that does not play any chemical role in the electrolysis.

inbreeding *See* ARTIFICIAL SELECTION.

incandescence The radiation of visible light from an object as a result of its high temperature. *See also* BLACK-BODY RADIATION.

incident ray An incoming ray of light.

incisor In mammals, a sharp TOOTH at the front of the mouth that is used for biting. Rodents such as rats have large incisors that continually grow and are adapted for gnawing. There are eight incisors in humans.

incus *See* EAR OSSICLE.

indehiscent (*adj.*) Describing a fruit that does not open spontaneously to shed its seeds, for example nuts.

independent variable *See* VARIABLE.

index *See* EXPONENT.

indicator A substance that produces a colour change in response to some chemical change, such as an increase in ACIDITY. The best known indicator is LITMUS. Indicators can be used to detect the END-POINT of a reaction in a TITRATION. *See also* UNIVERSAL INDICATOR.

indicator species A plant or animal with known ecological requirements whose presence in an area provides some information about the environmental conditions in that area. For example, some plants prefer acid soil and some alkaline soil, so their presence or not indicates the soil type. Some LICHENS are sensitive to levels of sulphur dioxide in the air, so their absence could be indicative of air pollution.

indium (In) The element with ATOMIC NUMBER 49; RELATIVE ATOMIC MASS 114.8; melting point 157°C; boiling point 2,080°C; RELATIVE DENSITY 7.3. Indium is a soft silvery metal. It is used as an abrasion-resistant coating on some metal parts and in the manufacture of some doped semiconductors (*see* DOPING).

induced charge Electric charge produced on an object as a result of the flow of electrons under the influence of a nearby charged object. The electric field produced will cause equal and opposite charges on opposite sides of an originally uncharged object. *See also* CHARGING BY INDUCTION.

induced fission NUCLEAR FISSION that occurs shortly after the nucleus has been struck by a neutron.

induced fit *See* LOCK-AND-KEY MECHANISM.

induced magnetism Temporary magnetism produced by the DOMAINS of a FERROMAGNETIC material being temporarily aligned by an external magnetic field.

inducible An ENZYME that is synthesized only when the SUBSTRATE is present. *Compare* CONSTITUTIVE.

inductance The property by which a change in the current flowing in an electric circuit produces an ELECTROMOTIVE FORCE either in the same circuit (SELF-INDUCTANCE) or in a neighbouring circuit with which it is magnetically linked (MUTUAL INDUCTANCE). *See also* ELECTROMAGNETIC INDUCTION, HENRY.

induction coil A device based on the TRANSFORMER principle used to produce high voltage pulses from a DIRECT CURRENT supply. Induction coils are used to generate the high voltages needed to produce sparks in PETROL ENGINE ignition systems. A steady current flows through a PRIMARY COIL containing a relatively small number of turns of thick wire wound around an iron core. A SECONDARY COIL containing many turns of thinner wire is wound on top of the primary. The current in the primary is broken by a device called a CONTACT BREAKER – a switch operated by the rotation of the engine. The very rapid change in MAGNETIC FLUX LINKAGE in the secondary when the primary current stops flowing produces a large voltage, which is used to produce the spark needed to ignite the petrol and air mixture in the engine.

induction motor A MOTOR that operates on ALTERNATING CURRENT supplies only. Induction motors use an arrangement of field coils to produce a rotating magnetic field pattern. The ARMATURE has no coils or connections to the supply, but EDDY CURRENTS set up within it produce forces tending to make the armature rotate with the rotating field pattern.

induction stroke In a PETROL ENGINE, the movement of the engine that draws fuel and air into the CYLINDER.

inductor A coil or any other circuit component with an INDUCTANCE.

inelastic collision A collision in which the total KINETIC ENERGY after the collision is less than that before the collision. An example of this is the case of two objects that coalesce (stick together), the kinetic energy lost being converted to heat energy. *Compare* ELASTIC, HYPERELASTIC. *See also* COEFFICIENT OF RESTITUTION.

inequality, *inequation* A mathematical statement that compares two numbers or algebraic expressions, describing one as being greater than (>) or less than (<) the other, or as 'greater than or equal to' or 'less than or equal to' (≥, ≤). The rules for handling inequalities are essentially the same as those for other EQUATIONS, but whenever an inequality is multiplied or divided by a negative number, or inverted, the sign of the inequality must be reversed.

inequation *See* INEQUALITY.

inert gas *See* NOBLE GAS.

inertia The resistance of an object to having its motion changed. According to Newton's second law, this depends on an object's mass. This idea of mass is sometimes called INERTIAL MASS to distinguish it from GRAVITATIONAL MASS. *See also* NEWTON'S LAWS OF MOTION.

inertial confinement A technique for obtaining the extreme temperatures needed for NUCLEAR FUSION by heating material so rapidly that it does not have time to expand before fusion begins. This is the method used in NUCLEAR WEAPONS but does not seem likely to be of much use for the peaceful use of nuclear fusion, where a continuous release of energy is required.

inertial mass MASS as a measure of resistance to change in motion. *Compare* GRAVITATIONAL MASS. *See also* NEWTON'S LAWS.

inertial navigation system (INS) A system used to enable aircraft and guided missiles to calculate their position by knowing their starting point and the duration of all accelerations in any direction. Such systems have the advantage that they do not rely on signals transmitted from satellites or ground-based stations.

inertial reference frame One of a series of co-ordinate systems, none of which seems to be accelerating when viewed from any other. In particular, co-ordinate systems that seem to be moving at a constant speed relative to the centre of mass of the observed Universe. *See also* FRAME OF REFERENCE.

infertility The inability of an organism to reproduce. Infertility can be due to the inability to produce GAMETES (no ova produced, or none or too few SPERM produced); blocked FALLOPIAN TUBES, so that sperm and ova cannot meet; or problems in maintaining the pregnancy, such as low hormone levels, recurrent spontaneous abortion, repeated unexplained early loss of embryos with a possible underlying immunological problem.

In humans it is now rare that problems of infertility cannot be overcome. The use of *IN VITRO* FERTILIZATION (IVF) has overcome the problem for many couples, particularly in cases of blocked fallopian tubes. Artificial insemination is another treatment that involves the artificial administration of sperm to a woman, either from her partner or from a donor (AID). Some of these treatments, especially AID and surrogacy, and the spare 'embryos' produced by IVF, raise difficult moral and ethical problems.

infinity A quantity that is so large that the largest imaginable number is not large enough to describe it. Infinity is represented in mathematics by the symbol ∞.

infinitesimal (*adj.*) A vanishingly small quantity, larger than zero but smaller than the smallest non-zero number.

inflammable (*adj.*) Able to burn.

inflammation In immunology, the local response to injury or infection, outwardly involving redness and swelling of the surrounding tissue. HISTAMINE is released by MAST CELLS and causes dilation of local blood vessels and an increase in permeability to allow through, for example, antibodies, FIBRINOGEN and NEUTROPHILS, which are needed for repair or recovery of the site. The presence of fibrinogen at the site of inflammation causes FIBRIN formation and clotting, to prevent haemorrhage and to trap any PATHOGENS to restrict the spread of infection. Neutrophils and later MONOCYTES (transforming into MACROPHAGES) pass through the blood capillaries and engulf pathogens and dead cells by PHAGOCYTOSIS, resulting in the accumulation of a yellowy liquid called pus (consisting of dead cells and bacteria) at the site of inflammation. *See also* ALLERGIC REACTION.

inflation In COSMOLOGY, a period of extremely rapid growth in the size of the Universe. *See* BIG BANG.

inflection A point on a curve where the GRADIENT is zero, but with the gradient just to each side of this point having the same sign.

inflorescence In plants, a term used for a specific arrangement of flowers on a single main stalk. Inflorescences can be divided according to their method of branching – cymose (definite) or racemose (indefinite). In a cymose inflorescence the tip of the main stalk produces a single flower and subsequent flowers arise on lower side branches. In a racemose inflorescence the main stalk increases in length by growing at its tip and flowers are borne along the whole length of the stalk, opening from below upwards. In a racemose inflorescence the oldest flowers are therefore found near the base. *See also* PEDICEL, PEDUNCLE.

infrared ELECTROMAGNETIC WAVES with wavelengths from about 1 mm to 7 x 10^{-7} m. Infrared radiation is emitted by all hot objects and can be detected by the heating effect it produces when absorbed by a blackened surface (*see* BOLOMETER, THERMOPILE). Infrared with a wavelength close to the visible part of the ELECTROMAGNETIC SPECTRUM (called near infrared) can also be detected by modified versions of photographic film and electronic devices used to detect visible light. The fact that warm objects produce more infrared than cold objects has led to the development of many military applications based on infrared cameras for night-time surveillance and heat-seeking missiles.

infrared spectroscopy The study of the INFRARED ABSORPTION SPECTRUM of organic molecules. This is a useful technique for determining molecular structure, as many COVALENT BONDS have resonant frequencies corresponding to the stretching or bending of the bonds that lie in the infrared region of the electromagnetic spectrum. Infrared radiation from a hot source is passed through the sample and then via a DIFFRACTION GRATING to a detector. The extent to which the radiation is absorbed is then represented as a plot of absorption against WAVENUMBER.

infrasound Sound of a frequency too low to be detected by the human ear.

inhibitor 1. A substance that slows down the rate of a chemical reaction or blocks it. *See also* ENZYME INHIBITION.

2. A material added to some glues, such as epoxy resins, to slow down the rate at which they harden.

inorganic (*adj.*) Describing compounds that contain no carbon or that are carbides or carbonates, or oxides or sulphides of carbon. Such substances are generally obtained from the Earth's crust or atmosphere. Inorganic chemistry is the study of the chemical reactions of the elements and the inorganic compounds they form.

in phase Describing two OSCILLATIONS that are exactly in step with one another.

input device Any device for entering information into a computer. The main input devices of PERSONAL COMPUTERS are the keyboard and mouse. Other devices are scanners, joysticks, touch-sensitive screens and microphones.

insect A member of the CLASS INSECTA of the PHYLUM ARTHROPODA.

Insecta A CLASS of the PHYLUM ARTHROPODA, consisting of the insects. It is a very diverse class of more than a million species. An insect's body is divided into three parts, the head, THORAX and ABDOMEN, with a pair of antennae on the head, three pairs of legs attached to the thorax and usually two pairs of wings (fore and hind), although not all insects are winged. The mouth parts can be adapted for biting and chewing, for example bees and locusts, or for piercing and sucking, for example butterflies and house flies. Some insects, for example lice and fleas, are PARASITES. An insect SENSORY SYSTEM is well developed; the antennae act as feelers and detect odours, there is one pair of compound eyes and many insects can detect sound.

The skeleton of insects is external and hard (except at the joints) and contains CHITIN. The thorax is responsible for locomotion, the legs being adapted to walk on land and in some, for example locusts, the hind legs are well developed to allow jumping. The wings are composed of two membranes with a strengthening framework of 'veins' between them. Differences in the wings form part of the classification of insects, for example the wings may be joined to operate as one pair, or hard as in beetles, or adapted to assist in balance. The abdomen is responsible for METABOLISM and reproduction. Respiratory gases are carried to and from the body muscles by TRACHEAE, which are fine tubes opening to the outside at pores called SPIRACLES. The sexes in insects are separate; the female produces eggs and FERTILIZATION is internal. Many insects mate once and then die. The life cycle of an insect often involves several changes in appearance and way of life, from young to adult, and is called METAMORPHOSIS.

Economically, insects are very important to humans; they can be serious pests that need to be controlled with insecticides, they carry diseases such as malaria, sleeping sickness, typhoid and dysentery, and they can cause damage to stored food, wood and clothing. Insects are also useful as pollinators of fruit and crops, as scavengers, in the study of genetics (e.g. the fruitfly *Drosophila*) and as a source of products such as honey and silk. The study of insects is called entomology.

insecticide A PESTICIDE that kills hazardous or unwanted insects.

insoluble (*adj.*) Describing a compound that will not dissolve (normally in water).

inspiration Breathing in. *See* BREATHING.

instruction set The complete set of MACHINE CODE instructions for a particular MICROPROCESSOR.

insulated gate field effect transistor *See* FIELD EFFECT TRANSISTOR.

insulator A material through which current or heat cannot flow. Except for graphite, all non-metals are electrical insulators in their solid form. Organic liquids are also electrical insulators, as are all gases unless ionized. *See also* BAND THEORY, CONDUCTOR, SEMICONDUCTOR.

insulin A protein HORMONE that is responsible for the regulation of blood sugar levels. Insulin is a 51 amino acid POLYPEPTIDE, made by the β-cells of the ISLETS OF LANGERHANS in the PANCREAS. Insulin is produced when blood sugar levels are high, for example after a meal, and has the effect of reducing the blood sugar levels. Insulin promotes uptake of free glucose by body cells (e.g. muscle and adipose cells) by altering the permeability of the CELL MEMBRANE to GLUCOSE. It also promotes uptake of amino acids by muscles and converts excess glucose (formed as a result of CARBOHYDRATE breakdown following a meal) into GLYCOGEN for storage in the liver. This is called GLYCOGENESIS. Deficiency in the production of insulin leads to a condition called DIABETES. Human insulin can now be produced from bacteria by GENETIC ENGINEERING techniques. *See also* GLUCAGON.

integer A whole number, positive or negative or zero. The numbers –1, 0 and 273 are all integers.

integral 1. In mathematics, a function that, when differentiated (*see* CALCULUS), gives a particular function. For example, if y and z are functions of x and $dy/dx = z$, then y is the integral of z with respect to x, represented as

$$\int z dx$$

On a graph, an integral is represented by the area between the curve that represents the function and the x-axis. An integral is generated by adding up the value of the function over a range of values of x.

Integrals may be either indefinite, with no specified range, or definite, taken between two specified limits. The indefinite integral of the function x^n is

$$x^{n+1}/(n+1) + c$$

where c is a constant, called the constant of integration, which can take on any value.

2. (*adj.*) Relating to an INTEGER.

integrated circuit, *silicon chip* A miniature electronic circuit made of a complex array of ACTIVE DEVICES produced on a single silicon wafer and designed to perform an electronic function. Most integrated circuits form elements in DIGITAL electronic circuits.

integrated pest management Combined strategies for the control of pests. There are five major strategies employed in integrated pest management: PESTICIDES, BIOLOGICAL CONTROL, cultural control (such as using agricultural practices to change the pest's habitat), breeding pest-resistant species and sterile-mating control.

Integrated Services Digital Network *See* ISDN.

integration In mathematics, the process of finding an INTEGRAL. This may be done analytically, following a set of algebraic procedures, or numerically, adding up the value of a function for a range of values of the independent variable. Numerical integration is often performed using a computer.

integrator A circuit that produces an output proportional to the integral of the input voltage over time. They are usually based on an OPERATIONAL AMPLIFIER using a CAPACITOR in the FEEDBACK circuit.

integument In SEED PLANTS, the protective layer surrounding the OVULE. Most flowering plants have two integuments, while GYMNOSPERMS have only one. The integuments later form the TESTA or seed coat.

intensity In physics, the power per unit area carried by a wave. The intensity of a wave is proportional to the square of its AMPLITUDE. The SI UNIT of intensity is the Wm^{-2} (watt per metre squared), but light waves may have their intensity quoted in LUX, while DECIBELS are often used for sound waves.

inter- (*prefix*) Indicates between or among, for example intercellular means between cells. *Compare* INTRA-.

interaction In general, any process in which one object exerts a force on another, or where an object changes its nature in some way. In particle physics, four distinct forms of interaction are known, sometimes called the FOUR FORCES OF NATURE. These are the STRONG NUCLEAR FORCE, the WEAK NUCLEAR FORCE, ELECTROMAGNETIC FORCE, and gravitation (see GRAVITY). *See also* ELECTROWEAK FORCE, GRAND UNIFIED THEORY, STANDARD MODEL.

interactive computing A computer processing system in which an operator must input data in order for the program to run to completion. *Compare* BATCH PROCESSING.

intercept The point at which a line or curve crosses a specified AXIS on a GRAPH.

intercostal muscle One of several muscles in TETRAPODS that allow the ribs to move up and down during breathing.

interface An ELECTRONIC system for gathering non-electronic information and feeding it into an electronic system, such as a COMPUTER, or for connecting two electronic systems together in such a way that information can be passed from one to another.

interference 1. In optics, the effect of two or more WAVES arriving at the same point at the same time. Interference is described as constructive if it leads to a greater AMPLITUDE and destructive if the amplitude is reduced. In light, interference is visible only if the interfering sources are coherent. If an incoherent light source is used, interference is only visible if a single beam of light is split in two – either by diffracting it though a narrow slit (interference by DIVISION OF WAVEFRONT), or by separating one beam into two by partial reflection (interference by DIVISION OF AMPLITUDE).

The effects of interference are studied using the PRINCIPLE OF SUPERPOSITION, which states that the effects of the interfering waves can be calculated by simply adding together (superposing) the effects that the individual waves would have on their own.

See also BEATS, DIFFRACTION, INTERFEROMETRY, PATH DIFFERENCE, THIN-FILM INTERFERENCE.

2. In TELECOMMUNICATIONS, an unwanted signal, either man-made or naturally occurring, which arrives at the receiving end of the system along with the desired signal. *See also* NOISE.

interferometer A device for splitting a beam of light into two, usually by partial reflection, and then recombining those beams after they have travelled along different paths. Observation of the INTERFERENCE patterns produced enables the difference in path length to be measured very precisely. *See also* INTERFEROMETRY.

interferometry A technique for combining light or radio waves received at two different points, taking due note of their relative PHASES. Interferometry is used for precise measurement of distances, accurate to a fraction of a wavelength of light.

In astronomy, radio interferometry has been used, combining the signals obtained by two or more separate RADIO TELESCOPES some distance away from one another. Interferometry involving radio telescopes in different continents has also provided direct evidence of CONTINENTAL DRIFT.

interferon *See* LYMPHOKINE.

interleukin *See* LYMPHOKINE.

intermetallic compound A compound formed between metals. Such compounds are ALLOYS in which the atoms of the metals are present in a simple ratio and form a regular lattice structure.

intermolecular bonding Weak INTERMOLECULAR FORCES, such as HYDROGEN BONDS, holding together molecules that are themselves covalently bound. At relatively low temperatures these bound structures can behave like single molecules, but at high temperatures they fall apart into the constituent molecules. Intermolecular bonding is particularly important in many biological processes. *See also* VAN DER WAAL'S BOND.

intermolecular forces The forces that act between one molecule and another in a substance. If these forces are strong enough compared to the energy of any thermal vibrations – in other words, if the substance is cold enough – the intermolecular forces will hold the substance together as a solid or liquid. At higher temperatures, or with weaker forces, the material will behave as a gas. In an IDEAL GAS there are no intermolecular forces.

All intermolecular forces are ELECTROSTATIC in origin and are the result of forces between the electrons and the nuclei of the molecules involved, governed by the rules of QUANTUM MECHANICS. In all cases, there is a link between the strength of the intermolecular forces and the melting point and also the amount of energy needed to melt the material (the LATENT HEAT). The way in which the intermolecular forces change with distance is linked to the stiffness of the resulting material, though most materials are much weaker than might be expected from the strengths of the individual intermolecular forces, owing to the existence of defects in the LATTICE structure, which also account for PLASTIC behaviour. Within the ELASTIC region, the extent to which the intermolecular forces vary linearly with separation between the molecules is related to the extent to which the material obeys HOOKE'S LAW.

See also HYDROGEN BONDING, IONIC SOLID, MACROMOLECULE, METAL, VAN DER WAAL'S FORCE.

internal combustion engine The most important engine of the 20th century, fuelled by petrol or diesel fuel. These fuels are both based on HYDROCARBONS, with petrol being made from molecules of smaller molecular mass than diesel. *See* DIESEL ENGINE, PETROL ENGINE. *See also* TURBOCHARGING.

internal energy The energy that atoms or molecules in a substance possess as a result of forces between themselves and the KINETIC ENERGY of their random thermal motion, as opposed to any bulk motion of the object, or external forces acting on the object. When an object is heated, provided no external WORK is done, the increase in the internal energy of the object is equal to the amount of heat energy supplied. In an IDEAL GAS, the internal energy is purely due to the kinetic energy of the molecules. *See also* FIRST LAW OF THERMODYNAMICS, JOULE'S LAW.

internal resistance The RESISTANCE a source of electrical energy appears to possess. When a battery or any other source of electrical energy is connected to a LOAD (some device that draws current from the supply of electricity), the voltage across the terminals of the supply will fall. The greater the load (i.e. the lower its resistance), the greater the fall in voltage. For simplicity in calculations, the supply can be treated as an ideal supply (one which can provide an unlimited current with no drop in voltage) in series with an internal resistance, across which there is a voltage (sometimes called the lost volts), which increases with the current drawn from the supply.

In applications where high currents are required, it is important to keep the internal resistance as low as possible. For example, the LEAD-ACID CELLS used in car batteries may be required to provide currents of several hundred amperes to start the car whilst the voltage across the terminals remains close to the level it would be with no load. *See also* MAXIMUM POWER THEOREM.

International Practical Temperature Scale (IPTS) A TEMPERATURE SCALE designed to conform as close as possible to thermodynamic temperature. The unit of temperature is the KELVIN (K). The latest version, devised in 1990, has 16 fixed points. The IPTS supersedes all other temperature scales for scientific purposes.

Internet The global computer network that links most of the world's computer networks. Services may be accessed from the individual networks, including ELECTRONIC MAIL; TELNET, a service that allows an Internet user to log on to another computer and use its applications as if it were on their own computer; and FTP (file transfer protocol) in which bulk data may be transmitted from one computer to another.

interphase The interval in the CELL CYCLE between nuclear divisions (usually in MITOSIS but also in MEIOSIS). Interphase is often called the 'resting phase' because the chromosomes are not visible, but it is actually a period of intense activity when the cell ORGANELLES are duplicated and the DNA content is doubled.

interpreter A piece of computer software that understands and executes programs written in a HIGH-LEVEL LANGUAGE. Unlike COMPILERS, interpreters execute the instructions directly, without the need for the intermediate step to translate them into MACHINE CODE. Interpreted programs run more slowly than compiled programs, but they have the advantage that when a change is made to the original source program it may be run immediately, without needing to be recompiled.

intersection 1. The point at which two lines or curves cross.

2. A SET whose elements are members of all of a number of specified sets. For example, if one set contains people with red hair and another contains those with green eyes, the intersection is the set of people with red hair and green eyes. The intersection of two sets is denoted by the symbol $\cap$.

interstellar (*adj.*) Between stars. The term is particularly used to describe large distances and also the low density gas and dust that fills the spaces between stars in a galaxy.

interstitial (*adj.*) Describing an atom located in the interstices, or spaces, between atoms in a regular crystal LATTICE. For example, in steel, carbon atoms occupy the interstices between the much larger iron atoms.

interstitial cell One of several cells in the human TESTIS situated in the spaces between the SEMINIFEROUS TUBULES. The interstitial cells secrete TESTOSTERONE and are themselves stimulated by LUTEINIZING HORMONE.

interstitial cell-stimulating hormone (ICSH) *See* LUTEINIZING HORMONE.

interstitial fluid *See* TISSUE FLUID.

interval A set of REAL NUMBERS lying between two specified limits, usually specified by a pair of INEQUALITIES. For example $-2 < x < 2$, means that x lies between -2 and 2.

intestinal juice An alkaline secretion produced by the wall of the SMALL INTESTINE when stimulated by the presence of food. Intestinal juice contains mineral ions that neutralize the acid CHYME from the stomach, and enzymes that aid digestion. These enzymes include PROTEASES to break down PEPTIDES into smaller peptides and amino acids. Other enzymes break down CARBOHYDRATES, for example AMYLASE, MALTASE, LACTASE and SUCRASE.

intestine *See* SMALL INTESTINE, LARGE INTESTINE.

intra- (*prefix*) Indicates in, within or inside of, for example, intracellular refers to within a cell. *Compare* INTER-.

intra-uterine device (IUD), ***coil*** A plastic or copper coil inserted into the UTERUS as a method of CONTRACEPTION. A number of devices are available, all of which prevent IMPLANTATION of a fertilized egg by causing inflammation of the uterine lining. Once in place an IUD provides continued protection that is 98 per cent effective, but there is a risk of infection that could lead to INFERTILITY. The IUD is not usually given to women who have not yet had a child.

intrinsic semiconductor A SEMICONDUCTOR with no DOPING and an equal number of electrons and HOLES.

inulin A POLYSACCHARIDE that consists of a chain of FRUCTOSE molecules. Inulin is a storage CARBOHYDRATE in many plants, for example the Dahlia. It is not found in animals.

invariant (*n., adj.*) In mathematics, a quantity that is unchanged during some transformation. For example, the area of a triangle is invariant under rotation.

inverse A mathematical operation that reverses the effect of some other operation. In group theory the inverse is the member of the group that combines with some other member to give the IDENTITY. If $f(x)$ and $g(x)$ are functions such that

$$f(g(x)) = x$$

for all x, then f is the inverse function of g.

inverse square law The behaviour of any quantity that radiates from a point source or the surface of a sphere with its strength falling in such a way that it is reduced by a factor of four when the distance from the source is doubled. ELECTRIC FIELDS and GRAVITATIONAL FIELDS, and the intensity of ELECTROMAGNETIC RADIATION all exhibit inverse square law behaviour.

inversion temperature The temperature below which a gas will cool in a FREE EXPANSION.

invertebrate An animal without a backbone, i.e. not a vertebrate. The majority of existing animal species are invertebrates, including SPONGES, FLATWORMS, ANNELIDS, ARTHROPODS, MOLLUSCS, ECHINODERMS and some invertebrate CHORDATES.

inverted (*adj.*) Upside down, particularly used of optical images formed by CONVERGING LENSES and mirrors.

inverter 1. (NOR gate) A LOGIC GATE with a single input and a single output that is the opposite of the input.

2. A device for producing an ALTERNATING CURRENT from a DIRECT CURRENT supply. The output is often converted to a higher voltage using a TRANSFORMER.

inverting amplifier An AMPLIFIER with a negative GAIN; that is, a positive input produces a negative output.

inverting input One of the inputs of a DIFFERENTIAL AMPLIFIER. Signals applied to this input are amplified with a negative GAIN.

in vitro (= in glass) A process or experiment that takes place outside the body of a living organism, for example in the laboratory. *Compare* IN VIVO.

in vitro* fertilization (IVF), *embryo transfer A technique where FERTILIZATION of an OVUM and SPERM is performed in a test-tube, to overcome problems of INFERTILITY. The procedure was first carried out by Edwards (1925–) and Steptoe (1913–88) in Cambridge, UK, and the first IVF baby was born in 1978. The technique involves removing an ovum (after administration of a fertility drug to increase production) and sperm and mixing them in suitable culture conditions outside the body and transferring fertilized eggs back to the mother. IVF is most suitable in cases of blocked FALLOPIAN TUBES but can be used for women who have problems maintaining the pregnancy, when a surrogate (substitute) mother can be used.

in vivo A process or experiment that takes place within a living cell or organism. *Compare* IN VITRO.

involuntary muscle, *unstriated muscle, smooth muscle* MUSCLE under the control of the AUTONOMIC NERVOUS SYSTEM.

Io One of the satellites of JUPITER. It is remarkable as the only place beyond the Earth where active volcanoes are observed. Io is heated by its interaction with Jupiter's strong magnetic field.

iodic acid Iodic(V) acid, HIO_3, and iodic(VII) acid or periodic acid, H_5IO_6. Iodic(V) acid is a pale yellow liquid, which decomposes on heating; RELATIVE DENSITY 4.6. It can be made by oxidizing iodine with hydrogen peroxide:

$$4H_2O_2 + I_2 \rightarrow 2HIO_3 + 4H_2O$$

Iodic(V) acid dissolves in water and is a strong acid and an OXIDIZING AGENT. Iodic(VII) acid is a white solid that decomposes on heating. It forms a weak acid in an aqueous solution.

iodide Any BINARY COMPOUND containing IODINE.

iodine (I) The element with ATOMIC NUMBER 53; RELATIVE ATOMIC MASS 126.9; sublimes at 183°C; RELATIVE DENSITY 4.9. Iodine is a HALOGEN and occurs as deep purple crystals. It is insoluble in water but soluble in organic solvents, such as ethanol or tetrachloromethane. Iodine is extracted as sodium iodide from sea water and used in the manufacture of silver iodide, an important light-sensitive chemical in the photographic industry. Iodine is concentrated by the body in the THYROID GLAND, which makes its radioactive ISOTOPES a dangerous part of NUCLEAR FALLOUT.

iodoform Common name for TRIIODOMETHANE.

iodoform test *See* TRIIODOMETHANE TEST.

iodopsin A light-sensitive pigment in the CONES of the eye. *See* RETINA.

ion An ATOM or MOLECULE that is not electrically neutral, having gained or lost ELECTRONS. An ion that has lost one or more electrons (cation) is positively charged, whereas one that has gained electrons (anion) is negatively charged. *See also* IONIC BOND, IONIC SOLID, IONIZATION, IONIZATION ENERGY, ZWITTERION.

ion exchange The exchange of ions between a solution and a solid material. When dissolved ions are washed past the solid, they are removed and replaced by a different ion. This process occurs in nature, particularly in the uptake of fertilizer by plants and is also used to remove magnesium and calcium ions from HARD WATER, replacing them with sodium ions. To reverse the process, and regenerate the ion exchange material, a concentrated solution of sodium ions, from dissolved sodium chloride, is passed over the exchange material. Similar techniques, based on synthetic POLYMERS can be used to remove all ANIONS and replace them with HYDROXIDE ions (OH^-) or to replace CATIONS with hydrogen ions (H^+). By using these together, water can be purified, producing deionized water.

ion exchange chromatography CHROMATOGRAPHY in which an ION EXCHANGE resin is used as the stationary phase. Competition between absorption rates for different ions means that different ions migrate through the chromatography column at different rates.

ionic (*adj.*) Describing a material, particularly a solid or a solution, that contains IONS or is held together by IONIC BONDS.

ionic bond A CHEMICAL BOND in which one or more electrons are transferred from one atom to another, creating a positive and negative ION. The attractive forces between these ions then holds the material together as a crystalline solid. These solids are generally hard, brittle materials with high melting points. They are usually soluble in water and insoluble in organic compounds. They always conduct electricity when molten or in solution.

Ionic bonds are generally formed between materials of highly differing ELECTRONEGATIVITY, thus the most electronegative and electropositive (*see* ELECTROPOSITIVITY) materials generally form ionic bonds. Although the removal of an electron requires an input of energy (*see* IONIZATION ENERGY), as may the formation of a negative ion (*see* ELECTRON AFFINITY), the energy released by the formation of a crystal from these ions more than offsets these requirements in substances that form ionic crystals. For example, when sodium reacts with chlorine to form sodium chloride, the formation of sodium ions from solid sodium requires 610 kJ mol^{-1}, whilst the formation of chlorine ions from molecular chlorine releases 223 kJ mol^{-1}. Thus the formation of the ions requires a net energy input of 387 kJ mol^{-1}. However, the formation of the crystal releases an energy of 779 kJ mol^{-1}, so the process as a whole is EXOTHERMIC.

Compare COVALENT BOND. *See also* FAJAN'S RULES, IONIC SOLID, SHELL.

ionic product of water The product of the CONCENTRATION of hydrogen and hydroxide ions in water, $[H^+][OH^-]$. This is a constant over all conditions of acidity and alkalinity and is equal to 10^{-14} mol^2 dm^{-6}.

ionic radius A measure of the effective size of an ION in an IONIC SOLID. The ionic radii of all ions increases with increasing ATOMIC NUMBER and is larger for ANIONS than for CATIONS in the same PERIOD. Typical ionic radii vary between about 0.03 nm (Be^{2+}) and 0.22 nm (Te^{2-}).

Ionic radii may be determined by finding the interatomic spacing in the crystal from its density and the masses of the atoms involved, or by X-RAY CRYSTALLOGRAPHY. This spacing is then allocated to the two types of ion in the crystal in a way that aims to give consistent results for all combinations of ions. The results are good enough to suggest that this technique is a valid one. However, it clearly contains some approximations arising from the different LATTICE structure of different crystals, the non-spherical distribution of charge in an ion, and the partially covalent nature of some bonds.

ionic solid A crystalline solid in which adjacent atoms gain and lose one or more electrons to form a regular LATTICE of IONS. Each ion is attracted to its nearest neighbours with the opposite charge. There is also a repulsive element in the interatomic force at short distance, caused by the effect of the PAULI EXCLUSION PRINCIPLE, which begins to promote electrons to higher ENERGY LEVELS as the atomic electron clouds of neighbouring atoms start to overlap. As a result of the strong nature of the ionic attractions, ionic materials tend to have melting and boiling points above room temperature and are usually fairly hard materials. *See also* COVALENT CRYSTAL.

ionic strength A measure of the effect of IONS in a solution on other ions in the solution. The ionic strength of an ion in solution is equal to its molar concentration multiplied by the square of its charge. The ionic strength of the solution overall is the sum of the ionic strength of the ions it contains.

ionization The process of creating IONS in a substance that previously contained neutral atoms or molecules, such as by a chemical reaction or IONIZING RADIATION. *See also* HEAT OF IONIZATION.

ionization energy, *ionization potential* The minimum energy needed to remove an electron infinitely far away from an atom, often specified either in ELECTRON-VOLTS or per mole of atoms. The first ionization energy relates to the removal of the first electron, the second ionization energy is the additional energy needed to remove a second electron and so on.

ionization potential *See* IONIZATION ENERGY.

ionized (*adj.*) Describing a material that contains IONS – for example as a result of chemical action as in acidified water, or IONIZING RADIATION.

ionizing radiation Any RADIATION that creates IONS in any matter through which it passes. Ionizing radiation may be a stream of high-energy particles (such as ELECTRONS, PROTONS or ALPHA PARTICLES) or short-wavelength ELECTROMAGNETIC RADIATION (ULTRAVIOLET radiation, X-RAYS or GAMMA RADIATION).

For all types of ionizing radiation, the level of radiation can be reduced by passing the radiation through a material containing as many electrons as possible, to provide the greatest number of opportunities for the radiation to lose energy by IONIZATION. Thus lead, which has a large number of electrons per unit volume, so is a dense material, is a much better absorber of ionizing radiation than aluminium.

See also BACKGROUND RADIATION, BECQUEREL, DECAY CONSTANT, RADIATION DETECTORS, RADIOCARBON DATING, TRACER TECHNIQUES.

ionosphere The outer layer of the ATMOSPHERE that is ionized by X-rays from the sun.

iridium (Ir) The element with ATOMIC NUMBER 77; RELATIVE ATOMIC MASS 192.2; melting point 2,410°C; boiling point 4,130°C. Iridium is a silvery TRANSITION METAL. It is used in the manufacture of hard alloys for wear-resistant surfaces, though its high cost limits this to fairly small bearings in precision machinery.

iris A coloured diaphragm of circular and radial muscles in the EYE that can alter the size of the PUPIL to control the amount of light entering the eye.

iron (Fe) The element with ATOMIC NUMBER 26; RELATIVE ATOMIC MASS 55.8; melting point 1,535°C; boiling point 2,750°C; RELATIVE DENSITY 7.9. Iron is the fourth most abundant element in the Earth's crust. It is extracted from its ores (mostly iron oxides) by REDUCTION by carbon in a BLAST FURNACE, for example:

$$2Fe_2O_3 + 3C \rightarrow 4Fe + 3CO_2$$

The resulting PIG IRON contains too much carbon to be used directly and is subsequently reacted, in an ELECTRIC ARC FURNACE or BESSEMER CONVERTER with oxygen to remove excess carbon, to make steel.

Iron is the most widely used metal. Its tendency to rust is offset by its high strength and low cost. Iron is a TRANSITION METAL that forms compounds of valency 2 (traditionally called ferric) and 3 (traditionally called ferrous).

iron chloride Iron(II) chloride, or ferric chloride, $FeCl_2$, and iron(III) chloride, or ferrous chloride, $FeCl_3$.

Iron(II) chloride is a greenish-yellow salt; melting point 670°C; decomposes on further heating; RELATIVE DENSITY 3.2. It can be made by heating iron in hydrogen chloride:

$$Fe + 2HCl \rightarrow FeCl_2 + H_2$$

Iron(III) chloride is a dark brown solid; melting point 306°C; decomposes on further heating; relative density 2.9. It can be formed by burning iron in chlorine:

$$2Fe + 3Cl_2 \rightarrow 2FeCl_3$$

Iron(III) chloride is soluble but hydrolyses partially to iron(III) hydroxide:

$$FeCl_3 + H_2O \rightarrow Fe(OH)_3 + 3HCl$$

iron losses In a TRANSFORMER, those energy losses attributable to the iron parts of the transformer. Iron losses are made up of energy losses due to HYSTERESIS and EDDY CURRENTS.

iron oxide Iron(II) oxide, FeO, iron(III) oxide, Fe_2O_3, and the mixed oxide, iron(II)–iron(III) oxide, Fe_3O_4.

Iron(II) oxide is a black powder; melting point 369°C; decomposes on further heating; RELATIVE DENSITY 5.7. It can be produced by reacting iron with ethanedioic acid to form iron(II) oxalate, which then decomposes to iron(II) oxide when heated in the absence of air:

$$Fe + (COOH)_2 \rightarrow Fe(COO)_2 + H_2$$

$$Fe(COO)_2 \rightarrow FeO + CO + CO_2$$

Iron(II) oxide is a base, dissolving in strong acids to form iron(II) salts, for example:

$$FeO + H_2SO_4 \rightarrow FeSO_4 + H_2O$$

Iron(III) oxide is a reddish brown solid; melting point 1,564°C; decomposes on further heating; relative density 5.2. It is formed naturally in rusting. It can also be formed by roasting iron in oxygen:

$$4Fe + 3O_2 \rightarrow 2Fe_2O_3$$

Iron(III) oxide is readily reduced, by carbon monoxide for example:

$$Fe_2O_3 + 3CO \rightarrow 2Fe + 3CO_2$$

Iron(II)–iron(III) oxide, which occurs naturally as the mineral magnetite, is a black ferrimagnetic solid. It contains iron atoms alternately in the +2 and +3 OXIDATION STATES, with their MAGNETIC MOMENTS in opposite directions.

iron sulphate Iron(II) sulphate $FeSO_4$ and iron(III) sulphate $Fe_2(SO_4)_3$.

Iron(II) sulphate generally occurs as the heptahydrate, $FeSO_4.7H_2O$, a bluish-green solid; RELATIVE DENSITY 1.9; melting point 64°C; decomposes on further heating. Iron(II) sulphate can be formed by the action of sulphuric acid on iron:

$$Fe + H_2SO_4 \rightarrow FeSO_4 + H_2$$

On heating, it decomposes to give iron(III) oxide:

$$2FeSO_4 \rightarrow Fe_2O_3 + SO_2 + SO_3$$

Iron(III) sulphate is a brownish yellow solid, which decomposes on heating; relative density 3.1. It can be formed by the OXIDATION of iron(III) sulphate, for example by the action of hydrogen peroxide:

$$2FeSO_4 + H_2SO_4 + H_2O_2 \rightarrow Fe_2(SO_4)3 + 2H_2O$$

irradiance The total power of ELECTROMAGNETIC RADIATION of all wavelengths arriving per unit area on a surface, measured in watts per square metre.

irradiation The process of bombarding a surface with RADIATION, particularly IONIZING RADIATION. Especially used in the context of using ionizing radiation to kill bacteria in foodstuffs.

irrational number A number that cannot be expressed exactly as a fraction, for example the square root of 2. *Compare* RATIONAL NUMBER.

irreversible (*adj.*) Describing a change that cannot happen in reverse, usually because to do so would violate the SECOND LAW OF THERMODYNAMICS.

ISDN (Integrated Services Digital Network) A high-BANDWIDTH digital communications network. It has the advantage over conventional telephone networks in that digital information can be transmitted directly between computers, etc., without the need for MODEMS.

isentropic (*adj.*) Describing a process that takes place without any change in ENTROPY, that is, a reversible process.

islets of Langerhans Groups of specialized cells in the PANCREAS that are concerned with regulation of blood sugar levels. There are two cell types: α-cells that secrete GLUCAGON and β-cells that secrete INSULIN.

isobar A line joining points of equal ATMOSPHERIC PRESSURE on a weather map. The CORIOLIS FORCE means that wind does not blow at right angles to the isobars, but more or less parallel to them.

isobutane *See* BUTANE.

isochronous (*adj.*) Describing an OSCILLATION whose period does not depend on its AMPLITUDE. *See also* SIMPLE HARMONIC MOTION.

isoelectronic (*adj.*) Describing a pair or group of compounds that have the same total number of VALENCE ELECTRONS. For example, nitrogen and carbon monoxide are isoelectronic as they both have a total of 10 valence electrons.

isoleucine An ESSENTIAL AMINO ACID, present in CASEIN and body tissue.

isomer A chemical compound that possesses the same composition and MOLECULAR MASS as another but differs in its physical or chemical properties. The differences are due to the structural arrangement of the constituent atoms. For example there are two organic compounds with the formula C_4H_{10} – butane $CH_3(CH_2)_2CH_3$ and methyl propane $CH_3CH(CH_3)CH_3$. These have the same MOLECULAR FORMULA but differ in their STRUCTURAL FORMULA and are therefore structural isomers:

$$CH_3 - CH_2 - CH_2 - CH_3 \qquad \text{butane}$$

$$CH_3 - \underset{\displaystyle CH_3}{CH} - CH_3 \qquad \text{methyl propane}$$

Structural isomers are distinct compounds with differing properties and can even belong to different classes of compounds.

Structural isomers differ in the order in which the atoms are joined together but some isomers are less obvious and are due to differences in their three-dimensional structure, seen only when models are constructed. Optical isomers are mirror images of each other and geometric isomers arise due to the different spatial arrangement of atoms around a central plane of symmetry.

See also GEOMETRIC ISOMERISM, OPTICAL ISOMERISM, TAUTOMERISM.

isomerase One of a group of enzymes that causes the rearrangement of groups within a molecule.

isometric growth In biology, where the growth of a given feature is the same as the growth of the entire organism. For example, the leaves of most plants exhibit isometric growth. *Compare* ALLOMETRIC GROWTH. *See also* GROWTH.

isomorphic (*adj.*) Describing two crystalline materials with the same LATTICE structure. Thus sodium chloride and potassium bromide are isomorphic, as they each form a crystalline structure in which each ANION is surrounded symmetrically by six CATIONS, resulting in a cubic structure described as FACE CENTRED CUBIC.

isoprene *See* 2-METHYLBUTADIENE.

isosceles (*adj.*) Describing a triangle that has two equal sides.

isostasy In GEOLOGY, the balance between the weight of a TECTONIC PLATE and the upthrust from the MAGMA in which it is floating. Changes in this balance cause continents to rise or sink.

isotactic polymer A POLYMER with a regular structure where all the substituted carbons (such as METHYL GROUPS) have the same stereoconfiguration and are on the same side of the carbon chain, if it is considered that the carbon atoms lie in the same plane. This arrangement gives a polymer great strength, for example in POLYPROPYLENE. *Compare* ATACTIC POLYMERS, SYNDIOTACTIC POLYMER.

isothermal (*adj.*) Describing a change that takes place at constant temperature.

isotonic (*adj.*) Describing a solution that has the same OSMOTIC PRESSURE as some reference solution.

isotope Any one of a series of atomic nuclei of the same element, each with the same number of PROTONS but different numbers of NEUTRONS, hence different masses. *See also* NUCLEUS, RELATIVE ATOMIC MASS.

isotropic (*adj.*) Describing a material, usually a crystalline solid, that has the same physical properties, such as thermal or electrical CONDUCTIVITY, regardless of the direction in which these properties are measured. *Compare* ANISOTROPIC.

iteration In mathematics, the process of solving an equation by a series of repeated calculations. The result from one calculation is used in the next calculation, with successive answers converging to give an increasingly accurate solution.

IUD *See* INTRA-UTERINE DEVICE.

IVF *See* *IN VITRO* FERTILIZATION.

J

J/ψ (J/psi) A MESON containing the charmed (*see* CHARM) QUARK and antiquark. Discovered independently by two experimental groups in 1974, it was the first evidence for the existence of charm. The J/ψ has a mass about 3.2 times that of the proton and a HALF-LIFE of around 10^{-20} s.

jejunum The part of the SMALL INTESTINE between the DUODENUM and the ILEUM. Its main function is the absorption of digested food.

jet engine An engine that burns fuel continuously to provide an outflow of high speed, high temperature gas. Jet engines are widely used to propel aircraft. Most modern jet engines are of the GAS TURBINE type, which are also used to propel ships and to generate electricity. *See also* PULSE-JET, RAMJET, TURBOPROP, HIGH-BYPASS ENGINE.

jet stream A particularly fast narrow current of wind that occurs at an altitude of 10 to 12 km. Wind speeds typically reach between 60 to 125 kmh^{-1}.

JFET (junction field effect transistor) *See* FIELD EFFECT TRANSISTOR.

joint In animals with skeletons, the point at which bones meet. Joints usually allow movement. Synovial joints allow free movement and consist of ball-and-socket joints (e.g. the hip and shoulder) or hinge joints (e.g. the elbow, knee, toes and fingers). In ball-and-socket joints, the round end of one bone fits into the hollow end of another bone and can turn in any direction. In this type of joint there is a layer of CARTILAGE for protection at the ends of the bones (called articular) and a fibrous covering called the synovial capsule around the joint. The inner layer of this is the synovial membrane, which secretes SYNOVIAL FLUID.

In hinge joints, the round end of one bone turns on the flat surface of another bone in one direction only. Other joints include gliding joints (e.g. between adjacent vertebrae and at the wrist and ankle), which allow one bone to move over the surface of another; pivot joints (e.g. between the first two vertebrae at the top of the VERTEBRAL COLUMN, allowing rotation of the head), which allow partial movement; and suture joints (such as those between the bones of the skull) which allow no movement. The bones of a joint are held together by LIGAMENTS.

joule (J) The SI UNIT of WORK and ENERGY. One joule of work is done when a force of one NEWTON moves through a distance of one metre in the direction of the force.

Joule's law For an IDEAL GAS, the INTERNAL ENERGY depends only on temperature, not on pressure or volume. REAL GASES show slight departures from Joule's law, most cooling as they expand, to compensate for work done against INTERMOLECULAR FORCES, though some gases, such as hydrogen, show an increase in temperature.

junction field effect transistor *See* FIELD EFFECT TRANSISTOR.

junction transistor A SEMICONDUCTOR device that exhibits the important property of GAIN – a small current can be used to control a much larger one. The commonest type of junction transistor is the NPN TRANSISTOR. It consists of a thin layer of lightly doped (*see* DOPING) P-TYPE SEMICONDUCTOR, called the BASE, sandwiched between two more heavily doped pieces of N-TYPE SEMICONDUCTOR material, the COLLECTOR and the EMITTER. The base-emitter junction is FORWARD BIASED and a current flows through it, mainly as a flow of electrons from the emitter to the base. Since the base is thin, many of these electrons diffuse through the base and enter the collector, which is held at a more positive potential than the base. Thus the current flowing into the base is small, but controls a much larger collector current. The PNP TRANSISTOR operates in the same way, but with all the polarities reversed. *See also* THERMAL RUNAWAY.

Jupiter The largest planet in the SOLAR SYSTEM. It has a mean diameter of 138,000 km, 11 times

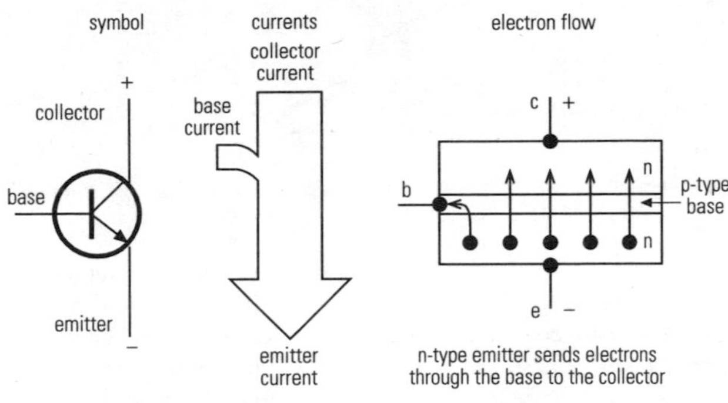

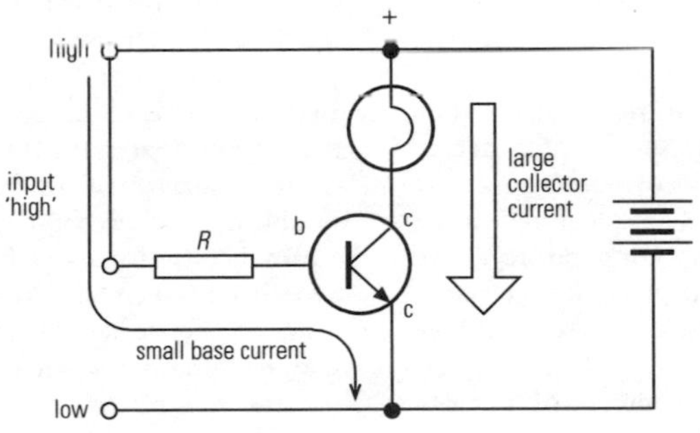

Junction transistor.

that of the Earth, and a mass of 1.9 x 10^{27} kg, 320 times the Earth's mass. Jupiter is composed mainly of hydrogen, so its average density is relatively low at 1,300 kg m^{-3}. Jupiter is the fifth planet in order of distance from the Sun, with a mean orbital radius of 5.2 AU (230,000,000 km). Jupiter takes 11.9 years to complete one orbit of the sun, yet rotates on its own axis in just 8 hours. Jupiter has no visible solid surface, but a complex atmosphere with active weather patterns is clearly visible. The most famous of these surface features is the GREAT RED SPOT. Numerous satellites are known and new small ones are routinely discovered. The four largest satellites, called the GALILEAN SATELLITES, are clearly visible with a small telescope. Jupiter has a faint RING SYSTEM, similar to that of SATURN but less developed.

Jurassic A PERIOD of geological time during the MESOZOIC ERA. It began after the TRIASSIC period (about 190 million years ago) and extended until the CRETACEOUS period (about 140 million years ago). Reptiles dominated in this period and birds evolved.

K

kaon *See* MESON.

karyotype The characteristics of the set of CHROMOSOMES of a given species. Karyotype describes the number, size and shape of the chromosomes. For example, the human karyotype consists of 46 chromosomes, arranged in 23 pairs.

Kastner-Kellner process The ELECTROLYSIS of BRINE in a MERCURY-CATHODE CELL to produce chlorine and sodium hydroxide.

kelvin (K) The SI UNIT of TEMPERATURE. The size of the kelvin is the same as the degree CELSIUS. The TRIPLE POINT of water is fixed at exactly 273.16 K: the kelvin is defined as the fraction 1/273.16 of the temperature above ABSOLUTE ZERO of the triple point of water. *See also* ABSOLUTE TEMPERATURE, TEMPERATURE SCALE.

Kelvin statement of the second law of thermodynamics No system can convert heat energy entirely to mechanical WORK – a certain amount of heat must be given out to the cooler surroundings. For example, the exhaust gas of an INTERNAL COMBUSTION ENGINE contains waste heat energy. If a HEAT ENGINE takes in heat at an ABSOLUTE TEMPERATURE T_1 and gives out waste heat at a lower temperature T_2, the maximum EFFICIENCY permissible by the SECOND LAW OF THERMODYNAMICS, even if there are no avoidable losses, such those due to friction, is $(1 - T_2/T_1)$.

Kepler's laws A set of three laws originally based on empirical observation of the SOLAR SYSTEM but now recognised as a consequence of the laws of MECHANICS and GRAVITY.

Kepler's first law sets out the arrangement of the Solar System, saying that all planets move in elliptical ORBITS, which have the Sun as a common FOCUS.

Kepler's second law is a consequence of the LAW OF CONSERVATION OF ANGULAR MOMENTUM, and the fact that the force of gravity acts along a line between the Sun and the planet. It states that the line joining the planet to the Sun sweeps out equal areas in equal intervals of time. Thus a planet moves fastest in its orbit when it is closest to the Sun.

Kepler's third law states that the square of the time taken to complete an orbit (called the period of the orbit) is proportional to the cube of the average radius of the orbit. Thus a planet close to the Sun, such as Mercury, orbits more quickly (88 days), than one further away (such as Jupiter, which has a period of 12 years). This law is a consequence of the fact that the force of gravity obeys an INVERSE SQUARE LAW; that is, the strength of gravity falls to one quarter of its original strength at double the distance from the Sun.

keratin In vertebrates, a fibrous sulphur-rich protein found in the skin that toughens the outer protective layer. Keratin is also found in hair, nails, hooves, feathers and horns.

kerosene A mixture of HYDROCARBONS of the ALKANE series obtained from PETROLEUM. It is a thin oil sold as a fuel under the common name of paraffin.

ketone, *alkanone* One of a group of organic compounds containing the CARBONYL GROUP (C=O) attached to two ALKYL GROUPS (R and R′) R-CO-R′ or RR′CO. The general formula of a ketone is $C_nH_{2n}O$, the same as an ALDEHYDE, but in aldehydes there is only one alkyl group and one hydrogen atom. The nomenclature of ketones follows that of the ALKANE with the same carbon skeleton, with the ending *-ane* being replaced by *-anone*. A number is placed before the *-one* in the name to indicate the position of the carbonyl carbon, for example pentan-3-one. Other examples include propanone (acetone) and phenylethanone. Ketones can be ALIPHATIC (usually liquids), AROMATIC (usually solids), CYCLIC or mixed. Ketones are formed by the OXIDATION of secondary ALCOHOLS:

$$\mathrm{RR'CHOH} \xrightarrow{\text{oxidized}} \mathrm{RR'C{=}O}$$

Ketones can be reduced back to secondary alcohols but are not themselves easily oxidized, unlike aldehydes, and this feature is used to distinguish them from aldehydes in FEHLING'S TEST or using TOLLEN'S REAGENT.

See also FRIEDEL–CRAFT'S REACTION, SCHIFF'S REAGENT.

ketose, *keto-sugar* A sugar having a KETONE group (C=O). *See also* MONOSACCHARIDE.

keto-sugar *See* KETOSE.

kidney In vertebrates, either one of a pair of major organs responsible for excretion of waste products and maintaining the balance of water and solutes in the body tissues and blood (osmoregulation; *see* OSMOSIS). The kidneys are located on the back wall of the ABDOMEN and are 7–10 cm long and 2.5–4 cm wide. Blood enters from the RENAL artery and leaves via the renal vein.

The functional unit of the kidney is called the NEPHRON, which is a filtering unit used to form URINE. There are over a million nephrons in a human kidney. The nephron consists of a knot of blood capillaries, called the GLOMERULUS, and a long tubule with several clearly defined regions. Some substances pass out of the blood through the glomerulus and enter the tubule as a filtrate. As the filtrate passes along the tubule, some of the substances are selectively reabsorbed, including water, and eventually urine is formed, which enters the URETER for excretion.

Two hormones are important in controlling osmoregulation by the kidney. ANTIDIURETIC HORMONE (ADH) controls the permeability of the walls of part of the tubule and so the degree to which water is drawn out from them. ALDOSTERONE is important in regulating the sodium ion concentration in the kidney. If one kidney is lost, the other usually becomes enlarged and takes over the function of both. If both kidneys are defective, DIALYSIS is needed to sustain life.

See also ADRENAL GLAND, BOWMAN'S CAPSULE, LOOP OF HENLE.

kilo- (k) A prefix before a unit indicating that the size of the unit is to be multiplied by 10^3, for example kilowatt (kW) is equivalent to one thousand watts.

kilogram The SI UNIT of mass. One kilogram is defined as the mass of the International prototype kilogram, a cylinder of platinum-iridium alloy kept at Sèvres, near Paris, France.

kinematics The study of objects moving with specified ACCELERATIONS, as opposed to DYNAMICS, which also studies the forces that produce those accelerations.

kinetic energy The energy of movement. The kinetic energy, E, of a mass m moving at a speed v is

$$E = {}^1/_2 mv^2$$

kinetics The measurement and study of the rates of chemical reactions and biological processes. In particular, the determination of the mechanisms of reactions, by studying the effects that factors such as temperature, pressure and CATALYSTS have on the rate. Kinetics looks at the sequence in which the events leading to a chemical reaction take place, and the intermediate products that are created. *See also* RATE OF REACTION.

kinetic theory The part of physics that explains the physical properties of matter in terms of the motion of its component atoms and molecules. The temperature of a body is dependant on the INTERNAL ENERGY, and therefore the velocity, of its molecules. If heat is supplied to a substance, such as a gas, the velocity of the particles increases and the temperature rises.

Kinetic theory explains the pressure of a gas in terms of the impacts of its molecules on to the walls of its container. If the temperature rises, the number of impacts per second increases, and the pressure increases. By making various assumptions about the molecules of an IDEAL GAS (such as negligible forces between molecules, the molecules themselves take up a negligible volume, etc.), kinetic theory gives rise to the GAS LAWS.

Kinetic theory also extends to liquids and solids and explains such physical phenomenon as THERMAL EXPANSION, CHANGES OF STATE and change of RESISTANCE in terms of the thermal motion of the particles.

See also BROWNIAN MOTION, STATISTICAL MECHANICS.

kingdom The major category in the CLASSIFICATION of organisms. The most common classification system today has five kingdoms, consisting of ANIMALIA (all ANIMALS), PLANTAE (all PLANTS), FUNGI (all fungi), PROTOCTISTA (including all ALGAE and PROTOZOA) (these four kingdoms consist of EUKARYOTES) and prokaryotae (all PROKARYOTES). Kingdoms are divided into phyla (*see* PHYLUM).

Kipp's apparatus A vessel used in laboratories for the production of gases from the reaction of a liquid (held in the upper part of the vessel) and a solid, held in the lower part. As gas pressure builds up in the apparatus, the liquid is forced back into the upper chamber, preventing further reaction. An example of the use of Kipp's apparatus is the preparation of hydrogen sulphide from the reaction of hydrochloric acid and iron sulphide.

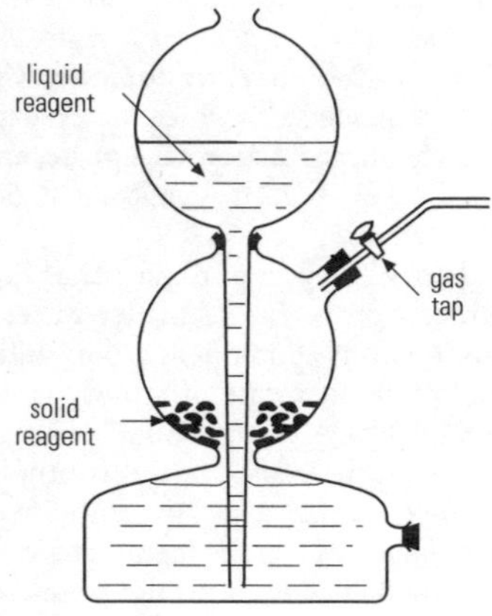

Kipp's apparatus.

Kirchhoff's laws A set of two laws that can be used to calculate the currents and voltages in a complex circuit.

Kirchhoff's first law states that at any junction in a circuit the total current entering the junction is equal to the total current leaving the junction. If currents entering the junction are taken as positive, whilst those leaving are negative, the sum of all the currents at any junction will be zero. This law is in effect a LAW OF CONSERVATION OF CHARGE: the charge per second entering the junction must equal the charge per second leaving the junction if charge is not to be created or destroyed. At any junction:

$$\Sigma I_{in} = \Sigma I_{out}$$

Kirchhoff's second law states that around any closed loop in a circuit, the sum of all the ELECTROMOTIVE FORCES (e.m.f.'s) in the loop must be the same as the sum of all the POTENTIAL DIFFERENCES (p.d.'s) across RESISTANCES in the loop. This is a consequence of the LAW OF CONSERVATION OF ENERGY, since the e.m.f.'s represent electrical energy gained per unit charge whilst the p.d.'s represent electrical energy lost per unit charge. Around any loop:

$$\Sigma E = \Sigma IR$$

for a loop containing e.m.f.'s E, currents I and resistances R.

Klinefelter's syndome *See* MUTATION.

klystron A device for the production of MICROWAVES. Electrons in a beam produced by THERMIONIC EMISSION from a hot filament have their speeds varied by a second ELECTRODE (the buncher). The faster electrons catch up with the slower ones and so bunches are formed. These produce a varying current at a collecting electrode (the catcher) and this is fed back to the buncher. The resulting oscillations have a frequency which depends on the voltage used to accelerate the electrons and the separation between the buncher and catcher. The buncher and catcher form cavities resonant (*see* RESONANT CAVITY) at the frequency of the electromagnetic radiation generated.

In a variation on this design, called the reflex magnetron, only a single cavity is used, with the electron beam being repelled back to the cavity by a negative electrode.

knocking A phenomenon occurring in PETROL ENGINES that reduces the power output. Knocking sounds result from explosions of unburned fuel-air mixture before it is ignited. The extent of knocking depends on the fuel composition and can be reduced by using fuel with a higher OCTANE NUMBER. Lead is used in petrol to reduce knocking but is being phased out due to its association with mental retardation in children.

Knudsen regime The situation in a gas at a very low pressure, where the MEAN FREE PATH of a molecule becomes large compared to the size of the apparatus, and the TRANSPORT COEFFICIENTS no longer correctly describe the behaviour of the gas. Similarly, a gas will no longer support sound waves once the pressure is low enough for the mean free path of the gas molecules to be comparable to the wavelength of the sound.

Kohlrausch's law The MOLAR CONDUCTIVITY of any IONIC solution at infinite dilution is the sum of the molar conductivity of the ions it

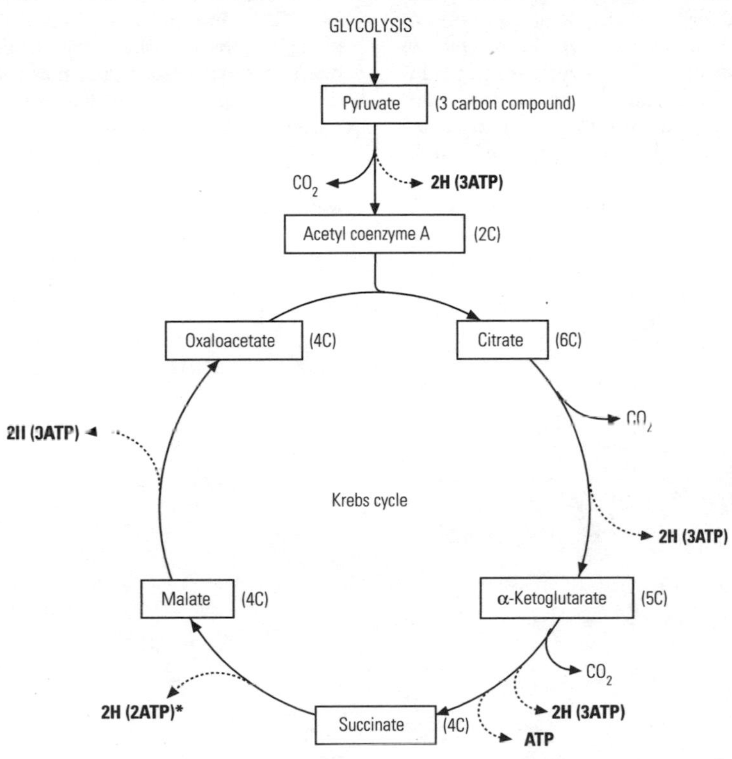

Summary of the Krebs cycle.

contains. This law simply expresses the idea that at low concentration, the ions in a solution carry charge independently of one another.

Krebs cycle, *TCA cycle, tricarboxylic acid cycle, citric acid cycle* The second stage of cellular RESPIRATION. In the Krebs cycle, the three-carbon compound pyruvate, formed from GLUCOSE during the first stage of GLYCOLYSIS, is converted, in the presence of oxygen, to carbon dioxide and hydrogen atoms. The Krebs cycle is named after its discoverer Hans Krebs (1900–81).

Before entering the cycle, pyruvate combines with a compound called COENZYME A to form acetyl coenzyme A, which then enters the Krebs cycle to combine with OXALOACETIC ACID. The cycle itself generates very little energy for the cell but the hydrogen atoms enter the ELECTRON TRANSPORT SYSTEM, where they later generate energy in the form of ATP. The carbon dioxide is removed as a waste product. The process is cyclic with the end-product, oxaloacetic acid, being the substrate for the next round.

In cells with mitochondria (*see* MITOCHONDRION), the enzymes needed for the Krebs cycle are found mostly in the mitochondrial matrices and so the reactions occur there. In PROKARYOTES the enzymes are free in the CYTOPLASM. The hydrogen atoms released are carried to the electron transport system by the carriers NAD or FAD – FAD yields less ATP than the NAD.

In addition to degrading pyruvate to carbon dioxide and providing hydrogen atoms for the electron transport system, the intermediate compounds of the Krebs cycle are needed to make other substances, such as FATTY ACIDS, AMINO ACIDS and CHLOROPHYLL, and so this cycle is considered to play a key role in a cell's biochemistry.

krypton (Kr) The element with ATOMIC NUMBER 36; RELATIVE ATOMIC MASS 83.8, melting point −157°C; boiling point −152°C. Krypton is a NOBLE GAS. Its low abundance in the atmosphere, compared to argon, means it is only used in special GAS DISCHARGE tubes.

Kupffer cell A MACROPHAGE found in the liver, forming part of the RETICULOENDOTHELIAL SYSTEM.

Kwashiorkor *See* DIET.

L

lachrymal gland, *tear gland* The gland in the eye of some vertebrates that secretes a watery substance that lubricates and cleanses the surface of the eye.

lactase An enzyme that breaks down LACTOSE into GLUCOSE and GALACTOSE.

lactate fermentation *See* FERMENTATION.

lactation In mammals, the production of milk by the MAMMARY GLANDS in response to the birth of their offspring. After birth of the FOETUS, milk production is stimulated by the hormone PROLACTIN, secreted by the PITUITARY GLAND. The sucking of the nipples of the mammary gland stimulates OXYTOCIN release from the pituitary, which causes contraction of the breast alveoli (*see* ALVEOLUS) resulting in the expression of milk (the 'let down' reflex).

The early milk produced is COLOSTRUM, which is an important source of antibodies (those that are too large to cross the placenta). Full milk produced after a few days contains fat, protein (CASEIN), sugar (LACTOSE), minerals and antibodies. Human milk (but not cows' milk) contains antibacterial antibodies and an iron-binding protein called lactoferrin that inhibits bacterial growth in the baby's intestine. Milk production can continue as long as the baby is suckling.

lacteal A small vessel in the VILLI of the SMALL INTESTINE through which fats are absorbed. *See also* ILEUM.

lactic acid, *2-hydroxypropanoic acid* ($CH_3CHOHCOOH$) A colourless, odourless organic liquid produced by LACTIC ACID BACTERIA during FERMENTATION. It is also produced by muscle cells when they are exercising vigorously and experience OXYGEN DEBT.

lactic acid bacteria A group of bacteria used in the dairy industry in the manufacture of cheese and yoghurt. When added to pasteurized milk (*see* PASTEURIZATION), lactic acid bacteria convert LACTOSE to LACTIC ACID. The fall in pH causes the milk to separate into solid curd and liquid whey. It is the curd that is used to make cheeses.

Other micro-organisms can be added to give variety to the ripening of cheese, for example, *Pencillium* is added to blue cheese, and the ripening of soft cheeses is helped by fungi growing on their surfaces. Lactic acid bacteria are also used in the manufacture of yoghurt; the starting material is pasteurized milk with the fat removed.

lactoferrin A protein present in human's milk that inhibits bacterial growth. *See* LACTATION.

lactose A DISACCHARIDE made from the combination of GLUCOSE and GALACTOSE. Lactose is found in mammalian milk (5 per cent in cows' milk) and is important to the suckling young.

lacuna In biology, a space or cavity, such as the spaces in BONE that contain the bone cells.

laevorotatory (*adj.*) Describing a form of OPTICAL ACTIVITY in which the plane of POLARIZATION of PLANE-POLARIZED light is rotated in an anticlockwise direction. It is also used to describe the ENANTIOMER of a compound exhibiting OPTICAL ISOMERISM that rotates plane-polarized light anticlockwise. This used to be denoted by the prefix *l-* but (–) is now used. This is not to be confused with the prefix L- used to indicate the configuration of CARBOHYDRATES and AMINO ACIDS.

lagging Thermal insulation, particularly around pipes or hotwater tanks to prevent loss of heat or damage by the water freezing.

lambda particle (λ) A neutral BARYON. The lambda particle is the lightest baryon to contain the STRANGE QUARK, with a mass about 1.2 times that of the proton and a relatively long HALF-LIFE of 2.6×10^{-10} s. *See also* SIGMA PARTICLE.

lamella Any thin layer, plate or membrane, especially the thin layers of tissue of which bone is formed.

lamina The blade of a leaf. *See* LEAF.

laminar flow In a fluid, smooth flow in which the motion of one part of the fluid is very similar to that in other nearby regions and does not change over time. *See also* DRAG.

laminated (*adj.*) Describing a construction made from many thin layers. *See also* EDDY CURRENT.

lance A water-cooled pipe used to carry a gas, usually oxygen, into a furnace. *See* OPEN HEARTH FURNACE, OXYGEN FURNACE.

lanthanides, *lanthanoids, rare earth elements* The 15 elements with ATOMIC NUMBERS from 57 (LANTHANUM) to 71 (LUTETIUM). Each has its outer electrons in 4f ORBITALS (except for lanthanum). They are grouped together in the PERIODIC TABLE because they have very similar chemical properties, which differ only slightly with atomic number.

lanthanoids *See* LANTHANIDES.

lanthanum (La) The element with ATOMIC NUMBER 57; RELATIVE ATOMIC MASS 138.9; melting point 918°C; boiling point 3,464°C; RELATIVE DENSITY 6.1. The first member of the LANTHANIDE series, it is a soft metallic element used in some alloys.

Large Electron Positron *See* LEP.

large intestine Part of the digestive tract between the SMALL INTESTINE and the ANUS. The large intestine is 1.5 m long and 6 cm in diameter and, like the small intestine, consists of muscular tubes with a folded inner lining increasing the surface area for absorption of food. It is linked to the small intestine by the CAECUM, which leads to the APPENDIX. The ileocaecal valve controls passage of food from the small intestine to the COLON. Water, minerals and vitamins in digestive secretions that have not been absorbed by the ILEUM are reabsorbed in the colon, and anything remaining is formed into FAECES. Excess calcium and iron salts are passed from the blood through the large intestine for removal with the faeces. Movement of faeces towards the RECTUM (for temporary storage) is aided by MUCUS secreted by the wall of the large intestine. The faeces are then expelled through the anus. *See also* DIGESTIVE SYSTEM.

Lamarckism A theory of EVOLUTION put forward by the French naturalist Jean-Baptiste Lamarck (1744–1829), which was never widely accepted. He suggested that useful characteristics acquired by an organism during its lifetime would be inherited by its offspring, while disuse of other characteristics would result in their eventual disappearance from the species. For example, he suggested that giraffes have long necks because they continually stretch to reach for high leaves. According to his theory, giraffes with longer necks through stretching would be able to pass this acquired characteristic onto their offspring. Although now discredited, Lamarck did influence the thinking of Charles Darwin (1809–82), who later proposed the still widely accepted theory of NATURAL SELECTION. *See also* DARWINSIM.

larva (*pl.* ***larvae***) A pre-adult form of many animals as they hatch from the egg and before they develop into sexually mature adults. Larvae may be very different from the adult, for example, tadpoles and caterpillars differ significantly from their adult forms. *See also* METAMORPHOSIS.

larynx, *voicebox* A cavity at the opening of the TRACHEA where it meets the PHARYNX. The larynx contains a series of LIGAMENTS called the vocal cords that vibrate producing sound waves. It is the Adam's Apple in humans. The complexity of the vocal cords varies; reptiles have none but in birds there is a structure called a syrinx with well-developed vocal cords.

laser Acronym for *l*ight *a*mplification by *s*timulated *e*mission of *r*adiation. A device for producing an intense parallel beam of COHERENT light.

A laser comprises a RESONANT CAVITY filled with some 'lasing' medium, with a mirror at either end, one of which is semi-reflecting. The atoms in the lasing medium are 'pumped' to an excited state by an external source of energy, such as flashlight, an electric current or another laser. This produces a POPULATION INVERSION, with more of the atoms in an EXCITED STATE than in the GROUND STATE. An electromagnetic STANDING WAVE set up in the cavity then causes STIMULATED EMISSION in which atoms move to a lower energy state, giving out coherent radiation. The radiation emitted by the atoms is reflected back and forth between the two mirrors, stimulating more emissions and so amplifying the amount of radiation. The radiation emerges from the semi-reflecting mirror as a powerful beam of light.

The lasing medium can be solid, liquid or gas, but infrared carbon dioxide lasers, red helium neon lasers and green argon ion lasers are amongst the most important technologically. Infrared semiconductor lasers are used to read COMPACT DISCS. Lasers are used as an alternative to scalpels in surgery, where they have the advantage that, since they cut tissue by heating it, there is less bleeding.

See also MASER.

laser-guided weapon A missile that finds its target by homing in on a laser beam aimed at the target by an air or ground-based observer.

laser printer A type of computer PRINTER that uses a LASER to remove charge from regions of a specially coated metal drum by the PHOTOELECTRIC EFFECT. This charge is then transferred to the page and used to attract ink powder, which is then melted onto the paper by a heated roller. This process is very similar to the operation of the PHOTOCOPIER.

lasso cell *See* CTENOPHORA.

latent heat The heat energy taken in or given out when a material changes its physical state at a constant temperature. Energy is taken in when a solid turns to a liquid, or a liquid to a gas, as energy is needed to overcome the forces holding the molecules together in a solid or liquid. When a gas condenses, or a liquid turns into a solid, energy is released as chemical bonds form holding the molecules together.

Latent heat can be measured in a CALORIMETER either by using a measured amount of electrical energy to bring about the change of state, or by a variation of the METHOD OF MIXTURES, in which a solid is placed in a liquid and melts. Since the substance will not start or end up at its melting point, the latter method also requires knowledge of the HEAT CAPACITY of the material.

See also SPECIFIC LATENT HEAT OF FUSION, SPECIFIC LATENT HEAT OF VAPORIZATION.

latitude A measure of position on the surface of the Earth, or any other planet. Latitude is measured in degrees north or south of the equator. *See also* LONGITUDE.

lattice An array of points arranged in space in some symmetrical pattern, such as the positions occupied by atoms in a crystal. *See also* ISOMORPHIC.

lattice energy The energy required to separate the molecules or ions of a crystal so they are infinitely far apart from one another. Often quoted for one mole of the material. *See also* BORN–HABER CYCLE.

lava Semi-molten rock, such as that ejected from a VOLCANO during an eruption. It is derived from MAGMA, and when reaching the surface it cools to form IGNEOUS rocks.

law of chemical equilibrium *See* EQUILIBRIUM CONSTANT.

law of conservation of angular momentum If there are no MOMENTS acting on an object its ANGULAR MOMENTUM will be constant. This law leads to Kepler's second law (*see* KEPLER'S LAWS) and explains why the rotational speed of a rotating skater increased if she pulls her arms in – as the MOMENT OF INERTIA is reduced with the mass being concentrated closer to the axis, the ANGULAR VELOCITY must increase to keep the angular momentum constant.

law of conservation of charge In any electric circuit or interaction between ELEMENTARY PARTICLES, the total charge remains unchanged. *See also* KIRCHHOFF'S LAWS.

law of conservation of energy In any closed system, where no energy can enter or leave, the total amount of energy remains constant. Thus energy can never be created or destroyed, though energy may be transferred from one form into another.

When energy is converted from one form to another, there is always a certain amount of heat energy produced as a by-product of the conversion process (*see* SECOND LAW OF THERMODYNAMICS). For example, burning fuel to produce mechanical energy in a motor car, or electricity in a power station, also results in the production of hot gases. This means that, as energy is converted into more useful forms, more and more of the energy is lost as low-grade heat.

See also HESS'S LAW, KIRCHHOFF'S LAWS.

law of conservation of momentum For any system in which no external forces act, the total MOMENTUM of the system is constant. In a system of particles interacting with each other, such as two particles colliding with one another, Newton's third law means that IMPULSES will occur in equal and opposite pairs. This means that the total change in momentum will be zero. This law is particularly useful in the study of collisions, which are divided into three classes: INELASTIC, ELASTIC and HYPERELASTIC.

For two objects of mass m and M, moving with velocities u and U respectively before collision and v and V after collision:

$$mu + MU = mv + MV$$

See also NEWTON'S LAWS OF MOTION.

law of constant composition *See* LAW OF CONSTANT PROPORTIONS.

law of constant proportions, *law of constant composition* A pure sample of any compound will always contain the same proportions of each of the constituent elements. If these masses are divided by the RELATIVE ATOMIC MASSES of the elements, the proportions will reduce to simple ratios. Thus sodium chloride contains 39 per cent sodium by mass and 61 per cent chlorine, and this is true for any pure sample of the material no matter how it is obtained. Once it is realized that the chlorine atoms are heavier than the sodium atoms, it can be calculated that this means that there is one atom of sodium for each atom of chlorine, and the formula NaCl can be deduced. This law, first put forward by John Dalton (1766–1844), was an important contribution to the fundamental ideas of chemistry (*see* DALTON'S ATOMIC THEORY).

law of equivalent proportions, *law of reciprocal proportions* If two elements, A and B, both form a compound with a third element C, then when A and B combine together they will do so in the same proportions in which they reacted with C. If A and B are oxygen and hydrogen for example, 12 g of carbon will combine with 4 g of hydrogen (to form methane) or 32 g of oxygen (to form carbon dioxide). When hydrogen and oxygen combine together to form water, 32 g of oxygen will combine with 4 g of hydrogen.

law of mass action The rate at which a chemical reaction takes place is dependent on the CONCENTRATION of the reagents. In a reaction where one MOLE of A reacts with one mole of B, the rate at which the reaction takes place is proportional to the concentrations of A and B, represented in equations by [A] and [B] respectively, so

$$\text{rate of reaction} = k[A][B]$$

where k is a constant, called the RATE CONSTANT. For a more complex reaction, involving different numbers of moles of the reagents, but still reacting in a single step,

$$xA + yB + zC \rightarrow \text{products}$$

where x, y and z are the numbers of moles involved,

$$\text{rate of reaction} = k[A]^x\,[B]^y\,[C]^z$$

The law of mass action applies only if the reacting molecules have free access to one another, so is not applicable to reactions involving solids, where only the atoms on the surface of the solid are available to react. It is for this reason that finely divided solids react more rapidly. *See also* RATE OF REACTION.

law of multiple proportions If two materials, A and B, react together to form more than one compound, the mass of A that combines with a fixed mass of B in the two compounds will be a simple ratio. For example, in water, H_2O, 2 g of hydrogen combines with 16 g of oxygen. In hydrogen peroxide, H_2O_2, the same 16 g of oxygen will combine with just 1 g of hydrogen – a simple ratio of 2:1.

law of octaves An early attempt at grouping the chemical elements into families with similar properties, based on an analogy with musical notes. Whilst some elements, notably the HALOGENS, were correctly classified, the idea was superseded by the more sophisticated classification of the PERIODIC TABLE.

law of reciprocal proportions *See* LAW OF EQUIVALENT PROPORTIONS.

lawrencium (Lr) The element with ATOMIC NUMBER 103. It does not occur naturally, and the longest lived ISOTOPE has a HALF-LIFE of only 3 minutes. Lawrencium was first synthesized by bombarding atoms of californium with high energy beryllium nuclei.

LCD *See* LIQUID CRYSTAL DISPLAY.

LDR *See* LIGHT-DEPENDENT RESISTOR.

leaching The process in which the passage of water or some other solvent through a porous solid material causes some part of that solid to be dissolved in the solvent. In particular, leaching is the washing away of substances out of the soil. This can be a result of DEFORESTATION, in which case useful nutrients are lost and soil fertility is reduced. Where fertilizers leach out of the soil, this can lead to POLLUTION of rivers, etc.

lead (Pb) The element with ATOMIC NUMBER 82; RELATIVE ATOMIC MASS 207.2; melting point 328°C; boiling point 1,740°C; RELATIVE DENSITY 11.4. Lead is a dense, soft metal, once used for covering roofs and making water pipes. Lead occurs widely in nature, mainly as lead sulphide in the ore galena. The ore is roasted to form the oxide, which is then reduced to metallic lead with carbon:

$$2PbS + 3O_2 \rightarrow 2PbO + 2SO_2$$

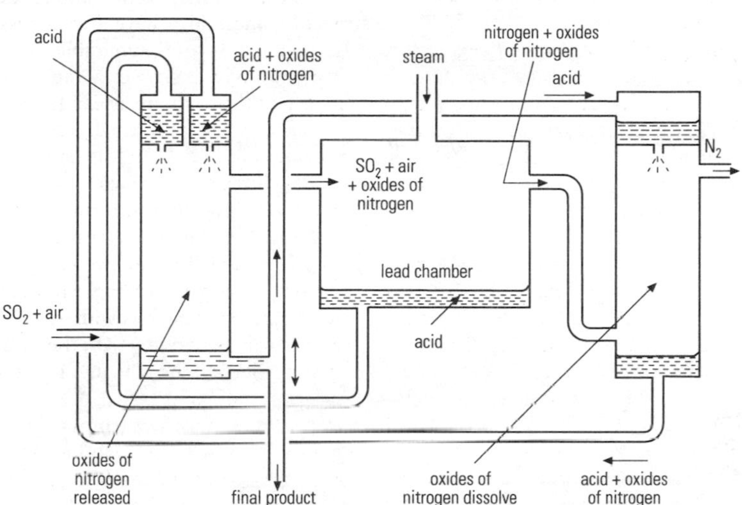

Lead chamber process.

$$2PbO + C \rightarrow 2Pb + CO_2$$

Lead is widely used in the ELECTRODES of LEAD-ACID CELLS and in tetraethyl lead as an additive to petrol, though concerns about exposure to lead in the environment (it is a cumulative poison) have meant tighter controls on its use. Its relatively low cost and high density also makes lead a common choice for protection against IONIZING RADIATION, such as X-rays and gamma radiation.

lead-acid cell A rechargeable electrochemical CELL with ELECTRODES of lead and lead sulphate and a sulphuric acid ELECTROLYTE, often used where high currents are needed, such as in motor vehicles.

lead carbonate ($PbCO_3$) A white solid, decomposes on heating, RELATIVE DENSITY 6.1. Lead carbonate is insoluble in water and can be made as a PRECIPITATE in the reaction between aqueous solutions of lead nitrate and ammonium carbonate:

$$PbNO_3 + NH_4CO_3 \rightarrow PbCO_3 + NH_4NO_3$$

Lead carbonate decomposes on heating to give lead oxide:

$$PbCO_3 \rightarrow PbO + CO_2$$

lead carbonate hydroxide, *basic lead carbonate, white lead* ($2PbCO_3.Pb(OH)_2$) A white powder; decomposes on heating; RELATIVE DENSITY 6.1. White lead was formerly used as a white pigment, but its high toxicity means it is no longer used. It reacts with hydrogen sulphide to form black lead sulphide:

$$2PbCO_3Pb(OH)_2 + 3H_2S \rightarrow 3PbS + 4H_2O + 2CO_2.$$

lead chamber process A process for the commercial manufacture of SULPHURIC ACID, which has largely been superseded by the CONTACT PROCESS. In the lead chamber process, sulphur dioxide, nitrogen dioxide, steam and oxygen are mixed in a lead chamber. The sulphur dioxide is oxidized to sulphur trioxide by the nitrogen dioxide:

$$SO_2 + NO_2 \rightarrow SO_3 + NO$$

The nitrogen oxide is then regenerated by the oxygen:

$$2NO + O_2 \rightarrow 2NO_2$$

The sulphur trioxide dissolves in the water to produce acid with a concentration up to 77 per cent of SATURATED. The YIELD of the reactions is increased by recycling unreacted FEEDSTOCK.

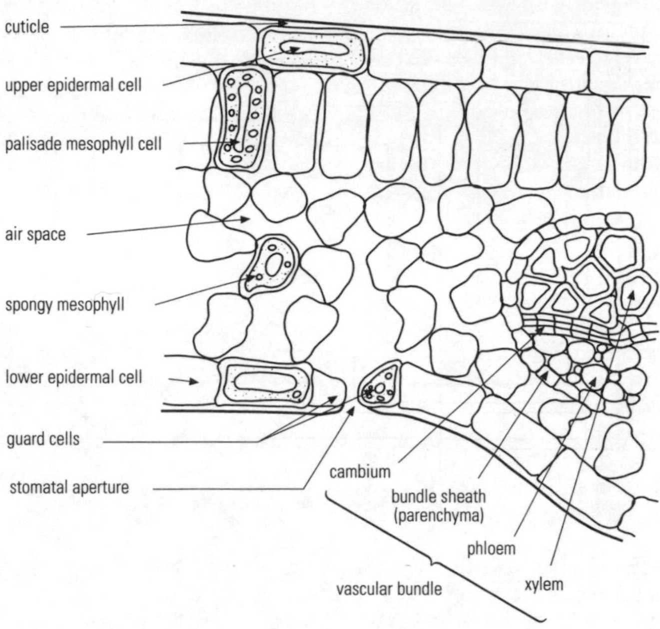

Structure of a dicotyledonous leaf (cross-section through the blade).

lead oxide Lead(II) oxide (or lead monoxide), PbO, or lead(IV) oxide (or lead dioxide), PbO_2.

Lead(II) oxide is a yellow solid; melting point 866°C; decomposes on further heating; RELATIVE DENSITY 9.5. Lead(II) oxide can be formed by heating lead nitrate:

$$2Pb(NO_3)_2 \rightarrow 2PbO + 4NO_2 + O_2$$

Lead oxide is AMPHOTERIC, dissolving readily in strong acids, for example:

$$PbO + 2HNO_3 \rightarrow Pb(NO_3)_2 + H_2O$$

but reacting with strong alkalis to form PLUMBATES, for example:

$$PbO + 2NaOH \rightarrow Na_2PbO_3 + H_2O$$

Lead(IV) oxide is a dark brown solid; decomposes on heating; relative density 9.4. Lead(IV)oxide can be made by the OXIDATION of lead(II) oxide, for example:

$$PbO + H_2O_2 \rightarrow PbO_2 + H_2O$$

Lead(IV) oxide is itself strongly oxidizing, and will reduce hydrochloric acid to give chlorine:

$$PbO_2 + 2HCl \rightarrow PbO + Cl_2 + H_2O$$

leaf A structure growing out from the stem of a plant, usually the main site of PHOTOSYNTHESIS. A leaf is made up of three parts: the sheath (or leaf base), the petiole (or stalk), and the lamina (or blade). A leaf blade is made up of MESOPHYLL cells, between the upper and lower EPIDERMIS, and a waxy CUTICLE that prevents water loss. The mesophyll cells contain numerous CHLOROPLASTS in which photosynthesis takes place.

Leaves have adapted many ways of ensuring they obtain sufficient light for photosynthesis: they have a large surface area to capture sunlight, they are thin and have a staggered arrangement on the plant to allow as much light through as possible, and chloroplasts are numerous and can move within the mesophyll to be in the best position for absorption of light. Leaves are also phototropic, moving towards the source of light (*see* TROPISM).

Evergreen leaves are called persistent, whereas those that fall off in the autumn are deciduous. A simple leaf, for example of a beech tree, is undivided, whereas a compound leaf, for example of a horse chestnut tree, is divided into several leaflets. Pinnate leaves are a form of compound leaves where the leaflets are arranged in rows either side of the midrib, for example in the ash.

leakage current A small current flowing through a material that is designed to be an insulator, such as the insulating layer in a CAPACITOR.

learning Behaviour in animals that is acquired and modified in response to experience and changes in the environment. Learning contrasts with innate or instinctive behaviour, which is inborn and usually cannot be modified. Forms of learning include HABITUATION, CONDITIONING and IMPRINTING. The highest form of learning is insight or intelligent behaviour, which involves the recall and adaptation of previous experiences to a new situation.

Le Chatelier's principle In a REVERSIBLE REACTION, any change in conditions (such as temperature, pressure or concentration of one of the reagents) will change the EQUILIBRIUM state of the reaction in such a way as to oppose the effects to the change. For example, in the formation of ammonia from nitrogen and hydrogen,

$$N_2 + 3H_2 \Leftrightarrow 2NH_3, \Delta H = -92.4 \text{ kJ mol}^{-1}$$

the equilibrium will shift to the right (more ammonia) if: (i) the pressure is increased (the formation of ammonia reduces the number of molecules present, tending to reduce the pressure; (ii) the concentration of nitrogen and hydrogen is increased or the concentration of ammonia reduced; or (iii) the temperature is reduced (the reaction is EXOTHERMIC, production of ammonia tends to increase the temperature). *See also* CHEMICAL EQUILIBRIUM.

lecithin A PHOSPHOLIPID that is an important component of CELL MEMBRANES of plants and animals.

lectin Any one of a group of proteins and GLYCOPROTEINS that can cause AGGLUTINATION of cells by cross-linking cell-surface CARBOHYDRATES and other ANTIGENS.

LED *See* LIGHT-EMITTING DIODE.

Lee's disc A method of measuring the THERMAL CONDUCTIVITY of a poor conductor. The sample is shaped into a thin disc and sandwiched between two metal plates, each drilled with a hole for a thermometer. The THERMAL RESISTANCE of these plates is much less than that of the sample, so the thermometers effectively enable the temperature at each surface of the sample to be measured. One plate is heated whilst the other looses heat by CONVECTION. Once a steady state has been established, the rate of heat flow can be found by removing the heat source and measuring the rate of temperature drop of the other metal plate. Knowing the mass and SPECIFIC HEAT CAPACITY of the lower plate, it is possible to find the rate at which heat must have been flowing through the sample to maintain this plate at a constant temperature.

legume A fruit plant with a pod that opens on both sides to release its dry seeds, for example in peas, beans and clovers. Legumes are members of the family Leguminosae. Their roots have ROOT NODULES that contain nitrogen-fixing bacteria, which improve soil fertility by fixing atmospheric nitrogen (*see* NITROGEN FIXATION). Legumes are therefore important in agriculture. The edible seeds are called pulses.

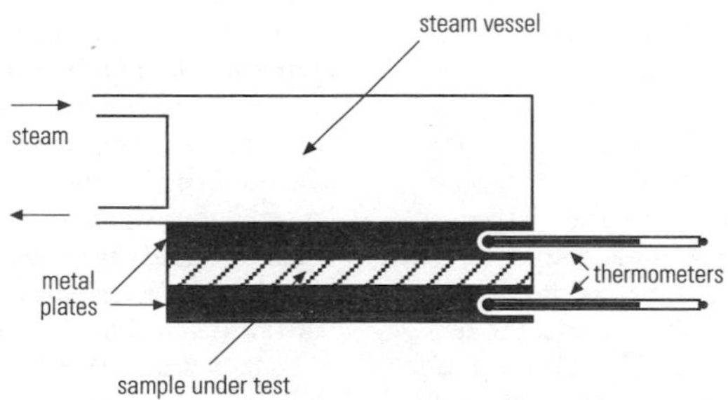

Lee's disc.

Leguminosae A family of flowering plants having LEGUMES as fruit and ROOT NODULES.

Lennard–Jones 6-12 potential A form of interatomic POTENTIAL function used to model the effects of VAN DER WAAL'S FORCES.

lens In optics, a curved piece of glass designed to refract light rays to form an image (*see* REFRACTION). The curved surfaces of a lens are described as CONCAVE if they curve inwards or CONVEX if they curve outwards. A lens with two concave surfaces will be diverging whilst a lens with two convex surfaces will be converging. Single lenses form images that are not perfect, but suffer form a number of defects called ABERRATIONS. Lenses that are used in optical instruments such as cameras or microscopes are almost always COMPOUND LENSES made up of several separate lenses made from different materials and with different curvatures to minimize the effect of these aberrations. *See also* CONVERGING LENS, DIVERGING LENS.

lenticel One of several small pores found in the trunk of trees or outer layer of woody stems through which gas exchange occurs. Lenticels consist of regions of loosely packed cells with air spaces between them, and are a site of some water loss (*see* TRANSPIRATION).

Lenz's law A law concerning the direction of the current produced by any process involving ELECTROMAGNETIC INDUCTION. It states that any such current will be in a direction that opposes the change that created it.

LEP (Large Electron Positron) Currently the world's largest SYNCHROTRON. Located at CERN, it is 27 km in circumference and accelerates electrons and positrons to 50 GeV for COLLIDING BEAM EXPERIMENTS.

lepton Any of a class of ELEMENTARY PARTICLES not affected by the STRONG NUCLEAR FORCE. Three charged leptons are known, the ELECTRON, the MUON and the TAU LEPTON, in order of increasing mass and decreasing HALF-LIFE (the electron is stable). For each of these particles there is also a NEUTRINO, a light (probably massless) particle carrying no charge, but the same QUANTUM NUMBER as the charged lepton (such as ELECTRON NUMBER, muon number, etc.). This quantum number is conserved in all interactions. Thus, for example, the decay of a muon produces an electron, a muon neutrino and an electron antineutrino, so the total muon number remains at 1 whilst the electron and the electron antineutrino have opposite electron number. *See also* STANDARD MODEL.

leucine An AMINO ACID essential in nutrition. It is produced in several ways, in particular by the digestion of proteins by ENZYMES in the PANCREAS.

leucocyte *See* WHITE BLOOD CELL.

leucocytosis An abnormal increase in the number of WHITE BLOOD CELLS in the blood.

leucopenia An abnormal reduction in the number of WHITE BLOOD CELLS in the blood.

leucoplast A PLASTID that is a colourless food storage body found in the cells of plant tissue not normally exposed to light. Leucoplasts include the STARCH-storing amyloplasts found in roots of many plants, oil-storing elaioplasts and protein-storing aleuroplasts.

lever A simple MACHINE in which a support rotates about a PIVOT, or FULCRUM. For a lever,

$$\text{effort force} \times \text{effort to pivot distance} = \text{load force} \times \text{load to pivot distance.}$$

See also EFFORT, LOAD.

Lewis acid A term used to describe substances that are electron acceptors. Thus Lewis acids include conventional ACIDS but also some additional substances, such as $AlCl_3$, which can accept electrons from a chloride ion to form the $(AlCl_4)^-$ ion.

Lewis base A term used to describe substances that have available ELECTRONS and are thus electron donors. These include conventional BASES but also some additional substances, for example ammonia.

LH *See* LUTEINIZING HORMONE.

lichen A member of the KINGDOM FUNGI that is a symbiotic association (*see* SYMBIOSIS) between an ALGA and a FUNGUS.

Liebig condenser A CONDENSER in which a straight glass tube is surrounded by a glass envelope through which cold water is passed.

ligament A band of strong, flexible CONNECTIVE TISSUE attaching two bones at a joint and restricting movement of the joint to prevent dislocation. Yellow elastic ligaments consist mostly of elastic fibres and are more extensible than ligaments made of COLLAGEN fibres.

ligand A molecule or ion that is attached to a metal ion in the formation of a complex (*see* COMPLEX ION). Water is the most common ligand, and many metal ions are HYDRATED, for example $[Cu(H_2O)]^{2+}$. The cyanide ion, CN^-, is another example, forming complexes with iron, such as $[Fe(CN)_6]^{3-}$. Some larger ligand molecules can attach themselves to an ion at more than one point and are described as bidentate, as opposed to the more common monodentate ligands, which attach themselves to a metal ion at only one point.

ligase One of a group of enzymes that forms bonds between two molecules using energy from the breakdown of ATP. DNA ligase in particular is very useful in GENETIC ENGINEERING because it joins DNA chains specifically cut by RESTRICTION ENDONUCLEASES.

light ELECTROMAGNETIC RADIATION of a wavelength to which the human eye is sensitive (VISIBLE LIGHT). The eye is sensitive to wavelengths from about 7×10^{-7} m (red) to 4×10^{-7} m (violet). The term is also commonly used for wavelengths outside this narrow range but where the properties being exploited are similar to those of visible light. Longer wavelengths are described as INFRARED and shorter wavelengths as ULTRAVIOLET. *See also* PHOTON, SPEED OF LIGHT, WAVE-PARTICLE DUALITY.

light-dependent resistor (LDR) A device whose RESISTANCE changes on exposure to light. An example of such a material is cadmium sulphide, which is an insulator in darkness: exposure to light releases FREE ELECTRONS, enabling a current to flow. LDRs are sometimes used in circuits that are required to react to changes in light level, though PHOTODIODES respond more quickly to changes in illumination, and can be made sensitive to a greater range of wavelengths of light.

light-emitting diode (LED) A SEMICONDUCTOR device that converts electrical energy into visible or infrared light. It comprises a DIODE in which the energy loss for CHARGE CARRIERS crossing the DEPLETION LAYER is large enough for them to produce a PHOTON of light. Thus light is produced when the diode is FORWARD BIASED. By making packages of LED's with suitable shaped segments, displays that form letters or numbers can be manufactured.

light filter *See* FILTER.

light gate An electronic timing device that measures the speed of a moving object by finding the time for which a light beam is obstructed by a card of known length attached to the object.

lightning The flash of light seen from a spark within a CUMULONIMBUS cloud, between two such clouds, or between a cloud and the ground. These sparks occur as a result of t he buildup of electrostatic charges (*see* STATIC ELECTRICITY) from friction between rain and hail in the turbulent core of the cloud. The thunder that accompanies a flash of lightning is produced by the expansion of the heated air.

A spark of lightning generated within a cloud cannot not usually been seen directly from the ground, but instead produces a general illumination of the sky. This effect is called sheet lightning. Forked lightning is the direct observation of a spark between one cloud and another or between the cloud and the ground.

See also CHARGING BY FRICTION, WEATHER SYSTEMS.

lightning conductor A metal rod connected between the top of a tall buildings and the ground that protects the building from damage by LIGHTNING. The rod builds up a large enough charge in the presence of a thundercloud to spray ions into the cloud, neutralizing it and reducing the chance of a lightning strike. If a lighting strike does occur, the lightning conductor will provide a low resistance path to Earth. Without this, a large amount of current would flow through the building, with the heat produced causing a great deal of damage.

light pollution In astronomy, man-made light sources scattering off particles in the atmosphere and making astronomical observations more difficult.

light year A unit of distance used in astronomy, the distance travelled by light in one year, equal to 9.46×10^{12} km.

lignin A naturally occurring chemical substance found in the CELL WALLS of plants, which is composed of rings of carbon atoms joined in a chain. Lignin provides strength to the plant, and is difficult to digest so provides protection from attack by many organisms. Lignin is found in all wood and is therefore of commercial importance. There are some regions of the cell wall where PLASMODESMATA are present and lignin is not deposited, so that water and dissolved minerals can pass between cells. *See also* SCLERENCHYMA.

lime A term used in the common names of many calcium compounds, for example, limestone which is calcium carbonate, quicklime (calcium oxide), slaked lime (calcium hydroxide). *See also* LIME WATER.

limestone A SEDIMENTARY rock, similar to chalk but harder and more compact, formed by the deposition of the shells of marine animals, which are fused into rock by the pressure of further layers. Limestone is composed mainly of calcium carbonate and sometimes magnesium carbonate. The porous and slightly soluble nature of limestone is responsible for HARD WATER in limestone area and for the formation of STALACTITES and STALAGMITES in limestone caves.

lime water Calcium hydroxide solution. In the presence of carbon dioxide, calcium carbonate is produced as a PRECIPITATE, though only in small quantities since calcium hydroxide is not very soluble. This can be used as a test for carbon dioxide, which turns the lime water 'milky':

$$Ca(OH)_2 + CO_2 \rightarrow CaCO_3 + H_2O$$

limit 1. One of the two ends of the range over which a sum or INTEGRAL is evaluated.

2. The value to which some function tends more and more closely as the independent variable moves closer and closer to a particular value, with the function not being defined for that value. For example, the limit of the function $(\sin x)/x$ is 1 as x tends to 0.

3. The value to which a sum or other SERIES approaches more and more closely as more and more terms in the series are taken into account.

limiting friction The maximum amount of STATIC FRICTION that is produced just before two objects start to slide over one another, often slightly greater than the DYNAMIC FRICTION.

line of action Of a force, a line that runs in the same direction as the force and passes through the point at which the force acts.

line of best fit A line drawn on a graph on which some data points are also plotted so as to minimize the sum of the squares of the distances of the points from the line.

line of force An imaginary line in an ELECTRIC FIELD or MAGNETIC FIELD that allows the direction of the associated force to be visualized.

line spectrum An EMISSION SPECTRUM or ABSORPTION SPECTRUM in which the wavelengths absorbed or emitted form a number of separate very narrow ranges or lines. Each line in a spectrum represents a closely defined wavelength, corresponding to the PHOTON energy equal to the difference in energy between the ENERGY LEVELS that produced the line.

linear (*adj.*) Relating to a line, in particular describing a relationship that would be shown by a straight line on a graph, with a given change in one quantity always causing a fixed change in some other quantity.

linear accelerator A PARTICLE ACCELERATOR in which charged particles are accelerated through a series of metal cylinders called DRIFT TUBES. Alternate drift tubes are connected to opposite sides of an ALTERNATING CURRENT supply. The frequency of the supply and the lengths of successive tubes are arranged so that the particles being accelerated, typically electrons, are in the field-free region inside the tubes whilst the polarity of the supply reverses, so they always experience an accelerating field.

linear equation An equation representing a function that contains only the first power of the independent variables. It gives a straight line when plotted on a graph. Linear equations have the form:

$$y = mx + c$$

where y is the dependent variable, x the independent variable, m the GRADIENT of the line, and c the x-INTERCEPT.

linear expansivity *See* THERMAL EXPANSIVITY.

line transect *See* TRANSECT.

linkage In genetics, the association of two or more GENES occurring on the same CHROMOSOME that therefore tend to be inherited together. Not all genes together on a chromosome will be inherited together because RECOMBINATION processes can occur, such as CROSSING-OVER of chromosome pairs, which shuffles the genetic material. Genes closer together on a chromosome are more likely to remain linked and be inherited together. SEX LINKAGE refers to genes carried on the sex chromosome that may have nothing to do with the sex of an organism but are linked to it by their position. *See also* MENDEL'S LAWS.

linoleic acid An unsaturated FATTY ACID abundant in many plant fats and oils.

Linz-Donawitz process The process for the production of STEEL in an OXYGEN FURNACE.

lipase One of a group of enzymes responsible for breaking down fats into FATTY ACIDS and GLYCEROL during digestion. Lipase is made by the PANCREAS and requires a slightly alkaline environment to function.

lipid Any one of a large group of organic compounds that are the major constituents in plant and animal fats, waxes and oils. Lipids are all ESTERS of FATTY ACIDS, and are soluble in alcohol but not water.

The most common alcohol with which fatty acids react to form lipids is GLYCEROL. The lipids formed in this way are called GLYCERIDES and can be mono, di or tri, depending on how many hydroxyl (OH) groups from glycerol have combined with the fatty acids, for example triglycerides have three fatty acids attached to glycerol. The properties of a lipid largely depend on the fatty acids present, because the alcohol is usually the same (glycerol). PHOSPHOLIPIDS are lipids containing glycerol in which one of the fatty acids is replaced by phosphoric acid.

Lipids provide an energy store in plants and animals, and in the form of fat give some protection and insulation to animals and their internal organs. Lipids also provide waterproofing for plants and animals, as oily secretions and waxes formed by the combination of fatty acids with an alcohol other than glycerol. The emulsion test (which gives a cloudy solution when lipid and alcohol mix) or the Sudan III test (a red dye that detects fats and oils) detect lipids in a solution.

lipoprotein A complex of LIPIDS and protein that is a structural component of all cell membranes. CHOLESTEROL is transported in the blood within such a complex, either free or esterified (*see* ESTER) to FATTY ACIDS.

liquation A process for extracting one substance from a mixture by heating until the material with the lower melting point melts and can be poured away.

liquefaction The process of turning a gas into a liquid.

liquid The state of matter in which a material is able to flow freely and take up the shape of its container, but where there is a distinct boundary between the material and its surroundings. Liquids are materials in which the molecules are closely spaced (like solids they are hard to compress, and they have similar densities to solids), but unlike most solids the molecules are randomly arranged. The forces between the molecules in a liquid are weak enough to allow liquids to flow, so they will take up the shape of any container in which they are placed, but the forces are strong enough to prevent the molecules from moving off into space, so a liquid has a fixed volume and will not expand to fill a volume.

liquid crystal Any liquid made from molecules that tend to line up under the influence of INTERMOLECULAR FORCES, producing long range ordering of a type more commonly associated with solid crystalline materials. Under the influence of an electric field, this ordering can extend throughout the whole of the liquid. *See also* LIQUID CRYSTAL DISPLAY.

liquid crystal display (LCD) An electronic display device that uses a LIQUID CRYSTAL material in which all the molecules can be aligned by an

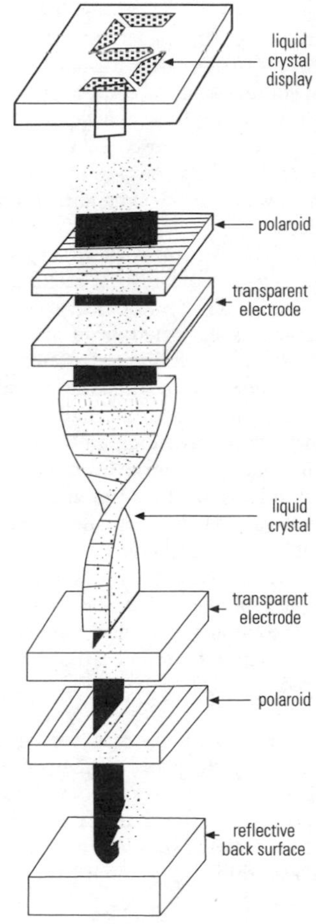

Liquid crystal display.

electric field (such materials are described as NEMATIC). The molecules of the liquid crystal are optically active, rotating the plane of polarization of light. By placing a piece of POLAROID in front of a suitable thickness of liquid crystal, areas of a display can be made which are clear in the absence of an electric field, but appear dark when an electric field is produced by applying a voltage to suitably shaped ELECTRODES.

liquid drop model A model of the atomic NUCLEUS that attempts to calculate the NUCLEAR BINDING ENERGY and other properties of the nucleus by treating it as a drop of liquid with a fixed DENSITY, SURFACE TENSION, SPECIFIC LATENT HEAT OF VAPORIZATION and so on. This model is most appropriate to nuclei containing large numbers of NUCLEONS. *See also* SHELL MODEL.

liquidus In a SOLID SOLUTION, the line on a graph of temperature against composition above which the substance is entirely liquid.

liquor A general term used in the chemical industry to describe a solution of some useful material in water, usually at a fairly high concentration.

lithium (Li) The element with ATOMIC NUMBER 3; RELATIVE ATOMIC MASS 6.9; melting point 180°C; boiling point 1,340°C; RELATIVE DENSITY 0.5. Lithium is extracted from its ores by ELECTROLYSIS. It is the lightest of the ALKALI METALS, the small Li^+ ion making it markedly less reactive than the other alkali metals. Lithium is used increasingly in the manufacture of high energy batteries for use with cameras and similar small domestic appliances.

lithium carbonate (Li_2CO_3) A white solid; melting point 735°C; decomposes on further heating; RELATIVE DENSITY 2.1. Lithium carbonate is only slightly soluble and can be obtained as a precipitate, for example by the reaction between lithium sulphate and sodium carbonate:

$$Li_2SO_4 + Na_2CO_3 \rightarrow Li_2CO_3 + Na_2SO_4$$

Lithium carbonate is used as a drug in the treatment of some depressive illnesses.

lithium hydride (LiH) A white solid; melting point 686°C; decomposes on further heating; RELATIVE DENSITY 0.82. Lithium hydride can be formed by burning lithium in hydrogen:

$$2Li + H_2 \rightarrow 2LiH$$

Lithium hydride is a largely IONIC compound, containing the H^- ion, with the unusual property of liberating hydrogen at the ANODE in ELECTROLYSIS. Lithium hydride reacts violently with water:

$$LiH + H_2O \rightarrow LiOH + H_2$$

lithium oxide (Li_2O) A white solid; melting point 1,200°C; decomposes on further heating. Unlike the oxides of other ALKALI METALS, it is stable in the presence of water. It is used as a lubricant and as a flux for welding.

lithosere A series of plant SUCCESSIONS originating on a bare rock surface.

lithosphere The region of solid rock forming the Earth's CRUST.

litmus A naturally occurring vegetable dye that can be used as an INDICATOR. A solution of litmus is red under ACIDIC conditions but blue in ALKALINE solutions.

litmus paper Paper soaked with LITMUS, used to test for ACIDIC or ALKALINE solutions.

litre (l) A unit of volume, correctly called the decimetre cubed (dm^3) in the SI system. One litre is equal to 10^{-3} m^3.

live That part of a mains electricity supply that is at a high ELECTRIC POTENTIAL relative to EARTH.

liver In vertebrates, a large organ derived from the ENDODERM that makes up 3–5 per cent of the body weight (2 kg). The liver is located in the upper ABDOMEN, below the DIAPHRAGM to which it is attached. It has many functions concerned with DIGESTION and maintaining a constant internal environment.

Oxygenated blood from the AORTA enters the liver via the HEPATIC artery, and blood leaves via the hepatic vein. The hepatic portal vein supplies blood from the INTESTINE that is rich in soluble digested food. Under the light microscope, the liver can be seen to be made up of lobules, the functional unit of which is called the acinus. The hepatic artery and hepatic portal vein combine to form channels called sinusoids, where the blood mixes and allows exchange of materials between the blood and liver cells lining the sinusoids. These liver cells are called HEPATOCYTES and they secrete the BILE fluid that assists in the break down and absorption of fats in the small intestine. Kupffer cells are also found lining the sinusoids; these are phagocytic cells (*see* PHAGOCYTOSIS) important in removing foreign bodies entering from the intestine and destroying worn-out or damaged blood cells.

The liver also converts excess GLUCOSE to GLYCOGEN for storage until needed, a process called GLYCOGENESIS. The liver plays a key role in the maintenance of blood glucose levels and can return glucose to the blood when needed by the break down of glycogen (GLYCOGENOLYSIS) or by converting protein and fats into glucose (GLUCONEOGENESIS). *See also* PANCREAS.

The liver also breaks down excess amino acids by DEAMINATION and can convert one amino acid to another by TRANSAMINATION to replace deficient non-essential amino acids. Other functions include the synthesis of many important PLASMA PROTEINS (such as ALBUMIN and FIBRINOGEN), storage of vitamins and minerals (e.g. iron, potassium, copper and zinc), break down of hormones, removal of toxins (such as nicotine and alcohol by converting to safer chemicals), storage of blood and production of heat (under control of the HYPOTHALAMUS).

liverwort A member of the CLASS HEPATICAE.

lixivation The process of separating one material from others by using a stream of water or some other solvent in which the desired material is dissolved, flowing out of the mixture.

load 1. In electricity and electronics, any device to which electrical power is supplied.

2. In mechanics, an object on which WORK is done by some machine, such as a pulley system, or the force acting on that object.

local area network (**LAN**) A data NETWORK that links computers in the same room or building. Typical local area networks are ETHERNETS or TOKEN RING NETWORKS.

lock-and-key mechanism A proposed mechanism of ENZYME action. The three-dimensional structure of the enzyme provides a specific ACTIVE SITE that can be compared to a lock, while the SUBSTRATE molecule it acts upon is the key. According to this theory only the correct substrate will fit the active site of an enzyme. An enzyme–substrate complex forms that is then a different shape to the substrate alone, and it therefore falls away from the enzyme leaving it free to attach to another substrate molecule. Modern interpretations of this theory suggest that the AMINO ACIDS that form the active site alter their relative three-dimensional positions as the substrate binds. This is an induced fit.

locus 1. (*mathematics*) A series of points all satisfying a certain equation, and forming a continuous line. For example, a circle can be thought of as the locus of points that are all a certain distance (the RADIUS) away from a single fixed point (the centre of the circle).

2. (*biology*) The position of a GENE within a DNA molecule. The position of a given locus can be determined by CHROMOSOME MAPPING. *See also* ALLELE.

logarithm The logarithm of a number to a given BASE is that number which, when the base is raised to the power of the logarithm, will give the original number. Thus if n is the logarithm to base x of y,

$$n = \log_x y$$

then x raised to the power n will give y,

$$y = x^n$$

Logarithms are usually taken to base 10 (common logarithms) or to base e (NATURAL LOGARITHMS).

logic circuit An electronic system handling DIGITAL signals and containing one or more LOGIC GATES.

logic gate The basic building block of any DIGITAL electronic system. A logic gate gives an output that is high or low depending on the state of one or more inputs. The simplest logic gate is the INVERTER or NOT gate, where the output is high if the input is low and vice versa. An AND GATE has an output only if all the inputs are high, whilst an OR GATE has a high output if any of its inputs is high. Particularly important is the NAND GATE (NAND = not and), where the output is high unless all the inputs are high. This gate is simple to manufacture and can be used as a building block for many more complex devices.

lone pair A pair of electrons in a single ORBITAL in the VALENCE SHELL of an atom. Such a pair do not take part in the usual covalent bonding (*see* COVALENT BOND) of the atom, though they can form a CO-ORDINATE BOND. The presence of lone pairs has important implications for the STEREOCHEMISTRY of an element. Ammonia, NH_3, for example, forms a TRIGONAL PYRAMIDAL structure, which could be regarded as TETRAHEDRAL with the lone pair forming one point in the TETRAHEDRON with hydrogen atoms at the other three. On the other hand, the ammonium ion, NH_4^+, is a tetrahedral structure with the lone pair forming a co-ordinate bond with a fourth hydrogen atom.

longitude A measure of position on the surface of the Earth, or any other planet. Longitude is measured in degrees east or west of the prime meridian; a line running from pole to pole and passing through a specified point. On Earth, the prime meridian is fixed as passing through Greenwich in the UK.

longitudinal wave A WAVE in which the motion of the particles that make up the wave is to and fro along the direction in which the wave is travelling. Such a wave can be thought of as a series of compressions and rarefactions; areas where the density of the particles is alternately higher and lower than normal. SOUND is an example of this kind of wave motion. *Compare* TRANSVERSE WAVE.

long period A PERIOD in the PERIODIC TABLE that includes TRANSITION METALS; that is, periods 4, 5, 6, 7 and 8. *See also* SHORT PERIOD.

long wave RADIO WAVES with a wavelength longer than about 600 m.

loop of Henle The part of the KIDNEY that is responsible for reabsorbing water, sugars and other useful minerals from the glomerular filtrate (*see* GLOMERULUS). The loop forms part of a long tubule within the functional unit called the NEPHRON, of which there are over a million in the kidney.

The loops of Henle are located in the inner region of the kidney, the MEDULLA. A loop is connected at one side by the proximal convoluted tubule to the BOWMAN'S CAPSULE, and at the other side by the distal convoluted tubule to a collecting duct. Much reabsorption is carried out by the proximal convoluted tubule and the loop of Henle concentrates the filtrate further. The degree to which the loop can concentrate the filtrate passing through it is related to the length of the loop, for example it is very long in the desert rat, where it is an advantage for URINE to be very concentrated.

A COUNTERCURRENT SYSTEM operates in the loop of Henle, in which sodium and chloride ions are removed by ACTIVE TRANSPORT from the wider, less permeable ascending limb into the interstitial spaces between the limbs of the loop and into the associated, parallel blood capillary, the vasa recta. The concentration of tissue fluid and blood is greater at the apex of the loop (in the medulla). This high concentration causes water to be drawn out of the more permeable descending limb of the loop by OSMOSIS and enter the vasa recta.

Lorentz contraction The apparent reduction in length of a moving object compared to its length at rest, according to the SPECIAL THEORY OF RELATIVITY. If an observer at rest with respect to an object observes a length l_0, then an observer moving at a speed v relative to the first will observe length l with

$$l = l_0\sqrt{(1 - v^2/c^2)}$$

where c is the SPEED OF LIGHT.

Lorentz transformation The mathematical transformation needed to move from co-ordinates based on one INERTIAL REFERENCE FRAME to another in the SPECIAL THEORY OF RELATIVITY.

Love wave In SEISMOLOGY, the transverse surface wave that travels along the upper part of the crust in an EARTHQUAKE. *See* SEISMIC WAVE.

low In DIGITAL electronics, a voltage level or other signal corresponding to a BINARY 0.

lowering of vapour pressure The reduction in the SATURATED VAPOUR PRESSURE of a solvent, depending on the concentration of dissolved material. Provided the dissolved material does not itself have a significant vapour pressure, the reduction in vapour pressure is an approximately COLLIGATIVE PROPERTY, depending far more on the molar concentration of the dissolved material than its chemical nature.

lowest common denominator The smallest INTEGER that can be used as a DENOMINATOR for all members of a set of specified fractions, or terms in an equation. Thus for the fractions $^1/_2$, $^1/_3$ and $^1/_6$, the lowest common denominator is 6 since they can all be expressed as fractions with 6 in the denominator, $^3/_6$, $^2/_6$ and $^1/_6$.

lowest common multiple The smallest INTEGER that has all of a set of specified integers as FACTORS. Thus the lowest common multiple of 4, 5 and 10 is 20.

low-grade heat Heat energy that has been used to raise the temperature of a large quantity of matter by a small amount and thus is of little use for further energy transformations. *See also* HEAT ENGINE.

low-level language A computer PROGRAMMING LANGUAGE in which each instruction corresponds to a specific MACHINE CODE instruction. A low-level language is designed for a particular computer and needs a knowledge of the internal working of that computer.

Lowry–Brønsted theory *See* ACID, BASE.

lumen 1. (*physics*) (lm) The SI UNIT of LUMINOUS FLUX. One lumen is equal to the amount of

VISIBLE LIGHT energy falling on one square metre at a distance of one metre from a point source of LUMINOUS INTENSITY of one CANDELA.

2. (*biology*) An inner, open space within tubular structures, such as blood vessels and the digestive system. In plants it is the space remaining within the CELL WALL of a cell that has lost its contents.

luminance The amount of VISIBLE LIGHT given off by a luminous surface, per second per square metre of its area, measured in CANDELA per square metre.

luminescence The emission of light from an object as a result of something other than high temperature, such as FLUORESCENCE or PHOSPHORESCENCE. *See also* THERMOLUMINESCENCE.

luminosity The total amount of light energy produced per second, such as by a star. *See also* APPARENT LUMINOSITY.

luminous flux The total amount of VISIBLE LIGHT arriving at a given surface per second. The SI unit of luminous flux is the LUMEN.

luminous intensity The amount of VISIBLE LIGHT given off per second by a light source. The SI UNIT of luminous intensity is the CANDELA.

lung A large, delicate spongy organ in which gaseous exchange takes place. Humans and most other TETRAPODS have two lungs in the THORAX, which are protected by the ribcage. The bulk of the lung tissue is made up of millions of sac-like structures called alveoli (see ALVEOLUS) that consist of thin, moist, membranous sheets over which blood passes, so that oxygen can enter the blood stream and carbon dioxide waste can leave. BREATHING causes air to enter and leave the lungs. Each lung is made airtight by two pleural membranes (the PLEURA), which form the PLEURAL CAVITY between them and into which pleural fluid enters that lubricates the lungs to allow for expansion. Pleurisy is an infection of the pleural cavity. *See also* BREATHING, RESPIRATORY SYSTEM.

lustre The characteristic reflective texture of metals, caused by the reflection of light by their surfaces.

luteinizing hormone (LH), ***interstitial cell-stimulating hormone,*** (ICSH) A GLYCOPROTEIN produced by the PITUITARY GLAND. In females, LH stimulates OVULATION (release of an OVUM from the OVARY), transforms the ruptured GRAAFIAN FOLLICLE into the CORPUS LUTEUM and controls the levels of OESTROGEN and PROGESTERONE if pregnancy follows. In males, LH stimulates the INTERSTITIAL CELLS of the testes (*see* TESTIS) to produce the hormone TESTOSTERONE. LH is under the control of the HYPOTHALAMUS, which releases gonadotrophic-releasing factor that stimulates LH but is itself inhibited by the testosterone it stimulates.

luteotrophic hormone (LTH) *See* PROLACTIN.

luteotrophin *See* PROLACTIN.

lutetium (Lu) The element with ATOMIC NUMBER 71; RELATIVE ATOMIC MASS 175.0; melting point 1,663°C; boiling point 3,402°C; RELATIVE DENSITY 9.8. It is the last member of the LANTHANIDE series of metals and has some catalytic properties that are used in the PETROCHEMICAL INDUSTRY.

lux The SI UNIT of ILLUMINANCE. One lux is an illuminance of one LUMEN per square metre, or the illuminance at a distance of one metre from a light source of LUMINOUS INTENSITY of one CANDELA.

lyase Any one of a group of enzymes that causes the addition or removal of a chemical group other than by HYDROLYSIS.

Lycopodophyta A PHYLUM of the PLANT KINGDOM consisting of the club mosses. Club mosses are small, evergreen plants with trailing or upright stems that have small spirally arranged leaves. They have sporangia (*see* SPORANGIUM) that are usually in cones. Some species are HOMOSPOROUS and others are HETEROSPOROUS.

lymph The fluid carried by the LYMPHATIC SYSTEM. It is a clear liquid made of tissue fluid and WHITE BLOOD CELLS (mostly LYMPHOCYTES).

lymphatic *See* LYMPH VESSEL.

lymphatic duct *See* LYMPH VESSEL.

lymphatic system, *lymphoid system* In vertebrates, a series of vessels carrying LYMPH throughout the body, and the organs associated with the production and storage of LYMPHOCYTES. The lymphatic system forms a second CIRCULATORY SYSTEM (the other being blood) that provides nutrients and WHITE BLOOD CELLS to body tissues and drains waste products away from body tissues.

Lymph capillaries are the smallest vessels and are found in all body tissues except nervous tissue. They are similar to blood capillaries but their walls are more permeable and allow even bacteria through. Lymph capillaries merge to form larger tubes called LYMPH VESSELS that pass lymph into large veins near to the heart. Along the lymphatics are a number of LYMPH NODES that are responsible for removing foreign bodies from lymph and are important in the IMMUNE RESPONSE. Lymph nodes, along with the SPLEEN, parts of the digestive system and respiratory and urinogenital regions, form secondary lymphoid tissue that are the sites of migration of lymphocytes from their site of production. The THYMUS, embryonic liver and adult BONE MARROW form the primary lymphoid tissue, where lymphocytes are produced. Some vertebrates, for example amphibians, have a lymph heart that pumps lymph around the body, but mammals rely on muscular contractions (of skeletal muscles pushing on lymph vessels), inspiratory movements (drawing lymph up to the THORAX) and the HYDROSTATIC PRESSURE of tissue fluid leaving the blood capillaries to push lymph along the lymphatic system.

See also CIRCULATORY SYSTEM.

lymph node In mammals, a small mass of tissue found at a number of sites along the major LYMPH VESSELS. Lymph nodes are a component of secondary lymphoid tissue (*see* LYMPHATIC SYSTEM). LYMPH passes through the nodes and is filtered by the engulfing action of MACROPHAGES that are contained in the lymph nodes, to remove foreign bodies such as BACTERIA. Lymph nodes contain many LYMPHOCYTES and are therefore important in the IMMUNE RESPONSE. Each node has its own blood supply. Larger lymph nodes include tonsils (in the throat of humans) and adenoids (at the back of the nose) and are often mistaken for glands that swell in response to an infection. The SPLEEN, found near the stomach in vertebrates, is the largest mass of lymphoid tissue.

lymphocyte In vertebrates, a WHITE BLOOD CELL with a large nucleus that is found in blood and LYMPH and is involved in the IMMUNE RESPONSE. There are two types of lymphocytes, called B CELLS and T CELLS, both of which are formed in BONE MARROW and later settle in the SPLEEN or a LYMPH NODE. B cells are responsible for the production of ANTIBODY to fight invasion by a foreign ANTIGEN, and also for the production of memory cells by which repeat infections are more quickly eliminated. T cells are involved in CELL-MEDIATED IMMUNITY. There is some overlap of their functions. *See also* LYMPHATIC SYSTEM.

lymphoid system *See* LYMPHATIC SYSTEM.

lymphokine A soluble substance produced by a LYMPHOCYTE that is involved in communication between cells of the IMMUNE RESPONSE. Lymphokines produced by T CELLS are often called interleukins, although other cells can make these too. Interleukin-1 (IL-1) activates many aspects of INFLAMMATION. IL-2 has been used clinically in the treatment of some CANCERS, and IL-3 stimulates growth of some immune cells.

Interferon is another example of a lymphokine and is made in cells infected by a VIRUS, from which it enters the bloodstream and then enters other uninfected cells, so making them immune to attack by the virus. There are three types of interferon: α, β and γ. Interferon-α has been used clinically in the treatment of leukaemia.

lymphoma A TUMOUR made of tissues of the LYMPHATIC SYSTEM, for example in Hodgkin's disease.

lymph vessel, *lymphatic* Any vessel of the LYMPHATIC SYSTEM that is like a vein with a non-return valve so that circulation is in one direction only (towards the heart). Lymph vessels from the right side of the head, THORAX and arm combine to form the right lymphatic duct, and the rest of the vessels drain into the thoracic duct. Both sides then pass lymph into large veins near the heart. Along the lymphatics are a number of LYMPH NODES.

lyophilic (*adj.*) (= solvent loving) Describing a SOL or a solid in a sol where the INTERMOLECULAR FORCES between the solid particles and the suspending liquid are strong. Such colloidal particles may break down further to form true solutions and do not coagulate readily (*see* COAGULATION). *Compare* LYOPHOBIC.

lyophobic (*adj.*) (= solvent-hating) Describing a SOL or a solid in a sol where the INTERMOLECULAR FORCES between the solid particles and the suspending particles are weak. Such materials tend not to disperse easily once coagulated (*see* COAGULATION). *Compare* LYOPHILIC.

lysin An ANTIBODY that chemically destroys foreign substances.

lysine An AMINO ACID essential for growth. It is produced by the HYDROLYSIS of certain proteins.

lysis The destruction of a cell by rupturing its CELL MEMBRANE or CELL WALL. *See also* LYSOSOME.

lysosome (*Lysis* = splitting, *soma* = body) A membrane-bound eukaryotic (*see* EUKARYOTE) cell ORGANELLE, formed by the GOLGI APPARATUS. Lysosomes are 0.2–0.5 µm in diameter and are similar to spherical MITOCHONDRIA but lack an internal structure. They contain a large number of enzymes (mostly HYDROLASES) in acid solution that are involved in autolysis, the complete cell breakdown after a cell's death, and in digesting worn-out organelles. A lysosome that has not been involved in hydrolytic activity is referred to as a primary lysosome, and the fusion of primary lysosomes with each other or material for digestion results in secondary lysosomes that are larger (*see* PHAGOSOME).

Lysosomes are abundant in WHITE BLOOD CELLS, where the enzymes attack ingested bacteria. In these cases material is digested within the lysosome, but enzymes can also be released outside the cell to break down other cells in a controlled selective way, for example during METAMORPHOSIS. Similar structures are seen in plants and fungi.

M

machine A device for transferring energy from one form to another, or enabling energy to be used more effectively.

An important class of machine involves only MECHANICAL ENERGY. Such machines, called simple machines, are usually based on moving a small FORCE through a large distance in order to make a larger force move through a smaller distance. The force applied to a machine is called the effort, and this force is converted into one that acts at the other end of the machine on the load. Since the load moves through a shorter distance than the effort, it also moves more slowly. This is expressed by the VELOCITY RATIO of the machine – the ratio of distance moved by the effort to distance moved by the load.

The MECHANICAL ADVANTAGE is the ratio of load to effort, and the purpose of simple machines is generally to exert a force on the load that is greater than the effort. If there were no loss of energy, as a result of FRICTION for example, the machine would be completely efficient and the mechanical advantage would be equal to the velocity ratio. There is always some energy loss, however, making the mechanical advantage smaller than the velocity ratio by a factor called the EFFICIENCY of the machine. This is the amount of useful work done by the machine divided by the total amount of energy supplied.

In general, the efficiency of machines varies depending on the type of energy conversion being performed; ELECTRIC MOTORS, for example, can be as much as 95 per cent efficient. Since most waste energy ends up as heat, heaters are usually close to 100 per cent efficient, though the heat may not be produced where it is needed – in a gas boiler for example, much heat will be lost in the hot gases released to the surrounding atmosphere.

See also LEVER, PULLEY.

machine code In computing, a set of instructions in BINARY code that a MICROPROCESSOR can understand and execute directly, without the need for a translator. Each microprocessor has its own machine code. Most computer programs are written in an easier to use HIGH-LEVEL LANGUAGE, which must be translated into machine code by a COMPILER or INTERPRETER so that it may be executed by the central processing unit.

Mach number The speed of an object, particularly an aircraft, expressed as a ratio to the SPEED OF SOUND in the surrounding air. Thus Mach 2 is equivalent to twice the speed of sound, etc.

MACHO (Massive Compact Halo Object) An invisible massive object in orbit around a galaxy. It is a possible form of missing matter in the Universe. *See* COSMOLOGY.

macroevolution *See* EVOLUTION.

macromolecule A very large MOLECULE. Examples of substances with macromolecules are POLYMERS, PROTEINS and HAEMOGLOBIN.

The term macromolecule may also be applied to a crystal structure in which the atoms are held together by COVALENT BONDS, so that the whole crystal is effectively a single giant molecule. Diamond and silicon dioxide are examples of such a structure, which are both tetrahedral in shape.

See also COVALENT CRYSTAL.

macrophage In vertebrates, a type of WHITE BLOOD CELL with many functions, mostly concerned with the removal of foreign or dead cells or debris (after infection or injury). Like PHAGOCYTES, macrophages can ingest particles (foreign or food) from their surroundings, which are subsequently broken down or destroyed, but macrophages are larger than other phagocytes, with a longer life span, and are found all over the body. In the liver (as KUPFFER CELLS) and SPLEEN, macrophages ingest worn-out blood cells. In the lungs they ingest dust and other inhaled particles, and in LYMPH NODES and the LYMPHATIC SYSTEM they are concerned with fighting infection and the IMMUNE RESPONSE. They are important in the immune response as ANTIGEN-PRESENTING CELLS. This system of circulating tissue macrophages forms the RETICULOENDOTHELIAL SYSTEM. *See also* MAJOR HISTOCOMPATIBILITY COMPLEX, OPSONIN, T CELL.

macrophage-activating cell *See* T CELL.

macroscopic (*adj.*) Visible to the naked eye. In science, a macroscopic state is a description of the behaviour of large-scale features of a system, such as temperature and pressure, as opposed to that of the individual atoms or molecules from which a system in made up.

mad cow disease *See* BOVINE SPONGIFORM ENCEPHALOPATHY.

magenta A bluish-red colour.

magic numbers In nuclear physics, a number of PROTONS or NEUTRONS in a NUCLEUS marking the end of a periodic variation of certain properties of the nucleus, such as binding energy per nucleon (*see* NUCLEAR BINDING ENERGY). There are unusually large numbers of stable ISOTOPES of elements with magic ATOMIC NUMBERS. Doubly-magic nuclei with magic numbers of protons and neutrons have very high binding energies. The magic numbers are 2, 8, 20, 28, 50, 82 and 126.

magma Molten rock, such as that which forms the Earth's MANTLE.

magnesium (Mg) The element with ATOMIC NUMBER 12; RELATIVE ATOMIC MASS 24.3; melting point 651°C; boiling point 1,107°C; RELATIVE DENSITY 1.7. Magnesium is an ALKALINE EARTH metal, widespread in nature. It is extracted by the ELECTROLYSIS of fused (molten) magnesium chloride. The metal burns with an intense white flame, and is used in flares to illuminate large areas at night. Magnesium's low density makes it a useful ingredient in some high strength, low density alloys used in the aerospace industry. It is also present in CHLOROPHYLL, so magnesium is a vital element for plant growth. *See also* THERMIT PROCESS.

magnesium carbonate ($MgCo_3$) A white solid; decomposes on heating; RELATIVE DENSITY 2.6. Magnesium carbonate often occurs in nature alongside calcium carbonate, for example in the mineral dolomite, $CaCO_3.MgCO_3$. On heating, magnesium carbonate decomposes to give magnesium oxide:

$$MgCO_3 \rightarrow MgO + CO_2$$

magnesium chloride ($MgCl_2$) A white solid; melting point 714°C; boiling point 1,412°C; RELATIVE DENSITY 2.3. Magnesium chloride can be formed by burning magnesium in chlorine:

$$Mg + Cl_2 \rightarrow MgCl_2$$

ANHYDROUS magnesium chloride is DELIQUESCENT, and the salt frequently occurs as the hexahydrate $MgCl_2.6H_2O$. On heating, this hydrolyses (*see* HYDROLYSIS):

$$MgCl_2 + H_2O \rightarrow MgO + 2HCl$$

magnesium hydroxide ($Mg(OH)_2$) A white solid; decomposes on heating; RELATIVE DENSITY 2.4. Magnesium hydroxide can be formed by the reaction between magnesium and steam:

$$Mg + 2H_2O \rightarrow Mg(OH)_2 + H_2$$

It decomposes on heating:

$$Mg(OH)_2 \rightarrow MgO + H_2O$$

Magnesium hydroxide is insoluble in water, but a suspension is used in medicines, such as milk of magnesia, to neutralize excess stomach acid:

$$Mg (OH)_2 + 2HCl \rightarrow MgCl_2 + 2H_2O$$

magnesium oxide (MgO) A white solid; melting point 3,800°C; decomposes on further heating; RELATIVE DENSITY 3.6. Magnesium oxide can be made by burning magnesium in air:

$$2Mg + O_2 \rightarrow 2MgO$$

Magnesium oxide is used as a REFRACTORY material, and as a hard antireflective coating on the glass surfaces of some lenses.

magnesium sulphate ($MgSO_4$) A white solid; decomposes on heating; RELATIVE DENSITY 2.7. Magnesium sulphate is soluble and occurs naturally in a HYDRATED form as the mineral Epsom salt, widely used in the processing of cloth and leather. Magnesium sulphate decomposes on heating to give magnesium oxide:

$$2MgSO_4 \rightarrow 2MgO + 2SO_2 + O_2$$

magnet Any object that is surrounded by a MAGNETIC FIELD and attracts or repels other magnets. Magnets also attract unmagnetized pieces of iron, nickel or cobalt and some alloys containing these elements (such materials are described as FERROMAGNETIC). This is the result of the magnet temporarily magnetizing the ferromagnetic material; a phenomenon known as INDUCED MAGNETISM.

Magnets are of two kinds: ELECTROMAGNETS and PERMANENT MAGNETS. Electromagnets are SOLENOIDS, usually with an iron core, which rely on an electric current for their magnetism and so can be turned on and off. In a permanent magnet, the magnetism is present continuously, without the need for a current. Permanent magnets can be made in many shapes, but the most common are bar magnets, with a POLE at each end of a bar, horseshoe magnets, a bar bent into an arc of a circle with a pole at each end of the arc, and slab magnets, flat pieces of magnetic material with a pole on each of the two large faces.

magnetic bottle Any pattern of MAGNETIC FIELDS that can be used for the MAGNETIC CONFINEMENT of a PLASMA.

magnetic circuit A closed path of MAGNETIC FIELD LINES around a CORE and through an air gap. If a small air gap is made in an otherwise closed core, producing a horseshoe ELECTROMAGNET, the effect will be to greatly reduce the FLUX through the core. The air gap has the same effect on the flux as a resistance on the flow of electric current. This effect is called RELUCTANCE. *See also* MAGNETOMOTIVE FORCE.

magnetic compass A small MAGNET, pivoted so it can freely move in a horizontal direction, used to find direction. The magnet (called the compass needle) aligns itself with the MAGNETIC FIELD OF THE EARTH, so that it points to the MAGNETIC NORTH. A scale under the needle shows the points of the compass, for use in navigation. *See also* PLOTTING COMPASS.

magnetic confinement A technique for containing a PLASMA by the use of MAGNETIC FIELDS whilst the plasma is heated to the very high temperatures required for NUCLEAR FUSION. *See also* TOKAMAK.

magnetic dipole The simplest form of magnet, having a single north-seeking POLE and a single south-seeking pole. The strength of a magnetic dipole is measured by its MAGNETIC MOMENT. *See also* DIPOLE, MAGNETIC MONOPOLE, PARAMAGNETISM.

magnetic field The field of force surrounding a magnet or a current-carrying conductor. A magnetic field will exert a force on a moving charge or a TORQUE on a MAGNETIC DIPOLE. A small permanent magnet, such as a PLOTTING COMPASS, will tend to turn to point in the direction of the field. The SI UNIT of magnet field is the TESLA.

The direction of the force on a current in a magnetic field is given by FLEMING'S LEFT-HAND RULE. The size of the force is proportional to the field strength, to the current and to the length of the current-carrying conductor in the field. It is also proportional to the sine of the angle between the current and the field: if the field and the current are in the same direction there will be no force; if the two are at right angles, the force will be a maximum. The force F on a wire of length l, carrying a current I at an angle θ to a magnetic field B is given by

$$F = BIl\sin\theta$$

Since a current is comprised of moving charges, the force on an electric current can be seen as a force on the individual moving CHARGE CARRIERS. The force on a moving charge is proportional to the strength of the magnetic field, the size of the charge and the speed at which it is moving. The force F on a charge q moving at a speed v at right angles to a field B is given by

$$F = Bqv$$

An electric current will produce a magnetic field. For a straight wire, the field strength is proportional to the current and inversely proportional to the distance from the wire. The MAGNETIC FIELD LINES form circles around the wire, the direction of which can be found using the RIGHT-HAND GRIP RULE. The magnetic field B at a distance r from a long straight wire carrying a current I is given by

$$B = \mu_0 I/2\pi r$$

where μ_0 is the PERMEABILITY of free space.

See also MAGNETIC FIELD OF THE EARTH, MAGNETIC FLUX, SUSCEPTIBILITY.

magnetic field lines, *magnetic flux lines* Lines that show the direction of a MAGNETIC FIELD at each point. This is the direction in which a PLOTTING COMPASS would point if it were placed at that point. The closer the lines the stronger the field at that point.

magnetic field of the Earth The Earth, and many other planets, produce their own MAGNETIC FIELDS, believed to be due to electric currents within their molten metallic CORES. This means that a freely suspended BAR MAGNET will tend to point in one direction. The POLE that points roughly north is called the north-seeking pole whilst the pole which points roughly south is the south-seeking pole. *See also* MAGNETIC NORTH.

magnetic flux (ϕ) A measure of the total amount of MAGNETIC FIELD passing through a given area. The magnetic flux is the magnetic field multiplied by the area over which the flux is being found and the COSINE of the angle between the field and the direction perpendicular to the plane of the area. If a coil has an area A with a magnetic field B passing through this area at an angle θ to the perpendicular to the circuit, and the coil contains N turns, the magnetic flux ϕ will be

$$\phi = BA\cos\theta$$

and the MAGNETIC FLUX LINKAGE ϕ_n will be

$$\phi_n = BAN\cos\theta$$

See also MAGNETOMOTIVE FORCE.

magnetic flux lines *See* MAGNETIC FIELD LINES.

magnetic flux linkage The MAGNETIC FLUX multiplied by the number of times this flux passes through the circuit.

magnetic force The force produced by a MAGNETIC FIELD on a magnet, an electric current or a moving charge. For a current I at an angle α to a magnetic field B, the magnetic force is

$$F = BIl\sin\alpha$$

where l is the length over which the current flows in the field. For a charge q moving at speed v at an angle α to the field, the force is

$$F = Bqv\sin\alpha$$

magnetic moment A measure of the strength of a MAGNETIC DIPOLE. The SI UNIT of magnetic moment is the AMPERE metre squared (Am^2), and is equivalent to the strength of field produced at large distances from a loop carrying a current of one ampere and enclosing an area of one square metre. Alternatively, this is the strength of magnetic dipole which will experience a TORQUE of one NEWTON metre (1 Nm) when placed at right angles to a magnetic field of one TESLA. *See also* MAGNETON.

magnetic monopole A hypothetical source of MAGNETIC FIELD, a north-seeking POLE without a south-seeking pole or vice versa. Despite extensive searches, no magnetic monopoles have been found. *See also* MAGNETIC DIPOLE.

magnetic north The direction in which a MAGNETIC COMPASS points. It is not quite the direction to the north pole because the Earth's magnetic poles do not quite coincide with its geographical poles – the points about which it rotates. The position of the Earth's magnetic poles also changes with time and the MAGNETIC FIELD OF THE EARTH has completely changed direction many times over its lifetime. *See also* MAGNETIC VARIATION.

magnetic resonance imaging (MRI) An imaging technique increasingly used in medicine. The patient is placed in a magnetic field, which aligns the SPINS of the protons in hydrogen atoms throughout the patient. A pulsed electromagnetic field destroys this alignment, and the signals produced as the protons realign themselves have frequencies characteristic of the chemical environment of the individual protons. TOMOGRAPHY techniques are used to produce a picture of the body in a series of slices. Unlike X-RAYS, MRI is non-ionizing (*see* IONIZING RADIATION) and can produce clear images of soft tissues. *See also* NUCLEAR MAGNETIC RESONANCE.

magnetic variation The angle between true north and MAGNETIC NORTH. Magnetic variation varies with time and with position on the surface of the Earth.

magnetism The collective term for all the effects resulting from the presence of MAGNETIC FIELDS. Magnetic fields are produced by electric currents and by many ELEMENTARY PARTICLES, including the electron. The magnetic effects of electrons are responsible for magnetic properties of several elements. *See also* DIAMAGNETISM, ELECTROMAGNETIC INDUCTION, FERROMAGNETIC, INDUCED MAGNETISM, MAGNET, MAGNETIC FORCE, MAGNETOMOTIVE FORCE, PARAMAGNETISM, PERMANENT MAGNET.

magnetite A mineral form of iron oxide, containing the mixed oxide, iron(II)–iron(III) oxide, Fe_3O_4.

magnetohydrodynamics (MHD) The study of electrically conducting fluids. MHD covers both molten metals and the properties of PLASMAS, such as those in a NUCLEAR FUSION reactor and the interaction of the SOLAR WIND with planetary magnetic fields.

magnetometer A device for measuring a MAGNETIC FIELD, particularly that of the Earth. Local variations of the Earth's magnetic field are sometimes indicative of disturbance of the underlying soil, so magnetometers are used by archaeologists as a surveying tool.

magnetomotive force The influence that creates the MAGNETIC FLUX in a MAGNETIC CIRCUIT. For a SOLENOID, this is equal to the product of the current in the coil and the number of turns (sometimes called the number of AMPERE-TURNS).

magneton A unit of MAGNETIC MOMENT used in atomic and nuclear physics. The Bohr magneton is 9.27 x 10^{-24}Am^2. The nuclear magneton is 5.05 x 10^{-27} Am^2.

magnetosphere The region in space that is influenced by the MAGNETIC FIELD OF THE EARTH.

magnetron A device for generating MICROWAVES. It comprises a filament surrounded by several ANODES, each forming a cavity resonant (*see* RESONANCE) at the frequency of microwaves being generated, and placed in a MAGNETIC FIELD. Electrons are emitted by the filament by the process of THERMIONIC EMISSION, accelerated towards the anodes and deflected by the magnetic field so that they move in a spiral path. Interactions between the electrons and the electric fields at the anodes set up oscillations at the RESONANT FREQUENCY of the cavities. The magnetron is often operated in pulses to generate microwaves for RADAR transmitters.

magnification The extent to which an image is larger than the original object, in a microscope or telescope for example.

magnification = image size/object size

In the case of a telescope, the term is used to denote angular magnification:

angular magnification = angle subtended by image/angle subtended by object

magnifying glass A CONVERGING LENS used with the object viewed less than one FOCAL LENGTH away from the lens, producing an enlarged VIRTUAL IMAGE.

magnitude Of a VECTOR, the size of the vector quantity, regardless of direction. If the COMPONENTS of a vector in the *x*, *y* and *z* directions are *a*, *b* and *c*, the magnitude of the vector is $\sqrt{(a^2 + b^2 + c^2)}$.

mainframe A large computer that performs large-scale operations, typically for commercial companies. They have a large number of PERIPHERAL DEVICES, a large BACKING STORE and a very fast CENTRAL PROCESSING UNIT. They are being superseded by high-powered MINICOMPUTERS and PERSONAL COMPUTERS.

main sequence star Any star lying on the main sequence of the HERTZSPRUNG–RUSSELL DIAGRAM, the diagonal band running from top left (hot bright stars) to bottom right (cool dim stars).

Stars are formed by the gravitational collapse of clouds of interstellar gas, which is made up of about 75 per cent hydrogen and 25 per cent helium. As the cloud collapses, provided it has a mass greater than about 0.08 SOLAR MASSes, it will heat up to a temperature at which further collapse is opposed by the release of energy from NUCLEAR FUSION, in which hydrogen is converted to helium. The chain of reactions by which this takes place is called the PP CHAIN. During the time when these reactions are taking place the star remains in more or less the same position on the Hertzsprung–Russell diagram, on the main sequence. The time spent as a main sequence star accounts for about 90 per cent of a star's lifetime, which explains why most of the stars observed are on the main sequence. Once the hydrogen in the core has all been fused the star moves away from the main sequence. *See also* RED GIANT, WHITE DWARF.

major histocompatibility complex (MHC) A mammalian GENE complex encoding ANTIGENS on the surface of most body cells that are unique to the individual. The MHC was originally identified as the region encoding antigens important in graft rejection (*see* TRANSPLANTATION).

The MHC regulates T CELL activity in such a way that foreign antigens are only recognized when they are in association with MHC molecules. There are two classes of MHC molecules: class I molecules are found in low levels on most body cells, but particularly on T cells, and class II molecules are found mostly on B CELLS and MACROPHAGES.

malaria An infectious disease in the tropics caused by the protozoan PARASITE *Plasmodium* carried by the *Anopheles* mosquito. The disease is characterized by a periodic high fever and enlarged SPLEEN and affects about 200 million people a year, more than any other organism in the tropics. There are four protozoa of the genus *Plasmodium* that can cause malaria, the most dangerous being *P. falciparum*.

The life cycle of the parasite is complex and involves an asexual stage in the liver and

RED BLOOD CELLS of humans and a sexual stage beginning in humans and continuing in the mosquito. When humans are bitten by an infected mosquito, the parasite enters the blood in the form of sporozoites (a circle-shaped form of parasite). These enter the liver, where they quickly divide to produce merozoites, which can reinfect other liver cells or enter red blood cells. They divide further and eventually cause the cell to rupture, releasing merozoites into the blood. The red blood cells rupture in synchrony every 2–3 days, causing the characteristic high fever. Some of the merozoites develop into GAMETES and remain dormant until a mosquito bites the human and takes the gamete into its stomach, where the gametes are fertilized and the ZYGOTE buries into the stomach wall. MEIOSIS occurs to produce large numbers of sporozoites that are released into the mosquito's body cavity and eventually to the SALIVARY GLANDS, to pass back to another human victim. The mosquito is the secondary host or VECTOR.

No vaccine is available, mainly due to the ANTIGENIC VARIATION shown by the parasite. Mosquitoes are largely resistant to insecticides, but synthetic drugs are used in the prevention or treatment of the disease. Drugs are most effective on the sporozoite and gamete stages, but relapses can occur due to the reservoir of merozoites in the liver. Some protection from malaria is provided by SICKLE-CELL DISEASE, which alters the shape of red blood cells making them harder to infect.

malate A SALT or ESTER of MALIC ACID.

malic acid An organic acid found particularly in apples but also other fruit. Malic acid is present in all living cells and is an intermediate of the KREBS CYCLE.

malignant (*adj.*) Of a TUMOUR, uncontrollable, rapidly spreading, or resistant to treatment. *See* CANCER.

malleable (*adj.*) Able to be beaten into a new shape without breaking. Malleability is an important property of many metals.

malleus *See* EAR OSSICLE.

Malpighian layer, *germinative layer* The deepest layer of skin EPIDERMIS that consists of actively dividing cells that move up through the layers of the epidermis. The Malpighian layer extends into the DERMIS, and gets its food and oxygen from capillaries in the dermis because there are no blood vessels in the epidermis. The brown pigment MELANIN is produced by the Malpighian layer. *See also* HAIR.

maltase An enzyme that breaks down MALTOSE into its constituent GLUCOSE molecules.

Maltese cross tube A device used to demonstrate the effects of THERMIONIC EMISSION. It comprises an electrically heated filament in an evacuated glass vessel, which also contains a hollow metal cylinder and a metal cross. The end of the tube opposite the filament is coated with a PHOSPHOR. The metal cylinder and the cross are made positive and the filament negative with a POTENTIAL DIFFERENCE of several thousand volts. The screen glows where it is struck by the electrons from the filament but there is a 'shadow' on the screen showing that electrons travel in straight lines and cannot pass through the metal of the cross.

maltose A DISACCHARIDE made from the combination of two GLUCOSE molecules. Maltose is a major constituent of malt and is therefore important in the manufacture of beer and whisky.

mammal A member of the vertebrate CLASS MAMMALIA in the PHYLUM CHORDATA.

Mammalia A CLASS of vertebrates characterized by the possession of MAMMARY GLANDS in the female, in addition to the presence of body hair (although in variable amounts), lungs, a lower jaw consisting of two bones, a middle ear consisting of three bones, and no nucleus in the RED BLOOD CELLS. There are three groups: (i) the placental mammals (the most advanced), with young that develop inside their mother within a uterus nourished by a placenta; (ii) MARSUPIALS; and (iii) MONOTREMES (the least evolved). Placental mammals have dominated the globe and are found in all parts of the world.

Mammals are very varied in size and shape, from a shrew to a blue whale, and live in very varied habitats, for example lions, monkeys and humans on land, and whales in the sea. All mammals are HOMOIOTHERMS (warm-blooded; able to maintain a constant body temperature), and all are heterodonts (have different types of teeth). There are 4,000 species of mammals, including the ORDERS of primates (humans, apes), rodents (rats, mice), carnivores (cats, dogs), lagomorphs (rabbits) and cetaceans (whales, dolphins).

mammary gland In female mammals, a milk-producing gland that is only active after the birth of their young. Mammary glands are formed on the chest or ABDOMEN, and the number varies from 2 to 20, according to species. Mammary glands develop during puberty from EPIDERMIS, and during PREGNANCY the hormones OESTROGEN and PROGESTERONE cause the development of milk glands that form a series of branching ducts ending in secretory alveoli (sacs; *see* ALVEOLUS). After birth of the foetus, milk production (LACTATION) is stimulated by the hormone PROLACTIN secreted by the PITUITARY GLAND. The sucking of the nipples of the mammary gland stimulates OXYTOCIN release from the pituitary, which results in the expression of milk. Milk production can continue as long as the baby suckles.

mandible 1. In ARTHROPODS, a part of the mouthparts that is involved in biting and crushing food.

2. In vertebrates, the lower jaw.

manganate(VI) Any salt containing the manganate(VI) ion, MnO_4^-. Salts containing this ion are dark green in colour. The ion can be converted to MANGANATE(VII) by manganese(IV) oxide in alkaline conditions:

$$MnO_4^- + MnO_2 + 2OH^- \rightarrow 2MnO_4^{2-} + H_2O$$

manganate(VII), ***permanganate*** Any salt containing the manganate(VII) ion, MnO_4^{2-}. Salts with this ion are deep purple in colour. Potassium manganate(VII), K_2MNO_4, is widely used as an OXIDIZING AGENT. Manganate(VII) ions can be converted to manganate(VI) ions in acid solutions:

$$3MnO_4^{2-} + 4H^+ \rightarrow 2MnO_4^- + MnO_2 + 2H_2O$$

manganese (Mn) The element with ATOMIC NUMBER 25; RELATIVE ATOMIC MASS 54.9; melting point 1,244°C; boiling point 2,040°C; RELATIVE DENSITY 7.4. Manganese occurs in nature as manganese oxide, and the metal can be extracted by heating the oxide with powdered magnesium:

$$MnO_2 + 2Mg \rightarrow Mn + 2MgO$$

Manganese is a TRANSITION METAL with some catalytic properties (particularly in manganese oxide) and is also used in the manufacture of some steels.

manganese(IV) oxide (MnO_2) A black solid; decomposes on heating. Manganese dioxide occurs naturally and is a strong OXIDIZING AGENT. It also has catalytic properties: the presence of manganese(IV) oxide will greatly increase the rate at which hydrogen peroxide, H_2O_2, decomposes. Manganese(IV) oxide is widely used in ZINC-CARBON CELLS, where it acts as a depolarizer, removing spent reagents from the ELECTROLYTE.

manganese-alkaline cell A common type of electrochemical CELL, non-rechargeable and more expensive than ZINC-CARBON CELLS, but longer lasting.

manometer An instrument for measuring pressure, comprising a U-shaped glass tube filled with a liquid. The difference in the height of the liquid in the two arms of the tube is proportional to the pressure difference between the two ends of the manometer, one of which is usually left open to ATMOSPHERIC PRESSURE.

mantle The layer of molten rock surrounding the Earth's CORE on which CONTINENTAL PLATES float, forming the Earth's CRUST.

mapping Any mathematical process in which members of one SET are linked with members of another set. If a single member of one set is mapped uniquely onto a single member of the other set, the mapping is described as a one-to-one mapping.

maramus *See* DIET.

marble A METAMORPHIC rock formed by the action of heat and pressure, on LIMESTONE. Small amounts of impurities such as iron and copper carbonates, give some marbles their characteristic colours and patterning.

maritime (*adj.*) Describing an AIR MASS that has travelled mostly over water.

mark, release, recapture In biology, a technique used to estimate the size of a population of mobile animals that can be tagged or marked in some way. A known number of animals are captured, tagged or marked and released into the population again. At a later date a given number of animals are collected at random and the percentage that are marked is recorded. Population size is calculated on the assumption that the percentage of marked animals in the second capture is the same as the percentage of marked animals in the population as a whole. The marked animals must have enough time to distribute themselves evenly throughout the population.

A problem with this method is that marked animals may be more likely to predation if the marking makes them more conspicuous. Also, it does not take into account the death or migration of marked animals.

Examples of this type of sampling are the marking of arthropods by means of a dab of non-toxic paint on their backs, and the attachment of ear tags to mammals and rings to the legs of birds.

See also QUADRAT, TRANSECT.

Markownikoff's rule During the ELECTROPHILIC ADDITION of hydrogen HALIDES (such as HBr) to an unsymmetrical ALKENE, the hydrogen goes to the end of the DOUBLE BOND that already has the greatest number of hydrogen atoms and the halide to the end with the least hydrogen atoms. For example, the addition of hydrogen bromide to propene ($CH_3CH{=}CH_2$) yields $CH_3CHBrCH_3$ instead of the alternative $CH_3CH_2CH_2Br$. This occurs because of the greater stability of the intermediate CARBONIUM ION formed during the production of the former. Sometimes the alternative product is formed, which is termed anti-Markownikoff addition.

Mars The fourth planet in order from the Sun, with an orbital radius of 1.52 AU (228 million km). It is rather smaller than the Earth, with a diameter of 6,780 km (0.53 times that of the Earth), and a mass of 6.4×10^{23} kg (0.11 times that of the Earth). Mars orbits the Sun every 1.9 years and rotates on its own axis every 24.6 hours.

As a result of its small size, Mars has only a very thin atmosphere. This atmosphere supports some weather effects, such as dust storms and OROGRAPHIC cloud, but has not produced significant erosion of the surface. Whilst the surface of the planet contains no liquid water, there is evidence for fluid flow at some time in the past, though it may have been short lived. In 1996 it was announced that some evidence had been found for the existence of primitive life forms at some stage in the planet's history.

marsupial A mammal whose young is born at an early stage of development and matures further within a pouch on the mother's body. Examples of marsupials include kangaroos and koalas. Marsupials are found particularly in Australasia and South America.

maser Acronym for *m*icrowave *a*mplification by *s*timulated *e*mission of *r*adiation. A device similar to a LASER, but operating in the MICROWAVE region of the ELECTROMAGNETIC SPECTRUM. Masers are sources of microwaves with very accurate frequencies and as such are used in many high-precision clocks.

mass A measure of the total amount of MATTER in an object, expressed either in terms of the resistance of an object to having its motion changed (INERTIAL MASS) or the effect of a gravitational field on the object (GRAVITATIONAL MASS). The SI UNIT of mass is the KILOGRAM. *See also* REST MASS.

mass defect The difference between the mass of an atomic NUCLEUS and the mass of the NEUTRONS and PROTONS from which it is made. A consequence of the SPECIAL THEORY OF RELATIVITY is that a loss of energy will result in an equivalent loss of mass, so that when nuclei were formed in stars, and energy was released, mass was lost. *See also* NUCLEAR BINDING ENERGY.

mass-energy equation The equation

$$E = mc^2$$

where E is energy, m mass and c the speed of light in a vacuum. This equation, an important consequence of the SPECIAL THEORY OF RELATIVITY, states that an object with more energy will appear more massive. *See also* EQUIVALENCE OF MASS AND ENERGY, REST MASS ENERGY.

Massive Compact Halo Object *See* MACHO.

mass number, *nucleon number* The total number of NUCLEONS (NEUTRONS and PROTONS) in a particular atomic NUCLEUS.

mass spectrometer An instrument for the measurement of the mass of atoms, molecules or fragments of a molecule, and the relative abundance of each mass present. A sample of material is IONIZED by bombardment with an electron beam. If the sample is molecular, the molecule will also be broken into fragments. The charged particles are accelerated in an electric field. The ions then enter a VELOCITY SELECTOR, a region in which electric and magnetic fields are applied at right angles to one another to produce opposing forces on the ions. These forces balance out only for particles moving at one particular speed, which then enter the next region of the device, where there is a magnetic field that deflects particles according to their charges and masses. A detector produces a reading of abundance against mass/charge ratio.

As well as measuring the relative abundance of ISOTOPES, the mass spectrometer is an important tool in determining the structure of organic molecules. In a KETONE or ALDEHYDE, for example, a fragment corresponding to the OH grouping (17 amu) would not be found, but in ALCOHOLS it will be present.

mass spectroscopy The use of a MASS SPECTROMETER to determine the structure of a compound or as a tool in QUALITATIVE ANALYSIS.

mast cell A cell similar to the blood BASOPHIL but containing different granules in its CYTOPLASM. Mast cells are often found in CONNECTIVE TISSUE. The numbers of mast cells often increase in certain allergies and they are involved in the IMMUNE RESPONSE to parasites. *See also* HISTAMINE.

matrix 1. (*mathematics*) A mathematical entity consisting of an array of elements arranged in rows (horizontal) and columns (vertical), each element in the matrix being a number or variable. A matrix is described as being m x n, where m is the number of columns and n is the number of rows. Matrices of the same size can be added and subtracted, whilst an $m \times n$ matrix can be multiplied with an $n \times p$ matrix to give an $m \times p$ matrix. Matrices can be used to represent related pieces of information, and matrix algebra can be used to solve SIMULTANEOUS EQUATIONS and to represent TRANSFORMATIONS such as ROTATIONS. *See also* DETERMINANT.

2. (*biology*) *See* EXTRACELLULAR MATRIX.

matter The collective term for all ATOMS. Matter is any substance with mass and which is not ANTIMATTER. In particle physics, matter is defined as all substances with a positive BARYON or LEPTON NUMBER.

maximum The largest of a set of numbers, or the largest value taken by a function. A local maximum is a value of a function that is larger than the value of that function for values of independent variable adjacent to that giving the maximum value, but which may be exceeded when the independent variable takes on other, very different, values.

maximum and minimum thermometer A thermometer for recording the highest and lowest temperatures reached since the thermometer was last reset. Traditionally, such thermometers use the expansion of alcohol to push a mercury thread around a U-shaped CAPILLARY TUBE. Steel markers are pushed along the tube by the mercury, but are left behind when the mercury moves away from them. These markers can be repositioned using a magnet. Modern maximum and minimum thermometers are electronic.

maximum power theorem In electricity, the POWER delivered to a LOAD is a maximum when the RESISTANCE of the load is equal to INTERNAL RESISTANCE of the power supply. If the load resistance is too large only a small current will flow, if the resistance is too small there will be too great a drop in the voltage across the load. In cases where it is important to deliver as much power to the load as possible, its resistance should be matched to the internal resistance of the supply.

maxwell Unit of magnetic flux in the C.G.S. SYSTEM. One maxwell is equal to 10^{-8} WEBER.

Maxwell–Boltzmann distribution A description of the range of energies possessed by molecules in a system in which those molecules are able to exchange energy freely with one another to maintain THERMAL EQUILIBRIUM. In the Maxwell–Boltzmann distribution the number of particles with an energy greater than E is proportional to $e^{-E/kT}$, where T is the ABSOLUTE TEMPERATURE and k is the BOLTZMANN CONSTANT. This quantity is called the BOLTZMANN FACTOR, and if E is the ACTIVATION ENERGY, the rate at which an ACTIVATION PROCESS takes place will depend on this factor. The Maxwell–Boltzmann distribution can also be used to describe range of speeds of molecules in a gas.

The situation is a little more complicated with FERMIONS, particularly at low temperatures, due to the PAULI EXCLUSION PRINCIPLE, which prevents all the particles from occupying the lowest energy states. The FREE ELECTRONS in a metal, for example, have far higher speeds than might be expected.

See also FERMI DISTRIBUTION.

Maxwell's equations A set of four equations describing the interaction of CURRENTS and CHARGES with ELECTRIC and MAGNETIC FIELDS. The first equation is essentially GAUSS' LAW. The second equation describes the creation of magnetic fields by currents and by changing electric fields. The third equation records the absence of MAGNETIC MONOPOLES, and the fourth is essentially FARADAY'S LAW OF ELECTROMAGNETIC INDUCTION. Taken collectively, they led to the discovery of ELECTROMAGNETIC WAVES and the recognition that light is an electromagnetic wave. The SPEED OF LIGHT in a vacuum, c, is related to the PERMITTIVITY of free space, ε_0, and the PERMEABILITY of free space, μ_0, by the equation

$$c^2 = 1/(\varepsilon_0\mu_0)$$

mean 1. ***arithmetic mean*** The sum of a set of numbers divided by the number of elements in the set. *See also* STANDARD DEVIATION.

2. ***geometric mean*** The *n*th ROOT of the product of all the values in a set, where *n* is the number of values in the set.

mean free path The average distance that a molecule in a fluid, usually a gas, travels before colliding with another molecule. This depends on the number of molecules per metre cubed and on the size of the two molecules, as measured by the COLLISION CROSS-SECTION. In liquids the mean free path is similar to the size of a molecule, and the molecules cannot be thought of as moving freely, as they can in gases.

For a gas with n molecules per metre cubed, each having a collision cross-section of σ, the mean free path is λ, where

$$\lambda = (\sqrt{2}n\sigma)^{-1}$$

See also TRANSPORT COEFFICIENT, KNUDSEN REGIME.

mean free time The average time for which a particle in a fluid, usually a gas, travels between collisions.

mean solar day *See* DAY.

mechanical advantage The factor by which the FORCE applied to a LOAD by a machine is greater than the EFFORT force:

$$\text{Mechanical advantage} = \text{load force}/\text{effort force}$$

mechanical energy A collective term for the forms of energy studied in MECHANICS: KINETIC ENERGY, GRAVITATIONAL POTENTIAL ENERGY and ELASTIC potential energy.

mechanical equivalent of heat A conversion factor, now obsolete in the SI system of units, for converting units of HEAT, measured in CALORIES, into units of WORK, measured in ERGS. It has the value 4.185 x 10^7 ergs/calorie.

mechanics The branch of physics that deals with the effect of FORCES on objects or structures, particularly gravitational and CONTACT FORCES. *See also* DYNAMICS, STATICS.

mechanoreceptor A RECEPTOR cell that detects pressure changes, gravity and vibrations (sound). *See also* SENSE ORGAN, EAR.

median The value that is midway between the largest and the smallest values found in a set of data. *See also* AVERAGE.

medical imaging The name given to a range of techniques that are used to obtain images of the inside of a human body (usually to diagnose illness) without resorting to surgery. Techniques used are often based on the transmission of X-rays, the reflection of ULTRASOUND and on NUCLEAR MAGNETIC RESONANCE. *See also* ENDOSCOPE, MAGNETIC RESONANCE IMAGING, RADIOGRAPH, TOMOGRAPHY, ULTRASOUND IMAGING.

medium wave RADIO WAVES with a wavelength from 200 m to 600 m.

medulla 1. In animals, the central part of some organs such as the ADRENAL GLAND, brain or kidney.

2. In plants, the central part of a stem or root, usually consisting of PARENCHYMA. *See also* CORTEX.

medulla oblongata Part of the vertebrate HINDBRAIN. The medulla oblongata is an enlarged region connected to the SPINAL CORD, and contains centres controlling respiration, blood pressure, the rate and strength of the heartbeat, in addition to other activities, such as swallowing, coughing, taste and touch. *See also* BRAIN.

medusa The free-swimming structural form in the life cycle of the CNIDARIA.

mega- (M) A prefix indicating that the size of a unit is to be multiplied by 10^6. For instance, one megawatt (MW) is equal to one million WATTS.

meiosis, ***reductive division*** A process of cell division in which each daughter cell formed contains half the number of CHROMOSOMES compared with the parent cell. In animals, it occurs in the formation of GAMETES as part of SEXUAL REPRODUCTION, so that when female and male gametes fuse during FERTILIZATION the resulting cells regain a full set of chromosomes, instead of doubling their total number of chromosomes. In plants, meiosis is part of SPORE formation.

The stages of meiosis are similar to MITOSIS, but two divisions occur (I and II):

(i) In prophase I the chromosomes become visible and, unlike in mitosis, associate

Stages of meiosis.

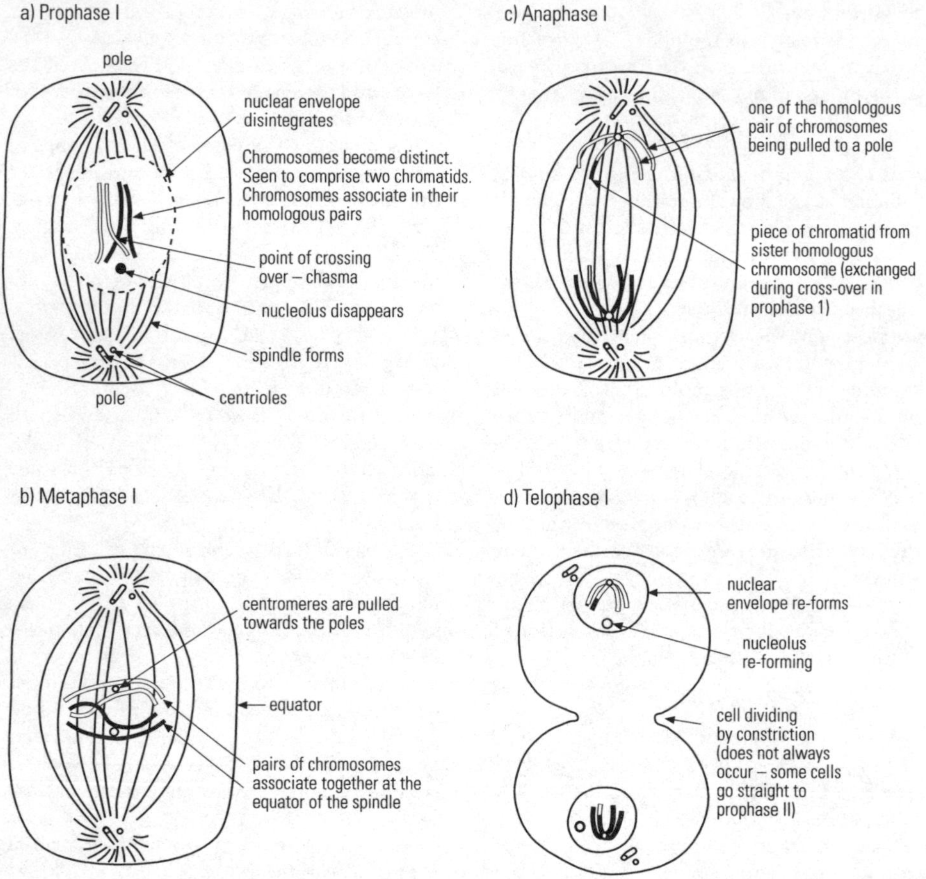

a) Prophase I

pole
nuclear envelope disintegrates
Chromosomes become distinct. Seen to comprise two chromatids. Chromosomes associate in their homologous pairs
point of crossing over – chasma
nucleolus disappears
spindle forms
pole
centrioles

c) Anaphase I

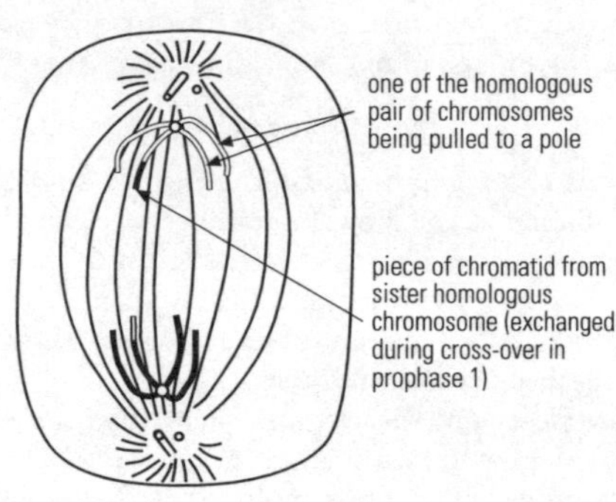

b) Metaphase I

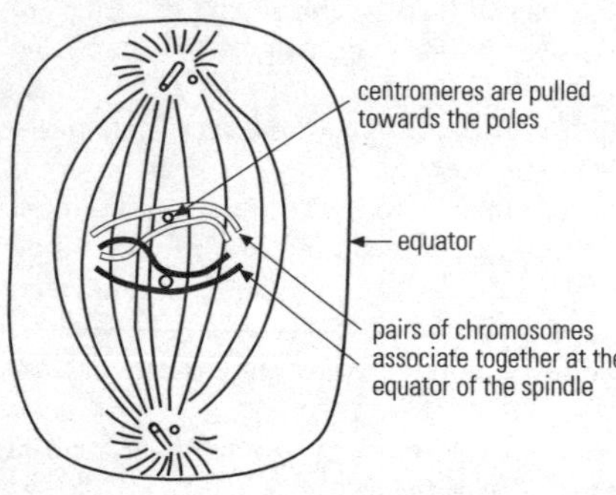

d) Telophase I

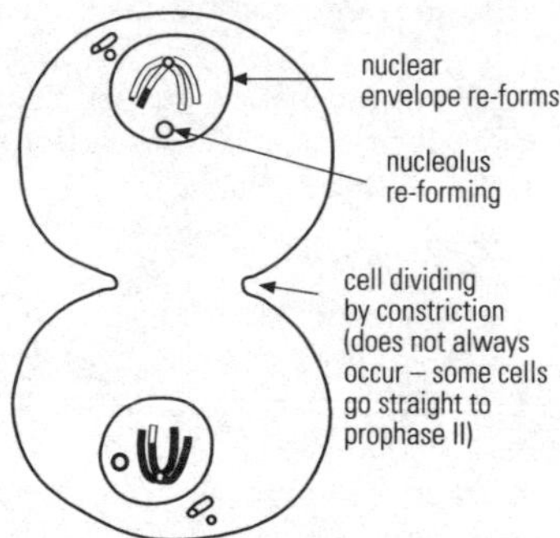

in their homologous (like) pairs (*see* CHROMOSOME). CROSSING-OVER may occur. The nuclear membrane disintegrates and a system of protein tubules called the spindle forms.

(ii) In metaphase I the pairs of chromosomes associate together at the equator of the spindle, but orientate towards opposite poles of the cell (this orientation is random relative to other pairs).

(iii) In anaphase I the spindle contracts and pulls the chromosome pairs apart to opposite poles.

(iv) In telophase I a nuclear membrane forms around each group, the spindle disappears and the cell may divide (to yield two cells each with a NUCLEUS and half the number of chromosomes) and enter INTERPHASE (but with no DNA replication; *see* MITOSIS). Some cells go straight to prophase II.

(v) In prophase II the nuclear membrane disintegrates and the spindle forms at right angles to that of the first mitotic division.

(vi) In metaphase II, anaphase II and telophase II the chromosomes arrange themselves (not in pairs) on the equator of the spindle, the CHROMATIDS are pulled apart to opposite poles, and nuclear membranes are reformed as in a typical mitotic division. The end result is four cells each with a nucleus and half the number of chromosomes.

e) Prophase II

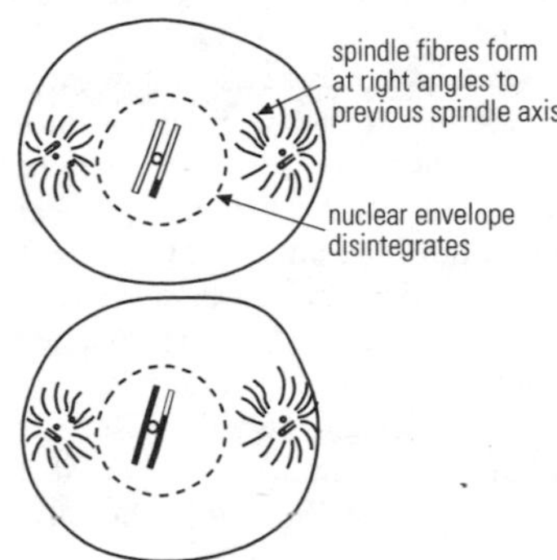

f) Metaphase II

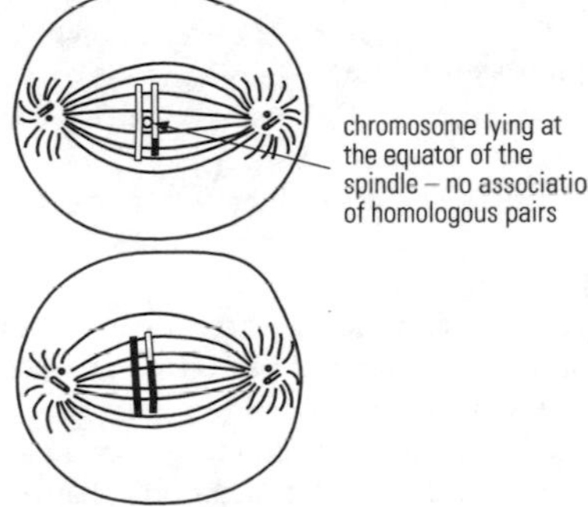

g) Anaphase II

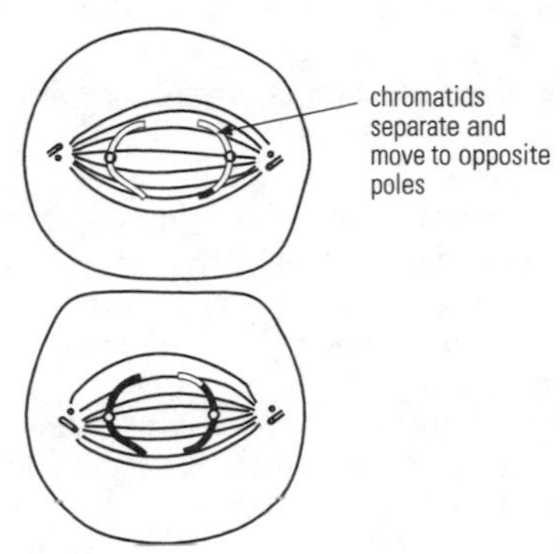

h) Telophase II

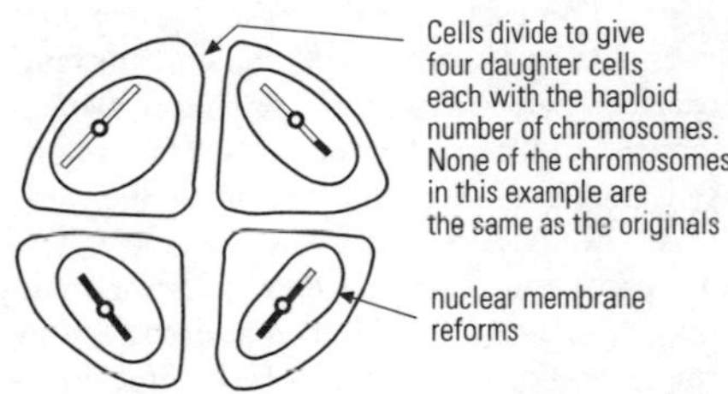

The main purpose of meiosis is to yield offspring that are different to their parent so that they can adapt to a changing or new environment. The genetic variability comes from the fusion of two gametes to give a different set of chromosomes to that of each parent. The crossing-over that occurs in prophase I and the random orientation of chromosome pairs during metaphase I provide further genetic variety.

M.E.K. *See* BUTANONE.

melamine A THERMOSETTING PLASTIC that is heat resistant, scratch resistant and difficult to break, so has many household uses.

melanin A brown pigment, produced by the MALPIGHIAN LAYER of EPIDERMIS, that determines skin, hair and eye colour and absorbs ultraviolet light, thereby protecting tissues beneath from sunlight. The amount of melanin in skin depends on genetic and environmental factors. *See also* MELATONIN.

melatonin A hormone-like substance secreted by the PINEAL GLAND. Production of melatonin is inhibited by light, so its levels are greatest during the night. In this way the pineal gland keeps track of changes in day length. Melatonin helps regulate certain seasonal changes in animals, such as the reproductive cycle of seasonally breeding animals. It also controls skin colour changes in certain animals by triggering the aggregation of the pigment MELANIN.

melting point The temperature at which a solid turns into a liquid, or vice versa. More technically, the melting point is the one temperature for a given pressure at which the solid and liquid can exist in equilibrium together. Most materials increase in volume, though only slightly, when they melt, and the melting point increases with increasing pressure. An important exception to this is water, which takes up less volume as a liquid than as a solid, thus water tends to melt under pressure.

membrane In biology, a layer surrounding cells or ORGANELLES that is made of LIPIDS and proteins. The membrane controls passage of molecules into and out of the cell or organelle. *See also* CELL MEMBRANE.

membrane potential In a living cell, the POTENTIAL DIFFERENCE across the CELL MEMBRANES. *See* NERVE IMPULSE.

memory The part of a computer system that stores data and programs permanently or temporarily. There are two types of memory: main memory and secondary memory. Main memory is held in semiconductor chips in the computer and is either read-only memory (*see* ROM, EROM, PROM) or RAM. Retrieval from main memory is usually fast, but chips are relatively expensive. Secondary memory include magnetic disks and tape and other media in which data can be stored permanently. These memory devices are cheap but retrieval is slow. *See also* BACKING STORE, BUBBLE MEMORY, IMMEDIATE ACCESS MEMORY.

memory cell *See* B CELL.

menarche *See* MENSTRUAL CYCLE.

Mendel's laws The original laws of inheritance set out by an Austrian monk, Gregor Mendel (1822–84), who studied the transmission of different features of the garden pea from parents to offspring. Although at that time GENES and DNA had not been discovered, much of what Mendel found forms the basis of our understanding of genetics.

Mendel's first law is segregation: an individual possesses a pair of ALLELES, one from each parent, that separate during MEIOSIS and go to different GAMETES so that they pass to different offspring, without blending. The offspring will be like one parent in a particular characteristic, for example eye colour, and not a mixture of both parental characteristics. This is also referred to as monohybrid inheritance: the inheritance of a single characteristic only. Which genetic feature is outwardly expressed depends on which allele is DOMINANT.

Mendel's second law is independent assortment: each member of an allelic pair can combine randomly with each of another pair. When considering the simultaneous inheritance of two characteristics (dihybrid inheritance), the presence of one does not affect the inheritance pattern of the other. This law does not hold true for linked genes (*see* LINKAGE), which tend to be inherited together.

Based on Mendel's laws, for simple genetics (excluding, for example, linkage and MUTATION) it is possible to predict the ratios of characteristics among offspring where the characteristics of the parents are known. This is called Mendelism.

See also DIHYBRID CROSS, MONOHYBRID CROSS.

mendelevium (Md) The element with ATOMIC NUMBER 101. Several isotopes are known, but the longest lived (mendelevium–256) has a HALF-LIFE of just 1.3 hours, too short for the material to be of any practical use.

Mendelism The theory of inheritance originally proposed by Gregor Mendel (1822–84). *See* MENDEL'S LAWS.

meninges In vertebrates, three membranous coverings enclosing and protecting the BRAIN and SPINAL CORD. The meninges consist of an innermost pia mater, a middle arachnoid and an outermost dura mater membrane. Between the arachnoid and the pia mater membranes is a space filled with CEREBROSPINAL FLUID, a clear, colourless solution of GLUCOSE and mineral ions and a few WHITE BLOOD CELLS but no protein. The fluid supplies nutrients and acts as a shock-absorber for the CENTRAL NERVOUS SYSTEM.

meniscus The name given to the curved shape of a liquid surface in a tube, or any similarly shaped object, such as a lens with one CONVEX and one CONCAVE surface. A meniscus lens may be either converging or diverging, depending on which surface is the more strongly curved. *See also* CAPILLARY EFFECT, CONVERGING LENS, DIVERGING LENS.

menopause The time in a woman's life when her reproductive capacity ends (usually between the age of 45 and 50 years). The changing levels of the hormones OESTROGEN

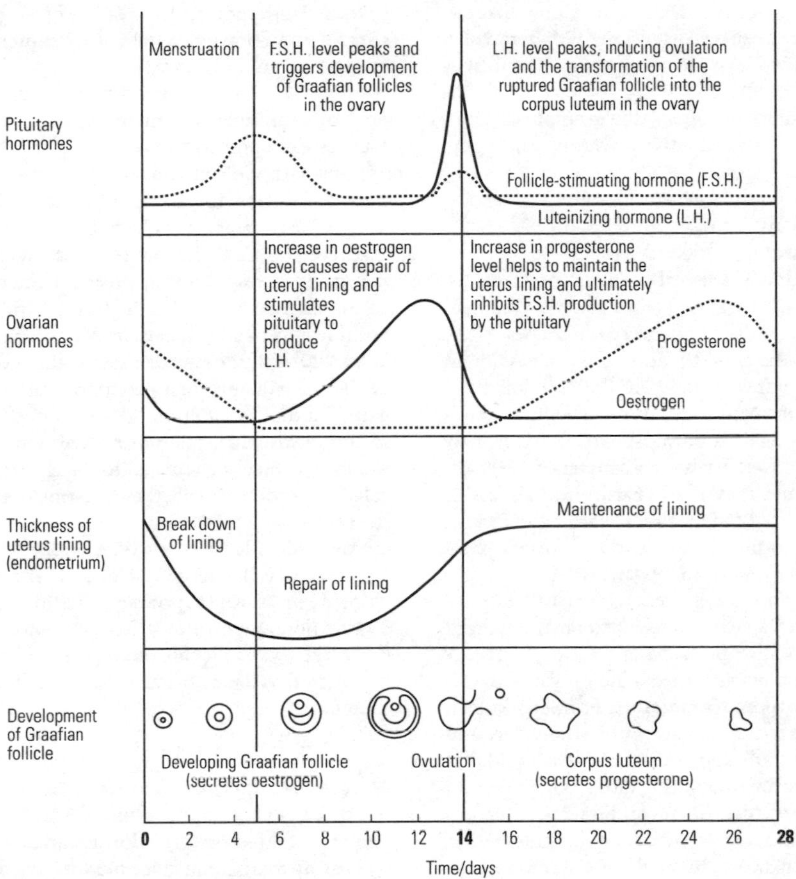

Summary of the menstrual cycle.

and PROGESTERONE can cause some symptoms associated with the menopause, such as hot flushes and osteoporosis (thinning of the bones), although often the menopause passes uneventfully. In more severe cases of post-menopausal problems, HORMONE-REPLACEMENT THERAPY (HRT) administered to replace some of the hormones and so relieve some of the symptoms.

menstrual cycle The cycle that occurs in female mammals of reproductive age to prepare the body for pregnancy. In humans, the start of the cycle is at puberty (at an average age of 12 years) and lasts until the MENOPAUSE (at an average age of 45–50 years). The first cycle is called menarche.

Each cycle lasts about 28 days, during which a GRAAFIAN FOLLICLE matures in the OVARY and releases its OVUM (ovulation) at about day 14. The CORPUS LUTEUM then develops (*see* OVARY) and produces PROGESTERONE. This causes the lining of the UTERUS to thicken and fill with blood vessels in preparation for IMPLANTATION of a fertilized egg. If no FERTILIZATION occurs then the corpus luteum dies and the uterine lining is shed, causing the loss of blood for about 5 days which is menstruation. The cycle is controlled by hormones including OESTROGEN and progesterone. It is the changing levels of these hormones that can cause pre-menstrual tension (PMT) and also menopausal symptoms, such as hot flushes and OSTEOPOROSIS.

In many mammals, for example cats and dogs, the reproductive cycle is called the OESTROUS CYCLE.

See also ENDOMETRIUM, HORMONE-REPLACEMENT THERAPY, PILL.

menstruation *See* MENSTRUAL CYCLE.

mensuration The use of geometry or CALCULUS, particularly INTEGRATION, to find the area, volume or CENTROID of geometrical shapes.

Mercury The closest planet to the sun, with an orbital radius of 0.39 AU (58 million km). Similar in many ways to our Moon, it is too small to retain an atmosphere and is heavily cratered. Mercury has a diameter of 4,900 km (0.38 times that of the Earth) and a mass of 3.3×10^{23} kg (0.055 times that of the Earth). Mercury orbits the Sun in 88 days and rotates about its own axis in 59 days.

mercury (Hg) The element with ATOMIC NUMBER 80; RELATIVE ATOMIC MASS 200.6; melting point –39°C; boiling point 357°C; RELATIVE DENSITY 13.6. Mercury is a metal, the only one that is liquid at room temperature. Formerly known as quicksilver, it is used in thermometers and in many applications that require a liquid that is electrically conducting or has a high density (10 times that of most liquids). Mercury occurs in nature as mercury sulphide, in the ore cinnabar. Mercury sulphide decomposes to mercury on heating:

$$HgS \rightarrow Hg + S$$

Mercury exhibits two valencies (*see* VALENCY). MONOVALENT compounds contain the mercury(I) ion, Hg_2^{2+}, while mercury(II) compounds containing the Hg^{2+} ion are more common. Mercury is a cumulative poison and concern has been expressed about high levels of mercury in the FOOD CHAINS of some industrialized areas.

mercury barometer A BAROMETER in which a tube filled with mercury and closed at the upper end is placed vertically, with its lower end in a trough of mercury open to the atmosphere. If the tube is long enough, the pressure of the air will not be sufficient to support the weight of the mercury and a vacuum (called a Toricellian vacuum) will appear at the top end of the tube. The height of the mercury column that can be supported is a measure of the pressure of the atmosphere, which is sometimes quoted as millimetres or inches of mercury (mmHg or inHg). The pressure that will support one millimetre of mercury is called one TORR. *See also* FORTIN BAROMETER.

mercury-cathode cell A device for the extraction of chlorine and sodium hydroxide from BRINE. The brine is electrolysed in a vessel with graphite ANODES at the top (from which chlorine gas is released) and a stream of mercury acting as the CATHODE at the base of the cell. Sodium released in the ELECTROLYSIS forms an AMALGAM with the cathode and is carried out of the cell. Water is then passed over the amalgam, and the sodium reacts, producing sodium hydroxide solution and hydrogen. The mercury is then recirculated to the electrolytic cell. This process is called the Kastner-Kellner process.

mercury chloride Mercury(I) chloride, Hg_2Cl_2, and mercury(II) chloride, $HgCl_2$. Mercury(I) chloride is a white powder; melting point 302°C; boiling point 384°C; RELATIVE DENSITY 7.0. Mercury(I) chloride can be made by heating mercury(II) chloride with an excess of mercury:

$$HgCl_2 + Hg \rightarrow Hg_2Cl_2$$

Mercury(II) chloride is a white powder; melting point 276°C; boiling point 303°C; relative density 5.4. Mercury(II) chloride can be made by heating mercury in chlorine:

$$Hg + Cl_2 \rightarrow HgCl_2$$

mercury oxide (HgO) A red or orange powder, which decomposes on heating. Mercury oxide can be formed by heating mercury in oxygen:

$$2Hg + O_2 \rightarrow 2HgO$$

At higher temperatures, this reaction is reversed and mercury oxide decomposes to mercury and oxygen.

mercury sulphide (HgS) A red or black solid. Mercury sulphide occurs naturally as the ore cinnabar, an important source of mercury. It decomposes on heating:

$$HgS \rightarrow Hg + S$$

meristem In plant tissue, a region of actively dividing cells that produces new tissue. Apical meristems are found at the growing tips of stems or roots and cause an increase in their length. Lateral meristems, for example the CAMBIUM, cause an increase in girth. The rigid cell wall of plant cells restricts their ability to grow, which is why meristems, which are actively growing immature cells, are needed. Meristem cultures involve growing shoots in nutrient mediums into new plants, which is used in plant propagation.

merocrine gland Any EXOCRINE GLAND that remains intact after secreting its product, for example SALIVARY GLANDS. *See also* APOCRINE GLAND, HOLOCRINE GLAND.

merozoite A stage in the life cycle of protozoans from the PHYLUM APICOMPLEXA, such as the *Plasmodium* parasite. *See* MALARIA.

mesh (*vb.*) Of GEARS, to meet together so the movement of one gear moves the other.

mesocarp The thick fleshy layer in the wall of a fruit, underneath the outer EXOCARP.

mesoderm The middle GERM LAYER of a developing animal EMBRYO. The mesoderm forms a NOTOCHORD that later becomes the VERTEBRAL COLUMN in vertebrates, and also gives rise to muscle, kidneys, the heart, blood cells, the reproductive system, eyes and connective tissue. The mesoderm is absent from primitive animals.

meson Any one of a family of unstable HADRONS, made up of a QUARK and an antiquark. They exist as positive, negative and neutral particles and they include the kaon, the PI-MESON and the psi particle.

mesophyll The layer of tissue in a LEAF blade between the upper and lower EPIDERMIS. The mesophyll comprises two layers. The palisade mesophyll is just below the upper epidermis, and below this is the spongy mesophyll.

The palisade layer consists of closely packed columnar cells containing numerous CHLOROPLASTS, in which PHOTOSYNTHESIS takes place (during daylight). The palisade cells are adapted to receive the components of photosynthesis: water from the nearby XYLEM, carbon dioxide from the air through the stomata (*see* STOMA), and sunlight, the passage of which is maximized by the vertical arrangement of the cells (with fewer cross-walls).

The spongy mesophyll consists of loosely arranged cells of a regular shape with fewer chloroplasts but more spaces for rapid DIFFUSION of gases (entering with the air through the stomata).

mesosome An infolding of the CELL MEMBRANE of bacteria that contains the ELECTRON TRANSPORT SYSTEM. Mesosomes are thought to function similarly to MITOCHONDRIA.

mesosphere A layer in the upper ATMOSPHERE in which temperature falls with increasing height.

Mesozoic The ERA of geological time that extended from the end of the PALAEOZOIC era (about 225 million years ago) to the start of the CENOZOIC era (about 65 million years ago). It encompasses the TRIASSIC, JURASSIC and CRETACEOUS PERIODS.

messenger RNA (mRNA) A type of RNA that acts as a template for PROTEIN SYNTHESIS in a cell. Messenger RNA is a single-stranded molecule thousands of NUCLEOTIDES long, formed into a helix, and represents less than 5 per cent of the total RNA in a cell. It is made in the nucleus and is a mirror copy of part of one of the DNA strands (*see* TRANSCRIPTION). From the nucleus, it enters the CYTOPLASM and associates with the RIBOSOMES, where it is subsequently involved in the production of a polypeptide (*see* TRANSLATION). Messenger RNA is very unstable. *See also* TRANSFER RNA, START CODON, STOP CODON.

meta In organic chemistry, a prefix used in disubstituted derivatives of BENZENE which indicates that the substituted groups are at positions 1 and 3 in the ring, for example 1,3-dichlorobenzene (*meta*-dichlorobenzene):

The use of the term *meta* in other situations, for example metaphosphoric acid, does not have the same meaning and is always written in full. *See also* *ORTHO*, *PARA*.

metabolic rate *See* METABOLISM.

metabolism The chemical processes occurring within a living organism. Metabolism is a continual process of building up of body tissue (anabolism) and breaking down of living tissue into energy and waste products (catabolism). The control of metabolism is complex, involving HORMONES and ENZYMES.

Metabolic rate refers loosely to the metabolic activity of an organism, measured by the respiratory rate (oxygen is used as a guide to metabolic activity). *See also* METABOLITE, THYROID GLAND.

metabolite A substance required for or produced by METABOLISM. Primary metabolites are involved in essential processes, whereas secondary metabolites are required for or produced by non-essential reactions. Secondary metabolites can be important in defence and are characteristic of a particular organism. For example fungi produce toxic secondary metabolites and plants produce some that make them less palatable.

metal Any of a class of elements that are typically lustrous, MALLEABLE, DUCTILE solids (mercury is a liquid) that are good conductors of heat and electricity. Metal ions replace the hydrogen in an ACID to form a SALT, and combine with the HYDROXIDE (OH^-) ion to form a BASE. Metals form ALLOYS with each other.

About 75 per cent of the known elements are classed as metals. The ALKALI METALS and ALKALINE EARTHS are generally soft silvery reactive metals that form positive ions. The TRANSITION METALS are harder, less reactive metals, and they form COMPLEX IONS.

The structure of a metal is typically a rigid lattice of positive ions, through which the outermost electrons, the VALENCE ELECTRONS, are free to move. The forces between the ions in the metal lattice are repulsive at short distances, but at larger separations the electrons screen the repulsion between ions, resulting in an attractive force that is strong enough to give most metals melting points well above room temperature.

The presence of FREE ELECTRONS in a metallic lattice make metals good CONDUCTORS of heat and electricity. The electrical CONDUCTIVITY allows metals to reflect electromagnetic radiation, including light, resulting in their characteristic shiny appearance. The interatomic forces in metals are generally weaker then those in IONIC SOLIDS, so metals are relatively soft, and in their pure forms are usually malleable and ductile.

See also ALLOY, METALLIC BONDING, THERMAL CONDUCTION.

metallic bonding The bonding that holds together atoms in a METAL. The overlapping of the ORBITALS in a metallic solid produces an energy band that is effectively a DELOCALIZED ORBITAL, so that VALENCE ELECTRONS can move freely from one atom to the next, allowing the metal to conduct heat and electricity. A simple model of a metallic structure is of a LATTICE of positive ions surrounded by a 'sea' of electrons. The relatively weak nature of metallic bonding compared to ionic bonding (*see* IONIC BOND) accounts for the softness and low melting points of metals when compared to IONIC SOLIDS.

metallic radius A measure of the effective size of a metal atom, defined as half the distance between adjacent atoms in a metal. Metallic radii increase with ATOMIC NUMBER within a given GROUP in the PERIODIC TABLE.

The metallic radii of the ALKALI METALS are somewhat larger than for other metals of similar atomic number as a result of the lower packing fraction achieved by the BODY CENTRED CUBIC lattice structure adopted by the alkali metals. Typical values lie between 0.1 nm and 0.3 nm.

metalloid, ***semi-metal*** Any element that has some of the properties of a METAL, but which is not completely metallic. Many of them are SEMICONDUCTORS, or form compounds that are semiconductors. Typical examples are ARSENIC, BORON, GERMANIUM, SILICON and TELLURIUM.

metallurgy The study of the properties of METALS, and more usually their ALLOYS, particularly in respect of their engineering properties.

metameric segmentation The division of the body of an animal into similar segments. *See also* ANNELIDA.

metamorphic (*adj.*) Describing a rock that was originally SEDIMENTARY in nature but has had its structure altered by subsequent heat and pressure.

metamorphosis A stage in the life cycle of most insects and amphibians and some fish, during which the body changes dramatically from one form to another. The changes involve major tissue reorganization, such as moulting (outer skin is shed, allowing the new soft body to alter its size and shape), and changes in GENE EXPRESSION. These changes are under hormonal control and occur relatively rapidly. The change from a tadpole to a frog is an example of metamorphosis.

Higher insects undergo complete metamorphosis, in which the LARVAE bear no resemblance to the adults. The larvae develop into PUPAE, which is a resting stage where no food is taken in and during which the organs and tissues of the young change into those of the adult. An example is the change of a caterpillar to a pupa, called a chrysalis, and then to a butterfly. Lower insects experience incomplete metamorphosis, in which the young develop through a series of stages resembling the adult, to the mature adult.

metaphase A stage in MITOSIS and MEIOSIS.

metastable equilibrium Any state that is a state of STABLE EQUILIBRIUM for small displacements but which is unstable for larger displacements, such as a ball lying in a small hollow in an upturned bowl.

metastable state A state of some physical system that is unstable, but in which the system will remain for an unusually long time, or a time sufficiently long for the state to be regarded as stable. SUPERHEATED water is an example of a metastable state, and some ENERGY LEVELS in atomic nuclei are also metastable.

metastasis A secondary TUMOUR caused by cells from the primary tumour being shed and travelling through the blood or LYMPHATIC SYSTEM to another part of the body.

metaxylem *See* XYLEM.

meteorology The study of the Earth's ATMOSPHERE, especially WEATHER SYSTEMS.

methanal, *formaldehyde* (HCHO) An ALDEHYDE that is a gas at ordinary temperatures with a pungent odour. It is used dissolved in water as formalin (a 37 per cent solution), which is a preservative for biological specimens. It is also a strong disinfectant (as a vapour or a spray) and can be used to sterilize surgical instruments. Other uses include the manufacture of foam, dyes, plastics and resins.

methane (CH_4) A colourless, odourless gas; boiling point –164°C; melting point –184°C. It is the simplest HYDROCARBON and the main component of NATURAL GAS. It reacts explosively with air, since it is odourless, this makes its presence ('firedamp') in coal mines a particular hazard. It is emitted by decaying vegetable matter and is therefore found in marshlands (as marsh gas) and is given off during SEWAGE DISPOSAL. Methane contributes to the GREENHOUSE EFFECT.

methanoic acid, *formic acid* (HCOOH) A colourless liquid CARBOXYLIC ACID that fumes slightly and has an unpleasant odour; boiling point 100.5°C; melting point 8.4°C. It occurs in ants, stinging nettles, sweat and urine. It is prepared by heating ethanedioic acid (oxalic acid) with GLYCEROL. It is used in textile dyeing and leather tanning.

methanol, *methyl alcohol* (CH_3OH) A simple colourless liquid ALCOHOL with a pleasant odour; boiling point 64.5°C. It is highly poisonous, causing blindness. It is usually made from coal or NATURAL GAS but can also be made by the dry DISTILLATION of wood, hence it is known as wood spirit. Methanol is used in the manufacture of methanal, MTBE and many other organic compounds, and as a solvent.

methionine An ESSENTIAL AMINO ACID containing sulphur, present in many proteins. *See also* START CODON.

method of mixtures A way of measuring HEAT CAPACITIES. A substance is heated to a measured temperature and then mixed with a second substance at a lower temperature. The mixture will be at a temperature between those of the two substances before they were mixed. If there are no heat losses, the heat energy given out by the hotter substance as it cools down will equal the heat energy taken in by the cooler substance as it heats up. If the heat capacity of one of the substances is known, that of the other can then be calculated.

methoxyethane An ETHER produced by a SUBSTITUTION REACTION of BROMOETHANE. *See* WILLIAMSON ETHER SYNTHESIS.

methyl alcohol *See* METHANOL.

methylated spirit ETHANOL containing METHANOL, making it undrinkable. It is used for industrial purposes as industrial methylated spirit.

methylbenzene, *toluene* ($C_6H_5CH_3$) An AROMATIC compound consisting of a BENZENE RING with a CH_3 (methyl) group attached. It is a colourless liquid derived from PETROLEUM; boiling point of 111°C; melting point –95°C. It is used in the manufacture of the explosive TNT, as a solvent and aircraft fuel and in the manufacture of PHENOL.

2-methylbutadiene, *isoprene* ($CH_2{=}C(CH_3)$-$CH{=}CH_2$, C_5H_8) A liquid obtained from PETROLEUM that is used in the manufacture of synthetic RUBBERS. Its boiling point is 34°C. Isoprene is polymerized by ZIEGLER-NATTA CATALYSTS.

methyl ethyl ketone *See* BUTANONE.

methyl orange, *4-dimethylamino-4′-azobenzene sodium sulphonate* ($C_{14}H_{14}N_3NaO_3S$) An organic dye used as an INDICATOR, particularly in TITRATIONS involving weak bases, and as a biological stain in the preparation of microscope slides. It is red below PH 3.1 and changes to yellow above pH 4.4. *See also* AZO COMPOUND.

methyl t-butyl ether *See* MTBE.

metre The BASE UNIT of length in the SI system (*see* SI UNIT). The metre was originally defined as the length of a standard metal bar, then in terms of a certain number of wavelengths of light, but now defined as the distance travelled by light in a vacuum in 1/(299,792,458) seconds.

metric system Any system of measurements based on the METRE as the unit of length, the GRAM as the unit of MASS and the SECOND as the unit of time, or on some multiple of these units. In science, the system almost universally used is the SI system of units, which defines units for all physical quantities, derived from seven BASE UNITS. *See* SI UNITS.

metric ton, *tonne* A unit of MASS. One tonne is equal to 1,000 kg.

metrology The science of measurement, in particular measuring quantities very precisely.

MHC *See* MAJOR HISTOCOMPATIBILITY COMPLEX.

micelle A small group of molecules loosely clumped together, in a COLLOID for example.

Michelson–Morley experiment An experiment, first performed in 1881, set up to measure the velocity of the Earth through the ether (the hypothetical medium that was thought to fill all space). The ether was thought to support the propagation of ELECTROMAGNETIC RADIATION through space, and the results of the Michelson–Morley experiment were to have led to a calculation of the velocity of light.

In the experiment, light waves were made to interfere with each other (*see* INTERFERENCE) after they had travelled along two paths set at right angles to one another. The interference pattern was unchanged when the whole experiment was rotated, suggesting that the motion of the Earth through space did not alter the speed of the light waves. This lead to an abandonment of the ether theory, and ELECTROMAGNETIC WAVES are now known to be able to travel through free space. The abandonment of the ether hypothesis also led indirectly to the formulation of the SPECIAL THEORY OF RELATIVITY.

micro- (μ) A prefix indicating that a unit is to be multiplied by 10^{-6}. For instance, the microampere (μA) is equal to one millionth of an AMPERE.

microbiology The study of MICRO-ORGANISMS. Microbiology has many applications, for example in medicine, industry and genetic engineering.

microbody *See* PEROXISOME.

microcomputer *See* PERSONAL COMPUTER.

microcontroller A MICROPROCESSOR with MEMORY built into a single INTEGRATED CIRCUIT. Some of the memory will be RAM for short-term storage of data whilst most will be ROM in which the sequence of instructions (PROGRAM) will be stored. Microcontrollers are increasingly used in the control of domestic appliances, such as washing machines or dishwashers.

microevolution *See* EVOLUTION.

microglial cell A type of small GLIAL CELL with an irregular shape. They occur more frequently in GREY MATTER than WHITE MATTER. Microglial cells may be PHAGOCYTES.

microhabitat A HABITAT with very specific living conditions, such as a fallen log in a forest.

micrometer An instrument for the accurate measurement of small objects (typically up to 10 cm). Many forms of micrometer exist, but typically the object to be measured is placed between a fixed surface and a moving surface, which can be advanced by a screw thread until it is just touching the object. The screw thread is machined accurately to a known pitch and provided with a scale which enables fractions of one rotation to be measured accurately. Accuracies of 10^{-6} m are possible using a simple micrometer equipped with a VERNIER scale.

micro-organism An organism too small to be seen by the naked eye and seen only under a microscope. Micro-organisms include VIRUSES, BACTERIA, PROTOZOA, YEASTS and some ALGAE. Micro-organism is a general term that is not significant for classification purposes.

microprocessor A complex INTEGRATED CIRCUIT that is capable of performing a large number of logical functions on BINARY numbers. The microprocessor contains an ARITHMETIC LOGIC UNIT (ALU), which performs various operations on numbers that are stored in SHIFT REGISTERS, with the output also being stored in a REGISTER, usually called the ACCUMULATOR. Instructions can be given to the microprocessor to perform various operations on the numbers in the registers or to move data between the registers and memory connected to the microprocessor.

micropyle In SEED PLANTS, a small hole at one end of the OVULE where the protective INTEGUMENT(S) is (are) absent. During FERTILIZATION the POLLEN TUBE containing the male GAMETE passes through the micropyle and so gains access to the egg cell within the ovule. In a mature SEED, the micropyle is seen as a small pore in the seed coat through which water enters at the start of GERMINATION. *See also* DOUBLE FERTILIZATION.

microscope An optical device that uses a system of lenses to magnify objects too small to be seen in fine detail with the naked eye. In 1665 Robert Hooke (1635–1703) was the first to record microscopic examination of cells in cork, and Anton van Leeuwenhoek (1632–1723) recorded bacteria in 1683.

A SIMPLE MICROSCOPE has a single lens but limited powers of magnification, whereas a COMPOUND MICROSCOPE uses two lenses and light passes from an object through the first lens (objective) to produce a magnified image that is then magnified further by the second lens (eyepiece). The total magnification is the product of the magnification of each lens and is maximally 1,500–2,000 times in a light microscope, achieved by oil-immersion objective lenses. The RESOLVING POWER of a light microscope is limited to two points 0.2 μm apart. The thinner the material being observed, the greater the clarity of image.

Preserved or fixed tissues are usually embedded in paraffin wax and thin sections (3–20 μm) are cut using an instrument called a microtome, but even thinner sections can be obtained by freezing the tissue in liquid nitrogen and cutting sections using a cryostat at –200° C. Various staining methods are used to enhance the image seen and to highlight specific cells or materials.

See also ELECTRON MICROSCOPE, FLUORESCENCE MICROSCOPY, PHASE CONTRAST MICROSCOPE.

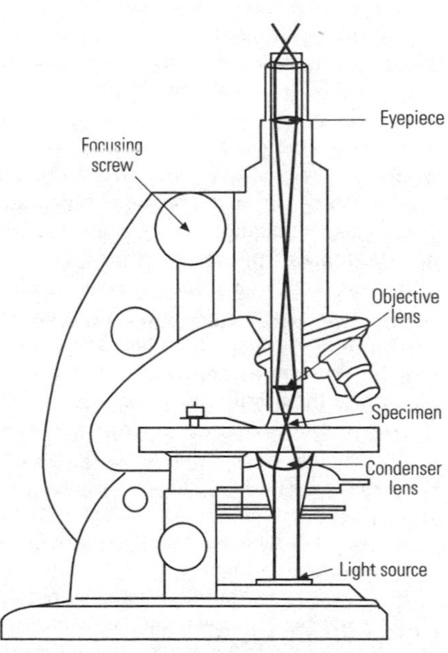

Light microscope.

microscopy The study or use of MICROSCOPES.

microtome A machine for cutting thin sections of tissue (frozen or embedded in paraffin wax) to be used in a light MICROSCOPE. The sections are 3–20 μm thick. A steel knife is used.

microtubule A hollow cylinder (about 25 nm in diameter), made of protein filaments (thread-like structures), that is an essential component of the CYTOSKELETON in almost all eukaryotic cells (*see* EUKARYOTE). Microtubules are also vital components of CILIA, FLAGELLA, and the SPINDLES formed in MITOSIS and MEIOSIS.

microvilli (*sing. **microvillus***) Minute finger-like projections on the surface of many eukaryotic cells (*see* EUKARYOTE), particularly EPITHELIA cells concerned with absorption, for example in the SMALL INTESTINE. In size each microvillus is about 1 μm long and 0.1 μm in diameter. Microvilli can extend and retract and form a BRUSH BORDER, increasing the surface area over which absorption can take place.

microwave ELECTROMAGNETIC RADIATION with a wavelength in the range 30 cm to 1 mm. The short wavelength means that microwaves can be formed into narrow beams using parabolic reflector dishes of a reasonable size without too much DIFFRACTION, thus they tend to be used for point-to-point communication rather than broadcasting. Microwaves are also often used in DBS (direct broadcasting satellite) systems. Their ability to form narrow beams with an aerial of reasonable size is also put to use in RADAR systems.

Microwave frequencies coincide with the RESONANT FREQUENCIES of some covalently bonded molecules (*see* COVALENT BOND). In particular, waves of 12.6 cm wavelength are resonantly absorbed by water molecules. This effect is used in microwave ovens, where microwave energy is resonantly absorbed and converted to heat by water in food.

See also KLYSTRON, MAGNETRON.

microwave spectroscopy The study of the absorption spectra (*see* ABSORPTION SPECTRUM) of MICROWAVES by gases, typically using microwaves with a wavelength of a few millimetres. Since these waves have frequencies that excite the natural frequencies of rotation of simple covalently bonded molecules (*see* COVALENT BOND), it is possible to study certain properties of these bonds, such as their natural length, and their angle to one another.

midbrain The smallest of three regions of the human brain, which links the FOREBRAIN and HINDBRAIN. This region contains the visual and auditory centres, for example controlling movement of the head to fix on an object or sound. *See also* PINEAL GLAND.

migration 1. (*biology*) The seasonal movement of certain animals, mostly birds and fish, to distant lands for breeding or feeding. For example, some birds fly south during the winter months and return to their breeding ground in the spring. In some species, for example locusts, the return journey is made by the next generation. Sometimes whole species migrate over a period of many years. The precise method of migration is unclear.

2. (*physics*) The tendency of charged particles to move under the influence of an electric field, in ELECTROPHORESIS for example.

3. (*chemistry*) The movement of a particular structural feature, such as a DOUBLE BOND, from one part of a molecule to another, particularly in a reaction where the molecule is otherwise unchanged.

mild steel An alloy of iron with a few per cent of carbon. Mild steel is cheap and easy to machine and to weld, but rusts quickly.

milk The fluid secreted by the MAMMARY GLANDS of the female mammals, to provide nourishment for their offspring. *See* LACTATION.

milk teeth *See* DECIDUOUS TEETH.

Milky Way Our own GALAXY, which appears as a band of stars across the sky. Study of the distribution of stars in the Milky Way shows that the SOLAR SYSTEM is located close to the plane of the galaxy, in one of the spiral arms.

mill In the chemical industry, a machine for breaking large lumps of material into smaller pieces, such as coal for burning in a furnace. Also used in a wider context to describe an industrial plant where a particular material is produced by chemical processes, for example a steel mill.

milli- (m) A prefix indicating a unit is to be multiplied by 10^{-3}. For instance, one milliampere (mA) is equal to one thousandth of an AMPERE.

millibar (mb) A unit of PRESSURE commonly used in METEOROLOGY. One millibar is equivalent to 100 PASCAL.

Millikan's oil drop experiment An experiment first performed by Robert Millikan (1868–1953) in 1909, in which the charge of an oil drop is measured by suspending it in an electric field. The charges of all the drops were found to be multiples of a single charge, the charge on the electron, thus showing the quantized (*see* QUANTUM) nature of electric charge.

mineral A naturally occurring non-living substance that has a characteristic chemical composition or contains a certain compound. Rocks are made up of mixtures of minerals. Technically, a mineral is often crystalline, having essentially a single chemical composition. More generally, a mineral is any material of commercial value that can be extracted from the Earth's crust, and under this definition includes petroleum oil and natural gas.

mineral acid An ACID obtained from MINERALS and used in chemical processes to produce other commercially important substances. The chief mineral acids are hydrochloric acid, sulphuric acid and nitric acid.

mineralocorticoid Any one of a group of CORTICOSTEROID hormones, secreted by the CORTEX of the ADRENAL GLAND, that is concerned with the METABOLISM of minerals. An example is ALDOSTERONE.

mineralogy The branch of GEOLOGY concerned with MINERALS.

mineral oil Any OIL derived from a MINERAL, particularly one derived from PETROLEUM.

minicomputer A multiuser computer whose size and capabilities lie between that of a PERSONAL COMPUTER and MAINFRAME.

minimum The smallest of a set of numbers, or the smallest value taken by a function. A local minimum is a value of a function that is smaller than the value of that function for values of independent variable adjacent to that giving the minimum value, but which may be exceeded when the independent variable takes on other, very different, values.

mini-pill *See* PILL.

minor planet *See* ASTEROID.

Miocene The fourth EPOCH of the TERTIARY PERIOD. It began after the OLIGOCENE epoch (about 26 million years ago) and extended until the PLIOCENE epoch (about 7 million years ago). The Miocene epoch saw the evolution of hoofed mammals and the spreading of grasslands.

mirage The illusion of water seen when looking at a hot surface, such as desert sand, at a shallow angle. The air just above the surface is heated by the surface, so is less dense and has a lower REFRACTIVE INDEX than the cooler air further from the ground. This leads to TOTAL INTERNAL REFLECTION of light from the sky as it approaches the surface. The unstable nature of the hot layer, resulting from CONVECTION, gives the reflected light an unsteady appearance, similar to light from the sky reflected in a pool of water. Unlike real water, the mirage remains at a fixed angle to the observer, so the mirage recedes as the observer approaches it.

mirror A device used to reflect some form of wave, usually visible light. Mirrors used for the REFLECTION of light are usually made from a thin layer of metal (usually aluminium) deposited on a layer of glass. For precision use, the metal is placed on the front surface of the glass (such mirrors are called front-silvered). Mirrors with the reflective coating behind the glass (back-silvered) are less easily damaged but suffer from multiple reflections off the front of the glass surface. *See* also CURVED MIRROR, PLANE MIRROR.

miscible (*adj.*) Describing two liquids that can mix together to form a single liquid. Most organic liquids are miscible with one another, but not with water. Water and ethanol are miscible. Liquids are most likely to be miscible if the degree of polarity is similar (*see* POLARIZATION). *See also* IMMISCIBLE.

missing mass Mass that is not observed in the Universe, but which must exist if the density is equal to the CRITICAL DENSITY. *See* COSMOLOGY.

mist Droplets of water formed by condensation in the atmosphere close to ground level, but with the visibility remaining above 1,000 m. *See also* FOG, WEATHER SYSTEMS.

mitochondria (*sing.* ***mitochondrion***) Rod-like (1–10 μm in length and 0.25–1 μm in width) or spherical membrane-bound bodies (ORGANELLES) within the CYTOPLASM of eukaryotic cells (*see* EUKARYOTE). Mitochondria occur in varying numbers and contain the enzymes responsible for energy production, in the form of ATP, during aerobic RESPIRATION. It is thought that mitochondria are derived from free-living bacteria that once invaded larger cells to become symbiotic (a mutually beneficial association; *see* SYMBIOSIS). Each mitochondrion contains its own mitochondrial DNA, and division of existing mitochondria yields new ones. There is a double outer membrane, the inner component of which is folded inwards and projects into the matrix as cristae. These are covered in particles associated with OXIDATIVE PHOSPHORYLATION. PLASTIDS of plants may be related to mitochondria.

mitosis A process in which a cell nucleus divides, usually prior to CELL DIVISION, to create two new cells each with the same number of CHROMOSOMES as the parent cell.

Mitosis can be divided into a number of phases: prophase, metaphase, anaphase and telophase. During prophase, the chromosomes become untangled and visible and a system of protein tubules called the spindle forms, from one pole of the cell to another pole, and the nuclear membrane disintegrates. During the subsequent phases the two strands (CHROMATIDS) that constitute chromosomes are separated and move to opposite poles of the cell. Unlike during MEIOSIS, homologous (like) chromosome pairs do not associate during mitosis. A new nuclear envelope forms around each group of chromatids, they become invisible again and the spindle disintegrates. The cell itself can divide after nuclear division but does not always (so multinucleate cells exist). It is during a further phase, INTERPHASE, that the DNA content of the cell doubles and cell ORGANELLES double, so that the two daughter cells formed contain the same number of chromosomes as the parent itself.

Only a specialized group of cells in plants can undergo mitosis (*see* MERISTEM), but most animal cells can. During growth or repair of

Stages of mitosis.

a) Interphase

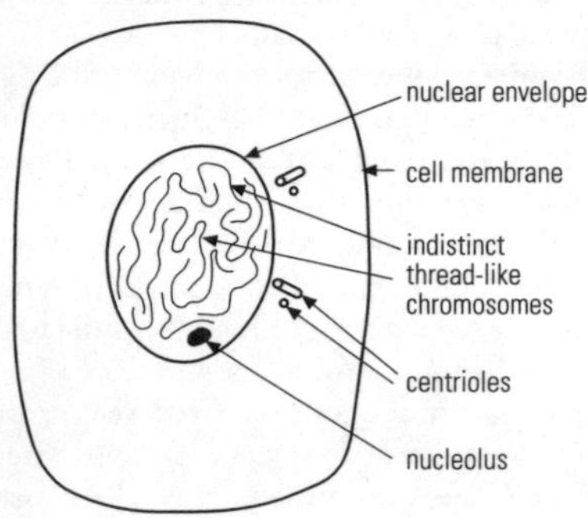

b) Prophase

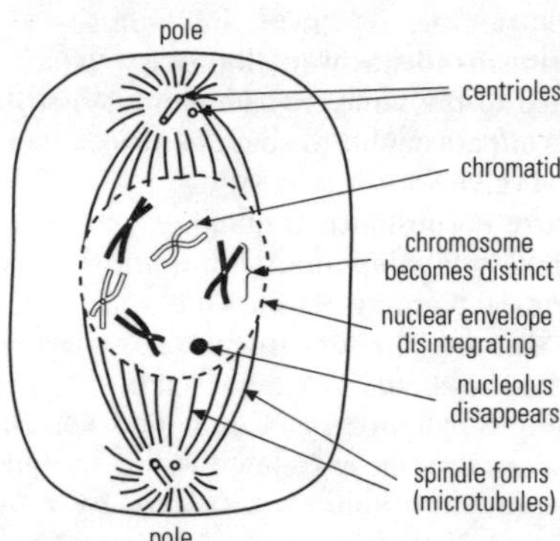

c) Metaphase

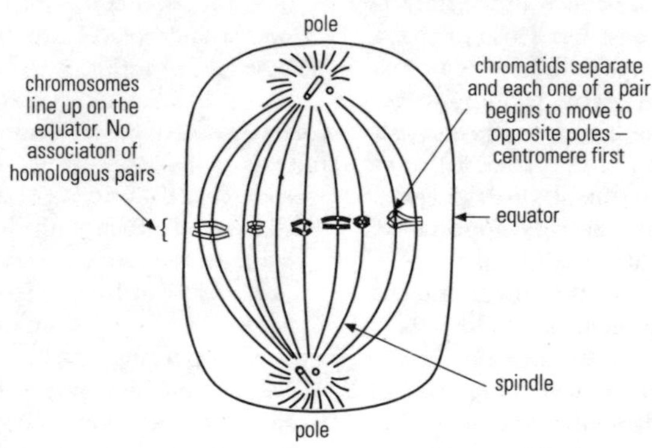

d) Anaphase

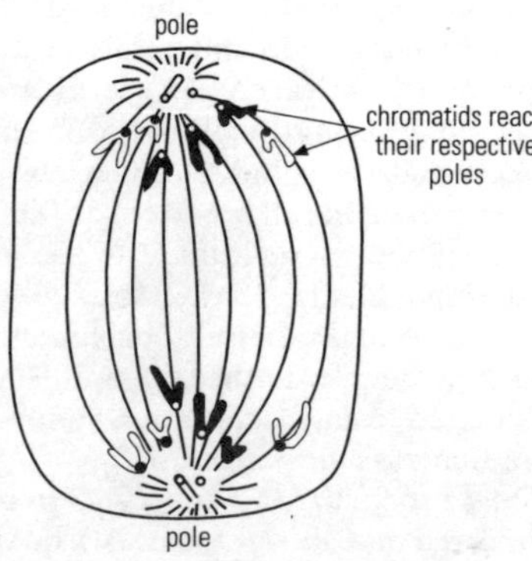

e) Telophase

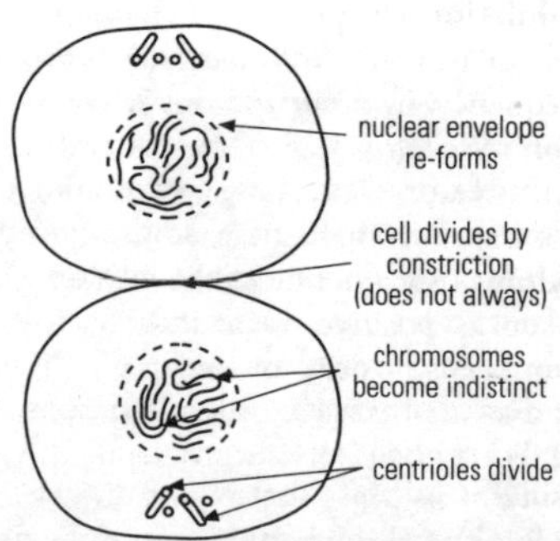

tissue, cell division is by mitosis because the new cells must be the same as the existing cells. ASEXUAL REPRODUCTION by mitosis yields organisms identical to the parent, which are suited to the same environment as their parents but are unable to adapt to new challenges.

mitral valve *See* BICUSPID VALVE.

mixture A substance containing two or more elements or compounds, but where there is no chemical bonding between the constituents of the mixture. The constituents can be separated without a chemical reaction taking place. For example, in a mixture of iron and sulphur, the iron can be removed using a magnet, but once the iron and sulphur have combined to form iron sulphide (a compound) they cannot be separated by physical means. *See also* CLATHRATE, COLLOID.

mnemonic In computing, a short sequence of letters used in an ASSEMBLY LANGUAGE and designed to correspond to a MACHINE CODE instruction.

mode The value in a set of data that occurs most often, or the range for which the frequency is greatest.

modem Acronym for *mod*ulator–*dem*odulator. A device for encoding and decoding digital information in such away that DUPLEX communication can be achieved between two computers connected by a telephone line or radio link.

moderator A material that absorbs energy from fast-moving neutrons in a NUCLEAR REACTOR, without absorbing the neutrons themselves. Neutrons slowed in this way are called thermal neutrons. They are more likely to cause fission in uranium–235 nuclei and less likely to be absorbed by uranium–238. DEUTERIUM and carbon–12 (as graphite) are often used as moderators. *See also* NUCLEAR FISSION.

modulation The process of changing some feature of a CARRIER WAVE, such as its amplitude or frequency, in order to convey some information. *See also* AMPLITUDE MODULATION, FREQUENCY MODULATION, PHASE MODULATION, PULSE CODE MODULATION, SIDEBAND.

modulus In mathematics the value of a number, taken as positive, regardless of whether the number is positive or negative. The modulus of a REAL NUMBER x is denoted by the symbol $|x|$. For a COMPLEX NUMBER, $a+ib$, the modulus is $\sqrt{(a^2 + b^2)}$.

modulus of rigidity bulk modulus A measure of the elasticity of a material subjected to a shearing force (*see* SHEAR). It is equal to the tangential force per unit area divided by the angular deformation in radians.

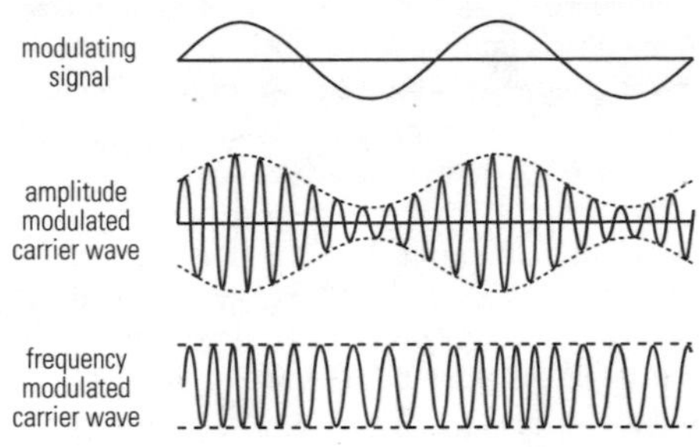

Modulation.

molality The CONCENTRATION of a solution, measured in terms of the number of MOLES of SOLUTE dissolved in one kilogram of SOLVENT.

molar conductivity The CONDUCTIVITY of a solution divided by its CONCENTRATION, in order to allow the contribution to the conductivity of each MOLE of SOLUTE ions to be determined. In a strong ELECTROLYTE (one that is fully IONIZED) the molar conductivity is essentially independent of concentration, except at very high concentrations. In a weak (only partially ionized) electrolyte, the molar conductivity rises only gradually with dilution, the material being fully ionized at very low concentrations. *See also* KOHLRAUSCH'S LAW.

molar gas constant (R) The constant of proportionality linking the ABSOLUTE TEMPERATURE of one MOLE of gas to the product of its pressure and volume. R is equal to 8.3 $JK^{-1}mol^{-1}$.
See IDEAL GAS EQUATION.

molar heat capacity The HEAT CAPACITY of a piece of material containing one MOLE of molecules. For an IDEAL GAS the molar heat capacity at constant pressure is greater than that at constant volume by R, the MOLAR GAS CONSTANT. For a material with a molar heat capacity at constant pressure of C_P and a molar heat capacity at constant volume of C_V:

$$C_P - C_V = R$$

See also DULONG AND PETIT'S LAW, RATIO OF SPECIFIC HEATS.

molarity The CONCENTRATION of a SOLUTION measured in MOLES of SOLUTE per decimetre cubed of solution. Molarity is often denoted by the letter M, so a 0.1 M solution will contain 0.1 moles of solute molecules in 1 dm^3 of solution.

molar volume The volume occupied by one MOLE of a given material. All IDEAL GASES have the same molar volume at a given temperature and pressure. At STANDARD TEMPERATURE AND PRESSURE this is 22.4 dm^3.

molecular biology The study of the molecular components of biological systems. Molecular biology particularly refers to the study of NUCLEIC ACIDS and PROTEINS.

molecular distillation A low pressure DISTILLATION technique in which molecules escaping from a liquid are condensed onto a cold condensing surface, after a passing through a short distance in a vacuum. This technique is used at low temperatures to distil molecules that decompose on heating, and also to distil molecules that react with oxygen in the air.

molecular formula A chemical FORMULA that gives the number of atoms present in a molecule. For example, ethane has a molecular formula of C_2H_6, though its EMPIRICAL FORMULA is CH_3. The molecular formula is determined by knowing the empirical formula and the RELATIVE MOLECULAR MASS. *See also* STRUCTURAL FORMULA.

molecular mass *See* RELATIVE MOLECULAR MASS.

molecular orbital An ORBITAL formed by the overlapping of orbitals between two atoms that are linked together by a COVALENT BOND. It is the reduction in energy brought about by the formation of one of the possible molecular orbitals, as compared to the two separate atomic orbitals, that makes the bond stable. *See also* ANTIBONDING ORBITAL, BONDING ORBITAL, DELOCALIZED ORBITAL, PI-BOND, SIGMA-BOND.

molecular sieve *See* SEMIPERMEABLE MEMBRANE.

molecular weight An obsolete term for RELATIVE MOLECULAR MASS.

molecule The smallest part of a chemical COMPOUND that can exist without it losing its chemical identity. Molecules are made of one or more ATOMS held together by IONIC or COVALENT BONDS. *See also* MACROMOLECULE.

mole fraction The proportion of a specified element or compound in a mixture, expressed in terms of the number of molecules present. Thus a mixture of 16 g of oxygen (O_2) with 4 g of helium (He) may contain 80 per cent oxygen by mass, but the mole fractions are 50 per cent each as there are two moles of gas molecules present: one mole of O_2 and one of He.

mollusc A member of the PHYLUM MOLLUSCA.

Mollusca The second largest PHYLUM in the animal KINGDOM, consisting of about 100,000 species, including snails, slugs, squids, octopuses, oysters, mussels and winkles. The members are all invertebrates whose bodies are divided into a head, muscular foot and a visceral mass. The body is soft with no internal skeleton and no limbs. In many species it is covered by a hard shell. Most molluscs are marine but some are freshwater and a few are terrestrial. The shells are variable and can be, for example, univalve, as in the snail, or bivalve, as in the mussel, as well as other forms. In squid, the shell is internal. Reproduction is sexual, eggs are laid, and many are HERMAPHRODITES.

Molluscs are important to humans as a food source and for pearls. Snails are the intermediate hosts for a number of parasitic diseases of humans and livestock, for example schistosomiasis (*see* TREMATODA). *See also* GASTROPODA, PELECYPODA.

molybdenum (Mo) The element with ATOMIC NUMBER 42; RELATIVE ATOMIC MASS 94.9; melting point 2,610°C; boiling point 5,560°C; RELATIVE DENSITY 10.2. Molybdenum is a TRANSITION METAL, used in the manufacture of hard steels.

moment The turning effect of a FORCE. The magnitude of a moment is equal to the magnitude of the force multiplied by the shortest distance between the axis and the line along which the force acts. If an object is in EQUILIBRIUM, the sum of the moments of all the forces about any point will be zero.

moment of inertia A quantity that measures the distribution of MASS within an object. The moment of inertia of a body is the ratio of the MOMENT acting on the body to the ANGULAR ACCELERATION caused by that moment. For a solid object, which can be thought of as made up of a series of points of mass m_i each located at a distance r_i from the axis, the moment of inertia I is given by

$$I = \Sigma m_i r_i^2$$

The ANGULAR MOMENTUM is $I\omega$, and the

rotational KINETIC ENERGY $^1/_2 I\omega^2$, where ω is the ANGULAR VELOCITY. *See also* ROTATIONAL DYNAMICS.

momentum The MASS of an object multiplied by its VELOCITY. Momentum is a measure of how easy or difficult it is to stop an object. To change the momentum of an object, an IMPULSE must be exerted equal to the change in momentum required. The impulse is the FORCE acting on the object multiplied by the time for which that force acts. For an object of mass m moving with velocity v, the momentum p will be

$$p = mv$$

See also LAW OF CONSERVATION OF MOMENTUM.

monatomic (*adj.*) Describing a elemental substance, such as a NOBLE GAS, whose atoms behave as separate entities and do not form molecules.

Mond gas *See* SEMI-WATER GAS.

monobasic acid Any ACID with a BASICITY of 1.

monochromatic (*adj.*) Describing ELECTROMAGNETIC RADIATION, in particular VISIBLE LIGHT, of a single wavelength. The term monochromatic is also used to describe a beam of particles where all particles have the same energy.

monoclinic (*adj.*) Describing a CRYSTAL structure where the UNIT CELL has two sets of faces at right angles to one another and the third is at some other angle. The size and shape of the unit cell are characterized by three lengths and one angle.

monoclonal antibody (MAb) An ANTIBODY produced in the laboratory by fusing an antibody-producing B CELL (from mice immunized with an ANTIGEN) with MYELOMA tumour cells, which do not exhibit the usual regulation of growth and CELL DIVISION. The resulting cell is called a hybridoma, and continues to divide and produce antibodies indefinitely. This allows large quantities of antibody with a single specificity to be made.

The technique was invented in 1975 by César Milstein (1927–) at Cambridge University, UK, and has been a major breakthrough in science. Monoclonal antibodies can be linked to markers to locate specific antigen targets, for example sources of disease, or specific cell types, or linked to CYTOTOXIC drugs which (if the antibody is directed at tumour cell antigens) could locate and hopefully destroy the tumour. One widely used technique using monoclonal antibodies is the ENZYME-LINKED IMMUNOSORBANT ASSAY.

monocotyledon In flowering plants, the possession of one COTYLEDON or seed leaf in the embryo. It is in contrast to DICOTYLEDONS, which are flowering plants with two cotyledons. Most monocotyledons are small plants, such as orchids, lilies, grasses and cereals, but some are large, such as palms. The leaves are usually narrow with parallel veins and smooth edges, the flower parts are grouped in threes, and the VASCULAR BUNDLES (XYLEM, PHLOEM) are arranged irregularly. POLLINATION is usually by wind. In most monocotyledons the cotyledons remain below ground following GERMINATION and are called hypogeal.

monocyte A type of vertebrate LEUCOCYTE that can differentiate into a MACROPHAGE.

monodentate *See* LIGAND.

monoecious Plants that have separate male and female flowers on the same individual plant. This arrangement favours cross-fertilization. *Compare* DIOECIOUS, HERMAPHRODITE.

monoestrus *See* OESTROUS CYCLE.

monoethylamine *See* ETHYLAMINE.

monohybrid cross In genetics, a cross between two animals or plants that are genetically identical except for one GENE. The one gene could be, for example, for seed colour, with one individual HOMOZYGOUS for the DOMINANT ALLELE (where the seed is green) and one homozygous for the RECESSIVE allele (where the seed is yellow). The offspring of such a cross (F_1 GENERATION) will all have green seeds; they are monohybrids (HYBRIDS for one gene only) and are identical to one another and resemble one parent. The recessive allele for yellow seeds is hidden but will be expressed in the next, (F_2) generation. In this F_2 GENERATION there will be on average three plants with green seeds and one plant with yellow seeds – a ratio of 3:1. In reality there are too many genetic differences to see this simple inheritance. *See also* DIHYBRID CROSS, MENDEL'S LAWS.

monohydric (*adj.*) Describing a compound with one HYDROXYL GROUP.

monomer A simple chemical compound that, under suitable conditions, can join with other identical monomers to form a long chain POLYMER.

monosaccharide A single sugar with the general formula $(CH_2O)_n$ that cannot be split into smaller CARBOHYDRATE units. When n is 3, the sugar is called a triose sugar, when *n* is 5 it is a pentose sugar, and when *n* is 6 it is hexose sugar. Monosaccharides are either aldoses (aldo-sugars), which have an ALDEHYDE group (CHO), or ketoses (keto-sugars), which have a KETONE group (C=O). Both GLUCOSE and FRUCTOSE have the formula $C_6H_{12}O_6$ but glucose is an aldose and fructose is a ketose, so their properties are different. Both can easily form ring structures; glucose usually has a six-sided pyranose ring and fructose a five-sided furanose ring, although both can form either ring structure. Most carbohydrates can form ISOMERS. Monosaccharides are sweet, soluble crystalline molecules (*see* CARBOHYDRATE). *See also* CARBOHYDRATE, DISACCHARIDE, POLYSACCHARIDE.

monotreme The least evolved mammal, where the young hatch from an egg outside the mother's body and are then nourished with milk. Only a few species of monotreme still exist, for example the platypus, because they have been displaced by other, more advanced species.

monotropy That form of ALLOTROPY in which there is only one stable form, with the other forms being unstable at all temperatures. Over time, the unstable form will always tend to change into the stable form. The two forms of phosphorous, red phosphorous and white phosphorous are an example of this. The white form will gradually turn into the red form. *See also* ENANTIOTROPY.

monovalent (*adj.*) Having a VALENCY of 1.

Moon The only natural satellite of the Earth. It orbits the Earth once every 27.3 days at an average distance of 384,400 km. It has a mean diameter of 3,476 km and a mass of about 0.012 that of the Earth. The Moon has no atmosphere or surface water.

mordant A colourless chemical with which a cloth is treated before being dyed. The mordant is absorbed into the fibres of the cloth and then forms a coloured complex with the dye molecules.

morning after pill *See* PILL.

moss A member of the CLASS MUSCI.

motherboard A computer's central circuit board that holds the MICROPROCESSOR, MEMORY and other main components of the computer. It provides the BUS that links all the computer's components.

mother liquor The concentrated solution from which crystals are formed or which remains after a CRYSTALLIZATION process. *See also* FRACTIONAL CRYSTALLIZATION.

motor Any device that converts ELECTRICAL ENERGY into mechanical WORK, usually in the form of rotational motion. *See also* D.C. MOTOR, INDUCTION MOTOR.

motor effect The force produced by a current-carrying wire in a magnetic field.

motor nerve A NERVE that carries an impulse away from the CENTRAL NERVOUS SYSTEM to an EFFECTOR muscle or gland. It is part of the EFFECTOR SYSTEM and causes voluntary and involuntary actions.

motor system *See* EFFECTOR SYSTEM.

mould A general name for superficial growth of a fungus on foodstuffs such as fruit or bread.

mouse A small hand-held control used to manipulate a pointer on a computer screen to select instructions from a list (called a menu), or to manipulate data. The mouse operates by sensors detecting the movement of a ball on the underside of the mouse which rotates as it is moved. This information is then fed to the computer along the wire which forms the 'tail' of the mouse.

mouth, *buccal cavity* The cavity forming the entrance to the DIGESTIVE SYSTEM. In mammals, it is enclosed by the jaws, cheeks and palate. Digestion of food begins in the mouth, where it is chewed (mastication) and mixed with SALIVA. The TONGUE is a muscular structure, attached to the floor of the mouth, which contains nerves and tastebuds. It aids the chewing process and pushes chewed food to the back of the mouth and into the PHARYNX. A number of reflexes exist to ensure that food goes down the OESOPHAGUS and air down the TRACHEA. *See also* EPIGLOTTIS, RESPIRATORY SYSTEM.

mouthpart An appendage around the mouth of ARTHROPODS that is adapted for feeding. Mouthparts are found in pairs, the number varying between groups, for example CRUSTACEANS have three pairs. In insects, the mouthparts can be adapted for piercing, sucking, biting or chewing. *See also* MANDIBLE.

moving-coil galvanometer An AMMETER in which the current flows through a coil suspended in a magnetic field. The result is a turning effect, which is opposed by a pair of flat coiled springs, called hairsprings. The hairsprings are also used to carry the current in and out of the coil. The coil rotates until the turning effect produced by the current is balanced by the turning effect of the springs. A pointer, or mirror with a beam of light shining onto it, is attached to the coil so the amount of motion, and thus the size of the current can be read from a graduated scale.

MRI *See* MAGNETIC RESONANCE IMAGING.

MTBE, *methyl t-butyl ether* An additive in petrol, used instead of lead to improve the OCTANE NUMBER.

mucin A sticky, jelly-like GLYCOPROTEIN that provides lubrication, intercellular bonding or binding of, for example, food (*see* SALIVA).

mucous membrane A thin layer of EPITHELIUM that lines all animal body cavities and canals that come into contact with the air. In particular, mucous membranes are found in the DIGESTIVE SYSTEM, RESPIRATORY SYSTEM and URINO GENITAL SYSTEM. The epithelium usually contains GOBLET CELLS that secrete MUCUS.

mucus A slimy secretion, containing MUCINS, that is produced by GOBLET CELLS of MUCOUS MEMBRANES in various parts of the body. In the RESPIRATORY SYSTEM, mucus helps to trap airborne particles for expulsion. In the DIGESTIVE SYSTEM, mucus helps to lubricate the food and protect the stomach from attack by digestive enzymes.

multicellular (*adj.*) Organisms or their parts that consist of more than one cell. *Compare* ACELLULAR, UNICELLULAR.

multinucleate cell A CELL with more than one NUCLEUS. *See also* CELL DIVISION.

multiple A number that is obtained by multiplying a specified number, usually an integer, by an integer. For example 12 is a multiple of 3, because 4 x 3 = 12.

multiplexing Any technique for sending more than one set of data down a single TELECOMMUNICATIONS link.

multiplier A resistance connected in series with a VOLTMETER to enable it to read higher voltages. The use of a multiplier has the advantage of increasing the resistance of the voltmeter.

multitasking In computing, running several programs, or different parts of the same program, simultaneously. The CENTRAL PROCESSING UNIT is shared and programs with calculations ready to run use the central processing unit while others wait for input or output.

mu-meson *See* MUON.

muon, *mu-meson* A charged LEPTON with a mass about 210 times that of the electron, to which it decays with a HALF-LIFE of about 2×10^{-6} s.

Musci A CLASS of the PHYLUM BRYOPHYTA consisting of the MOSSES. Mosses are small, non-flowering plants with no true roots. The sexual organs are at the tips of the leaves (*see* ANTHERIDIUM, ARCHEGONIUM). They thrive in damp conditions and show ALTERNATION OF GENERATIONS.

muscle Animal cells or fibres, derived from embryonic MESODERM, that have the ability to contract, causing movement of joints and other body movements. There are three types: voluntary, involuntary and cardiac.

Voluntary muscle (also called striated, striped, skeletal muscle) is activated by MOTOR NERVES under voluntary control and is concerned with locomotion and joint movement. It is composed of large, long fibres that consist of multinucleate cells, or syncytia (*see* SYNCYTIUM), with many nuclei held together by CONNECTIVE TISSUE and surrounded by a membrane. Within the SARCOPLASM of each fibre are longitudinal MYOFIBRILS, each with a distinctive pattern of bands caused by the distribution of the proteins ACTIN and MYOSIN. The bands form repeating units called sarcomeres.

Involuntary muscle (also called unstriated, smooth muscle) consists of spindle-shaped cells arranged in sheets or bundles bound by connective tissue and is under the control of motor nerves from the AUTONOMIC NERVOUS SYSTEM (involuntary nervous system). It is characterized by its ability to contract slowly and rhythmically over a long period of time. Therefore, involuntary muscle is important in the DIGESTIVE SYSTEM, by allowing PERISTALSIS, in the walls of blood vessels and tubes of the URINO-GENITAL SYSTEM.

CARDIAC MUSCLE is a specialized muscle found only in the heart, and is under involuntary control.

See also MUSCULAR CONTRACTION.

muscular contraction The response of MUSCLE cells to stimulation, causing a force in one direction. The muscle may shorten during a contraction or remain the same length.

The contraction of voluntary, or striated, muscle is understood best and involves the 'sliding filament theory', although the contraction of involuntary and cardiac muscle is thought to be similar. Muscle fibres consist of several MYOFIBRILS, which themselves consist of thin and thick filaments of the proteins ACTIN and MYOSIN respectively. These form a distinctive banding pattern across the myofibril (which may be seen with a microscope) which alters during a contraction, showing that the filaments slide past one another.

In order for contraction of voluntary muscle to occur, a number of conditions must be met. Firstly, the muscle needs stimulation by an impulse from a MOTOR NERVE. Secondly, the actin and myosin filaments must make contact to form a complex called ACTOMYOSIN. This complex can only be formed in the presence of calcium ions. Where the nerve meets the muscle is the NEUROMUSCULAR JUNCTION, and there

Sliding filament theory of muscular contraction.

are many of these spread throughout a muscle to ensure rapid contraction of all the fibres simultaneously. Each NERVE IMPULSE releases a jet of ACETYLCHOLINE, which diffuses across to the outer membrane of the muscle and depolarizes it, generating an ACTION POTENTIAL. Calcium ions are then released from the SARCOPLASMIC RETICULUM, where they are stored, where this comes into contact with infoldings of the muscle fibre's outer membrane, called

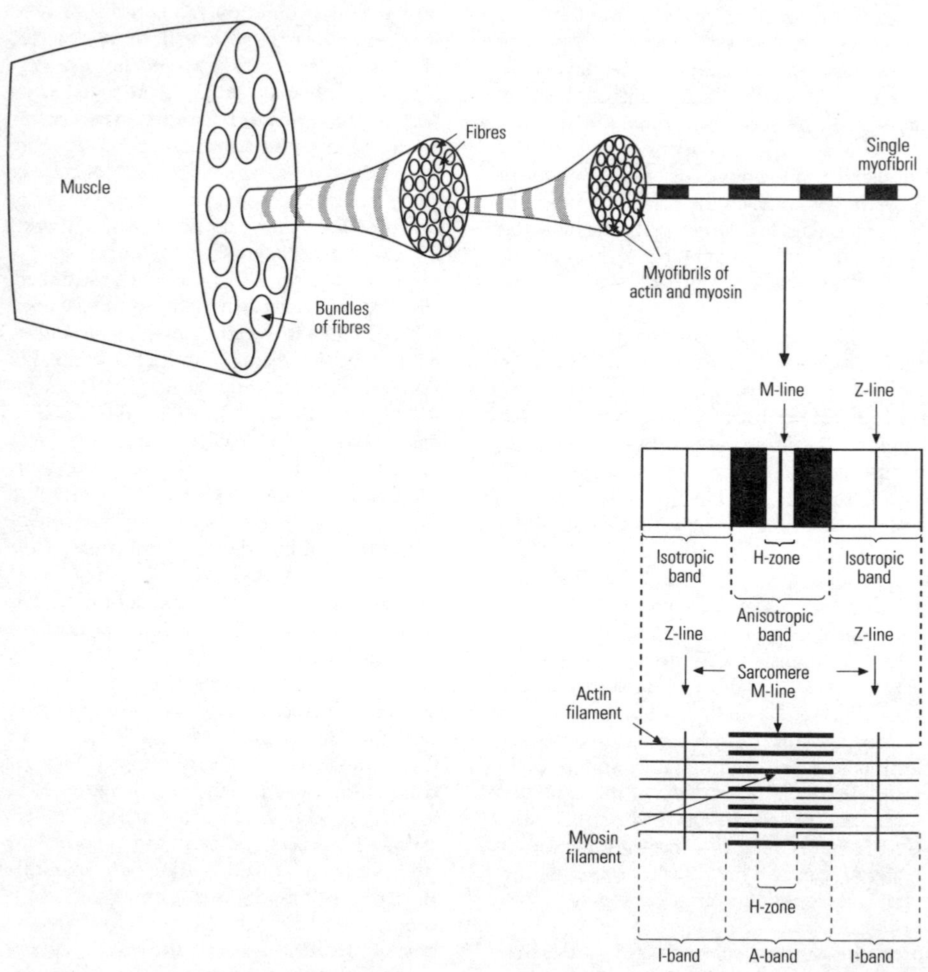

Detailed structure of muscle.

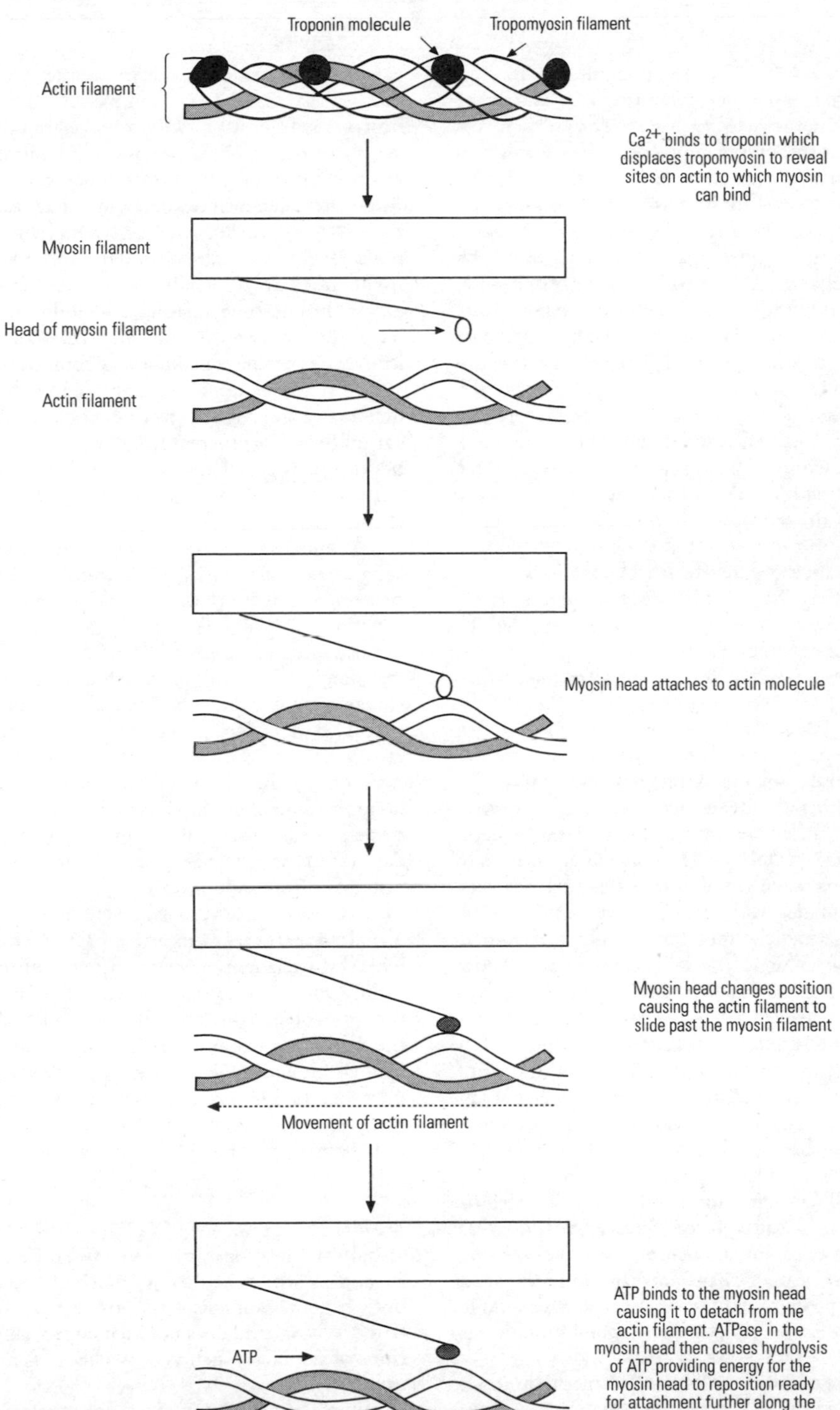
Troponin molecule
Tropomyosin filament
Actin filament
Ca2+ binds to troponin which displaces tropomyosin to reveal sites on actin to which myosin can bind
Myosin filament
Head of myosin filament
Actin filament
Myosin head attaches to actin molecule
Myosin head changes position causing the actin filament to slide past the myosin filament
Movement of actin filament
ATP
ATP binds to the myosin head causing it to detach from the actin filament. ATPase in the myosin head then causes hydrolysis of ATP providing energy for the myosin head to reposition ready for attachment further along the actin filament

transverse tubules. The calcium ions bind to protein molecules called troponin, which are attached to actin filaments. This in turn displaces tropomyosin molecules, also attached to actin, to reveal sites on the latter to which myosin can bind, forming an actomyosin complex. The myosin heads bind to the actin filament (following the HYDROLYSIS of ATP by the enzyme ATPase within the myosin heads), forming bridges, and the heads change position, so causing the actin filament to slide past the stationary myosin filament. This is the muscle contracting. ATP then binds to the myosin head, causing it to detach from the actin filament and reposition ready for reattachment further along the filament. After contraction, calcium ions are pumped back into the sarcoplasmic reticulum and tropomyosin once again hides the myosin binding sites on the actin filament. The muscle is now relaxed. The GLYCOGEN store found in muscles provides the supply of glucose needed to regenerate the ATP.

mutagen A factor, for example ultraviolet radiation, IONIZING RADIATION (such as X-rays, ALPHA PARTICLES, BETA PARTICLES and GAMMA RADIATION) and chemicals, that increases the natural spontaneous rate of MUTATION.

mutation An alteration of a GENE in an organism caused by an alteration of the genetic material (DNA or RNA). A mutation usually occurs when DNA is replicated and mistakes are made. Most mutations are undesirable. Any beneficial mutations would be favoured by NATURAL SELECTION. There is a natural, low spontaneous rate of mutation but certain factors called MUTAGENS increase this rate.

Most mutations occur in body cells (not in the formation of GAMETES) and are therefore not passed on. The mutation may be as small as the omission, insertion or substitution of a single base (*see* NUCLEOTIDES) in the DNA, which is called a point mutation. Although small, a point mutation can have serious effects because it can create STOP CODONS (nonsense mutations) or alter the reading frame of the DNA (frame shift mutations) so that protein synthesis is affected. An example of a disease resulting from a point mutation is SICKLE-CELL DISEASE.

Mutations can be much larger than those at the gene level, affecting CHROMOSOME structure or number. Whole sets of chromosomes can be duplicated or one chromosome can be deleted or added. An example of this in humans is in DOWN'S SYNDROME, where there are three copies of chromosome 21. Similar examples involving the sex chromosomes are Klinefelter's syndrome, where individuals have the GENOTYPE XXY, XXXY or XXXXY and a male PHENOTYPE with some female development, and Turner's syndrome, where there is an X chromosome missing and individuals have the genotype XO and a small, sexually immature female phenotype. Chromosome structure can be mutated during MEIOSIS by a deletion of a portion, or by inversion, translocation (which is different from CROSSING-OVER because it occurs between non-homologous chromosomes) or duplication of a portion of chromosome.

A mutation does not always affect the organism because it may not be translated into protein, and if it is it may be within a non-functional part of the protein. This would be a neutral mutation, but these can be important and they build up with time. In the laboratory, mutagenesis is a process that can be used to modify existing gene products.

Predictions can be made about the likelihood of offspring having a particular condition, taking account of the family genetic history, and some defects, particularly chromosomal, can be detected in early pregnancy by examining foetal cells taken from AMNIOTIC FLUID.

See also ONCOGENE, POLYMORPHISM.

mutual inductance A measure of the effect whereby a changing current in one coil will induce an ELECTROMOTIVE FORCE in another coil. It is equal to the induced e.m.f. divided by the rate of change of current. The SI UNIT of mutual inductance is the HENRY. If two coils have a mutual inductance M, a rate of change of current $\mathrm{d}I/\mathrm{d}t$ in one will produce an e.m.f. in the other of

$$E = M\mathrm{d}I/\mathrm{d}t$$

See also LENZ'S LAW.

mutualism A relationship between two different species in which neither partner suffers, or both benefit. Mutualism is another term for true SYMBIOSIS, and does not include the variations of symbiosis such as COMMENSALISM and PARASITISM.

mycelium The main body of a fungus that consists of a mass of thread-like HYPHAE.

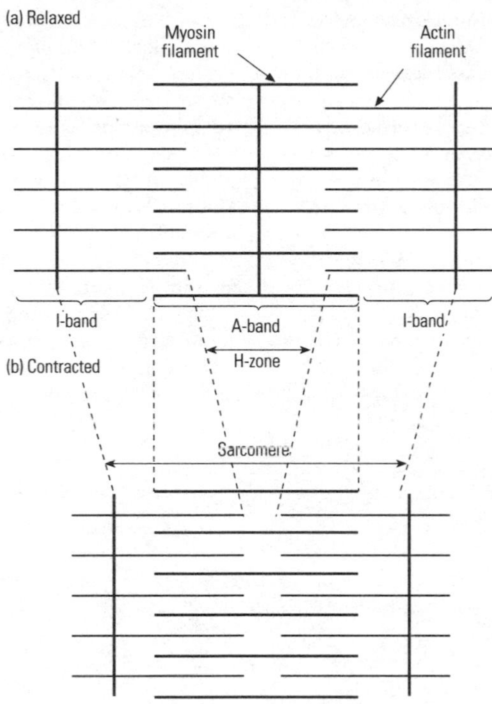

Changes in the myofibril banding pattern during muscular contraction.

mycorrhiza Structures formed by the association of a fungus with a root of a higher plant, such as a pine, oak, beech or birch tree.

myelin sheath An insulating layer around some nerve AXONS that speeds up the passage of NERVE IMPULSES. The myelin sheath is made up of fats and proteins contained within many layers of membrane laid down by specialized SCHWANN CELLS. The myelin sheath is discontinuous and is absent at intervals, called nodes of Ranvier, along the axons. A nerve impulse travels faster along these axons because the myelin sheath acts as an electrical insulator and the impulse therefore jumps from node to node.

myeloma A malignant TUMOUR of BONE MARROW. *See also* MONOCLONAL ANTIBODY.

myofibril A structural component of striated (voluntary) MUSCLE fibres. There are several myofibrils to each fibre and they have a distinctive banding pattern due to the distribution of the proteins ACTIN and MYOSIN. Actin is made of thin filaments, which form a light (isotropic) I-band across the muscle fibril. Myosin is made of thick filaments, and where these overlap with actin filaments a darker (anisotropic) A-band is seen. These bands alternate across the myofibril, causing its striated appearance. Within each light band is a central Z-line, and within each dark band is a lighter region called the H-zone, which may have a central dark M-line. During MUSCULAR CONTRACTION, the actin filaments slide over the myosin filaments as they are pulled together. This has the effect of shortening the I-band and H-zone while the A-band remains unchanged.

myogenic (*adj.*) Originating in or forming MUSCLE tissue. The contractions of the CARDIAC MUSCLE are said to be myogenic since they are stimulated within the heart itself.

myoglobin A globular PROTEIN found in vertebrate muscle that is closely related to

HAEMOGLOBIN and binds oxygen. Myoglobin has a single HAEM group (human haemoglobin has four) and a greater affinity for oxygen than haemoglobin. Myoglobin stores oxygen until it is needed in situations of extreme exertion, when the blood oxygen supply from haemoglobin is not sufficient to keep up with demands of muscle cells. Myoglobin is a red colour and is responsible for the coloration of meat.

During really vigorous exercise, if sufficient oxygen cannot be provided, muscle cells switch from AEROBIC RESPIRATION to ANAEROBIC RESPIRATION, which requires no oxygen but causes a build up of LACTIC ACID in the muscle, which is felt as cramp. This physiological state is known as OXYGEN DEBT. When exercising is stopped, the lactic acid is broken down and the oxygen debt is paid off.

See also RESPIRATORY PIGMENT.

myopia *See* SHORT-SIGHTEDNESS.

myosin A protein found in most eukaryotic cells (*see* EUKARYOTE). Two classes of myosin exist; both consisting of a head and tail region. Myosin I is involved in cell locomotion, and myosin II involved in MUSCULAR CONTRACTION. Myosin II forms the thick filaments of muscle myofibrils, along with the thin filaments of ACTIN. The head regions of myosin contain actin-binding sites and it is the interaction of myosin with actin to form actomyosin that is important in muscular contraction.

N

NAD (nicotinamide adenine dinucleotide) A COENZYME derived from the vitamin nicotinic acid (*see* VITAMIN B), which is an electron carrier in the ELECTRON TRANSPORT SYSTEM and in the KREBS CYCLE in RESPIRATION. When reduced NAD receives a hydrogen atom it becomes NADH, which carries the electrons. In its phosphorylated form NAD is NADP.

NADP (nicotinamide adenine dinucleotide phosphate) The phosphorylated form of NAD (*see* PHOSPHORYLATION). NADP is important as an electron carrier in PHOTOSYNTHESIS. When reduced, NADP receives a hydrogen atom to become NADPH. NADP is not as abundant in animal cells as NAD.

NAND gate (NAND = not and) A LOGIC GATE in which the output is not HIGH unless all the inputs are high. This gate is simple to manufacture and can be used as a building block for many more complex devices.

nano- (n) A prefix indicating that a unit is to be multiplied by 10^{-9}. For instance, one nanometre (nm) is one billionth of a metre.

nanotechnology The technology of building machines, such as electric motors, and eventually whole robots, on a very small scale. Nanotechnology devices are typically only a few nanometres in size, and approach the limits that can be achieved given the sizes of individual atoms. Although promising for the future, nanotechnology is still very much at the experimental stage.

narcotic One of a group of strong pain-killing and sleep-inducing drugs. Examples include morphine, heroin, codeine, alcohols and barbiturates. These drugs have the problem of inducing dependency and are therefore used under strict control.

nasopharynx The upper part of the PHARYNX.

nastic movement A plant movement in response to an external stimulus that is unrelated to the direction of the stimulus (unlike TROPISM). The movement can be due to growth or changes in TURGOR pressure. Examples of stimuli are light (PHOTONASTY), temperature (THERMONASTY) and chemicals (CHEMONASTY). *See also* THIGMONASTY.

native (*adj.*) Describing a MINERAL that contains essentially a single element not combined chemically. Only relatively unreactive materials occur in this form, such as copper, gold and sulphur.

natural frequency The FREQUENCY at which an oscillating system will oscillate if it is displaced from EQUILIBRIUM and then allowed to oscillate without any further external forces. *See also* FORCED OSCILLATION, RESONANCE.

natural gas A mixture of gases often found in pockets above PETROLEUM in the Earth's CRUST and which is one of the three main FOSSIL FUELS. It consists of HYDROCARBONS, mainly methane with some ethane, butane and propane.

natural killer cell (NK cell) In higher vertebrates, a type of WHITE BLOOD CELL that is important in recognizing alterations to the surface of virally infected and cancerous cells and then killing them.

natural logarithm A LOGARITHM to BASE e. The natural logarithm of x is denoted by the symbol $\ln x$.

natural number Any positive whole number, such as 1, 2, 3 etc.

natural selection The process by which individuals within a POPULATION, with characteristics that favour survival in their environment, reproduce more efficiently and so are selected (non-randomly) in preference to individuals without those characteristics. The frequency of a favourable characteristic therefore increases in a population, and the frequency of an undesirable one decreases. The process of natural selection was recognized in 1858 by Charles Darwin (1809–1882) and Alfred Wallace (1823–1913).

It is now known that natural selection is a slow process relying upon random gene MUTATIONS (some of which are favourable and passed on to future generations) and on the genetic RECOMBINATION resulting from SEXUAL

REPRODUCTION. It takes many generations for a particular trait, for example beak shape, limb shape and fur coat, to become an adaptation, and at any one time there will be a range of individuals with respect to any one character. The effect of the environment on this selective gene transmission is called selection pressure. Sometimes more than one distinct form of a characteristic can coexist within a population (*see* POLYMORPHISM).

The process of natural selection can cause problems for humans, for example the development of ANTIBIOTIC RESISTANCE by certain bacteria. Over many generations, through the inheritance of many favoured VARIATIONS, natural selection will eventually lead to the formation of new species (although other factors affect evolution).

See also ADAPTIVE RADIATION, ARTIFICIAL SELECTION, EVOLUTION, GENETIC DRIFT.

naval *See* UMBILICAL CORD.

neap tide The TIDE produced when the gravitational effects of the Moon and Sun oppose each other, producing the smallest rise and fall in tide level.

nectar A sugary liquid secreted by some plants that attracts insects, birds or other animals to the flower for POLLINATION. Nectar consists of sugars, amino acids and other nutrients. Bees use nectar to make honey.

nectary A specialized GLAND, near the base of some flowers, that produces NECTAR.

negative In physics, the name given to one of the two types of electric charges. Electrons are negatively charged particles. Objects that are negatively charged have more electrons than protons.

negative feedback 1. (*electronics*) FEEDBACK that is out of PHASE with the original signal, thus tending to reduce the GAIN of an AMPLIFIER. Negative feedback usually has the advantages of increasing the stability of an amplifier against external changes, such as temperature or supply voltage fluctuations. It also reduces DISTORTION.

2. (*biology*) Where the end-product of a pathway inhibits the ENZYME at the start of the pathway, as occurs in many metabolic reactions. *See also* HOMEOSTASIS.

nekton The swimming animals of the PELAGIC zone of a mass of water, for example whales and fish. *See also* PLANKTON.

nematic (*adj.*) Describing a material containing long molecules that can be aligned by the application of an electric field. *See* LIQUID CRYSTAL.

Nematoda A PHYLUM consisting of worms that have an unsegmented, cylindrical body with two openings to the gut (mouth and anus) and lack CILIA or FLAGELLA. There are 10,000 known species, which are thought to represent only about 2 per cent of the phylum, living in a variety of habitats.

The triploblastic (three-layered) body is pointed at both ends and has a tough outer CUTICLE of protein. Most nematodes are free-living but some are PARASITES. Nematodes move by a series of longitudinal contractions down their body. Food is pumped into the intestine by the PHARYNX and is varied, including algae, animals and organic debris. Reproduction is sexual; male and female individuals are separate and FERTILIZATION is internal. Parasitic nematodes may have a complex life cycle with several hosts, for example roundworms, threadworms and eel worms cause diseases of humans, such as elephantiasis, and attack plant roots such as the potato.

Compare FLATWORM.

nematode A member of the PHYLUM NEMATODA.

neo-Darwinism *See* EVOLUTION.

neodymium (Nd) The element with ATOMIC NUMBER 60; RELATIVE ATOMIC MASS 144.2; melting point 1,016°C; boiling point 3,068°C; RELATIVE DENSITY 7.0. Neodymium is a pale yellow metal of the LANTHANIDE series. Its SALTS are pink and sometimes used to colour glass and ceramics. Neodymium salts are also used in some lasers.

neon (Ne) The element with ATOMIC NUMBER 10; RELATIVE ATOMIC MASS 20.2; melting point –249°C; boiling point –246°C. Neon is an inert gas that occurs in very small amounts in air. It is obtained by FRACTIONAL DISTILLATION of liquid air. The bright orange-red colour of its GAS DISCHARGE has led to its widespread use in decorative lighting and advertising signs.

neon lamp A light source using a GAS DISCHARGE through neon. It gives a bright orange-red light, commonly used in advertising signs.

neoplasm *See* TUMOUR.

nephron The functional unit of the KIDNEY. A nephron is a filtering unit forming URINE. There are over a million nephrons in the human kidney.

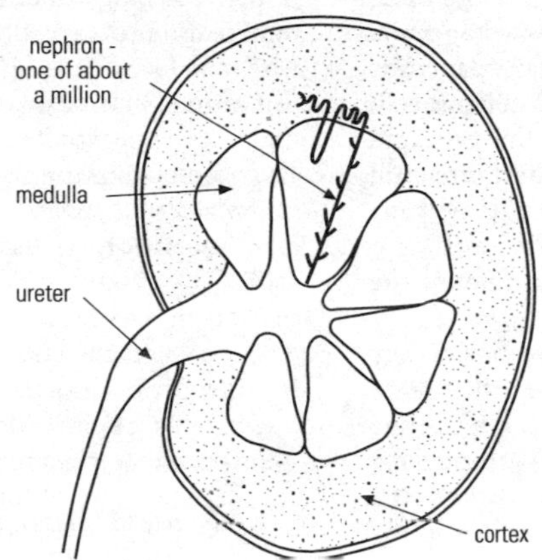

Above: Mammalian kidney showing the position of a nephron.
Below: Regions of the nephron.

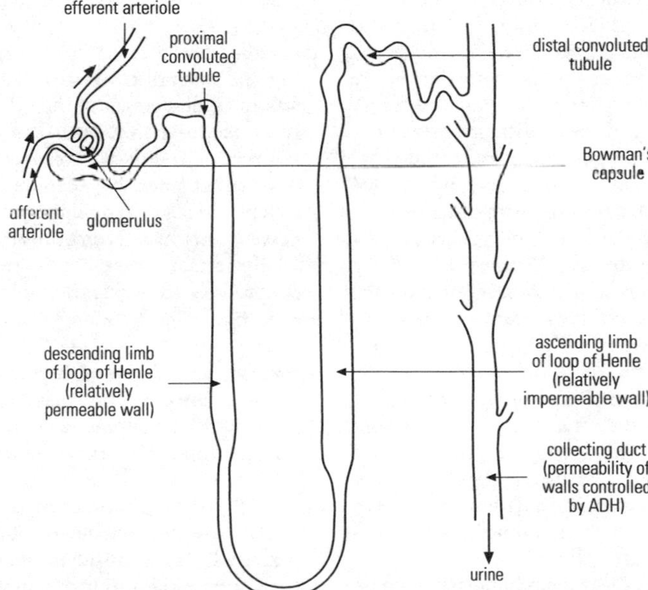

A nephron is made up of a tight knot of blood capillaries called the GLOMERULUS, which is surrounded by a cup-shaped structure called the BOWMAN'S CAPSULE. This capsule forms part of a long tubule that has several clearly defined regions. The tubule extends from the Bowman's capsule along the proximal convoluted tubule, to a long narrow collecting tubule called the LOOP OF HENLE. The distal convoluted tubule links the loop at the other side to a collecting duct. Blood containing waste materials passes over the Bowman's capsule and useful minerals and water are reabsorbed back into the blood. Over 80 per cent of reabsorption occurs within the proximal convoluted tubule. The cells here are well suited to reabsorption because they have MICROVILLI, to increase the surface area, and contain many MITOCHONDRIA, to provide the ATP needed for ACTIVE TRANSPORT. Food substances, water and sodium are reabsorbed by the proximal convoluted tubule. Further reabsorption and concentration of the filtrate occurs in the loop of Henle. The remaining fluid passes through the distal convoluted tubule to the collecting duct and then to the URETER as urine.

The distal convoluted tubule controls the PH of the blood and urine by adjusting excretion and retention of hydrogen (H^+) and hydrogen carbonate (HCO^{3-}) ions. ANTIDIURETIC HORMONE controls the permeability of the walls of the distal convoluted tubule and the collecting duct, and hence the degree to which water is drawn out from them by OSMOSIS, which occurs due to the high salt concentration in the inner (MEDULLA) region of the kidney.

Neptune The eighth most distant planet from the Sun, with an orbital radius of 30.1 AU (4,500 million km). It is the smallest and most distant of the GAS GIANTS, with a diameter of 48,000 km (3.8 times that of the Earth) and a mass of 1.0×10^{26} kg (17 times that of the Earth). Neptune takes 165 years to orbit the Sun, and rotates on its own axis in 16 days.

Little was known about Neptune until it was visited by the VOYAGER spacecraft in 1989. Images taken then discovered six new satellites, bringing the total to eight, and a faint RING SYSTEM. The atmosphere is composed largely of methane and hydrogen. Strong winds were detected, and a dark spot similar to Jupiter's GREAT RED SPOT, though this may be a temporary feature.

Historically, Neptune is important because its existence was predicted in 1846, ahead of its discovery, by calculations based on unexplained disturbances in the orbits of other planets.

neptunium (Np) The element with ATOMIC NUMBER 93. The longest lived ISOTOPE (neptunium–237) has a HALF-LIFE of 2 million years. Neptunium is produced in many nuclear reactors from the BETA DECAY of uranium–239, which is formed when the common isotope uranium–238 is struck by a neutron. The beta decay of neptunium produces plutonium. Chemically, neptunium is a reactive member of the ACTINIDE series of metals.

Nernst equation An equation that enables a REDOX POTENTIAL under non-standard conditions to be calculated from STANDARD REDOX POTENTIALS. If the standard redox potential is $E^{\ominus}$, then if the ABSOLUTE TEMPERATURE is T and the number of electrons transferred per reaction is z, the new redox potential E will be

$$E = E^{\ominus} + (kT/ze) \ln([\text{O.A.}]/[\text{R.A.}])$$

where e is the charge on an electron, k is the BOLTZMANN CONSTANT, and [O.A.] and [R.A.] are the concentrations of the oxidized and reduced (*see* OXIDATION, REDUCTION) forms of the reagents, respectively.

Nernst heat theorem Any chemical change involving pure crystalline materials, and taking place at ABSOLUTE ZERO will involve no change in ENTROPY.

nerve A bundle of NEURONES and GLIAL CELLS together with their associated CONNECTIVE TISSUE and blood vessels surrounded by a connective tissue sheath. Nerves can contain either sensory or motor neurones or a mixture of both (mixed nerves). Each neurone conducts independently of its neighbour. Nerves are of varying lengths but can be nearly as long as the whole animal.

nerve cell *See* NEURONE.

nerve fibre The AXON of a NEURONE and its MYELIN SHEATH if present.

nerve impulse A wave of chemical and electrical changes, affecting the membrane of a NEURONE, which passes along NERVE FIBRES to relay information rapidly between different parts of the body.

All living cells have a POTENTIAL DIFFERENCE across their membranes (the membrane potential) that is caused by the distribution of four ions, sodium (Na^+), potassium (K^+), chloride (Cl^-) and organic ANIONS (COO^-). In a normal state the membrane of a neurone is negatively charged inside with respect to the outside (resting potential) and the membrane is said to be polarized (*see* POLARIZE). Following appropriate stimulation above a threshold value, sodium and potassium ions move across the membrane by DIFFUSION or CATION PUMPS, and the membrane becomes positively charged. This is called an ACTION POTENTIAL and the membrane is said to be depolarized. This is an all or none response. Once an action potential has been received by the nerve cell body, it travels quickly along the AXON to a SYNAPSE, which is linked either to other neurones or EFFECTOR cells (e.g. muscles). Here a NEUROTRANSMITTER is released that diffuses across the synapse and stimulates either an impulse through another nerve cell or the action of an effector cell. Very quickly the membrane returns to its resting potential and is thus repolarized.

The speed of transmission of an impulse can vary from 1 to 100 $cm s^{-1}$. There is a refractory period after an action potential, which separates a second impulse from the first.

nerve net A network of NERVE cells that forms a primitive type of NERVOUS SYSTEM in some simple invertebrates

nervous system All of the cells forming nervous tissue, the NEURONES, NERVES, RECEPTORS and GLIAL CELLS, that detect and relay information about an animal's internal and external environment and co-ordinate a response. All animals have a nervous system derived from embryonic ECTODERM, although the nervous system can vary from being very simple, as in the 'nerve net' of jellyfish and sponges, to very complex, as in humans.

The nervous system of humans can be divided into a number of parts. (i) The SENSORY SYSTEM is formed from receptors collecting information from the internal and external environment and the sensory neurones carrying this information to the CENTRAL NERVOUS SYSTEM, where it is processed. (ii) The central nervous system (CNS) receives information from sensory neurones, interprets and sends messages to motor neurones to stimulate the appropriate action. In vertebrates, the CNS consists of a BRAIN and SPINAL CORD enclosed and protected by the spinal column surrounded by three membranous coverings called the MENINGES. (iii) The EFFECTOR SYSTEM transmits the information received from the CNS to effectors, which act upon it.

See also AUTONOMIC NERVOUS SYSTEM, REFLEX.

network In computing, a collection of computers linked together so they can communicate with each and share PERIPHERAL DEVICES. *See* ETHERNET, INTERNET, LOCAL AREA NETWORK, PACKET SWITCHING, TOKEN RING NETWORK, WIDE AREA NETWORK.

neural network An artificial network of processors that tries to mimic the transmission of impulses in the human brain. A neural network may be an electronic or optical construction or a computer simulation.

Small processors (nodes) are connected together in a manner rather like the network of NEURONES (nerve cells) in the brain. Each node sends a signal or remains silent, depending on the signal it receives from other nodes. The strength of the output depends on that of the input, and each signal can be given a weight. Neural networks can be taught to recognize complex patterns by adjusting the weight of the signals.

Neural networks are used in research in artificial intelligence, and some have been taught to recognize handwriting, to search for patterns in stock market movements and to recognize and distinguish between complex objects, such as faces.

neurilemma The outer covering of a NERVE FIBRE.

neuroendocrine cell *See* NEUROSECRETORY CELL.

neurogenic (*adj.*) Originating in or stimulated by the NERVOUS SYSTEM or NERVE IMPULSES.

neuromuscular junction The point at which a MOTOR NERVE meets a MUSCLE and a SYNAPSE forms. The muscle membrane under the nerve is called the endplate. A NERVE IMPULSE releases a jet of ACETYLCHOLINE, which depolarizes the endplate and generates an ACTION POTENTIAL that travels along the muscle, causing it to contract (*see* MUSCULAR CONTRACTION).

neurone, *nerve cell* A major cell type of the NERVOUS SYSTEM specialized to transmit information rapidly between different parts of the body, in the form of NERVE IMPULSES. The cell body of a neurone consists of a nucleus and cytoplasm and one or more short projections called DENDRITES (branching from a DENDRON), which conduct impulses towards the neurone. A longer extension (usually only one) called the AXON conducts impulses away from the cell body of the neurone to the SYNAPSE, which links with another nerve cell or an EFFECTOR cell such as muscle. Sensory neurones carry impulses towards the CENTRAL NERVOUS SYSTEM; relay neurones are found within the central nervous system and connect with motor neurones, which carry impulses away from the central nervous system to an EFFECTOR cell (muscles or glands). Neurones are bundled together by CONNECTIVE TISSUE which, with associated blood vessels, forms NERVES.

neurosecretory cell, *neuroendocrine cell* Any cell of the NERVOUS SYSTEM that can both conduct NERVE IMPULSES and secrete HORMONES. Neurosecretory cells are found, for example, in the HYPOTHALAMUS. *See also* ENDOCRINE SYSTEM.

neurotransmitter A chemical of low MOLECULAR MASS that is released at a SYNAPSE and transmits impulses between NEURONES or between neurones and EFFECTORS. Neurotransmitters are stored in vesicles located at the end of a nerve AXON and are released upon arrival of a NERVE IMPULSE. The neurotransmitter then diffuses across the synapse to RECEPTORS on the postsynaptic cell (*see* SYNAPSE). Neurotransmitters can be excitatory, generating an ACTION POTENTIAL, or they can be inhibitory. Neurotransmitters are inactivated, often by enzymes, to ensure that impulses do not merge at the synapse.

About 50 neurotransmitters are known, including ACETYLCHOLINE, NORADRENALINE, ADRENALINE, ENDORPHINS and ENCEPHALINS. A number of drugs exist that can mimic neurotransmitters, for example amphetamines mimic the action of noradrenaline; nicotine mimics natural neurotransmitters. Other drugs affect the release of neurotransmitters, for example caffeine increases release, whereas beta-blockers inhibit release.

neutral 1. (*adj.*) Describing an object with no CHARGE or that contains equal amounts of POSITIVE and NEGATIVE charge.

2. In a mains electricity system, a wire that remains close to EARTH potential throughout the a.c. cycle (*see* ALTERNATING CURRENT).

neutral equilibrium A state where the displacement of a system produces no forces on it, for example a ball lying on a flat surface.

neutralization The process of an ACID reacting with a BASE to form a SALT plus water, for example:

$$HCl + KOH \rightarrow KCl + H_2O$$

Such reactions are all based on the formation of a water molecule from a hydrogen ion and a hydroxyl ion:

$$H^+ + OH^- \rightarrow H_2O$$

neutral mutation *See* MUTATION.

neutrino Any one of a family of three light (probably massless) neutral LEPTONS. They are the electron neutrino, the muon neutrino and the tau neutrino. Each carries the same ELECTRON NUMBER, muon number, etc., as whichever charged lepton it is associated with. Neutrinos have no charge, probably zero rest mass, and move at the speed of light. Each neutrino has an associated antineutrino, the most familiar of which, the anti particle of the electron neutrino, is emitted from the nucleus during BETA DECAY.

Neutrinos interact only via the WEAK NUCLEAR FORCE, so even in solid rock, they have a MEAN FREE PATH of many LIGHT YEARS. Despite this, neutrino interactions have been observed, including those from beta decay in NUCLEAR REACTORS, from the sun and from SUPERNOVA explosions.

neutron The neutral particle found in the nuclei of all elements except hydrogen. It is slightly more massive than the PROTON. Outside the nucleus, the neutron has a mean life of about 12 minutes, before decaying into a proton, an electron and an antineutrino (*see* NEUTRINO). *See also* ATOM, HADRON, ISOTOPE, QUARK, MASS NUMBER.

neutron degeneracy pressure The mechanism that supports NEUTRON STARS against further gravitational collapse. It is caused by the PAULI EXCLUSION PRINCIPLE acting on NEUTRONS in much the same way as it produces ELECTRON DEGENERACY PRESSURE. The densities at which neutron degeneracy pressure acts are those associated with the atomic nucleus, about 10^{15} times the density of ordinary solid matter.

neutron diffraction A technique similar to X-RAY DIFFRACTION, but using low energy NEUTRONS, which have a DE BROGLIE WAVELENGTH similar to the X-rays they replace. As the neutrons are scattered by the nuclei rather than by the electrons of the crystal, this technique can be used to obtain further information about crystal structures. *See also* WAVE NATURE OF PARTICLES.

neutron star An extremely dense collapsed star. If the active core of a star (which is about 10 per cent of its total mass) exceeds 1.4 solar masses, the NUCLEAR FUSION goes beyond iron and energy starts to be absorbed by further fusion rather than released. This accelerates rather than opposes further collapse and the core shrinks rapidly. Protons and electrons fuse to form neutrons and the collapse is only halted by NEUTRON DEGENERACY PRESSURE with a density roughly 10^{15} times that of ordinary matter. The outer layers rebound off this very rigid core and the resulting explosion, in which the star suddenly increases its brightness many millions of times, is called a SUPERNOVA. The remnant of the core is called a neutron star. Neutron stars are often observed as PULSARS.

neutrophil A type of GRANULOCYTE (blood cell) with cytoplasmic granules that do not take up acid or basic dyes. The majority of granulocytes are of this type.

newton The derived SI UNIT of FORCE. Force is equal to the mass times the acceleration, so one newton is defined as the force that will make a mass of one kilogram accelerate at one metre per second per second.

Newtonian mechanics MECHANICS based on NEWTON'S LAWS OF MOTION, and not taking account of the effects of QUANTUM MECHANICS or RELATIVITY. Newtonian mechanics generally works well provided the systems being considered are large compared to the size of an atom and move at speeds small compared to the speed of light.

Newton's law of gravitation The gravitational attraction between two objects is proportional to their masses and inversely proportional to the square of the distance between their respective centres of mass. The gravitational force F between two masses m and M separated by a distance r is

$$F = GMm/r^2$$

where G is the GRAVITATIONAL CONSTANT.

Newton's laws of motion Three laws that together sum up the behaviour of objects under the influence of FORCES acting upon them. These laws were first put forward by Isaac Newton (1642–1727) in his famous work the *Principia* (1687). They formed the basis of the study of motion until the early years of the 20th century, when the SPECIAL THEORY OF RELATIVITY suggested that Newton's laws are an approximation to the truth, and are most valid for particles moving at speeds much less than the speed of light.

Newton's first law states that an object will remain at rest, or continue to move in a straight line at a steady speed, unless it is acted upon by an unbalanced external FORCE. In other words if the vector sum of the forces acting on an object is zero, it will not accelerate.

Newton's second law states that the rate of change of MOMENTUM of a body is directly proportional to the FORCE on it. If the object has a constant MASS, this is equivalent to saying that the force is directly proportional to the mass times the ACCELERATION. This introduces the idea of INERTIAL MASS as a measure of the amount of matter in an object: the greater the mass the less the motion of the object will be changed by a given force. In other words, mass is a measure of the INERTIA of a body, or its resistance to having its motion changed. For an object of mass m experiencing a force F, the acceleration will be a, with

$$F = ma$$

Newton's third law is usually stated as 'for every action there is an equal and opposite reaction', i.e. if object A exerts a force (the action) on object B , then object B will exert an equal but opposite force (the reaction) on object A.

Newton's rings A THIN-FILM INTERFERENCE effect seen when a CONVEX lens is placed on a flat glass plate, producing circular interference fringes. The effect was first described in 1704 by Isaac Newton (1642–1727). It arises from INTERFERENCE between light which is partially reflected from the bottom of the lens and the top of the glass plate.

niacin *See* VITAMIN B.

Nicad *See* NICKEL-CADMIUM CELL.

nickel (Ni) The element with ATOMIC NUMBER 28; RELATIVE ATOMIC MASS 58.7; melting point 1,450°C; boiling point 2,840°C; RELATIVE DENSITY 8.9. Nickel is a hard metal, and is fairly unreactive, which has led to its use as an electrolytically deposited coating on some steel items (*see* ELECTROPLATING). It is also used in the manufacture of NICKEL-CADMIUM CELLS. Nickel is widely used as a catalyst in a number of reactions in the petrochemical industry, such as HYDROGENATION. Nickel salts often have a characteristic pale green colour. Along with iron, nickel is the major constituent of the Earth's core and is also found in some METEORITES.

nickel-cadmium cell (Nicad) A high capacity rechargeable electrochemical CELL, often used to power portable electrical appliances.

nickel carbonyl ($Ni(CO)_4$) A colourless liquid; melting point –25°C; boiling point 43°C. Nickel carbonyl is a CO-ORDINATION COMPOUND, formed by passing carbon monoxide gas over the powdered metal. The reaction is reversed at higher temperatures leading to the decomposition of the compound.

nickel oxide Nickel(II) oxide, NiO, or nickel(III) oxide (nickel peroxide), Ni_2O_3. Nickel(II) oxide is a green solid; decomposes on heating; RELATIVE DENSITY 6.6. It can be made by heating nickel nitrate in the absence of air:

$$2Ni(NO_3)_2 \rightarrow 2NiO + 4NO_2 + O_2$$

Nickel(III) oxide is a grey solid; decomposes on heating; relative density 4.8. It can be made by burning nickel in air, or by heating nickel(II) oxide in air:

$$4Ni + 3O_2 \rightarrow 2Ni_2O_3$$

$$4NiO + O_2 \rightarrow 2Ni_2O_3$$

nicotinamide adenine dinucleotide *See* NAD.

nicotinamide adenine dinucleotide phosphate *See* NADP.

nicotinic acid *See* VITAMIN B.

niobium (Nb) The element with ATOMIC NUMBER 41; RELATIVE ATOMIC MASS 92.9; melting point 2,468°C; boiling point 4,742°C; RELATIVE DENSITY 8.6. Niobium is a TRANSITION METAL, used in the manufacture of some special steels and in niobium-tin alloy. This alloy is the conductor in many superconducting magnets (*see* SUPERCONDUCTIVITY), as it retains its superconductivity up to the relatively high temperature of 22 K and is able to remain superconducting in quite strong magnetic fields.

nitrate Any salt containing the nitrate ion, NO_3^-. Nitrates form stable crystals containing WATER OF CRYSTALLIZATION. Most nitrates are soluble and many are important fertilizers since the nitrate ion is an important source of fixed nitrogen (*see* NITROGEN FIXATION). *See also* NITROGEN CYCLE.

nitration In organic chemistry, the introduction of a nitro-group (NO_2) to ARENES. This is achieved using a mixture of concentrated nitric and sulphuric acids, which produces the nitronium ion (NO_2^+), a strong ELECTROPHILE. The reactions are ELECTROPHILIC SUBSTITUTIONS. Nitration is an important step in industry in the production of a wide range of compounds. For example the nitronium ion reacts with benzene to give nitrobenzene.

nitric acid (HNO_3) An important MINERAL ACID; melting point –42°C; boiling point 83°C; RELATIVE DENSITY 1.5. Nitric acid is manufactured by the action of concentrated sulphuric acid on NITRATES, for example:

$$KNO_3 + H_2SO_4 \rightarrow KHSO_4 + HNO_3$$

This produces concentrated nitric acid as a gas. An alternative process is the OXIDATION of ammonia by excess air with a platinum-rhodium catalyst:

$$NH_3 + 2O_2 \rightarrow HNO_3 + H_2O$$

In practice this reaction proceeds via nitrogen oxide, NO, some of which is oxidized to nitrogen dioxide, NO_2, which dissolves in water to form nitric acid, with the remaining NO being recycled. The acid formed in this process is dilute, but can be concentrated by DISTILLATION up to 68.5 per cent. Nitric acid is an important FEEDSTOCK in the fertilizer, dyestuff and explosives industries.

nitric oxide *See* NITROGEN MONOXIDE.

nitride Any BINARY COMPOUND containing NITROGEN. The ALKALI METALS and ALKALINE EARTHS form nitrides on heating. These contain the ion N^{3-}, but are fairly unstable. They are hydrolysed (*see* HYDROLYSIS), for example, to form the metal HYDROXIDE and ammonia:

$$Na_3N + 3H_2O \rightarrow 3NaOH + NH_3$$

nitrification The process occurring in the soil by which ammonia (from UREA, URINE and the break down of protein by AMMONIFICATION) is oxidized by bacteria to form NITRATES. The free-living chemosynthetic (*see* CHEMOSYNTHESIS) bacterium *Nitrosomonas* oxidizes ammonium ions (NH_4^+) to NITRITES (NO_2^-), which are toxic but quickly oxidized to nitrates (NO_3^-) by *Nitrobacter*. These processes release energy, which the bacteria use for their own respiratory processes. Nitrification is reduced if soil temperature and pH are low. Because nitrates are soluble, they can easily leach out of the soil, causing nitrogen deficiency, so artificial fertilizers are often added to prevent growth limitation. *See also* NITROGEN CYCLE, DENITRIFICATION.

nitrile An organic compound containing a carbon-nitrogen TRIPLE BOND, an organic CYANIDE. The general formula is RCN, where R is a HYDROCARBON group, for example CH_3CN, ethanenitrile. Nitriles are considered to be CARBOXYLIC ACID derivatives. Nitriles are formed by dehydration (*see* DEHYDRATE) when AMIDES are heated with phosphorus(V) oxide:

$$CH_3CONH_2 \xrightarrow[-H_2O]{P_2O_5} CH_3C{\equiv}N$$

or by the reaction of a HALOGENOALKANE with sodium/potassium cyanide in ethanol:

$$C_2H_5Br + CN^- \rightarrow C_2H_5CN + Br^-$$

When reacted with acid or alkali, nitriles are hydrolysed to the corresponding carboxylic acid. They are reduced to primary AMINES. Nitriles are useful as intermediates in the synthesis of certain organic compounds, particularly in providing a way of adding an extra carbon atom to a chain. They have other uses such as insect repellents, weed control and fuel additives.

nitrite Any salt containing the nitrite ion, NO_2^-. Nitrites are easily oxidized (*see* OXIDATION) to NITRATES.

nitrobenzene ($C_6H_5NO_2$) A colourless liquid; boiling point 211°C; melting point of 6°C.

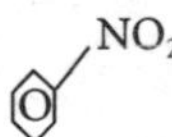

Nitrobenzene is produced by reacting benzene with a mixture of sulphuric and nitric acids. The NITRONIUM ION is an intermediate in this process. Most of the nitrobenzene manufactured is used in the dyestuffs industry, either as nitrobenzene or reduced to PHENYLAMINE.

nitrogen (N) The element with ATOMIC NUMBER 7; RELATIVE ATOMIC MASS 14.0; melting point –210°C; boiling point –196°C. Nitrogen gas makes up 78 per cent of the atmosphere, but is chemically fairly unreactive due to the TRIPLE BOND in the N_2 molecule ($N{\equiv}N$).

Nitrogen is an essential element for life, being present in all proteins and nucleic acids. Some bacteria are able to 'fix' nitrogen from the air and incorporate it into the growth of certain plants (*see* NITROGEN CYCLE, NITROGEN FIXATION). NITRATES are often used as a fertilizer since this is a form of nitrogen that can be readily used by many organisms.

Nitrogen is obtained commercially from the FRACTIONAL DISTILLATION of liquefied air and is used to manufacture AMMONIA (*see* HABER PROCESS). The gas is also used to provide an inert environment in welding and metallurgy. Liquid nitrogen is commonly used to cool other objects to low temperatures.

nitrogen cycle (*See also* diagram on following page.) The circulation of nitrogen, mostly by living organisms, through the ECOSYSTEM. Nitrogen is an essential mineral for all organisms because it is used to make proteins and other organic compounds. Although the atmosphere is 78 per cent nitrogen, this cannot be readily used by most organisms (*see* NITROGEN FIXATION). Plants obtain nitrogen from NITRATES in the soil, by absorption through the roots, and convert them to proteins. The proteins are passed to HERBIVORES and CARNIVORES in the FOOD CHAIN, and nitrogen is eventually returned to the soil as excrement or when organisms die.

There are several important groups of bacteria involved in the processes of the nitrogen cycle. The free-living chemosynthetic (*see* CHEMOSYNTHESIS) bacteria *Nitrosomonas* and *Nitrobacter* are important in NITRIFICATION, which is the process by which ammonia is oxidized to form nitrates. Bacteria that can utilize atmospheric nitrogen by nitrogen fixation to make nitrogenous compounds are called nitrogen-fixing bacteria.

Anaerobic bacteria, such as *Pseudomonas denitrificans*, and *Thiobacillus denitrificans* are also important because they convert nitrates in the soil back to atmospheric nitrogen in the process of DENITRIFICATION. DECOMPOSERS (organisms capable of feeding on excrement and other dead organisms) are crucial in the nitrogen cycle for breaking down proteins, amino acids and other nitrogenous compounds to form ammonia, ammonium ions (by AMMONIFICATION) and amines (by putrefaction), which are used by the nitrifying bacteria.

nitrogen dioxide *See* DINITROGEN TETROXIDE.

nitrogen fixation The process by which atmospheric nitrogen is converted to nitrogenous compounds by the action of nitrogen-fixing bacteria. These bacteria can be free-living bacteria, such as *Azotobacter* and *Clostridium*, or CYANOBACTERIA, such as *Nostoc*. They convert atmospheric nitrogen to ammonia, which they use to make amino acids.

Nitrogen fixation is also carried out by symbiotic bacteria such as *Rhizobium* (*see*

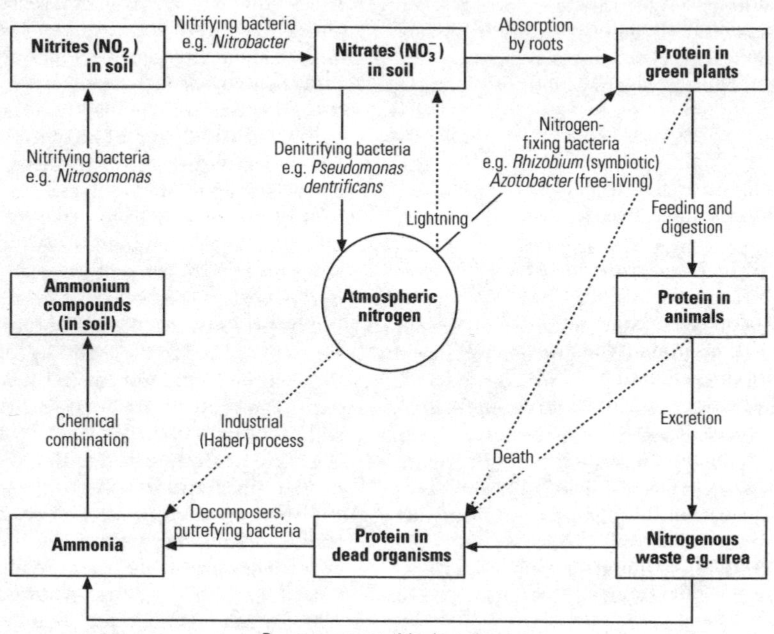

A summary of the nitrogen cycle.

SYMBIOSIS) that live in specialized ROOT NODULES on the roots of leguminous plants, for example beans and peas, and a few non-leguminous plants. LEGUMES are therefore important crops for improving soil fertility.

See also NITROGEN CYCLE.

nitrogen hydride *See* AMMONIA.

nitrogen monoxide (NO) A colourless gas; melting point –164°C; boiling point –152°C. Nitrogen monoxide can be prepared by the reaction of sodium nitrite with sulphuric acid and iron(II) sulphate:

$$2NaNO_2 + 2H_2SO_4 + 2FeSO_4 \rightarrow 2NO + Na_2SO_4 + Fe_2(SO_4)_3 + 2H_2O$$

Nitrogen monoxide is readily oxidized (*see* OXIDATION) to dinitrogen tetroxide:

$$2NO + O_2 \rightarrow N_2O_4$$

nitroglycerine ($C_3H_5(ONO_2)_3$) A colourless, odourless oil produced by treating GLYCEROL with a mixture of nitric and sulphuric acids. It is a powerful explosive used in the preparation of dynamite. Nitroglycerine is also used in the treatment of the chest complaint angina.

nitro-group The NO_2 group.

nitronium ion (NO_2^+) A strongly electrophilic ion (*see* ELECTROPHILE) produced by the reaction of concentrated nitric and sulphuricacids:

$$HONO_2 + 2(HO)_2SO_2 \rightarrow NO_2+ + H_3O+ + 2HOSO_3^-$$

The nitronium ion is important in NITRATION reactions.

nitrous acid (HNO_2) A weak acid, formed by the action of strong acids on nitrites, for example:

$$Ba(NO_2)_2 + H_2SO_4 \rightarrow 2HNO_2 + BaSO_4$$

Nitrous acid is stable only as a gas or a dilute aqueous solution. The aqueous acid dissociates on heating to give nitric acid and nitrogen monoxide:

$$3HNO_2 \rightarrow HNO_3 + 2NO + H_2O$$

nitrous oxide *See* DINITROGEN OXIDE.

NK cell *See* NATURAL KILLER CELL.

NMR An abbreviation for NUCLEAR MAGNETIC RESONANCE.

nobelium (No) The element with ATOMIC NUMBER 102. It does not occur naturally, since the longest lived ISOTOPE (nobelium–259) has a HALF-LIFE of only 3 minutes.

noble (adj.) An archaic term describing a metal or gas that is unreactive. *See* NOBLE GAS, NOBLE METAL.

noble gas, *inert gas, rare gas* Any of the elements HELIUM, NEON, ARGON, KRYPTON, XENON and RADON that occupy GROUP 18 (formerly group VIII or group 0) of the PERIODIC TABLE. They are characterized by having full VALENCE SHELLS of electrons. Thus they do not readily form IONS or COVALENT BONDS, though some will form compounds with fluorine.

Helium was discovered first in the spectroscopic analysis of the Sun. Apart from hydrogen, it is by far the most abundant element in the universe. It is, however, relatively rare on Earth as it is does not combine chemically to form a solid and as a gas the mean speed of its atoms is high enough for the gas to escape quite rapidly from the Earth's atmosphere. Some NATURAL GAS sources contain significant amounts of helium, believed to be a result of ALPHA PARTICLE production in nearby uranium deposits.

The other noble gases (except RADON) were discovered by spectroscopic analysis of the products of FRACTIONAL DISTILLATION of liquid air. Argon is by far the most abundant and is used to provide an inert atmosphere for example, in light bulbs, and in welding easily oxidized materials such as aluminium.

See also SHELL.

noble metal An archaic term for a metal that does not react with the common MINERAL ACIDS. Gold and platinum are examples. Such metals are used in jewellry as they do not TARNISH, and are also used in electrical contacts as they resist OXIDATION. *See also* BASE METAL.

node 1. (*mathematics*) A point at which two or more lines meet, or where a curve intersects itself.

2. (*physics*) A point of minimum AMPLITUDE in a STANDING WAVE. The distance from one node to the next is half the WAVELENGTH of the wave.

3. (*biology*) A swelling or lump. In zoology, an example is a LYMPH NODE; in botany, an example is a region on a plant stem from which leaves develop.

node of Ranvier *See* MYELIN SHEATH.

noise Any unwanted sound and, by extension, any unwanted signal, such as in an electronic system. Noise that covers a broad spread of frequencies is called white noise.

non-disjunction The failure of one or more pairs of CHROMOSOMES to separate at MEIOSIS. *See also* DOWN'S SYNDROME.

non-Euclidean space *See* EUCLIDEAN SPACE.

non-inertial frame of reference Any one of two or more CO-ORDINATE systems where an observer stationary with respect to one of the co-ordinate systems would see an observer stationary with respect to any of the other co-ordinate systems as accelerating.

non-inverting amplifier An AMPLIFIER with a positive GAIN.

non-inverting input One of the inputs of a DIFFERENTIAL AMPLIFIER. Signals applied to this input will be amplified with a positive GAIN.

nonmetal Any element that is not a METAL. Nonmetals are poor conductors of electricity and tend to have low melting points; many of them are gases at room temperature. Nonmetals are electronegative (*see* ELECTRONEGATIVITY) and tend to form negative ions or COVALENT BONDS. The OXIDES of nonmetals generally dissolve in water to form acids.

non-reducing sugar A sugar that cannot act as a REDUCING AGENT in solution, as indicated by a negative BENEDICT'S TEST or FEHLING'S TEST. *See also* CARBOHYDRATE.

non-renewable (*adj.*) A term used to describe FOSSIL FUELS and other energy sources that are being consumed at a rate which far exceeds the production of new reserves. *See also* RENEWABLE RESOURCE.

non-return valve *See* VALVE.

nonsense codon *See* STOP CODON.

nonsense mutation *See* MUTATION.

nonviscous flow In HYDRODYNAMICS and AERODYNAMICS, an approximation in which the effects of VISCOSITY are neglected. Nonviscous flow is assumed in BERNOULLI'S THEOREM. *See also* SUPERFLUIDITY.

nonvolatile memory Computer MEMORY in which data stored in the memory is not lost when the computer is switched off. *See also* BUBBLE MEMORY.

noradrenaline, *norepinephrine* A HORMONE and NEUROTRANSMITTER that is, like ADRENALINE, derived from the amino acid tyrosine. It is secreted by the MEDULLA of the ADRENAL GLAND and by modified NEURONES of the sympathetic NERVOUS SYSTEM (*see* AUTONOMIC NERVOUS SYSTEM). Noradrenaline maintains arousal in the brain, for example in response to external stress, dreaming and emotion.

norepinephrine *See* NORADRENALINE.

NOR gate (NOR = NOT OR) A LOGIC GATE that implements the NOR function. A NOR gate has two or more inputs and a single output that is HIGH only if all the inputs are LOW.

normal An imaginary line at right angles to a surface.

normal distribution, *Gaussian curve* In statistics, a bell-shaped curve obtained when the frequency distribution for a characteristic that shows continuous variation is plotted on a graph. It is the PROBABILITY distribution for random events centred on a MEAN $<x>$ and having a STANDARD DEVIATION σ, such that the probability of finding a value x is proportional to $\exp[(x - <x>)^2/\sigma]$. An example is the height of individuals in a population; most individuals are of intermediate height, with a few at each extreme. *See also* AVERAGE.

normal reaction The COMPONENT of the force of contact between the two surfaces that is at right angles to those surfaces.

normal salt A SALT formed when all the hydrogen available in an ACID (*see* ACIDIC HYDROGEN) has been replaced by metal ions. An example is sodium sulphate, Na_2SO_4, compared to sodium hydrogensulphate, $NaHSO_4$, which is an ACIDIC SALT. *See also* BASIC SALT.

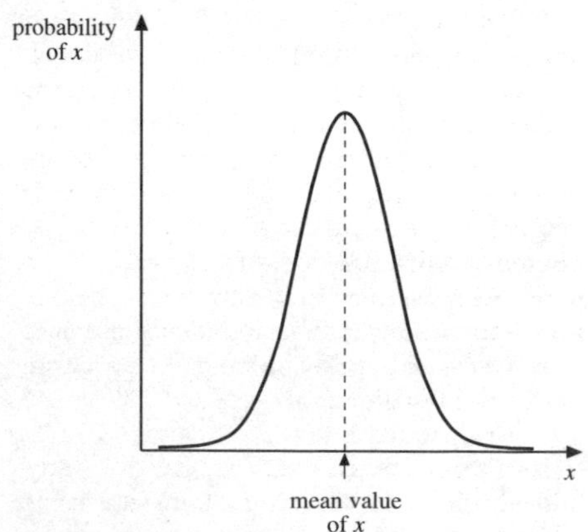

Normal distribution.

normal solution An obsolete term for a solution that has a CONCENTRATION such that one decimetre cubed of the solution contains, or will react with, one MOLE of hydrogen atoms. Thus a normal solution of hydrochloric acid, H_2SO_4, or copper sulphate, $CuSO_4$ will have a concentration of 0.5 mol dm^{-3}, whilst a normal solution of hydrochloric acid, HCl, will have a concentration of 1 mol dm^{-3}. The normality of a solution is often represented by the letter N, thus 2N indicates a solution of twice normal concentration.

norplant A long-lasting contraceptive preparation (*see* CONTRACEPTION) that is surgically placed under the skin and releases female hormones for up to 5 years.

northern lights *See* AURORA.

nose The SENSE ORGAN for smell and an opening of the respiratory tract. There are numerous olfactory RECEPTORS detecting smell that are found in the roof of the nasal cavity within the MUCOUS MEMBRANE lining the whole cavity. This membrane moistens and warms the air entering the nose and traps dirt. Small hairs inside the nostrils also prevent the entry of foreign objects. A septum of CARTILAGE divides the external part of the nose.

NOT gate *See* INVERTER.

notochord A stiff, but flexible, rod of tissue that exists between the DIGESTIVE SYSTEM and CENTRAL NERVOUS SYSTEM of all CHORDATES at some stage in their life. In most vertebrates the notochord occurs in the embryo and is replaced in adults by the VERTEBRAL COLUMN.

npn transistor A JUNCTION TRANSISTOR in which the EMITTER and COLLECTOR are made of N-TYPE SEMICONDUCTOR, whilst the BASE is made of P-TYPE SEMICONDUCTOR.

n-type semiconductor A SEMICONDUCTOR in which the charge is carried predominantly by electrons rather than HOLES.

nucellus A mass of cells within the OVULE of a SEED PLANT. The nucellus is completely surrounded by the INTEGUMENT except for a small hole (MICROPYLE) at the tip. The cells of the nucellus differentiate and divide to form the EMBRYO SAC and the egg cell (the female GAMETE). The nucellus also provides nutrition for the developing ovule.

nuclear binding energy The energy released when an atomic NUCLEUS is formed from its constituent PROTONS and NEUTRONS. The binding energy is largest for the largest nuclei, but the binding energy per NUCLEON is greatest for iron–56. *See also* MASS DEFECT.

nuclear energy The energy released in processes involving the atomic NUCLEUS, particularly radioactive decay (*see* RADIOACTIVITY), NUCLEAR FUSION and NUCLEAR FISSION.

nuclear fallout The radioactive material produced after the NUCLEAR FISSION of material in a NUCLEAR WEAPON. Many of the elements produced are chemically active and may be incorporated into the FOOD CHAIN. *See also* NUCLEAR WINTER.

nuclear fission The splitting of a heavy atomic NUCLEUS into two smaller nuclei, with the emission of energy. Some nuclei show SPONTANEOUS FISSION, a form of radioactive decay (*see* RADIOACTIVITY), which takes place as an alternative to ALPHA DECAY. A more useful process is INDUCED FISSION, in which a nucleus, described as FISSILE, breaks up after absorbing a slow moving NEUTRON. The most important fissile ISOTOPES are uranium–238 and plutonium–239.

Because lighter nuclei contain a smaller proportion of neutrons than heavy nuclei, a number of neutrons, usually two or three, are produced in each fission. Lighter nuclei are also more tightly bound than heavy nuclei (*see* NUCLEAR BINDING ENERGY) and thus some energy is released in the process. Most of this energy is in the form of the KINETIC ENERGY of the neutrons.

In a NUCLEAR REACTOR, the energy of the neutrons is absorbed when they collide with a material called a MODERATOR, which contains light nuclei that do not absorb neutrons. The moderator becomes heated by the energy absorbed from the neutrons which, now moving more slowly, can go on to produce further fission and release more neutrons. The result is a CHAIN REACTION, where each step produces neutrons that can go on to produce further neutrons. In its uncontrolled form, this is the principle of the ATOMIC WEAPONS that were used against the Japanese by the Americans at the end of the Second World War.

See also THERMONUCLEAR REACTION.

nuclear fusion The bringing together of two light atomic nuclei to form a heavier one. Since the heavier nuclei, up to iron–56, are more tightly bound (*see* NUCLEAR BINDING ENERGY), energy will be released in this process. However, nuclei are positively charged so must be very fast moving to overcome the ELECTROSTATIC repulsion that tends to keep them apart. Thus nuclear fusion can only take place at extremely high temperatures. The right conditions exist for nuclear fusion in the cores of stars, and the fusion of hydrogen to helium is the mechanism by which the Sun produces the energy necessary for life on Earth. This fusion process involves the conversion of protons to neutrons, and therefore involves the WEAK NUCLEAR FORCE, so it happens much too slowly to be suitable as a source of energy for use on Earth.

In a HYDROGEN BOMB extreme temperatures are produced by triggering a NUCLEAR FISSION reaction. This fission explosion then triggers the fusion reaction, which releases energy so quickly that most of the nuclei have fused before the material of the bomb has time to expand. This method of triggering fusion is called INERTIAL CONFINEMENT.

A more likely method for the peaceful use of nuclear fusion is MAGNETIC CONFINEMENT. At the temperatures needed for fusion, matter is in the form of a PLASMA, a gas of electrons and nuclei. As this gas is made up of charged particles, they can be confined by magnetic fields in what is known as a magnetic bottle. Various arrangements of magnetic fields have been tried, but the most successful seems to be the TOKAMAK.

In the D-T REACTION, much of the energy released by the fusion is carried by the neutron, which, being uncharged, will escape the magnetic field and deposit its energy in the surroundings of the machine. One proposal is to surround the tokamak with a blanket of lithium – the neutron will react with the lithium nuclei to produce further TRITIUM, thus 'breeding' new fuel. The energy released by the neutron would be used to heat water to drive a steam TURBINE to generate electricity in the conventional manner.

See also COLD FUSION.

nuclear magnetic resonance (NMR) A technique widely used in organic chemistry to identify organic molecules and to determine their structures. The technique can also be used to examine living organs without destroying them, which has revolutionized the diagnosis of disease.

Some of the nuclei in a molecule possess the property of SPIN, which can be in one of two orientations. If a magnetic field is applied to the molecule these differences in spin cause a splitting of the nuclear ENERGY LEVELS. The molecule is then subjected to an additional weak, oscillating magnetic field. At a precise frequency the nuclear magnets resonate and it is this that is recorded and amplified.

The resonance frequencies of a particular element depend on its environment. Thus using NMR it is possible, for example, to detect three different types of hydrogen atoms in ETHANOL – those in CH_3, CH_2 and OH groups. In addition, information about molecular structure can be obtained by the coupling of nuclei.

See also MAGNETIC RESONANCE IMAGING.

nuclear magneton A unit of MAGNETIC MOMENT equal to 5.05×10^{-27} Am^2. *See* MAGNETON.

nuclear physics The branch of physics that concerns itself with the atomic NUCLEUS, in particular the ENERGY LEVELS of NUCLEONS within the nucleus, collisions between nuclei and the processes of RADIOACTIVITY.

nuclear reaction Any process in which a NUCLEUS is changed in some way. In particular, a process triggered by a nucleus being struck by some incoming particle, often a PROTON, NEUTRON or PHOTON of GAMMA RADIATION. Nuclear reactions differ from CHEMICAL REACTIONS in that the energies involved are often a significant fraction of the REST MASS energies of the nuclei, so the total rest mass present before and after the reaction may be significantly different. *See also* EQUIVALENCE OF MASS AND ENERGY, NUCLEAR FISSION, NUCLEAR FUSION, THERMONUCLEAR REACTION.

nuclear reactor A sealed vessel containing FISSILE material together with CONTROL RODS and a MODERATOR and some means of extracting the heat produced in the moderator as it absorbs the KINETIC ENERGY of the neutrons released in NUCLEAR FISSION.

A CHAIN REACTION is usually controlled by using control rods made of a neutron-absorbing material such as boron or cadmium. If the nuclear reactor is cooled by passing a suitable fluid through it, the heat extracted can be used to make steam for an electrical generator – this is the basis of the nuclear power station. The energy released by each fission is many million times greater than the energy released in any single chemical reaction, thus the energy yield for a given mass of fuel is far greater than could be obtained from conventional fuels such as coal or oil.

The nuclei produced by the fission process tend to have too many neutrons to be stable so are often radioactive. In the case of a NUCLEAR WEAPON this material is called NUCLEAR FALLOUT, in a power station it is the RADIOACTIVE WASTE.

See also FAST BREEDER REACTOR.

nuclear weapon Any device that uses the energy released from NUCLEAR FUSION (more common nowadays) or NUCLEAR FISSION, for destructive purposes. Nuclear weapons were originally designed as bombs to be dropped from aircraft, but now also exist in the form of missiles and artillery shells. *See also* NUCLEAR FALLOUT, NUCLEAR WINTER.

nuclear winter A hypothetical period of prolonged cold weather following a war in which widespread use is made of NUCLEAR WEAPONS. It has been suggested that such a war would put sufficient dust into the atmosphere to have significant effects on the climate.

nuclease A general term for an ENZYME that degrades NUCLEIC ACID. *See* RESTRICTION ENDONUCLEASE.

nucleic acid The complex organic acid present in the cells of all organisms that is responsible for their genetic make-up. The two types of nucleic acid are DNA and RNA, and each is made of long chains of NUCLEOTIDES.

nucleoid The region of a prokaryotic cell (*see* PROKARYOTE) that contains the genetic material in the form of CHROMOSOMES or PLASMIDS. In contrast to a NUCLEUS, a nucleoid is not bounded by a membrane.

nucleolus (*pl.* ***nucleoli***) A spherical body within the nucleus of a non-dividing eukaryotic cell (*see* EUKARYOTE). There may be one or more nucleoli present, and they contain RNA and protein and are concerned with the synthesis of RIBOSOMES. Nucleoli stain with basic dyes and their size reflects their level of activity.

nucleon A particle found in an atomic NUCLEUS: a NEUTRON or a PROTON.

nucleon number *See* MASS NUMBER.

nucleophile (Greek = nucleus loving) Any molecule or ion that forms a new COVALENT BOND by donating or sharing its electrons, such as the hydroxyl ANION (OH^-).

Nucleophiles are often negatively charged but do not need to be. For example, water is a nucleophile because of the LONE PAIR of electrons on the oxygen atom. A nucleophile is a LEWIS BASE.

nucleophilic substitution A SUBSTITUTION REACTION in which an atom or group in a molecule is replaced by a NUCLEOPHILE (which provides the electrons to form the new COVALENT BOND). The general reaction can be considered to be as follows:

$$R–X + :NUC \rightarrow R–NUC + X^-$$

where R is a HYDROCARBON, metal or METALLOID group and the nucleophile, NUC, and X can be a variety of organic or inorganic ANIONS.

There are two types of nucleophilic reactions: Sn1 (substitution, nucleophilic, unimolecular) and Sn2 (substitution, nucleophilic, bimolecular) reactions. In Sn1 reactions the R-X bond is broken first and the nucleophile then enters the reaction to provide the electrons to form the R-NUC bond. In Sn2 reactions the nucleophile forms the R-NUC bond at the same time as the R-X bond breaks (hence the reaction is bimolecular).

Nucleophilic substitution reactions are used to introduce various FUNCTIONAL GROUPS into organic molecules, for example into HALOGENOALKANES. In particular, primary halogenoalkanes usually react by nucleophilic substitution reactions.

nucleoside The organic base and PENTOSE sugar part of a NUCLEOTIDE, i.e. a nucleotide without the PHOSPHATE group. In RNA the sugar is RIBOSE and the common nucleosides are adenosine, guanosine, cytidine and uridine. In DNA the sugar is DEOXYRIBOSE and the nucleosides found are the same as RNA except that uridine is replaced by thymidine.

nucleotide The constituent unit of the nucleic acids DNA and RNA, which itself consists of an organic base, a PENTOSE sugar (RIBOSE $C_5H_{10}O_5$ or DEOXYRIBOSE $C_5H_{10}O_4$) and a PHOSPHATE group. There are five organic bases that are either PURINES (adenine and guanine), which have double rings (one with six sides, one with five), or PYRIMIDINES (cytosine and thymine or uracil), which have two single, six-sided rings. The organic bases are abbreviated to A, G, C, T or U, and the order in which they are placed in the nucleic acid strand contains the GENETIC CODE. The three components of a nucleotide join together in a CONDENSATION REACTION. Links then form similarly between the sugar and phosphate groups of two or more nucleotides, to form dinucleotides or polynucleotides. Although the main role of nucleotides is in the formation of nucleic acids, they are also found in other molecules, for example AMP, ADP, ATP, NAD, NADP and FAD.

nucleus (*pl.* ***nuclei***) **1.** (*physics*) The positively charged massive centre of an ATOM. It is made up of particles called PROTONS, which are positively charged, having a charge equal in size to the negative charge on an ELECTRON, and NEUTRONS, which have no charge and have slightly more mass than the protons. Protons and neutrons are collectively called NUCLEONS. The neutrons and protons are both far more massive than the electrons that surround the nucleus, but the nucleus is far smaller than the atom itself. A typical atom is 10^{-10} m in diameter, whilst a nucleus is 10^{-14} m across.

The number of protons in the nucleus determines the number of electrons needed to produce a neutral atom. It is the arrangement of these electrons that determines the chemical properties of an ELEMENT. Thus different numbers of protons in the nucleus produce atoms of different elements.

The number of neutrons in a nucleus has no affect on the chemical properties, but does affect the mass of the atom. The number of protons in a nucleus is called the ATOMIC NUMBER (Z). The total number of nucleons is called the MASS NUMBER (A). A NUCLIDE may therefore be represented by the notation $^{A}_{Z}X$ where X is the element.

See also LIQUID DROP MODEL, RUTHERFORD SCATTERING EXPERIMENT, SHELL MODEL.

2. (*biology*) A central dense body surrounded by a nuclear membrane or envelope, found in almost all eukaryotic cells (human RED BLOOD CELLS have no nucleus; *see* EUKARYOTE). They contain the genetic material of the cell in the form of CHROMOSOMES within a liquid nuclear sap. The nucleus controls all the activities of the cell, including cell division by MITOSIS or MEIOSIS. The nuclear membrane is a double membrane containing many pores, which allow large molecules such as RNA to pass between the nucleus and the CYTOPLASM. Within the nucleus of a non-dividing cell one or more nucleoli (*see* NUCLEOLUS) are present, which are concerned with the synthesis of RIBOSOMES.

nuclide An atomic NUCLEUS identified as having a particular number of NEUTRONS and PROTONS. Thus nuclei of different ISOTOPES or different elements are different nuclides.

numeral A written symbol representing a single digit. In the decimal system, the numerals are 0, 1, 2, ...9. In the hexadecimal system, A, B,... F are also numerals.

numerator The number, or algebraic variable which is divided by the DENOMINATOR in a fraction. Thus in *a/b, a* is the numerator.

nut A dry, INDEHISCENT (does not split open) fruit with a single seed surrounded by a hard woody wall (PERICARP), for example the hazelnut and chestnut. A nut is formed from more than one CARPEL but only one seed develops, the others abort (*compare* ACHENE). The term nut is used to describe not only true nuts, such as hazelnut and acorns, but also hard-shelled fruits, for example almonds and walnuts, which are really DRUPES, and seeds, for example Brazil nuts and peanuts.

Nutrasweet *See* ASPARTAME.

nutrition The processes by which living things take in food and use it. There are two types of nutrition: AUTOTROPHIC NUTRITION and heterotrophic nutrition (*see* HETEROTROPH).

nylon A synthetic POLYAMIDE similar in structure to protein. It was first synthesized as an alternative to SILK and is more elastic and stronger than silk. There is a range of nylons, the most common being nylon–6 and nylon–6,6. Like proteins, nylon fibres contain NH and CO groups but in the latter there are different carbon chains between these groups. In nylon–6 the repeating unit is $-NH(CH_2)_5CO$, the number thus referring to the number of carbon atoms in this unit. Nylon–6,6 is a CONDENSATION POLYMER made from two different molecules, a diamine $H_2N(CH_2)_6NH_2$ and a diacid $HOCO(CH_2)_4COOH$. Nylon has many uses, in hosiery, carpets, textiles, medical sutures and moulded plastics.

nylon-6,6	nylon-6
|	|
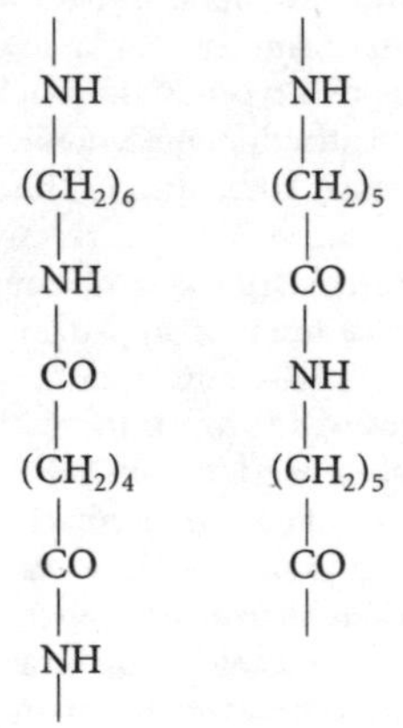 NH	NH
|	|
$(CH_2)_6$	$(CH_2)_5$
|	|
NH	CO
|	|
CO	NH
|	|
$(CH_2)_4$	$(CH_2)_5$
|	|
CO	CO
|	|
NH	
|	

O

obesity *See* DIET.

objective *See* OBJECT LENS.

object lens, *objective* In a REFRACTING TELESCOPE, microscope or binoculars, the lens that collects light from the object being viewed.

object-oriented programming An approach to computer programming based on software 'objects' whose capabilities mirror those of the real world. For example, a bank account may be an 'object' that would store within itself itself its own balance and would know how to respond to messages asking it to make a withdrawal or deposit.

object program The MACHINE CODE version of a program that has been assembled or compiled. *See* ASSEMBLER, COMPILER.

obligate parasite A PARASITE that cannot live independently of its host.

obtuse angle Any angle between 90 and 180 degrees.

occlusion 1. The occurrence of small pockets of liquid or vapour in a crystalline material.

2. The ADSORPTION of gas molecules, especially hydrogen into INTERSTITIAL sites in a metallic LATTICE. The occlusion of hydrogen by palladium has been proposed as a safe way of storing hydrogen as a fuel for road vehicles, and was also proposed as a mechanism for COLD FUSION.

oceanography The study of the physical, chemical, geological and biological features of the ocean.

octahedral (*adj.*) Having the shape of an octahedron; that is, a figure with eight triangular faces, each side having the same length. In chemistry, the term is used to describe a molecule or RADICAL in which a central atom is surrounded by six others, held by COVALENT BONDS or HYDROGEN BONDS. Common examples include the hexacynanoferrate(III) ion, $Fe(CN)_6^{3-}$, and many hydrated TRANSITION METAL ions, such as $Cu(H_2O)_6^{2+}$.

octahedron A POLYHEDRON with eight plane faces.

octane (C_8H_{18}) The eighth member of the ALKANE series, of which there are 18 possible ISOMERS. They are found in PETROLEUM and have boiling points between 99°C and 125°C. 2,2,4-trimethylpentane (iso-octane) is the most important isomer, being a colourless liquid with a boiling point of 99°C. It has anti-knock properties (*see* KNOCKING) and is used as a standard in determining the OCTANE NUMBER of petrol.

octane number, *octane rating* A numerical representation of the ability of PETROL to resist KNOCKING. This is measured by comparing the ease with which a petrol mixture burns in comparison to iso-octane (2,2,4-trimethylpentane) in a blend with heptane. Pure heptane is given a value of 0 and pure iso-octane a value of 100. Thus petrol with an octane value of 97 burns like a mixture of 97:3 iso-octane:heptane, by volume under standard conditions. Petrol with a higher value burns faster than that with a lower value.

octane rating *See* OCTANE NUMBER.

octave An interval in PITCH corresponding to a doubling in frequency of the sound.

octet A group of eight electrons in the outermost shell, the stable configuration of an atom or ion; that is, a full set of s- and p-ORBITALS.

octet rule The broad principle that elements tend to react in ways that leave them with an octet of VALENCE ELECTRONS.

oersted (Oe) Unit of MAGNETIC FIELD strength in the C.G.S. SYSTEM. One oersted is equal to 79.58 Am^{-1}.

oesophagus The tube carrying food from the MOUTH to the STOMACH. It is a muscular tube, 23 cm long in humans, that begins in the lower part of the PHARYNX. The oesophagus contains glands in its lining that secrete MUCUS to lubricate the food. *See also* DIGESTIVE SYSTEM.

oestrogen A STEROID HORMONE produced by the GRAAFIAN FOLLICLE in the OVARY of mammals. Oestrogen actually refers to a group of hormones, including synthetic ones; the main synthetic oestrogen used in humans is oestradiol. Oestrogens cause the development of secondary sexual characteristics, such as breasts, fat deposition and help prepare the UTERUS for pregnancy and maintain the pregnancy if it follows. Oestrogens stimulate production of LUTEINIZING HORMONE and repair the uterine lining following MENSTRUATION if pregnancy does not follow.

Synthetic oestrogens are a major component of the contraceptive PILL, and are thought to be responsible for the side-effects. It has been suggested recently that environmental levels of oestrogens, for example in plastics, are too high and could be responsible for feminization of males and a possible reduction in male fertility.

oestrous cycle The hormonal cycle occurring in many non-human mammals that is equivalent to the MENSTRUAL CYCLE of humans. The oestrous cycle is characterized by an oestrous phase (in oestrus or 'in heat'), during which OVULATION occurs and the female's sexual desire is heightened so that mating is more likely to occur. There may be long periods between consecutive oestrous phases, with several oestrous phases each year (polyoestrus) or a single oestrous period each year (monoestrus). Oestrus often occurs at a time that will favour the survival of the offspring.

oestrus *See* OESTROUS CYCLE.

ogive A graph showing how the CUMULATIVE FREQUENCY of a set of data increases over the range of the data. For example, in a statistical analysis of the heights of individuals in a population, the ogive will be the graph of the total number of individuals having a height less than the specified value. For a NORMAL DISTRIBUTION, the ogive climbs slowly at first, then more rapidly before levelling out as it approaches 100 per cent of the population.

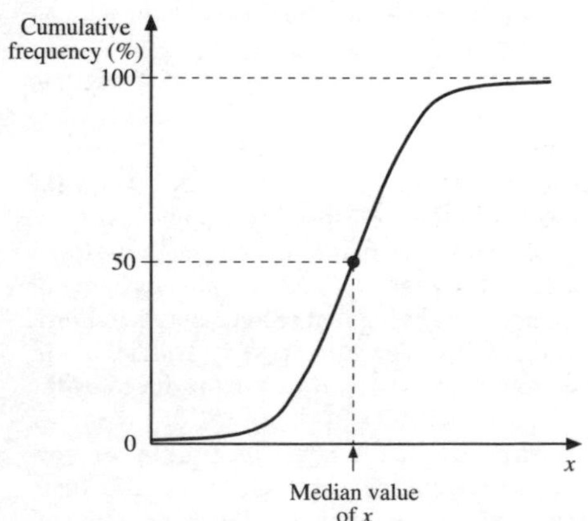

Ogive.

ohm (Ω) The SI UNIT of RESISTANCE. One ohm is the resistance of a conductor that requires a POTENTIAL DIFFERENCE of one volt across it to make a current of one AMPERE flow.

ohmmeter A device for measuring resistance. A simple ohmmeter contains a source of ELECTROMOTIVE FORCE in series with a MOVING-COIL GALVANOMETER and a VARIABLE RESISTOR, which is adjusted so the galvanometer reads full scale when there is no other resistance in the circuit. Any additional resistance reduces the current, leading to a non-linear scale reading in the opposite sense to a normal scale. More modern instruments provide a direct digital read-out.

ohmic (*adj.*) Obeying OHM'S LAW.

Ohm's law For some materials, under constant physical conditions, the CURRENT flowing is proportional to the POTENTIAL DIFFERENCE, i.e. the RESISTANCE is constant. Many conductors, particularly metals, follow this law provided they are kept at a constant temperature. Such material are said to be ohmic. Non-ohmic conductors have a resistance that varies with the current flowing through them.

oil One of many types of naturally occurring HYDROCARBONS, which can be solid (FATS, WAXES) or liquid. Oils are flammable and usually insoluble in water. Mineral oils are those obtained from refining PETROLEUM and are used as fuels and lubricants. Essential oils are those obtained from plants which possess pleasant odours used in perfumes and flavourings. Fixed oils are LIPIDS found in animals and plants, such as fish and nuts, and are used, for example, in foods, soaps and paints.

oil-drop experiment *See* MILLIKAN'S OIL DROP EXPERIMENT.

olefin The common name for ALKENE.

oleic acid An unsaturated FATTY ACID that is abundant in many animal and plant fats.

oleum Concentrated SULPHURIC ACID containing dissolved sulphur trioxide, SO_3.

olfaction The sense of smell or the action of smelling.

olfactory (*adj.*) Relating to smell.

olfactory receptor *See* SENSE ORGAN.

Oligocene The third EPOCH of the TERTIARY PERIOD. It began after the EOCENE epoch (about 38 million years ago) and extended until the MIOCENE epoch (about 26 million years ago). During the Oligocene epoch mammals continued to evolve, and some modern mammals, such as the pig and the rhinoceros first made their appearance at this time.

Oligochaeta A CLASS of the PHYLUM ANNELIDA that includes the earthworm (*Lumbricus*). Earthworms move by contraction and relaxation of muscles with the aid of chaetae (bristles), by which they anchor themselves. They have no distinct head and are HERMAPHRODITE. The earthworm is of particular economic importance to humans because it contributes to soil formation and improvement, by improving aeration and drainage, mixing vegetation and soil, and neutralizing acid soil with their gut secretions. *Compare* POLYCHAETA and HIRUDINEA. *See also* ANNELIDA.

oligodendrocyte A type of GLIAL CELL within the CENTRAL NERVOUS SYSTEM that deposits the MYELIN SHEATH.

omega-minus particle A BARYON containing three STRANGE QUARKS. It has a mass 1.8 times that of the proton and a HALF-LIFE of 8.2×10^{-9} s, decaying by the WEAK NUCLEAR FORCE. Discovered in 1964, the omega-minus particle was an important success for the model of particle physics, which eventually led to the discovery of quarks.

ommatidium (*pl.* ***ommatidia***) One of many units of the compound eye of an insect. Each ommatidium consists of a CORNEA, a lens, a CONE and a group of RECEPTOR cells with light-sensitive pigments linked to a NERVE FIBRE.

omnivore An animal, for example humans, apes and ants, that eats both plants and animal meat in its diet and has gut bacteria that help digestion of a range of substances.

oncogene A tumour-inducing GENE identified in DNA that arises from MUTATIONS in normal genes. Cellular oncogenes are sometimes called proto-oncogenes. Oncogenes usually originate from a host cell and are carried by a virus, often a RETROVIRUS. They may undergo mutation and cause cancer in host cells when they become infected by the virus. Most viruses that are known to be capable of transforming a normal cell to a tumour cell are found to have oncogenes that are inserted into the host cell DNA and induce abnormal growth and division (in several different ways). Viral oncogenes are generally associated with cancer, but chemicals and radiation can activate cellular oncogenes to cause cancer as well.

on-line In computing, working and communicating with another computer via a NETWORK.

ontogeny The whole course of an organism's development, from fertilized egg to maturity. *Compare* PHYLOGENY.

oocyte In animals, an immature female GAMETE that gives rise to an OVUM. *See* OVARY.

oogenesis In animals, the process of ova production. *See* OVARY.

oogonium In animals, an immature female GAMETE that gives rise to OOCYTES. *See* OVARY.

Oomycete A member of the PHYLUM OOMYCOTA.

Oomycota A PHYLUM of the KINGDOM PROTOCTISTA, consisting of water moulds and related organisms. Oomycetes have HYPHAE but are distinct from fungi and they possess FLAGELLA. Oomycetes can be SAPROTROPHS or PARASITES and they can reproduce asexually (by ZOOSPORES) or sexually. The best example is *Phytophthora infestans,* the cause of potato blight and the Irish famine of 1845.

Oort cloud A region around the edge of the SOLAR SYSTEM, from where COMETS are believed to originate.

op amp *See* OPERATIONAL AMPLIFIER.

opaque (*adj.*) Describing a material which does not permit light (or some other specified form of electromagnetic radiation) to pass through it. *Compare* TRANSPARENT, TRANSLUCENT.

open-cast mining The extraction of a mineral from the surface of the Earth, as opposed to DEEP MINING where a vertical shaft has to be drilled to reach the mineral. *See also* DRIFT MINING.

open circuit (*n., adj.*) A circuit in which there is some break, either introduced deliberately – such as by a switch – or as the result of a fault.

open hearth furnace A device, now largely obsolete, for the manufacture of steel from PIG IRON. The pig iron is melted by burning a mixture of air, gas and oil above it in a shallow furnace (the 'hearth'). Oxygen is injected into the top of the furnace through water-cooled pipes called lances, and this OXIDIZES impurities in the steel, in particular carbon. The carbon may also be oxidized by the inclusion of HAEMATITE, an ore containing iron oxide. LIMESTONE is added to form a SLAG with the less volatile impurities. The molten steel is then removed from the furnace in large ladles, the slag is skimmed off, and the steel poured into moulds. The exhaust gas is used to heat a number of bricks arranged in an open lattice pattern. The gas flow is periodically reversed so that two sets of bricks, one at each end of the furnace, are alternately used to pre-heat the incoming gas and heated by the outgoing gases. The open hearth furnace is a slow process compared to the OXYGEN FURNACE and the ELECTRIC ARC FURNACE, which have largely replaced it. *See also* BESSEMER CONVERTER.

open loop gain The GAIN of an AMPLIFIER, particularly an OPERATIONAL AMPLIFIER, which has no FEEDBACK.

operant conditioning *See* CONDITIONING.

operating system The program that provides basic capabilities of a computer. An operating system's functions include managing the printers, drivers and other input and output devices; managing the reading and writing of FILES; loading and running programs; organizing the memory; providing an interface to networks; backing up data; monitoring and adjusting performance; etc. *See also* DOS, UNIX.

operation A mathematical manipulation applied to a number or pair of numbers resulting in a new number according to pre-determined rules, for instance multiplication.

operational amplifier (*abbrev. **op amp***) The key building-block of many ANALOGUE electronic circuits. An operational amplifier is a type of INTEGRATED CIRCUIT that forms a high GAIN DIFFERENTIAL AMPLIFIER, with an OPEN LOOP GAIN of typically several million. The operational amplifier also has a very high input resistance, so virtually no current flows into either input. The output voltage cannot exceed the supply voltage, so in many applications, the two inputs are at virtually the same ELECTRIC POTENTIAL. The NON-INVERTING INPUT is often connected to EARTH (zero potential), so the INVERTING INPUT is close to earth potential, and forms what is called a VIRTUAL EARTH, a point in a circuit that although not connected to earth, can always be taken as being at earth potential.

Most operational amplifier circuits also involve NEGATIVE FEEDBACK, in which some of the output is fed back to the inverting input. This reduces the gain of the circuit but produces a system with properties that do not depend on the detailed performance of the operational amplifier itself, and so are not affected by changes caused by changes in temperature, ageing, etc.

See also AMPLIFIER.

operculum 1. In botany, a covering over the SPORANGIUM of some fungi and mosses that opens to allow the release of mature SPORES.

2. In zoology, a bony flap covering the GILL slits of bony fish, which protects the gills and aids their ventilation.

operon A group of adjacent GENES on a CHROMOSOME that act together to produce a POLYPEPTIDE or ENZYME. Operons were discovered in bacteria by the French biochemists Jacob (1920–) and Monod (1910–76) in 1961.

An operon is made up of structural genes (CISTRONS) that are responsible for the production of the polypeptide, and operator genes that regulate the switching on and off of the structural genes. Another gene, further away from the group, called the regulator gene, codes for a protein called the repressor, which can bind to the operator gene and prevent it switching on the structural genes. Operons are found mostly in PROKARYOTES and are less common in EUKARYOTES, where regulatory mechanisms are more complex.

See also GENE EXPRESSION.

opiate A term referring to drugs derived from OPIUM. The term opiate is also used to refer to natural pain-relieving chemicals produced by the body, such as ENDORPHINS and ENCEPHALINS.

opium A drug that is obtained from unripe seeds of the opium poppy *Papaver somniferum*. It contains a number of ALKALOIDS including the pain-killing substances morphine and codeine, and also a highly poisonous substance called thebaine. Morphine is a strong but addictive pain-killing drug and heroin is an even stronger synthetic derivative of morphine.

opsonin A protein, often an ANTIBODY, that coats the surface of foreign substances, making them more vulnerable to ingestion by MACROPHAGES.

optic nerve A large nerve between the eye and the brain. It carries information from the sensory cells in the RETINA to the visual centres in the brain. The optic nerve develops as part of the brain wall.

optical axis *See* PRINCIPAL AXIS.

optical activity The ability of certain substances to rotate the plane of polarized light (*see* POLARIZATION). An optically active substance may be a crystal, liquid or solution. Optical activity arises from a lack of symmetry in the three-dimensional structure of the molecules (or crystal lattices) concerned. It gives rise to optical ISOMERS, which differ in the direction in which they rotate the plane of polarized light. Optically active substances are either LAEVOROTATORY or DEXTROROTATORY, according to the direction of rotation. *See also* OPTICAL ISOMERISM, POLARIMETER.

optical binary *See* BINARY SYSTEM.

optical centre The point at which the PRINCIPAL AXIS meets a lens. It is the thickest point in a CONVERGING LENS; the thinnest point in a DIVERGING LENS.

optical character recognition (**OCR**) A computer technique for recognizing letters on a printed page. This enables printed text fed into a computer by an OPTICAL SCANNER to be converted into a form which can be manipulated further, by a word processor for example. Whilst such systems work well with simple text those presently available have problems with more complex pages, such as those containing a mixture of sizes of type, or diagrams.

optical fibre *See* FIBRE OPTIC.

optical isomerism ISOMERS that are mirror images of one another. Unlike structural isomers, optical isomers are only obvious when models are constructed. They arise due to asymmetry of the molecule and differ in the direction in which they rotate the plane of polarized light (*see* POLARIZATION). The two mirror images cannot be superimposed on one another and are related in the same way as the right hand is to the left hand. This is termed chirality and the molecules are said to be CHIRAL.

The two isomers are called enantiomers. A pair of enantiomers are identical in all respects except for their orientation in space. Their physical and chemical properties are the same except for the direction in which they rotate the plane of a beam of polarized light when passed through solutions containing each enantiomer. One enantiomer rotates the plane of polarized light to the right (dextrorotatory; denoted +) and the other rotates it an equal amount to the left (laevorotatory; denoted –). Although most chemical reactions of two enantiomers are identical they do differ in their reaction with other chiral molecules.

Many naturally occurring compounds are chiral (for example amino acids) but usually only one enantiomer is found in nature.

See also GEOMETRIC ISOMERISM, RACEMIC MIXTURE.

optical scanner A device for viewing a paper document and converting the page into BINARY data to be processed by a computer. *See also* OPTICAL CHARACTER RECOGNITION.

optics The study of visible light. Optics is usually divided into ray, or geometrical, optics, which deals with the phenomena of REFRACTION and REFLECTION, and wave optics, which deals with DIFFRACTION and INTERFERENCE.

orbit The path taken by an object, called a SATELLITE, which is moving under the influence of the GRAVITY of some larger object, such as a star or planet. If the total energy of the orbiting satellite is positive, it will have enough KINETIC ENERGY to escape completely from the gravitational influence and the orbit will be open, with the satellite curving around the source of gravity before travelling on into space. If the total energy is negative, so the GRAVITATIONAL POTENTIAL ENERGY more than outweighs the kinetic energy, the orbit will be closed and the satellite will trace out a closed path, either an ELLIPSE (a flattened circle) or a circle.

Orbital motion was first studied by Kepler, who put forward three laws to describe the orbits of planets around the Sun (*see* KEPLER'S LAWS). Kepler did not fully understand the force of gravity that was responsible for these orbits; this was explained by Newton, who studied the orbit of the Moon, realizing that, like an apple falling from a tree, it was in FREE FALL. By calculating the CENTRIPETAL ACCELERATION of the Moon, Newton was able to find how the force of gravity varied with distance.

See also CIRCUMPOLAR ORBIT, GEOSTATIONARY ORBIT.

orbital A region of space in an ATOM or MOLECULE that can be occupied by a maximum of two electrons. An orbital is often thought of as the volume that encloses the space within which the probability of finding an electron is greater than a specified figure, since in principle, electron WAVEFUNCTIONS extend out infinitely far from the atom.

Two electrons in a single orbital often form a stable configuration called a LONE PAIR, whilst a single electron is often active in the formation of chemical bonds. Orbitals may

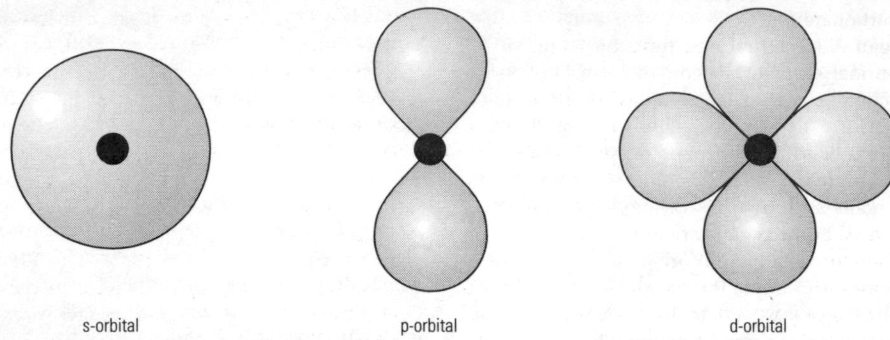

Electron orbitals.

overlap and share electrons between atoms to form a COVALENT BOND, or an orbital may gain or lose an electron so that the atom or molecule forms an ION.

Atomic orbitals are labelled by a number in the sequence 1, 2, 3 etc., called the PRINCIPAL QUANTUM NUMBER, with higher numbers indicating higher energies. The principal quantum number indicates the size of the orbital. They are also labelled by a letter, s, p, d, f, called the SUBSIDIARY QUANTUM NUMBER. This denotes the shape of the orbital and the ANGULAR MOMENTUM of an electron in that orbital. For a principal quantum number of 1, only an s-orbital exists. For principal quantum number of 2 there is one S-ORBITAL and three P-ORBITALS. With a principal quantum number of 3 there is a single s-orbital, three p-orbitals and five D-ORBITALS, and so forth. The energy of the orbitals, and hence the sequence in which they are filled by electrons is basically

1s, 2s, 2p, 3s, 3p, 4s, 3d, 4p, 5s, 4d, 5p,
6s, 4f, 6p, 7s, 5f

The order in which electrons go into these orbitals to form neutral atoms is responsible for the structure of the PERIODIC TABLE.

See also HUND'S RULE OF MAXIMUM MULTIPLICITY, HYBRID ORBITAL, MOLECULAR ORBITAL, PAULI EXCLUSION PRINCIPLE, SHELL.

order 1. (*mathematics*) In an equation containing DERIVATIVES of a function, the largest number of times any function is differentiated (*see* CALCULUS). Thus an equation in which some quantity is differentiated twice, but no quantity is differentiated more than twice, is called a second order differential equation.

2. (*biology*) One of the subdivisions of CLASS in the CLASSIFICATION of organisms. The names of orders for birds and fish end in '*-formes*', for mammals, amphibians and reptiles '*-a*' and for fungi and plants '*-ales*'. Orders consist of groups of FAMILIES.

3. (*chemistry*) A measure of the extent to which the rate of a reaction depends on the CONCENTRATION of the reagents involved. In the reaction A + B → products, the rate of reaction is proportional to $[A]^m[B]^n$, where [A] is the molar concentration of A and so on. The reaction is said to be *m*th order in A and *n*th order in B, with an overall order of $m + n$. *See also* RATE OF REACTION.

4. (*physics*) The value of the number n in DIFFRACTION and DIFFRACTION GRATING equations, such as

$$d\sin\theta = n\lambda$$

where d is the width of the diffraction slits, θ is the angle of the diffraction and λ is the wavelength of light).

ordinary ray *See* BIREFRINGENCE.

Ordovician A PERIOD of geological time during the PALAEOZOIC ERA. It began after the CAMBRIAN period (about 440 million years ago) and extended until the SILURIAN period (about 139 million years ago). At this time all life was confined to the sea, and the first fish were beginning to evolve.

ore Any naturally occurring rock from which a metal can be extracted, such as iron ore, which is usually iron(III) oxide, Fe_2O_3, or iron(II) carbonate, $FeCO_3$.

organ A structural and functional unit of an animal or plant. An organ consists of more than one type of TISSUE and is co-ordinated to perform usually one main function. Organs often work together in organ systems, for example the DIGESTIVE SYSTEM which includes organs such as the STOMACH, LIVER and PANCREAS. Some organs are also GLANDS and can therefore belong to more than one organ system, for example the pancreas is also part of the ENDOCRINE SYSTEM. Examples of organs in plants include roots, stems and leaves.

organelle A discrete structure found in living cells that is specialized for a particular function. Organelles include the NUCLEUS, which contains the DNA, and cytoplasmic organelles such as MITOCHONDRIA, RIBOSOMES, CHLOROPLASTS, LYSOSOMES, ENDOPLASMIC RETICULUM, and GOLGI APPARATUS. *See also* PLASTID.

organic A term that was used in the late eighteenth century to refer to compounds obtained from living material in contrast to those derived from minerals (which were termed INORGANIC). Now 'organic' refers to all compounds containing both carbon and hydrogen, except the HYDROGENCARBONATES.

Organic compounds are more numerous than inorganic compounds and form the basis of life. Although many organic compounds are made only by living organisms, such as PROTEINS, CARBOHYDRATES, it is now possible to manufacture many synthetically. Many organic compounds are derived from PETROLEUM, the remains of microscopic marine organisms.

Organic compounds may consist only of carbon and hydrogen – HYDROCARBONS – linked by COVALENT BONDS, although frequently they also contain oxygen, nitrogen, sulphur, phosphorus or HALOGENS similarly linked to the carbon atoms. This contrasts to the IONIC BONDING typical of many inorganic compounds.

Organic chemistry is the study of organic compounds. Examples of organic compounds include ALKANES, ALKENES, ALKYNES, ALCOHOLS, ALDEHYDES, KETONES, ETHERS, ESTERS, AMINES, AMIDES, AMINO ACIDS, proteins, carbohydrates.

organism A living individual plant or animal. *See also* MICRO-ORGANISM.

organ of Corti *See* COCHLEA.

organogenesis *See* EMBRYONIC DEVELOPMENT.

organometallic compounds Organic compounds that are linked to metal or metal-like atoms, such as $NaCH_3$. Vitamin B_{12} is the only naturally occurring organometallic compound. The compounds are used in organic chemistry as GRIGNARD REAGENTS and as ZIEGLER-NATTA CATALYSTS.

OR gate A LOGIC GATE with two or more inputs and an output that is high if any of the inputs is high.

origin In a GRAPH, or any system of CO-ORDINATES, the point at which the axes intersect, where the value of all the co-ordinates is zero.

origin of life The beginning of life on Earth about 4,000 million years ago. There are a number of theories regarding how life began. The steady-state theory suggests that the Earth and its life forms have always existed. The spontaneous generation theory (abiogenesis) suggests that life arose from non-living matter on a number of separate occasions. This does not seem to occur now, but could have played a role in the origin of life. The creation theory suggests that God created the Earth and all its life forms. The cosmozoan theory suggests that life began elsewhere in the universe and later arrived on earth. The biochemical evolution theory suggests that simple molecules were formed that combined into complex ones and then into cells.

The biochemical EVOLUTION theory is the one most widely accepted by scientists today. It is thought that simple molecules, such as amino acids and MONOSACCHARIDES, were formed by the combination of gases present in the Earth's early atmosphere (carbon dioxide, methane, hydrogen, ammonia and water), the energy for which came from ultraviolet radiation from the sun and electrical energy from lightning storms. Such molecules probably floated on the surface of the oceans as a 'primeval soup', and when concentrated enough formed POLYMERS, such as STARCH, and eventually developed into cells.

See also EVOLUTION.

ornithine cycle *See* UREA.

ornithology The study of birds.

ornithophily POLLINATION of flowers by birds. Flowers adapted for ornithophily, for example tropical plants, are often unscented, because birds do not respond well to smell, and instead are large and brightly coloured (red or orange) with large amounts of NECTAR.

orographic (*adj.*) Describing the formation of cloud and sometimes rain when moist air is forced upwards by a land mass. Orographic effects account for the higher than average rainfall on the side of any hills that face the PREVAILING WIND, and the dry area, called the RAIN SHADOW on the lee (downwind) side of the prevailing wind.

ortho In organic chemistry, a prefix used in disubstituted derivatives of BENZENE which indicates that the substituted groups are at positions 1 and 2 in the ring, such as 2-nitrophenol (*ortho*-nitrophenol, *o*-nitrophenol).

OH
NO_2

The use of the term *ortho* in other situations, for example orthocarbonates does not have the same meaning and is always written in full. *See also* META, PARA.

orthorhombic (*adj.*) Describing a crystal structure in which the UNIT CELL has all its faces at right angles to one another, with all the faces rectangular, but none square, so there are three different lengths characterizing the size of the unit cell.

ortho-sulpho benzimide *See* SACCHARIN.

oscillation A motion that repeats itself at regular intervals. The time taken for one complete oscillation is called the period, whilst the number of oscillations in one second is called the frequency. The unit of frequency is the HERTZ (Hz).

In order to oscillate, a system must have a position of STABLE EQUILIBRIUM, so that when displaced from the equilibrium position, the system will experience a force tending to return it to equilibrium. The mass of the system means that it will overshoot its equilibrium position and move away from equilibrium in the opposite direction and so on.

For many systems it is the case, at least to some degree of approximation, that the restoring force tending to return the system to equilibrium is directly proportional to the displacement from the equilibrium position. The motion is said to be ISOCHRONOUS, i.e. the period does not depend on the amplitude. In this case the system is said to exhibit SIMPLE HARMONIC MOTION, with the displacement from equilibrium varying with the SINE or COSINE function.

In any oscillation there is an interchange between KINETIC ENERGY and POTENTIAL ENERGY – GRAVITATIONAL POTENTIAL in the case of the SIMPLE PENDULUM and ELASTIC potential in the SPRING PENDULUM. In each case the kinetic energy is a maximum and the potential energy a minimum as the pendulum passes through its equilibrium position, and the kinetic energy is zero and the potential energy a maximum at the extremes of the oscillation. The total energy is constant throughout the oscillation.

For simple harmonic motion with a displacement x from equilibrium, moving with a velocity v and an acceleration a:

$$a = -\omega^2 x$$

where ω is the ANGULAR FREQUENCY. If the motion is started by releasing the system from rest:

$$x = A\cos\omega t$$

where A is the amplitude:

$$v = -\omega A\sin\omega t$$

In simple harmonic motion, the maximum velocity is ωA and is reached as the system passes through its equilibrium position:

$$a = -\omega^2 A\cos\omega t$$

the maximum acceleration is $\omega^2 A$

$$v = \pm\omega(A^2 - x^2)^-$$

See also DAMPING, FORCED OSCILLATION, RESONANCE.

oscillator A circuit for generating ALTERNATING CURRENT, particularly at high frequency for the generation of a CARRIER WAVE in a radio system. The frequency of the oscillator is fixed by some resonant device (*see* RESONANCE), which may be a PIEZOELECTRIC quartz crystal with a mechanical resonance at the desired frequency, or a tuned circuit containing an INDUCTANCE and a CAPACITANCE. In the latter case, the current in the inductance and the capacitance will be exactly out of PHASE, and at one particular frequency they will also be equal in magnitude. Thus the circuit will appear to have a very high resistance at a single frequency. An AMPLIFIER circuit with POSITIVE FEEDBACK is used, so some of the output is fed back and further amplified. Such a circuit will oscillate if it has sufficient GAIN, the tuned circuit can be used to fix the frequency at which this feedback will be greatest. *See also* CRYSTAL OSCILLATOR.

oscilloscope *See* CATHODE RAY OSCILLOSCOPE.

osmiridium A naturally occurring alloy containing mostly osmium (typically 20 to 50 per cent) and iridium (typically 50 per cent), together with varying quantities of other elements.

osmium (Os) The element with ATOMIC NUMBER 76; RELATIVE ATOMIC MASS 190.2; melting point 3,045°C; boiling point 5,027°C; RELATIVE DENSITY 22.6. Osmium is a transition metal, and the densest of all the elements, with a density more than three times that of iron. Osmium is a hard bluish metal used in some steels and has some catalytic properties.

osmoregulation *See* OSMOSIS.

osmosis The movement of a liquid SOLVENT, usually water, from a less concentrated SOLUTION to a more concentrated solution through a SEMIPERMEABLE MEMBRANE (one that is permeable in both directions to water but varying in PERMEABILITY to the SOLUTE) until the two concentrations are equal or isotonic.

Osmosis is a passive process requiring no energy. If external pressure is applied to the more concentrated solution, osmosis is prevented and this provides a measure of the OSMOTIC PRESSURE or osmotic potential of the more concentrated solution, which is measured in pascals (Pa). The osmotic pressure is greater the more concentrated the solution.

In animal cells, the less dilute solution is called hypotonic, and the more concentrated solution is called hypertonic. The passage of water by osmosis will occur across a semi permeable membrane from any solution of weaker osmotic pressure to one of higher osmotic pressure, regardless of whether the dissolved substance on both sides of the membrane is the same or not.

Osmosis is vital in controlling the distribution of water in living organisms, for example in the transport of water from the roots up to the stems of plants, and in maintaining a constant water balance (osmoregulation) to prevent the concentration of salts becoming too high or too low and affecting vital functions. Fish have a protective mechanism to counteract osmosis. Without such a mechanism saltwater fish would lose fluid by osmosis and freshwater fish would gain excess fluid.

See also REVERSE OSMOSIS.

osmotic potential *See* OSMOTIC PRESSURE.

osmotic pressure, *osmotic potential* The pressure difference that can occur across a SEMIPERMEABLE MEMBRANE as a result of OSMOSIS. It is defined as the pressure that needs to be applied across a semipermeable membrane to prevent osmosis. KINETIC THEORY shows that the osmotic pressure is proportional to the CONCENTRATION of the SOLUTE and to the ABSOLUTE TEMPERATURE of the liquid.

ossification The process by which BONE develops from CARTILAGE. Specialized cells called osteoblasts secrete EXTRACELLULAR MATRIX onto the surface of existing cartilage, and calcium phosphate crystals deposit within the matrix to form bone. Osteoblasts that become included within the bone structure during its development become osteocytes (bone cells) and cease to divide and form bone matrix.

Osteichthyes A CLASS of VERTEBRATES consisting of the bony fish (which have a bony skeleton), for example the salmon, herring and stickleback. Osteichthyes forms the largest class in the PHYLUM CHORDATA. *Compare* CHONDRICHTHYES.

osteoblast A specialized cell responsible for the formation of BONE by the process of OSSIFICATION.

osteoclast A multinucleate cell in BONE that allows remodelling of bone shape during growth by breaking down the calcified matrix under the regulation of parathyroid hormone (*see* THYROID GLAND).

osteocyte A bone cell. *See* OSSIFICATION.

osteoporosis A human condition in which bones become brittle and weak and are easily broken. Osteoporosis occurs in elderly people, particularly in women past the MENOPAUSE, and is possibly linked to a drop in OESTROGEN levels.

Ostwald's dilution law For a solution in which a substance AB only weakly DISSOCIATES into ions A^+ and B^-, the degree of IONIZATION, provided it is small, is inversely proportional to the square root of the concentration. This result follows from the application of CHEMICAL EQUILIBRIUM laws to the system of ions and molecules. If the degree of dissociation is α and the number of molecules present before dissociation is n in a volume V, then

$$\alpha = (KV/n)^{1/2}$$

where K is a constant.

otolith *See* SEMI-CIRCULAR CANAL.

Otto cycle *See* FOUR STROKE CYCLE.

outbreeding *See* ARTIFICIAL SELECTION.

out of phase Describing two OSCILLATIONS that are not exactly in step with one another or that are exactly out of step with one another.

output device Any device for displaying the information processed by a computer. The main output devices of a PERSONAL COMPUTER are the VISUAL DISPLAY UNIT and PRINTER.

oval window *See* EAR OSSICLE.

ovary 1. In animals, the female GONAD that produces the OVUM. In humans, there are two ovaries, about 25 by 35 mm, in the lower ABDOMEN situated close to the FALLOPIAN TUBES. Each month, a mature ovum is released from the ovary, a process called OVULATION, and is either fertilized (*see* FERTILIZATION) or lost as part of the menstrual flow (*see* MENSTRUAL CYCLE).

The outer layer of the ovary contains germinal EPITHELIAL cells that begin to divide to form ova while in the FOETUS, before birth. The process of ova production is called oogenesis, and begins with oogonia dividing by MITOSIS to form primary oocytes, which then divide by MEIOSIS. Oogenesis stops in the newborn at the primary oocyte stage. At this stage many oocytes are organized in structures called primary follicles. The newborn female contains all the genetic information it needs to provide for its offspring.

At puberty a hormone produced by the PITUITARY GLAND, called FOLLICLE-STIMULATING HORMONE, restarts oocyte development within the follicle. At any one time there will be oocytes and follicles at different stages of maturity in the ovary. A GRAAFIAN FOLLICLE is the largest, mature stage and contains the secondary oocyte (produced by meiotic division of a primary oocyte) that is released as the mature ovum. After the ovum has been released, the Graafian follicle remains as a CORPUS LUTEUM that secretes the steroid hormone PROGESTERONE. The ovaries also produce the steroid hormone OESTROGEN.

2. For plant ovaries, *see* CARPEL.

overdamped (*adj.*) Describing a system that could oscillate if allowed to move freely, but where the level of DAMPING is too large to permit this. *See also* UNDERDAMPED.

overfishing Fishing at a rate that exceeds the sustainable yield and causes the depletion of fishing stock. Overfishing is a result of modern fishing methods, which use huge factory ships and specialized equipment to locate shoals of fish. In the North Sea, cod and haddock have suffered through overfishing and herring are almost extinct. Overfishing is a serious problem in the developing world because stocks used by local people have been depleted.

There are now restrictions on fishing, although these are a cause of much controversy between countries. The mesh size of fishing nets is regulated to allow smaller, immature fish to fall through so that they can reach sexual maturity and breed to replenish the stocks.

overtone Frequencies of STANDING WAVES higher than the FUNDAMENTAL frequency for the system in question. The next highest frequency after the fundamental is called the first overtone, the next frequency above this is the second overtone, etc. With a standing wave on a stretched string, there is a NODE at each end, and the overtones will include all the HARMONICS.

In a pipe that is open at only one end, such as in a wind instrument, there is a node at one end and an ANTINODE at the other end. The node in air pressure is at the open end, but the node in air displacement is at the closed end. In each case there is an antinode at the opposite end. Such a pipe will produce overtones containing only odd harmonics and thus a sound which is musically different from that produced by a string. A pipe that is open at both ends, however, will produce all the harmonics.

oviduct *See* FALLOPIAN TUBE.

ovipary Reproduction by which the female lays eggs that develop outside her body. Ovipary is the most common form of reproduction. *See also* OVOVIVIPARY, VIVIPARY.

ovovivipary Reproduction by which fertilized eggs develop within the female but gain no nutrition from her, for example in fish, reptiles and many insects. *See also* OVIPARY, VIVIPARY.

ovulation In female animals the release of a mature OVUM (egg cell) from the OVARY. Ovulation is under hormonal control as part of the MENSTRUAL CYCLE, and occurs monthly in women from puberty until the MENOPAUSE.

ovule In SEED PLANTS the structure containing the female GAMETE that develops into the SEED after fertilization (*see* DOUBLE FERTILIZATION). The ovule consists of an EMBRYO SAC containing several nuclei, one of which is the egg nucleus or female gamete, surrounded by nutritive tissue called the NUCELLUS. There are one or two protective layers called the INTEGUMENTS that develop into the TESTA after fertilization. In flowering plants (ANGIOSPERMS) the ovule is within an ovary (*see* CARPEL) fixed to the wall by a stalk, but in conifers (CONIFEROPHYTA) it is not enclosed by an ovary and is found on an ovule-bearing scale within a CONE.

ovum (*pl. **ova***) The female GAMETE before FERTILIZATION. In higher organisms, such as humans, it is called an egg and is produced in the OVARY. It is larger than the male SPERM and non-motile, and consists of a nucleus, a large CYTOPLASM containing yolk grains as a food source, and a thick outer membrane surrounded by cumulus cells, which aid movement of the ovum toward the UTERUS and provide nutrients. Once fertilized, the ovum is called a ZYGOTE. In some species, such as birds, the egg is covered by a shell.

In plants, the ovum is called an egg cell (*see* OVULE).

oxalic acid *See* ETHANEDIOIC ACID.

oxaloacetic acid A colourless, crystalline acid that combines with ACETYL COENZYME A at the beginning and end of the KREBS CYCLE. It is also produced in C_4 PLANTS in the light-independent stage of PHOTOSYNTHESIS. *See also* CALVIN CYCLE.

oxidase *See* OXIDOREDUCTASE.

oxidation In general, any chemical reaction that increases the OXIDATION STATE, of an element, so the element concerned is now in a state where it has lost extra electrons (or where extra electrons could be imagined to have been lost, if COVALENT BONDS are treated as IONIC). In particular, oxidation is the addition of oxygen or the removal of hydrogen from an element, or the removal of electrons to form a positive ION (or an ion more positive than was originally present).

The simplest examples of oxidation are those in which oxygen is added, for example when magnesium is oxidized when it burns in air:

$$2Mg + O_2 \rightarrow 2MgO$$

This example also involves the removal of electrons from the magnesium as magnesium oxide is ionic:

$$Mg \rightarrow Mg^{2+} + 2e^-$$

and an increase of the oxidation state of the magnesium, from 0 to +2.

See also OXIDIZING AGENT, REDOX REDUCTION, REDUCTION.

oxidation number *See* OXIDATION STATE.

oxidation state, *oxidation number* The number of electrons that are lost, or that can be imagined to be lost, by an element in forming a particular compound, ion or COMPLEX ION. If the compound is IONIC, the oxidation state is simply the charge on the ions, or adds up to this charge if the ion is a FREE RADICAL. In COVALENT BONDS, the electrons involved in the bond are imagined to be wholly attached to the more electronegative of the two elements (*see* ELECTRONEGATIVITY). If the two elements have equal electronegativity, the electrons are imagined to be shared equally. These rules mean that for any neutral compound, the oxidation states must add up to zero.

Fluorine always has an oxidation state of –1 and the oxidation state of oxygen is –2, except in PEROXIDES, where it is –1. The oxidation state of hydrogen is +1 except in metallic HYDRIDES, where it is –1. ALKALI METALS all have an oxidation state of +1 and ALKALINE EARTHS of +2. HALOGENS usually have oxidation states of –1, though +7 is sometimes found, such as in the ClO_3^- ion. Sulphur exhibits a wide range of oxidation states, though apart from –2 in H_2S, only +4 and +6 are common.

Many of the TRANSITION METALS can form ions or complexes with different oxidation states. For example, iron can form Fe^{2+} and Fe^{3+} ions with oxidation states of +2 and +3. In the case of the transition metals, the oxidation state is often shown by inserting the oxidation number, shown as a Roman numeral, in the chemical name.

oxidative phosphorylation The process by which ATP is formed through AEROBIC RESPIRATION, by the transfer of electrons in the ELECTRON TRANSPORT SYSTEM, which results in the oxidation of hydrogen atoms. Oxidative phosphorylation occurs in the MITOCHONDRIA, within particles found on the folded inner layer of the double membrane.

oxide Any BINARY COMPOUND containing oxygen. Both metals and non metals form oxides, many by direct combustion in oxygen, for example, magnesium burns in oxygen to form magnesium oxide:

$$2Mg + O_2 \rightarrow 2MgO$$

and sulphur burns to form sulphur dioxide:

$$S + O_2 \rightarrow SO_2$$

The oxides of metals are BASIC, or AMPHOTERIC in the case of some TRANSITION METALS, and have high melting points. The oxides of non-metals have lower melting points (many are gases at room temperature) and are ACIDIC.

oxidize (*vb.*) To bring about an OXIDATION process.

oxidizing agent A material that readily brings about an OXIDATION, itself being reduced (*see* REDUCTION). Oxygen is an obvious example. Hydrogen peroxide, which is itself reduced to water, and potassium manganate(VII), reduced to manganese(II) oxide, are other examples.

oxidoreductase One of a group of ENZYMES, for example, dehydrogenases and oxidases, that transfer O and H atoms between substances in OXIDATION/REDUCTION reactions.

oxoacid An acid containing a non-metallic atom bound to one or more oxygen atoms. Sulphuric acid, H_2SO_4, and nitric acid, HNO_3, are two examples.

2-oxopropanoic acid *See* PYRUVIC ACID.

oxygen (O) The element with ATOMIC NUMBER 8; RELATIVE ATOMIC MASS 16.0; melting point –214°C; boiling point –183°C. Oxygen makes up 20 per cent of the atmosphere, mostly as the molecule O_2 though some occurs as OZONE, O_3 (*see* OZONE LAYER). It is the most common element in the Earth's CRUST, occurring in SILICATES, SILICA and CARBONATES. It is very reactive and will combine with all elements apart from the NOBLE GASES and fluorine.

Oxygen is vital to organisms that carry out AEROBIC RESPIRATION. These organisms absorb oxygen directly from the atmosphere or make use of dissolved oxygen in water. Oxygen also supports COMBUSTION – many materials will burn very rapidly in pure oxygen. Oxygen is extracted from liquefied air by FRACTIONAL DISTILLATION and in its liquid form is a common OXIDIZING AGENT in rocket propulsion.

oxygen debt The oxygen needed to break down LACTIC ACID produced by a fatigued muscle during anaerobic RESPIRATION. During vigorous exercise, if the lungs cannot supply enough oxygen to a muscle, the cells will switch from aerobic to anaerobic respiration in order to produce energy. This results in a build up of lactic acid, which causes cramp. After resting, the body will use extra oxygen (obtained by the automatic panting response) to break down the lactic acid, so paying off the oxygen debt. *See also* FERMENTATION.

oxygen furnace A device used in modern steel making. The furnace is a steel vessel with a REFRACTORY lining. Molten PIG IRON is introduced into the furnace and oxygen blown onto the surface through a water-cooled LANCE. This OXIDIZES many impurities, particularly carbon. Those impurities that are less VOLATILE form a SLAG with LIMESTONE, which has been added to the furnace. Other elements may be added to alter the composition of the steel. Once the steel is ready, the furnace is tipped onto its side and the steel tapped off into moulds. This process is called the Linz-Donawitz process. *See also* ELECTRIC ARC FURNACE.

oxyhaemoglobin *See* HAEMOGLOBIN.

oxytocin A PEPTIDE hormone, produced by the PITUITARY GLAND of birds and mammals, that is involved in the contraction of the smooth muscle of the UTERUS during birth and, in mammals, the muscular contractions causing expression of milk from the MAMMARY GLANDS during LACTATION.

ozone (O_3) A colourless gas with a distinctive odour; melting point –192°C; boiling point –112°C. Ozone can be made by the action of an electric discharge on oxygen:

$$3O_2 \rightarrow 2O_3$$

Ozone is produced in the upper atmosphere and plays an important part in protecting the Earth's surface from ultraviolet radiation. It is a powerful OXIDIZING AGENT. *See also* OZONE LAYER.

ozone hole An area of lower than usual ozone concentration in the OZONE LAYER above the Earth's poles. *See* ATMOSPHERE.

ozone layer A protective layer consisting of the gas ozone, O_3, 15–40 km above the Earth's surface. It is is formed by the effect of ultraviolet (UV) radiation on oxygen molecules. UV light splits oxygen (O_2) molecules into two atoms, one of which then combines with oxygen to create ozone.

The ozone layer prevents harmful UV radiation reaching the Earth's surface, but in recent years it has become clear that the layer is being damaged by human activities. *See also* CHLOROFLUOROCARBON.

P

pacemaker, *sinoatrial node, SA node* In vertebrates, a group of muscle cells in the wall of the heart that controls the basic rate of contractions of the HEART. The pacemaker is found in the wall of the right ATRIUM. The cells of the pacemaker spontaneously and rhythmically contract, and this rate can be adjusted according to demand by the AUTONOMIC NERVOUS SYSTEM.

An artificial pacemaker can be implanted in a person whose heart beats irregularly. This operates by stimulating the heart muscles with minute electric shocks at regular intervals to restore the normal heartbeat.

See also PURKINJE FIBRES.

packet switching A method of transmitting data over a computer NETWORK in which the data is sent in discrete packets of information rather than a steady stream of BITS and BYTES. The connection between the two machines is closed when no data is flowing between them.

Palaeocene The first EPOCH of the TERTIARY PERIOD which extended about 11 million years until the EOCENE epoch (about 54 million years ago). The Palaeocene epoch saw the evolution of hoofed mammals and the spreading of grasslands.

palaeontology The study of FOSSILS. Palaeontology looks at the structure and evolution of extinct organisms and their environment, as can be determined from their fossil remains. It also has applications in geology, as fossils help date the rock strata in which they are found and help identify potential sources of FOSSIL FUELS.

Palaeozoic The ERA of geological time that lasted for 345 million years, from the end of the PRECAMBRIAN to the start of the MESOZOIC era (about 225 million years ago). It is the first era of PHANEROZOIC TIME. It is subdivided into the Lower Palaeozoic, which encompasses the CAMBRIAN, ORDOVICIAN and SILURIAN periods, and the Upper Palaeozoic, which comprises the DEVONIAN, CARBONIFEROUS and PERMIAN periods.

palisade mesophyll *See* MESOPHYLL.

palladium (Pd) The element with ATOMIC NUMBER 46; RELATIVE ATOMIC MASS 106.4; melting point 1,551°C; boiling point 3,140°C; RELATIVE DENSITY 12.3. Palladium is a TRANSITION METAL with important catalytic properties. It is also notable in that it can absorb as much as 1,000 times its own volume of hydrogen by the process of OCCLUSION.

palmitic acid A saturated FATTY ACID that occurs in many plant and animal fats.

palynology *See* POLLEN.

pancreas A GLAND in all vertebrates (except some fish) that is part of the DIGESTIVE SYSTEM and located near the DUODENUM. In humans the pancreas is behind and below the stomach and is about 18 cm long. The pancreas has two major roles, as both an EXOCRINE GLAND and an ENDOCRINE GLAND.

As an endocrine gland, specialized groups of cells within the pancreas called the ISLETS OF LANGERHANS secrete the hormones INSULIN and GLUCAGON directly into the blood, which act antagonistically to regulate blood sugar levels.

As an exocrine gland the pancreas produces an alkaline PANCREATIC JUICE, when stimulated by the hormones SECRETIN and CHOLECYSTOKININ, which is secreted into the duodenum where it aids digestion.

pancreatic juice A neutral secretion produced by the PANCREAS, when stimulated by the hormones SECRETIN and CHOLECYSTOKININ made by the SMALL INTESTINE. Pancreatic juice is secreted into the DUODENUM, where it aids digestion. The juice contains AMYLASES, PROTEASES, LIPASES and NUCLEASES. In order for these enzymes to function correctly, mineral ions (such as sodium hydrogencarbonate) present in the pancreatic juice neutralize the acid mix coming from the stomach. The enzymes are made by acinar cells in the tubules of the pancreas. *See also* SECRETIN.

pandemic *See* DISEASE.

Paneth cell *See* CRYPT OF LIEBERKÜHN.

Pangaea A single large land mass that is believed to have broken up into the TECTONIC PLATES that form the continents we observe today.

pantothenic acid *See* VITAMIN B.

paper chromatography CHROMATOGRAPHY in which the stationary phase is absorbent paper. A spot of the mixture to be separated is placed near the base of a paper strip, the bottom edge of which is placed in a solvent. The solvent moves up the paper by the CAPILLARY EFFECT, and the different components in the mixture are carried along at different rates. Colourless components may be made visible by 'developing' the CHROMATOGRAM – spraying it with some material that reacts to make the components visible. Colourless components that fluoresce (*see* FLUORESCENCE) may be viewed in ultraviolet light.

para In organic chemistry, a prefix used in disubstituted derivatives of BENZENE which indicates that the substituted groups are at positions 1 and 4 in the ring. An example is 1,4-nitrophenol (*para*-nitrophenol, *p*-nitrophenol):

The use of the term *para* in other situations, for example, paraformaldehyde, has no structural significance and is always written in full. *See also* META, ORTHO.

parabola A CONIC SECTION formed when the plane intersecting the cone is parallel to a straight line drawn on the surface of the cone. A parabola can also be represented by any equation of the form

$$y = ax^2 + bx + c$$

where *a*, *b* and *c* are constants.

parabolic mirror A MIRROR with a reflecting surface having a shape with a cross-section that is part of a PARABOLA (this shape is called a PARABOLOID of revolution). *See* CURVED MIRROR.

paraboloid A three-dimensional surface that has a parabolic (*see* PARABOLA) cross-section in one direction and a circular cross-section in the perpendicular direction.

paracetamol, *4-acetamidophenol* ($C_8H_9NO_2$) A popular pain-killing drug that also reduces fever. It is less damaging to the stomach than ASPIRIN, but liver damage can be caused by excess doses.

paraffin *See* ALKANE.

parallel (*adj.*) **1.** (*mathematics*) Describing two lines or planes, such that the closest distance from a point on one line or plane to the other line or plane is always the same, regardless of which point is chosen.

2. (*physics*) Describing electric devices that are connected in such a way that current can return to the power supply by flowing through one device or another. The voltage across each element in a parallel circuit will be the same, whilst the total current flowing will be equal to the sum of the currents flowing in the individual elements.

parallel processing A computer development that involves having several circuits working on different parts of a problem at the same time. *See also* SERIAL LOGIC.

parallelogram A geometrical figure with four straight sides, with pairs of sides parallel to one another and with parallel sides having equal lengths. Unlike a rectangle, the two pairs of sides are not at right angles to one another. The area of a parallelogram is equal to one half the length of one of the sides multiplied by the perpendicular distance between the two sides having this length.

paramagnetism A weak form of magnetism found in some elements and molecules (such as O_2), causing these materials to have a RELATIVE PERMEABILITY slightly greater than 1. A paramagnetic material will align parallel to any applied magnetic field.

Paramagnetic materials contain electrons with a total ANGULAR MOMENTUM that is not zero. This causes each atom or molecule to behave as a MAGNETIC DIPOLE and these dipoles will line up in an external magnetic field. The alignment is destroyed by thermal vibrations, so paramagnetism generally decreases with increasing temperature.

parameter A quantity in a mathematical equation on which the other variables depend, but which is kept constant while the other variables are being investigated. The calculation may later be repeated with a different value for the parameter. *See also* PARAMETRIC EQUATION.

parametric equation A mathematical equation in which a number of variables are expressed in terms of one or more PARAMETERS. Thus for a particle moving in a circle, the position of the particle may be expressed in terms of the parameter *t*, with

$$x = r\cos \omega t,\ y = r\sin\omega t$$

Calculating *x* and *y* for just one value of *t* gives a single point on the circle. By using all values of *t*, the whole of the path is plotted.

paramylon *See* EUGLENOPHYTA.

parapodium In POLYCHAETA, any of the paired fleshy appendages with numerous chaetae (bristles), used for locomotion.

parasite An organism living on another organism (the host) and dependent on it for nourishment at the expense of the host. A parasitic organism must spend some time on the host (so a fly bite is not a parasitic relationship). Parasites that live inside the host, for example the liver tapeworm, are endoparasites, and parasites that live outside the host, for example fleas and lice, are ectoparasites.

Obligate parasites cannot survive without their host, while facultative parasites can, and partial parasites, for example mistletoe, can photosynthesize (*see* PHOTOSYNTHESIS) and also be parasitic. It is crucial for the parasite to be able to reproduce in a host. The host a parasite reproduces in (or lives in when it becomes sexually mature) is a primary (definitive) host. A secondary (intermediate) host is one in which the parasite does not reproduce but it used for some stages of its development. Parasites often have complex life cycles with one or two hosts interspersed with periods of free-living. The host can mount an IMMUNE RESPONSE to a parasite, but parasites are very resistant to immune attacks.

Examples of parasites include *Phytophthora infestans,* which causes potato blight (causing serious crop loss), the malarial parasite *Plasmodium* (*see* MALARIA) and parasitic flatworms (tapeworms and flukes). Liverfluke causes epidemics of 'liver rot' in sheep and cattle. The pork tapeworm *Taenia solium* uses the pig as its intermediate host and then attaches to the intestinal mucous membrane of humans as its primary host. Its eggs are passed out in the host's faeces. Parasites cause a great deal of damage, as diseases in humans and domestic animals and as a cause of crop damage.

parasitism An association between two different species in which one partner (the PARASITE) gains considerably at the expense of the other (the host). Parasitism can be considered as a variation of SYMBIOSIS.

parasympathetic nervous system *See* AUTONOMIC NERVOUS SYSTEM.

parathormone *See* PARATHYROID GLAND.

parathyroid gland Either one of a pair of ENDOCRINE GLANDS found near to or embedded within the THYROID GLAND. Parathyroid glands produce parathormone, which regulates the level of calcium in the blood to ensure that it is high enough to permit normal muscle and nerve activity. Another hormone called calcitonin works antagonistically with parathormone – calcitonin reduces levels of calcium ions in the blood, and together they maintain the correct balance.

parenchyma A simple (one cell-type only) plant tissue, consisting of loosely packed, almost spherical, cells with thin CELLULOSE walls that provide the main packing tissue within a plant. Parenchyma cells may also store food, and intercellular spaces allow gas exchange to occur. Photosynthetic parenchyma is called CHLORENCHYMA and is found in some stems and the MESOPHYLL of leaves (*see* LEAF). EPIDERMIS is a specialized parenchyma. *See also* GUARD CELL, STOMA.

parity The property of being even or odd. In computing, parity is a scheme for detecting and correcting errors. Each piece of data must have an odd or even number of BITS. If the data should have an even number of bits and comes out with an odd number, then an extra bit, called the parity bit, is added. This scheme helps detect all one-bit errors.

parotid gland A large SALIVARY GLAND.

parsec A unit of distance used in astronomy. One parsec is the distance of a star that makes an angle of one arc second (1/3,600 degrees) between the Earth and the Sun. One parsec is equivalent 3.09×10^{13} km.

parthenogenesis The process by which a female egg (OVUM) develops without being fertilized by a male GAMETE. It is a modified form of SEXUAL REPRODUCTION, although because the offspring are from a single parent it is ASEXUAL REPRODUCTION. In some cases FERTILIZATION with male SPERM is the stimulus for parthenogenesis, even though the male chromosomes are not taken into the nucleus. Parthenogenesis can be induced artificially, for example in rabbits, by stimulating the egg. Some plants, for example dandelions, and certain fish reproduce by parthenogenesis, and others use it at some stage during their life cycle, for example aphids use it in the summer to build up large population numbers. In the honey bee, unfertilized eggs develop into male bees and fertilized eggs develop into female bees.

partial parasite An organism that is capable of PHOTOSYNTHESIS but is also parasitic. *See* PARASITE.

partial pressure In a mixture of gases, that part of overall pressure that can be attributed to the presence of one specified gas in the mixture; that is, the pressure that it would exert if it were alone. *See also* DALTON'S LAW OF PARTIAL PRESSURE.

particle accelerator A device designed to accelerate charged particles to high energies so that they can be fired at targets or made to collide with particle beams moving in the opposite direction. An electric field is used to accelerate the particles. In modern machines the particles will pass through this electric field many times over. Magnetic fields are used to focus the particle beam, and in some machines to steer the beam around a circular path. *See also* COLLIDING BEAM EXPERIMENTS, CYCLOTRON, LINEAR ACCELERATOR, SYNCHROTRON, SYNCHROCYCLOTRON.

particle detector A device used in particle physics to detect and track the paths of SUBATOMIC PARTICLES. They are used to identify new particles produced when fast-moving electrons or protons hit a target or an oncoming beam of POSITRONS or antiprotons. Many such experiments use a combination of detectors combined with shielding through which only certain types of particle (such as MUONS) can pass. A common feature of most detectors is that they operate in a magnetic field so the CHARGE and MOMENTUM of the particles can be deduced from the curvature of the tracks they leave in the detector.

See also BUBBLE CHAMBER, CLOUD CHAMBER, DRIFT CHAMBER, PARTICLE ACCELERATOR, SPARK CHAMBER.

particle physics The study of ELEMENTARY PARTICLES. *See also* GRAND UNIFIED THEORY, QUANTUM THEORY, STANDARD MODEL.

partition constant A quantity used to determine the feasibility of SOLVENT EXTRACTION. For any given material, the partition constant is determined by dissolving some of the material in a mixture of SOLVENTS. The partition constant is the ratio of the concentration in solvent B to the concentration in solvent A. It is largely independent of the amount of material used and the amount of solvent present, provided neither concentration approaches saturation. If the partition constant is large, solvent extraction can be successfully used with solvent B extracting the material from solvent A. *See also* HENRY'S LAW.

parturition In mammals, the birth of a FOETUS at the end of a full-term PREGNANCY. The precise control of the onset of labour and the subsequent birth is not fully understood, but the foetus seems to play a role by stimulating release of PROSTAGLANDIN from the UTERUS, which causes it to contract. Synthetic prostaglandins can be used to induce labour. The PEPTIDE hormone OXYTOCIN also plays a part in parturition.

pascal (Pa) The SI UNIT of PRESSURE. One pascal is equivalent to a pressure of one NEWTON per metre squared.

Pascal A computer PROGRAMMING LANGUAGE that is quite simple and therefore is good as a teaching program. However, it has limited use in applications outside teaching.

Pascal's law In any confined fluid, pressure is transmitted uniformly throughout the fluid and acts at right angles on any surface in contact with that fluid. *See also* HYDROSTATIC PRESSURE.

Pascal's triangle A triangular array of numbers. Each number is the sum of the two numbers directly above it.

1
1 1
1 2 1
1 3 3 1
1 4 6 4 1
1 5 10 10 5 1

The numbers in Pascal's triangle are also the BINOMIAL COEFFICIENTS, with the *r*th element in the *n*th line (counting the top line as line zero) being the binomial coefficient ^nC_r.

passive device An electronic device, such as a RESISTOR, CAPACITOR or INDUCTOR, that cannot amplify a current. The behaviour of passive devices can be described fully by simple mathematical equations. *See also* ACTIVE DEVICE.

passive immunity *See* IMMUNITY.

pasteurization A process used to kill some disease-causing bacteria present in food without spoiling the food. The process was discovered by Louis Pasteur (1822–1895) in the 1850s. He first used it on wine and beer by heating it to slow down the multiplication of bacteria and hence the souring process. The most well-known pasteurized food product is pasteurized milk, which is heated to 72°C for 15 seconds and rapidly cooled to 10°C, which kills the organisms causing tuberculosis, typhoid, diphtheria and dysentery. Some beneficial bacteria are also killed, which reduces the nutritional value of milk.

path difference The difference in the length of two paths by which light beams travel from a source to a point where they overlap and interfere (*see* INTERFERENCE). Path difference is often expressed as a number of wavelengths of the light in the material concerned (this is sometimes called an optical path difference). One important example of interference is the case where two waves start in PHASE and interfere after travelling along different paths. For CONSTRUCTIVE INTERFERENCE the path difference must be either zero or a whole number of wavelengths. For DESTRUCTIVE INTERFERENCE, the path difference must be a whole number of wavelengths plus an odd half-wavelength.

pathogen A disease-causing agent, usually a MICRO-ORGANISM.

Pauli exclusion principle A consequence of QUANTUM THEORY, which states that no two fermions may occupy the same quantum state. This has applications in the arrangements of electrons in atoms, where each ORBITAL contains only two electrons, each in one of the two possible SPIN states, described as spin up and spin down, depending on the direction of the spin ANGULAR MOMENTUM relative to some reference direction.

Pauling electronegativity A commonly used scale for comparing the ELECTRONEGATIVITY of elements. Pauling measured electronegativities from differences in the energies of COVALENT BONDS and deduced a scale in which the most electronegative element, fluorine, is allocated an electronegativity of 4.0. The least electronegative elements, the ALKALI METALS have electronegativities of about 0.9 on this scale.

p-block element Any chemical element in GROUPS 13, 14, 15, 16, 17 and 18 (formerly groups III to VIII) of the PERIODIC TABLE; that is, an element in which the outer electrons are in P-ORBITALS.

p.d. *See* POTENTIAL DIFFERENCE.

peat A brownish compact mass of partly decayed vegetation, saturated with water, that has accumulated in wetland areas. It is generally the first stage in the formation of COAL. It is used as a fertilizer and, when dried, as fuel.

pectoral fin *See* FIN.

pectoral girdle, *shoulder girdle* In vertebrates, the skeletal support, within the THORAX, for the trunk that consists of the bones and muscles necessary for attachment of the upper limbs or fins. The pectoral girdle is attached to the sternum (breast bone) at the front of the chest. The scapula (shoulder blade) and clavicle (collar bone) form part of the pectoral girdle.

pedicel In botany, a small stalk of an individual flower of an INFLORESCENCE.

peduncle In botany, the main stalk of an INFLORESCENCE, or a stalk bearing a single flower. It contains XYLEM and PHLOEM to provide nourishment for the flower.

pelagic (*adj.*) Describing animals and plants that live in the mass of water of the sea or a lake, in contrast to the bottom of the water (benthos). Pelagic organisms can be divided into PLANKTON and NEKTON.

Pelecypoda A CLASS of the PHYLUM MOLLUSCA (the bivalves), including mussels. Bivalves have shells that are in two halves and they have an indistinct head with no tentacles. They are filter feeders that trap organic matter from the water around them in CILIA located in GILLS.

pellicle The outer layer of protein that protects and maintains the shape of certain unicellular organisms, such as *Euglena*, of the phylum EUGLENOPHYTA.

pelvic fin *See* FIN.

pelvic girdle, *hip girdle* In vertebrates, a set of bones in the PELVIS providing support and a site for muscle attachment to enable movement of the legs, hindlimbs or back fins.

pelvis In vertebrates, the lower region of the ABDOMEN that is bounded by a set of bones called the PELVIC GIRDLE (hip girdle) to which the lower limbs and the muscles needed for their movement are attached.

pendulum A system in which a point-like mass exhibits oscillatory motion. *See* SIMPLE PENDULUM.

penicillin The first ANTIBIOTIC, discovered in 1929 by Alexander Fleming (1881–1955), that now forms a family of antibiotics obtained from moulds of the genus *Penicillium*. It can also be produced synthetically.

penis The male sex organ, used to insert sperm into the female VAGINA during sexual intercourse (*see* SEXUAL REPRODUCTION). The penis contains erectile tissue and a good blood supply that allows it to remain stiff when filled with blood (upon sexual stimulation). In most mammals (but not humans) the penis is stiffened by a bone. The penis itself consists of spongy tissue covered by an elastic skin and the end is expanded to form the sensitive region (the glans penis), which is protected by a retractable skin called the foreskin.

pentadactyl limb The basic limb found in mammals, birds, reptiles and amphibians, consisting of an upper limb with one bone, a lower limb with two bones and a hand/foot with five digits (fingers/toes). This basic pattern has been modified by loss or fusion of bones to suit the lifestyle of different vertebrate species; sometimes it has been greatly altered, for example in birds and bats for flying and in whales for swimming.

pentahydrate A HYDRATE containing five parts water to one part of the compound. For example, sodium thiosulphate forms the pentahydrate $Na_2S_2O_3.5H_2O$.

pentane (C_5H_{12}) The fifth member of the ALKANE series. There are three possible ISOMERS of pentane. *n*-Pentane has a boiling point of 38°C. The pentanes are all inflammable liquids extracted from petroleum.

pentanol, *amyl alcohol* ($C_5H_{11}OH$) The fifth member in the series of ALCOHOLS. There are eight isomers of this formula. 1-Pentanol has a boiling point of 137°C. They are all oily liquids that do not mix with water.

pentose A MONOSACCHARIDE containing five carbon atoms in the molecule. Examples are RIBOSE and DEOXYRIBOSE, the sugar components in NUCLEIC ACIDS.

pepo A type of berry with a hard exterior, for example cucumber fruit.

pepsin A PROTEASE enzyme found in the stomach that breaks down proteins during digestion. Pepsin is secreted as its precursor, pepsinogen, which is activated by the stomach acidity to form pepsin.

pepsinogen *See* PEPSIN.

peptide Two or more AMINO ACIDS joined together by a peptide bond (–CO-NH–) that forms between the amino group (NH_2) of one amino acid and the carboxyl group (COOH) of another amino acid with the loss of a water molecule. This is called a CONDENSATION REACTION. A long chain peptide (more than three amino acids) is called a polypeptide, which can then fold or twist to form a PROTEIN. Some peptides are HORMONES, for example OXYTOCIN and ANTI-DIURETIC HORMONE.

percentage A fraction expressed with a denominator of one hundred, denoted by the symbol %. Thus 5% is a fraction of 5/100 or 0.05.

perchlorate Any salt containing the chlorate(VII) ion, ClO_4^-.

perchlorous acid *See* CHLORIC ACID.

perennation The ability of plants to survive for more than one year by means of underground storage organs such as a CORM, RHIZOME, BULB, ROOT or TUBER.

perennial plant A plant that continues to grow year after year. In woody perennials, such as trees, the stems remain above ground during winter and new growth starts from these in the spring. In herbaceous perennials, the above-ground growth dies in the winter and new growth begins from an underground storage organ such as a RHIZOME or CORM. *Compare* ANNUAL PLANT, BIENNIAL PLANT.

perianth A collective term for the outer layers of a flower, consisting of the PETALS and SEPALS. Petals form the inner layer of the perianth. Collectively, the petals are called the corolla and their function is to attract pollinators such as insects or birds. The sepals form the outer layer of the perianth. Collectively, the sepals are known as the calyx. Sepals are usually green and are a site of PHOTOSYNTHESIS. The main function of sepals is to protect the other floral parts when the flower is a bud.

In most DICOTYLEDONS the perianth is composed of two whorls, the calyx and corolla, but in MONOCOTYLEDONS these two are indistinguishable and are called tepals. Dicotyledons usually have sepals and petals in groups of five, and monocotyledons usually have them in groups of three.

pericarp The wall of a fruit. In fleshy fruits such as berries, the pericarp is composed of three layers, a tough outer exocarp, a middle mesocarp that is often fleshy, and an inner endocarp that surrounds the seeds. In fruits such as the nut, the pericarp is dry and hard.

pericline A dome-shaped arrangement of rock STRATA. If this contains a layer of porous rock surrounded by impermeable rock, oil or natural gas may be trapped in the porous rock.

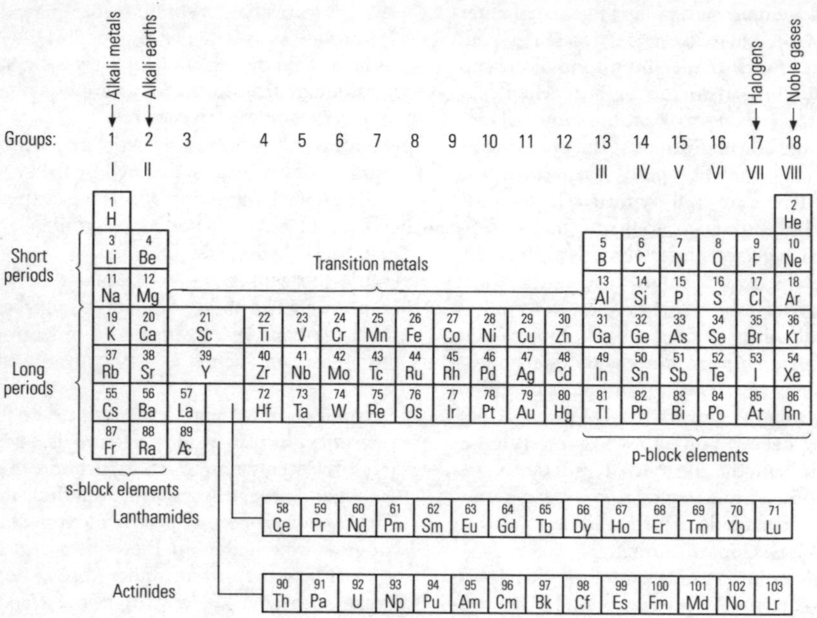

The periodic table of the elements.

pericycle *See* ROOT.

perigee For an object in orbit around the Earth, the position in the orbit that is closest to the Earth.

perihelion For an object in orbit around the Sun, the position in the orbit that is closest to the Sun.

perilymph Fluid found in the inner EAR. *See* COCHLEA.

period 1. (*physics*) The time taken to complete an ORBIT or OSCILLATION.

2. (*chemistry*) A horizontal row of elements in the PERIODIC TABLE, containing a sequence of elements of consecutive atomic number and steadily changing chemical properties. *See also* LONG PERIOD, SHORT PERIOD.

3. (*geology*) A division of geological time, a subdivision of an ERA, which is further subdivided into EPOCHS.

periodic acid *See* IODIC ACID.

periodic function Any mathematical function that takes on the same value repeatedly for regular intervals of the independent variable. In particular the trigonometric functions, SINE, COSINE, TANGENT etc., have a period of 360° (or 2π radians). So

$$\sin(\theta+360°) = \sin\theta$$

for any θ.

periodic law The principle that certain physical and chemical properties of the elements, such as IONIZATION ENERGY, reactivity, melting point and boiling point, vary in a periodic way with ATOMIC NUMBER. It was this recognition that led to the construction of the PERIODIC TABLE.

periodic table An arrangement of the ELEMENTS in order of increasing ATOMIC NUMBER that brings out the similarities between them. Atomic number increases along the rows of the table (called PERIODS) and the table is arranged so that elements with similar properties are placed in the same column or GROUP. Thus the left-hand column of the table, group 1 (formerly group I), contains the ALKALI METALS, lithium, sodium, potassium etc., each of which is a metal that reacts violently with water to form a HYDROXIDE containing a singly charged positive metal ion. The next column, group 2 (formerly group II), contains the ALKALINE EARTHS, metals that are reactive enough to form oxides on exposure to air, but which react only slowly with water. To the right of the table are group 17 (formerly group VII) and group 18 (formerly group VIII). Group 17 contains the HALOGENS, reactive elements that tend to form singly charged negative ions, whilst group 18 contains the NOBLE GASES, with a full outer shell of electrons, which are the least reactive of the elements.

In the middle of the table are the TRANSITION METALS, which show only slight changes with increasing atomic number, as electrons fill the D ORBITALS.

The periodic table also contains two long series of elements, the LANTHANIDES and the ACTINIDES. Within each of these series the chemical properties are generally very similar, with each element differing only in the number of electrons in f-orbitals.

See also SHELL.

periosteum A tough fibrous membrane surrounding bones, to which muscles and tendons attach.

peripheral device Any device that attaches to a computer, such as a printer, scanner, tape drive, external hard drive, modem, etc.

peripheral nervous system (PNS) The combined SENSORY SYSTEM and EFFECTOR SYSTEM.

periscope A device that enables objects to be viewed over or around some obstacle preventing direct viewing – such as the ocean surface in the case of a submarine periscope. A pair of totally internally reflecting prisms (*see* TOTAL INTERNAL REFLECTION) are used to bend the light through 90° at the top and bottom of the periscope tube.

peristalsis Involuntary waves of movement, caused by contraction of smooth muscle, that pass along tubular organs, for example in the intestines and earthworms. Peristalsis helps to move material along the tube.

peritoneal cavity The body cavity enclosing the DIGESTIVE SYSTEM, surrounded by an EPITHELIAL membrane called the peritoneum and containing peritoneal fluid.

peritoneal fluid The fluid contained in the PERITONEAL CAVITY.

peritoneum The SEROUS MEMBRANE lining the PERITONEAL CAVITY.

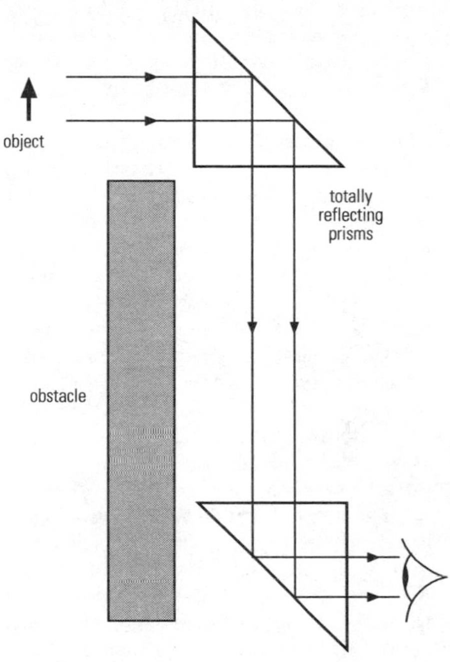

Periscope.

permanent gas A gas that cannot be turned into a liquid by the action of pressure alone. Oxygen and nitrogen are examples. *See* GAS.

permanent hardness HARDNESS in water that cannot be removed by boiling. It is caused principally by dissolved calcium sulphate and, to a lesser extent, magnesium sulphate. *See* WATER SOFTENING. *See also* HARD WATER, TEMPORARY HARDNESS.

permanent magnet A MAGNET that retains its magnetism indefinitely. The atoms in a permanent magnet each behave like a small BAR MAGNET, due to the imbalance in the rotational motion of their electrons – effectively an electric current within the atom. In FERROMAGNETIC materials these atoms are all aligned in a single direction over a region of space called a DOMAIN. In an unmagnetized piece of a ferromagnetic material, neighbouring domains are magnetized in different directions, so the total magnetic effect is zero. Under the influence of an external magnetic field, the domains which are aligned in the same direction as the field tend to grow, producing INDUCED MAGNETISM.

Once all the domains are aligned the material cannot be magnetized any further – it is then described as saturated. In some materials, such as pure iron, that are described as magnetically soft, the domains do not retain their alignment when the applied field is removed.

The amount of magnetism remaining when the applied field is removed is called the REMANENCE, whilst the amount of reverse magnetic field needed to demagnetize a magnetized material is called the COERCIVE FORCE. Magnetically soft materials are used in ELECTROMAGNETS and TRANSFORMER cores, where the direction of magnetization must constantly be reversed without any unnecessary waste of energy. The alignment of domains can be destroyed by heating – the temperature at which a material loses its ferromagnetic properties is called the CURIE POINT.

The removal of any permanent magnetism, called DEGAUSSING, can also be achieved by moving the material through a SOLENOID in which an ALTERNATING CURRENT is flowing. As the sample moves past the solenoid, it is magnetized first in one direction then the other, with successive magnetizations becoming weaker and weaker.

permanganate The old name for MANGANATE(VII).

permeability (μ) A measure of the degree to which a material may be magnetized. The absolute permeability μ is the ratio of MAGNETIC FLUX density B induced in a medium to the MAGNETIC FIELD strength H of the external field inducing it:

$$\mu = B/H$$

The RELATIVE PERMEABILITY μ_r is the ratio of the magnetic flux density of a material to that induced in free space μ_0 by the same magnetic field strength:

$$\mu r = \mu/\mu_0$$

The quantity μ_0 is known as the permeability of free space and has the value $4\pi \times 10^{-7}$ HENRY per metre. *See also* AMPERE.

Permian The last PERIOD of the PALAEOZOIC ERA, which began about 280 million years ago at the end of the CARBONIFEROUS period and lasted about 55 million years. The era was characterized by widespread deserts. Amphibians and reptiles were the dominant land animals and GYMNOSPERMS replaced ferns, clubmosses and horsetails as the dominant plants.

permittivity (ε) The ratio of the ELECTRIC FLUX density induced a material to the external ELECTRIC FIELD strength inducing it. The permittivity of free space ε_o is a fundamental constant that measures the strength of the ELECTROSTATIC force in a vacuum. It is given in the equation:

$$F = Q_1Q_2/4\pi\varepsilon_o r^2$$

where F is the force between two charges Q_1 and Q_2 separated by a distance r in a vacuum. This equation is a statement of COULOMB'S LAW. ε_o has the value 8.85×10^{-12} $\mathrm{Fm^{-1}}$. *See also* RELATIVE PERMITTIVITY.

permutation Any one of the possible ways in which a number of items can be chosen from a set. In counting permutations, the order in which the objects are taken is significant, so in picking three letters from the alphabet, *abc*, and *acb* are different permutations. In selecting n objects, from a group of N, the number of permutations possible is

$$N!/(N-n)!$$

if no object can be selected more than once. *Compare* COMBINATION.

permutite Any naturally occurring mineral, such as ZEOLITE, which is used as an ION EXCHANGE medium in WATER-SOFTENING processes. Modern synthetic materials have largely replaced these materials.

pernicious anaemia A deficiency of VITAMIN B_{12} where RED BLOOD CELLS are malformed, resulting in bruising and slow recovery from minor injuries.

peroxide An OXIDE containing the O_2^{2-} ion. They are all strong OXIDIZING AGENTS, such as sodium peroxide, Na_2O_2.

peroxisome, *microbody* A small, spherical, membrane-bound body within a eukaryotic cell (*see* EUKARYOTE) that contains the enzyme CATALASE. Peroxisomes have very little internal structure other than being granular. Catalase breaks down hydrogen peroxide, which is a toxic by-product of several biochemical reactions. Peroxisomes are found particularly in cells that are metabolically active, such as in the liver.

peroxosulphuric(VI) acid (H_2SO_5) A colourless crystalline solid, which decomposes on heating.

Peroxosulphuric(VI) acid can be made by the OXIDATION of concentrated sulphuric acid by hydrogen peroxide:

$$H_2SO_4 + H_2O_2 \rightarrow H_2SO_5 + H_2O$$

The reaction is reversed on exposure to excess water or on heating. Peroxosulphuric(VI) acid is a powerful OXIDIZING AGENT.

perpendicular (*n., adj.*) A line at an angle of 90° (at right angles) to a specified plane or line. The point at which the perpendicular meets the line or plane is sometimes called the foot of the perpendicular.

perpendicular bisector A line that is at right angles to a line joining two points and which divides that line into two sections of equal length.

perpetual motion The state of some hypothetical machine which, once set in motion, would continue to move for ever without any further energy input.

The existence of friction and other dissipative forces make perpetual motion impossible on a macroscopic scale, though electrons in atoms, for example, do represent a form of perpetual motion. Systems such as these are sometimes described as showing perpetual movement rather than perpetual motion, since the original idea was that perpetual motion machines might be used as sources of energy. Once the LAW OF CONSERVATION OF ENERGY was understood, interest in such devices declined.

personal computer, *microcomputer* A computer designed to be used by one person only at a time. In contrast to MAINFRAMES and MINICOMPUTERS, which have networks of terminals that allow a number of people to work at the same time.

Personal computers are either designed to fit on a person's desktop, or are portable. A number of personal computers can be linked together in a NETWORK so that users can share programs and data.

Perspex (trade name for ***poly(methylmethacrylate)***, ***poly(methyl 2-methylpropenoate)***) A clear, tough plastic with many uses, for example in advertising signs, protective screens and aeroplane canopies. It is an addition POLYMER containing carbon-carbon bonds similar to POLY(ETHENE) and POLYVINYL CHLORIDE but more rigid due to its structure-hindering movement of carbon chains.

pest Any living organism that has a detrimental effect on human welfare, profit or convenience. Pest does not include organisms that are the direct cause of disease. Most pests damage crops, livestock or buildings and may also spread disease. *See also* PESTICIDE.

pesticide A poisonous chemical, used to kill PESTS that may be hazardous to health or simply unwanted by humans. Pesticides are named after the organisms they kill: insecticides kill insects, fungicides kill fungi, herbicides kill plants considered to be weeds, and rodenticides kill rodents. Pesticides are used in farming and gardening but because of their toxic nature they can cause a pollution problem. A good pesticide is one that is specific to the organism to which it is directed, is easily broken down to harmless substances so it does not persist in the environment, and does not accumulate within an organism to be passed along the FOOD CHAIN. *See also* BIOMAGNIFICATION.

petal Part of a flower that forms the inner layer of the PERIANTH. The collective term for petals is the corolla. The main function of petals is to attract pollinators, such as insects and birds, and they are therefore often brightly coloured, large and scented and may produce a sweet NECTAR at their base. Petals are derived from modified leaves and may be absent or small in wind-pollinated plants.

petiole The stalk of a leaf. *See* LEAF.

petrochemical industry Those industries that use PETROLEUM or NATURAL GAS as the raw materials for the manufacture of a wide range of products, such as detergents, fertilizers, paints. Most petroleum is used in the manufacture of fuel and it is only the remaining 10 per cent that forms the basis of the petrochemical industry.

petrol, *gasoline* (*US*) A fuel used in motor vehicles which consists of a mixture of HYDROCARBONS obtained from PETROLEUM. Leaded petrol contains tetraethyl lead and dibromoethane to improve the engine performance but the lead from exhaust fumes enters the atmosphere as simple lead compounds which are thought to cause mental impairment in young children. For this reason unleaded petrol was introduced and is gradually taking over in the UK, although there is a suggestion that the fumes from this may contain some CARCINOGENS. Unleaded petrol has a lower OCTANE NUMBER than leaded petrol.

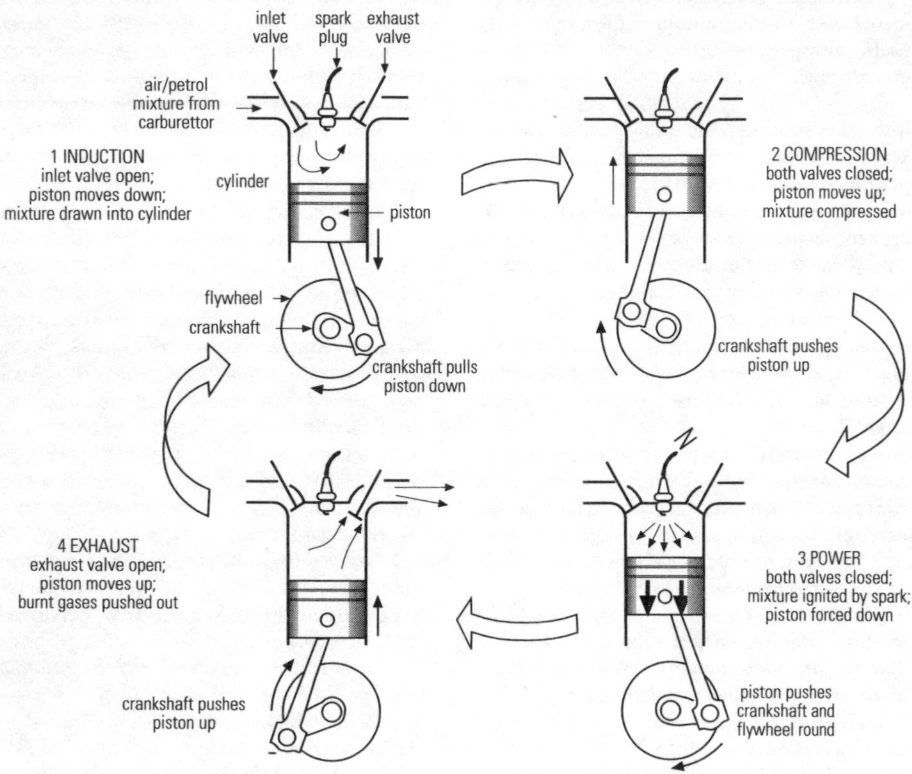

Cycle of operation of the petrol engine.

petrol engine An INTERNAL COMBUSTION ENGINE burning PETROL. Petrol engines are mostly used for smaller vehicles, such as cars, small vans and light aircraft.

A mixture of petrol and air is drawn into the engine by a PISTON moving down a CYLINDER during the INDUCTION STROKE. It is then compressed, and thus heated, as the piston moves back up the cylinder during the COMPRESSION STROKE. During this time the fuel vaporizes, and as the piston reaches the top of the stroke it is ignited by a spark plug, to which a high voltage from an INDUCTION COIL is connected. The expanding, burning gases force the piston back down the cylinder for the POWER STROKE, and as the piston returns up the cylinder, the burnt gases are pushed out of the cylinder in the EXHAUST STROKE. The cycle then repeats. This sequence is called the four-stroke cycle, since the piston moves up and down the cylinder four times for each load of fuel burnt. Many engines contain several cylinders, typically four for a car engine, to produce a more even supply of power.

See also CATALYTIC CONVERTER, COMPRESSION RATIO, DIESEL ENGINE, KNOCKING, PRE-IGNITION, TURBOCHARGING, TWO-STROKE CYCLE.

petroleum, *crude oil* A natural mineral oil, one of the three FOSSIL FUELS that have formed from the decayed remains of marine micro-organisms millions of years ago. Petroleum is found in layers of porous rock such as limestone and sandstone, often below a pocket of NATURAL GAS, which keeps it under pressure. Thus when a pipe from an oil well is forced into an oil pocket sometimes the oil rises up the borehole under natural pressure, although often pumping is necessary to extract it.

Petroleum consists of HYDROCARBONS mixed with varying amounts of oxygen, sulphur, nitrogen or other elements. It is a green/brown flammable liquid. Crude petroleum is separated by FRACTIONAL DISTILLATION into other products, including fuel oil, petrol, kerosene, diesel, lubricating oil, paraffin wax, petroleum jelly and asphalt. The larger ALKANES obtained from fractional distillation are broken down further by a process called CRACKING into lower alkanes, which are more in demand as motor or aviation fuel.

About 90 per cent of the world's oil supply is used as fuel and the rest is used in the manufacture of many everyday products, such as detergents, plastics, insecticides, fertilizers, drugs, synthetic fibres, paints, toiletries. The burning of petroleum fuel is a major cause of environmental pollution, as are the oil spillages occurring during transportation.

petrology The branch of GEOLOGY that deals with rocks, in particular their formation and chemical and physical structure.

pH A measure of the acidity or alkalinity of a solution. pH 7 indicates a neutral solution, numbers smaller than this indicate ACIDIC solutions, whilst larger numbers indicate ALKALINE solutions. The pH value actually indicates the concentration of hydrogen ions in the solution. If $[H^+]$ is the CONCENTRATION of hydrogen ions in mol dm^{-3}:

$$pH = -\log[H^+]$$

Phaeophyta A PHYLUM from the KINGDOM PROTOCTISTA that consists of brown ALGAE. Brown algae are diverse, and most are marine. They include all the larger seaweeds and some smaller ones. No brown alga is unicellular. Brown algae contain CHLOROPHYLL a and c and other pigments, which give them their colour. Both ASEXUAL REPRODUCTION and SEXUAL REPRODUCTION can occur.

phage *See* BACTERIOPHAGE.

phagocyte A type of WHITE BLOOD CELL that can ingest particles (foreign micro-organisms or food) from its surroundings to form internal VACUOLES consisting of the particle surrounded by the CELL MEMBRANE (PHAGOCYTOSIS). The vacuoles may fuse with LYSOSOMES and the particle is then destroyed or broken down. MACROPHAGES are phagocytic but are larger than other phagocytes and have a longer life span.

phagocytosis The process by which cells obtain particles too large to enter by DIFFUSION or ACTIVE TRANSPORT. The cell invaginates to form a cup-shaped depression that contains the particle. This is then pinched off to form a VACUOLE (or PHAGOSOME), and a LYSOSOME fuses with it to become a heterophagosome (a type of secondary lysosome; *see* PHAGOSOME). Enzymes are then released to break down the particle and useful elements are absorbed by the cell. This process occurs in *Amoeba* as a means of feeding and in a few specialized cells (PHAGOCYTES) such as WHITE BLOOD CELLS and MACROPHAGES, which ingest harmful bacteria or other invading bodies. *See also* PINOCYTOSIS, POTOCYTOSIS.

phagosome In biology, a type of secondary LYSOSOME. Heterophagosomes result from the fusion of primary lysosomes with substances entering the cell for digestion, for example bacteria. Such substances are then digested by the enzymes contained within the lysosome. Some of the products pass out through the lysosomal membrane and others remain as residual bodies.

Autophagosomes result from the fusion of a primary lysosome with cell ORGANELLES wrapped in a membrane (probably originating from smooth ENDOPLASMIC RETICULUM). This may occur during cell reorganization or to remove worn-out organelles.

Phanerozoic The division of geological time comprising the PALAEOZOIC, the MESOZOIC and CENOZOIC (the current) ERAS. It began about 570 million years ago, after the PRECAMBRIAN. Rock strata from the Phanerozoic may be identified by the existence of fossils, whereas there are relatively few fossils dating from the Precambrian.

pharynx The cavity in the throat at the back of the mouth. The upper part of the pharynx (nasopharynx) is an airway but the rest is for the passage of food. Air enters from the nostrils and into the pharynx for passage, then to the TRACHEA (windpipe). Food is pushed to the pharynx by the tongue for passage into the OESOPHAGUS, leading on to the stomach. To ensure that food goes down the oesophagus and air down the trachea, a number of reflexes exist. The EPIGLOTTIS closes the opening to the trachea and a soft palate closes the opening to the nasal cavity. In addition to the trachea and oesophagus meeting at the pharynx, the

EUSTACHIAN TUBES from each middle ear also enter the pharynx. *See also* RESPIRATORY SYSTEM.

phase 1. A measure of the stage that an OSCILLATION has reached at a given instant, particularly when comparing two oscillations. Phase is usually expressed as an angle, with the complete oscillation being represented by 360°, or 2π radians. Two oscillations moving together have a phase difference of 0 and are said to be in phase, whilst two oscillations exactly out of step with one another, so they are always moving in opposite directions, are said to be exactly out of phase, or to have a phase difference of 180° or π radians.

2. Any one of the different arrangements in which the molecules of a certain substance may exist, as a gas, liquid, or as one or more solid ALLOTROPES.

phase contrast microscopy A modification of a light MICROSCOPE that utilizes the different REFRACTIVE INDEX of features within an object so that a transparent object can be seen in detail without the need for a coloured stain.

phase diagram A graph of temperature against pressure that shows changes in melting and boiling points with pressure. Three lines on this diagram indicate the combinations of pressure and temperature at which two states can exist together in equilibrium. Where the three lines meet is the TRIPLE POINT. At pressures below that of the triple point, the liquid state does not exist and a solid when heated will sublime (turn from solid to gas). Carbon dioxide at normal (atmospheric) pressures is an example of this: solid carbon dioxide (called dry ice) will turn directly into a gas.

phase modulation A method of transmitting information in which the relative phase of a CARRIER WAVE is varied with the amplitude of the signal. *See also* MODULATION.

phase rule A rule that determines the number of DEGREES OF FREEDOM for any system containing one or more PHASES in equilibrium. If the number of phases present is P, the number of degrees of freedom is F and the number of chemically distinct components present in the system is C, then

$$F + P = C + 2$$

phase velocity In a travelling wave, the speed at which a point of a given PHASE, for example a crest in a TRANSVERSE WAVE, travels through a material. It is related to the frequency and wavelength of the wave by the equation

$$\text{speed} = \text{frequency} \times \text{wavelength}$$

In many materials the speed of the wave does not depend on the frequency or the wavelength but only on the material through which the wave is travelling. The stiffer the material the faster the wave will travel as the motion of one part of the material is passed on more effectively to neighbouring regions. Waves also travel more rapidly in less dense materials, as the smaller mass accelerates more rapidly. Materials in which the speed of the wave does depend on the wavelength of the wave are called DISPERSIVE.

For a solid of YOUNG'S MODULUS E and density ρ, the speed of the wave is given by

$$v = \sqrt{(E/\rho)}$$

For sound travelling in a gas at a pressure p and density ρ, with a RATIO OF SPECIFIC HEATS γ,

$$v = \sqrt{(\gamma p/\rho)}$$

For a transverse wave on a string stretched by a tension T, and with a mass per unit length of μ,

$$v = \sqrt{(T/\mu)}$$

phenol, *carbolic acid* (C_6H_5OH) The simplest member of a group of AROMATIC chemical compounds (the phenols) characterized by a HYDROXYL GROUP (OH) being attached to an aromatic ring. Phenol itself consists of a hydroxyl group attached to a BENZENE RING:

Na^+O^-

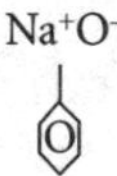

Pure phenol forms colourless crystals with a strong odour; melting point 43°C; boiling point 183°C. Phenol is insoluble in water at room temperature but MISCIBLE with water in all proportions at 84°C. Phenol is weakly acidic but is caustic and toxic by absorption through the skin. There are a number of processes by which phenol can be manufactured, for example, by the CUMENE PROCESS. Other phenols can be made by warming aqueous solutions of DIAZONIUM SALTS.

Since phenols possess a hydroxyl group they have many reactions in common with

ALCOHOLS. For example they form ESTERS with ACID CHLORIDES or ACID ANHYDRIDES. The carbon-oxygen bond is not easily broken so it is not possible to substitute a HALOGEN group in a phenol in the same way that HALOGENOALKANES are formed from alcohols. They are much more acidic than alcohols but less so than CARBOXYLIC ACIDS. A solution of phenol in alkali gives the PHENOXIDE ion.

Phenol is used in the manufacture of THERMOSETTING PLASTICS, dyes, explosives and pharmaceuticals.

phenol-formaldehyde resin RESIN formed by the reaction of PHENOL with FORMALDEHYDE. It is a THERMOSETTING PLASTIC, used for mouldings.

phenotype The actual visible traits of an organism, for example eye or hair colour, that is determined by the GENOTYPE and the effects of dominance (*see* ALLELE) and environmental factors. *See also* VARIATION.

phenoxide The ion formed when PHENOL is placed in an alkali solution, for example.

phenylalanine An ESSENTIAL AMINO ACID, normally converted to the amino acid TYROSINE in the body.

phenylamine, *aniline* ($C_6H_5NH_2$) A colourless, oily liquid with a distinctive odour; melting point –6.2°C; boiling point 184°C. It is a simple AROMATIC compound consisting of a BENZENE RING with an NH_2 group attached to the ring. Phenylamine is highly poisonous. On contact with air it turns brown. It is made by the reduction of NITROBENZENE. Phenylamine has an important use in the rubber industry and in the manufacture of dyes and drugs.

phenylethene, *styrene* (C_8H_8) An organic liquid made from BENZENE and ETHENE; boiling point 146°C. It is used to manufacture POLYMERS such as POLYSTYRENE and the synthetic rubber SBR.

phenyl group (C_6H_5) A group derived from BENZENE with the removal of a hydrogen atom.

pheromone A chemical signal (such as an odour), considered to be a 'social hormone', that operates in a hormone-like manner between individuals of a species rather than within an individual. Pheromones are often used to attract mates, for example female silk moths can attract male mates many kilometres away. Ants produce pheromones that warn other ants of danger, and bees produce several pheromones each with its own effects.

phloem A compound plant tissue (made of a number of cell types) with the main function of conducting sugars and other food materials from the leaves to other parts of the plant. Phloem consists of PARENCHYMA cells, sclereids and fibres (*see* SCLERENCHYMA), and specialized 'sieve elements' with associated companion cells. The sieve elements are long, thin-walled cells joined end to end, whose cross-walls partially break down to form 'sieve tubes' with 'sieve plates'. This results in the formation of large pores in the cross-wall, which allow the continuous passage of nutrients. The sieve tube elements are living but have no nucleus and very little CYTOPLASM, so depend on adjacent companion cells to sustain them. Phloem is usually found associated with XYLEM in structures (or vein) called VASCULAR BUNDLES. *See also* TRANSLOCATION.

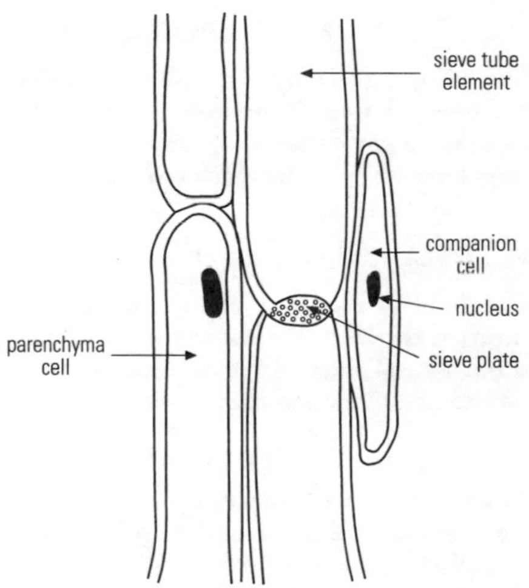

Longitudinal section of the phloem.

phon A unit of loudness which, unlike the DECIBEL, takes into account the differing response of the ear to sounds of differing frequencies. The loudness of a sound measured in phons is the same as the loudness in decibels of a 1 kHz tone perceived as being of equal loudness to the sound being measured.

phosgene *See* CARBONYL CHLORIDE.

phosphate Any salt containing the phosphate ion, PO_4^{3-}. Phosphates can be made by the reaction of metals with phosphoric acid, for example:

$$3Mg + 2H_3PO_4 \rightarrow Mg_3(PO_4)_2 + 3H_2$$

Phosphates, particularly ammonium phosphate, are widely used as fertilizers. Phosphates are extensively used as BUFFER SOLUTIONS.

phosphine (PH_3) A colourless gas; melting point –133°C; boiling point –88°C. Phosphine is similar to, but less reactive than, ammonia. It can form stable complexes (*see* COMPLEX ION) with TRANSITION METAL ions, which makes it highly toxic. Phosphine can be manufactured by the reaction between phosphorus and strong alkalis, for example:

$$8P + 15NaOH + 3H_2O \rightarrow 7PH_3 + 5Na_3PO_4$$

Phosphine is soluble in water, and will react with some acids to produce phosphinium salts, for example:

$$PH_3 + HCl \rightarrow PH_4Cl$$

phosphinic acid (H_3PO_3) A colourless solid; melting point 74°C; decomposes on further heating; RELATIVE DENSITY 1.7. Phosphinic acid may be made by hydrolysing (*see* HYDROLYSIS) phosphorus(III) chloride with water:

$$PCl_3 + 3H_2O \rightarrow H_3PO_3 + 3HCl$$

It is only dibasic, since one of the hydrogen atoms is covalently bonded to the phosphorus atom. Phosphinic acid decomposes on heating to give phosphine and phosphoric(V) acid:

$$4H_3PO_3 \rightarrow PH_3 + 3H_3PO_4$$

phosphoenolpyruvate (PEP) A METABOLITE that plays an important role in the light-independent stage of the PHOTOSYNTHESIS of C_4 PLANTS. *See* CALVIN CYCLE.

phospholipid A LIPID based on GLYCEROL and comprising a PHOSPHATE group, with one or more FATTY ACIDS. Phospholipids are found throughout living systems because they are found in CELL MEMBRANES. The phosphate end is HYDROPHILIC (water-loving) and the other end is HYDROPHOBIC (water-repelling). *See also* FLUID MOSAIC MODEL.

phosphor Any material that, when struck by electrons, converts some of the KINETIC ENERGY of the electrons into visible light.

phosphor bronze An alloy of copper with up to 10 per cent tin and up to 1 per cent phosphorus. The alloy is hard-wearing and corrosion resistant, and is frequently used for metal parts that must be strong and will be exposed to water, for example in ships' propellers.

phosphorescence The production of visible light as a result of a chemical reaction, electron bombardment or other process in which electrons move from one ENERGY LEVEL to a lower energy level. Also, the emission of visible light some time after light of a shorter wavelength (e.g. ultraviolet) has been absorbed. *Compare* FLUORESCENCE.

phosphoric(V) acid (H_3PO_4) A white solid; melting point 42°C; decomposes on further heating; RELATIVE DENSITY 1.8. Phosphoric acid is formed by burning phosphorus in a mixture of air and steam:

$$4P + 6H_2O + 5O_2 \rightarrow 4H_3PO_4$$

Phosphoric acid is highly soluble in water. It is a weak tribasic acid (*see* BASICITY), and readily forms both ACIDIC SALTS and NORMAL SALTS.

phosphorus (P) The element with ATOMIC NUMBER 15; RELATIVE ATOMIC MASS 31.0; melting point 44°C (white allotrope); boiling point 280°C (white allotrope), RELATIVE DENSITY 1.8 (white), 2.2 (red). It occurs in several ALLOTROPES, the most common of which are white phosphorus, which is toxic and highly reactive, and red phosphorus, which is less reactive and non toxic. White phosphorus ignites spontaneously on exposure to air.

Phosphorus is essential for life. It is required for the formation of NUCLEIC ACIDS and certain energy-carrying molecules (*see* ATP). Phosphorus is also an important constituent of bones and teeth. PHOSPHATES are used as fertilizers. *See also* PHOSPHORUS CYCLE.

phosphorus chloride Phosphorus(III) chloride (phosphorus trichloride), PCl_3, and phosphorus(V) chloride (phosphorus pentachloride), PCl_5.

Phosphorus(III) chloride is a colourless liquid; melting point –112°C; boiling point 76°C; RELATIVE DENSITY 1.6. It is produced by burning excess phosphorous in chlorine:

$$2P + 3Cl_2 \rightarrow 2PCl_3$$

Phosphorus(IV) chloride is a pale yellow solid; sublimes at 160°C; relative density 3.6. It is produced by heating phosphorus(III) chloride in chlorine:

$$PCl_3 + Cl_2 \rightarrow PCl_5$$

Both chlorides are hydrolysed (*see* HYDROLYSIS) by water:

$$PCl_3 + 3H_2O \rightarrow H_3PO_3 + 3HCl$$

$$PCl_5 + 4H_2O \rightarrow H_3PO_4 + 5HCl$$

Both of the above reactions are violent enough for the chloride to fume in moist air.

Phosphorus(III) chloride is an important material for the manufacture of organic phosphorus compounds. Phosphorus(IV) chloride is readily reduced to phosphorus(III) chloride and is used to introduce chlorine into organic molecules.

phosphorus cycle The recycling of phosphorous throughout the ECOSYSTEM. A wide range of important biological chemicals contain phosphorus, including NUCLEOTIDES, ATP and proteins. The main source of phosphorous is from rocks, which release it into the ecosystem through erosion. Plants absorb dissolved PHOSPHATES, which are passed to other animals in the FOOD CHAIN. Phosphates are recycled through plant and animal waste and decay upon death, including bones and shells, which can themselves be eroded to provide more dissolved phosphates or be deposited in rocks.

phosphorus oxide Phosphorus(III) oxide, P_4O_6, and phosphorus(V) oxide (phosphorus pentoxide), P_4O_{10}.

Phosphorus(III) oxide is a white waxy solid; melting point 24°C; boiling point 174°C. It can be formed by burning phosphorus in a limited supply of oxygen:

$$2P + 3O_2 \rightarrow P_2O_6$$

On heating it loses phosphorus and forms a polymeric form of phosphorus oxide. Phosphorus(III) oxide dissolves in water to give phosphonic acid:

$$P_4O_6 + 6H_2O \rightarrow 4H_3PO_3$$

Phosphorus(V) oxide is a white DELIQUESCENT powder; sublimes at 360°C; RELATIVE DENSITY 2.4. It can be formed by burning phosphorus in an excess of oxygen:

$$4P + 5O_2 \rightarrow P_4O_{10}$$

Phosphorus(V) oxide dissolves in water to produce phosphoric(V) acid:

$$P_4O_{10} + 6H_2O \rightarrow 4H_3PO_4$$

phosphorylation The transfer of a PHOSPHATE group by the enzyme phosphorylase, often from inorganic phosphate ions, to an organic compound. Phosphorylation can be photophosphorylation (using light energy; *see* PHOTOSYNTHESIS) or OXIDATIVE PHOSPHORYLATION (occurring during cellular RESPIRATION in aerobic cells).

photoautotroph *See* AUTOTROPH.

photocathode A negative ELECTRODE designed to release electrons when struck by light. *See* PHOTOELECTRIC CELL, PHOTOELECTRIC EFFECT.

photocell *See* PHOTOELECTRIC CELL.

photochemistry The study of reactions that are affected by light or ultraviolet radiation. PHOTONS of light can excite certain molecules, or break them into FREE RADICALS, allowing a reaction to proceed which would otherwise have too high an ACTIVATION ENERGY. For example the reaction between hydrogen and chlorine:

$$H_2 + Cl_2 \rightarrow 2HCl$$

cannot proceed at normal temperatures unless ultraviolet light is used to DISSOCIATE the chlorine molecules into free atoms.

photochromism The property of some DYES that undergo a reversible colour change on exposure to electromagnetic radiation, especially visible light. Photochromic materials are used in some spectacles, which darken on exposure to sunlight.

photoconductive (*adj.*) Describing a material, such as cadmium sulphide, that conducts electricity when light falls upon it. The CONDUCTIVITY of the material increases with the intensity of the light. The action of light creates ELECTRON-HOLE PAIRS in an otherwise insulating material.

photoconductive cell *See* PHOTOELECTRIC CELL.

photocopier A machine for copying printed or written documents. A light illuminates the document to be copied and a lens forms an image of this document on a charged drum with a PHOTOCONDUCTIVE surface. The action of the photoconductive layer causes electric charge to escape from the drum in those regions which are to be white on the final document. Ink, in the form of a dry powder

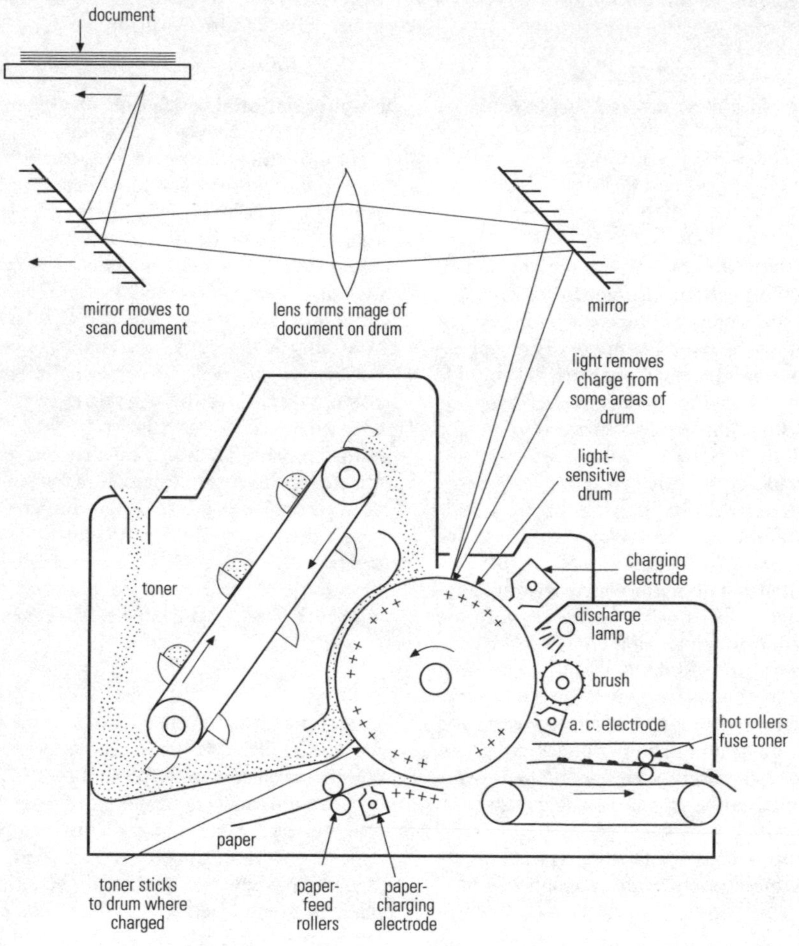

Photocopier.

known as toner, sticks to the charged portions of the drum. A piece of paper is given an electrostatic charge (*see* STATIC ELECTRICITY) and then rolled onto the drum, transferring ink from the drum to the paper. The paper then passes though a heated roller, which melts the ink, allowing it to soak into the paper to form a permanent image.

photodiode A REVERSE BIASED PN JUNCTION DIODE in which incoming PHOTONS create ELECTRON-HOLE PAIRS, allowing a current to flow. The size of the current is dependent on the intensity of the light. Photodiodes are used to measure the intensity of light or as switches to detect light.

photoelasticity The property of many transparent plastics, which exhibit BIREFRINGENCE to an extent dependent on the degree of STRESS to which the material is subjected. By building models of a complex structure in a photoelastic material and viewing this in polarized light (*see* POLARIZATION), the effects of stresses on the structure can be made visible.

photoelectric cell, *photocell* One of several devices for detecting light and other forms of electromagnetic radiation. A photoemissive cell consists of a negative ELECTRODE (the photocathode) and a positive-collecting electrode (ANODE) in a vacuum. PHOTONS striking the photocathode liberate electrons, in a process known as photoelectricity (*see* PHOTOELECTRIC EFFECT). The electrons are attracted to the anode, and the resulting electric current is a measure of the light intensity.

In a photovoltaic cell, a potential difference is set up between two layers as a result of irradiation by light (*see* SOLAR CELL). In a photoconductive cell, the conductivity of a SEMICONDUCTOR increases on exposure to light.

photoelectric effect The emission of electrons from the surface of a metal on exposure to electromagnetic radiation. Electrons are emitted only if the wavelength is below some minimum which depends on the metal used. If light of too long a wavelength is used, the effect is not observed no matter how bright the light.

The particle-like nature of light first became apparent from studies of the photoelectric effect. It contradicts the wave model of light, which holds that the energy in a light beam can be increased either by increasing the intensity or by having a shorter wavelength.

Einstein's explanation of the photoelectric effect (1901) suggested that light comes in quanta (*see* QUANTUM), or individual particles, called PHOTONS. Each electron gains the energy needed to escape by absorbing the energy of one photon. The energy of each photon is related to the frequency, f, of the light by

$$E = hf$$

where E is the photon energy and h is a FUNDAMENTAL CONSTANT called PLANCK'S CONSTANT. The minimum energy needed for an electron to escape from the surface of a metal is called the WORK FUNCTION, and the photoelectric effect cannot take place for a given metal if the photon frequency is so low (i.e. the wavelength so long) that the photon energy is smaller than the work function.

The theory is further justified by experiments in which the energy of the ejected electrons is measured by collecting them on a plate held at a potential negative to the metal plate from which they are produced. The maximum POTENTIAL DIFFERENCE that can exist with photoelectrons still arriving is called the stopping potential and is proportional to the maximum energy with which photoelectrons are released. The photoelectric effect is the basis of many light-detecting devices.

See also CCD, PHOTOELECTRIC CELL, PHOTOMULTIPLIER, PHOTOVOLTAIC EFFECT.

photoemission The release of electrons from a metal ELECTRODE by the PHOTOELECTRIC EFFECT.

photoemissive cell *See* PHOTOELECTRIC CELL.

photographic film A light-sensitive surface that relies on a chemical reaction caused by light striking grains of a silver halide, which is made visible by chemical reactions when the film is developed.

photoionization The IONIZATION of a substance, usually a gas, by ELECTROMAGNETIC RADIATION. The PHOTONS involved must have an energy greater than or equal to the first IONIZATION ENERGY of the material involved, hence this effect is usually restricted to ultraviolet light and X-rays. Photoionization of the upper atmosphere by radiation from the Sun is responsible for the creation of the IONOSPHERE.

photolysis The light-dependent stage of PHOTOSYNTHESIS.

photometer Any device for measuring light intensity. Modern photometers generally comprise a SOLAR CELL in conjunction with a digital VOLTMETER calibrated to give a direct reading of light intensity.

photometry The science of measuring light intensity.

photomultiplier A sensitive light-detecting device based on the PHOTOELECTRIC EFFECT. Light strikes a PHOTOCATHODE and releases electrons. A series of collecting ELECTRODES then multiplies the number of electrons. The first electrode is sufficiently positive that electrons reach it with sufficient energy to knock out further electrons, in a process known as SECONDARY EMISSION. This is repeated several times, with each collecting electrode being more positive than the previous one, so that a single photon releasing a single electron will result in many thousands of electrons being produced. In this way, individual photons can be detected with high efficiencies.

photon The QUANTUM OF ELECTROMAGNETIC RADIATION, having an energy related to the frequency, f, of the light by

$$E = hf$$

where E is the photon energy and h is PLANCK'S

CONSTANT. *See also* PHOTOELECTRIC EFFECT, QUANTUM ELECTRODYNAMICS, WAVE-PARTICLE DUALITY.

photonasty The NASTIC MOVEMENT of plants in response to light. For example, many flowers open in the day and close at night.

photoperiodism The mechanism in plants by which they can respond to changes in day length and therefore regulate some of their activities. Photoperiodism is regulated by the light-sensitive pigment PHYTOCHROME.

The ratio of light and dark controls the flowering of many plants, which fall into three groups. (i) In long-day plants (LDP), flowering is triggered by a dark period of less than a critical length. In other words, flowering occurs mostly in the summer. Examples include the radish, clover and petunia. (ii) In short-day plants (SDP), flowering is triggered when the dark period exceeds the critical length; that is, mostly in the winter, for example the chrysanthemum. (iii) In day-neutral plants, flowering is unaffected by the length of day, for example the cucumber and begonia. The categories are not absolute and some plants need a combination of long and short days before flowering. The stimulus is detected by the leaves and thought to be transmitted by a growth substance (as yet unidentified) called 'florigen'.

Another factor affecting flowering is vernalization, when exposure of the plant to a period of low temperatures stimulates flowering. Vernalization is thought to be affected by a growth substance, vernalin, that is a GIBBERELLIN. This, together with photoperiodism, ensures that flowering occurs at specific times of the year. Both can be controlled artificially, with commercial implications.

Animals are also affected by photoperiodism (*see* BIORHYTHM).

photophosphorylation PHOSPHORYLATION that uses light energy.

photoreceptor A receptor cell that detects light and some forms of electromagnetic radiation. *See* CONE, EYE, ROD, SENSE ORGAN.

photorespiration A very active type of RESPIRATION that occurs outside the MITOCHONDRIA in plants, when light intensity is high, carbon dioxide levels low and oxygen levels high. Photorespiration involves the reoxidation of some of the CARBOHYDRATES formed during PHOTOSYNTHESIS to carbon dioxide. No energy is generated, so the process seems wasteful and reduces the efficiency of photosynthesis. It only occurs in C_3 PLANTS (temperate plants; about 85 per cent of plants) and not in C_4 PLANTS (tropical plants).

photosensitive (*adj.*) Describing a device or substance that is stimulated to action by, or reacts in the presence of, light or some other form of electromagnetic radiation.

photosynthesis The use of light energy by green plants to convert carbon dioxide from the air, and water to CARBOHYDRATES and oxygen. Photosynthesis is a form of AUTOTROPHIC NUTRITION or self-feeding and all animals depend on it because it is the means by which basic food (sugar) is created and oxygen is released for use by aerobic organisms. Photosynthesis occurs in the CHLOROPLASTS of higher plants and ALGAE by means of various light-trapping pigments, the most common of which is CHLOROPHYLL. There are two stages to photosynthesis: the light-dependent stage (photolysis) and the light-independent stage (formerly called the dark reaction).

The light-dependent stage occurs in the grana of the chloroplast, and involves the splitting of water by light to yield oxygen, hydrogen ions (protons) and electrons with the simultaneous conversion of ADP to ATP by PHOTOPHOSPHORYLATION, so converting light energy to chemical energy. This can be non-cyclic photophosphorylation, in which electrons are lost from the chlorophyll to be passed into the light-independent stage and are replaced by electrons from a water molecule (with the production of ATP). Thus the same electrons are not recycled through the chlorophyll. Cyclic photophosphorylation also occurs in which electrons from the chlorophyll are returned to it via an electron carrier system. NADPH is also formed during the light-dependent stage by the reduction of NADP to NADPH and H^+.

Electrons and protons from the light-dependent stage (NADPH and H^+) are used in the light-independent stage (along with the ATP generated) to convert carbon dioxide from the air (entering by DIFFUSION through the STOMATA and eventually into the chloroplast STROMA) into carbohydrates. The light-independent stage occurs in the stroma of chloroplasts and is also called the CALVIN CYCLE. This latter stage differs between C_3 PLANTS and

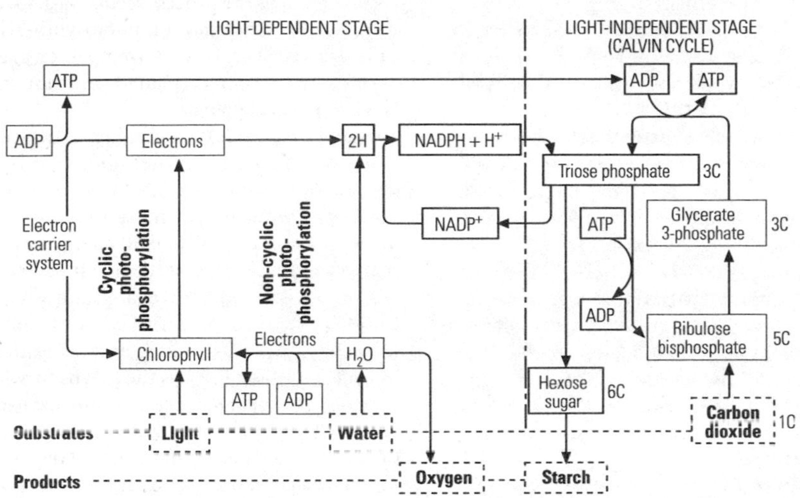

Summary of photosynthesis.

C$_4$ PLANTS. Most of the products of photosynthesis are derived from intermediates of the Calvin cycle.

A plant is capable of synthesizing all the organic materials necessary for its survival, which includes carbohydrates, LIPIDS and PROTEINS. A number of factors affect photosynthesis: light intensity, carbon dioxide concentration, temperature (the light-independent stage is temperature-dependent), and chlorophyll and oxygen concentrations. Under certain conditions PHOTORESPIRATION occurs in some plants at the expense of photosynthesis. As light intensity increases, so does the rate of photosynthesis. The point at which as much carbon dioxide is being used in photosynthesis as is evolved from RESPIRATION is called the compensation point. Beyond this the rate of photosynthesis increases until light saturation is reached – that is, where an increase in light intensity has no effect on photosynthesis.

Compare CHEMOSYNTHESIS.

phototaxis *See* TAXIS.

phototropism The directional growth of a plant in response to an external light stimulus. Shoots are positively phototropic and bend towards a light source; some roots are negatively phototropic and bend away from a light source; leaves position themselves at right angles to light and are called diaphototropic. Phototropism can be explained in terms of AUXIN, a PLANT GROWTH SUBSTANCE. Auxin moves away from the light source, causing cell expansion and therefore bending of a shoot towards the light. At higher concentration of auxins can inhibit growth, and roots are more sensitive to auxin than shoots, so the same concentration that causes positive phototropism in shoots will cause inhibition of growth in roots (on the side away from the light) therefore causing the root to bend away from the light.

See also TROPISM.

photovoltaic cell *See* SOLAR CELL.

photovoltaic effect A PHOTOELECTRIC EFFECT in which a POTENTIAL DIFFERENCE is created between two layers as a result of light falling on the boundary between the two.

phylogeny The evolutionary development of an organism or species. *Compare* ONTOGENY.

phylum (*pl. phyla*) The major division within a KINGDOM. An example is the phylum CHORDATA, which includes mammals, birds, reptiles, amphibians, fish and tunicates. In plant classification the term 'division' is sometimes used instead of phylum. The subdivisions of phyla are CLASSES. *See* CLASSIFICATION.

physical binary *See* BINARY.

physical chemistry The branch of chemistry that deals with the effect of chemical reactions on the physical properties of the materials involved, such as their melting points or electrical CONDUCTIVITY.

physics The branch of science that deals with MATTER, ENERGY and their interactions. Physics attempts to find laws, usually mathematical in form, that accurately describe a wide range of phenomena throughout the universe.

physiology The study of the functional processes and activities that occur within living organisms. In higher organisms physiology includes the study of interactions between cells, tissues and organs. Although usually used in relation to animals, physiology can also refer to the processes occurring within plants.

physisorption The ADSORPTION of a gas by a solid in which the molecules of the adsorbed gas are held on the surface of the adsorbing solid by VAN DER WAALS' FORCES.

phytochrome A light-sensitive protein pigment that, by absorbing light of different wavelengths, enables a plant to detect changes in day-length, so controlling PHOTOPERIODISM. There are two forms of phytochrome: phytochrome 660 (P_{660}), which absorbs red light with a wavelength of 660 nm; and phytochrome 730 (P_{730}), which absorbs light at the far red end of the ELECTROMAGNETIC SPECTRUM, at wavelength 730 nm. These two forms are interconvertable and even short exposure to light at the appropriate wavelength causes a switch. Daylight and darkness also have an effect; during daylight P_{660} is converted to P_{730} and in the dark P_{730} is converted to P_{660} (although this is slower).

Phytochrome is distributed in small amounts all over the plants but mostly at the growing tips. The two forms have different (usually ANTAGONISTIC) effects on plants. P_{730} stimulates GERMINATION of some seeds, causes an increase in leaf area, stimulates flowering in long day plants (*see* PHOTOPERIODISM) and induces formation of other plant pigments. P_{660} inhibits these effects but stimulates flowering in short-day plants. The ratio of the two forms can be important in breaking DORMANCY of some seeds.

phytomenadione *See* VITAMIN K.

phytoplankton The plant constituent of PLANKTON.

pi (π) An IRRATIONAL NUMBER equal to the circumference of any circle divided by its diameter. Pi is equal to 3.1416 (to 4 decimal places).

pia mater The innermost of the three membranes that cover the BRAIN and SPINAL CORD. *See* MENINGES.

pi-bond, π-bond A COVALENT BOND that has an electron distribution that is not symmetrical about the line joining the two atoms. Covalent DOUBLE BONDS and TRIPLE BONDS often consist of a SIGMA-BOND surrounded by one or two pi-bonds.

pico- (p) A prefix placed in front of a unit to denote that the size of that unit is to be multiplied by 10^{-12}. For instance a picofarad (pF) is one million millionth of a FARAD.

pie chart A method for displaying data using a circle as the whole sample, divided into segments. The angles of the segments show the percentage of the whole represented by each portion. It provides a clear, simple representation of proportions but does not give precise information.

pi electrons, π *electrons* ELECTRONS associated with a PI-BOND.

piezoelectric The effect by which a POTENTIAL DIFFERENCE (p.d.) appears between the faces of certain materials when they are subject to STRESS. The reverse effect also occurs, with the material deforming in response to an applied p.d. This effect is used in many electronic TRANSDUCERS. *See also* CRYSTAL OSCILLATOR.

pig iron The metal removed from the bottom of a BLAST FURNACE. It has too high a carbon content (3–5 per cent) to be used directly. Pig iron is refined and converted into CAST IRON and STEEL.

pigment In biology, any substance, for example HAEMOGLOBIN and CHLOROPHYLL, that gives a characteristic colour to animal or plant tissue.

piliferous layer The layer of cells on the surface of a plant ROOT that produces the ROOT HAIRS.

pill A general term for a variety of contraceptive tablets (*see* CONTRACEPTION) taken orally that use the female hormones to interfere with the MENSTRUAL CYCLE. The combined pill contains synthetic hormones that mimic OESTROGEN and PROGESTERONE produced in pregnancy. It prevents OVULATION and alters the CERVIX, making it hostile to SPERM. It is the most effective form of birth control (99 per cent) but has some side-effects, for example thrombosis due mainly to the oestrogens, headaches and high blood pressure. The mini-pill works in a similar way but contains progesterone only and prevents ovulation and the IMPLANTATION of a fertilized egg; it has fewer side-effects but a slightly lower success rate.

The morning-after pill can be taken up to 72 hours after unprotected intercourse and prevents implantation of a fertilized egg, but contains high levels of hormones and is not for regular use.

pilus (*pl. **pili***) A small hair-like outgrowth on the surface of many bacteria that assist them in attaching to certain surfaces. DNA is exchanged or donated between bacteria through a pilus. *See also* CONJUGATION.

pi-meson, *pion* (π^+, π^-, π^o) The lightest MESON, with a mass 0.15 times the proton mass, and the one with the longest HALF-LIFE. The positive and negative pi-mesons have half-lives of 2.6 x 10^{-8} s. The half-life of the neutral pion (sometimes called pi-zero) is far shorter, 8 x 10^{-17} s.

pineal body *See* PINEAL GLAND.

pineal gland, *pineal body, epiphysis* A small mass of nervous tissue within the vertebrate MIDBRAIN. The nervous connection to the brain is lost but a nerve supply comes from the SYMPATHETIC NERVOUS SYSTEM. The pineal gland secretes two products that are linked to body rhythms, a hormone-like substance called melatonin (which is inhibited by light) and a NEUROTRANSMITTER called serotonin. Serotonin causes general inhibition of activity (opposes NORADRENALINE).

In lower vertebrates, such as amphibia and some snakes, the pineal gland develops a basic lens and retina to form an eye (so it seems to be derived from an eye) that is situated on top of the head. In fish, it detects the surrounding light and controls colour change involved in camouflaging. In birds, it detects changes in length of daylight and stimulates breeding behaviour as spring approaches.

See also SENSE ORGAN.

pinion A small GEAR, the smaller of a pair of gears. *See also* RACK AND PINION.

pinna 1. In mammals, part of the outer EAR.

2. In botany, the primary division of a pinnate LEAF.

pinnate leaf A type of compound LEAF.

pinocytosis A process similar to PHAGOCYTOSIS that is used by living cells for the uptake of liquids not solids.

pion *See* PI-MESON.

pipette A graduated glass tube for delivering a measured quantity of a liquid REAGENT. It consists of a glass stem with a bulb about half-way up the stem. Two lines are marked on either side of the bulb, the volume of liquid specified will have been delivered by the pipette when the liquid level has moved from one line to the other.

piston A round metal plate, usually with a flat surface, which moves up and down in a CYLINDER in response to pressure changes. The piston is a key part in the INTERNAL COMBUSTION ENGINE and in many HYDRAULIC and PNEUMATIC systems.

pitch 1. A black, sticky substance derived from the DESTRUCTIVE DISTILLATION of COAL TAR or wood. It is a liquid when hot but solid when cold. Its uses include roofing, paving and waterproofing, for example, wooden ships.

2. A musical description related to the FREQUENCY of a sound; a doubling in frequency represents an increase in pitch of one OCTAVE.

pitchblende The principle ore of uranium, consisting mainly of uranium(IV) oxide, UO_2. Pitchblende is also a source of radium, thorium and polonium.

pith The soft, spongy tissue in the centre of stems or roots of VASCULAR PLANTS. Pith is usually PARENCHYMA tissue. Pith also refers to the white, fibrous tissue between the rind and pulp of fruits such as oranges.

pitot head A forward-facing open tube fitted to an aircraft as part of a PITOT-STATIC SYSTEM.

pitot-static system In aircraft, a system used to measure the difference between the pressure of the air striking the front of the aircraft and the static pressure of the surrounding air, in order to measure the speed of the aircraft through the air.

pituitary gland, *hypophysis* A small ENDOCRINE GLAND found in the centre of the vertebrate brain that secretes a number of hormones that have control over the activities of the other ENDOCRINE GLANDS. The pituitary gland is itself largely under the control of the HYPOTHALAMUS, to which the POSTERIOR pituitary is attached.

The posterior pituitary is of nervous origin and communicates with the hypothalamus via nerves. It stores and secretes (under hypothalamic control) ANTI-DIURETIC HORMONE (ADH) and OXYTOCIN (made by the hypothalamus), concerned, respectively, with the salt/water balance and the functioning of the MAMMARY GLANDS and UTERUS.

The other region of the pituitary is the ANTERIOR pituitary, which produces five hormones called trophic hormones (which affect other endocrine glands) and a non-trophic hormone, GROWTH HORMONE, which controls body METABOLISM and growth generally. The trophic hormones are THYROID-STIMULATING HORMONE, ADRENOCORTICOTROPHIC HORMONE, FOLLICLE-STIMULATING HORMONE, LUTEINIZING HORMONE and PROLACTIN.

See also THYROID.

pivot A fixed point about which an object, such as a LEVER, can rotate. To reduce the effects of friction, some form of lubrication or BEARING is often used at a pivot.

pixel Any one of the large number of small single elements from which an image is formed on the screen of a VISUAL DISPLAY UNIT or other computer image-forming system.

pK A measure of the strength of an ACID or BASE. For an acid, is the extent to which the acid DISSOCIATES in aqueous solution, to release hydrogen ions. Likewise for a base, it is the extent to which the base dissociates to release HYDROXIDE ions.

$$\mathrm{pK} = -\log_{10} K$$

where K is the DISSOCIATION CONSTANT for the acid.

placenta 1. The structure in most mammals that attaches the developing FOETUS to the UTERUS and which links the blood supply of the mother and baby so that exchange of oxygen, nutrients and waste products can sustain the foetus throughout PREGNANCY. The placenta is connected to the foetus by an UMBILICAL CORD.

The blood supply of the mother and baby are separate (and therefore AGGLUTINATION, or clumping, is prevented) and exchange of nutrients, for example GLUCOSE, amino acids, vitamins, fats and oxygen, and waste products, for example UREA and carbon dioxide, occurs between the capillaries of mother and baby by DIFFUSION. Some maternal antibodies can cross the placenta, and provide protection against some diseases during the early months of life, but some PATHOGENS can also cross, for example rubella (the German measles virus), causing physical and mental damage to the baby, or HIV causing AIDS. Many organisms do not cross the placenta, however, and it allows considerable protection against disease. The placenta also prevents many chemicals or high levels of maternal hormones entering the foetal blood but some, for example alcohol, nicotine, caffeine and heroine can cause damage to the foetus. The placenta also has an essential immunological role in the protection of the foetus.

Once fully developed, at 3 months, the placenta takes over the role of hormone production necessary to maintain the pregnancy from the earlier TROPHOBLAST. At this time it also begins to secrete OESTROGEN, PROGESTERONE, HUMAN CHORIONIC GONADOTROPHIN and human placental lactogen. Many other proteins are secreted by the placenta that are of considerable interest but the function of many is as yet unknown. The placenta is shed at birth as the afterbirth.

See also ENDOMETRIUM.

2. The part of the ovary of a flowering plant to which the OVULES are attached. *See* CARPEL.

Planck's constant (h) A constant that determines the scale of the effects of WAVE-PARTICLE DUALITY. Since h is small, the effects of QUANTUM MECHANICS only become apparent at short distances or over short times. $h = 6.6 \times 10^{-34}$ Js.

See also PHOTOELECTRIC EFFECT.

plane A flat surface in space. Any vector normal to a particular plane always has the same direction, and any straight line joining any two points on the plane lies entirely in the surface.

plane mirror A flat MIRROR. When objects are viewed by looking at light reflected from a plane mirror, a VIRTUAL IMAGE is seen, which is as far behind the mirror as the object was in front of the mirror, with the line joining object and image being at 90° to the plane of the mirror.

plane-polarized (*adj.*) Describing an ELECTROMAGNETIC WAVE in which the electric and magnetic fields each oscillate in a single plane, perpendicular to one another and to the direction of wave propagation. *See* POLARIZATION. *Compare* CIRCULARLY POLARIZED.

planet A large, roughly spherical, celestial object composed mainly of rock or gas and in orbit around the Sun. The smaller planets, including the Earth, are composed mostly of rock and are called TERRESTRIAL PLANETS. The larger plants are made mostly of gas and are called

GAS GIANTS. The periods of the orbits of the planets increase with distance from the sun (*see* KEPLER'S LAWS), ranging from 88 days for Mercury to 240 years for Pluto.

planetary nebula A stage in the life cycle of some stars in which the outer layers of a RED GIANT are blown off into space over a period of typically one million years, revealing a WHITE DWARF at the core of the star.

planetary satellite *See* SATELLITE.

plankton A small organism living on the surface of fresh or salt water that is an important food source for larger animals such as fish and whales. The plants (ALGAE) of plankton are called phytoplankton, and they are non-motile, simple organisms of varying sizes that move with the water currents. The animals of the plankton are called zooplankton, and they feed on phytoplankton and are themselves food for larger fishes. Zooplankton are mostly able to move by FLAGELLAE. *See also* NEKTON, PELAGIC.

plant An immobile multicellular, eukaryotic (*see* EUKARYOTE) organism capable of PHOTOSYNTHESIS. Plants are the primary producers in all FOOD CHAINS and therefore all animal life is dependent on them. They contain CELLULOSE in their cell walls. Plants make up the kingdom Plantae.

One common method of plant classification is into the PHYLA: BRYOPHYTA (mosses and liverworts); LYCOPODOPHYTA (club mosses); SPHENOPHYTA (horsetails), FILICINOPHYTA (FERNS); CONIFEROPHYTA (conifers); and ANGIOSPERMOPHYTA (angiosperms; flowering plants). The study of plants is called botany.

See also VASCULAR PLANT, SEED PLANT.

Plantae The KINGDOM consisting of PLANTS.

plant growth substance, *plant hormone* A chemical substance (natural or artificial) that modifies plant growth. Plant growth substances are not produced by a particular area of the plant (in contrast to animal HORMONES) and may affect different areas of growth or a related process (for example, leaf fall, flowering or fruit ripening). There are five groups of growth substances: AUXINS, GIBBERELLINS, CYTOKININS, ABSCISSIC ACID (inhibitor) and ETHENE.

plant hormone *See* PLANT GROWTH SUBSTANCE.

plasma 1. (*physics*) A highly IONIZED gas in which the number of FREE ELECTRONS approximately equals the number of positive IONS. As a whole, a plasma is electrically neutral. They are produced at such high temperatures that collisions between particles are violent enough to produce almost complete IONIZATION.

Most of the matter in the Universe exists as a plasma, which is sometimes described as the fourth state of matter. Plasmas occur in stars and in interstellar space, and are induced in thermonuclear reactors and in GAS DISCHARGE tubes.

2. (*biology*) The liquid part of BLOOD (without the cells) that is a watery yellowish fluid containing many dissolved sugars, fats and proteins, including antibodies, hormones and salts as well as the blood-clotting substances (*see* BLOOD CLOTTING CASCADE). Plasma is obtained by centrifugation of blood to sediment the cells.

plasma cell *See* B CELL.

plasma membrane *See* CELL MEMBRANE.

plasma protein One of many PROTEINS found in blood PLASMA. Examples include ALBUMIN, GLOBULIN and blood-clotting factors (PROTHROMBIN, FIBRINOGEN).

plasmid A small circular loop of DNA found in bacterial cells that is separate from the bacterial chromosome but replicates with it. Usually only one copy is present but several can be found. Plasmids can confer useful properties onto the cell, such as ANTIBIOTIC RESISTANCE. If a plasmid is used as a carrier for foreign DNA, it is called a vector (vectors can also be BACTERIOPHAGES). By growing cultures of the bacteria containing the plasmid with its foreign DNA fragment, many copies of the foreign DNA can be obtained. Plasmids are an invaluable tool in GENETIC ENGINEERING.

plasmodesma (*pl. **plasmodesmata***) A fine cytoplasmic tube (30–60 nm in diameter) that is present in the CELL WALL of a plant, at intervals or in regions. There is no LIGNIN at these sites and so 'pits' are formed, through which water and dissolved minerals can pass from one cell to another.

plasmolysis The shrinkage of a cell CYTOPLASM away from the cell wall as a result of water loss by OSMOSIS. Water leaves the cell VACUOLE, which causes the cytoplasm to shrink without affecting the overall cell shape or size. The cell is then called 'flaccid'. Plasmolysis can be induced in the laboratory by placing plant cells in a saline or sugar solution, so water leaves the cells by osmosis. It is unlikely to occur naturally except in extreme conditions. *See also* TURGOR.

plaster of Paris *See* CALCIUM SULPHATE.

plastic 1. (*n.*) Any synthetic organic POLYMER that is liquid at some stage in its manufacture. Plastics are classified by their behaviour on heating. THERMOPLASTICS soften on heating and harden as they cool, for example POLYSTYRENE, POLYVINYL CHLORIDE. THERMOSETTING PLASTICS remain rigid in their final shape and do not soften on heating, for example RESINS, MELAMINE, POLYESTER, BAKELITE.

During manufacture plastics are moulded and shaped while they are in the heat-softening stage and then cooled for thermoplastics, or heated further for thermosetting plastics to yield the final product. The processing of plastics can produce a wide variety of materials from rigid and inflexible to soft and bendy. They can be extruded to make pipes and rods. Plastics can be strengthened by the addition of CARBON FIBRES for use in aircraft or engineering.

Due to the difficulties in disposal of plastics a number of BIODEGRADABLE plastics are now in use, including polyhydroxybutyrate (PHB), which is made from sugar and digested by micro-organisms in the soil.

2. (*adj.*) Describing a material or process in which the material is deformed by a force and does not return to its original shape once the force is removed.

plasticizers Organic liquids of high RELATIVE MOLECULAR MASS that are added to PLASTICS to increase their flexibility. Plasticizers have an important use in POLYVINYL CHLORIDE (PVC). PVC undergoes dramatic changes in the presence of a plasticizer, from the rigid substance used to make window frames (unplasticized uPVC) to the flexible plastic used in shoes.

plastid The general name for a cell ORGANELLE of plants and ALGAE, containing a series of internal membranes bounded by a double membrane. Plastids contain a small amount of DNA, and some RIBOSOMES of the prokaryotic type (*see* PROKARYOTE) may be present. They develop by division of existing plastids and there can be one or more per cell of various shapes. Plastids mature into CHLOROPLASTS or CHROMOPLASTS, which contain coloured pigments, or LEUCOPLASTS, which are colourless.

platelet, *thrombocyte* In vertebrates, a small cell-like fragment that buds from larger cells in the BONE MARROW and is involved in blood clotting following injury. Platelets adhere to injured ENDOTHELIAL cells (lining blood vessels), and aggregate releasing a number of factors needed for prevention of bleeding and repair of the damaged tissue. In the BLOOD-CLOTTING CASCADE, platelets release an enzyme called thrombokinase which converts the inactive enzyme prothrombin into the active thrombin. This in turn converts the soluble plasma protein FIBRINOGEN into the insoluble FIBRIN, which forms a clot over the wound. *See also* BLOOD.

plate tectonics The study of the motion of TECTONIC PLATES.

platinum (Pt) The element with ATOMIC NUMBER 78; RELATIVE ATOMIC MASS 195.1; melting point 1,772°C; boiling point 3,800°C; RELATIVE DENSITY 21.3. Platinum is a very unreactive metal and is used to make INACTIVE ELECTRODES for ELECTROLYSIS. It also acts as a catalyst in a wide range of reactions.

Platonic solid Any of the five regular POLYHEDRONS – the cube, dodecahedron, icosahedron, octahedron, tetrahedron.

Platyhelminthes A PHYLUM consisting of the flatworms, which are free-living or parasitic (*see* PARASITE) invertebrates of great economic importance. Flatworms have a head region and a flattened shape showing bilateral symmetry. Their body walls are triploblastic (three layers), and there is one opening to the gut. Many are HERMAPHRODITES that can self-fertilize, but usually cross-fertilization occurs. The life cycle of parasitic flatworms can be complex and often involves several intermediate hosts (*see* PARASITE).

Examples are the liverfluke (*Fascida hepatica*) of sheep and cattle (the intermediate host is a snail) that causes liver rot; the blood fluke (*Schistosoma*) of humans (the intermediate host is a snail) that causes schistosomiasis (bilharzia), damaging the lungs and liver; the pork tape worm (*Taenia solium*) that infects humans (the intermediate host is the pig), causing anaemia, diarrhoea, weight loss, intestinal pain and blockage.

Some flatworms move by secreting MUCUS and gliding over this by means of CILIA, some produce muscular contractions down the length of their body, but parasitic flatworms are usually attached by hooks or suckers to their host. Flatworms are CARNIVORES that either trap prey in the mucus they produce, or suck blood and cells into their PHARYNX. Tapeworms absorb pre-digested food from their host by diffusion through their body surface (*see* CESTODA, TREMATODA, TURBELLARIA).

Pleistocene The first EPOCH of the QUATERNARY PERIOD. It began after the PLIOCENE epoch (about 2 million years ago) and extended until the HOLOCENE epoch (about 10,000 years ago). The Pleistocene is also known as the Ice Age, and this time saw the advance of GLACIERS into much of North America and Northern Europe. There were about four glacial advances, separated by interglacial periods of warmer weather. During the Pleistocene humans evolved into modern *Homo sapiens.*

pleura *See* LUNG.

pleural cavity *See* LUNG.

pleurisy An infection of the pleural cavity. *See* LUNG.

Pliocene The last EPOCH of the TERTIARY PERIOD. It extended for about 5 million years, after the MIOCENE epoch (about 7 million years ago). The forerunner of humans made their appearance in this time.

plotting compasses A small MAGNETIC COMPASS used for determining the direction of a MAGNETIC FIELD in the laboratory.

plumbate Any compound containing the plumbate ion, PbO_3^{2-}. Plumbates are formed by the reaction between lead(IV) oxide and concentrated strong alkalis, for example:

$$PbO_2 + 2NaOH \rightarrow Na_2PbO_3 + H_2O$$

plumule The part of the embryo of a SEED PLANT that develops into the shoot and bears the first true leaves (*see* GERMINATION).

Pluto The ninth, and last, planet from the Sun, with an orbital radius of 39.4 AU (5.9 billion km). Relatively little is known about Pluto, which was not discovered until 1930, except that it is small, and probably composed of rock, ice and frozen methane, with a mass of 6.6×10^{23} kg (0.11 times that of the Earth), and a diameter of 2,700 km (0.21 times that of the Earth). It has a satellite called Charon, which is almost as large as the planet itself. Pluto has an orbital period of 248 years and rotates about its own axis every 6.9 days.

plutonium (Pu) The element with ATOMIC NUMBER 94; melting point 641°C; boiling point 3,232°C. It is chemically reactive and highly toxic. It has 13 ISOTOPES, all radioactive. The most important isotope (plutonium–239) has a HALF-LIFE of 24,000 years. Plutonium–239 is FISSILE and an important fuel source for NUCLEAR POWER. It is produced after uranium–238 captures a neutron and then BETA DECAYS twice. Plutonium is produced in FAST BREEDER REACTORS.

pneumatic (*adj.*) Describing a system in which air is used to transmit forces from one place to another or to drive a machine. Because no sparks are produced, pneumatic tools are often used in places where there is a danger of explosion, such as coal mines.

pneumatophore An erect plant root that rises above the soil surface, for example in swamps, to aid gas exchange.

pn junction diode A simple electronic device, produced by forming a junction between P-TYPE and N-TYPE SEMICONDUCTORS. Electrons from the n-type material, and HOLES from the p-type, diffuse across the boundary and cancel each other out, forming a DEPLETION LAYER in which there are very few CHARGE CARRIERS. When the p-type material is made negative and the n-type positive, the diode is said to be reverse biased – the holes in the p-type material and the electrons in the n-type layer are pulled away from the junction and no current will flow. If the polarity is reversed, with the p-type material positive and the n-type negative, charge carriers are attracted into and across the depletion layer and a current flows. The diode is then said to be forward biased.

Diodes can be used for RECTIFICATION, the conversion of alternating to direct current, and for this purpose a BRIDGE RECTIFIER circuit is often used. They are also used for DEMODULATION in radio circuits. If the junction is made sufficiently large, a reverse biased diode will

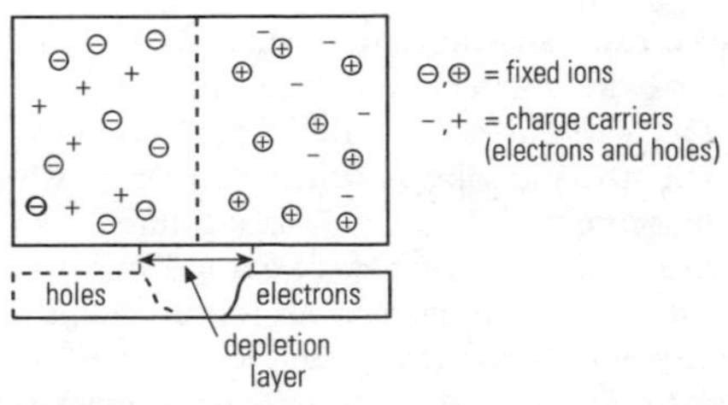

pn junction diode.

have a significant CAPACITANCE, which can be used in a TUNED CIRCUIT. Such diodes are called varicap diodes.

pnp transistor A JUNCTION TRANSISTOR in which the EMITTER and COLLECTOR are made of P-TYPE SEMICONDUCTOR, whilst the BASE is N-TYPE.

PNS *See* PERIPHERAL NERVOUS SYSTEM.

poikilotherm, *ectotherm* An animal that exhibits POIKILOTHERMY. A poikilotherm is also known as a cold-blooded animal.

poikilothermy, *ectothermy* (*poikilo* = various, *thermo* = heat) A condition where the body temperature of an animal fluctuates depending on the external temperature. Poikilothermy is characteristic of all animals except birds and mammals, which maintain a constant body temperature (HOMEOTHERMY). Invertebrates, fish, amphibians and reptiles are all poikilotherms. Despite being called cold-blooded, poikilotherms can have body temperatures as high as homeotherms (or higher). Poikilotherms have behavioural means of temperature control, such as cooling in water, shivering, lying in the sun and sheltering.

point In mathematics, a position in space, on a graph for example, usually defined by its CO-ORDINATES, and having no physical extent, so it can be fully represented by a single set of co-ordinates rather than by a range of co-ordinates.

point mutation The omission, insertion or substitution of a single base (*see* NUCLEOTIDES) in DNA. *See* MUTATION.

Poiseulle's equation An equation for finding the rate of volume flow of a viscous fluid in LAMINAR FLOW down a tube. If the VISCOSITY of the fluid is η and the tube has radius r and length l, the volume flow per second, V, caused by a pressure difference p between the ends of the tube is

$$V = \pi p r^4 / 8 l \eta$$

poison 1. A substance that damages the health of a living organism.

2. (*chemistry*) A material that, when present in very small quantities, can interfere with the action of a CATALYST. Many common poisons in this sense are also toxic to living creatures as they interfere with the functions of enzymes. Arsenic is an example.

polar (*adj.*) **1.** Relating to the POLE. A polar AIR MASS is one that has come from regions close to the one of the Earth's poles.

2. (*chemistry*) Describing a molecule or a substance, particularly a SOLVENT, that contains POLAR BONDS. Many IONIC solids are soluble in polar solvents, including water and ethanol.

polar bond A COVALENT BOND in which the electrons spend a higher proportion of their time closer to one atom in the bond than the other, thus one atom in effect carries a partial negative charge whilst the other has an equal positive charge. These are sometimes indicated on diagrams by δ+ and δ–. The degree of polarity in a bond depends on the difference in ELECTRONEGATIVITY between the two elements forming the bond. Typical examples are hydrogen chloride, HCl, and water H_2O, each of which have polar bonds with the hydrogen atom forming the positive end of the bond.

The extreme version of a polar bond, with

$$\underset{\delta+}{H} - \underset{\delta-}{Cl}$$

$$\overset{\delta+}{H} \text{ bonded to } \underset{\delta-}{O} \text{ and } \underset{\delta-}{O}$$

the electrons spending all their time attached to one atom, is an IONIC BOND. Polar bonds tend to lead to a greater attraction between molecules and materials with higher boiling points than would be expected from VAN DER WAALS' BONDING.

See also HYDROGEN BONDING.

polar co-ordinates CO-ORDINATES that in two-dimensions define the location of a point in terms of its distance from the ORIGIN and the angle between the direction to the point and a fixed direction. These co-ordinates are generally denoted as r and θ and if the x-axis is taken as the $\theta = 0$ direction, they are related to CARTESIAN CO-ORDINATES by the equations:

$$x = r\cos\theta \text{ and } y = r\sin\theta$$

In three dimensions, cylindrical polar co-ordinates or spherical polar co-ordinates may be used. Cylindrical polar co-ordinates use a second distance, z to measure the position of the point above or below the plane of the origin. Spherical polar co-ordinates measure the distance r of the point from the origin together with a polar angle θ, above or below the plane perpendicular to an axis called the polar axis, and an azimuthal angle ϕ found by projecting the point into the plane perpendicular to the polar axis.

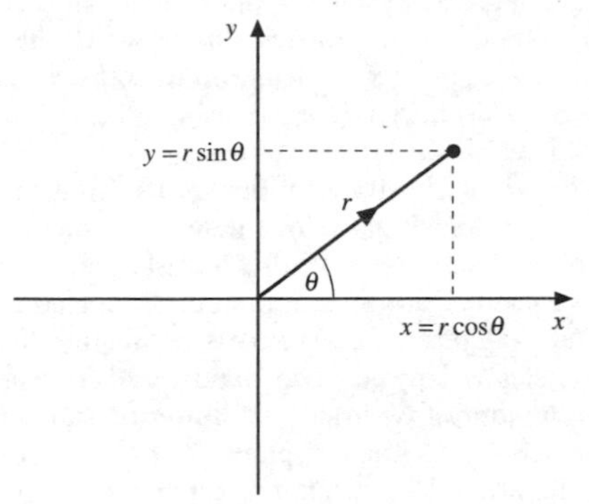

Polar co-ordinates.

polarimeter A device for analysing polarized light (*see* POLARIZATION), particularly for measuring the OPTICAL ACTIVITY of a solution. The sample to be analysed is placed between two pieces of POLAROID. The first (the polarizer) polarizes the light and the second (the analyser) can be rotated until no light passes through the polarimeter. The degree of rotation required to prevent any light passing through the polarimeter is a measure of the optical activity of the sample.

polarizability A measure of the ease with which an ION can have its electron distribution distorted by a neighbouring ion of the opposite charge in an IONIC SOLID. A high value of polarizability indicates a tendency to form a polar COVALENT BOND rather than an IONIC BOND. Larger ions are more polarizable than small ones, which means that ANIONS tend to have higher polarizabilities. TRANSITION METAL ions tend to have relatively high polarizabilities due to the more widely spread nature of the D-ORBITALS. Polarizability is measured on an arbitrary scale, with typical values ranging from 0.03 (Li^+) to 10 (S^{2-}).

polarization 1. (*physics*) The direction of motion of the material through which a TRANSVERSE WAVE is travelling. In the case of an ELECTROMAGNETIC WAVE, the direction of polarization is the direction of the electric field, since it is this, rather than the magnetic field, that is responsible for most of the physical effects of the wave. Electromagnetic waves may be PLANE-POLARIZED, with the electric field in a single direction, or CIRCULARLY POLARIZED, in which case the electric field direction follows a circular path, which may be either left- or right-handed, as seen by an observer facing in the direction in which the wave is travelling. A circularly polarized wave will have its direction of polarization reversed on reflection. Radio aerials transmit and receive waves that are plane-polarized in the direction of the wire elements of the aerial. *See also* BIREFRINGENT, BREWSTER'S LAW, OPTICAL ACTIVITY, POLARIZATION BY REFLECTION, POLAROID.

2. (*chemistry*) A measure of the extent to which molecules have been polarized (*see* POLARIZE) by an electric field.

polarization by reflection Unpolarized light reflected from a non-metallic surface will become partially polarized, with the electric field predominantly in the direction parallel to the reflecting surface. This reflected light is completely polarized when the ANGLE OF INCIDENCE is equal to the BREWSTER ANGLE for the material, at which point the reflected and refracted light (*see* REFRACTION) are at right angles. *See also* BIREFRINGENT, BREWSTER'S LAW.

polarize (*vb.*) **1.** Of light, to transmit, reflect or scatter one direction of POLARIZATION in unpolarized light more strongly than another.

2. Of an ELECTRIC FIELD, to pull a molecule or atom away from symmetry so the centres of positive and negative charge no longer coincide, or to align POLAR MOLECULES.

polar molecule A molecule in which the electrons in a COVALENT BOND are not evenly distributed, so they have a higher probability of being found at one end of the molecule than the other. *See also* HYDROGEN BONDING, POLAR BOND.

polar nucleus Either one of two nuclei within the OVULE of a flowering plant that migrates from each pole of the EMBRYO SAC towards the centre. Polar nuclei fuse with one of the male nuclei carried to the ovule by the POLLEN TUBE, and form the PRIMARY ENDOSPERM NUCLEUS. This is part of DOUBLE FERTILIZATION.

Polaroid Trade name for a transparent plastic material that plane-polarizes light passing through it (*see* POLARIZATION). The plastic contains long-chain molecules along which electrons can travel, absorbing the energy from electromagnetic radiation. During manufacture, the plastic is stretched, aligning the molecules. Light that is polarized perpendicular to the direction of alignment will be transmitted through the Polaroid, whilst light polarized parallel to the molecules is absorbed. Polaroid filters are used in some sunglasses and are sometimes fitted to cameras to reduce the amount of reflected (rather than scattered) light and so reduce glare.

pole 1. The point on the surface of an object, a planet for example, at which the axis of rotation passes through that surface.

2. The region on the surface of a magnet at which the MAGNETIC FIELD LINES enter and leave the magnet.

pole and barn paradox A RELATIVISTIC PARADOX in which a man runs into a barn carrying a pole which is the same length as the barn. According to an observer at rest in the barn, the pole is now shorter than the barn (LORENTZ CONTRACTION) so it would be possible to shut the doors at each end of the barn and trap the pole in the barn. According to the runner, the barn is now shorter than the pole, so it would not be possible to trap the pole. The resolution of this paradox comes from the idea that the shock wave of the leading edge of the pole running into the barn door cannot travel along the pole faster than the speed of light, thus the pole cannot be perfectly rigid, and so the runner can, just, be trapped in the barn.

pollen The grain of SEED PLANTS carrying the male GAMETE. Pollen is produced in POLLEN SACS of the ANTHERS of flowering plants and within male cones in CONIFEROPHYTA. Pollen grains are usually yellow with a hard outer wall and can be light, for wind dispersal, or heavier, larger, sticky and spiny for insect dispersal. Pollen grains are extremely resistant to decay, which makes them a useful tool for gathering information on the abundance of past species. This is termed pollen analysis or palynology. *See also* DOUBLE FERTILIZATION, POLLEN TUBE.

pollen analysis *See* POLLEN.

pollen sac A structure within the ANTHER of flowering plants containing POLLEN. There are usually four pollen sacs (two in each lobe of the anther).

pollen tube An outgrowth from a POLLEN grain when it lands on a STIGMA of a flowering plant and begins to germinate. The pollen tube transports two male GAMETES to the female gamete (egg nucleus) to allow DOUBLE FERTILIZATION. The tube passes through the STYLE and towards the OVULE, where it enters through the MICROPYLE. It penetrates the EMBRYO SAC and releases the male gametes, one of which then fuses with the egg nucleus and the other with two polar nuclei (*see* POLAR NUCLEUS). The pollen tube then disintegrates.

pollination The transfer of POLLEN from the male to female parts of plants. In flowering plants (ANGIOSPERMS) this transfer is from ANTHERS to STIGMAS, and in CONIFERS it is from male to female cones. In cross-pollination the transfer is between two plants, and in self-pollination it is within one flower or another flower on the same plant. Although self-pollination does occur, it is less favoured than cross-pollination because it reduces the variability of the population. To reduce self-pollination, the STAMEN and stigma may mature at different times. If the stamen develops first, the plant is termed protandrous; in protogyny plants the stigma and OVULES develop first.

Transfer of pollen can be by wind, water, insects, birds, bats or other small animals, and flowers are usually adapted for one form of pollination. Pollination is not the same as FERTILIZATION, which occurs after pollination (*see* DOUBLE FERTILIZATION).

See also ANEMOPHILY, ENTOMOPHILY, ORNITHOPHILY.

pollution Contamination of the environment by the by-products of human activity, mostly from industrial and agricultural processes. Air pollution is largely due to the burning of FOSSIL FUELS in the home, industry or the combustion engine of vehicles. Such pollutants include smoke (tiny particles of carbon), sulphur dioxide, carbon dioxide and carbon monoxide, nitrogen oxides and lead, particularly from car exhaust emissions. These contribute to the GREENHOUSE EFFECT and ACID RAIN. Other important air pollutants are CHLOROFLUOROCARBONS, which contribute to the thinning of the OZONE LAYER.

Water pollution can be the result of rain polluted by contaminants in the air, but is also due to the release of toxic chemicals, for example copper, zinc, lead, mercury and cyanide, into rivers and seas, killing fish and plant species. Oil pollution of water (usually accidental) is localized, but devastating to seabirds, shellfish and seaweed. Oil spillages can be degraded by some micro-organisms. Sewage is treated (*see* SEWAGE DISPOSAL) but still contains large amounts of PHOSPHATES from washing powders and detergents that remain a problem as a source of water pollution.

Land (terrestrial) pollution comes from the dumping of solid wastes, for example slag heaps from ore digging, metal refining and coal mining, which are unsightly and often cannot sustain any vegetation. Domestic rubbish is usually disposed of by burning; some plastics are now being used that can be degraded by micro-organisms to avoid the dangerous gases given off by burning. Other land pollutants are pesticides and noise.

Radioactive pollution can be from medical waste, televisions, watches, waste from the nuclear power industry, or from testing nuclear weapons. Disposal of NUCLEAR WASTE has been at sea or by burial on land, but decay can take thousands of years and presents problems of safety, pollution and security.

See also EUTROPHICATION, GLOBAL WARMING, LIGHT POLLUTION.

polonium (Po) The element with ATOMIC NUMBER 84; RELATIVE ATOMIC MASS 210.0; melting point 254°C; boiling point 962°C. Polonium has no stable ISOTOPES, but occurs in nature from the decay of longer lived heavy radioactive elements. The longest lived isotope (polonium–209) has a HALF-LIFE of 100 years, which makes it far more radioactive than most other naturally occurring elements. It has been used as a heat source for radioactive electrical generators. Polonium was the first element to be discovered as a radioactive decay product, and was extracted from pitchblende, a uranium ore by Pierre and Marie Curie in 1898.

polyamide Any polymerized AMIDE, such as NYLON.

Polychaeta A class of the PHYLUM ANNELIDA, including the lugworm (*Arenicola*) and ragworm (*Nereis*). Polychaetes are marine, free-swimming and can move slowly on structures called 'parapodia', with which they can burrow. They are clearly segmented, with a distinct head, and they possess many chaetae (bristles). Polychaetes have separate sexes and FERTILIZATION is external. *Compare* HIRUDINEA, OLIGOCHAETA.

Polychaete A member of the PHYLUM POLYCHAETA.

poly(chloroethene) *See* POLYVINYL CHLORIDE.

polycrystalline (*adj.*) Describing any solid material that occurs in pieces with no regular shape, but on examination can be seen to be made of many small CRYSTALS. Within each crystal the molecules are arranged regularly, but the direction of alignment of this structure varies from one small crystal to the next, and the boundaries between one crystal and the next (GRAIN BOUNDARIES) are often irregular.

polyester A synthetic THERMOSETTING PLASTIC, formed by the condensation POLYMERIZATION of POLYHYDRIC alcohols with DIBASIC acids. Polyesters are used in the manufacture of synthetic fibres in textiles such as terylene and Dacron. They can be reinforced by the addition of glass fibres and then used in car bodies and boats.

poly(ethene), ***poly(ethylene),*** trade name ***polythene*** An addition POLYMER of ETHENE (C_2H_4). It is a tough, white THERMOPLASTIC, repeatedly softened on heating. It is widely used, for example, in bottles, toys, packaging, pipes and electric cable.

Polythene may be manufactured by high pressure POLYMERIZATION of ethene gas with trace amounts of oxygen. An unpaired electron from the oxygen molecule initiates the polymerization by attacking an ethene molecule. This produces a RADICAL that attacks another ethene molecule and so on until 1 to 10 thousand ethene molecules are joined together, then the reaction is terminated. This method yields low-density polythene.

In another method, high-density polythene is made using catalysts at low pressure. This method was first used by Karl Ziegler (1898–1973) and the catalyst is suitably termed ZIEGLER-NATTA CATALYST. This yields polythene which is more rigid at low temperatures and softer at higher temperatures.

poly(ethylene) *See* POLY(ETHENE).

polygene Any one of a group of GENES that interact together to have an effect on a phenotypic characteristic (*see* PHENOTYPE). Individually, the genes have little effect but combined their effects are marked. Polygenes affect characteristics that show continuous variation, such as the height and weight of an organism. An individual organism receives a range of genes from any polygenic complex because of the random assortment of genes during MEIOSIS. This usually ensures that the characteristic is intermediate, for example neither very tall nor very small. A characteristic that is determined by polygenes is called a polygenic character.

polygon A closed two-dimensional figure having straight sides. The type of polygon is defined by the number of sides: 3 sides is a triangle; 4 a quadrilateral; 5 a pentagon; 6 a hexagon; etc. If a polygon has sides all of the same length it is described as regular.

polyhedron A closed three-dimensional figure having at least four faces, each of which is flat. In a regular polyhedron, all sides have the same length. There are only five such regular polyhedrons, called the Platonic solids. These are the tetrahedron (four triangular faces), the cube (six square faces) the octahedron (eight triangular faces), the dodecahedron (12 faces each in the shape of a pentagon) and the icosahedron (20 triangular faces).

polyhydric Of ALCOHOLS, those containing more than one HYDROXYL GROUP. Dihydric alcohols contain two hydroxyl groups and trihydric contain three.

polymer A large molecule made up of two or more similar or identical repeated MONOMERS joined together by POLYMERIZATION to form a chain or branching matrix. There are many naturally occurring polymers, including PROTEINS, POLYSACCHARIDES, NUCLEIC ACIDS and also many synthetic polymers including poly(ethene), NYLON, POLYSTYRENE, POLYVINYL CHLORIDE (PVC).

Addition polymers are those in which the polymer chain is built up by the simple addition of identical monomers. Condensation polymers are those in which the chain can be built up of two or more different monomers with the loss of a small molecule, such as water, during the joining of the monomers. Biochemical polymers are usually condensation polymers, whereas many synthetic polymers, including polystyrene and PVC, are addition polymers.

See also ATACTIC POLYMER, ISOTACTIC POLYMER, RUBBER, SYNDIOTACTIC POLYMER.

polymerase An enzyme, for example DNA polymerase and RNA polymerase, that joins MONOMERS together to form POLYMERS.

polymerase chain reaction *See* GENE AMPLIFICATION.

polymerization The chemical process resulting in the joining together of MONOMERS to form POLYMERS. This can be addition polymerization in which a polymer chain is built up by the addition of identical monomers to each other, for example POLYSTYRENE, POLYVINYL CHLORIDE (PVC), PERSPEX. The repeating units in an addition polymer are identical to the initial reacting monomers.

The other type of polymerization is condensation polymerization, in which the joining of monomers results in the loss of a small molecule, usually water. Condensation polymerization can involve two or more different monomers (co-polymerization). In contrast to addition polymers the repeating units in condensation polymers are not identical to the reacting monomers. Examples of the latter include NYLON and POLYESTER.

See also ZIEGLER-NATTA CATALYST.

poly(methyl 2-methylpropenoate) *See* PERSPEX.

poly(methylmethacrylate) *See* PERSPEX.

polymorphism 1. (*biology*) In genetics, the existence of more than one distinct form of a particular characteristic within a species population that cannot be explained by MUTATION. Examples include the BLOOD GROUP SYSTEM in humans and different colour forms of the peppered moth (which depends on their environment).

2. (*chemistry*) The existence of an element or compound in more than one crystalline form. For example, sulphur forms rhombic crystals below 96°C and monoclinic crystals at higher temperatures, up to its melting point of 119°C. The distinction between polymorphism and ALLOTROPY is that polymorphism relates only to crystalline solids.

polymorphonuclear leucocyte *See* GRANULOCYTE.

polynomial A mathematical expression containing three or more terms. The degree of a polynomial is the largest power that it contains. Thus

$$ax^3 + bx^2 + cx + d$$

is a third-degree polynomial.

polynucleotide A long-chain molecule of NUCLEOTIDES.

polyoestrus *See* OESTROUS CYCLE.

polyp The sedentary structural form in the life cycle of the CNIDARIA.

polypeptide A PEPTIDE consisting of three or more AMINO ACIDS. Polypeptide chains can then fold or twist to form a PROTEIN.

poly(phenylethene) *See* POLYSTYRENE.

polyploid A nucleus, cell or organism that has three or more sets of CHROMOSOMES. Polyploidy is rare in animals but common in plants. Polyploidy can arise spontaneously, for example if diploid GAMETES (with two sets of chromosomes each) self-fertilize or if chromosome numbers double after fertilization. Polyploidy can also be induced by treatment with a chemical called colchicine. This prevents spindle formation during MEIOSIS and therefore chromosomes cannot separate. Some polyploids are sterile, depending on whether HOMOLOGOUS chromosomes can pair during meiosis. Sterile polyploids can still undergo ASEXUAL REPRODUCTION.

An example of polyploidy in plants is wheat, where varieties with four or six sets of chromosomes exist. Many flowering plants are polyploid, including bananas, potatoes, apples and sugar cane. Polyploid varieties usually have an advantage, such as larger fruit.

poly(propene) *See* POLYPROPYLENE.

polypropylene, *poly(propene)* A POLYMER of propene which is rigid, very strong and resistant to abrasion. It is used in moulded furniture and can be hinged for use in make-up compacts for example. Polypropylene can also be spun into fibres to make ropes or carpets.

polysaccharide A CARBOHYDRATE consisting of a variable number of MONOSACCHARIDES joined together in chains that can be branched or not and can fold for easy storage. Polysaccharides can be broken down into their constituent DISACCHARIDES or monosaccharides for use by an organism. In chains, they are insoluble. Examples include CELLULOSE, CHITIN, STARCH and GLYCOGEN.

polystyrene, *poly(phenylethene)* A THERMOPLASTIC, like POLY(ETHENE) an addition POLYMER but more rigid due to the presence of BENZENE RINGS. It is widely used in an expanded form produced by impregnating polystyrene with pentane and heating in steam where it is blown into a solid foam. This forms a light, white material which is used in packaging and for insulation.

polytetrafluoroethene (PTFE) (trade name ***Teflon***) An addition POLYMER made from the MONOMER tetrafluoroethene (CF_2CF_2). It is a THERMOSETTING PLASTIC with a high melting point and is used for non-stick surfaces on kitchen pans.

polythene The trade name for POLY(ETHENE).

polyunsaturate A FATTY ACID with two or more DOUBLE BONDS.

polyurethane A major THERMOSETTING PLASTIC used as a liquid in paints or varnish and as a foam in upholstery. It is a condensation POLYMER consisting of the MONOMER –NHCOOH–.

polyvinyl chloride, *PVC, poly(chloroethene)* A THERMOPLASTIC, like POLY(ETHENE) an addition POLYMER. It is rigid and tough but can be made flexible by the addition of PLASTICIZERS. Unplasticized it is used in window frames (uPVC) and record discs. Plasticized PVC is widely used for shoes, luggage, flooring and toys.

pome A type of PSEUDOCARP, for example apples and pears.

pons A broad band of nerve fibres that links the MEDULLA OBLONGATA to the MIDBRAIN.

population A group of animals of one species living together in a particular area. The actual size of a population is determined by the balance between the birth rate and the death rate, although other factors such as immigration and emigration also influence population size. The human population is increasing exponentially and it was suggested as far back as 1798 by Thomas Malthus (1766–1834) that future food supplies would not support our population size; this problem is still true today. *Compare* COMMUNITY. *See also* DENSITY-DEPENDENCE.

population inversion A state in which more atoms are in an EXCITED STATE than in some lower energy state, such as the GROUND STATE. *See also* LASER.

p-orbital The second lowest energy ORBITAL for a given PRINCIPAL QUANTUM NUMBER. p-Orbitals exist only for principal quantum numbers of two or greater. There are three p-orbitals for a given principal quantum number, each formed of two lobes with each of the three orbitals at right angles to the other two.

Porifera A PHYLUM consisting of the sponges. Sponges are simple invertebrate animals that are usually marine and possess no organs or tissues. The body is hollow but is lined by outer contractile cells and inner choanocytes (collar cells) that bear FLAGELLA, the movement of which causes water to circulate, therefore providing a constant supply of food particles and oxygen. Between these two cell types are amoebocytes, which store food and give rise to reproductive cells. ASEXUAL REPRODUCTION or

SEXUAL REPRODUCTION can occur in sponges; they are HERMAPHRODITE but cross-fertilization does occur. There are many openings all over the body wall that can be strengthened with protein, silica or calcium carbonate.

porous (*adj.*) Describing any material that contains many small cracks or holes, able to absorb water, air or some other fluid.

port In computing, a socket that allows a computer to be connected to a PERIPHERAL DEVICE or a NETWORK.

position vector A VECTOR whose length and direction are equal to the distance and direction of a specified point from the ORIGIN.

positive In physics, the name given to one of the two types of electric CHARGE. PROTONS are positively charged. Objects that are positively charged normally reach that state by losing electrons rather than by gaining protons.

positive feedback FEEDBACK that is IN PHASE with the original signal. This generally has the effect of destabilizing a circuit and may lead to OSCILLATION. *See also* NEGATIVE FEEDBACK.

positron The ANTIPARTICLE of the ELECTRON. The positron is a stable ELEMENTARY PARTICLE with the same mass as an electron, but the opposite charge and LEPTON number. Although positrons are stable in isolation, when a positron meets an electron, they will annihilate one another, producing two or more gamma rays.

posterior (*adj.*) In biology, the rear (hind) end of the body of an organism. In humans it is the back side. Posterior also refers to the back of a structure such as an ORGAN or GLAND. In plants, it refers to buds or flowers that are nearest to the main stem. *Compare* ANTERIOR.

post-synaptic neurone *See* SYNAPSE.

potash *See* POTASSIUM CARBONATE.

potash alum *See* ALUM.

potassium (K) The element with ATOMIC NUMBER 19; RELATIVE ATOMIC MASS 39.1; melting point 64°C; boiling point 774°C; RELATIVE DENSITY 0.9. Potassium is a highly reactive ALKALI METAL, and its compounds give a characteristic violet colour when ionized in a flame. It is widespread in nature and slightly radioactive due to a small concentration (0.12 per cent) of the long-lived radioactive ISOTOPE potassium–40, which has a HALF-LIFE of 1.25×10^9 years.

Potassium is essential to all living organisms. In animals it aids in the transmission of nerve impulses, while in plants it is required for growth.

potassium bromide (KBr) A white solid; melting point 734°C; boiling point 1,435°C; RELATIVE DENSITY 2.7. Potassium bromide may be manufactured by the direct reaction between bromine and potassium:

$$2K + Br_2 \rightarrow 2KBr$$

It is widely used in the photographic industry for the manufacture of silver bromide, used in light-sensitive emulsions.

potassium carbonate, *potash* (K_2CO_3) A white solid; melting point 891°C; decomposes on further heating; RELATIVE DENSITY 2.4. Potassium carbonate is an important naturally occurring source of potassium. It is used in the manufacture of soaps and in textile processing.

potassium chlorate ($KClO_3$) A white powder; melting point 360°C; decomposes on further heating; RELATIVE DENSITY 2.3. Potassium chlorate is produced around the ANODE in the ELECTROLYSIS of concentrated potassium chloride:

$$KCl + 3H_2O \rightarrow KClO_3 + 3H_2$$

It decomposes on heating, releasing oxygen:

$$2KClO_3 \rightarrow 2KCl + 3O_2$$

This reaction has led to the use of potassium chlorate in some explosives. It is also a powerful OXIDIZING AGENT, a feature exploited in its use as a weedkiller and disinfectant.

potassium chloride (KCl) A white powder; sublimes at 1,700°C; RELATIVE DENSITY 2.0. Potassium chloride occurs naturally, and can be extracted from brine by FRACTIONAL CRYSTALLIZATION. It is used as a raw material for the manufacture of other potassium salts, particularly potassium chromate.

potassium chromium sulphate, *chrome alum* ($K_2SO_4.Cr_2(SO_4)_3.12H_2O$) A crystalline DOUBLE SALT; melting point 89°C; loses water on further heating; RELATIVE DENSITY 1.8. Potassium chromium sulphate is deep red in colour and can be made by crystallizing a solution containing equal concentrations of potassium and chromium sulphates.

potassium dichromate ($K_2Cr_2O_7$) An orange solid; melting point 396°C; decomposes on further heating; RELATIVE DENSITY 2.7. Like all dichromates it is stable only in acidic solution. In the presence of alkali, chromates are produced, for example:

$$K_2Cr_2O_7 + 2NaOH \rightarrow K_2CrO_4 + Na_2CrO_4 + H_2O$$

Potassium dichromate is used as a laboratory OXIDIZING AGENT for certain analytical techniques. It is also used as a reagent for chromium ELECTROPLATING.

potassium hydrogencarbonate ($KHCO_3$) A white solid; decomposes on heating; RELATIVE DENSITY 2.2. Potassium hydrogencarbonate can be produced by passing carbon dioxide through potassium carbonate solution:

$$K_2CO_3 + CO_2 + H_2O \rightarrow 2KHCO_3$$

Potassium hydrogencarbonate reacts with acids to produce carbon dioxide, a reaction that is exploited in some cooking processes and in certain types of fire extinguisher, for example:

$$KHCO_3 + CH_3COOH \rightarrow CH_3COOK + H_2O + CO_2$$

Potassium hydrogencarbonate decomposes on heating:

$$KHCO_3 \rightarrow KOH + CO_2$$

potassium hydroxide, *caustic potash* (KOH) A white solid; melting point 360°C; boiling point 1,320°C; RELATIVE DENSITY 2.0. Potassium hydroxide can be prepared by reacting potassium with water:

$$2K + 2H_2O \rightarrow 2KOH + H_2$$

Potassium hydroxide is used in many situations where a concentrated alkaline solution is required, but is highly damaging to body tissues, so must be handled with particular care – even small splashes in the eye can cause permanent damage.

potassium iodide (KI) A white solid; melting point 686°C; boiling point 1,330°C; RELATIVE DENSITY 3.1. Potassium iodide can be made by the reaction between iodine and potassium hydroxide:

$$6KOH + 3I_2 \rightarrow KIO_3 + 5KI + 3H_2O$$

The potassium iodate so formed can be converted to potassium iodide by heating:

$$2KIO_3 \rightarrow 2KI + 3O_2$$

Potassium iodide is widely used in the manufacture of silver iodide, an important compound in light-sensitive emulsions.

potassium nitrate, *saltpetre* (KNO_3) A colourless solid; melting point 334°C; decomposes on further heating; RELATIVE DENSITY 2.1. Potassium nitrate occurs naturally, and is used as a fertilizer and as an OXIDIZING AGENT, especially in some explosives, such as gun powder.

potassium sulphate (K_2SO_4) A colourless solid; melting point 1,062°C; decomposes on further heating; RELATIVE DENSITY 2.7. Potassium sulphate occurs in nature. It is manufactured by the reaction between potassium chloride and concentrated sulphuric acid:

$$2KCl + H_2SO_4 \rightarrow K_2SO_4 + 2HCl$$

It is used as a fertilizer and in the manufacture of cement.

potential difference (p.d.) A more technical term for VOLTAGE, the p.d. between two points being the amount of energy converted from electrical energy to other forms when one COULOMB of charge flows between the two points.

$$\text{Potential difference} = \frac{\text{energy transformed}}{\text{charge flow}}.$$

potential divider A pair of RESISTORS with a supply voltage connected across both of the resistors so that a fraction of this voltage appears across each of the resistors. This arrangement is used in volume controls in radios and AMPLIFIERS. For a potential divider with resistors R_1 and R_2 and an input voltage V_{in} and an output voltage V_{out}

$$V_{out} = V_{in}R_2/(R_1 + R_2)$$

potential energy The ENERGY possessed by an object or system as a result of its position or state, such as a stretched spring or a mass in a GRAVITATIONAL FIELD. It is the amount of work done by the object or system moving from a state at which it is said to have no potential energy to a higher state. *See also* GRAVITATIONAL POTENTIAL ENERGY.

potentiometer An alternative name for a VARIABLE RESISTOR, particularly when used as an adjustable POTENTIAL DIVIDER.

potocytosis A process used by cells, similar to PHAGOCYTOSIS, in which small molecules and ions can be concentrated before entering the CYTOSOL at special CELL MEMBRANE sites called caveolae.

pound (lb) A unit of mass, now obsolete in science but still in everyday use in the US. One pound is approximately 0.454 kilograms.

poundal A unit of force, now obsolete in science, equal to the pull of gravity on Earth on a mass of one pound. One poundal is approximately 4.5 NEWTONS.

powder coating A way of giving a plastic surface to a metal object, preventing corrosion, in a way similar to the application of paint. The object to be coated is given an ELECTROSTATIC charge and then sprayed with an oppositely charged powder. The opposite charges ensure that the powder sticks to the object. The object is then heated to melt the powder into a solid layer.

power 1. (*mathematics*) The number of times a quantity must be multiplied by itself in an algebraic expression, thus a^5, which means $a \times a \times a \times a \times a$, is referred to as 'the fifth power of a' or 'a to the power (of) 5' or just 'a to the fifth'. If the power is 2 or 3, the number is said to be 'squared' or 'cubed' respectively.

2. (*physics*) The rate at which WORK is done. The SI UNIT of power is the WATT.

$$\text{power} = \text{energy/time} = \text{force} \times \text{speed}$$

power series *See* SERIES.

power stroke The stage in the operation of a PETROL ENGINE where fuel burns and expands to drive the engine.

pp chain The NUCLEAR FUSION reactions by which a MAIN SEQUENCE STAR fuses hydrogen into helium. The details are complicated, but the first step in the chain involves two protons fusing to form a hydrogen–2 nucleus, with the release of a POSITRON and a NEUTRINO. Various steps are then possible, but most commonly the hydrogen–2 nucleus fuses with another proton to form a helium–3 nucleus with the emission of a gamma ray. Two helium–3 nuclei then fuse and release two protons to give a helium–4 nucleus.

praseodymium (Pr) The element with ATOMIC NUMBER 59; RELATIVE ATOMIC MASS 140.9; melting point 934°C; boiling point 3,512°C; RELATIVE DENSITY 6.8. Praseodymium is LANTHANIDE metal. Its IONIC compounds have a bright green colour and are used to colour glass and ceramics.

Precambrian The division of geological time that preceded the PHANEROZOIC and lasted from 4.5 billion years ago (when the Earth was first formed) to 570 million years ago. Fossils from this time are very rare.

precession The slow change in alignment of the axis of a more rapidly spinning object, such as a GYROSCOPE, under the influence of an external force.

precipitate A solid or solid-phase separated from a solution, which may subsequently settle. In chemistry, a precipitate is a finely divided powder suspended in a liquid, which is formed when a reaction between two soluble salts in solution produce an insoluble salt.

precipitation 1. (*meteorology*) The fall of water from the atmosphere, either as RAIN, SLEET, SNOW or HAIL.

2. (*chemistry*) The formation of a PRECIPITATE.

precipitin An ANTIBODY that combines with a soluble ANTIGEN to form an antigen–antibody complex, which then precipitates and can be removed (or separated if used experimentally).

precursor Any compound produced in some intermediate step in a series of chemical reactions, and which is then involved in further reactions leading to the desired final product.

predation A relationship in an ecological COMMUNITY in which one animal species (the predator) captures, kills and feeds on another (the prey). The level of predation influences the size of POPULATIONS. *See also* FOOD CHAIN.

pregnancy The duration of EMBRYO development within the UTERUS of mammals, beginning with conception and ending at birth. The duration varies between species, for example it lasts 40 weeks in humans, 18–22 months in elephants, and 60 days in cats.

Some of the early signs of pregnancy in humans are cessation of menstruation, tenderness or enlargement of breasts, and the detection of the hormone HUMAN CHORIONIC GONADOTROPHIN in the urine, which forms the basis of pregnancy testing. Levels of the steroid hormones OESTROGEN and PROGESTERONE prepare for and maintain the pregnancy, and there are changes during pregnancy in the circulating levels of many other constituents of blood. In humans and other primates the pregnancy length seems to be determined by the FOETUS, although the precise control of the onset of labour and birth is not fully understood (*see* PARTURITION).

One in five pregnancies fail, usually very early and often unnoticed (except by a late period) but sometimes later as a miscarriage. These failures are often due to an abnormality in the foetus. Foetal death after 24 weeks is called a stillbirth, when the baby is born dead.

Factors such as drugs, alcohol and certain PATHOGENS can affect foetal development (*see* PLACENTA).

The term GESTATION is more commonly used for pregnancy in animals other than humans.

See also BLASTOCYST, ECTOPIC PREGNANCY, EMBRYO, EMBRYONIC DEVELOPMENT, IMPLANTATION.

pre-ignition The condition in a PETROL ENGINE where the fuel starts to burn before the spark has been generated in the SPARK PLUG. Until the 1980s, lead-based compounds were routinely added to petrol to enable the petrol-air mixture to be compressed to a greater extent without igniting. Concerns over the high levels of toxic lead compounds in the air of some large cities have lead to the introduction of unleaded petrol, which has no such additives, but can only be burnt in engines designed to compress the fuel a little less before igniting it. *See also* COMPRESSION RATIO.

premolar In mammals, a large TOOTH behind the INCISOR teeth that has two cusps (ridges) and a single or double root, and is used for grinding food. Humans have eight premolars and they are present in the milk teeth (*see* TOOTH).

preon A hypothetical ELEMENTARY PARTICLE from which QUARKS and LEPTONS are made, postulated to explain the large number of different types of such particles. No workable model of particle physics based on preons has yet been constructed and there is no experimental evidence for their existence.

pressure The FORCE acting on each square metre of area. The unit of pressure is the PASCAL (Pa), this being a pressure of one NEWTON per square metre.

The amount of pressure acting on an object is related to the amount it is deformed or damaged, thus shoes with small heels do more damage to a soft floor than larger heels worn by a person of the same weight. Knives are made with sharp edges, and nails with sharp points, to minimize the contact area and make the pressure produced as large as possible. Padded seats are more comfortable than hard ones, because they deform to produce a large contact area and less pressure.

On a molecular level, the pressure of a fluid can be thought of in terms of the pressure exerted on a surface by the molecules of the fluid. The pressure is equal to the average change in MOMENTUM per molecular collision multiplied by the average number of collisions per second per square metre. Because there are a very large number of collisions, they are felt as a constant pressure, rather than a series of separate impacts.

See also BAROMETER, BOURDON GAUGE, DALTON'S LAW OF PARTIAL PRESSURE, DYNAMIC PRESSURE, HYDROSTATIC PRESSURE, HYDROSTATICS, KINETIC THEORY.

pressure law A GAS LAW used to define the IDEAL GAS TEMPERATURE SCALE. It states that, for a fixed mass of gas held in a constant volume, the pressure is proportional to the ABSOLUTE TEMPERATURE, i.e. the pressure divided by the temperature is a constant. For a fixed mass of an ideal gas with a pressure p and absolute temperature T held at constant volume,

$$p/T = \text{constant}$$

See also BOYLE'S LAW, CHARLES' LAW, CONSTANT VOLUME GAS THERMOMETER, IDEAL GAS EQUATION.

pressure relief valve *See* VALVE.

pre-synaptic neurone *See* SYNAPSE.

prevailing wind The most common wind direction at a given location. *See also* AIR MASS.

Prévost's theory of exchanges The absorption and emission of THERMAL RADIATION from a body is equal when it is in EQUILIBRIUM with its surroundings. In this state, the temperature of the body remains constant. If the body and its surroundings are at different temperatures there is a net flow of energy. *See also* HEAT, THERMAL EQUILIBRIUM.

primary cell Former name for a non-rechargeable electrochemical CELL.

primary coil In a TRANSFORMER or INDUCTION COIL, the coil that is connected to the power supply.

primary colour In optics, red, green or blue, the three colours of light that when mixed together in equal proportions produce white light. When mixed in other proportions they can produce any other colour (except black).

primary endosperm nucleus A nucleus within the EMBRYO SAC of a flowering plant that results from the fusion of one of the male nuclei carried by the POLLEN TUBE with the two polar nuclei (*see* POLAR NUCLEUS) present in the embryo sac. The primary endosperm nucleus is TRIPLOID and it divides further to form the ENDOSPERM. *See also* DOUBLE FERTILIZATION.

primate A member of the ORDER Primates in the CLASS MAMMALIA, including apes, monkeys and humans (called anthropoids) and lemurs, bushbabies, lorises and tarsiers (called prosimians). Features of primates include a large brain relative to the body, forward-directed eyes with good colour vision, five digits on limbs, nails instead of claws, opposable thumbs and big toes, gripping hands and feet, good mobility of limbs and shoulder joints, only two MAMMARY GLANDS, young that are usually born singularly and are nourished by a PLACENTA. Many of these characteristics are adaptations to a climbing mode of life in trees.

prime factor A FACTOR that is also a PRIME NUMBER. Thus the prime factors of 20 are 1, 2, and 5.

prime number A number that has no FACTORS other than 1 and the number itself. 1 is also a prime number by definition. The first few prime numbers are 1, 2, 3, 5, 7, 11, 13, 17. Apart from 2 all prime numbers are odd. The discovery of large prime numbers is an important aspect of some techniques for coding data so it cannot be understood by those who do not have access to the code.

primitive cell *See* UNIT CELL.

primordium In botany, a cell or group of cells that is immature but will develop into a specific structure, for example a leaf primordium.

principal axis, *optical axis* The line joining the centre of a LENS or MIRROR to the PRINCIPAL FOCUS, at right angles to the FOCAL PLANE. It is the line along which a ray of light can travel through an optical system without any change of direction, except for 180° reflection at a mirror.

principal focus A point associated with a LENS or CURVED MIRROR, through which all rays originally parallel to the PRINCIPAL AXIS will pass after REFRACTION or REFLECTION at the lens or mirror. For a DIVERGING LENS or a CONVEX mirror, this is a VIRTUAL FOCUS – the rays diverge from the principal focus as if they had originated from an imaginary point on the other side of the lens or mirror.

principal quantum number The QUANTUM NUMBER used to label an ORBITAL to give a broad indication of its energy. Thus in describing a 2s orbital, 2 is the principal quantum number.

principle of equivalence One of the starting points of the GENERAL THEORY OF RELATIVITY, which states that there is no difference between a system that is undergoing a steady acceleration and one that is in a constant GRAVITATIONAL FIELD. Thus an observer in a lift who suddenly finds himself weightless will not know whether the lift cable has broken and the lift has entered FREE FALL or if the Earth's gravity has mysteriously been turned off.

This equivalence is a consequence of two ways of thinking about MASS. Mass can be thought of as the resistance to acceleration, INERTIAL MASS in Newton's second law (*see* NEWTON'S LAWS OF MOTION) and also as the quantity that controls the size of gravitational interactions in NEWTON'S LAW OF GRAVITATION – this is GRAVITATIONAL MASS. The equivalence of these two masses leads to the idea that all objects fall with the same acceleration in a given gravitational field, and thus to the principal of equivalence.

principle of least time *See* FERMAT'S PRINCIPLE.

principle of superposition The total displacement of two or more WAVES arriving at a point is equal to the sum of the displacements of the individual waves. Each wave leaves the point of superposition unaltered. *See* INTERFERENCE.

printer A device, normally connected to the output of a computer, for producing text or diagrams on paper. Printers come in various forms. A DOT-MATRIX PRINTER forms characters by having an array of fine wires which press against an inked ribbon. In a DAISYWHEEL printer characters are selected from a rotating wheel and pressed against an inked ribbon. This is slower than a dot-matrix printer and can produce fewer characters, but produces higher quality. BUBBLE-JET PRINTERS use a jet of fine ink droplets that are deflected electrostatically. *See also* LASER PRINTER.

prion An infectious protein fragment thought to be responsible for BOVINE SPONGIFORM ENCEPHALOPATHY in cattle, SCRAPIE in sheep and CREUTZFELDT–JAKOB DISEASE in humans.

prism 1. A three-dimensional shape formed by extending a two-dimensional shape, especially a triangle, into the third dimension, so it has two faces that are POLYGONS with all the others being rectangular. The volume of a prism is equal to the length of its parallel sides multiplied by the area of the face perpendicular to these sides.

2. A triangular block of glass or some other transparent material used to refract light (*see* REFRACTION) in optical systems.

probability A mathematical measure of how likely an event is. If an event cannot happen, it is given a probability of 0, if it is certain, the probability is 1. If an event is expected to happen n times out of N, the probability is n/N. If an event has a probability p, the probability of it not happening is $1 - p$. For two events that are independent, with probabilities p and q, the probability of both happening is pq. For two alternative events, one or the other of which can happen, but not both, if the probabilities for each are p and q, the probability that one or the other will occur is $p + q$.

probability amplitude A quantity used in QUANTUM MECHANICS to find the probability of a system being in a particular state or a particle in a particular position. Probability amplitudes obey the PRINCIPLE OF SUPERPOSITION and so can interfere constructively or destructively (*see* INTERFERENCE), giving wave-like properties to the particles they describe. The probability of finding a particular state is the square of the probability amplitude. *See also* WAVE-PARTICLE DUALITY.

procambium Tissue derived from the MERISTEM in VASCULAR PLANTS that gives rise to the VASCULAR BUNDLES. The procambium consists of elongated cells that are grouped into strands just behind the growing points of stems and roots.

producer gas The gas formed by passing air over hot COKE. The coke is partially oxidized (*see* OXIDATION) by the oxygen, forming a mixture of nitrogen and carbon monoxide. This reaction heats the coke and is sometimes used in conjunction with the production of WATER GAS. *See also* SEMI-WATER GAS.

productivity In biology, the quantity of carbon compounds formed by the primary producers (plants) in a FOOD CHAIN during PHOTOSYNTHESIS, which can be used by the consumers (animals). The gross productivity is the total quantity of carbon compounds produced. The net productivity is the gross productivity minus the quantity of carbon compounds used by the plants themselves in their RESPIRATION. Thus the net productivity represents the amount of food available to consumers in an ECOSYSTEM.

progesterone In mammals, a STEROID HORMONE that regulates the MENSTRUAL CYCLE and prepares the UTERUS for PREGNANCY. Progesterone is secreted by the CORPUS LUTEUM in the OVARY and if FERTILIZATION of an OVUM occurs, this provides the progesterone needed for the maintenance of early pregnancy until the PLACENTA takes over at 3 months. If fertilization does not occur the corpus luteum regresses and dies, and levels of progesterone fall dramatically, causing menstruation. Progesterone also inhibits production of FOLLICLE-STIMULATING HORMONE.

program A set of instructions for a COMPUTER. Modern programs are increasingly sophisticated and allow the operator to perform complex tasks without being at all aware of the calculations taking place within the machine. Programs can be divided into APPLICATIONS PROGRAMS, which are designed to allow a user to perform a task in the simplest way possible, and SYSTEMS PROGRAMS, which perform tasks essential to the computer's operation, but generally without the user being aware of these processes taking place.

The most common types of applications programs are word processors and spreadsheets. Word-processing programs are used to enter text that can then be altered and checked for spelling before being fed to a printer, which produces a paper version of the text (called a hard copy in computer jargon). Spreadsheets perform numerical manipulations on data entered in rows and columns – totalling all the figures in a column for example. *See also* ASSEMBLER, COMPILER, INTERPRETER.

programming language A set of instructions and rules for their use that can be put together in a sequence to allow a computer to perform a certain task. Languages are defined as high-level if they do not require the programmer to concern himself with the operation of the computer itself, and low–level if they are require the operator to think in terms of the way the computer functions, with instructions relating directly to use of certain sections of the computer memory etc.

High-level languages are generally easy to learn and can be used on a wide variety of computers. Low-level languages are often limited to a small number of machines since the detailed structure of a computer will vary from one machine to another. In either case, the instructions will have be converted into MACHINE CODE, a set of BINARY instructions to the CENTRAL PROCESSING UNIT before the program instructions can be carried out. *See also*

ASSEMBLY LANGUAGE, BASIC, C, COBOL, FORTRAN, OBJECT-ORIENTED PROGRAMMING, PASCAL.

progression A SERIES of algebraic expressions added together, each one differing from the previous one in some way, typically either by the addition of a certain number (an ARITHMETIC PROGRESSION) or by multiplication by a certain number (a GEOMETRIC PROGRESSION).

progressive wave, *travelling wave* A WAVE that transmits ENERGY in one direction through a material, as opposed to a STANDING WAVE.

projectile Any object that is launched and then allowed to move freely under the influence of GRAVITY. *See* FREE FALL. *See also* BALLISTICS.

projector A device designed to produce an enlarged image on a screen. In a motion picture projector the object is a transparent piece of photographic film. Because the image is enlarged, the light falling on any one part of the screen is likely to be rather dim, so the light falling on the object being projected is usually made as bright as possible using CONDENSERS.

Prokaryotae The KINGDOM consisting of PROKARYOTES.

prokaryote (*Pro* = before, *karyote* = nucleus) A simple organism in which the genetic material DNA is not contained within a nucleus and no membrane-bound ORGANELLES exist. Prokaryotes are thought to be the first forms of life on Earth. BACTERIA and CYANOBACTERIA are prokaryotes; all other organisms are EUKARYOTES.

Prokaryotes are all unicellular but can be found in filaments and clusters. The DNA of prokaryotes forms a coiled structure called a nucleoid; often more than one nucleoid exists within a single cell because nucleoids replicate faster than the cell divides. There are no MITOCHONDRIA or CHLOROPLASTS in prokaryotic cells, but there are structures thought to function similarly – MESOSOMES and CHROMATOPHORES. RIBOSOMES in prokaryotes are smaller than in eukaryotes and there is no MITOSIS or MEIOSIS.

prolactin, *luteotrophic hormone, luteotrophin* In vertebrates a protein hormone secreted by the PITUITARY GLAND that stimulates LACTATION (milk production) and promotes secretion of PROGESTERONE by the CORPUS LUTEUM.

PROM (programmable read-only memory) A MEMORY device consisting of an INTEGRATED CIRCUIT that is programmed after manufacture and can hold the data permanently.

promethium (Pm) The element with ATOMIC NUMBER 61; RELATIVE ATOMIC MASS 145.0; melting point 1,042°C; boiling point 3,000°C; RELATIVE DENSITY 7.3. Promethium has only radioactive ISOTOPES, all of which have HALF-LIVES of 20 years or less. Promethium–147 occurs in nature in minute amounts from the SPONTANEOUS FISSION of uranium.

promoter In chemistry, any substance that, when present in very small quantities, can increase the effectiveness of a CATALYST.

proof A series of mathematical steps designed to illustrate the truth of a mathematical statement, by showing that it can be deduced from fundamental AXIOMS.

propagate (*vb.*) **1.** (*physics*) Of a WAVE, to travel through a medium, which in the case of ELECTROMAGNETIC WAVES may be empty space.

2. (*biology*) To breed (of animals) or multiply (of plants, by cuttings, grafts, etc.).

propan-1, 2, 3-triol *See* GLYCEROL.

propane (C_3H_8) A colourless gas with an odour, of the ALKANE series of HYDROCARBONS; boiling point –45°C; melting point –190°C. It is a component of NATURAL GAS and can also be made from reduction of propene. Propane is used as a fuel and a refrigerant.

propanol, *propyl alcohol* (C_3H_7OH) The third member in the series of ALCOHOLS, a colourless liquid which is a mixture of the ISOMERS propan-1-ol ($CH_3CH_2CH_2OH$) and propan-2-ol ($CH_3CHOHCH_3$). The boiling points are 97°C and 82°C respectively. Propanol is used in perfumery.

propanone, *acetone* (CH_3COCH_3) A colourless, flammable liquid; boiling point 56°C. It is miscible with water and widely used as nail varnish remover. Propanone is a KETONE with a characteristic odour. It is manufactured by the CUMENE PROCESS.

propene, *propylene* (C_3H_6,) A colourless organic gas; boiling point –48°C; melting point –185°C. It is second member of the ALKENE series of HYDROCARBONS. Propene is made from petroleum by CRACKING. It is widely used in the manufacture of resins and plastics, such as POLYPROPYLENE.

propenenitrile, *acrylonitrile* ($CH_2{=}CHCN$) An organic liquid; boiling point 78°C. It readily POLYMERIZES and is used in the manufacture of synthetic RUBBER.

propenoic acid, *acrylic acid* ($CH_2{=}CHCOOH$) A colourless liquid made from the oxidation of the ALDEHYDE propanal; boiling point 141°C; melting point 13°C. Propenoic acid can be polymerized (*see* POLYMERIZATION) to produce important synthetic POLYMERS used as artificial fibres (ACRYLICS), acrylic artist's paint, adhesives. The derivative methyl propenoate (methyl acrylate) is polymerized to produce glass-like RESINS for use as lenses, dentures and transparent parts. Another derivative is polymeric methyl methyl-acrylate, better known under its trade name of PERSPEX.

prophase The first stage of MITOSIS and MEIOSIS.

proportional (*adj.*) Describing two quantities that vary together in such a way that the ratio of one to the other is a constant, shown by the symbol $\propto$. If $y \propto x$ then y/x will be a constant, called the constant of proportionality. If x and y are plotted on a graph, the result will be a straight line passing through the ORIGIN, with a gradient equal to the constant of proportionality.

prop root A form of ADVENTITIOUS ROOT that grows from the lower part of a plant stem or trunk to the ground to provide extra support, for example in maize and some woody plants.

propyl alcohol *See* PROPANOL.

propylene *See* PROPENE.

prorennin The inactive precursor of RENNIN.

prostaglandin In mammals, any one of a group of complex FATTY ACIDS synthesized continuously by most nucleated cells. Prostaglandins act in a similar way to HORMONES as chemical messengers between cells; they can be released directly into the blood but usually only act locally. A rich source of prostaglandins is SEMEN, where they were discovered, but they are made all over the body and may act as an intermediary between a hormone binding to its RECEPTOR on the target cell and the activation of the second messenger. Their effects include stimulating the contraction of smooth muscle (for example the UTERUS during PARTURITION), regulation of the production of stomach acid, modifying other hormonal activity, assisting in blood clotting by causing PLATELET aggregation, and being responsible for INFLAMMATION following injury or infection.

In excess, prostaglandins may be involved in causing inflammatory disorders such as arthritis. Pain-relieving drugs, such as aspirin, act by inhibiting prostaglandins. Prostaglandins are of great potential importance in the alteration of blood pressure and broncodilation and constriction.

prostate gland In male mammals, a GLAND at the base of the BLADDER that opens into the URETHRA. The prostate gland secretes an alkaline fluid that forms up to a third of the volume of SEMEN. In humans it is a rich source of PROSTAGLANDINS. In later life in humans, the prostate gland often enlarges to block the urethra and has to be removed. *See also* SEXUAL REPRODUCTION.

prosthetic group A non-protein group that firmly attaches to a PROTEIN or ENZYME to create a functional complex, in contrast to a COENZYME. Examples include HAEMOGLOBIN, which contains iron as its prosthetic group, and GLYCOPROTEINS, which contain CARBOHYDRATES as the prosthetic groups. Metal ions such as Zn^{2+}, K^+ and Na^+ are often prosthetic groups for enzymes, providing a charge needed in an active site. *See also* COFACTOR.

protactinium (Pa) The element with ATOMIC NUMBER 91. Only radioactive ISOTOPES are known, the longest lived of which (protactinium–231) has a HALF-LIFE of 34,000 years. It exists in minute amounts in uranium ores, produced by the decay of uranium–235.

protandrous (*adj.*) Describing a plant whose STAMENS develop before the CARPELS, ensuring that self-pollination does not occur. *See* POLLINATION.

protease A general term for an ENZYME, for example TRYPSIN and PEPSIN, that digests PROTEINS to PEPTIDES.

protein Any one of a group of complex organic compounds, essential to all living organisms, that have a large RELATIVE MOLECULAR MASS and consist of AMINO ACIDS linked together. Proteins always contain carbon, hydrogen, oxygen and nitrogen, usually sulphur and sometimes phosphorus.

The amino acids link together to form PEPTIDES or POLYPEPTIDES, and the sequence of amino acids in a polypeptide chain is its primary structure. There is an amino end (NH_2) and carboxyl end (COOH) to any protein molecule. The ultimate shape of the protein molecule depends on the types of bonds that exist within it, and the shape of the polypeptide chain is called the secondary structure. HYDROGEN BONDS can form that often result in a polypeptide chain coiling into an ALPHA HELIX or BETA-PLEATED SHEETS. Other types of bonding can occur, including IONIC BONDS and HYDROPHOBIC interactions between non-polar R groups (*see* AMINO ACIDS), which cause folding of the protein to shield these groups from water. Together these bonds cause folding and twisting of the constituent polypeptide chains of a protein into a three-dimensional structure called the tertiary structure. A large complex protein molecule has many polypeptide chains combined and incorporates non-protein groups (*see* PROSTHETIC GROUP, CO-FACTOR) vital to its function into its structure, which is then referred to as the quaternary structure. The shape of a protein is crucial to its functioning, for example in providing ENZYME-binding sites.

There are a limitless number of proteins and, unlike CARBOHYDRATES, they vary from species to species. Their functions are numerous. Fibrous proteins, for example COLLAGEN, provide a structural role, while globular proteins, for example enzymes, provide a metabolic role. Conjugated proteins incorporate non-protein groups into their structure that play a vital role in their functioning, for example the haem group in HAEMOGLOBIN.

Proteins can be denatured, for example by heat, which breaks up the three-dimensional structure and prevents it from functioning. Protein is an essential requirement of the human diet to provide energy; it is not usually stored in the body so needs to be included regularly in the diet (60 g per day).

See also PROTEIN SYNTHESIS.

protein synthesis The manufacture of all the PROTEINS needed by an organism. Protein synthesis is ultimately controlled by the DNA contained within the cell nucleus. A copy of the genetic code contained within the DNA is carried, in the form of RNA, to the cell cytoplasm, where it determines which AMINO ACIDS are to be linked in which order. This in turn determines the type of protein made.

Animals obtain the ESSENTIAL AMINO ACIDS from their diet and synthesize the others. Plants synthesize their own amino acids from NITRATES in the soil and CARBOHYDRATE products from, for example, the KREBS CYCLE. The amino acids combine with a specific TRANSFER RNA (tRNA) by which they are carried to the RIBOSOMES, all in the cell cytoplasm. In the ribosomes, the amino acids meet MESSENGER RNA (mRNA), which has been synthesized in the cell nucleus by a process called TRANSCRIPTION. Ribosomes then attach to one end of the mRNA (where there is a START CODON) and attract a specific amino acid carried by the tRNA in a process called TRANSLATION. In this way amino acids are linked in a specific order to form POLYPEPTIDES. Protein synthesis ends when a STOP CODON occurs on the mRNA. The polypeptides are then assembled into proteins.

prothrombin The precursor of the enzyme THROMBIN. *See also* BLOOD-CLOTTING CASCADE.

protocol In computing, the agreed set of operational procedures that allows data to be transferred between devices.

protoctist A member of the KINGDOM PROTOCTISTA.

Protoctista A CLASSIFICATION KINGDOM made of single-celled eukaryotic (*see* EUKARYOTE) organisms with varied characteristics, including PROTOZOA, slime mould and ALGAE. Some, but not all, protoctists feed by PHOTOSYNTHESIS, most have SEXUAL REPRODUCTION and some have FLAGELLA. Important protoctists include *Plasmodium* (the cause of MALARIA), *Trypanosoma* (the cause of sleeping sickness), AMOEBA, *Phytophthora infestans* (the cause of potato blight and the Irish famine of 1845) and *Euglena*. *See also* OOMYCOTA, EUGLENOPHYTA).

protogynous (*adj.*) Describing a plant whose CARPELS develop before the STAMENS, ensuring that self-pollination does not occur. *See* POLLINATION.

proton The positively charged ELEMENTARY PARTICLE found in the NUCLEUS of an ATOM. The number of protons in a nucleus is called the ATOMIC NUMBER and fixes the number of ELECTRONS needed to produce a neutral atom, which in turn determines the chemical properties of the element.

The mass of a proton is 1.6×10^{-27} kg. It has a charge of 1.6×10^{-19} C, equal but opposite to the charge on an electron.

See also HADRON, QUARK

proton decay The decay of nuclear matter, particularly PROTONS into LEPTONS, predicted by GRAND UNIFIED THEORIES, but not confirmed experimentally. If the proton is unstable, its lifetime must be many orders of magnitude greater than the age of the Universe.

protoplasm The CYTOPLASM and NUCLEUS of a CELL, including all the structures in the cytoplasm but excluding the CELL WALL, if present.

protoxylem *See* XYLEM.

protozoa (*sing. **protozoan***) A collective term for a group of four phyla (*see* PHYLUM) of the KINGDOM PROTOCTISTA, consisting of single-celled organisms without rigid cell walls. Protozoa are all aquatic, eukaryotic (*see* EUKARYOTE) organisms that lack CHLOROPHYLL. Most can be free-living or parasitic (*see* parasite) but some, for example *Plasmodium* (the cause of MALARIA), are solely parasitic. The protozoa are divided into phyla mostly by their mode of locomotion. Some protozoa move by CILIA, for example *Paramecium*, some byFLAGELLA, for example *Trypanosoma* (the cause of sleeping sickness), and some by pseudopodia (*see* PSEUDOPODIUM), for example AMOEBA. Reproduction is commonly by BINARY FISSION. Protozoa form a large proportion of PLANKTON and are therefore an important food for fish and other aquatic animals. *See also* APICOMPLEXA, CILIOPHORA, RHIZOPODA, ZOOMASTIGINA.

provirus A viral GENOME that is integrated into the CHROMOSOME of a host cell and remains latent there for long periods of time before being expressed. *See also* RETROVIRUS, VIRUS.

proximal convoluted tubule In the KIDNEY, the part of the NEPHRON between the BOWMAN'S CAPSULE and the LOOP OF HENLE along which useful minerals and water are reabsorbed back into the blood.

Prussian blue A blue pigment, the DOUBLE SALT potassium iron hexacyanoferrate, $KFe[Fe(CN)_6]$.

pseudocarp A soft fruit that results from parts other than the OVARY being incorporated into the fruit structure. An example is the strawberry, which develops from the RECEPTACLE (flower stalk) and the true fruits are the pips on the outer surface. Other examples are apples and pears, in which the outer skin and fleshy tissue develop from the receptacle after FERTILIZATION and the CARPELS (true fruits) form the core that surrounds the seeds. This grouped pseudocarp is called POME. The pineapple is a multiple pseudocarp that is composed of fleshy tissue from the receptacles of many flowers.

pseudopodium (*pl. **pseudopodia***) An extension of PROTOPLASM that forms in some cells, for example AMOEBA and MACROPHAGES, and is used as a method of locomotion (amoeboid movement) or for engulfing food or foreign bodies.

psi particle *See* J/ψ.

Pteridophyta A division of the PLANT KINGDOM that is no longer used. It included the FERNS, CLUB MOSSES and HORSETAILS, which are now classified into phyla (*see* PHYLUM) of their own: FILICINOPHYTA, LYCOPODOPHYTA and SPHENOPHYTA, respectively.

PTFE *See* POLYTETRAFLUOROETHENE.

ptyalin An AMYLASE found in the SALIVA of some mammals, including humans.

p-type semiconductor A SEMICONDUCTOR in which the charge is predominantly carried by HOLES rather than electrons.

puddling The process of producing WROUGHT IRON in a REVERBERATORY FURNACE using a mixture of PIG IRON and haematite ore, which comprises mostly iron(III) oxide. Oxygen from the haematite oxidizes the carbon in the pig iron (*see* OXIDATION) and produces almost pure iron.

pulley A wheel with a groove in it around which there is a rope supporting a LOAD, or a belt transmitting a load to another pulley. It is possible for an EFFORT to be applied to a rope which loops around several pulleys, so that the TENSION is used several times over in supporting the load. If two wheels of different sizes are used, the different MOMENTS of the tension in a belt around the two pulleys produces a machine with a VELOCITY RATIO that is equal to the inverse ratio of the pulley sizes.

pulmonary (*adj.*) Relating to the lungs.

pulsar (*See also* diagram on following page.) A STAR that flashes with a well defined period, typically ranging from a few hundred times per second to once every few seconds. Pulsars are now recognized as NEUTRON STARS, and the flashing is caused by the interaction between the radiation from the neutron star and its intense magnetic field. This causes radiation to be emitted only along the magnetic poles of the neutron star. If the magnetic poles are not aligned with the axis of rotation of the star, a beam of radiation will sweep through space and an observer will see a pulse whenever this points towards his telescope.

pulse 1. (*physics*) A wave or electrical signal of short duration.

2. (*biology*) In vertebrates, a rhythmic throbbing that can be felt where ARTERIES are close to the surface, for example in the wrist. The pulse is caused by the contractions of the heart muscle forcing blood into the AORTA,

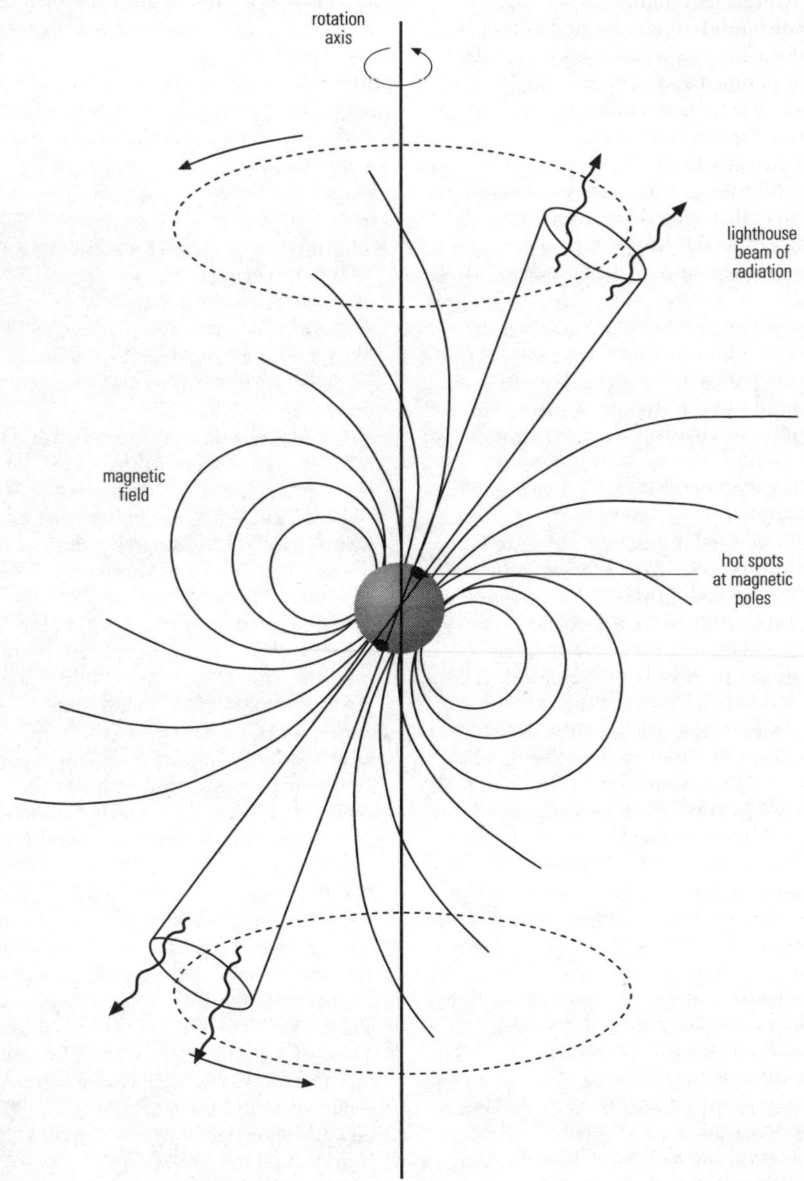
rotation axis
lighthouse beam of radiation
magnetic field
hot spots at magnetic poles

Pulsar.

which results in a sudden increase in pressure in the arteries that, because of their elastic walls, causes a swelling or throbbing to pass through them. An average human pulse rate is about 70 beats per minute.

pulse-code modulation A method of transmitting information in which the AMPLITUDE of a signal is sampled and converted into a BINARY code, which is transmitted as a series of on and off pulses. *See also* MODULATION.

pulse-jet An early form of JET ENGINE that operated by admitting air through a series of slats which were then closed by the pressure of the burning gas. As this left the rear of the engine, the slats opened again to admit a fresh supply of air.

pump A device for moving or changing the pressure of a fluid. A pump normally consists of a chamber together with two valves through which fluid can flow in one direction only, or some other mechanism for ensuring that the fluid can flow in one direction only. Fluid enters the chamber through one of the valves. The volume of the chamber is then reduced and fluid expelled through the other valve.

pumped storage An energy storage scheme that uses electricity at times of low demand to pump water from a low-level reservoir to one at a higher level. At times of high electricity demand, the process is reversed to generate electricity. Although some energy is inevitably lost in the process, it has economic benefits as there is no way of storing surplus electrical energy in sufficient quantities and it enables the most cost-efficient generating plant to be used more continuously.

pupa, *chrysalis* The stage between LARVA and adult in INSECTS undergoing METAMORPHOSIS. The protective covering made by the larva to surround it during its pupal stage is called a cocoon.

pupil An aperture at the centre of the IRIS of the EYE. The size of the pupil can be adjusted by the muscles of the iris to control the amount of light entering the eye.

purine A type of organic base occurring in NUCLEIC ACIDS and NUCLEOTIDES that consists of a double ring, one with six sides and one with five. Adenine and guanine are the most common.

Purkinje fibres Specialized fibres of CARDIAC MUSCLE that conduct waves of excitation from the PACEMAKER (in the wall of the right ATRIUM) to the apex of the VENTRICLES. The Purkinje fibres form a network throughout the wall of the ventricles and ensure simultaneous contractions of both ventricles.

pus The yellowy fluid consisting of dead cells and bacteria that accumulates at the site of an INFLAMMATION.

putrefaction The largely anaerobic (*see* ANAEROBE) decomposition of organic matter by micro-organisms. Foul-smelling AMINES are often produced.

PVC *See* POLYVINYL CHLORIDE.

p-wave In SEISMOLOGY, the LONGITUDINAL wave that arrives first after an EARTHQUAKE. *See* SEISMIC WAVE.

pyloric sphincter A ring of muscles at the exit of the STOMACH that relax and contract to allow food to leave the stomach.

pyramid A three-dimensional geometrical shape having a rectangular (usually square) base, and three triangular sides (usually EQUILATERAL) meeting at a point called the apex. The volume of a pyramid is $Ah/3$, where A is the area of the base, and h is the height of the apex above the base.

pyramid of biomass A diagrammatic representation of the dry mass of the organisms at each level of a FOOD CHAIN. Pyramids of biomass overcome some of the problems of a PYRAMID OF NUMBERS, but they are difficult to achieve. They are largely dependent on the time span used, for example the biomass of crops in summer can be very different to the biomass in winter. *See also* PYRAMID OF ENERGY.

pyramid of energy A diagrammatic representation of the energy used at each level in a FOOD CHAIN. A pyramid of energy is determined over a given time period, which provides an accurate picture of the food chain but is difficult to obtain. *See also* PYRAMID OF BIOMASS, PYRAMID OF NUMBERS.

pyramid of numbers A diagrammatic representation of the numbers of individuals at each level in a FOOD CHAIN. Each level is shown as a horizontal bar. Usually the numbers of primary consumers at the bottom of the pyramid exceeds the number of secondary consumers at the top of the pyramid, and so the overall shape is of a classic pyramid. However, the pyramid of numbers does not take into account the size of an organism, so a tree counts as one organism, as does an insect, which sometimes results in an inverted pyramid or bulging in the middle. *See also* PYRAMID OF BIOMASS, PYRAMID OF ENERGY.

pyrenoid A region of small protein bodies embedded in the CHLOROPLASTS of many algae that is concerned with converting the early products of PHOTOSYNTHESIS into compounds for storage.

pyridoxine *See* VITAMIN B.

pyrimidine A type of organic base occurring in NUCLEIC ACIDS and NUCLEOTIDES that consists of two single, six-sided rings. The most common are cytosine, thymine and uracil.

pyrites A mineral SULPHIDE of some metals, having a cubic crystal form. In particular iron pyrites, which is iron(II) sulphide, FeS, sometimes called fools' gold as its metallic lustre has led to it being mistaken for NATIVE gold.

pyroelectricity The production of a charge on the surface of certain materials as a result of electrical POLARIZATION produced by a change in temperature.

pyrolysis The DECOMPOSITION of a compound (usually organic) under the action of heat.

pyrometer Any device for measuring high temperatures, especially one that operates by analysing visible light or infrared radiation given off by a hot object. One simple form of pyrometer works by viewing the hot object behind the filament of a light bulb. The current through the filament is adjusted until the filament disappears from view, being neither brighter nor less bright than the object being viewed. At this point the filament and the hot object are at the same temperature. A VARIABLE RESISTOR controlling the filament temperature is calibrated to read temperature directly.

pyrometry The measurement of high temperatures. *See* PYROMETER.

pyruvate Any salt or ester of PYRUVIC ACID.

pyruvic acid, *2-oxopropanoic acid* ($CH_3COCOOH$) A colourless, pleasant-smelling organic acid; melting point 13°C. It is an important product in the metabolism of carbohydrates and proteins.

Pythagoras' theorem For a right-angled triangle, the length of the longest side (the hypotenuse) is related to the lengths of the other two sides by

$$c^2 = a^2 + b^2$$

where a, b and c are the lengths of the sides, with c being the hypotenuse.

Q

Q factor A number that describes the amount of DAMPING in an oscillating system. The Q factor is 2π times the total energy in the OSCILLATION divided by the energy lost to heat due to the DAMPING forces in one oscillation. The larger the Q factor, the longer a system can continue to oscillate on its own. *See also* RESONANCE.

QCD *See* QUANTUM CHROMODYNAMICS.

QED *See* QUANTUM ELECTRODYNAMICS.

quadrant One quarter of a circle, or some other angular range of 90°.

quadrat A folding square frame (usually 1 m^2) placed on the ground and used to study numbers and types of plants and animals within it, as a representation of an area as a whole. Sampling can be random or systematic (placed at regular intervals). Usually a number of quadrats are studied in order to reach a valid conclusion about the distribution, type and number of species within an area.

A point quadrat or point frame consists of two vertical poles with a horizontal bar across the top of them. A long pin is placed through holes at regular intervals along the horizontal bar and every species touched by the pin is recorded. The point quadrat is useful where there is dense vegetation because it samples at different levels.

Quadrats are not reliable for more mobile animals that move away when disturbed. To gain information on population sizes of such animals, MARK methods are used.

See also TRANSECT.

quadratic equation A mathematical equation of the form

$$ax^2 + bx + c = 0$$

or any equation that can be rewritten in this form, containing no power of x higher than the second. A quadratic equation can be solved using the formula

$$x = [-b \pm \sqrt{(b^2 - 4ac)}]/2a$$

that is, these are the values of x for which the equation is satisfied. If the quantity $(b^2 - 4ac)$, called the discriminant, is positive, the equation will have two real solutions. If the discriminant is zero, there will only be a single solution. If the discriminant is negative, there will be no REAL NUMBERS that satisfy the equation, but two COMPLEX NUMBERS will satisfy it.

quadrilateral Any closed geometrical shape that has four sides. If all the sides are equal in length, the shape will be a RHOMBUS, or a SQUARE if they are also parallel. If one pair of sides is parallel, the shape is a TRAPEZIUM. If there are two pairs of parallel sides the shape is a PARALLELOGRAM, unless the sides are also at right angles to one another, in which case it is a RECTANGLE.

qualitative analysis The branch of chemical ANALYSIS that aims to discover which elements or compounds are present in a sample, without measuring the amount of the materials present. A wide range of techniques can be used, from simple chemical tests for certain ions, to sophisticated SPECTROSCOPY and MASS SPECTROSCOPY techniques.

quantitative analysis The branch of chemical ANALYSIS that uses techniques enabling the quantity of a certain element or compound present in a sample to be measured. The chief techniques of quantitative analysis are GRAVIMETRIC ANALYSIS, in which a component, such as a PRECIPITATE, is separated from the sample and weighed, and VOLUMETRIC ANALYSIS, in which the volume of material that can react with the sample in a particular way is measured, for instance by TITRATION.

quantized Describing a quantity that occurs only in fixed amounts, often multiples of a basic unit (or QUANTUM). For example, electric CHARGE always occurs in multiples of the electron charge.

quantum (*pl. **quanta***) A particle, or the group of waves associated with such a particle, which has a fixed (quantized) value of some quantity. Thus, for example, the PHOTON is a quantum of ELECTROMAGNETIC RADIATION with a fixed amount of energy. The term quantum also refers to the minimum amount by which certain properties, such as the energy or ANGULAR MOMENTUM, of a system can change.

quantum chromodynamics (QCD) The QUANTUM THEORY of the STRONG NUCLEAR FORCE. QCD explains that the force between two QUARKS is carried by particles called GLUONS and increases with their separation until there is sufficient energy to create a quark-antiquark pair. Thus quarks are only seen in COLOUR neutral states made up of three quarks (BARYONS), three antiquarks (an antibaryon) or a quark and an antiquark (a MESON).

quantum electrodynamics (QED) A description of the interactions between charged particles, combining the principles of QUANTUM THEORY and the SPECIAL THEORY OF RELATIVITY. QED explains the forces between charged particles in terms of the exchange of PHOTONS. The ideas of QED were subsequently applied to the STRONG NUCLEAR FORCE and the WEAK NUCLEAR FORCE to produce the STANDARD MODEL of particle physics.

quantum mechanics A system of mechanics developed from QUANTUM THEORY. Quantum mechanics explains the properties of atoms, molecules and ELECTROMAGNETIC RADIATION.

quantum number A number that represents the value of some quantity such as CHARGE, which is conserved in certain types of interactions and which is only found in whole number multiples of some basic quantity. *See also* PROBABILITY AMPLITUDE.

quantum physics *See* QUANTUM THEORY.

quantum theory, *quantum physics* The theory that energy is absorbed or released in discrete, indivisible units called quanta. According to quantum theory, there is no real distinction between effects traditionally described in terms of WAVES, such as light, and objects that are more usually thought of as particles, such as ELECTRONS. This double nature of waves and particles is referred to as WAVE-PARTICLE DUALITY. The strange nature of quantum theory has given rise to a number of GEDANKENEXPERIMENTS (thought experiments) carried out with idealized apparatus, to illustrate some of the ideas. *See also* BAND THEORY, ENERGY LEVEL, GAUGE BOSON, HEISENBERG'S UNCERTAINTY PRINCIPLE, HYDROGEN SPECTRUM, PAULI EXCLUSION PRINCIPLE, PHOTOELECTRIC EFFECT, QUANTUM CHROMODYNAMICS, QUANTUM ELECTRODYNAMICS, SPIN, VIRTUAL PARTICLES, WAVE NATURE OF PARTICLES.

quark Any member of the fundamental family of particles from which all HADRONS, including the PROTON and NEUTRON are made. Quarks are held together in hadrons by the STRONG NUCLEAR FORCE and come in six varieties called FLAVOURS. Protons and neutrons are composed of up and down quarks. Heavier quarks called strange, charm, top and bottom quarks are also known. Each flavour of quark has a corresponding antiparticle.

Each quark has charge of $+^2/_3$ or $-^1/_3$ in units of the electron charge. Each also carries BARYON NUMBER (a quantity conserved in all interactions) $^1/_3$. The heavier quarks carry quantum numbers called strangeness, charm, etc., which are conserved in interactions involving the STRONG NUCLEAR FORCE but not in those involving the WEAK NUCLEAR FORCE. Hadrons containing these quarks have relatively long HALF-LIVES, the times involved (typically 10^{-10} s) are short by conventional standards, but long compared with the 10^{-23} s lifetime typical of hadrons that decay by way of the strong force.

See also ASSOCIATED PRODUCTION, QUANTUM CHROMODYNAMICS, STANDARD MODEL.

quarter The result of dividing some specified quantity or number into four equal parts.

quartz The most abundant mineral in the Earth's CRUST, containing mostly silicon dioxide. In its purest form, quartz is colourless, but small amounts of impurities produce coloured gemstones, such as amethyst. Quartz is PIEZOELECTRIC, and mechanically resonant quartz crystals are used in OSCILLATORS.

quasar Contraction of quasi-stellar object. An object that appears as a star even when viewed through a powerful telescope, but which has an unusually high RED-SHIFT. It is now believed that many quasars are in fact extremely distant galaxies. See also DOPPLER EFFECT, HUBBLE'S LAW

Quaternary The current PERIOD of geological time, which began with the PLEISTOCENE EPOCH, about 2 million years ago. It followed the TERTIARY period. During the Quaternary, humans came to dominate the Earth.

quenching A process in which a material is heated and the cooled suddenly, usually using oil or water. Then result is that the large numbers of DISLOCATIONS produced by thermal vibrations at high temperatures become 'frozen' in place and tangled with one another, so are unable to move through the metal. The result is a material that is very hard, but brittle. *See also* ANNEALING.

quicklime *See* CALCIUM OXIDE.

quotient The result of dividing one number by another. For example, the quotient of 20 and 5 is 4. An intelligence quotient is obtained by dividing a child's mental age, determined in a specified way, by the true age of the child, multiplied by one hundred.

R

racemic mixture A 1:1 mixture of the two ENANTIOMERS of a compound exhibiting OPTICAL ISOMERISM. A racemic mixture does not rotate PLANE-POLARIZED LIGHT since one enantiomer rotates it in one direction and the other equally in the opposite direction. A racemic compound has the prefix (+/–) denoting the presence of equal amounts of the (+) and (–) enantiomers.

rack and pinion A device for converting rotational motion into linear, side to side motion, such as in the steering mechanism of a car. The rotation of the steering wheel rotates a small gear (the pinion) which meshes with a toothed bar (the rack), which moves the car's wheels from side to side.

rad 1. (*mathematics*) Abbreviation for RADIAN.
2. (*physics*) A former unit of absorbed DOSE of IONIZING RADIATION. One rad is equivalent to 10^{-2} GRAY.

radar A system based on the reflection of MICROWAVES for the detection of objects that reflect the microwaves, particularly ships and aircraft. Radar systems use a rotating AERIAL to transmit a series of microwave pulses. A receiver then picks up any reflected signal. The time taken for the pulse to return enables the distance to the reflecting object to be calculated and the direction of the aerial at the time the signal is received provides information on the direction of the reflecting object. *See also* SONAR.

radial symmetry A type of structure of an organ or organism in which a cut through the centre in any direction produces two halves that are mirror images of each other. *Compare* BILATERAL SYMMETRY.

radian (*abbrev.* ***rad***) The SI UNIT of plane ANGLE. The length of an arc which subtends an angle θ is equal to the radius of the arc multiplied by the angle in radians. Whenever an angle (as opposed to a trigonometrical function of the angle) appears in an equation, it is essential that the angle be expressed in radians. One radian is equal to 57.3°, a complete circle is 2π radians.

radiance The total amount of ELECTROMAGNETIC RADIATION emitted by an object, per square metre of receiving area as measured at a distance of one metre from the object.

radiant energy RADIATION, particularly ELECTROMAGNETIC RADIATION, emitted by an object. *See* BLACK-BODY RADIATION.

radiation In general, the emission of rays, waves or particles from a source, with intensity falling off according to an INVERSE SQUARE LAW. In particular, the term radiation used for IONIZING RADIATION and ELECTROMAGNETIC RADIATION. *See also* BACKGROUND RADIATION, BLACK-BODY RADIATION, IRRADIATION, RADIOACTIVITY, THERMAL RADIATION.

radiation detectors Any device for detecting IONIZING RADIATION by the IONIZATION it produces. Types of radiation detectors include the spark counter, Geiger–Müller tube and solid-state detector. In each case the ionization produced by the particle occurs in a region where there is an ELECTRIC FIELD.

In the case of a spark counter or Geiger–Müller tube, two ELECTRODES in a gas are used to produce an electric field that is strong enough for the ions to be accelerated and gain sufficient energy to create further ions when they collide with other gas molecules. The spark counter operates in air at ATMOSPHERIC PRESSURE and produces a spark that can be seen and heard.

In the Geiger–Müller tube, lower pressures are used so the ions accelerate for longer before colliding with other molecules. In this way, a lower voltage can be used and the detector is more sensitive. Only ALPHA PARTICLES are sufficiently ionizing to be detected by spark counters; a Geiger–Müller tube will detect BETA PARTICLES and GAMMA RADIATION but can only detect alpha particles if the window through which the radiation must pass to reach the low pressure gas of the tube is sufficiently thin. The pulse of current that is produced when an ionizing particle enters a Geiger–Müller tube is counted electronically and can be fed to a loudspeaker to produce a clicking sound.

In a solid-state detector, which is essentially a REVERSE BIASED PN JUNCTION DIODE, electrons and HOLES are produced allowing a pulse of current to flow in proportion to the amount of energy deposited in the detector by the ionizing radiation.

radiation fog FOG formed by the rapid cooling of air close to ground that radiates its heat away into space on a clear night. *See also* ADVECTION FOG.

radical A group of atoms, such as CH_3 or NH_2 that frequently occur together and pass unchanged through many chemical reactions, being held together by strong COVALENT BONDS. *See also* FREE RADICAL.

radicle The part of the embryo in a SEED PLANT that becomes the primary ROOT. Once GERMINATION begins the radicle appears first and its apical MERISTEM pushes through the soil and may develop into the entire root system. *See also* ADVENTITIOUS ROOT.

radio The transmission of information, particularly speech or music by use of RADIO WAVES. To do this, a CARRIER WAVE is used, a radio wave of some specified frequency, generated by an electronic circuit called an OSCILLATOR.

To send any information, the carrier wave must be modulated, or changed in some way. The simplest way to do this is simply to switch it on and off in accordance with some agreed code, such as Morse code. More sophisticated communication systems involve changing the amplitude (AMPLITUDE MODULATION, AM) or frequency (FREQUENCY MODULATION, FM) of the carrier wave in accordance with the ANALOGUE or DIGITAL information that is to be transmitted.

At the receiving end, a radio receiver needs firstly to select the required signal from the many radio waves arriving, and a TUNED CIRCUIT is used for this. The system then must extract the information conveyed by the carrier wave. This process is called demodulation. The received information can then be converted back to its original form.

See also LONG WAVE, RADIO, SHORT WAVE, UHF, VHF.

radioactive series, *decay series* A series of NUCLIDES, all but the last radioactive, where the decay of the first nuclide gives rise to the second, which decays into the third and so on. In general, the first member of such a series has a much longer HALF-LIFE than the others, so that a sample of this nuclide will create the other nuclides in the series with an equilibrium between the rates of production and decay. Thus the amounts of each of the members of the decay series, apart from the final product, will be proportional to their half-lives. *See also* RADIOACTIVITY.

radioactive waste The radioactive material produced in the NUCLEAR FISSION of material in a NUCLEAR REACTOR. This material is extracted from the spent fuel and has to be stored carefully until its radioactivity has fallen to a safe level – for some ISOTOPES in the waste, this takes several thousand years.

radioactivity The spontaneous decay of unstable atomic nuclei with the emission of IONIZING RADIATION, either ALPHA PARTICLES, BETA PARTICLES or GAMMA RADIATION. Radioactivity occurs spontaneously in many naturally-occurring radioisotopes without any external influence, and may also be induced in certain unstable nuclei by bombarding with NEUTRONS or other particles.

Radioactivity can be harmful to living tissue because of the damage done to living cells by the ionizing radiation. In particular, the formation of FREE RADICALS in the vicinity of the DNA in a cell can lead to MUTATIONS that may cause cancer. However, radioactivity can also be used to kill cancerous cells: a cell is most vulnerable to genetic damage when it is dividing, and since cancer cells divide more rapidly than healthy cells, they are more easily killed. To ensure that healthy tissue does not receive a dose which may create mutations leading to further cancers, the radiation source is either implanted in the patient or is in the form of a beam aimed at the patient from several directions, overlapping to form a large dose at the location of the tumour.

See also ACTIVITY, ALPHA DECAY, BACKGROUND RADIATION, BETA DECAY, DECAY CONSTANT, GAMMA RADIATION, HALF-LIFE, RADIOACTIVE SERIES, RADIOACTIVE WASTE, TRACER TECHNIQUE.

radiocarbon dating An important example of the use of the decrease in activity of a radioactive ISOTOPE to find the age of some objects. All living organisms extract carbon from their surroundings. Some of this carbon will be the unstable isotope carbon–14, which is produced in the atmosphere by the interaction of

COSMIC RADIATION with nitrogen nuclei. When the organism dies, it stops taking in carbon from its surroundings, and since carbon–14 has a HALF-LIFE of about 5,700 years, measurement of the proportion of carbon–14 compared to stable carbon–12 enables the age to be determined. One technique is to measure the level of radioactivity, but since this is very small, greater precision can be achieved and smaller samples are needed if a MASS SPECTROMETER is used to detect the nuclei.

Radiocarbon dating assumes that the proportion of carbon–14 in the atmosphere has remained constant. This is not quite true, as has been shown by the comparison of radiocarbon dates with those obtained by other methods, such as examining the growth rings of trees – a technique called dendrochronology.

radiochemistry The study of the effects of RADIOACTIVITY on chemistry. It includes the use of radioactive ISOTOPES to label certain atoms in a molecule – so the path of a reaction can be followed more clearly – and radiolysis, the effect of IONIZING RADIATION on chemical reactions.

radiograph An X-RAY shadow picture, often used in medical diagnosis, taken by placing the patient between an X-RAY TUBE and a suitable detector, such as a fluorescent screen or photographic film. The amount of X-radiation absorbed depends on the density of the tissue through which the X-rays pass. Since bone absorbs far more radiation than soft tissue, radiographs can be used to examine bone structures. In other cases, a CONTRAST ENHANCING MEDIUM, such as barium sulphate, may be introduced into the patient – as a BARIUM MEAL to examine the stomach for example, or injected into an artery to produce an image of the arteries (arteriogram). A related form of imaging involves the GAMMA CAMERA. *See also* MEDICAL IMAGING.

radioisotope Any radioactive ISOTOPE.

radio telescope A radio receiver connected to a large dish AERIAL and used to detect radio waves from space. Since radio waves have a far longer wavelength than visible light, they are diffracted more (*see* DIFFRACTION) so a radio telescope will produce an image with much less RESOLUTION than a visual telescope, though it may be larger. *See also* INTERFEROMETRY.

radiotherapy The treatment of disease by RADIATION, from an X-ray machine or a radioactive source. Radiation reduces the activity of dividing cells and is used particularly in the treatment of cancer. Special high-energy X-ray machines are often used for radiotherapy that are more powerful than the low-voltage machines usually used for taking X-rays. There can be unpleasant side-effects from radiotherapy, such as hair loss and nausea. Radioactive substances can be administered to a patient to have a more local effect, for example the use of radioactive iodine in the treatment of thyroid disease.

radio waves ELECTROMAGNETIC WAVES with a wavelength greater than about 1 mm. They are produced by oscillating electric charge. Radio waves are used in communications – in radar, television and RADIO broadcasting. Radio waves are emitted by stars and detected by RADIO TELESCOPES.

Radio waves can be generated by producing oscillating currents in electric circuits and feeding this current into a wire or pattern of wires called an AERIAL or antenna. The oscillating charges in the aerial set up an oscillating electromagnetic field which then spreads out into space as an electromagnetic wave. A similar aerial can be used to detect the wave, with either the electric or magnetic field producing a current in the aerial. A RESONANT CIRCUIT can be used to separate currents of different frequencies so that the receiving system can be tuned to respond to signals at one frequency only.

Although there is no theoretical upper limit on the wavelength of a radio waves, wavelengths longer than 10 km are of little practical use.

radium (Ra) The element with ATOMIC NUMBER 88; RELATIVE ATOMIC MASS 226.0; melting point 700°C; boiling point 1,140°C; RELATIVE DENSITY 5.1 Only radioactive ISOTOPES are known, with HALF-LIVES up to 1,600 years (radium–226). Radium was one of the first radioactive elements to be discovered and extracted from pitchblende, a uranium ore in which is produced by the ALPHA DECAY of uranium, via thorium. Radium was the source of the ALPHA PARTICLES used in the RUTHERFORD SCATTERING experiment and was widely used in luminous paints until the cancer risks of IONIZING RADIATION became known.

radius (*pl.* ***radii***) A line joining any point on a circle to the centre of the circle; the length of such a line. The radius of a circle is one half its diameter.

radon (Rn) The element with ATOMIC NUMBER 86; RELATIVE ATOMIC MASS 222.0; melting point –72°C; boiling point –62°C. Radon is a naturally occurring RADIOACTIVE element produced by the ALPHA DECAY of radium. It is an inert gas and thus can easily be separated chemically from the elements that produce it. The longest lived ISOTOPE (radon–222) has a HALF-LIFE of just over 3 days.

Concern has been expressed about the level of radon in some mines and in homes situated over some rock formations that may release significant amounts of radon into the atmosphere.

radula The tongue of MOLLUSCS, consisting of a horny strip bearing rows of teeth and used for rasping food.

rain Droplets of water that fall to the Earth's surface from clouds. Droplets are formed when water vapour in the cloud condenses on CONDENSATION NUCLEI in the atmosphere. When they grow too large to be supported by any upward movement of air into the cloud, they fall to the ground as rain.

rainbow An arc of colours, seen when water droplets, usually from falling rain, are illuminated with the Sun roughly behind the observer. The effect is caused by sunlight being refracted (*see* REFRACTION) on entering each raindrop, undergoing TOTAL INTERNAL REFLECTION at the rear of the drop, and then being refracted again on leaving the drop. DISPERSION causes the different colours in the sunlight to be refracted at slightly different angles.

rainforest A dense forest found in wet climates. Over half the tropical rainforests (where it is wet and hot) occur in Central and South America, and the rest in Africa and south-east Asia. Temperate rainforests also exist, in New Zealand for example. Rainforests are of global importance because they contain a great diversity of species, including many useful plants (used for example as medicines, oils, resins and beverages), and provide much of the oxygen needed for plant and animal respiration. They also help regulate global weather patterns.

More than half of the world's rainforests have been destroyed (and not replaced) for timber or cleared to use the land for agricultural purposes. This is called DEFORESTATION, which can lead to DESERTIFICATION. Despite the destruction, rainforests still house most of our growing wood and many of the Earth's species of plants and animals.

rain shadow A dry area on the side of a hill or mountain that faces away from the PREVAILING WIND. OROGRAPHIC uplifting of incoming moist air causes rain on the upwind side of the hills, leaving dry air on the downwind side.

RAM (**random access memory**) Computer memory for storing and retrieving data, essentially an array of many thousands of BISTABLES that can each be set to a HIGH or LOW state in which they then remain until reset. The disadvantage of RAM is that it loses the information stored once the power supply is removed. *See also* DRAM, SRAM.

ramjet A type of JET ENGINE in which the air is forced into the engines by the motion of the aircraft through the air. Ramjets have been used in some high-speed experimental aircraft, but as they can only be started when already moving at high speed, they are unsuitable for routine use.

random (*adj.*) Describing a quantity or outcome that cannot be predicted beforehand, but only expressed as a PROBABILITY. For example, when two dice are thrown, it is impossible to predict the total score, but the probabilities of various scores can be calculated.

random access memory *See* RAM.

range The set of values that a FUNCTION can take on. For example, the range of the function $\sin x$ is $-1 \leq \sin x \leq 1$.

Raoult's Law In any mixture of liquids the SATURATED VAPOUR PRESSURE above the liquid is equal to the sum of the vapour pressure of each liquid multiplied by its MOLE FRACTION. Thus for a solution of a VOLATILE liquid in a non-volatile solvent, the major contribution to the vapour pressure will be proportional to the concentration of the volatile liquid. The truth of this law can be seen by applying KINETIC THEORY to the escape of molecules from the surface of the mixture, but in practice Raoult's law is only approximately true due to interactions between the molecules in the mixture.

rare earth elements *See* LANTHANIDES.

rarefaction A reduction in density, as between compressions when a sound wave travels through a gas.

rare gas *See* NOBLE GAS.

raster A scanning pattern in which an electron beam covers the whole face of a CATHODE RAY TUBE in a series of lines. *See* TELEVISION.

rate constant The constant of proportionality in the LAW OF MASS ACTION. For all reactions, the rate constant may be affected by temperature and CATALYSTS. For AQUEOUS reactions it is virtually independent of pressure, but in many reactions involving gases it is pressure dependent. *See* LE CHATELIER'S PRINCIPLE.

rate-determining step In a chemical reaction that proceeds via a number of steps, the slowest step, which effectively limits the rate of the overall reaction. In a reaction A→B→C, the reaction B→C may be limited by the rate at which B is produced by the first stage in the reaction, which is then called the rate-determining step. Alternatively, the reaction A→B may go to completion almost immediately, but B→C proceed only slowly, in which case this is the rate-limiting step in the production of A from C.

When considering techniques to increase the rate of a reaction, for example in an industrial process, it is most important to increase the rate of the rate-determining step. *See also* RATE OF REACTION.

rate of change The amount by which a quantity changes for a fixed change in some other quantity. In particular the change in a time-dependent quantity per second of time. *See also* SPEED, VELOCITY.

rate of reaction The rate at which a chemical reaction takes place, usually expressed in moles of product produced per second from one decimetre cubed of reacting material. The rate of reaction may be altered by the presence of a CATALYST, and will also depend on the concentrations of reagents and products present and on the physical conditions.

See also ACTIVATION ENERGY, ARRHENIUS' EQUATION, KINETICS, LAW OF MASS ACTION, LE CHATELIER'S PRINCIPLE, RATE CONSTANT, RATE DETERMINING STEP.

ratio A description of the relative size of two quantities, often expressed as a pair of integers (a simple ratio). For example if two quantities are in the ratio 3:1 it means that the first quantity is three times the second one. If the ratio is 3:2, it means that the first quantity divided by 3 is equal to the second quantity divided by two, i.e the first quantity is 3/2 times the second one.

rational number A number that can be expressed as a fraction containing only integers. *Compare* IRRATIONAL NUMBER, SURD.

ratio of specific heats The HEAT CAPACITY of a substance at constant pressure divided by the heat capacity at constant volume. For air, this ratio, usually given the symbol γ, is about 1.4. For a material with a MOLAR HEAT CAPACITY at constant pressure of C_P and a molar heat capacity at constant volume of C_V

$$\gamma = C_P / C_V$$

ray An infinitely narrow, parallel-sided beam of RADIATION, such as light, or a line drawn to represent such a beam. The term is also loosely used to denote radiation of any type.

Rayleigh's criterion An EMPIRICAL rule for discovering the RESOLUTION of a telescope, microscope or DIFFRACTION GRATING. Rayleigh's criterion states that two objects of equal brightness can just be resolved as separate if the central maximum of the DIFFRACTION pattern from one object falls in the same place as the first minimum of the diffraction pattern from the other object.

Rayleigh wave In SEISMOLOGY, the surface wave that produces a rolling wave-like motion of the surface in an EARTHQUAKE. *See* SEISMIC WAVE.

rayon The general name for a group of textiles and artificial fibres made from CELLULOSE. The cellulose used is usually obtained from wood fibre, made into a solution and filaments regenerated from this. Viscose and acetate (*see* ETHANOATE) are types of rayon.

reactance That part of the IMPEDANCE of a circuit which is due to its CAPACITANCE or INDUCTANCE. The total impedance Z of a circuit containing a CAPACITOR or INDUCTOR is given by

$$Z^2 = R^2 + X^2$$

where R is the RESISTANCE and X is the reactance of the capacitor or inductor. The reactance is the peak value of the POTENTIAL DIFFERENCE across the inductor or capacitor, divided by the peak value of the current through it (equivalently, the ROOT MEAN SQUARE (r.m.s.) voltage divided by the r.m.s. current). There is a PHASE difference between the voltage and the current of 90°, with the current leading the voltage for a capacitance and the voltage leading the current for an inductance.

The SI UNIT of reactance is the OHM. For an ANGULAR FREQUENCY ω, the reactance of a capacitance C is

$$X = \omega C$$

and for an inductance L it is

$$X = 1/\omega L$$

reaction *See* ACTION AND REACTION.

reaction profile A diagrammatic way of describing a chemical process that proceeds via some intermediate state, usually of higher energy, i.e. a process having an ACTIVATION ENERGY. The energy of the reagents is shown as a line at the left of the diagram, with the energy of the products shown as a line to the right, lower than the reagents line if the reaction is EXOTHERMIC. A 'hill' between them represents the energy of the intermediate state, which must be reached before the reaction can proceed. The effect of any CATALYST is to reduce the height of the barrier. An increase in temperature increases the number of reagent molecules with sufficient energy to cross the barrier and therefore increases the rate of the reaction. *See also* ACTIVATION PROCESS.

reactivity In chemistry, the ease with which a particular element takes part in chemical reactions and the rate at which such reactions proceed. For example sodium is said to be a reactive metal – it reacts violently with cold water and will spontaneously burn in air. Zinc is a less reactive metal – it reacts with water only at high temperatures and will only burn if ignited in a pure oxygen atmosphere.

In the case of these reactions, the reactivity can be related to the relative ease with which sodium forms IONS compared to zinc. This lower IONIZATION ENERGY makes sodium more reactive. In turn, this stems from the fact that sodium has only a single VALENCE ELECTRON in a 3s ORBITAL, where it is almost completely screened from the nuclear charge. In zinc there are two 4s electrons, each incompletely screened from the nuclear charge by the 3d electrons.

See also REACTIVITY SERIES.

reactivity series A list of elements, usually metals, in order of REACTIVITY. The details of a reactivity series may differ depending on which reaction is being considered, but the broad features are the same. Generally speaking, the group 1 elements (ALKALI METALS, formerly group I) are the most reactive of the metals, followed by the group 2 elements (ALKALINE EARTHS, formerly group II) followed by aluminium, zinc, iron and copper. Within groups 1 and 2 the reactivity increases with increasing ATOMIC NUMBER, thus potassium is more reactive than sodium and calcium is more reactive than magnesium. These trends are due to the decreasing IONIZATION ENERGY caused by the increasingly effective screening of the nuclear charge from the valence electrons.

Amongst the HALOGENS, the lighter elements are the most reactive, with chlorine being more reactive than bromine for example. This is a consequence of the fact that these elements generally gain electrons in reactions unlike metals, which tend to lose them. The less complete screening in the lighter elements increases the ELECTRON AFFINITY of these elements, making them more reactive.

See also ELECTROCHEMICAL SERIES.

reactor Any vessel in which some kind of reaction takes place, in particular a vessel for containing NUCLEAR FISSION. *See also* NUCLEAR REACTOR.

read only memory See ROM.

reagent Any substance that takes part in a CHEMICAL REACTION.

real gas Any gas that shows significant departures from the behaviour predicted by the IDEAL GAS EQUATION. The behaviour of such gases can be modelled fairly well by VAN DER WAAL'S EQUATION, which takes account of the volume taken up by the gas molecules themselves and of the attractive forces between the molecules (VAN DER WAAL'S FORCES).

If the density of the gas is high enough and the temperature low enough, van der Waal's equation predicts a region where volume would increase with increasing pressure. This is unrealistic, and this region is in fact the area where the gas would exist as a SATURATED VAPOUR along with some liquid. The temperature at which this is first seen is called the critical temperature – above this temperature there is no distinction between the liquid and gas states.

See also IDEAL GAS.

real image An IMAGE in which the light rays from any given point on the object actually do meet at the corresponding point on the image.

If a screen is placed at in the place where a real image is formed, the rays will be scattered into the eyes of the observer and a representation of the object will be seen. *See also* VIRTUAL IMAGE.

real number A number that contains no imaginary part. In other words, any number that is not a COMPLEX NUMBER.

real time In computing, a computer system that responds to events as they happen and has strict limits on how much time it takes to process and deliver a computation.

receptacle The end of a flower stalk (PEDUNCLE) to which the flower parts are attached. The receptacle is often rounded but can be flattened or cup-shaped.

receptor 1. A cell that is specialized to receive a particular stimulus and convert this energy into a NERVE IMPULSE. The stimulus may be chemical, electrical, mechanical (movement, pressure) or temperature or light. *See also* SENSE ORGAN.

2. A PROTEIN in a MEMBRANE forming a receptor site. This can be a channel-linked receptor associated with passage of an ion, for example, sodium, across the membrane. Other membrane receptor sites are associated with ENDOCYTOSIS, for example Fc receptors on MACROPHAGES that bind to an ANTIBODY once it has bound to a PATHOGEN, and therefore permit endocytosis of the pathogen. GROWTH FACTOR receptors, for example for INSULIN, are linked to EFFECTOR molecules such as ENZYMES and binding to these receptors sets off a cascade of intercellular signals that control many aspects of the cell cycle, differentiation and GENE EXPRESSION.

3. An intracellular receptor called a nuclear receptor by which some HORMONES operate. Nuclear receptors are proteins that after binding with the hormone can then bind to a specific region of DNA to inhibit or stimulate TRANSCRIPTION of particular genes. STEROID HORMONES operate in this way.

recessive In genetics, an ALLELE that is masked by a DOMINANT allele and therefore is not expressed in the HETEROZYGOUS form. For example, the allele for blue eyes is recessive and the allele for brown eyes is dominant. A heterozygous individual with one blue and one brown allele will have brown eyes. A recessive allele will only be expressed in the HOMOZYGOUS form.

rechargeable (*adj.*) Describing an electrochemical CELL in which the chemical reactions can be reversed by forcing a current through the cell against its ELECTROMOTIVE FORCE, enabling the cell to be re-used.

reciprocal 1 divided by the quantity concerned. For instance the reciprocal of x is $1/x$, which may be written x^{-1}.

recombinant DNA DNA in which the NUCLEOTIDE sequence has been altered by its combination with another fragment of DNA. The nucleotide sequence can be altered by incorporation of, or exchange with, the new fragment of DNA. This can be a natural event occurring during MEIOSIS as a result of RECOMBINATION, or it can be a result of deliberate manipulation of genes. Recombinant DNA technology refers to deliberate manipulation. *See also* GENETIC ENGINEERING.

recombination 1. (*biology*) A process in genetics in which genetic material is rearranged, resulting in genetic VARIATION of the offspring. This is is evolutionarily advantageous. One of the recombination processes is CROSSING-OVER of some CHROMOSOME pairs during MEIOSIS, which results in segments of chromosomes exchanging. The new combinations are called recombinants. Another recombination process is random reassortment of the chromosomes when pairs are split to go into GAMETES (the gametes receive only one of each pair of chromosomes).

2. (*chemistry*) Any chemical process in which charged particles come together to form a neutral molecule or atom. For example:

$$H^+ + Cl^- \rightarrow HCl$$

or

$$Cu^{2+} + 2e^- \rightarrow Cu$$

recombination frequency *See* CROSS-OVER VALUE.

recrystallization A process for obtaining a pure sample of a crystalline material. The material is produced in solution and then crystallized. The crystals are then dissolved again and recrystallized, possibly several times.

rectangle A geometrical figure having four straight sides at right angles to one another. If the lengths of the sides of a rectangle are a and b, the area is ab.

rectangular hyperbola *See* HYPERBOLA.

rectification The process of converting ALTERNATING CURRENT to DIRECT CURRENT.

rectifier Any device designed to convert ALTERNATING CURRENT to DIRECT CURRENT. Nowadays a rectifier usually comprises a PN JUNCTION DIODE or an array of such diodes forming a BRIDGE RECTIFIER.

rectify (*vb.*) To convert ALTERNATING CURRENT to DIRECT CURRENT.

rectum The final part of the DIGESTIVE SYSTEM in animals in which FAECES are stored before their elimination via the ANUS to the outside.

rectus muscle A muscle that is of equal width or depth throughout its length. *See also* EYE.

red blood cell, *erythrocyte* Any one of the cells carried in blood PLASMA that form almost half of the blood by volume and the majority by number. Human red blood cells are biconcave, 7 μm in diameter, have no nucleus, are made by red BONE MARROW, stored by the SPLEEN, and carry oxygen around the body complexed to HAEMOGLOBIN, which gives the cells their red colour.

Red blood cells have a life of about 4 months and are therefore constantly being destroyed in the liver and spleen and being replaced. There are about six million red blood cells per millilitre of adult human blood.

See also WHITE BLOOD CELL.

red giant A large cool STAR. Since the surface area is large, the LUMINOSITY is high despite the relatively low temperature. Red giants lie to the top right of the HERTZSPRUNG–RUSSELL DIAGRAM.

Stars with masses greater than about 0.5 SOLAR MASSES are able to fuse helium and some are able to fuse further elements. In the case of a star of about 8 solar masses or more, this fusion will continue to produce heavier elements such as silicon and iron. During this stage of its lifetime the star's outer layers swell and cool. Eventually the outer layers are thrown off in a cloud of gas called a PLANETARY NEBULA revealing a WHITE DWARF at the core.

See also CEPHEID VARIABLE.

red lead *See* DILEAD(II) LEAD(IV) OXIDE.

redox couple A pair of sets of reagents, one of which represents an oxidized form and the other a reduced form of the same set of elements (*see* OXIDATION, REDUCTION). For example:

$$Fe^{3+} + e^- \rightarrow Fe^{2+}$$

is a redox couple. By convention, the more oxidized form (the OXIDIZING AGENT) is written on the left of the equation, so the more positive the REDOX POTENTIAL the more likely it is to proceed from left to right.

redox half-cell A device for measuring or comparing REDOX POTENTIALS. In its simplest form, a pair of HALF-CELLS comprise two vessels each containing an INACTIVE ELECTRODE (such as a platinum electrode). Each cell contains the reagents for a different redox process. The two cells are linked by a conducting bridge, typically a piece of POROUS material soaked in aqueous potassium chloride solution. The bridge provides an electrical connection without allowing any extra chemical reaction. The POTENTIAL DIFFERENCE between the two electrodes is the difference in redox potentials for the two REDOX COUPLES. The redox potential for hydrogen,

$$2H^+ + 2e^- \rightarrow H_2$$

is assigned the value zero.

redox potential An ELECTRODE POTENTIAL deduced for a reaction that does not necessarily involve a metal entering an ionic solution. The redox potential is the POTENTIAL DIFFERENCE, measured relative to a standard HYDROGEN ELECTRODE, needed to bring about a specified OXIDATION (at an ANODE), or REDUCTION (at a CATHODE). Such potentials can be deduced from known STANDARD ELECTRODE POTENTIALS in a REDOX HALF-CELL.

The more positive the redox potential for a REDOX COUPLE, the more likely it is to proceed in the direction of reduction, whilst the lower (less positive or more negative) redox potentials indicate a tendency to oxidation. When a reaction involves a redox process, it will proceed in the direction that involves the redox couple with the more positive redox couple being reduced. Thus in the reaction

$$2Na + Cl_2 \rightarrow Na^+ + Cl^-$$

the redox potential for

$$Na^+ + e^- \rightarrow Na$$

is –2.71V, whilst for

$$Cl_2 + 2e^- \rightarrow 2Cl^-$$

it is +1.36. Thus this reaction proceeds in the direction that oxidizes sodium and reduces chlorine.

See also NERNST EQUATION, REDOX REACTION, STANDARD REDOX POTENTIAL.

redox reaction A reaction in which one material is reduced whilst another is oxidized. Since REDUCTION and OXIDATION are reverse processes they always occur together and any reaction can be regarded as a redox reaction. Thus to consider a process in these terms is really a particular way of analysing it, particularly the analysis of a reaction in terms of REDOX POTENTIALS.

red-shift The apparent increase in wavelength of an ELECTROMAGNETIC WAVE by the DOPPLER EFFECT. This effect may be seen in the light reaching us from distant galaxies, leading to the theory that the Universe is expanding and that it began in a BIG BANG. *See also* HUBBLE'S LAW.

reducing agent A material that brings about a REDUCTION, being itself oxidized (*see* OXIDATION). Hydrogen is an important reducing agent.

reducing sugar A SUGAR that can act as a REDUCING AGENT in solution, as indicated by a positive BENEDICT'S TEST or FEHLING'S TEST. This depends on the presence of a free ALDEHYDE or KETONE group. Most MONOSACCHARIDES are reducing sugars as are most DISACCHARIDES except SUCROSE. *See also* CARBOHYDRATE.

reduction The opposite process to OXIDATION. It is the reduction of the OXIDATION STATE of an element, typically by the addition of hydrogen, the removal of oxygen or the addition of electrons to form a negative (or less positive) ion. For example, the reaction between copper oxide and hydrogen involves the reduction of the copper:

$$CuO + H_2 \rightarrow Cu + H_2O$$

The copper has oxygen removed and is converted from ionic Cu^{2+} to neutral copper. This represents a change in oxidation state from +2 to 0. Chlorine is reduced when it is burnt with hydrogen to form hydrogen chloride, which in aqueous solution involves the formation of the Cl^- ion. *See also* CATHODIC REDUCTION, REDOX REACTION, REDUCING AGENT.

reductive division *See* MEIOSIS.

reed relay A pair of contacts made of a FERROMAGNETIC material and enclosed in a glass tube. In a magnetic field, INDUCED MAGNETISM holds the contacts together; when the field is removed they spring apart. A reed relay can be operated with an ELECTROMAGNET or by bringing up a PERMANENT MAGNET. *See also* RELAY.

refine (*vb.*) To increase the purity of a material by some chemical process. For example, copper can be refined by ELECTROLYSIS. In this process the impure copper is made the ANODE of an electrochemical CELL containing copper(II) sulphate as an ELECTROLYTE. Pure copper is deposited on the CATHODE whilst impurities form a sludge that sinks to the bottom of the electrolyte.

Many organic compounds can be refined by FRACTIONAL DISTILLATION, which takes advantage of differences in boiling point.

See also FRACTIONAL CRYSTALLIZATION, RECRYSTALLIZATION.

reflectance, *reflectivity* The proportion of the ELECTROMAGNETIC RADIATION falling on a surface that is reflected by that surface. A perfectly reflecting surface will have a reflectance of 1, whilst a perfectly black surface will have a reflectance of 0.

reflecting telescope A TELESCOPE in which light is gathered and focused to form an image by a CURVED MIRROR. *See also* REFRACTING TELESCOPE.

reflection 1. (*mathematics*) A mathematical operation that maps every point to a point called its image, such that the point and its image equally distant from a specified line (the line of reflection). The line of reflection forms the PERPENDICULAR BISECTOR of any line that joins a point to its image.

2. (*physics*) The process by which a wave strikes the boundary between one medium and another, and leaves the boundary travelling in a new direction but through the same medium as it started in.

A beam of light reflected from a smooth surface such as a mirror leaves in a single direction. This is called regular or specular reflection. If the surface is less smooth the light will be scattered, being reflected in many different directions.

The direction at which light strikes a reflecting surface is called the angle of incidence, this angle being measured from an imaginary line called the normal, which is at right angles to the surface. The angle at which the light leaves the reflecting surface, again measured from the normal is called the angle of reflection.

The behaviour of light striking a mirror is described by the laws of reflection. These state (i) the angle of reflection is equal to the angle of incidence and (ii) the incident ray (the ray of light arriving at the mirror), the reflected ray (the ray leaving the mirror), and the normal all lie in the same plane.

See also FERMAT'S PRINCIPLE, TOTAL INTERNAL REFLECTION.

reflectivity *See* REFLECTANCE.

reflex An automatic, involuntary response in animals to a particular stimulus that is under the control of the NERVOUS SYSTEM but only involves a few NEURONES. In humans, examples of reflexes include the knee-jerk reflex, withdrawal of a hand or foot from a painful stimulus, breathing, dilation or constriction of the eye iris in response to light, sneezing, coughing, regulation of the heart rate.

A reflex arc is the pathway of neurones involved in a reflex action and may consist simply of a stimulus sending a NERVE IMPULSE along a sensory neurone to a single SYNAPSE (monosynaptic) in the spinal cord with an effector neurone causing a response. In vertebrates, however, there is often an intermediate (relay) neurone involved in transmitting impulses. Reflex arcs with two or more synapses are termed polysynaptic. If a reflex arc is restricted to the spinal cord and does not involve the brain, it is a spinal reflex.

Involuntary reflexes are important, especially in primitive animals, to avoid danger. Involuntary responses can be modified by experience (conditioned reflexes), where the first stimulus causing a simple reflex action becomes associated with a second stimulant that involves transmission of nerve impulses to the brain. An example of this was demonstrated by Ivan Pavlov (1849–1936), who conditioned dogs to salivate in response to the sound of a bell that indicated that food was to be delivered.

reflex angle An angle greater than 180° but less than 360°.

reflex arc The pathway of NEURONES involved in a REFLEX action.

reflex magnetron *See* KLYSTRON.

reflux A chemical reaction in which the reagents are heated and the vapour collected, condensed and returned to the reaction vessel.

refracted ray The ray of light leaving a boundary between two transparent materials, the light having passed from one material into the other. *See* REFRACTION.

refracting telescope A TELESCOPE in which light is gathered and focused to form an image by a lens. *See also* REFLECTING TELESCOPE.

refraction The process by which a wave, such as visible light, changes speed as it passes from one medium to another. Unless the wave meets the boundary between the two materials at right angles, one side of the WAVEFRONTS will meet the boundary before the other side and will thus experience the change of speed sooner, the result being a change in the direction of the wave. If a wave passes into a medium where it is slowed down (generally a denser medium) the wave will be refracted in such a way that it travels more closely aligned with the direction of the normal (the imaginary line at right angles to the boundary between the two surfaces). If the wave passes into a medium in which it travels more quickly (generally a less dense material), it will be refracted so as to diverge from the direction of normal.

The angle at which the incoming ray of light (called the incident ray) hits the boundary between the two materials is measured from the normal and is called the angle of incidence. The angle at which the outgoing ray (the refracted ray) leaves the boundary is called the angle of refraction and is also measured from the normal. The amount of refraction is given by SNELL'S LAW.

See also APPARENT DEPTH, BIREFRINGENCE, FERMAT'S PRINCIPLE, REFRACTIVE INDEX, REFRACTIVITY.

refractive index In REFRACTION, the ratio of the sine of the ANGLE OF INCIDENCE to the sine of the ANGLE OF REFRACTION. It is also equal to the ratio of the speeds of the wave in the two materials. When a refractive index is quoted for a single medium, the light is taken as entering that medium from a vacuum. Since the refractive index varies with wavelength, the wavelength is taken as that for yellow light (589 nm) unless otherwise stated. *See also* REFRACTIVITY, SNELL'S LAW.

refractivity A measure of the extent to which a medium refracts light.

$$\text{refractivity} = \text{refractive index} - 1$$

See also REFRACTIVE INDEX.

refractometer Any device for measuring REFRACTIVITY.

refractory (*adj.*) Describing a material that can be heated to a high temperature without losing its mechanical strength. Refractory materials are used for lining furnaces and kilns.

regelation The process of the refreezing of ice after it has been melted by the application of pressure, rather than by any increase in temperature. This effect is commonly demonstrated in an experiment in which a weighted copper wire passes through a block of ice, the pressure of the wire melting the ice beneath it, with the ice re-freezing after the wire has passed. The LATENT HEAT of fusion released by the freezing ice is conducted through the thickness of the copper wire and melts the next layer of ice.

register The part of a MICROPROCESSOR that holds the program and data currently being worked on. The size of the register determines how much data can be worked on at once and therefore helps to determine the size and capacity of the microprocessor. *See also* SHIFT REGISTER.

regular (*adj.*) Of reflection, *see* SPECULAR.

reheat system, *afterburner* A system used in military GAS TURBINE engines to increase power by burning extra fuel behind the turbines.

reinforced concrete A COMPOSITE MATERIAL that consists of a mesh of steel wires or rods encased in concrete. The concrete is strong under compression (i.e. when squashed), but tends to crack under tension (i.e. when stretched). This is counteracted by the steel, which is strong under tension but would buckle when compressed unless much thicker bars were used.

relative atomic mass, *atomic weight* The mass of an ATOM measured in ATOMIC MASS UNITS; that is, on a scale where a single atom of carbon–12 has a mass of 12 exactly. One MOLE of atoms will have a mass in grams that is equal to the relative atomic mass of the atom in atomic mass units.

In the case of an element that occurs with several ISOTOPES, the relative atomic mass is normally given as an average of the isotopes, weighted by their natural abundances. For example, chlorine occurs with two isotopes with relative atomic masses of approximately 35 and 37. However, the lighter isotope is roughly three times more common than the heavier one, so any compound made from a naturally occurring chlorine sample will have this ratio of isotopes. Thus the relative atomic mass of chlorine is 35.5.

relative density (r.d.), ***specific gravity*** The DENSITY of a substance at a specified temperature, divided by the maximum density of water (its density at 4°C). Since the density of water is close to 1 gcm^{-3}, the relative density is close to the density in grams per centimetre cubed. If its relative density is less than 1, a substance will float on water; if its density is greater than 1, it will sink.

The relative density of a gas is usually quoted relative to dry air – both at the same temperature and pressure.

relative humidity HUMIDITY expressed as a fraction of the maximum amount of water vapour that air can hold at that temperature. *See also* SATURATED.

relative molecular mass (rmm), ***molecular weight*** The mass of a MOLECULE measured in ATOMIC MASS UNITS. The relative molecular mass of a molecule is equal to the sum of the RELATIVE ATOMIC MASSES of the atoms from which the molecule is composed. *See also* DUMAS' METHOD.

relative permeability The factor by which a material changes the strength of any MAGNETIC FIELD in which it is placed. It is the ratio of the magnetic field in a toroidal (doughnut shaped) SOLENOID with a core made of a specified material to the field in a similar solenoid with no core, ignoring the effects of saturation and hysteresis. *See also* PERMEABILITY, SUSCEPTIBILITY.

relative permittivity The amount by which the CAPACITANCE of a CAPACITOR is increased by the introduction of a DIELECTRIC into the space between the plates. *See also* PERMITTIVITY.

relative velocity The VELOCITY of an object as seen from another moving object. *See also* GALILEAN RELATIVITY.

relativistic (*adj.*) Travelling at speeds close to the SPEED OF LIGHT. *See* SPECIAL THEORY OF RELATIVITY.

relativistic paradox A paradox, or apparently nonsensical result, arising from the SPECIAL THEORY OF RELATIVITY. All such paradoxes can be resolved by careful consideration of the physics involved. *See* POLE AND BARN PARADOX, TWIN PARADOX.

relativity A blanket term covering the RELATIVITY PRINCIPLE, SPECIAL THEORY OF RELATIVITY and GENERAL THEORY OF RELATIVITY. *See also* GALILEAN RELATIVITY.

relativity principle The idea that there is no

observer who has a privileged viewpoint and can claim to be truly at rest. Thus a view of events seen by any observer is equally valid, and all velocities are relative to a given observer. This idea lead to the SPECIAL THEORY OF RELATIVITY.

relay An electrical switch operated by the current flowing in an ELECTROMAGNET. Relays are used to control circuits using the flow of current in a separate circuit. The currents and voltages controlled may be far larger than those in the electromagnet. Car starter motors, for example, require currents of several hundred AMPERES. A relay is used to operate them, with the current in the electromagnet, which is only a few amperes, controlled by a key operated switch. *See also* REED RELAY.

relay neurone *See* NEURONE.

relay station An installation, normally on a hilltop or high building, which receives RADIO signals from one direction, and amplifies and re-transmits them, increasing the coverage of a VHF transmitter.

reluctance In a MAGNETIC CIRCUIT, the MAGNETOMOTIVE FORCE divided by the MAGNETIC FLUX. It is analogous to RESISTANCE in electrical circuits. As with electrical resistance, the reluctance of any part of a magnetic circuit is proportional to its length, and inversely proportional to its cross-sectional area. *See* RELATIVE PERMEABILITY.

rem (Abbreviation for **Röntgen equivalent man**) A former unit for effective dose of IONIZING RADIATION. One rem is equivalent to 10^{-2} SIEVERT.

remainder The amount remaining when one NATURAL NUMBER is subtracted from another as many times as is possible without giving a negative result (this process of repeated subtraction is called division). For example, when 7 is divided by 3, the remainder is one.

remanence The amount of magnetism retained by a FERROMAGNETIC material when the magnetizing field is removed. *See also* PERMANENT MAGNET.

renal (*adj.*) Relating to the KIDNEY.

renal dialysis *See* DIALYSIS.

renewable resource Natural resources that can be replaced naturally in a reasonable amount of time, for example wood, soil, water and fish. Although they are renewable, the continued supply of such resources relies on their proper use and conservation by humans. Renewable energy is power obtained from a renewable source, for example SOLAR ENERGY, WAVE POWER, HYDROELECTRIC POWER, GEOTHERMAL ENERGY and wind. NON-RENEWABLE resources cannot be replaced, for example coal, oil and metal ores. Some resources, for example used in motor cars and tin cans, could be recycled but it is often uneconomical to do so.

renin An enzyme produced by the kidney in response to a drop in blood pressure. Renin stimulates ANGIOTENSIN.

rennin An enzyme that coagulates milk in the stomach and is therefore important to young animals in their digestion of milk. Rennin is secreted as the inactive prorennin, which is activated in the STOMACH to form rennin.

reproductive system The system of organs and tubes involved in SEXUAL REPRODUCTION.

reptile A member of the vertebrate CLASS REPTILIA.

Reptilia A CLASS of VERTEBRATES consisting of the reptiles. Reptiles lay hard-shelled eggs on land. The eggs are filled with yolk and fully formed young hatch from them. In some reptiles, for example lizards and snakes, the eggs remain inside the female and they give birth to live young. The skin of reptiles is covered with horny scales and they are POIKILOTHERMS, unable to maintain their body temperature. Reptiles usually have four legs and some live in water (but breathe air) and some live on land; they possess lungs. Their METABOLISM is slow and their teeth are all the same.

Many species of reptiles are now extinct, for example the dinosaurs, but surviving species include crocodiles, alligators, tortoises, lizards and snakes.

RES *See* RETICULOENDOTHELIAL SYSTEM.

resin ($(C_5H_8)_n$) A material with a high RELATIVE MOLECULAR MASS that softens at high temperatures. Natural resins are exuded from trees such as pines and firs and harden in air. Synthetic resins are produced by POLYMERIZATION and used in adhesives, plastics and varnishes. Examples include EPOXY RESINS and PHENOL-FORMALDEHYDE RESINS. Soft resins are used in ointments.

resistance A measure of the difficulty with which a CURRENT flows through a CONDUCTOR. It is equal to the VOLTAGE across the object whose resistance is being measured divided by the current through that object. The unit of resistance is the OHM.

Resistance = potential difference/current

When resistors are connected in SERIES, the total resistance is larger than any of the resistors in the circuit, it will be equal to the sum of the resistances in the circuit. When resistors are connected in PARALLEL, the total resistance is smaller than the resistance of the smallest resistor – connecting another resistor in parallel, no matter how large, can only make it easier for the current to flow. The total resistance is the reciprocal of the sum of the reciprocals of the individual resistors.

For resistances R_1, R_2, R_3 connected in series, the total resistance is R where

$$R = R_1 + R_2 + R_3$$

If these resistances are connected in parallel,

$$1/R = 1/R_1 + 1/R_2 + 1/R_3$$

See also CONDUCTANCE, INTERNAL RESISTANCE, OHMMETER, RESISTIVITY, THERMAL RESISTANCE, WHEATSTONE BRIDGE.

resistance thermometer A thermometer that uses a WHEATSTONE BRIDGE circuit to measure temperature by the change in resistance of a length of platinum wire.

It is important to compensate for any changes in the resistance of the wires connecting the sensor to the rest of the bridge circuit. This is done by incorporating a pair of DUMMY LEADS, which run from the opposite arm of the bridge to the point where the sensor is located, but are connected together without being connected to the sensor. In this way, any changes in the resistance of the connecting wires, due to temperature changes for example, affect both arms of the bridge equally, so do not affect the balance of the bridge.

resistance wire Wire made of a metal alloy, deliberately designed to have a high RESISTANCE.

resistivity The RESISTANCE between opposite faces per one metre cube of material. Resistivity is measured in OHM metres. The resistance of a conductor is proportional to its length and inversely proportional to the cross-sectional area through which the current flows. It also depends on the material used and the temperature of the material, and these are the two factors accounted for by the resistivity of the material.

For a conductor of length l with cross-sectional area A and resistivity ρ, the resistance R is given by

$$R = \rho l/A$$

See also CONDUCTIVITY, THERMAL COEFFICIENT OF RESISTIVITY.

resistor A device designed to have a constant RESISTANCE, to control the current in a circuit. Small resistors are usually made from a film of metal or metal oxide on a ceramic former, coated with a layer of hard varnish to prevent moisture or dirt from altering the resistance. The value of the resistance is usually marked on the resistor by a series of coloured stripes.

Larger resistors, capable of dissipating more power are made from a coil of metal alloy wire called resistance wire, designed to have a larger resistance than normal wire. Resistors made in this way are called wire wound resistors.

See also VARIABLE RESISTOR.

resolution, *resolving power* The ability of an optical system, such as a camera, microscope telescope or SPECTROMETER to show fine detail in an image. It is usually measured in terms of the closest separation of two objects that can just be shown to be separate by the system under consideration. The resolution is inversely proportional to the WAVELENGTH of light being used. Thus the resolution of a light microscope is limited to two points 0.2 μm apart, compared to typically 1 nm for the ELECTRON MICROSCOPE.

The resolution of an instrument is ultimately limited by DIFFRACTION. The light rays must pass through an APERTURE or lens at some stage, and will be diffracted as they do so. Resolution is often calculated in terms of RAYLEIGH'S CRITERION.

resolve (*vb.*) The process of finding the COMPONENT of a vector in a specified direction, or a pair of directions, usually at right angles to one another. *See also* RESULTANT.

resolving power *See* RESOLUTION.

resonance The state of an oscillating system when its NATURAL FREQUENCY of OSCILLATION is close to the frequency of some periodically varying driving force (*see* FORCED OSCILLATION). Resonance can occur in electrical circuits, mechanical systems, atoms and molecules.

At resonance the amplitude of the oscillating motion is a maximum and the PHASE difference between the oscillation and the driving force is 90°, with the motion lagging behind the force by one quarter of an oscillation. The size and sharpness of the resonance depends on the Q FACTOR of the oscillating system: the larger this is, the greater the response at resonance and the more rapid the change in phase as the resonance is approached.

Resonance can be useful in that it can be used to select one frequency and reject others, in tuning a radio for example. Resonance can also be a problem in systems which are subject to vibration – such as a car driving over an uneven road. In such systems, DAMPING is usually applied to ensure that the oscillation of the system does not reach dangerous levels.

See also TUNED CIRCUIT.

resonance hybrid A molecule or set of COVALENT BONDS within a molecule that cannot be described in terms of simple single and double covalent bonds, but which is a quantum mechanical mixture of a number of possibilities, with the resonance hybrid having a lower energy.

The BENZENE molecule can be regarded as a resonance hybrid between two alternative structures in which single and double bonds alternate about the ring, and the carbon dioxide molecule is a resonance hybrid involving double bonds and CO-ORDINATE BONDS.

resonant cavity A hollow space, often cylindrical in shape, with a particular RESONANT FREQUENCY; particularly such a cavity with metal walls and resonant at a frequency in the MICROWAVE region of the ELECTROMAGNETIC SPECTRUM.

resonant circuit *See* TUNED CIRCUIT.

resonant frequency The FREQUENCY at which the response to a periodic driving force reaches a maximum. *See* RESONANCE.

respiration The biochemical processes occurring within cells to break down food molecules to release energy. There are three stages to respiration: GLYCOLYSIS, the KREBS CYCLE and the ELECTRON TRANSPORT SYSTEM.

Respiration is more correctly called 'internal cellular respiration' and contrasts with 'external respiration', which is the exchange of oxygen and carbon dioxide during BREATHING. Most food is converted into the sugar glucose, which is then converted into carbon dioxide and water with the release of energy. Fats and proteins can also be used as respiratory substrates (without being first converted to CARBOHYDRATES) but only during starvation (or dieting).

Respiration is usually aerobic (requiring oxygen) and occurs in MITOCHONDRIA of EUKARYOTES and MESOSOMES of PROKARYOTES. Some cells can function for short periods without oxygen, but most die. Some organisms, such as certain bacteria, yeast and parasites, are ANAEROBES and can use GLUCOSE to make energy without the use of oxygen. This is called anaerobic respiration and is less efficient than aerobic respiration, but because it produces alcohol and carbon dioxide it is of great use in the baking and brewing industry (*see* FERMENTATION).

In plants, there is a variable balance between respiration and PHOTOSYNTHESIS. The oxygen produced as a waste product of photosynthesis goes directly from the CHLOROPLAST, where it is made, to the mitochondria, where it can be used for respiration. This occurs during the dark in most plants (C_3 PLANTS). During daylight some plants engage in the apparently wasteful process of PHOTORESPIRATION. Carbon dioxide released by respiration can be used for photosynthesis.

respiratory chain *See* ELECTRON TRANSPORT SYSTEM.

respiratory pigment In most vertebrates and some invertebrates, one of a group of coloured proteins that can bind weakly to oxygen in the blood or other tissues, to increase the uptake, transport and unloading of oxygen. HAEMOGLOBIN is the main example of a respiratory pigment in blood, which contains iron and colours the blood red.

Other examples in blood are haemocyanin, which contains copper and is blue, and is found in the MOLLUSCS and CRUSTACEANS; haemoerythrin, which contains iron and is red/brown, and is found in ANNELIDS; chlorocruorin, which contains iron and is green, and is found in POLYCHAETES. MYOGLOBIN is found in muscles of all vertebrates but does not colour them.

Respiratory pigments are efficient because they have a high affinity for oxygen when there is a high concentration of oxygen (and so pick oxygen up easily), but a low affinity for oxygen when it is in low concentrations (thus release oxygen where it is needed). Respiratory pigments can also bind to carbon dioxide or to carbon monoxide (*see* HAEMOGLOBIN).

The term respiratory pigment can be used to refer to substances, such as CYTOCHROMES, involved in the ELECTRON TRANSPORT SYSTEM.

respiratory quotient (RQ) The ratio of the volume of carbon dioxide produced by an organism to oxygen consumed during aerobic RESPIRATION. The theoretical RQ value for CARBOHYDRATES as respiratory substrates is 1, for fats 0.7 and for proteins 0.8, although a mixture of these is normally utilized.

respiratory system The series of components involved in the exchange of gases (usually oxygen and carbon dioxide) by an organism. In humans, the respiratory system consists of two lungs each containing millions of air sacs called alveoli (*see* ALVEOLUS), where exchange of oxygen and carbon dioxide between air and the bloodstream takes place. The opening of the respiratory tract is considered to be the nose, through which air enters and passes to the mouth cavity, entering at the PHARYNX. The TRACHEA is the main airway running into the lungs, which branches into two tubes called bronchi (*see* BRONCHUS). The bronchi divide further into small tubes called bronchioles, and then into the air sacs or alveoli. Dust and other unwanted particles are cleared from the respiratory tract by fine hairs in the nose and by a sticky liquid called MUCUS secreted by the trachea and bronchus.

Birds, like mammals, have lungs in which gaseous exchange takes place. In fish, the respiratory system consists of a series of GILLS, and in insects gases enter and leave by SPIRACLES. In plants, the gases oxygen and carbon dioxide are involved in both respiration and PHOTOSYNTHESIS, and gas exchange occurs through pores in the leaves called stomata (*see* STOMA).

See also BREATHING.

rest energy *See* REST MASS ENERGY.

rest mass (m_0) The MASS of an object, especially an ELEMENTARY PARTICLE, when at rest relative to the observer measuring the mass. Since mass increases with velocity, this is an unambiguous way of specifying the mass of a particular type of elementary particle, regardless of its motion. *See also* REST MASS ENERGY, SPECIAL THEORY OF RELATIVITY.

rest mass energy The ENERGY an object has by virtue of its REST MASS. It is equal to m_0c^2, where m_0 is the rest mass, and c is the speed of light in a vacuum. Rest mass energies, expressed in ELECTRON-VOLTS, are often used in preference to masses in particle physics.

restriction endonuclease, *restriction enzyme* An enzyme derived from a bacterium that cuts a chain of DNA between specific NUCLEOTIDE base sequences. Many restriction endonuclease enzymes exist that are specific for different nucleotide sequences. They are all NUCLEASE enzymes. Any fragment of DNA produced in this way can be joined to other DNA by the use of another enzyme called DNA LIGASE. Hence manipulation of DNA is possible, and is used in GENETIC ENGINEERING.

restriction enzyme *See* RESTRICTION ENDONUCLEASE.

resultant In mathematics, the VECTOR obtained by adding a number of vectors together, particularly the vector sum of the forces acting on an object.

reticuloendothelial system (RES) The system of circulating tissue MACROPHAGES, including those in CONNECTIVE TISSUE (histiocytes), LIVER (Kupffer cells), SPLEEN, LUNGS, LYMPH NODES and LYMPHATIC SYSTEM.

retina In vertebrates and cephalopods (higher molluscs), a light-sensitive layer at the back of the EYE. In humans, the retina contains sensory cells called rods and cones, which convert the light energy they receive into NERVE IMPULSES that travel along the OPTIC NERVE to the brain. At the point where the optic nerve leaves the retina, there are no rods or cones and this is called the blind spot.

There are about 6 million cones per eye, mostly concentrated in a region on the retina called the fovea centralis. The cones are sensitive to colour and are used mostly for day vision. Because each cone has its own NEURONE connection to the brain, visual acuity is high compared to rods, which share neurones. There are more rods (about 120 million), which are distributed throughout the retina. Rods are mostly used for night vision and cannot distinguish colour.

The basic structure of rods and cones is similar apart from the shape. Each rod possesses thousands of vesicles containing a photosensitive pigment called rhodopsin. Exposure to light causes rhodopsin to split into its constituent parts, which generates an ACTION POTENTIAL resulting in a nerve impulse. The pigment is resynthesized by energy provided from MITOCHONDRIA, which are also present in the rod.

In cones a similar process occurs but the photosensitive pigment is iodopsin, which is less sensitive to light (so more is needed to initiate a nerve impulse).

Colour vision is explained by the trichromatic theory, in which it is thought that three forms of iodopsin exist occurring in three different types of cones. These respond to three different types of light, green, blue and red, and other colours are perceived by a combined stimulation of these.

retinol *See* VITAMIN A.

retort A vessel used for DISTILLATION, traditionally in the form of a glass bulb with a long stem.

retrovirus An important group of RNA viruses that can use the enzyme REVERSE TRANSCRIPTASE to make DNA from their single-stranded RNA. The DNA form of the virus is a PROVIRUS. The human immunodeficiency virus (HIV) causing AIDS is a retrovirus. Some human cancers are thought to be caused by retroviruses that can carry (and mutate) host cell genes capable of inducing CANCER in other cells. These genes are called ONCOGENES and most viruses known to cause cancer contain them, but some cellular oncogenes exist that can be activated by other non-viral factors. Many retroviruses are harmless and in fact the proviral DNA may be integrated permanently into the host cell DNA and passed through generations as an endogenous virus.

reverberation The repeated REFLECTION of sounds from the walls of an enclosed space.

reverberation time The time taken for the volume of echoing sound in a room, such as a concert hall, to fall by 60 DECIBELS.

reverberatory furnace A device for heating solid materials, particularly metal ores, in which a mixture of gas and air is burned above a hearth containing the material to be heated. Gas and air are fed in at one end and leave through a chimney, or flue, at the other. The furnace has a low roof, lined with a REFRACTORY material to direct as much heat as possible onto the contents of the hearth.

reverse biased Describing the state of a PN JUNCTION DIODE when a voltage is applied that tends to increase the size of the DEPLETION LAYER, i.e. with the P-TYPE SEMICONDUCTOR negative and the N-TYPE SEMICONDUCTOR positive. In this state, no current can flow. *See also* FORWARD BIASED.

reverse osmosis The passage of a solvent through a SEMIPERMEABLE MEMBRANE from a region of high SOLUTE concentration to one lower – that is, in the reverse direction to OSMOSIS. Reverse osmosis occurs on the application of a pressure greater than the OSMOTIC PRESSURE.

reverse transcriptase An enzyme that synthesizes single-stranded COPY DNA (cDNA) from a single-stranded RNA template. The enzyme is found in RETROVIRUSES, but is also useful experimentally in GENETIC ENGINEERING.

reversible reaction Any chemical reaction that can proceed in either direction. For example the reaction:

$$3H_2 + 2N_2 \Leftrightarrow 2NH_3$$

is said to be reversible as nitrogen and hydrogen can combine to form ammonia, or ammonia can dissociate to form nitrogen and hydrogen.

All reactions are to some extent reversible, but a reaction is considered reversible if the EQUILIBRIUM state, where the reaction proceeds in both directions at the same rate, is reached with comparable concentrations of the reagents and products. The position of the equilibrium will vary depending on the conditions of the reaction, such as temperature and pressure.

See also CHEMICAL EQUILIBRIUM, EQUILIBRIUM CONSTANT, FREE ENERGY, LE CHATELIER'S PRINCIPLE.

rhenium (Re) The element with ATOMIC NUMBER 75; RELATIVE ATOMIC MASS 186.2; melting point 3,180°C; boiling point 5,620°C; RELATIVE DENSITY 22.0. It is a TRANSITION METAL, whose hardness and high melting point have led to its use in some alloys.

rheology The study of the deformation and flow of matter, especially plastic solids and viscous liquids.

rheostat Former term for a VARIABLE RESISTOR, particularly when used to control the brightness of a lamp.

rhesus disease, *haemolytic disease of the newborn* A condition of newborn babies in which the foetal RED BLOOD CELLS are broken, down causing anaemia, heart failure and possible brain damage. Rhesus disease arises during pregnancy when a mother who does not have the RHESUS FACTOR in her blood (is rhesus negative) carries a foetus that is rhesus positive. The mother will produce ANTIBODIES to the rhesus ANTIGEN if foetal blood crosses the placenta, and these can pass across the placenta causing rhesus disease. Complete blood transfusion with rhesus negative blood is then necessary.

A first child is usually unaffected because foetal blood only passes into the mother towards the end of pregnancy or at birth and the antibody response is slow to build up, but it does continue to build up after birth. To prevent the antibodies affecting subsequent rhesus-positive foetuses, the mother is given anti-rhesus GLOBULIN just after the first pregnancy to prevent the formation of antibodies.

rhesus factor, *antigen D* In humans, a protein on the surface of RED BLOOD CELLS involved in the rhesus BLOOD GROUP SYSTEM. The factor was first identified in rhesus monkeys, hence its name. Most individuals (75–85 per cent) possess the factor and are called rhesus positive (Rh+), while those who do not carry the factor are rhesus negative (Rh–). Anti-rhesus ANTIBODIES are produced by Rh– people after exposure to the rhesus ANTIGEN (the antibodies are not naturally occurring). The rhesus system can cause a problem during pregnancy as a Rh– mother carrying a Rh+ baby will produce antibodies to the rhesus antigen if foetal blood crosses the placenta. The antibodies can pass across the placenta causing RHESUS DISEASE.

rhesus system *See* BLOOD GROUP SYSTEM.

rhizoid A simple hair-like outgrowth on a plant that serves as a ROOT because it is able to absorb water and nutrients. Rhizoids are often found on BRYOPHYTES, some FUNGI and ALGAE.

rhizome An underground horizontal, branching stem, protected by scaly leaves. An example is the iris. The stem grows horizontally year after year, and buds form between the scaly leaves and develop into new vertical shoots. It is a structure of VEGETATIVE REPRODUCTION.

Rhizopod A member of the PHYLUM RHIZOPODA.

Rhizopoda A PHYLUM from the KINGDOM PROTOCTISTA that consists of PROTOZOA. The members (rhizopods) move and trap food by means of PSEUDOPODIA. Reproduction is generally by BINARY FISSION but can be by the formation of GAMETES. Rhizopods can be free-living or PARASITES, and are an irregular shape. An example is *Amoeba*.

rhodium (Rh) The element with ATOMIC NUMBER 45; RELATIVE ATOMIC MASS 102.9; melting point 1,966°C; boiling point 3,727°C; RELATIVE DENSITY 12.4. Rhodium is chemically fairly unreactive, and is sometimes used with platinum as a material for THERMOCOUPLES. Rhodium salts have a characteristic red colour.

Rhodophyta A PHYLUM from the KINGDOM PROTOCTISTA that consists of red ALGAE. Red algae are mostly marine; they contain CHLOROPHYLLS a and d and other pigments, which give them their colour. They are characterized by a complete absence of FLAGELLA at any stage.

rhodopsin A light-sensitive pigment found in the RODS of the eye. *See* RETINA.

rhombus A geometrical shape that has four straight sides of equal length, with opposite sides parallel to one another, but with adjacent sides not at 90°.

rib A long, narrow, curved bone of vertebrates extending from the spine. In humans, there are 12 pairs, joined at the back to the vertebrae of the spine. The ribs form the ribcage, which can move to allow chest expansion by movement of INTERCOSTAL MUSCLES between the ribs. In humans, the upper seven ribs are joined by CARTILAGE to the breast bone (sternum) at the front of the THORAX, the next three are joined to each other at the ends by cartilage, and the last two (floating ribs) are not attached at the front. In fish and reptiles, the ribs extend along most of the spine, but in mammals they are confined to the thorax where they provide protection for the lungs and heart.

ribcage *See* RIB.

riboflavin *See* VITAMIN B.

ribonuclease *See* RNASE.

ribonucleic acid *See* RNA.

ribose ($C_5H_{10}O_5$) A PENTOSE sugar that is a component of RNA.

ribosomal RNA (rRNA) A type of RNA that is a large molecule, making up more than half the mass of a cell's total RNA and more than half the mass of the RIBOSOMES. Ribosomal RNA can be either a single or double helix and its base sequence is similar in all organisms.

ribosome Any one of many granules of PROTEIN and RIBOSOMAL RNA (rRNA) in a cell that are the site of PROTEIN SYNTHESIS. Ribosomes are associated with ENDOPLASMIC RETICULUM (as rough ER) in eukaryotic cells (*see* EUKARYOTE), and are present in the cytoplasm of both eukaryotic and prokaryotic cells (*see* PROKARYOTE).

Ribosomes of prokaryotic cells are slightly smaller than those of eukaryotic cells. Ribosomes are often linked together to form chains of polyribosomes or polysomes.

ribozyme RNA molecules that can act as ENZYMES as well as PROTEIN.

ribulose bisphosphate A carbon dioxide acceptor in the CALVIN CYCLE.

Richter scale A logarithmic scale used to measure the magnitude of EARTHQUAKES. The magnitude is calculated as a function of the total energy released by the earthquake and expressed on a logarithmic scale of 0 to 10.

rickets A deficiency of VITAMIN D that causes a bone disorder caused by lack of calcium deposits, in which the bones do not harden and therefore bend out of shape. Rickets is also associated with kidney disease.

right angle An angle of 90°, one quarter of a circle. The angle between vertical and horizontal.

right-angled triangle Any triangle with one angle a RIGHT ANGLE. *See also* PYTHAGORAS' THEOREM.

right hand grip rule A rule for finding the direction of a MAGNETIC FIELD around a current-carrying wire or SOLENOID. For a wire, imagine gripping the wire with the right hand, with the thumb pointing in the direction of CONVENTIONAL CURRENT flow; the fingers then wrap around the wire in the same direction as the field lines. For a solenoid, the fingers are made to curl around like the conventional current and the thumb gives the field direction.

ring system In astronomy, a set of flat discs found around all the GAS GIANTS made of small particles in ORBIT in bands around the equator of the planet. JUPITER'S rings are too faint to be seen from Earth and URANUS' and NEPTUNE'S rings are too dark to be observed against the dark background of the sky, but SATURN'S rings have been known almost since the invention of the telescope.

RISC (Reduced Instruction Set Computing) An approach to the design of MICROPROCESSORS, in which the instruction set contains relatively few and much simpler functions, compared to the older CISC microprocessors. Programs written for RISC chips are much longer than those for CISC chips, because the RISC chips need more of their simple instructions to do the same work. However, since they run simple instructions very, very fast, rather than complex instructions more slowly, RISC chips are much faster.

RNA (ribonucleic acid) A NUCLEIC ACID associated mainly with the synthesis of proteins from DNA. RNA is found in the nucleus and cytoplasm of cells. It is usually a single-stranded chain of NUCLEOTIDES synthesized from DNA by the formation of BASE PAIRS. The organic bases in RNA are adenine, guanine, cytosine and uracil (which replaces the thymine of DNA) and a PENTOSE sugar that is always RIBOSE. There are three main forms of RNA, all concerned with PROTEIN SYNTHESIS. These are RIBOSOMAL RNA (rRNA), TRANSFER RNA (tRNA) and MESSENGER RNA (mRNA). In some viruses, for example RETROVIRUSES, RNA can make up the hereditary material, instead of DNA.

RNA polymerase *See* POLYMERASE.

RNAse (ribonuclease) One of many enzymes that hydrolyse RNA by breaking down the sugar–phosphate bonds. *See also* RESTRICTION ENDONUCLEASE.

roasting The process of heating an ore in air. This can be used to extract some impurities such as sulphur and sulphur dioxide.

rod A type of light-sensitive cell found in the RETINA of most vertebrates. Rods are mostly used for night vision and cannot distinguish colour. *See* RETINA.

rodent A member of the order Rodentia in the CLASS MAMMALIA. It includes most of the species of mammals. Rodents are characterized by the presence of a single pair of incisors at the front of the upper and lower jaw that continue to grow as they are worn down. They are PLACENTAL MAMMALS and can be OMNIVORES (plant or meat eating) or HERBIVORES (plant eating). Examples include rats, mice, squirrels, beavers, porcupines and guinea pigs.

rodenticide A PESTICIDE that kills RODENTS.

roentgen *See* RÖNTGEN.

roller bearing *See* BEARING.

ROM (read only memory) A data storage system into which data can only be placed once. Once the data is stored the memory is 'blown' and the state of each unit of memory is fixed. *See also* EROM, PROM.

röntgen, *roentgen* Former unit for DOSE of IONIZING RADIATION. One röntgen is approximately 8.7 x 10^{-7} GRAY.

root 1. (*biology*) The part of a plant that is usually underground, that anchors the plant and absorbs water and dissolved mineral ions. Roots usually grow downwards and towards water (*see* TROPISM) but some plants produce aerial roots that absorb moisture from the atmosphere or provide further anchorage. The root has a central column of vascular tissue (XYLEM and PHLOEM) surrounded by a single-celled ring of endodermis. Inside the endodermis is a layer of PARENCHYMA called the pericycle, from which lateral roots originate. The bulk of the root is made of cortex (parenchyma) storing starch, which is surrounded by the outer exodermis and a one-cell thick epidermal layer with ROOT HAIRS. The end of the root is protected by a root cap, where the actively dividing cells push through the soil (*see* MERISTEM).

Water and dissolved mineral ions are absorbed into the root hairs, through the cortex, endodermis and pericycle into the xylem, to travel upwards through the roots, stems and to the leaves by TRANSLOCATION. The roots of certain plants can form a symbiotic (*see* SYMBIOSIS) relationship with some bacteria (*see* ROOT NODULE). *See also* ADVENTITIOUS ROOT, CONTRACTILE ROOT, FIBROUS ROOT, PNEUMATOPHORE, TAP ROOT.

2. (*mathematics*) A solution to a mathematical equation, a value of an independent variable for which a function takes on a particular value, often zero. In particular, a root is the value of x for which $x^2 = y$ (called the square root of y), or $x^3 = y$ (the cube root of y).

root hair A delicate extension from the epidermal cells (*see* EPIDERMIS) on the surface of a plant ROOT. The layer of cells producing the root hairs is called the piliferous layer. Root hairs are found near the tip of young roots and increase the surface area for absorption of water and mineral ions. They function for only a few weeks and are continually replaced nearer the growing tip. *See also* TRANSLOCATION.

root mean square (r.m.s.) A form of average in which a quantity is squared, the mean value found and then the square root taken. If an ALTERNATING CURRENT flows through a resistor, the r.m.s. values of the voltage and current are the ones that would produce the same heating effect as that voltage or current flowing continuously. For a SINUSOIDAL current or voltage, the root mean square value is 0.707 times the peak value.

root nodule A swelling on the ROOTS of leguminous plants (LEGUMES), such as beans, peas and clover, that is caused by infection with nitrogen-fixing bacteria such as *Rhizobium* (*see* NITROGEN FIXATION). A symbiotic relationship exists: the bacteria convert atmospheric nitrogen into nitrates that the plant can use, independently of soil nitrates, and in return the bacteria get a source of nutrition. The nodule develops like a lateral root and during the early stages of development central cells fill up with the bacteria enclosed in a membrane (thus the bacteria remain extracellular). Neither the legume nor the bacteria alone can fix nitrogen, and it is thought that the site of fixation is this membrane separating the two. Leguminous plants are very important in improving soil fertility because they can use atmospheric nitrogen, and without them the level of nitrates in the soil would not be sufficient to support vegetation cover. *See also* NITROGEN CYCLE.

root tuber *See* TUBER.

rotation A geometrical operation that maps every point onto another point in such a way that the distance of the point from a fixed point, called the centre of rotation, remains fixed, but the direction of the line between the centre of rotation and the point is rotated by a specified angle.

rotational dynamics The study of the way in which an object rotates, described in terms an angle between a point on the object and a fixed direction. The line around which the object rotates is called the AXIS.

The rotational motion of a body is described in terms of its ANGULAR VELOCITY, the rate of change of its angular position with time. If there is a force acting on the system along a line that does not cross the axis, the angular velocity of the object will change. The rate of change of angular velocity with time, called the ANGULAR ACCELERATION, is proportional to the MOMENT of the force. The amount of angular acceleration also depends on the mass of the object and the way in which that mass is distributed; the acceleration will be less for massive objects or for objects where the mass is far from the axis. These two factors are taken account of in the MOMENT OF INERTIA.

For a point of mass m moving at a speed v at a distance r from the axis, with the motion at right angles to the line joining the object and the axis

$$\omega = v/r$$

where ω is the angular velocity.

roughage, *dietary fibre* Plant material, mainly CELLULOSE from plant cell walls, that passes through the human digestive tract almost unchanged (it is not digested). The function of roughage is to provide bulk to the other material passing through the digestive system, stimulate PERISTALSIS, and ease movement of food throughout the digestive system. Roughage is important for healthy living and too little can lead to intestinal disorders and constipation.

RQ *See* RESPIRATORY QUOTIENT.

rRNA *See* RIBOSOMAL RNA.

rubber A natural or synthetic POLYMER that has elasticity at room temperature. Natural rubber comes mostly from the tree *Hevea brasilienis* where it exudes from groves cut in the trunk. It is then coagulated with methanoic or ethanoic acid to form a solid rubber. It is treated before use by a process called VULCANIZATION. This yields a harder, stronger end-product that is less affected by external temperatures than untreated rubber. Most of the natural supply of rubber comes from Malaysia. There is a large variety of synthetic rubbers that are cheaper than natural rubber and can be adapted to special purposes. For example SBR, styrene-butadiene rubber, is used for car tyres, shoe soles and can be blended with natural rubber.

rubidium (Rb) The element with ATOMIC NUMBER 37; RELATIVE ATOMIC MASS 85.5; melting point 38°C; boiling point 688°C; RELATIVE DENSITY 1.5. Rubidium is an ALKALI METAL, highly reactive and combining violently with water or oxygen in the atmosphere.

rumen The first of four chambers that form the stomachs of RUMINANTS.

ruminant An even-toed mammal with a complex DIGESTIVE SYSTEM, including several stomach chambers, for the digestion of plant food. Cattle, goats, deer and giraffes all have a four-chambered stomach, while camels have three chambers. Food is stored and initial digestion carried out in the first chamber, called the rumen. This involves the enzyme cellulase (*see* CELLULOSE), which is provided by symbiotic bacteria. The food is then returned to the mouth for further chewing before being passed to the next stomach.

runner A vegetatively reproducing structure similar to a STOLON, but a number of stems develop from the parent plant and travel across the soil surface, periodically producing ADVENTITIOUS ROOTS, and new plants grow at these points. Examples include the strawberry and buttercup.

rust (*n., vb.*) The powdery red material formed when iron CORRODES. The chemical nature of rust is complex, but it is essentially a mixture of HYDRATED iron(III) oxide, Fe_2O_3 with iron(III) oxyhydroxide, FeO(OH).

ruthenium (Ru) The element with ATOMIC NUMBER 44; RELATIVE ATOMIC MASS 101.1; melting point 2,310°C; boiling point 3,900°C; RELATIVE DENSITY 12.4. Ruthenium is a TRANSITION METAL. It often occurs in association with platinum. It is used in some platinum alloys and as a catalyst.

Rutherford–Bohr atom A model of the ATOM, incorporating the idea of an atomic NUCLEUS and the idea that ELECTRONS only occupy certain ENERGY LEVELS with QUANTIZED values of ANGULAR MOMENTUM. *See also* BOHR THEORY, HYDROGEN SPECTRUM, RUTHERFORD SCATTERING EXPERIMENT.

rutherfordium The name proposed for the element with ATOMIC NUMBER 104, also called unnilquadium. It has only radioactive ISOTOPES, with very short HALF-LIVES. It is synthesized by bombarding heavy nuclei with fast moving ions. Its discovery was first claimed by a Russian group in 1964, but its existence was not independently confirmed until 1967. Little is known about its physical or chemical properties.

Rutherford scattering experiment The experiment performed in 1909 by Geiger and Marsden that showed the existence of the atomic nucleus. ALPHA PARTICLES from a radium source were fired in a vacuum at a thin sheet of gold foil and the emerging particles were detected. Most particles passed straight through the foil, but a few were deflected through very large angles, suggesting that they had experienced a large force from a massive object.

Rydberg constant The constant *R* in the RYDBERG EQUATION, equal to 1.10×10^7 m^{-1}. The rutherford–bohr model of the atom showed that this constant is related to other

FUNDAMENTAL CONSTANTS by the relationship

$$R = me^4/8\varepsilon_0^2h^3c$$

where m is the mass of an electron, e the charge on an electron, ε_0 the PERMITTIVITY of free space, h PLANCK'S CONSTANT and c the SPEED OF LIGHT.

Rydberg equation An equation which gives the wavelengths λ, of the lines in the EMISSION SPECTRUM of the hydrogen atom.

$$1/\lambda = R\,(1/n^2 - 1/m^2)$$

where R is the RYDBERG CONSTANT and n and m are positive integers, with m greater than n.

S

saccharide *See* MONOSACCHARIDE, DISACCHARIDE and POLYSACCHARIDE.

saccharin, *ortho-sulpho benzimide* ($C_7H_5NO_3S$) An artificial sweetener derived from AMINES. It is a white solid 500 times sweeter than sugar but with a bitter aftertaste and potentially carcinogenic (*see* CARCINOGEN). It has therefore largely been replaced by other sweetening agents.

sacculus *See* SEMI-CIRCULAR CANAL.

sacrificial cathode A piece of fairly reactive metal, such as zinc, attached to a steel structure exposed to moisture. The sacrificial cathode CORRODES in preference to the steel. *See* SACRIFICIAL CORROSION.

sacrificial corrosion A CORROSION process in which two pieces of metal are in electrical contact. Surrounding moisture sets up an electrochemical cell (*see* CELL) between the two metals and the one with the more negative ELECTRODE POTENTIAL (that is, the more reactive metal), called a SACRIFICIAL CATHODE, is attacked by corrosion far more rapidly than the other metal.

Zinc is used as a sacrificial cathode to prevent iron from corroding. In steel ships for example, a piece of zinc is attached to the outside of the hull and replaced as it is eroded by corrosion. *See also* GALVANISING.

safety valve *See* VALVE.

salicylic acid, *2-hydroxybenzoic acid* (HOC_6H_4COOH) An organic acid that occurs as white, crystalline needles when pure; melting point 159°C. At 200°C it decomposes to phenol and carbon dioxide. Salicylic acid is the active ingredient in ASPIRIN and is also used in DYES. It occurs naturally in willow tree bark and oil of wintergreen.

saliva A fluid produced by the SALIVARY GLANDS that helps digestion of food in the mouth. Saliva is over 99 per cent water, and contains an enzyme called salivary AMYLASE that converts STARCH to sugar. It also contains mineral ions to maintain the correct pH for the enzyme, and a sticky MUCIN that helps bind food and make it easier to swallow.

salivary amylase *See* AMYLASE.

salivary gland Any EXOCRINE GLAND in the mouth that produces SALIVA. There are three pairs of salivary glands: sublingual, submaxillary and parotid. *See also* MEROCRINE GLAND.

Salmonella A GENUS of BACTERIA that cause typhoid and paratyphoid fevers and salmonella food poisoning. The bacteria are carried in food (particularly poultry) and water and can be passed on by human carriers in the unhygenic preparation of food. Domestic pets can also be carriers. Food poisoning can be severe but is usually not fatal. Vaccination against typhoid fever is available.

salt Any compound produced by a reaction in which some or all of the hydrogen in an ACID (*see* ACIDIC HYDROGEN) is replaced by a metal or other positive ion. An example is the action of sulphuric(VI) acid on magnesium. Hydrogen is displaced from the acid to produce magnesium sulphate, a salt:

$$Mg + H_2SO_4 \rightarrow MgSO_4 + H_2$$

Salts are also formed when an acid reacts with a base, for example:

$$2HCl + Ca(OH)_2 \rightarrow CaCl_2 + 2H_2O$$

In general,

$$\text{acid} + \text{base} \rightarrow \text{salt} + \text{water}$$

See also ACIDIC SALT, BASIC SALT, NORMAL SALT.

Salting out The process of removing an organic substance from solution in water or ethanol by the addition of a concentrated sodium chloride solution. This increases the POLAR nature of the solvent and reduces the solubility of non-polar compounds.

saltpetre *See* POTASSIUM NITRATE.

samarium (Sm) The element with ATOMIC NUMBER 62; RELATIVE ATOMIC MASS 150.3; melting point 1,073°C; boiling point 1,791°C; RELATIVE DENSITY 7.5. Samarium is one of the more widely occurring of the LANTHANIDE metals. It is used for its catalytic properties and in some FERROMAGNETIC alloys.

sample In statistics, a selection from a larger group that have been measured or studied in greater detail.

Sandmeyer reaction *See* DIAZONIUM SALTS.

sandstone A SEDIMENTARY rock formed by the deposition of layers of sand, fused into rock by the pressure from further layers. Sandstone is porous and permeable and used primarily as a building material.

SA node *See* PACEMAKER.

saponification A process used in the manufacture of SOAP. It consists of the HYDROLYSIS of an ESTER by treatment with a strong ALKALI. The ester is split to yield the ALCOHOL from which the ester was derived and a salt of the constituent FATTY ACID.

saprophyte *See* SAPROTROPH.

saprotroph, ***saprophyte*** An organism that feeds on dead organisms or excrement. Saprotrophs cannot make food for themselves and are a type of HETEROTROPH. Most saprotrophs are bacteria or fungi and they absorb material through their cell walls, some directly and others after breakdown by enzymes that they release onto the food material to cause extracellular digestion.

Saprotrophs play an essential role in recycling elements in dead organisms, making them available again to living organisms, but they can also cause spoilage of food or other materials at great economic loss. Some are useful to humans because of the by-products in their break down of organic material, for example they are used in brewing, baking and yoghurt production.

Compare DETRITIVORE. *See also* DECOMPOSER.

sarcolemma *See* CARDIAC MUSCLE.

sarcoma A malignant TUMOUR of CONNECTIVE TISSUE, for example of bone, cartilage or blood.

sarcomere The repeating unit of cross-striations in a MUSCLE fibre.

sarcoplasm The CYTOPLASM of MUSCLE fibres.

sarcoplasmic reticulum A system of internal membranes found within MUSCLE fibres that controls calcium ion concentration and is important in MUSCULAR CONTRACTION.

satellite Any body in ORBIT around another body. The term usually refers to a natural or artificial satellite that moves in a closed orbit around a planet under the influence of the GRAVITATIONAL FIELD of that planet.

Large numbers of natural satellites have been discovered around the larger planets, whilst the Earth and Pluto are exceptional in having satellites with sizes which are a substantial fraction of the parent planet. The Earth's Moon, the four large satellites of JUPITER (called the GALILEAN SATELLITES) and Titan, the largest satellite of SATURN, are all comparable in size to small planets, but most of the others are far smaller.

Since the 1960s there has been an increasing use of artificial satellites. These are mainly communications satellites, for relaying telephone, radio and television signals around the globe; navigation satellites; military satellites; astronomy satellites, including telescopes that monitor the radio, infrared, ultraviolet and gamma ray emissions from the Solar System and distant galaxies; weather-monitoring satellites; and land resources satellites, used for routine mapmaking and for geological, agricultural and oceanographic purposes. Two special types of orbit are often used, the GEOSTATIONARY ORBIT and the CIRCUMPOLAR ORBIT.

saturated (*adj.*) **1.** Describing a SOLUTION or VAPOUR that can hold no more dissolved or evaporated substance of a specified type. *See also* SOLUBILITY PRODUCT.

2. Describing a MAGNETIC material in which all the DOMAINS are aligned, so the magnetism is as strong as possible.

saturated compound Any organic compound in which the carbon atoms are linked by single COVALENT BONDS only, such as ethane and other ALKANES. They therefore can only react further by SUBSTITUTION REACTIONS. *Compare* UNSATURATED COMPOUNDS.

saturated vapour The state of a VAPOUR whose PARTIAL PRESSURE is equal to its SATURATED VAPOUR PRESSURE, the maximum density that the vapour can have at that temperature. If a saturated vapour is cooled, the liquid will condense.

saturated vapour pressure The PRESSURE in the VAPOUR above a liquid at which the molecules leave and re-enter the liquid at the same rate. The saturated vapour pressure increases with temperature. Once the saturated vapour pressure reaches the pressure of the atmosphere above the liquid, bubbles can form and the liquid will be at its BOILING POINT. *See also* RAOULT'S LAW, SUPERSATURATED.

Saturn The sixth planet in order from the Sun, with an orbital radius of 9.53 AU (1.4 billion km). Saturn is a GAS GIANT like JUPITER, and is the second largest planet in the SOLAR SYSTEM, with a diameter of 110,000 km (8.5 times that of the Earth), and a mass of 5.7 x 10^{26} kg (95 times that of the Earth). Saturn is most famous for its complex RING SYSTEM. It has an atmosphere of hydrogen and methane, similar to Jupiter's. Eighteen satellites are known, of which the largest, TITAN, is large enough to retain its own atmosphere. Saturn orbits the sun every 29 years and rotates on its own axis every 10 hours.

s-block element Any element in GROUPS 1 or 2 of the PERIODIC TABLE, with the outer electrons in an S-ORBITAL.

SBR (styrene-butadiene rubber) An artificial RUBBER that is a COPOLYMER made from BUTADIENE combined with PHENYLETHENE (styrene). It is rubbery in texture similar to natural rubber and is the most important synthetic rubber.

scalar A quantity that has no associated direction such as mass and temperature. *Compare* VECTOR, which is specified by its magnitude and direction.

scale A series of marks along the axis of a graph, or on some other diagram, to show how distance on the diagram relates to the true size of the quantity being represented. For maps and plans, the scale is also often represented as a ratio. A scale of 1:50,000 on a map, for example, indicates that one unit of distance on the map represents 50,000 units of distance on the ground.

scalene (*adj.*) In mathematics, describing a TRIANGLE with sides of different lengths.

scandium (Sc) The element with ATOMIC NUMBER 21; RELATIVE ATOMIC MASS 45.0; melting point 1,540°C; boiling point 2,850°C; RELATIVE DENSITY 3.0. Scandium has some use as a catalyst and in the manufacture of REFRACTORY ceramics.

scanning electron microscope A type of ELECTRON MICROSCOPE developed in the 1960s, in which a fine beam of electrons passes over the surface of the specimen and some electrons are absorbed and others are reflected. Secondary electrons may also be emitted by the specimen, and these and the reflected electrons are amplified to form an image showing the three-dimensional exterior of the specimen on a screen. The RESOLUTION is not as good as in the TRANSMISSION ELECTRON MICROSCOPE (about 10 nm) and the overall magnification is 10–200,000 times. *See also* SCANNING TRANSMISSION ELECTRON MICROSCOPE, SCANNING TUNNELLING MICROSCOPE.

scanning transmission electron microscope (STEM) A type of ELECTRON MICROSCOPE that combines features of both the TRANSMISSION ELECTRON MICROSCOPE and the SCANNING ELECTRON MICROSCOPE to produce a magnification of 90 million times. A fine beam of electrons moves over the specimen but a thin slice is used so that the electrons also pass through it. A computer interprets the electrical signal formed by reflected electrons and those penetrating the specimen, to form an image on a screen.

scanning tunnelling microscope A type of ELECTRON MICROSCOPE invented in 1981 that magnifies an image by passing a small tungsten probe over the surface of the specimen. The tip of the probe may be as fine as one atom, and it moves very close to the specimen so that electrons jump (or tunnel) between the specimen and probe. The magnitude of the electron flow depends on how close the probe is to the specimen, and therefore the contours of the surface can be determined and transmitted to a screen. Images of 100 million times can be obtained, which means individual atoms can be resolved.

scapula In humans, the shoulder blade. *See* PECTORAL GIRDLE.

scatter diagram A GRAPH on which individual data points, each consisting of a measured value for each of two different quantities, are plotted to see how closely a variation in one

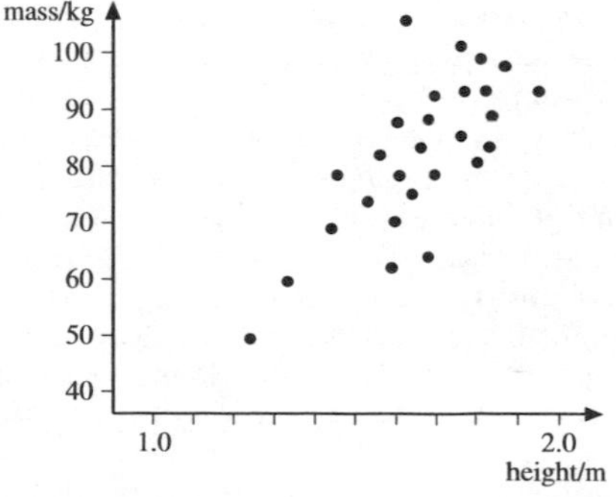

Scatter diagram.

quantity is related to a variation in the other. If the two quantities are completely dependent on one another, the points will lie on a single line or curve. If there is no relationship, they will be scattered randomly over the graph.

Schiff's bases Weakly basic compounds prepared from the reaction of AROMATIC AMINES with ALIPHATIC or aromatic ALDEHYDES and KETONES.

Schiff's reagent A solution that distinguishes ALDEHYDES and KETONES. ALIPHATIC aldehydes and aldose sugars (*see* MONOSACCHARIDES) cause the colourless solution to change to magenta. In AROMATIC aldehydes and aliphatic ketones the colour develops more slowly. Aromatic ketones do not react.

Schrödinger's cat A thought experiment designed to illustrate the strange behaviour of WAVEFUNCTIONS and the philosophical problems with the interpretation of QUANTUM THEORY in PROBABILITY terms. A cat is sealed in a box and is either killed or allowed to survive with equal probability depending on the outcome of some quantum mechanical process, such as the decay of a single RADIOACTIVE particle within one HALF-LIFE. The outcome is not known until the box is opened, so for an external observer the cat has a wavefunction that is a mixture of alive and dead until the observation is made, but alive or dead thereafter. This raises the question of when and how the wavefunction changes and the artificial boundary between an experiment and its observer.

Schrödinger's equation An equation that provides information about the WAVE NATURE OF PARTICLES and describes the behaviour of a quantity called the WAVEFUNCTION. The wavefunction represents a PROBABILITY AMPLITUDE, the square of which is a measure of the probability of finding the particle at a given point.

Schwann cell A specialized GLIAL CELL that is responsible for the formation of the MYELIN SHEATH surrounding some nerve AXONS.

scintillation counter A PARTICLE DETECTOR based on materials, called scintillators, that are particularly efficient at converting the energy deposited by charged particles into visible light. This light can than be channelled along an OPTICAL FIBRE and detected by a PHOTOMULTIPLIER.

scintillator A material that is particularly efficient at converting the energy deposited by IONIZING RADIATION into visible light.

sclera *See* SCLEROTIC COAT.

sclerotic coat, *sclera* The white of the EYE; the tough outer layer of the eye.

sclereid *See* SCLERENCHYMA.

sclerenchyma Plant tissue consisting of thick-walled cells providing strength and support. When mature, the sclerenchyma cells die and the cell contents are lost, leaving only the wall. The CELL WALL contains large deposits of LIGNIN that provides extra strength. Lignin is difficult to digest and so provides protection from attack by many organisms. There are some regions of the cell wall where PLASMODESMATA are present and lignin is not deposited, so forming pits, through which water and dissolved minerals pass from one cell to another. Some sclerenchyma cells are spherical (sclereids) and are found singularly or in small clusters in hard shells of fruit, seed coats, bark and stem CORTEX. Other cells are elongated (fibres) and are found in bundles providing the main supporting tissue of mature stems. Sclerenchyma cells are associated with the vascular tissue (*see* PHLOEM, XYLEM).

SCP *See* SINGLE CELL PROTEIN.

scrapie A disease of sheep affecting their CENTRAL NERVOUS SYSTEM, causing progressive degeneration. Scrapie is characterized by the spongy appearance of the brain after death, due to the presence of numerous holes in the tissue.

The disease is thought to be caused by a 'prion', which is a protein fragment capable of self-replication in animal cells. It is possible that the causal agent switches on a gene that encodes the prion protein.

See also BOVINE SPONGIFORM ENCEPHALOPATHY, CREUTZFELDT– JAKOB DISEASE.

screen 1. A surface onto which an optical image is projected, designed to scatter the received light into the eyes of those viewing the image.

2. A conducting layer designed to protect a device from the influence of nearby electromagnetic fields.

scrotum In male mammals, a skin sac that contains the TESTES and keeps the SPERM cooler than body temperature. The muscles in the wall of the scrotum can contract and relax to bring the sac closer to or further from the ABDOMEN as the temperature varies.

scrubber A vessel in which water is passed through a mixture of gasses to remove a soluble component from the gas mixture. In the production of COAL GAS, for example, water is used in a scrubber to remove ammonia, which is often present as an impurity.

scurvy A deficiency of VITAMIN C in which an abnormal type of COLLAGEN is made, causing weakness of muscles and joints, skin sores and ulcers, bleeding of teeth, gums and other organs due to burst capillaries, and dry skin and hair.

Scyphozoa A CLASS of the PHYLUM CNIDARIA, including the jellyfish.

sea breeze A wind blowing from the sea to the land driven by the CONVECTION CURRENTS produced as the land heats up more quickly than the sea during the day. *See also* WEATHER SYSTEMS.

seam A layer of a particular MINERAL, particularly coal, that is thick enough to be exploited by mining. A coal seam may have a thickness from a few metres to several hundred metres. Seams are found in SEDIMENTARY rock and tend to be roughly horizontal, unless subsequent folding has brought them to the surface, where they can sometimes be exploited by OPEN-CAST MINING.

Searle's bar A method of measuring the THERMAL CONDUCTIVITY of a good conductor by heating one end of a bar of the material whilst keeping the other end at a fixed temperature. The TEMPERATURE GRADIENT can be measured by inserting thermometers into holes drilled in the sample a known distance apart. The rate of heat flow can be found either by heating the bar electrically at a known rate, or by measuring the rate at which energy has to be removed from the cool end of the bar to maintain the temperature at a steady level. If the bar is cooled with water, for example, the heat flow can be found by measuring the flow rate of the cooling water and its temperature rise. To ensure that heat only flows through the sample under test and not to or from the surroundings, the whole experiment should be thermally insulated, or LAGGED.

sebaceous gland In mammals, a gland in the SKIN that produces an oily secretion called sebum, which gives skin and hair its water-resistance and lubrication. Over-secretion of sebum is a cause of acne.

sebum *See* SEBACEOUS GLAND.

secant A function of angle, the reciprocal of COSINE. In a right-angled triangle, the secant of an angle is equal to the hypotenuse of the triangle divided by the side of the triangle adjacent to the angle concerned.

second The SI UNIT of time. One second is defined as being equal to the time taken for 9,192,631,770 oscillations of the electromagnetic radiation produced in a transition between two specified ENERGY LEVELS of an atom of caesium–133.

second filial generation *See* F_2 GENERATION.

secondary cell Former term for a rechargeable CELL.

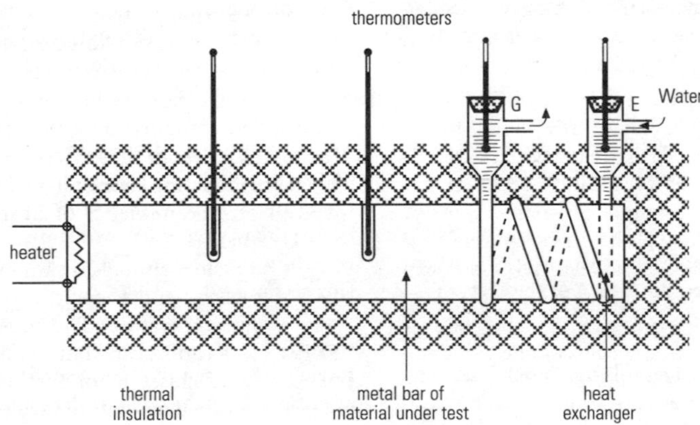

Searle's bar.

secondary coil In a TRANSFORMER or INDUCTION COIL, the coil that supplies energy to a LOAD.

secondary colour Any colour that can be made by mixing two PRIMARY COLOURS. The secondary colours are cyan (blue and green), magenta (blue and red) and yellow (green and red).

secondary emission The release of electrons from an ELECTRODE that is itself struck by high energy electrons. Generally this effect is undesirable, but it is used to advantage in the PHOTOMULTIPLIER.

secondary wavelet A small section of WAVEFRONT used to predict the position of the next wavefront in HUYGENS' CONSTRUCTION.

second law of thermodynamics Any change will bring about an increase in the total ENTROPY of the system if the change is irreversible, or produce no entropy change if it is reversible. Systems that appear to produce order out of chaos, such as living organisms, do not actually result in a net decrease in entropy, as they give out sufficient heat to increase the entropy of their surroundings by an amount that more than compensates for the decrease in entropy of the living organism itself. *See also* ARROW OF TIME, CLAUSIUS STATEMENT OF THE SECOND LAW OF THERMODYNAMICS, KELVIN STATEMENT OF THE SECOND LAW OF THERMODYNAMICS.

secretin A hormone secreted by the SMALL INTESTINE in response to acid CHYME from the STOMACH. Secretin inhibits production of GASTRIC JUICE and stimulates production of PANCREATIC JUICE and BILE. *See also* PANCREAS.

secretion The production and release of a substance, for example HORMONES and ENZYMES, from a cell or specialized GLAND that is useful to the organism. The substance produced can also be called a secretion. *Compare* EXCRETION.

secretory cell A general name for any cell that produces and releases a specific substance. *See also* SECRETION.

sector A piece of a circle, bounded by two radii and an ARC joining them. If the angle between the radii is θ°, then the area *A* of the sector is

$$A = \pi r^2\theta°/360$$

sedimentary (*adj.*) Describing a rock that has formed from the action of pressure forcing together small particles of other rocks that had been broken down by the forces of erosion.

Seebeck effect *See* THERMOELECTRIC EFFECT.

seed The reproductive structure containing the EMBRYO and food stores that develops from a fertilized OVULE in higher plants (ANGIOSPERMS and CONIFERS). The seed is protected by a hard, impermeable outer seed coat or TESTA. After fertilization (*see* DOUBLE FERTILIZATION) the diploid ZYGOTE divides by MITOSIS and develops into the embryo, and the food source for the growing embryo is the ENDOSPERM or the COTYLEDONS.

In flowering plants (ANGIOSPERMS), the seeds are enclosed within a FRUIT that protects them during development. In CONIFERS, the seed is unprotected. When the conditions (such as temperature and water) are favourable, the seed will germinate (*see* GERMINATION) and develop into a new plant. Until the conditions are right, the seed will remain in a state of DORMANCY.

seed crystal A small CRYSTAL added to a SUPERSATURATED solution to start the process of CRYSTALLIZATION. Once the seed crystal is added, ions leave the solution under the influence of the electric field surrounding the ions already in the lattice of the seed crystal.

seed leaf *See* COTYLEDON.

seed plant A plant bearing SEEDS. There are two phyla (*see* PHYLUM) of seed-bearing plants, the CONIFEROPHYTA and ANGIOSPERMOPHYTA. Together, these form the largest group of plants, including most of the Earth's land vegetation, and the structurally most complex plants.

seismic wave A wave travelling through the Earth or along its surface, such as those produce during an EARTHQUAKE. Seismic body waves, which travel through the Earth can be both longitudinal and transverse. The LONGITUDINAL WAVES travel faster than the TRANSVERSE WAVES and are called P-WAVES (for primary). P-waves are the first to arrive after an earthquake. The transverse body waves travel more slowly and cannot pass through the molten core, which will not support this type of wave motion. They are called S-WAVES (for secondary).

There are two types of seismic surface waves, called Love waves (L-waves) and Rayleigh waves, and it is these vibrations that are responsible for most of the damage caused by an earthquake. L-waves are low frequency transverse waves that move along the upper part of the crust. Rayleigh waves travel a little deeper into the crust, causing a rolling wave-like motion of the surface. Seismic waves are detected with an instrument called a SEISMOGRAPH.

seismograph An instrument for detecting EARTHQUAKES. Seismographs can also be used to detect the waves returning from man-made explosions, an important technique in geology. In particular they are used to detect reserves of oil or natural gas, which often become trapped in periclines (dome-shaped structures) of porous rock capped by impermeable (non-porous) rock.

seismology The study of EARTHQUAKES. Seismology gives a good deal of information about the structure of the Earth's crust as SEISMIC WAVES are reflected and refracted from various strata (layers) of rock, and sudden discontinuities (FAULTS) in these strata.

selection pressure *See* NATURAL SELECTION.

selenide Any BINARY COMPOUND containing selenium and a more electropositive element (*see* ELECTROPOSITIVITY). Hydrogen selenide is covalent (*see* COVALENT BOND), whilst the selenides of the ALKALI METALS have IONIC structures. Selenides of the TRANSITION METALS tend to be non-STOICHIOMETRIC compounds.

selenium (Se) The element with ATOMIC NUMBER 34; RELATIVE ATOMIC MASS 79.0; melting point 217°C; boiling point 685°C; RELATIVE DENSITY 4.8. Selenium is a METALLOID, and its ability to conduct electricity on exposure to light has led to its use in some PHOTOCONDUCTIVE CELLS.

self-inductance A measure of the effect in which a changing current in a coil causes an ELECTROMOTIVE FORCE (e.m.f.) in the same coil, tending (by LENZ'S LAW) to oppose the change in current. Such an e.m.f. is sometimes called a BACK E.M.F. The self inductance of a coil is equal to the induced e.m.f. divided by the rate of change of current which produces it. The SI UNIT of self inductance is the HENRY. A rate of change of current $\mathrm{d}I/\mathrm{d}t$ in a single coil with self-inductance L will produce a back e.m.f. of

$$E = L\mathrm{d}I/\mathrm{d}t$$

self-pollination The transfer of pollen from the male part to the female part of the same plant. *See* POLLINATION.

semen The fluid in which SPERM are carried during their ejaculation from the PENIS at copulation or mating. Semen is secreted by the SEMINAL VESICLES, PROSTATE GLAND and COWPER'S GLANDS. It contains alkaline chemicals (to neutralize the acidity of the vagina which would kill sperm), sugars (which nourish the sperm and help make them mobile), MUCUS (in which the sperm swim) and PROSTAGLANDINS (which cause muscular contractions of the UTERUS and OVIDUCTS, so helping the sperm reach the OVUM).

semi-circle One half of a circle, a SECTOR where the angle between the radii is 180°. The area of a semi-circle is $\pi r^2/2$.

semi-circular canal Any one of three tubes in the inner EAR concerned with balance. Semi-circular canals are linked to the COCHLEA by two sac-like parts, the utriculus and the sacculus, both of which contain a fluid called endolymph. If there is movement of the head, the endolymph displaces a gelatinous plate called the cupula that is found within the ampulla, a swollen region found in each canal. This movement of the cupula is detected by sensory hairs and transmitted to the brain by the auditory nerve so that the imbalance can be corrected. The utriculus and sacculus provide information on the position of the body relative to gravity, as well as movement due to acceleration or deceleration. This is achieved by deposits of calcium carbonate called otoliths, which are embedded in a jelly-like substance. These respond to vertical and lateral movements of the head and displace the sensory hairs, thereby transmitting a message to the brain.

semiconductor A material intermediate between a CONDUCTOR and an INSULATOR, which conducts electricity but not very well. By far the most important semiconductor is silicon, the material from which many electronic devices, particularly INTEGRATED CIRCUITS, are made.

Pure semiconductors do not conduct at all at very low temperatures, but thermal vibrations of the LATTICE, or the addition of certain impurities, can make the material conduct. The addition of impurities, a process called DOPING, must be carefully controlled if the semiconductor is to have predictable properties. In the case of silicon, the lattice is held together by each atom forming four COVALENT BONDS with its neighbours. In such a structure there are no free CHARGE CARRIERS. The addition of an impurity with five VALENCE

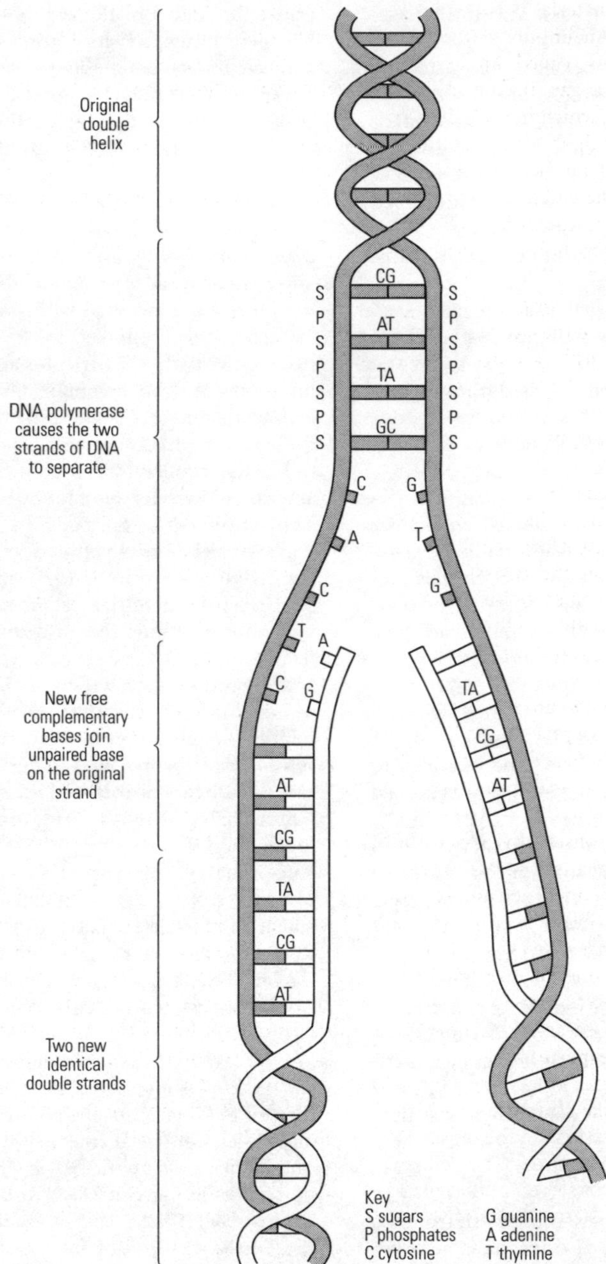

Semi-conservative replication of DNA.

ELECTRONS, called a donor impurity, can release a FREE ELECTRON that can carry charge through the material. Such a material is called an n-type semiconductor. An impurity with only three valence electrons, called an acceptor impurity, will create a gap in the electron structure of the lattice, which may be filled by an electron from a neighbouring bond, In effect, the shortage of an electron, called a hole, moves through the lattice like a positive charge carrier. A semiconductor in which current is carried mostly by holes is called a p-type semiconductor.

The action of thermal vibrations or light will release electrons from the lattice structure, producing ELECTRON-HOLE PAIRS, meaning that semiconductors conduct better at high temperatures, or on exposure to light.

See also BAND THEORY, INTRINSIC SEMICONDUCTOR.

semi-conservative replication The method by which DNA replicates under the control of the enzyme DNA POLYMERASE. During this process DNA polymerase causes the two strands of DNA to separate. The NUCLEOTIDE bases on each strand then join with new, free complementary bases and the strands rejoin. Part of the strand is rejoining while the remaining unpaired bases continue to attract their complimentary nucleotides. The result is four strands making two identical DNA molecules, each consisting of one original strand and one new strand.

Evidence for semi-conservative replication comes from the experiments of Meselsohn and Stahl, who followed the incorporation of the ISOTOPE nitrogen–15 (heavy nitrogen) into DNA of the bacteria *Escherichia coli*.

semi-lunar valve Any one of two crescent-shaped non-return VALVES, one in the AORTA and one in the PULMONARY artery, that prevent blood from re-entering the heart.

semi-metal *See* METALLOID.

seminal vesicle A sac-like structure extending from each of the VAS DEFERENS in the reproductive system of male mammals. The seminal vesicles produce a MUCUS secretion that forms part of the volume of SEMEN. *See also* SEXUAL REPRODUCTION.

seminiferous tubule One of many coiled tubes in the male TESTIS that are lined with germinal EPITHELIUM cells from which SPERM are produced. In between the tubules lie INTERSTITIAL CELLS that secrete the male hormone TESTOSTERONE. The seminiferous tubules merge to form the vasa efferentia, which are a group of small ducts through which mature sperm pass into the EPIDIDYMIS to be stored.

semipermeable membrane A material through which one type of molecule can pass but not another. Typically, such membranes are considered in processes such as OSMOSIS, where the SOLVENT molecules can pass through the membrane, but the SOLUTE cannot.

One explanation of this effect is that the solute molecules are too large too pass through the membrane, which acts as a 'molecular sieve'. However, some membranes still work with IONIC solutions where the solute ions are smaller than the solvent molecules, suggesting that this cannot be the explanation for every case.

See also CELL MEMBRANE, DIALYSIS.

semi-water gas, *Mond gas* A mixture of WATER GAS and PRODUCER GAS formed by mixing air and steam and passing them over hot COKE. If the balance of the gases is right, a reasonable fuel gas can be produced without having to supply any extra energy to maintain the temperature of the coke.

sense organ An organ containing nervous tissue that has specialized RECEPTORS for detecting specific stimuli and converting them into a NERVE IMPULSE, so that an animal can gain information about its surroundings. In humans, the main sense organs are the EYE (which detects light and colour), the EAR (which detects sound and balance), the NOSE (which detects smell), the TONGUE (which detects taste) and small sense organs in the SKIN (which detect temperature, pressure and pain).

Chemoreceptors are receptor cells that detect chemicals. They are involved in taste and smell and are found on the tongue and nose. Olfactory (smell) receptors are linked with taste and are found at the back of the nose. Photoreceptors detect light. Mechanoreceptors detect pressure changes, gravity and vibrations (sound) and there are many in the ear. Thermoreceptors detect temperature changes, for example in the skin. Electroreceptors detect electrical fields (which is important in fish).

Many senses, for example hearing and smell (important for detecting food, danger and mates), are more acutely developed in animals other than humans. Some species see outside the human SPECTRUM, for example insects can see in the ultraviolet range and snakes can see in the INFRARED range. Most mammals, however, cannot distinguish colours. The PINEAL GLAND detects light and keeps a track of daylength and seasons in some animals.

sensory neurone *See* SENSORY SYSTEM.

sensory system The part of the NERVOUS SYSTEM that consists of RECEPTORS collecting information from the internal and external environment, and sensory NEURONES carrying this information to the CENTRAL NERVOUS SYSTEM where it is processed.

sepal A part of a flower that forms the outer layer of the PERIANTH. The collective term for sepals is the calyx. Sepals are derived from modified leaves and are usually green and capable of PHOTOSYNTHESIS. Their main function is to protect the other floral parts when the flower is a bud. In some flowers the sepals are brightly coloured and assist in attracting insects.

septum (*pl. **septa***) Any dividing partition in a plant or animal.

sequence A set of linked algebraic terms, referred to by their position in the sequence, for example a_1, a_2 a_3 etc.

serial logic, *series logic* A DIGITAL electronic system in which only one element of a problem or calculation is dealt with at a time, with the result then being stored in memory for later use. The high speed at which MICROPROCESSORS work is due to the very short time they take to complete each step, but further gains have been achieved by moving to PARALLEL PROCESSING.

series 1. (*mathematics*) A sequence of mathematical quantities added together, each calculated by a set of rules determined by its position in the series. An example is a power series, where the *n*th term in the series contains the *n*th power of some variable. An example of a power series is

$$y = a_0 + a_1x + a_2x^2 + a_3x^3 + \ldots + a_nx^n,$$

where all the *a*'s are constants. *See also* PROGRESSION.

2. (*physics*) In an electric CIRCUIT, devices are said to be in series if they are connected one after another, so that all the CURRENT flowing through one device has to flow through the next one. The current is the same in each element and the VOLTAGE across the combination is equal to the sum of the voltages across the individual circuit elements.

series logic *See* SERIAL LOGIC.

series wound Describing a D.C. MOTOR in which the ARMATURE and the FIELD COILS are in SERIES.

serotonin A substance present in the brain, intestines and blood platelets that acts as a NEUROTRANSMITTER and also induces the constriction of blood vessels.

serous membrane A thin membrane lining a closed body cavity, such as the PLEURAL CAVITY and PERITONEAL CAVITY.

Sertoli cells The cells lining the SEMINIFEROUS TUBULES that protect and nourish immature SPERM cells. *See* TESTIS.

serum The fluid that remains after blood has been allowed to clot and the clot removed. Serum is a watery, yellowish liquid, the same as blood PLASMA except for the removal of the clotting substances. It contains many dissolved proteins, including ANTIBODIES, sugars, fats, HORMONES and SALTS.

Antiserum is blood serum containing antibodies to a specific ANTIGEN, and can be made artificially by administrating the required ANTIGEN. This can be useful in protection against some diseases (*see* VACCINATION).

set In mathematics, a group of objects linked together by some common property, such as the set of all REAL NUMBERS, or the set of all people with red hair. Sets are often denoted by curly brackets, thus the set of all people with red hair might be denoted by {red hair}. *See also* MAPPING, SUBSET, UNIVERSAL SET, VENN DIAGRAM.

severe combined immune deficiency *See* IMMUNODEFICIENCY.

sewage disposal The removal of any water-borne waste products passing through sewers, including human excrement, household waste and industrial waste. The sewage passes through sewers to a sewage works to be treated before its discharge into rivers and the sea. Untreated (or raw) sewage causes water POLLUTION and EUTROPHICATION.

In the treatment of sewage, large pieces of debris are first filtered out or broken down and heavier inorganic material (detritus and grit) is allowed to deposit out, which can then be dumped. The remaining sewage flows into large tanks for primary sedimentation, where sand, silt and organic material settles out, forming sludge that can be digested into simpler compounds and semi-dried to be used as fertilizer or dumped at sea. Methane gas is produced during the digestion of sludge, which can be used to generate electricity, so potentially making sewage treatment works self-sufficient. The remaining sewage, with most solid material removed, is called effluent and is removed for further break down of dissolved organic materials by a variety of aerobic micro-organisms, for example urea is broken down into nitrates by *Nitrosomonas* and *Nitrobacter* (*see* NITRIFICATION, AMMONIFICATION). These micro-organisms are allowed to settle out, forming HUMUS that is digested with the sludge. The remaining effluent is then safe to pass into rivers and seas.

Treatment of sewage successfully removes complex chemicals like DDT, organic compounds and most PATHOGENS, but some potential pollutants, such as phosphorus and copper, are only partially removed and small numbers of some pathogens may survive (for example those causing paratyphoid and dysentery).

sex determination The process by which the sex of an organism is determined. In humans, like many other species, sex determination is dependent on two sex CHROMOSOMES, the X-chromosome and the Y-chromosome. The Y-chromosome is shorter than the X-chromosome and lacks many of the GENES carried on the X-chromosome. A pair of identical X-chromosomes (XX: homogametic) produces a female, whereas a non-identical pair of an X- and Y- chromosome (XY: heterogametic) produces a male. Maleness is caused by a single gene (of 14 BASE PAIRS) on the Y-chromosome that is present but inactive on the X-chromosome.

Sex determination is different in some other species. In birds, fish, butterflies and reptiles the males are mostly XX and the females XY. In other insects the male has no Y chromosomes and is XO. In some fish and reptiles, environmental factors such as temperature can affect sex determination.

sex linkage The association of a particular characteristic with the sex of an organism. Sex-linked characteristics are carried by GENES present on the sex CHROMOSOMES (HETEROSOMES), which may have nothing to do with the sex of an organism but are linked to it by their position (*see* LINKAGE). The X-CHROMOSOME carries many such genes. Two examples of X-linked genes in humans are those causing HAEMOPHILIA and red/green colour blindness. Both genes are RECESSIVE and are expressed almost exclusively in males because the males usually have only one X and one Y chromosome (XY; *see* SEX DETERMINATION). If the recessive ALLELE is present on the X-chromosome then it is expressed because there is no possibility of it being masked by a DOMINANT allele on another X-chromosome. For haemophilia or red/green colour blindness to occur in females (usually XX), both X-chromosomes would have to carry the recessive allele, which is very rare. Usually one of the female X-chromosomes has the dominant allele that masks the recessive one. Such a female is called a carrier, because she can pass the recessive allele onto her offspring, where it may be expressed in a male.

Dominant mutant X-linked genes are unusual but can occur, for example the absence of incisor teeth, and these are more often seen in females because there is twice the chance of it occurring. Genes linked to the Y-chromosome are rare and will only occur in males, where they will be inherited whether they are dominant or recessive.

See also MENDEL'S LAWS.

sexual reproduction The production of offspring, that requires the union of two GAMETES usually from different individuals, that are genetically different from either of the parents. Most organisms (except BACTERIA and CYANOBACTERIA) show some form of sexual reproduction. The gametes are usually female ova (eggs; *see* OVUM) and male SPERM. Some primitive organisms, for example earthworms and sponges, are HERMAPHRODITES, with one individual producing both gametes, although cross-fertilization still often occurs. The genetic VARIATION provided by sexual reproduction allows for adaptation to changed environmental conditions (and therefore greater chance of survival; *compare* ASEXUAL REPRODUCTION). However, the rate of reproduction is slower since a partner has to be found before the process of mating (copulation or coitus) can begin.

In some animals, the breeding cycle is affected by seasonal changes such as a rise in temperature. In birds there is seasonal growth of the GONADS during the spring. These control

Human male reproductive system.

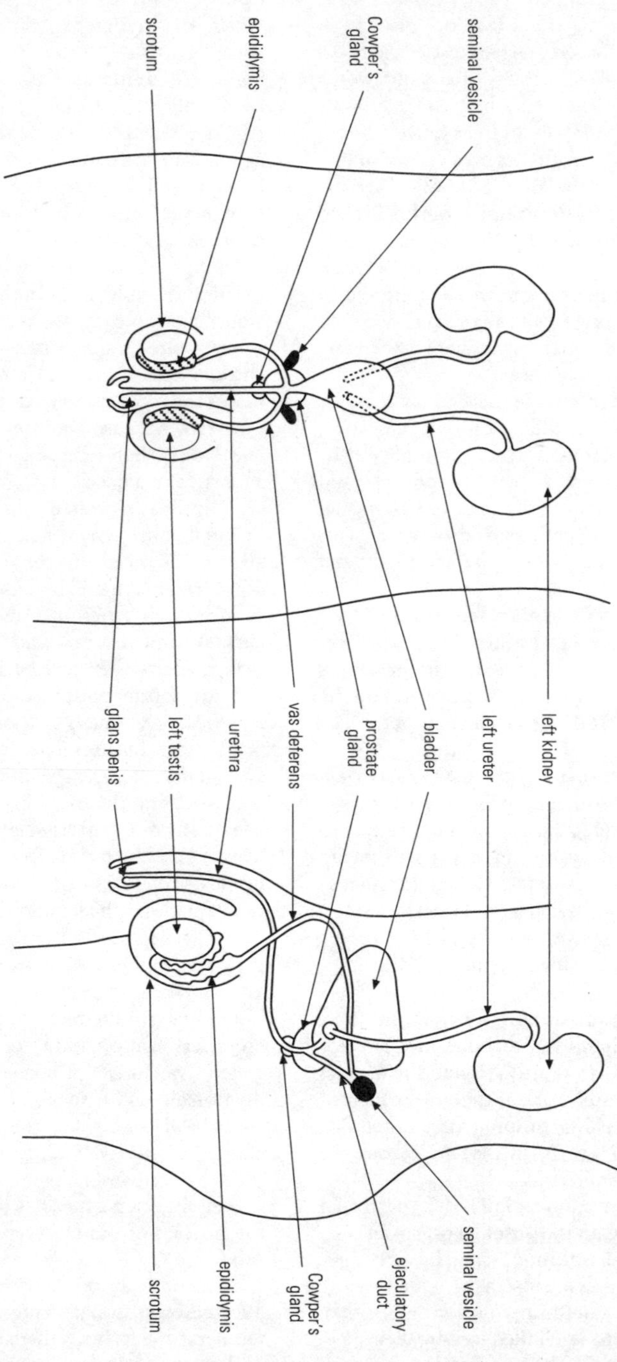

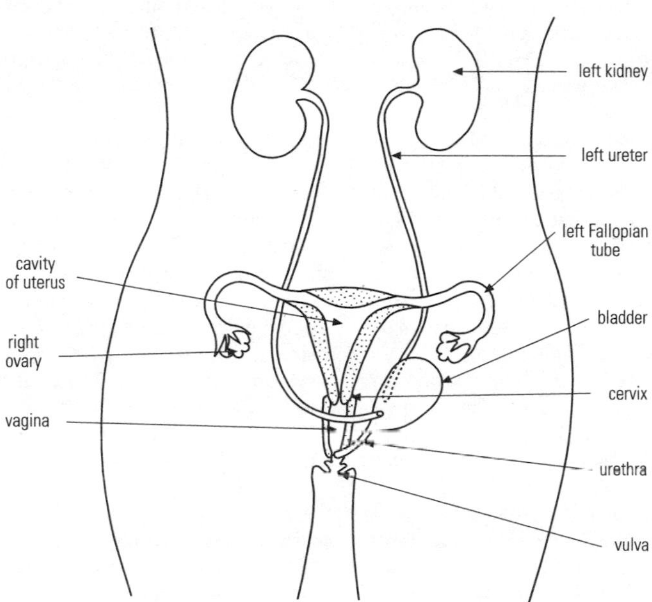

Human female reproductive system.

mechanisms optimize the chances of survival of the young. In some non-human mammals, the female is only receptive to breeding at certain times in her OESTROUS CYCLE. In these species courtship behaviour by the males is used to determine whether a female is receptive or not, judged by her behavioural response. In some species, for example cats and rabbits, the act of mating stimulates OVULATION.

Many species recognize a sexually mature individual by their development of secondary sexual characteristics, such as bright feathers in birds, growth of antlers in deer, the mane of a lion. In humans secondary sexual characteristics include the growth of pubic hair in both sexes, growth of facial hair and deepening of the voice in males. Human females develop breasts and wider hips.

In mammals, sexual stimulation causes an increase in the blood supply to the GENITALS of both male and females, in males causing the PENIS to become hard and erect. Stimulation of sensory cells at the tip of the penis cause muscular contractions that move the sperm from the VAS DEFERENS into the URETHRA, where they mix with the other constituents of SEMEN. The contractions of the penis cause ejaculation (expulsion of the semen) and sperm are propelled through the CERVIX and into the UTERUS, where they swim to the OVIDUCTS and if successful fertilize an egg (*see* FERTILIZATION). Ejaculation in humans is accompanied by a pleasurable feeling called orgasm, which the female can also experience by the contraction of an equivalent set of muscles.

See also ACROSOME REACTION, ASEXUAL REPRODUCTION, DOUBLE FERTILIZATION, FERTILIZATION, MENSTRUAL CYCLE, POLLINATION, PREGNANCY.

shadowmask In a CATHODE RAY TUBE for a COLOUR TELEVISION system, a metal plate with many small holes through which electrons can pass to reach only certain areas on the screen. Electrons starting at one of the three ELECTRON GUNS can only hit one type of PHOSPHOR on the screen.

shale A SEDIMENTARY rock formed by the deposition of layers of mud, fused into rock by pressure from further layers.

shear 1. (*physics*) The deformation of a material in which two parallel surfaces move by differing amounts in a direction parallel to the surfaces; that is, the two surfaces slide over one another. *See also* SHEAR STRAIN, SHEAR STRESS.

2. (*mathematics*) A geometrical operation that maps every point onto another point by a TRANSLATION parallel to a specified line, but with the size of the translation proportional to the distance of the point from the line.

shear modulus An ELASTIC MODULUS describing the stiffness of a material under a shearing force. It is equal to the SHEAR STRESS divided by the SHEAR STRAIN.

shear strain A measure of the deformation produced by a SHEAR. The relative distance moved by two parallel planes in the material undergoing a shear divided by the separation between those planes. *See also* SHEAR MODULUS.

shear stress A measure of the strength of a force producing a SHEAR. The size of the force divided by the surface area parallel to the force over which the force acts. *See also* SHEAR MODULUS.

sheath In plants, the base of a leaf. *See* LEAF.

shell A series of atomic ORBITALS of roughly similar energies. Completing the filling of one shell and starting to fill the next results in a sudden change in chemical properties, atomic size, IONIZATION ENERGY etc. This represents the start of a new PERIOD in the PERIODIC TABLE. The shells are labelled K, L, M, N, O, P. The K-shell is closest to the nucleus and can hold two electrons, the next shell is called the L-shell and holds a further eight electrons. The *n*th shell out from the nucleus can hold $2n^2$ electrons.

Elements with a full outermost shell of electrons are particularly stable; these are the NOBLE GASES. Elements that have a few electrons beyond a full shell will tend to lose them to form positive ions (for example sodium, which has 11 electrons, one beyond a full shell). Those with a few electrons too few, will tend to gain electrons to complete a shell, forming negative ions (for example fluorine with nine electrons, one short of a full shell). Those that are not close to a full shell tend to form COVALENT BONDS (carbon for example, with six electrons, four short of a full shell, will form four covalent bonds) or else exhibit more complex behaviour, for example the TRANSITION METALS show multiple VALENCY.

The shell model does not take full account of the energy differences within a single shell and so does not fully account for the behaviour of the elements of higher ATOMIC NUMBERS, where a full consideration of the orbitals (sometimes called sub-shells) must be used, though even for heavy elements it provides a useful description of their X-ray spectra (*see* SPECTRUM), which depend on the motion of electrons in the innermost shells.

shell model A model of the structure of the atomic nucleus that treats individual protons and neutrons as existing in ENERGY LEVELS, similar in pattern to those occupied by electrons in an atom. The shell model works well for small nuclei, but the greater complexity of the interactions in the nucleus mean that the model breaks down for larger nuclei, for which the LIQUID DROP MODEL is more often used.

shift register A series of BISTABLES used with their clock inputs activated together and the set and reset inputs of one bistable connected to the outputs of the next bistable in the line. Digits can be fed in one after another, and the clock pulses cause them to move along the register from one bistable to the next. This is the way in which numbers are stored in computers for short-term memory, and how data can be entered step by step, as in a pocket calculator. *See also* CLOCKED BISTABLE.

SHM *See* SIMPLE HARMONIC MOTION.

shock wave A sudden change in pressure in a fluid, propagating as a LONGITUDINAL WAVE, such as that produced by an aircraft travelling though air at a SUPERSONIC speed. Such a shock wave in air is heard as a SONIC BOOM.

short-circuit A circuit with a low resistance, such that a dangerously large current may flow, possibly limited only by the INTERNAL RESISTANCE of the power supply.

short period Any of the first three PERIODS in the PERIODIC TABLE, which do not contain any TRANSITION METALS. The short periods are periods 1, 2 and 3, and involve elements with electrons only in S-ORBITALS and P-ORBITALS.

short-sightedness, ***myopia*** A common defect of vision in which the eye lens is too strong for the size of the eyeball. Nearby objects can be seen clearly, but the sufferer cannot see distant objects clearly. It can be corrected using a DIVERGING LENS in front of the eye, effectively weakening the eye lens.

short wave RADIO WAVES with a wavelength from 200 m to 10 m. Short wave radio signals (and to some extent MEDIUM WAVE) can be reflected from an upper layer of the Earth's atmosphere called the IONOSPHERE, which is IONIZED by radiation from the Sun. Reflected radio waves can be used for long-distance communication, but the degree of reflection varies with the state of the ionosphere, which is in turn affected by the time of day, time of year and the activity of the Sun. Short wave communications systems are increasingly being replaced by systems based on communications satellites.

shoulder girdle *See* PECTORAL GIRDLE.

shower A sudden period of rain, often heavy and with large droplets, lasting only a few minutes and produced by CONVECTIVE activity within CUMULUS or CUMULONIMBUS clouds. *See also* WEATHER SYSTEMS.

shunt Former term for PARALLEL as applied to electric circuits. Still used to denote a RESISTOR connected in parallel with a meter to produce an AMMETER of lower sensitivity and lower RESISTANCE.

shunt-wound (*adj.*) Describing a D.C. MOTOR in which the ARMATURE and the FIELD COILS are in PARALLEL.

shutter A device designed to allow light to enter for only a short period of time. Used in a CAMERA to control the time for which light is able to reach the film.

sickle-cell disease An inherited blood disorder in which the RED BLOOD CELLS are sickle-shaped and very fragile, and are therefore easily lost from the circulation, causing anaemia and possibly death. Sickle-cell disease is most common in people of black African origin but is also found in people from north-east India and the east Mediterranean.

The disease is caused by a point MUTATION in HAEMOGLOBIN. Substitution of a single NUCLEOTIDE causes the wrong amino acids to be incorporated into two of the POLYPEPTIDE chains making up the haemoglobin molecule. In the HETEROZYGOUS state not all cells are sickle-shaped and the anaemia is less debilitating than in the HOMOZYGOUS state.

In regions of Africa where death from MALARIA is a threat, the heterozygous sickle-cell condition is an advantage because the sickle-shaped cell is less easily invaded by the malarial parasite.

sideband A range of FREQUENCIES on each side of the CARRIER WAVE frequency in any MODULATED radio wave. Sidebands are produced by beats formed between the carrier frequency and the modulating frequency and account for the BANDWIDTH needed to transmit a radio signal. In an amplitude modulated signal (*see* AMPLITUDE MODULATION) the sidebands extend on each side of the carrier frequency by an amount equal to the highest frequency present in the modulating signal. *See also* SINGLE SIDEBAND.

sidereal day *See* DAY.

siemens (S) The SI UNIT of CONDUCTANCE. An object with a conductance of one siemens has a RESISTANCE of one OHM.

Siemens-Martin process The process of producing steel from a mixture of steel scrap and haematite (an ore containing iron oxide) in an OPEN HEARTH FURNACE. The iron oxide oxidizes some of the carbon in the steel.

sieve plate *See* PHLOEM.

sievert (Sv) The SI UNIT of radiation DOSE EQUIVALENT. A dose equivalent of one sievert will have the same effect on living tissue as a dose of BETA PARTICLES depositing an energy of one JOULE per kilogram.

sieve tube *See* PHLOEM.

sigma-bond (**σ-bond**) A COVALENT BOND that has an electron distribution symmetrical about the line joining the bound atoms. Single covalent bonds are sigma bonds. *See also* PI-BOND.

sigma particle (Σ^+, Σ^0, Σ^-) Any one of a family of three BARYONS, one positive, one neutral and one negative, all containing a STRANGE QUARK and having masses around 1.3 times the mass of a PROTON. The charged particles decay via the WEAK NUCLEAR FORCE with a HALF-LIFE of 0.8×10^{-10} s for the positive particle and 1.5×10^{-10} s for the negative particle. The neutral SIGMA PARTICLE decays to the LAMBDA PARTICLE by the ELECTROMAGNETIC FORCE, with a half-life of just 5.8×10^{-20} s.

sign The expression of whether a number is greater or less than zero. Numbers greater than zero are called positive (+), numbers less than zero are negative (–).

signal Any form of energy, particularly an electrical voltage or a modulated electromagnetic wave, used to convey information.

significant figure A digit in a number that is important in determining the size of a number rather than simply locating the other digits in their correct places. In the number 0.015, the 1 and 5 are significant figures, but the zeros are not – they serve only to put the one and five in the correct place after the decimal point. In the number 2,300 the zeros may be significant, if the number is meant as distinct from 2,301, but are not significant if the number is only approximate, expressing a value between 2,250 and 2,350. The number of significant figures in a number is a measure of how accurately it claims to represent a particular quantity.

silane 1. (SiH_4) A colourless gas; melting point –184°C; boiling point –112°C. Silane can be produced by the reaction of magnesium silicide with acids:

$$Mg_2Si + 4HCl \rightarrow SiH_4 + 4HCl$$

Silane is a powerful REDUCING AGENT, and ignites spontaneously on contact with air:

$$SiH_4 + 2O_2 \rightarrow SiO_2 + 2H_2O$$

It is used in the semiconductor industry to deposit silicon in the manufacture of INTEGRATED CIRCUITS.

2. A generic term for any compound containing silicon and hydrogen, with the general formula Si_nH_{2n+2}, for example disilane, Si_2H_6. They are all highly unstable.

silica A common mineral, composed mostly of SILICON DIOXIDE, SiO_2. Sand is composed mostly of fine grains of silica.

silica gel A water-absorbent material made by drying a GEL of sodium silicate in water. The dried gel will absorb moisture from the air, and can be regenerated by heating. Cobalt chloride is often included with the gel as an INDICATOR, turning from blue to pink as the gel hydrates.

silicate Any compound of SILICA, SO_2, and one or more metal ions. Natural silicates form the major components of most rocks.

silicon (Si) The element with ATOMIC NUMBER 14; RELATIVE ATOMIC MASS 28.1; melting point 1,410°C; boiling point 2,355°C; RELATIVE DENSITY 2.3. Silicon is widespread in the Earth's crust and core. It is a METALLOID and a SEMICONDUCTOR.

The most important use of silicon is in the electronics industry, where relatively small amounts of highly pure silicon are used for the manufacture of INTEGRATED CIRCUITS (silicon chips). The production of large single crystal silicon wafers has been at the forefront of many recent developments in materials technology.

silicon carbide (SiC) A black solid; melting point 2,700°C; decomposes on further heating; RELATIVE DENSITY 3.2. Silicon carbide is made by heating silicon oxide with carbon:

$$SiO_2 + 2C \rightarrow SiC + CO_2$$

Silicon carbide is a very hard material and is widely used in abrasives.

silicon chip A colloquial term for INTEGRATED CIRCUIT.

silicon controlled rectifier *See* THYRISTOR.

silicon dioxide (SiO_2) A colourless glassy solid; melting point 1,610°C; boiling point 2,230°C; RELATIVE DENSITY 2.3. Various impure forms of silicon dioxide are extremely common in the Earth's crust. Silicon dioxide is widely used in the manufacture of glass.

silicones Organic POLYMERS derived from silicon. They are used in greases, sealing compounds, resins, synthetic rubber.

silk A natural protein fibre produced by modified SALIVARY GLANDS of silkworms and other ARTHROPODS.

Silurian A PERIOD of geological time during the PALAEOZOIC ERA. It extended for about 45 million years from the end of the ORDOVICIAN period to the DEVONIAN period (about 395 million years ago). The first fish evolved during this period, and at the end of the Silurian period, the first land plants appeared.

silver (Ag) The element with ATOMIC NUMBER 47; RELATIVE ATOMIC MASS 107.9; melting point 962°C; boiling point 2,212°C; RELATIVE DENSITY 10.5. Silver is easily worked and TARNISHES fairly slowly, which has lead to its widespread use in the manufacture of jewellry and other ornamental work. It is also a very good conductor of heat and electricity, having the lowest RESISTIVITY of any metal.

Silver HALIDES are decomposed by the action of light, and photographic developers can be used to amplify this change, enabling small amounts of light to control large changes. This chemistry is the basis of the photographic process.

silver bromide (AgBr) A pale yellow solid; melting point 432°C; decomposes on further heating; RELATIVE DENSITY 6.5. Silver bromide may be precipitated from other silver salts, for example:

$$AgNO_3 + KBr \rightarrow AgBr + KNO_3$$

Silver bromide is sensitive to light, and a suspension will slowly decompose to give a black precipitate of finely divided silver. It is less light sensitive than silver iodide, but more sensitive than silver chloride.

silver chloride (AgCl) A white solid; melting point 455°C; boiling point 1,350°C; RELATIVE DENSITY 5.6. Silver chloride can be precipitated from other silver salts, for example:

$$AgNO_3 + KCl \rightarrow AgCl + KNO_3$$

It decomposes slowly on exposure to light, a suspension of silver chloride slowly turning black due to the production of a fine precipitate of metallic silver. Silver chloride will dissolve readily in dilute ammonia due to the formation of the silver diamine complex, $[Ag(NH_3)_2]^+$.

silver iodide (AgI) A yellow solid; melting point 556°C; boiling point 1,506°C. Silver iodide can be precipitated from other silver salts, for example:

$$AgNO_3 + KI \rightarrow AgI + KNO_3$$

It is the most light sensitive of the silver halides, and rapidly decomposes on exposure to light, a fact that has led to its widespread use in photographic emulsions.

silver nitrate ($AgNO_3$) A colourless solid; melting point 212°C; decomposes on further heating; RELATIVE DENSITY 4.3. Silver nitrate is important as it is the only soluble salt of silver. It is widely used in the photographic industry to prepare silver iodide. It is also used in QUALITATIVE ANALYSIS to test for halides, forming a white or yellow precipitate, soluble in dilute ammonia in the case of a chloride, soluble in concentrated ammonia for a bromide, and insoluble in ammonia for an iodide.

simple harmonic motion (SHM) An oscillating motion of an object or system about a fixed point such that the acceleration of the object is always directed to the fixed point and is proportional to the displacement from the fixed point.

If acceleration is a, the displacement is x and the angular frequency is ω, then

$$a = -\omega^2 x$$

The – sign indicates that the acceleration is directed towards the fixed point. Simple harmonic motion is isochronous (period does not depend on the amplitude).

See also OSCILLATION.

simple microscope A CONVERGING LENS used as a magnifying glass to produce an enlarged VIRTUAL IMAGE. *See also* COMPOUND MICROSCOPE, MICROSCOPE.

simple pendulum A mass (called a bob) swinging on the end of a string or a pivoted rigid support. A simple pendulum is isochronous (period does not depend on amplitude) for small oscillations – the period increases for larger swings. The period of oscillation of a simple pendulum is also independent of the mass of the bob. Historically, the simple pendulum was used as the timekeeping element in many clocks, though PIEZOELECTRIC quartz crystals are now used.

For a simple pendulum of length l, the period is

$$T = 2\pi(l/g)^{1/2}$$

See also CONICAL PENDULUM, OSCILLATION.

simplex (*adj.*) Describing a system, such as a two-way radio link using a single frequency, in which communications cannot be sent and received at the same time. *Compare* DUPLEX.

simultaneous equation Any member of a set of equations that are all required to be true at the same time. If the number of different equations is equal to the number of independent variables (*see* FUNCTION) they contain, it may be possible to find a single set of values of the independent variables that satisfy all the equations. If the number of equations is less than the number of independent variables, there will be a range of values for which the equations are satisfied. If there are more equations than variables, it is generally not possible to find a set of values that satisfies all of them.

sine A function of angle. In a right-angled triangle, the sine of an angle is equal to the length of the side of the triangle opposite to the angle divided by the HYPOTENUSE.

sine wave A wave or oscillation the value of which varies smoothly with time in a way that can be described by the SINE function, such as SIMPLE HARMONIC MOTION.

single cell protein (SCP) A food source rich in proteins and vitamins that is manufactured by the large scale FERMENTATION of a variety of raw materials by different micro-organisms, then dried and sold as an animal or human food. As well as being a good healthy food SCP is theoretically economical to make because the raw materials used can be agricultural or industrial waste, alcohol, sugars or petroleum chemicals. The micro-organisms used are varied and include fungi, yeasts, algae and bacteria.

One example of an SCP is Pruteen, manufactured as an animal food. The manufacture of Pruteen uses methanol, which is a waste product of other processes, making it cheap to make, and the aerobic bacterium *Methylophilus methylotrophus* ferments this to yield the tasteless, cream-coloured Pruteen. A human SCP is mycoprotein manufactured by the action of the fungus *Fusarium graminearum* on flour waste.

The success of SCP has been limited because other food surpluses, for example butter and grain mountains, make it unnecessary and greater demand would make it less economical because not enough raw materials would be available through waste.

single circulation *See* CIRCULATORY SYSTEM.

single sideband (ssb) A form of amplitude modulated RADIO transmission in which the BANDWIDTH required is reduced by transmitting only one of the two SIDEBANDS produced when a CARRIER WAVE is amplitude modulated. *See also* AMPLITUDE MODULATION.

sinoatrial node *See* PACEMAKER.

sinusoid A channel or cavity in certain organs, such as the liver, where blood mixes and allows exchange of materials between the blood and the tissue it is supplying.

sinusoidal (*adj.*) Describing a SINE WAVE.

site of special scientific interest (SSSI) An area of land in the UK worthy of protection due to the animals, plants or geological features present there. *See also* CONSERVATION.

SI unit The unit of measurement in the internationally agreed METRIC SYSTEM (Système International). In any physical equation, if all quantities are substituted into the equation in SI units, the result will also be in SI units. All units in the SI system are expressed in terms of seven base units and two supplementary units. The base units are METRE (length), SECOND (time), KILOGRAM (mass), AMPERE (CURRENT), KELVIN (TEMPERATURE), MOLE (amount of substance) and CANDELA (LUMINOUS INTENSITY). The supplementary units are the RADIAN (ANGLE) and STERADIAN (SOLID ANGLE).

Any quantity that cannot be expressed directly in terms of one of these units can be expressed in terms of a derived unit, such as the metre per second for velocity. Eighteen of the derived units are given special names. The derived units with special names are NEWTON (FORCE), PASCAL (PRESSURE), JOULE (ENERGY), WATT (POWER), COULOMB (CHARGE), FARAD (CAPACITANCE), OHM (RESISTANCE), SIEMENS (CONDUCTANCE), VOLT (POTENTIAL DIFFERENCE), HERTZ (FREQUENCY), TESLA (MAGNETIC FIELD strength), WEBER (MAGNETIC FLUX), HENRY (INDUCTANCE), LUMEN (LUMINOUS FLUX), LUX (ILLUMINANCE), BECQUEREL (ACTIVITY), GRAY (radiation DOSE) and SIEVERT (radiation DOSE EQUIVALENT).

skeletal muscle *See* VOLUNTARY MUSCLE.

skeleton A rigid structure in animals that provides support and protects internal organs. A skeleton may be internal (endoskeleton), as in vertebrates, and made of BONE and CARTILAGE that provide attachment points for MUSCLES in addition to support and protection of vital organs (such as the skull and ribs). The axial skeleton refers to the head and body, while the appendicular skeleton refers to the limbs (including the hips and shoulder bones). In some vertebrates, for example sharks, the entire skeleton is made of cartilage to provide extra flexibility.

The skeleton can also be external (EXOSKELETON), as in ARTHROPODS, providing the same function as above. A third type of skeleton is a HYDROSTATIC SKELETON, as found in invertebrates such as earthworms, where the body cavity is full of fluid.

skin The external body covering of vertebrates that consists of an outer EPIDERMIS and an inner DERMIS. The epidermis consists of several layers, the surface of which is a tough waterproof layer of dead cells that provides protection and prevention of water loss. This is constantly worn away and replaced by the living cells of the underlying tissue. The deepest layer of epidermis is called the MALPIGHIAN LAYER, which determines skin colour and protects the tissues beneath from ultraviolet light. The dermis contains blood vessels, hair follicles, nerves, SEBACEOUS GLANDS and SWEAT GLANDS embedded in CONNECTIVE TISSUE. Beneath the dermis is an insulating layer of subcutaneous fat that is also a long-term food reserve. The study of skin and its disorders is called dermatology. *See also* SENSE ORGAN.

slab magnet A short fat BAR MAGNET, with the POLES on the larger faces.

slag Non-metallic impurities released in smelting metals. These impurities are often removed by the addition of a suitable material, such as calcium carbonate, which reacts to form a molten layer on top of the metal. The term slag particularly refers to impurities extracted from iron ore and produced as a waste product in a BLAST FURNACE. The waste material is mostly in the form of SILICATES and SILICA, which combine with the calcium in the calcium carbonate to form calcium silicate.

slaked lime *See* CALCIUM HYDROXIDE.

sleet SNOW that melts before it reaches the ground. *See also* WEATHER SYSTEMS.

sliding filament theory *See* MUSCULAR CONTRACTION.

slip plane In a metallic CRYSTAL, a surface along which the layers of atoms can move relatively easily. Thus a metal crystal is more DUCTILE when being pulled along a slip plane than at an angle to one. The POLYCRYSTALLINE nature of many metal samples makes the presence of slip planes less obvious than the CLEAVAGE PLANES in non-metallic crystals.

slip ring A metal ring attached to a rotating machine, such as the ARMATURE of an ALTERNATOR, to allow current to be carried in or out using brushes.

small intestine Part of the DIGESTIVE SYSTEM, located after the STOMACH. In humans the small intestine consists of a 6-m long, 4-cm diameter, coiled muscular tube situated in the lower abdomen. It is responsible for DIGESTION of food and absorption of the products. It consists of the DUODENUM, JEJUNUM and ILEUM. The duodenum is a short length of intestine where most digestion occurs. The ileum is concerned mostly with absorption and is specialized for this. The jejunum is not anatomically distinct from the ileum.

The inner lining of the walls of the small intestine are folded to increase their surface area, and consists of projections called VILLI, which contain smooth muscle fibres to allow repeated relaxation and contraction to help food and enzyme mixing. When food enters the duodenum, it mixes with BILE and PANCREATIC JUICES that help to neutralize the acid CHYME from the stomach and contain a variety of enzymes needed for the breakdown of proteins, fats, carbohydrates and nucleic acids. Further digestive juices are made by the intestinal wall itself. Some non-digestive enzymes are also made that activate other enzymes entering from the PANCREAS. The contents of the small intestine are passed along by PERISTALSIS.

See also INTESTINAL JUICES, SECRETIN.

smelt (*vb.*) To extract a metal from its ORE by heating in combination with a suitable REDUCING AGENT.

smoke detector A device for detecting small particles, particularly smoke, in the air. They are used in fire alarm systems. Many smoke detectors operate by measuring the IONIZATION current produced by a source of ALPHA PARTICLES. Smoke absorbs the alpha particles, reducing the ionization current.

smoothing The removal of changes in a DIRECT CURRENT supply. A RECTIFIER, for example, will provide a very unsteady voltage from an ALTERNATING CURRENT supply. A CAPACITOR can be used to even out these variations, charging during the peaks of the supply voltage and discharging to fill in the gaps.

smooth muscle *See* INVOLUNTARY MUSCLE.

Snell's law In REFRACTION, the SINE of the ANGLE OF INCIDENCE divided by the sine of the ANGLE OF REFRACTION is equal to a constant for a given pair of materials. This constant is called the REFRACTIVE INDEX:

$$\sin i / \sin r = n$$

where i is the angle of incidence, r is the angle of refraction and n is the refractive index of the material.

snow Frozen water, in a hexagonal crystal form, that falls to the Earth's surface from clouds when water condenses at temperatures below its freezing point. *See also* WEATHER SYSTEMS.

soap A cleansing agent consisting of a mixture of sodium salts of various FATTY ACIDS, usually sodium stearate, sodium oleate and sodium palmitate. It is made by the action of a strong ALKALI (for example sodium hydroxide or potassium hydroxide) on animal or vegetable fats, by the process of SAPONIFICATION.

soda A term now largely obsolete, used in the common names of many sodium compounds, such as caustic soda, which is sodium hydroxide, and washing soda (sodium carbonate).

soda water *See* CARBONIC ACID.

sodium (Na) The element with ATOMIC NUMBER 11; RELATIVE ATOMIC MASS 23.0; melting point

98°C; boiling point 892°C, RELATIVE DENSITY 1.0. Sodium is very common in the Earth's crust, occurring mostly as dissolved sodium chloride in sea water, from which it is extracted as sodium hydroxide by ELECTROLYSIS in a MERCURY-CATHODE CELL. Sodium is an ALKALI METAL, reacting with water and burning in air.

Sodium compounds produce a characteristic bright orange colour in a flame test, and this is also the colour of the GAS DISCHARGE in sodium vapour, which is widely used as an electric light source more efficient than filament lamps. The vast majority of sodium compounds are soluble.

sodium bicarbonate *See* SODIUM HYDROGENCARBONATE.

sodium carbonate (Na_2CO_3) A white solid; melting point 851°C; decomposes on further heating. Sodium carbonate is most frequently found as the decahydrate, $Na_2CO_3.10H_2O$, commonly known as washing soda, an efflorescent (*see* EFFLORESCENCE) colourless crystalline solid; RELATIVE DENSITY 1.4. Sodium carbonate is produced in the SOLVAY PROCESS. Sodium carbonate is widely used in water treatment as it removes dissolved calcium and magnesium salts from hard water. *See also* WATER SOFTENING.

sodium chlorate(V) ($NaClO_3$) A white solid; melting point 250°C; decomposes on further heating; RELATIVE DENSITY 2.5. Sodium chlorate is manufactured by the reaction between chlorine and concentrated sodium hydroxide:

$$6NaOH + 3Cl_2 \rightarrow NaClO_3 + 5NaCl + 3H_2O$$

It decomposes on heating:

$$2NaClO_3 \rightarrow 2NaCl + 3O_2$$

Sodium chlorate is an powerful OXIDIZING AGENT. It is used in matches and as a weedkiller.

sodium chloride, *common salt* (NaCl) A white solid; melting point 801°C; boiling point 1,413°C; RELATIVE DENSITY 2.2. Sodium chloride is the chief dissolved component of BRINE, from which it is extracted by evaporation. It is an essential component of the ELECTROLYTE structure of biological organisms, and is used commercially as the raw material for the manufacture of other sodium salts.

sodium cyanide (NaCN) A white solid; melting point 564°C; boiling point 1,496°C. Sodium cyanide is made by the reaction between hydrogen cyanide and sodium hydroxide:

$$HCN + NaOH \rightarrow NaCN + H_2O$$

Like most cyanides, sodium cyanide is highly toxic.

sodium dichromate ($Na_2Cr_2O_7$) An orange-red crystalline solid, which decomposes on heating. It is usually found as the dihydrate, $Na_2Cr_2O_7.2H_2O$; RELATIVE DENSITY 2.5. It is widely used as an OXIDIZING AGENT and as a MORDANT in dying. It is used in QUANTITATIVE ANALYSIS, though being slightly HYGROSCOPIC it is less suitable than the more expensive POTASSIUM DICHROMATE.

sodium fluoride (NaF) A white solid; melting point 993°C; boiling point 1,695°C; RELATIVE DENSITY 2.6. It can be manufactured by the reaction of sodium hydroxide with hydrogen fluoride:

$$NaOH + HF \rightarrow NaF + H_2O$$

It is used in the manufacture of ceramic glazes, and is introduced into drinking water at low concentrations, where it helps prevent tooth decay.

sodium hydrogencarbonate, *sodium bicarbonate* ($NaHCO_3$) A white solid; decomposes on heating; RELATIVE DENSITY 2.2. Sodium hydrogencarbonate can be prepared by passing carbon dioxide through sodium carbonate solution:

$$Na_2CO_3 + CO_2 + H_2O \rightarrow 2NaHCO_3$$

Sodium carbonate reacts with acids to liberate carbon dioxide, and is widely used as a remedy for excess stomach acid:

$$NaHCO_3 + HCl \rightarrow NaCl + H_2O + CO_2$$

It decomposes on heating, again releasing carbon dioxide:

$$NaHCO_3 \rightarrow NaOH + CO_2$$

This reaction is exploited in 'dry powder' fire extinguishers, which use sodium hydrogencarbonate propelled by carbon dioxide.

sodium hydroxide, *caustic soda* (NaOH) A white soapy solid; melting point 318°C; boiling point 1,390°C; RELATIVE DENSITY 2.1. Sodium hydroxide is produced as a by-product of the ELECTROLYSIS of BRINE in MERCURY-CATHODE CELLS. It is a strong alkali, DELIQUESCENT and widely used in the manufacture of soap. It is extremely corrosive to living tissue, and even small splashes in the eye can cause permanent damage.

sodium nitrate ($NaNO_3$) A white solid; melting point 306°C; decomposes on further heating; RELATIVE DENSITY 2.3. Sodium nitrate can be prepared by the reaction between sodium carbonate and nitric acid:

$$Na_2CO_3 + 2HNO_3 \rightarrow 2NaNO_3 + H_2O + CO_2.$$

Sodium nitrate decomposes on heating:

$$4NaNO_3 \rightarrow 2Na_2O + 2N_2O_4 + O_2$$

It is used as a fertilizer.

sodium peroxide (Na_2O_2) A creamy white solid; decomposes on heating; RELATIVE DENSITY 2.8. Sodium peroxide is formed by burning sodium in excess oxygen:

$$2Na + O_2 \rightarrow Na_2O_2$$

It decomposes in water to give sodium hydroxide and hydrogen peroxide:

$$Na_2O_2 + 2H_2O \rightarrow 2NaOH + H_2O_2$$

It is a powerful OXIDIZING AGENT, and releases oxygen on heating:

$$2Na_2O_2 \rightarrow 2Na_2O + O_2$$

sodium pump, *cation pump* A mechanism in the CELL MEMBRANE of most animal cells in which sodium ions (Na^+) are pumped out of the cell and potassium ions (K^+) are pumped into the cell by ACTIVE TRANSPORT, an energy-requiring process. The sodium pump is important in the regulation of cell volume, by altering the osmotic potential (*see* OSMOSIS) and establishing a resting potential (*see* NERVE IMPULSE).

sodium sulphate (Na_2SO_4) A white powder; melting point 888°C; decomposes on further heating; RELATIVE DENSITY 2.7. Sodium sulphate is formed in the reaction between sodium chloride and concentrated sulphuric acid:

$$NaCl + H_2SO_4 \rightarrow Na_2SO_4 + 2HCl$$

It is used in the manufacture of glass.

sodium sulphide (Na_2S) An orange solid; melting point 1,180°C; decomposes on further heating, RELATIVE DENSITY 1.9. Sodium sulphide can be formed by the REDUCTION of sodium sulphate with carbon:

$$Na_2SO_4 + 2C \rightarrow Na_2S + 2CO_2$$

It is hydrolysed by water (*see* HYDROLYSIS):

$$Na_2S + 2H_2O \rightarrow 2NaOH + H_2S$$

It is used in the paper industry to soften wood-pulp.

sodium sulphite (Na_2SO_3) A white solid; which decomposes on heating; RELATIVE DENSITY 2.6. Sodium sulphite is made by the reaction between sulphur dioxide and sodium carbonate:

$$Na_2CO_3 + SO_2 \rightarrow Na_2SO_3 + CO_2$$

It is a mild REDUCING AGENT and is used as a preservative in some foodstuffs. It liberates sulphur dioxide on heating, or on reaction with strong acids:

$$Na_2SO_3 \rightarrow Na_2O + SO_2$$

$$Na_2SO_3 + 2HCl \rightarrow 2NaCl + H_2O + SO_2$$

sodium thiosulphate, *hypo* ($Na_2S_2O_3$) A white solid; decomposes on heating; RELATIVE DENSITY 1.7. It is more frequently encountered as clear crystals of the pentahydrate, $Na_2S_2O_3.5H_2O$; melting point 42°C. Sodium thiosulphate is manufactured by the reaction between sulphur dioxide, sulphur and sodium hydroxide:

$$SO_2 + S + 2NaOH \rightarrow Na_2S_2O_3 + H_2O$$

Its chief use is in photographic processing, where its ability to dissolve unexposed silver halides from a photographic emulsion leads to its use as a photographic fixer.

soft (*adj.*) In physics, describing a FERROMAGNETIC material in which the alignment of the DOMAINS is easily altered, so the material is easily magnetized and demagnetized.

software A general term for computer programs, files and other associated documentation. *Contrast* HARDWARE.

soft water Water that is not HARD WATER; in other words, which does not contain dissolved calcium and magnesium SALTS.

soil Small particles of rock, formed by weathering, mixed with organic material overlying the bedrock of the Earth's surface. The grains of soil have water and air between them, and the texture and properties of soil determine which plants can grow in it. Different types of soil develop under different climatic and physical conditions, for example deep soils develop in warm, wet climates and in valleys, and shallow soils develop in cool dry areas and on slopes.

Soil consists of a number of layers. The top soil is the uppermost layer that contains most of the nutrients needed by plants, but it is susceptible to SOIL EROSION. Below the top soil lies the sub-soil, which does not provide many nutrients but is important in the drainage of water in the soil.

Clay soil has very small particles (mostly aluminium silicate) and water passes through slowly. Sandy soil has larger particles (mostly silicon dioxide) through which water can pass quickly. The best soils for plant growth are a mixture of sand and clay (loam soil). The HUMUS content of soil determines its fertility. The BIOTIC (living) element of soil plays an important role in soil texture and fertility (as well as in decomposition); burrowing animals, for example earthworms, improve aeration and drainage and assist humus breakdown.

The effects of the nature of soil on the ECOSYSTEM are collectively called edaphic factors.

soil erosion The wearing away or redistribution of SOIL caused by the action of water, wind, ice and human intervention. Erosion is a serious problem if left unchecked, as it may result in decline of the land, infertility and eventually in the formation of deserts. Destruction of forests and other vegetation and mechanization of farming causes unnecessary soil erosion.

soil profile The series of layers, or horizons, seen in a vertical section through SOIL to the rock beneath. The soil profile provides important information about the character of soils.

sol A COLLOID in which solid particles are suspended in a liquid. *See also* HYDROSOL, LYOPHILIC, LYOPHOBIC.

solar (*adj.*) Associated with the sun.

solar cell, *photovoltaic cell* A device used to convert light energy, usually from the Sun, into electricity. A solar cell typically consists of a glass or plastic plate with a very thin gold layer, through which light can pass, in contact with a thin layer of copper oxide on a copper base.

solar day *See* DAY.

solar energy Energy from the Sun. With the exception of NUCLEAR POWER and TIDAL POWER, all the energy currently available on Earth originated in the Sun. Many schemes have been proposed to convert energy from the Sun directly into heat or electricity. Direct conversion to electricity has found limited use in remote locations, such as providing the power for spacecraft. However, the technology is too expensive, and the amount of energy falling on each square metre too small, to make this energy form economically viable in competition with continued use of FOSSIL FUEL reserves. The use of solar energy for heating water is more widespread, but the temperatures reached are not high enough for this to be used as the sole energy source when heating water.

solar mass The mass of the sun, about 2×10^{30} kg. The masses of other stars are often expressed in terms of solar mass.

solar panel A BATTERY of SOLAR CELLS.

Solar System The name given to the Sun and the bodies held in ORBIT around it by its GRAVITY. The Sun appears to be a typical star, and there is some evidence that other stars may have similar systems. The bodies in orbit around the Sun include PLANETS, PLANETARY SATELLITES, ASTEROIDS and COMETS.

solar wind The stream of high-energy particles, mostly PROTONS and ELECTRONS, given off by the Sun. The interaction between the solar wind and the Earth's atmosphere and magnetic field is responsible for AURORAE and can influence the IONOSPHERE and the state of the Earth's magnetic field.

solenoid A coil of wire used to produce a MAGNETIC FIELD. The magnetic field around a single wire is very weak unless a large current is flowing. To produce a larger field, wire is often wound into a coil. Solenoids often have iron cores to further increase the field strength – the field from the current aligns the DOMAINS in the iron, producing a far stronger field. In a solenoid without an iron core, the field is proportional to the current flowing in the coil, and to the number of turns per metre length of solenoid. The field at the ends of a long solenoid is half that at the middle, and the pattern of field lines is rather like that produced by a BAR MAGNET. The field direction can be found using the RIGHT-HAND GRIP RULE.

The magnetic field B near the centre of a long solenoid of n turns per metre, carrying a current I is

$$B = \mu_0 nI$$

where μ_0 is the PERMEABILITY of free space.

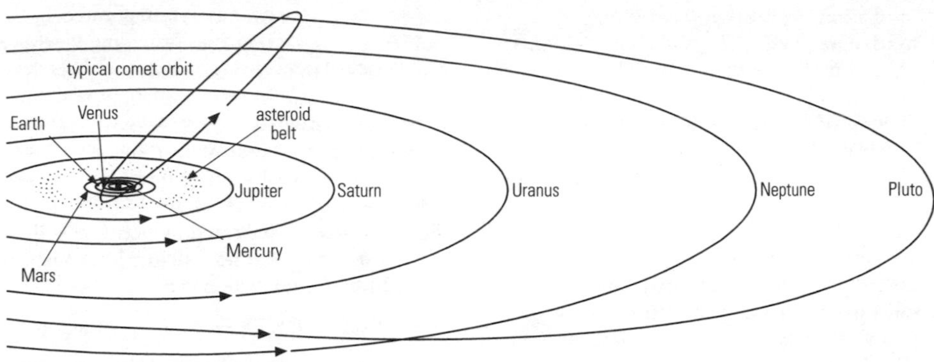

Above: Relative orbits of the planets in the Solar System.
Below: Relative sizes of the planets in the Solar System.

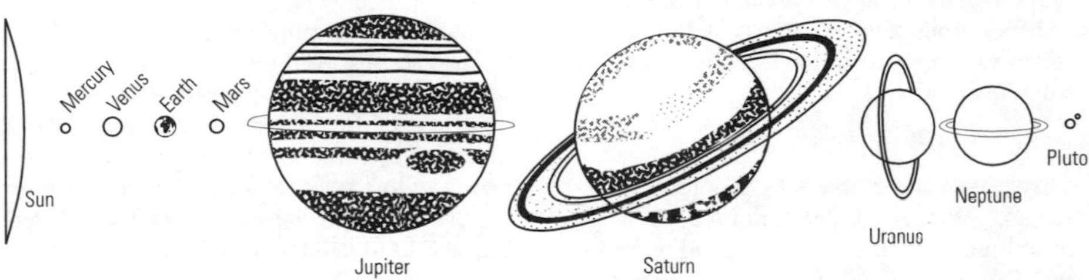

solid The state of matter in which a substance retains its shape. The molecules are closely packed, so a solid is not easily compressed, and rigidly held together. Solids may be either CRYSTALLINE, in which case the molecules are arranged in a regular LATTICE, or AMORPHOUS, in which there is no regular arrangement of the atoms. A crystalline solid has definite melting point at which it becomes liquid, whereas an amorphous solid becomes increasingly pliable over a range of temperatures until it assumes liquid properties.

Crystalline solids can occur as CRYSTALS, with the lattice ordering being maintained over long distances to produce pieces of material with symmetrical shapes reflecting the ordered nature of the lattice. This is particularly the case with ionic materials.

See also BAND THEORY, COVALENT CRYSTAL, IONIC SOLID, POLYCRYSTALLINE.

solid angle An angle in three dimensions, as at the point of a cone. Solid angle is defined as the ratio of an area of a section of a sphere to the square of its distance from the point where the angle is measured. Solid angles are measured in STERADIAN.

solid solution A solid MIXTURE in which the atoms, ions or molecules of the constituents are entirely intermixed, rather than appearing as small crystals of each type of material. Certain alloys, such as those formed between gold and silver, are solid solutions. When such a solution is heated, it does not have a single melting point, but melts over a range of temperatures. *See* LIQUIDUS, SOLIDUS.

solid-state (*adj.*) Describing an electronic system that does not use any THERMIONIC VALVES.

solid-state detector A REVERSE BIASED JUNCTION DIODE used to detect IONIZING RADIATION from the ELECTRON–HOLE PAIRS produced in the

DEPLETION LAYER by the passage of the ionizing radiation. *See* RADIATION DETECTORS.

solid-state physics The physics of solid materials, in particular their thermal, electrical and magnetic properties. The creation and investigation of new semiconducting materials, and their use for electronic devices, such as INTEGRATED CIRCUITS, is an important area of solid-state physics.

solidus In a SOLID SOLUTION, the line on a graph of temperature against composition below which the material is entirely solid.

soliton A pulse-like disturbance that is able to move through a DISPERSIVE material without itself spreading.

solubility A measure of the amount of material that can DISSOLVE in a given amount of SOLVENT. Usually measured in gram per decimetre cubed, or mole per decimetre cubed. The solubility of ionic salts in water generally increases with temperature, but gases become less soluble at high temperatures. *See also* HENRY'S LAW.

solubility product For a material A_xB_y that dissolves to form a saturated IONIC solution, the solubility product K_s is

$$K_s = [A^{m+}]^x[B^{n-}]^y$$

where the square brackets, [ ] denotes CONCENTRATION. This value is constant for a SATURATED solution of a given salt at a specified temperature. If the actual value of this concentration product is less than the solubility product, the solution will not be saturated. If the value of this concentration product exceeds the solubility product, the solid will start to be formed as a PRECIPITATE. This explains the COMMON ION EFFECT: to precipitate solid A_xB_y, the concentration of B ions that is required can be reduced by increasing the concentration of A ions and vice versa.

soluble (*adj.*) Describing a material that will DISSOLVE in a particular SOLVENT (usually water) to form a SOLUTION. Many, but by no means all, IONIC salts will dissolve in water, forming HYDRATED ions. The breaking up of the crystal LATTICE requires an input of energy that is obtained from the hydration process. Almost all salts of ALKALI METALS are soluble, as are many salts of the TRANSITION METALS. The ALKALINE EARTH metals form salts with far lower solubilities. Organic solids are generally insoluble in POLAR solvents such as water. *See also* HEAT OF SOLUTION.

solute A solid that is dissolved to form a SOLUTION. *See* SOLUBLE.

solution A liquid that comprises a SOLVENT and a dissolved solid or gas. *See also* DISSOLVE, HEAT OF SOLUTION, SATURATED, SOLUBILITY, SOLUBILITY PRODUCT, SOLUBLE.

solvation The process of forming a SOLUTION, particularly the breaking down of an IONIC SOLID as it dissolves in water.

Solvay process The reaction between sodium chloride solution and calcium carbonate to produce sodium carbonate:

$$2NaCl + CaCO_3 \rightarrow Na_2CO_3 + CaCl_2$$

This cannot be carried out directly since calcium carbonate is insoluble. Instead, sodium chloride solution (brine) is saturated with ammonia in a scrubber and trickled down a column called a SOLVAY TOWER, whilst the calcium carbonate is heated to form carbon dioxide:

$$CaCO_3 \rightarrow CaO + CO_2$$

A solution containing ammonium chloride is formed, whilst sodium hydrogencarbonate, which is less soluble, is formed as a PRECIPITATE. The rate of formation of the precipitate is increased by cooling the Solvay tower and by a COMMON ION EFFECT with the sodium ions in the brine. The precipitate is extracted by filtering, and heated to form sodium carbonate:

$$2NaHCO_3 \rightarrow Na_2CO_3 + H_2O + CO_2$$

The ammonium chloride is reacted with calcium oxide from the heated calcium carbonate, regenerating the ammonia and producing calcium chloride as a waste product:

$$CaO + 2NH_4Cl \rightarrow CaCl_2 + 2NH_3 + H_2O$$

Solvay tower A vessel used in the SOLVAY PROCESS. It comprises a tall column fitted with metal plates to enable a liquid flowing down the tower to react with a gas being passed up the tower.

solvent A liquid in which a substance will dissolve to form a SOLUTION. The term solvent particularly refers to VOLATILE organic liquids, such as ALKANES, in which a wide range of organic materials will dissolve.

solvent extraction A method of extracting a SOLUTE from a SOLUTION containing a mixture of materials by mixing the solution with a second solvent in which the material being extracted is more soluble. Organic molecules can be extracted from water by using an IMMISCIBLE organic solvent such as benzene. The PARTITION CONSTANT is used as a measure of how much more readily the material will dissolve in the new solvent than the original one.

somatic (*adj.*) Relating to the body. A somatic cell is any cell other than a GAMETE.

somatotrophin *See* GROWTH HORMONE.

sonar A system similar in principle to RADAR, but which locates objects using the reflection of SOUND. Sonar is used in ships to measure the depth of the water and to detect fish, submarines, etc.

sonic boom The sudden loud sound heard when a SHOCK WAVE from a SUPERSONIC aircraft reaches the ear.

s-orbital The lowest energy ORBITAL for a given PRINCIPAL QUANTUM NUMBER. There is only a single s-orbital for each principal quantum number and the electron WAVEFUNCTION for an s-orbital is spherical in shape.

sorption pump A device for creating or improving a vacuum by the absorption of gas molecules into a solid, such as charcoal or a ZEOLITE.

sound A LONGITUDINAL WAVE motion that can be heard by the human EAR. Sound waves need a supporting medium: they can travel through solids, liquids and gases, but not through a vacuum.

The human ear uses a thin layer of skin called the EARDRUM to convert the pressure changes in the air into the movement of small bones which then stimulate nerves producing the sensation of sound. The human ear will respond to sound waves in a frequency range of roughly 16 Hz to 15 kHz; the sensitivity is less towards the end of this range and the upper limit decreases with age. Other animals can hear higher frequency sounds and bats use sound waves to locate their prey, sending out pulses of high frequency sound and listening for the reflections. The reflection of a sound wave is called an echo.

See also INFRASOUND, SONAR, SPEED OF SOUND, ULTRASOUND.

source In electronics, the ELECTRODE from which CHARGE CARRIERS enter the CHANNEL in a FIELD EFFECT TRANSISTOR.

source code The language in which a computer program was written, as opposed to the MACHINE CODE, which the computer understands and executes directly.

source slit In a SPECTROMETER or SPECTROSCOPE, a narrow slit from which the light to be examined is diffracted. The SPECTRUM is then a series of images of this slit in different wavelengths. *See also* DIFFRACTION, SPECTROSCOPY.

space-time Space and time considered as a single entity, mathematically described by a VECTOR in four dimensions. By considering space and time together, it is possible to simplify the equations of the SPECIAL THEORY OF RELATIVITY, which treat both distances and times as dependent on the motion of the observer. The GENERAL THEORY OF RELATIVITY describes space-time as being distorted by the effects of GRAVITY.

spark The unstable flow of electric currents through gases at high pressures (greater than about 0.1 ATMOSPHERES), with the emission of light and sound. Sparks occur as a result of AVALANCHE BREAKDOWN.

spark chamber A PARTICLE DETECTOR which comprises a volume of gas through which parallel wires are run in a criss-cross arrangement, with alternate sets of wires at right angles to one another and connected to opposite sides of a high-voltage supply. Sparks between the wires form first along any IONIZATION trails left by fast-moving charged particles passing through the chamber. The spark trails can be photographed or the information fed to a computer by detecting the pulses of current in those wires that experienced a spark. *See also* DRIFT CHAMBER.

spark counter A RADIATION DETECTOR that detects IONIZING RADIATION by producing a spark from a high voltage supply in the air between two ELECTRODES through which the ionizing radiation passes.

spark plug A pair of ELECTRODES used to produce a SPARK that ignites the fuel/air mixture in a PETROL ENGINE. They are built to withstand high pressures and temperatures.

spatula A laboratory tool shaped like a small spoon, or having a flat surface, used to handle small amounts of powdered material

special theory of relativity Proposed by Albert Einstein (1879–1955) in 1905, this theory takes as its starting point the idea that there is no one viewpoint for any physical system that is more valid than any other. Thus any observer who is in an non-accelerating frame of reference (*see* INERTIAL REFERENCE FRAME), will agree with a second such observer about the laws of physics. Thus physics experiments performed in a train that is moving (but not accelerating) are just as valid, and should lead to the same conclusions as those performed by an observer at rest on the surface of the Earth, which is in any case moving through space.

When James Maxwell (1831–1879) put forward the electromagnetic theory of light in 1864, it became clear that light waves travelled at a constant speed irrespective of any relative motion between the source of the light and the observer measuring its speed. This was confirmed experimentally in the MICHELSON–MORLEY EXPERIMENT. Einstein took the incompatibility between the constant SPEED OF LIGHT and Newton's laws and made the bold step of devising a new system of mechanics in which NEWTON'S LAWS OF MOTION were only approximately valid, with the approximation becoming increasingly good at speeds much less than the speed of light (3×10^8 ms^{-1}). Objects travelling at speeds sufficiently close to the speed of light for the differences between relativity and NEWTONIAN MECHANICS to be important are said to be RELATIVISTIC.

Consequences of the special theory of relativity include the idea that distances and times between events are different for different observers; the result that nothing can travel faster than the speed of light; and that mass appears to increase for a moving particle. The increase in mass with KINETIC ENERGY turns out to be just one example of another idea that comes from the special theory of relativity, the EQUIVALENCE OF MASS AND ENERGY.

The predictions of the special theory of relativity have subsequently been verified experimentally with fast-moving FUNDAMENTAL PARTICLES; for example, PI-MESONS moving at speeds very close to the speed of light are able to reach the end of a tube that, according to Newtonian mechanics, is far too long for them to pass along without undergoing radioactive decay. As seen by an observer in the laboratory, this is explained by the fast-moving particles decaying more slowly (TIME DILATION). An observer moving along with the particles would see them decaying at the normal rate, but would see the tube as being shorter (LORENTZ CONTRACTION).

If an observer at rest with respect to an object observes a mass m_0, then an observer moving at a speed v relative to the first will observe mass m, length l and time t with

$$m = m_0/\sqrt{(1 - v^2/c^2)}$$

where c is the speed of light.

See also GALILEAN RELATIVITY, GENERAL THEORY OF RELATIVITY, QUANTUM ELECTRODYNAMICS, POLE AND BARN PARADOX, TWIN PARADOX.

speciation The emergence of new species during EVOLUTION. The cause of speciation is isolation of a group within a POPULATION. ADAPTIVE RADIATION is the result of geographical separation that prevents the separated groups interbreeding; the groups adapt differently to their surrounding and eventually a new species emerges. Isolation can also be due to physiological factors, such as incompatibility of genitalia, or behavioural factors, such as incompatibility of timing of breeding; the isolation of groups causes them to breed amongst themselves to survive, and therefore form new species. *See also* EVOLUTION.

species The lowest level in the CLASSIFICATION scheme. Different populations can exist within a species for example different breeds of dog. Members of the same species can all interbreed to produce fertile offspring. A native species has existed in a country since prehistoric times, whereas a naturalized species has been introduced to a country by humans and adapted to the new country. Exotic species need human intervention to survive.

About 1.4 million species have been identified. Species become extinct, for example through destruction of habitats, and new ones can form (*see* SPECIATION) by diversifying from other members of their group. There are more species in tropical regions than temperate ones. *See also* BINOMIAL NOMENCLATURE, GENUS.

specific charge The CHARGE carried by an object, especially a proton, electron or ion, divided by its mass.

specific gravity *See* RELATIVE DENSITY.

specific heat capacity The HEAT CAPACITY per unit mass of a substance. Specific heat capacity is measured in $Jkg^{-1}K^{-1}$.

$$\text{Energy flow} = \text{mass} \times \text{specific heat capacity} \times \text{temperature change}$$

The specific heat capacity of water is unusually large at 4200 $Jkg^{-1}K^{-1}$. This is due to the HYDROGEN BONDS in water, which absorb energy as the water is heated. The consequence of this is that water heats up and cools down more slowly than most other substances.

For solids and liquids it makes little difference whether the heat capacity is measured under conditions of constant pressure or constant volume, but gases expand substantially when they are heated under a constant pressure. The work then done in pushing back the atmosphere makes the specific heat capacity at constant pressure greater.

Specific heat capacities can be measured by finding the amount of heat needed to change the temperature of a certain quantity of the material by a measured amount. If the temperature of the container in which the specific heat capacity is being measured also changes during the experiment, it is important to take account of the energy involved in heating the container rather than the material under test. This is done by performing the measurement in a CALORIMETER.

See also CONSTANT FLOW METHOD, COOLING CORRECTION, EQUIPARTITION OF ENERGY, METHOD OF MIXTURES, MOLAR HEAT CAPACITY, RATIO OF SPECIFIC HEATS.

specific latent heat of fusion The amount of HEAT needed to turn one kilogram of solid material at its melting point into liquid at the same temperature.

specific latent heat of vaporization The energy needed to convert one kilogram of liquid at its boiling point to gas at the same temperature.

specific volume The reciprocal of DENSITY, the volume occupied by unit mass of a material, usually at a specified temperature and pressure.

spectacles LENSES worn in front of the eyes to correct defects of vision. *See* LONG-SIGHTEDNESS, PRESBYOPIA, SHORT-SIGHTEDNESS.

spectral class A classification of STARS according to their surface temperature. The temperature of a star determines the features of its BLACK BODY spectrum and of the LINE SPECTRUM which is seen superimposed on this. This enables stars to be classified in a sequence O, B, A, F, G, K, M, in order of decreasing temperature.

spectral line A narrow range of wavelengths present in an EMISSION SPECTRUM or absent from an ABSORPTION SPECTRUM. The wavelength of a spectral line corresponds to the energy of a transition between two ENERGY LEVELS in the atom or ion which produced the line. *See also* SPECTROSCOPY, SPECTRUM.

spectrometer An instrument for forming and recording a SPECTRUM. *See* SPECTROSCOPY.

spectroscope An instrument for forming a SPECTRUM and viewing it directly. *See* SPECTROSCOPY.

spectroscopic binary *See* BINARY STAR.

spectroscopy The study of the ELECTROMAGNETIC RADIATION produced by a sample, usually in the INFRARED, visible and ULTRAVIOLET regions of the ELECTROMAGNETIC SPECTRUM. Spectroscopy is a powerful tool for chemical analysis. INFRARED SPECTROSCOPY gives information about the chemical bonds in organic molecules, whilst visible and ULTRAVIOLET SPECTROSCOPY provides information about which elements are present and their IONIZATION states. Visible spectroscopy is also used in astronomy to provide information about the surface temperatures of stars and, via the DOPPLER EFFECT, about their motion.

A SPECTROMETER is a device used to produce a record of a SPECTRUM, whilst a SPECTROSCOPE enables a visible spectrum to be viewed directly. In either instrument, light from the sample is used to illuminate a slit. Light diffracting (*see* DIFFRACTION) from this slit (the SOURCE SLIT) is focused into a parallel beam by a CONVERGING LENS. The lens and slit assembly is called a collimator. The light is separated into its different WAVELENGTHS by a DIFFRACTION GRATING.

See also ATOMIC ABSORPTION SPECTROSCOPY, ATOMIC EMISSION SPECTROSCOPY, MICROWAVE SPECTROSCOPY.

spectrum (*pl.* ***spectra***) The arrangement of ELECTROMAGNETIC RADIATION into its constituent energies in order of wavelength or frequency. White light separated into its component wavelengths by a PRISM or DIFFRACTION GRATING gives a characteristic spectrum of coloured bands.

When a sample is heated, or bombarded with ions or electrons, or absorbs photons of electromagnetic radiation, it emits radiation of wavelengths characteristic to the sample. This type of spectrum is called an emission spectrum. If radiation of a continuous range of wavelengths is passed through a sample, the sample absorbs certain characteristic wavelengths. When the transmitted radiation is viewed by a SPECTROSCOPE, the absorbed wavelengths show up as dark bands or lines. This is called an absorption spectrum. A line spectrum is one in which only certain wavelengths appear, while a continuous spectrum is one in which all the wavelengths in a certain range appear.

The term 'spectrum' can also apply to any distribution of entities or properties arranged in order of increasing (or decreasing value). For example, a mass spectrum is an arrangement of molecules, ions or isotopes by mass (*see* MASS SPECTROSCOPY).

See also ELECTROMAGNETIC SPECTRUM, HYDROGEN SPECTRUM, RYDBERG EQUATION, SPECTRAL LINE, WAVE NATURE OF PARTICLES.

specular, *regular* (*adj.*) In physics, describing a REFLECTION in which light is all reflected in one direction rather than being scattered in all directions.

speed The distance an object travels divided by the time it takes to travel through that distance. If the direction of travel is specified, the quantity is a VELOCITY, which is a vector (a quantity with both size and direction), as opposed to a scalar (a quantity with no associated direction). *See also* SPEEDOMETER.

speed of light (*c*) All ELECTROMAGNETIC WAVES, including light, travel through empty space at the same speed, 3.00×10^8 ms^{-1}. In the SI system, the speed of light is now fixed by the definition of the metre as being 299,792,458 ms^{-1}.

The first measurement of the speed of light was made by Römer in 1674 and involved studying the orbits of the satellites of the planet Jupiter. These appear to be completed more frequently as the Earth moves nearer to Jupiter, and less frequently as it moves away, due to the differing times taken for the light to reach the Earth. Later methods involved chopping a light beam into a series of pulses, using a toothed wheel (Fizeau, 1849), or a rotating octagonal mirror (Michelson, 1878). In each of these two cases the pulsed light beam is reflected from a distant point and returns to the apparatus. It can only re-enter the optical system if the wheel or mirror has now moved far enough for the light pulse to pass through the next gap between the teeth, or to be reflected off the next face of the mirror. In this way, the speed of the light is related to the rotation of the wheel or mirror.

In modern methods of measuring the speed of electromagnetic waves, radio waves form a STANDING WAVE pattern and their frequency is measured by counting the waves over a known time interval and the wavelength from the separation of nodes in the standing wave pattern.

speed of sound The speed at which sound, and other LONGITUDINAL WAVES, pass through a specified medium, usually air. The speed of sound in air is 331 ms^{-1} (738 mph) at 0°C, but increases proportionally to the square root of the absolute temperature. The speed of sound in a gas is proportional to the ROOT MEAN SQUARE speed of the molecules, so sound travels faster in hydrogen and helium than in air, and more slowly in carbon dioxide. The speed of sound in LIQUIDS and SOLIDS is generally faster than in air.

In a gas, the speed of sound is $(\gamma p/\rho)^{1/2}$ where p is the pressure, γ the RATIO OF SPECIFIC HEATS and ρ the DENSITY. In a solid the speed is $(E/\rho)^{1/2}$, where E is the YOUNG MODULUS and ρ the density.

See also MACH NUMBER.

speedometer A speed-measuring instrument. The speedometer commonly used in cars exploits the effect of EDDY CURRENTS set up in a metal disc by a magnet rotating at the same rate as the wheels of the car. The DOPPLER EFFECT can also be used to measure the speed of objects ranging from cars in police speed-traps to stars (*see* HUBBLE'S LAW).

sperm, *spermatozoon* (*pl. spermatozoa*) The male GAMETE of animals that is produced in large numbers in the TESTIS. A sperm is much smaller than the female gamete (OVUM). It consists of a head (containing the nucleus), a middle region with many MITOCHONDRIA for energy, and a tail (FLAGELLUM) by which it moves. At the tip of the head is a specialized structure called an ACROSOME that contains enzymes ready for release when the sperm meets the ovum. These enzymes dissolve the hard outer membrane of the ovum, allowing the sperm head and ovum to fuse. When the sperm leave the body (*see* SEXUAL REPRODUCTION) there are about 500 million sperm in about 3 cm^2 of SEMEN. Before FERTILIZATION of the ovum can occur, the sperm must go through a final maturation stage in the female tract called 'capacitation'. *See also* ACROSOME REACTION, FERTILIZATION.

spermatid *See* TESTIS.

spermatocyte In animals, an immature male GAMETE that gives rise to a SPERM cell. *See* TESTIS.

spermatogenesis In animals, the formation and maturation of SPERM. *See* TESTIS.

spermatozoon *See* SPERM.

sperm duct *See* VAS DEFERENS.

Sphenophyta A PHYLUM of the plant KINGDOM consisting of the horsetails. There is only one surviving genus of this group, called *Equisetum*. Horsetails are found in damp conditions and consist of an underground RHIZOME that gives rise to jointed aerial stems with prominent nodes and small leaves whorled at the nodes. They are HOMOSPOROUS with SPORANGIA in CONES.

sphere A three-dimensional shape enclosed by a curved surface with all points on that surface at the same distance from a single point, the centre of the sphere. If the radius of the sphere (the distance of the surface from the centre) is r, then the area of the surface is $4\pi r^2$, whilst the volume is $4\pi r^3/3$.

spherical aberration *See* ABERRATION.

spherical co-ordinates *See* POLAR CO-ORDINATES.

spherical mirror A MIRROR with a reflecting surface that is part of a sphere. *See* CURVED MIRROR.

sphincter Any ring of smooth muscle in the wall of a tubular structure such as the ALIMENTARY CANAL. Sphincters contract and relax to control the opening of the tube, for example for the passage of food or FAECES. Examples include the pyloric and cardiac sphincters (*see* STOMACH) and the anal sphincter (*see* ANUS). Control of sphincters can sometimes be voluntary, for example of the anal sphincter.

spider A member of the CLASS Arachnida of the PHYLUM ARTHROPODA. A spider possesses eight legs and an unsegmented ABDOMEN connected by a thin 'waist' to the cephalothorax (head and THORAX merged). The eyes are usually simple. Many species of spider exude a liquid from the underside of the abdomen, which hardens in air to form the thread seen in webs; the spider traps prey in its web, injects substances into the prey to subdue it, and then digests and sucks up the juices.

There are about 30,000 species of spiders found all over the world, some of which produce poisonous toxins, for example the black widow and tarantula.

spin Attempts to produce a theory of QUANTUM MECHANICS that also incorporated the ideas of the SPECIAL THEORY OF RELATIVITY led to the discovery that some types of particle possess an inherent ANGULAR MOMENTUM rather as if they were spinning. The spin of a particle is always a whole number times $h/4\pi$, where h is PLANCK'S CONSTANT. Particles with half-integral spin (odd number times this basic unit) are called FERMIONS, of which the electron is an example. Particles with no spin or integral spin (an even number times $h/4\pi$), such as photons, are called BOSONS.

See also PAULI EXCLUSION PRINCIPLE, STERN-GERLACH EXPERIMENT.

spinal cord A major component of the CENTRAL NERVOUS SYSTEM (CNS), linking the PERIPHERAL NERVOUS SYSTEM (outside the CNS) to the BRAIN. The spinal cord consists of a cylinder of nervous tissue protected by VERTEBRAE, with a small central canal running through the centre containing CEREBROSPINAL FLUID. Surrounding this canal is GREY MATTER, usually forming an H-shape in cross-section, and around this WHITE MATTER. Surrounding all of this are membranes called the MENINGES. Pairs of spinal nerves extend along the length of the spinal cord; the uppermost of these is called the dorsal root (which only carries sensory nerves) and the lowest is called the ventral root (which only carries effector nerves). The swelling within the dorsal root, called the dorsal root GANGLION, is formed by cell bodies of sensory neurones. *See also* VERTEBRAL COLUMN.

spindle The group of protein fibres formed in a cell during MITOSIS and MEIOSIS. The spindles draw the pairs of CHROMOSOMES apart as the cell divides.

spine *See* VERTEBRAL COLUMN.

spiracle 1. In insects, a pore in their hard exterior through which oxygen enters the body and carbon dioxide leaves. Spiracles form a complex series of tubes called tracheae that divide to form smaller tubes called tracheoles, which then enter the tissues directly. Oxygen is carried in this tracheal system and not in the blood. This system is considered to be equivalent to the TRACHEA of vertebrates.

2. In many fish, the opening of the remains of the first GILL slit.

spirillum (*pl.* ***spirilla***) Any spiral-shaped bacterium. *See* BACTERIA.

spleen The largest mass of lymphoid tissue (*see* LYMPH NODE), found near the stomach in vertebrates. The spleen has blood circulating through it rather than LYMPH and is important in destroying worn-out blood cells and storing new RED BLOOD CELLS and LYMPHOCYTES to be pumped into the circulation when needed. *See also* RETICULOENDOTHELIAL SYSTEM.

split-ring commutator A metal ring, split into segments that are connected to the coils in a D.C. MOTOR or DYNAMO. As the commutator rotates with the coils, brushes carry current in and out through those segments of the commutator that are best placed to produce a turning effect in the desired direction (in a motor) or are generating the largest VOLTAGE (in a dynamo).

sponge A member of the PHYLUM PORIFERA.

spongy mesophyll *See* MESOPHYLL.

spontaneous combustion COMBUSTION that takes place without an external triggering factor, such as a flame. The heat needed to start the combustion process is generated internally by the reagents, usually by a slow OXIDATION process.

spontaneous emission The change of an atom from one ENERGY LEVEL to another with the emission of a PHOTON, taking place without the influence of any external electric field. *See also* STIMULATED EMISSION.

spontaneous fission NUCLEAR FISSION by radioactive decay, without having been struck by a neutron. *See also* INDUCED FISSION, RADIOACTIVITY.

spontaneous generation theory *See* ORIGIN OF LIFE.

sporangium (*pl.* ***sporangia***) In fungi and plants, the structure within which SPORES are produced.

spore The reproductive structure of many primitive plants, including FUNGI, FERNS and mosses (*see* BRYOPHYTA), and also some BACTERIA and ALGAE. Spores usually consist of a single cell and they can develop without the need for fusion with another cell. Spores can form sexually or asexually and are a means of very rapid increase in the size of a population. They are light and easily dispersed, and can remain dormant (*see* DORMANCY) until the conditions are favourable for their development. *See also* GERMINATION.

sporophyll A leaf bearing sporangia (*see* SPORANGIUM). Examples include the scale-like leaves of CONES of CONIFERS, and the STAMENS and CARPELS of flowering plants.

sporophyte One form of a plant that shows ALTERNATION OF GENERATIONS. The sporophyte is the DIPLOID generation that produces HAPLOID spores by MEIOSIS. *See also* GAMETOPHYTE.

sporozoan A member of the PHYLUM APICOMPLEXA.

sporozoite An infective stage in the life cycle of protozoans from the PHYLUM APICOMPLEXA, such as the *Plasmodium* parasite. *See* MALARIA.

sporulation The formation of SPORES (the reproductive body of many lower plants, such as mosses, ferns and fungi). Sporulation is a common means of ASEXUAL REPRODUCTION, whereby numerous small unicellular spores are produced (by MEIOSIS, so providing genetic VARIATION) that are easily dispersed and grow to new organisms when the conditions are suitable.

spreadsheet A program for the manipulation of mainly numerical data, particularly for financial purposes, such as accounting.

spring constant *See* HOOKE'S LAW.

spring pendulum A mass oscillating on the end of a spring. A spring pendulum is isochronous (period of oscillation is independent of amplitude) provided the spring obeys HOOKE'S LAW. For a spring pendulum with SPRING CONSTANT k, and a mass m, the period is

$$T = 2\pi(k/m)^{1/2}$$

spring tide The TIDE produced when the gravitational effects of the Moon and Sun combine, producing the greatest rise and fall in tide level.

SPS (Super Proton Synchrotron) Currently the largest proton SYNCHROTRON, located at CERN (European Centre for Nuclear Research) in Geneva. The SPS is 7 km in circumference and can accelerate electrons to 270 GeV. It can accelerate antiprotons to the same energy in the other direction, and was used in the COLLIDING BEAM EXPERIMENTS that led to the discovery of the W BOSON and Z BOSON in 1984.

square 1. A geometrical figure having four straight sides of equal length and at right angles to one another. If the length of the sides is a, the area of the square is a^2.

2. A number multiplied by itself, for example, the square of 3 is 9.

square wave An electrical signal that switches suddenly and regularly between two values, generally remaining at each of the two values for equal periods of time.

SRAM (static random access memory) Memory chips used by computers to provide IMMEDIATE ACCESS MEMORY. Like DRAM, BITS are stored as electric charges on a grid of tiny electric circuits. Unlike DRAM, the circuits hold the charges indefinitely, and SRAM does not have to be constantly read and rewritten. SRAM is faster and more reliable that DRAM, but also more expensive.

ssb *See* SINGLE SIDEBAND.

SSSI *See* SITE OF SPECIAL SCIENTIFIC INTEREST.

stabilizer 1. A negative CATALYST added to postpone an undesirable chemical reaction, such as in some explosives and glues.

2. A material added to one of the substances in a SOL, especially an EMULSION, to prevent it separating into two separate phases.

stable equilibrium An EQUILIBRIUM state where a small displacement from the equilibrium position will cause the system to create forces tending to return it to its equilibrium position. A simple example of this is a ball resting in the bottom of a bowl.

stalactite A downward-growing calcium carbonate column formed by the seepage and evaporation of water containing dissolved calcium compounds in limestone caves. *See also* STALAGMITE.

stalagmite An upward-growing calcium carbonate column formed by the dripping and evaporation of water containing dissolved calcium compounds in limestone caves. *See also* STALACTITE.

stamen The essential male reproductive structure in a flowering plant (ANGIOSPERM). The stamen surrounds the CARPELS and consists of a long stalk called the filament, with an ANTHER at its apex that produces the POLLEN grains that contain the male GAMETES. There are variable numbers of stamens and they can be in different positions, which is useful in classification of flowering plants. The collective name for the stamens is androecium.

standard atmosphere *See* ATMOSPHERE.

standard deviation In statistics, a measure of how much the values recorded depart from the MEAN. The standard deviation is the square root of the mean value of the square of the difference of each value from the mean. Thus if $<x>$ denotes the mean of x, the standard deviation, σ, is

$$\sigma = \sqrt{<(x - <x>)^2>}$$

See also NORMAL DISTRIBUTION.

standard electrode potential The POTENTIAL DIFFERENCE in an electrochemical CELL with a concentration of one MOLE of metal ions per decimetre cubed, and a temperature of 25°C, with an ANODE made of the same metal, measured relative to a HYDROGEN ELECTRODE. *See also* ELECTROCHEMICAL SERIES.

standard enthalpy of formation *See* STANDARD HEAT OF FORMATION.

standard form A way of expressing a number as a number with one digit before the decimal point multiplied by a power of 10. For example, 154.79 is written in standard form as 1.5479×10^2, and 0.01471 as 1.471×10^{-2}. Standard form is particularly useful in expressing numbers that are so large or so small that they would otherwise contain a large number of zeros.

standard heat of formation The HEAT OF FORMATION measured with all reagents and products at atmospheric pressure and a temperature of 25°C.

standard model In ELEMENTARY PARTICLE physics, the picture of the FOUR FORCES OF MATTER unified to produce the ELECTROWEAK FORCE, the STRONG NUCLEAR FORCE and GRAVITY, with these forces acting on three pairs of LEPTONS (electron, muon and tau lepton and their NEUTRINOS) and three pairs of QUARKS (up, down, charm, strange, top and bottom). This model has firm experimental support, whilst anything beyond this is widely regarded as unsupported by experimental evidence. *See also* HIGGS BOSON, QUANTUM THEORY, UNIFIED FIELD THEORY.

standard redox potential A REDOX POTENTIAL measured at one MOLE per decimetre cubed concentration of all reagents and at a temperature of 25°C. *See also* NERNST EQUATION.

standard temperature and pressure (s.t.p.) Conditions where the pressure is one ATMOSPHERE (1.01×10^5 Pa) and the temperature is 0°C.

standing wave An oscillating motion along the length of some object, such as a stretched string or air in a pipe. To support a standing wave, the object must be capable of supporting the motion of a PROGRESSIVE WAVE along its length and have ends at which this wave will be REFLECTED. The standing wave is formed by the superposition of waves travelling in both directions along the length of the system, each being reflected when it reaches the end. A standing wave can exist only at those frequencies where the wave is IN PHASE with itself after it has travelled along the system, been reflected at one end, travelled back along the system and been reflected at the other end. On reflection, there may be a 180° phase change. If this occurs at both ends, or at neither end, then standing waves will be supported if they have a wavelength such that the length of the system is a whole number of half-wavelengths. If there is a phase reversal on reflection at one end only, the wavelength must be such that the length of the system is a whole number of half-wavelengths plus an odd quarter wavelength.

Standing waves are also important in QUANTUM THEORY, as the ENERGY LEVELS in an atom are determined by the patterns of standing electron waves that can be supported within the electric field around the nucleus (*see* WAVE-PARTICLE DUALITY).

See also ANTINODE, FUNDAMENTAL, HARMONIC, OVERTONE, NODE.

Stanford Linear Accelerator Centre (SLAC) California, USA. The home of what is currently the largest LINEAR ACCELERATOR. This machine is 3 km long and can accelerate electrons to an energy of 30 GeV.

stannate Any compound containing the stannate ion, SnO_4^{2-}. Stannates are formed by the reaction between tin oxide and alkalis, for example:

$$SnO_2 + 2NaOH \rightarrow Na_2SnO_4 + H_2O$$

stapes *See* EAR OSSICLE.

star A self-luminous celestial body. A star is a large, roughly spherical mass of gas, mostly hydrogen and helium held together by its own gravity and releasing energy by NUCLEAR FUSION. The Sun is a fairly typical star, though stars with masses between 0.1 and 50 times the mass of the sun are fairly common.

See also BLACK HOLE, GALAXY, HERTZSPRUNG–RUSSELL DIAGRAM, MAIN SEQUENCE STAR, NEUTRON STAR, PP CHAIN, PULSAR, RED GIANT, SPECTRAL CLASS, STELLAR PARALLAX, SUPERNOVA, WEIN'S LAW, WHITE DWARF.

starch A complex POLYSACCHARIDE that is the main food reserve of green plants. Starch consists of hundreds of GLUCOSE polymers, some of which are straight chain molecules (amylose) and some branched molecules (amylopectin). The main dietary sources of starch for humans are cereals, potatoes and LEGUMES. Starch is used in many ways in industry, for example to stiffen paper and textiles, as a thickening agent in foodstuffs, and as glucose syrups (by HYDROLYSIS of starch). *See also* CARBOHYDRATE.

start codon The CODON AUG that is always present at one end of a MESSENGER RNA (mRNA) molecule and at which PROTEIN SYNTHESIS begins. The codon codes for the amino acid methionine, and the first RIBOSOME to be attracted to the mRNA in TRANSLATION attaches to this AUG codon. *See also* GENETIC CODE.

states of matter The three physical forms in which a substance can usually exist: SOLID, LIQUID and GAS. The different properties of these three states are explained by the KINETIC THEORY of matter. PLASMA is sometimes regarded as a fourth state of matter.

state symbol *See* CHEMICAL EQUATION.

static electricity Effects produced by the separation of opposite electric CHARGES when these charges are at rest. Such effects are described as electrostatic. An object with no overall charge is said to be electrically neutral. When two objects are rubbed together some electrons may be transferred from one object to the other, leaving one with a negative charge and the other with an equal positive charge. This process is called CHARGING BY FRICTION.

static equilibrium *See* EQUILIBRIUM.

static friction The FRICTION between two objects that are not moving relative to one another. *See also* LIMITING FRICTION.

statics The branch of mechanics that deals with objects at rest. For an object to remain at rest the vector sum of all the FORCES acting on it must be zero, and the total MOMENT produced by all the forces, taken about any point, must also be zero. These conditions for EQUILIBRIUM are the starting point for all statics problems. *See also* DYNAMICS.

stationary point A point at which the GRADIENT of a function is zero, but where the sign of the gradient on either side of the function has the same sign, so that this does not represent a local MAXIMUM or MINIMUM value for the function.

statistical mechanics A development of KINETIC THEORY that applies the laws of MECHANICS to systems containing large numbers of particles in order to calculate quantities such as TEMPERATURE and PRESSURE from the average values of the ENERGY and MOMENTUM of these particles. *See also* FERMI DISTRIBUTION, MAXWELL–BOLTZMANN DISTRIBUTION.

statistics The branch of mathematics that deals with the collection and interpretation of large numbers of results, and the prediction of the outcome of sampling processes, where a small number of measurements are taken on a much larger group. *See also* CHI-SQUARED TEST, NORMAL DISTRIBUTION, SAMPLING, STANDARD DEVIATION.

steady-state theory The cosmological theory that the universe has existed in a steady state throughout time and will continue to do so, with a constant average density. The theory does not explain certain observable events, and most cosmologists now favour the BIG BANG theory. *See also* COSMOLOGY, ORIGIN OF LIFE.

stearic acid A saturated FATTY ACID that occurs in many plant and animal fats.

steel An alloy in which IRON is the predominant component. *See also* BESSEMER CONVERTER, ELECTRIC ARC FURNACE, OPEN HEARTH FURNACE, OXYGEN FURNACE.

Stefan–Boltzmann constant, *Stefan's constant* (σ) The constant σ in STEFAN'S LAW, equal to $5.67 \times 10^{-8}\text{Wm}^{-2}\text{K}^{-4}$. Early work on the QUANTUM MECHANICS of BLACK-BODY RADIATION showed that the Stefan–Boltzmann constant could be expressed in terms of other fundamental constants by the relationship

$$\sigma = 2\pi^4 k^5/15h^3c^2$$

where k is the BOLTZMANN CONSTANT, h is PLANCK'S CONSTANT and c is the SPEED OF LIGHT.

Stefan-Boltzmann law *See* STEFAN'S LAW.

Stefan's constant *See* STEFAN–BOLTZMANN CONSTANT.

Stefan's law, *Stefan-Boltzmann law* That the total amount of BLACK-BODY RADIATION emitted per unit area from a BLACK BODY at an ABSOLUTE TEMPERATURE T is σT^4, where σ is the STEFAN–BOLTZMANN CONSTANT.

stellar parallax A method of measuring the distances of nearby stars. A nearby star will appear to change its position slightly against the background of more distant stars as the Earth moves around its orbit. This leads to a unit of distance called the PARSEC. Stellar parallax gives distances for only the nearer stars.

stem The main axis of a plant, which supports the leaves, buds and flowers. The stem contains vascular tissue (*see* VASCULAR BUNDLE), which conducts food, water and minerals between the roots and the leaves.

stem tuber *See* TUBER.

step-down Describing a TRANSFORMER that has a TURNS RATIO less than 1, so the voltage across the SECONDARY COIL is less than that across the PRIMARY COIL.

step-up Describing a TRANSFORMER with a TURNS RATIO greater than 1, so the voltage across the SECONDARY COIL is greater than that across the PRIMARY COIL.

steradian The unit of SOLID ANGLE. The size of the solid angle at the point of a cone is equal to the area of the section of a sphere at the base of the cone divided by the square of the distance of this surface from the point of the cone. A complete sphere encloses a solid angle of 4π steradian at its centre.

stereochemistry The branch of chemistry that concerns itself with the shapes of MOLECULES, particularly ORGANIC molecules.

stereoisomerism The existence of two or more ISOMERS that differ only in the spatial orientation of the atoms in the molecule. *See* GEOMETRIC ISOMERISM, OPTICAL ISOMERISM.

sterilization 1. The killing or removal of living organisms. Sterilization can be achieved by heat treatment (boiling), by chemical disinfectants, irradiation with gamma radiation and filtration. It is important in food processing, medicine and research.

2. A surgical operation to prevent reproduction. In males this is called vasectomy and the passage of sperm is blocked. In females the FALLOPIAN TUBES are tied to prevent FERTILIZATION. Sterilization can be used as a means of CONTRACEPTION.

Stern–Gerlach experiment The experiment in which SPIN was first observed experimentally. In this experiment a beam of neutral silver atoms was split into two when passed through an inhomogeneous (non-uniform) magnetic field.

sternum In humans, the breast bone. *See* PECTORAL GIRDLE.

steroid hormone Any one of a group of HORMONES, for example OESTROGEN, PROGESTERONE, TESTOSTERONE, CORTISONE and ALDOSTERONE, that are derived from LIPIDS. Steroid hormones have a complex structure consisting of four carbon rings. A large subgroup of steroids is the sterols, including CHOLESTEROL.

sterol A large subgroup of STEROID HORMONES of which CHOLESTEROL is a member.

stibnite An ore of antimony, containing mostly antimony sulphide.

stigma In flowers, a structure within the CARPEL (female reproductive structure) that is supported by the STYLE and is specialized for receiving POLLEN. The stigma often has hairs and produces a sticky secretion to attract pollen grains.

stimulated emission The emission of a PHOTON by an atom in the presence of ELECTROMAGNETIC RADIATION. When a photon of a certain energy interacts with an excited atom, the atom drops from its excited energy level to a lower one, with the emission of a photon of energy equal to the difference in energy between the two levels. For stimulated emission to take place, the interacting photon must have the same energy as the emitted photon. Stimulated emission in a RESONANT CAVITY is responsible for the operation of the LASER. *See also* SPONTANEOUS EMISSION.

stoichiometric (*adj.*) Describing a compound in which the proportions of the elements reacting to form the compound are always the same. Such compounds obey the LAW OF CONSTANT PROPORTIONS.

Some solids are effectively mixtures of more than one compound. These are sometimes stoichiometric, such as the mixed iron oxide Fe_3O_4, which is effectively a mixture of iron (II) oxide and iron (III) oxide, but many minerals occur as non-stoichiometric crystals.

stoichiometry The proportions of various ELEMENTS that are involved in the formation of a COMPOUND.

Stokes' law A law predicting the amount of DRAG produced by a spherical object of radius r, moving with LAMINAR FLOW through a liquid of VISCOSITY η at a speed v. Stokes' law states the viscous drag will be F, where

$$F = 6\pi\eta rv$$

stolon A long vertical stem, for example the blackberry, that reproduces vegetatively by bending over to touch the soil and developing ADVENTITIOUS ROOTS (roots growing from stems instead of other roots), at which point a new plant grows. A stolon is similar to a RUNNER.

stoma (*pl.* ***stomata***) One of many pores in the EPIDERMIS of plants that is the main route of water loss in a plant and the site of carbon dioxide and oxygen exchange. Stomata can be opened or closed by a pair of GUARD CELLS, which surround them. The guard cells change shape (by changes in turgidity) to adjust water loss. Factors affecting stomatal opening include levels of carbon dioxide, light, temperature, humidity, air currents and water availability. For example, the pores are open during cold weather but closed during hot weather to reduce evaporation of water. Stomata close at night, when PHOTOSYNTHESIS cannot take place, to reduce water loss. Stomata are also found in large numbers on the aerial parts of a plant. *See also* LEAF, TRANSPIRATION.

stomach A muscular cavity in animals, situated just below the DIAPHRAGM, that produces acids and enzymes to digest food entering it from the OESOPHAGUS, and then passes the food to the SMALL INTESTINE for further digestion and absorption.

The lining of the stomach is a folded layer called the gastric mucosa, embedded in which are gastric pits lined with secretory cells producing GASTRIC JUICE, in response to production of GASTRIN. Gastric juice is made up of mostly water mixed with hydrochloric acid, to provide the acid environment needed for the digestive enzymes to function. The enzymes PEPSIN and RENNIN are made by the gastric pit as inactive precursors, which are activated by the acidity of the stomach. Goblet cells in the stomach lining secrete MUCUS, which helps protect the lining from its own gastric juice and also helps lubricate movement of food. Cells of the stomach lining are replaced continually to avoid damage by the acidity.

Food is thoroughly mixed in the stomach until a creamy chyme forms, which is released slowly into the small intestine in manageable amounts. The cardiac and pyloric sphincters are two rings of muscles at the entrance and exit of the stomach, respectively, which can relax and contract to allow food to enter and leave. The stomach itself contracts periodically throughout digestion (to aid mixing and food movement).

Some HERBIVORES have stomachs with more than one chamber to allow CELLULOSE breakdown to occur without affecting the environment needed for the other microorganisms vital for digestion.

See also CROP, GIZZARD.

stone fruit *See* DRUPE.

stonewort Any member of the phylum CHAROPHYTA.

stop codon, *termination codon, nonsense codon* Any one of three CODONS that do not code for an AMINO ACID and cause PROTEIN SYNTHESIS to come to an end. The codons are UAA, UAG and UGA. When a stop codon is reached on a MESSENGER RNA (mRNA) molecule during TRANSLATION, the POLYPEPTIDE chain is released from the mRNA. *See also* GENETIC CODE, MUTATION.

stopping potential In the PHOTOELECTRIC EFFECT, the voltage needed to prevent all the electrons released from reaching a nearby negative ELECTRODE.

storage granule Insoluble material found in all cells as a store of food energy. STARCH granules are present in the CYTOPLASM of plant cells and in CHLOROPLASTS. GLYCOGEN granules are found in animal cells. Oil or LIPID droplets can be found in both plant and animal cells.

strain A measure of the extent to which an object has been deformed. The TENSILE strain in an object is the amount by which it has been extended divided by its original length.

strain energy *See* ELASTIC ENERGY.

strange A FLAVOUR of QUARK, with charge $-^1/_3$ in units of the electron charge.

strangeness A QUANTUM NUMBER carried by the strange QUARK. It is +1 for the strange quark and −1 for the strange antiquark, but zero for all other particles. Strangeness is conserved in all interactions except the WEAK NUCLEAR FORCE, thus strange particles decay into non-strange particles relatively slowly.

stratigraphy The branch of GEOLOGY that deals with the formation and arrangement of rock strata.

stratopause The boundary between the STRATOSPHERE and the MESOSPHERE. *See* ATMOSPHERE.

stratosphere The layer of the upper ATMOSPHERE in which temperature rises with increasing height due to the absorption of ultraviolet radiation.

stratum, (*pl. strata*) A layer, especially of rock in geology. *See also* STRATIGRAPHY.

stratum corneum *See* EPIDERMIS.

stratum granulosum *See* EPIDERMIS.

stratus Cloud that takes the form of an even layer, usually covering most or all of the sky. *See also* CUMULUS, WEATHER SYSTEMS.

streamlined (*adj.*) Having a smooth shape designed to minimize DRAG forces and minimize the amount of TURBULENT flow.

streptomycin An ANTIBIOTIC that affects PROTEIN SYNTHESIS.

stress The FORCE on a solid object divided by the cross-sectional area over which that force acts. Stress is a useful way of measuring the force on an object regardless of its size. The point at which a material will break, for example, can be expressed as a stress, regardless of the size of the sample being considered.

striated muscle *See* VOLUNTARY MUSCLE.

striped muscle *See* VOLUNTARY MUSCLE.

stroboscope A device that produces flashes of light at regular intervals, used in physics experiments and as a special effect in discotheques.

stroboscopic photograph A series of pictures recorded on a single piece of film by illuminating the moving object with light from a STROBOSCOPE.

stroma *See* CHLOROPLAST.

strong acid An ACID that is almost completely IONIZED when dissolved in water. An example is sulphuric acid, H_2SO_4, which forms H^+ ions (which associate with water molecules to form the oxonium ion, H_3O^+) and sulphate ions SO_4^{2-}. Note that a strong acid is not the same as a concentrated acid – indeed at high concentrations the level of IONIZATION of many acids falls.

strong nuclear force The force that holds QUARKS together inside PROTONS, NEUTRONS and other HADRONS. The strong nuclear force was first thought to be the force responsible for holding protons and neutrons together in the atomic nucleus. It has since been recognized that this effect is a left-over from the far stronger forces between quarks. The interactions between quarks are described by the exchange of particles called GLUONS, and a full QUANTUM MECHANICAL theory of the strong interaction has been produced. This theory is called QUANTUM CHROMODYNAMICS and the quality that quarks possess that has the same role as CHARGE in QUANTUM ELECTRODYNAMICS is called COLOUR, though it has nothing to do with colour in the visual sense.

strontium (Sr) The element with ATOMIC NUMBER 38; RELATIVE ATOMIC MASS 87.6°C; boiling point 1,300°C; RELATIVE DENSITY 2.5. Strontium is an ALKALINE EARTH metal. Strontium ISOTOPES are a common constituent of radioactive fallout. Since strontium is chemically similar to, but more reactive than calcium, there are concerns about radioactive strontium entering the FOOD CHAIN and being incorporated into the bone structure of those exposed to the fallout.

strontium carbonate ($SrCO_3$) A white solid; decomposes on heating; RELATIVE DENSITY 3.7. It can be made by passing carbon dioxide through strontium hydroxide solution:

$$Sr(OH)_2 + CO_2 \rightarrow SrCO_3 + H_2O$$

On heating, it decomposes to strontium oxide:

$$SrCO_3 \rightarrow SrO + CO_2$$

Strontium carbonate is used as a red colouring in pyrotechnic flares and fireworks, and as a PHOSPHOR in CATHODE RAY TUBES.

strontium oxide (SrO) A white solid; melting point 2,430°C; boiling point 3,000°C. It can be manufactured by heating strontium carbonate, or by burning strontium in oxygen:

$$2Sr + O_2 \rightarrow 2SrO$$

It is used as a raw material in the manufacture of other strontium salts.

structural formula A chemical formula that gives some indication of the way in which the atoms are arranged in the molecule. The two ISOMERS of dichloroethane for example, both have the MOLECULAR FORMULA $C_2H_4Cl_2$, but they can be written as different structural formulae, CH_2ClCH_2Cl or CH_3CHCl_2. *See also* EMPIRICAL FORMULA.

structural isomerism *See* ISOMER.

style In flowers, a structure at the centre of the CARPEL (the female reproductive structure). The style can be a long slender stalk that supports the STIGMA at the top in a position that is optimal for receiving POLLEN. In some flowers it is short or even absent.

styrene *See* PHENYLETHENE.

styrene-butadiene rubber *See* SBR.

subatomic particle Any LEPTON or HADRON, including in particular the PROTON, NEUTRON and ELECTRON, from which atoms are made.

subduction zone A place on the Earth's surface where one TECTONIC PLATE slides under another, with the lower plate re-melting whilst the upper plate may be forced up to form mountain ranges. Subduction zones are areas where EARTHQUAKES occur most frequently as the plates grate past one another.

sublime(*vb.*) To change directly from a solid to a gas without passing through a liquid phase.

sublimation The direct change from solid to gas without passing through a liquid phase when a material is heated. Solid carbon dioxide (dry ice) and iodine are two examples of materials that sublime rather than melt at ATMOSPHERIC PRESSURE.

sublingual gland Any SALIVARY GLAND situated beneath the tongue.

submarine An underwater vessel that can float or sink by adjusting its average density by flooding tanks with water or using compressed air to force that water out of the tanks. Submarines are used by the military, as they can move quickly and are difficult to detect since sea water conducts electricity and so is virtually opaque to electromagnetic radiation. SONAR systems can be used to detect submarines, but must be immersed in water to operate, so cannot be carried on fast aircraft or satellites.

submaxillary gland Either of a pair of SALIVARY GLANDS situated close to the lower jaw.

subset A SET whose members are all members of some larger set, e.g. the set of INTEGERS is a subset of the set of REAL NUMBERS.

subsidiary maximum In a DIFFRACTION pattern, a maximum of INTENSITY other than the central maximum.

subsidiary quantum number The QUANTUM NUMBER used to label an ORBITAL giving a measure of the ANGULAR MOMENTUM of an electron in the orbital, and describing the shape of the orbital. The subsidiary quantum number is usually expressed as a letter in the sequence s, p, d, f, from an early mistaken categorization of lines in SPECTRA as sharp, principal, diffuse and fine.

sub-soil *See* SOIL.

substitution The process of replacing an algebraic variable, such as in an equation, by a numerical value or by some other expression. Substituting $x = y + 2$ in the expression $2x + 5$ gives the expression $2(y + 2) + 5$.

substitution reaction A reaction in which one atom or group of atoms in a molecule is replaced by another. An example is the chlorination of methane:

$$CH_4 + Cl_2 \rightarrow CH_3Cl + HCl$$

where a hydrogen atom is replaced by a chlorine atom. *See also* ELECTROPHILIC SUBSTITUTION, NUCLEOPHILIC SUBSTITUTION.

substrate A substance that is acted upon by an ENZYME. Substrate is also the material on which micro-organisms grow, for example agar, or the surface to which cells in TISSUE CULTURE attach.

succession The series of changes occurring within a COMMUNITY, from its origin to its stable climax. The first colonization of an area is called primary succession, and recolonization of an established area destroyed, for example by fire or flood, is called secondary succession. A climax community, such as a deciduous oak woodland, eventually forms in which there is a balance between the species (both plant and animal) sustained in the area, and new varieties only replace established species. Usually one or two dominant plants and animals form the greatest biomass.

A series of plant successions developing in a particular area is called a sere. For example, a lithosere develops on bare rock surfaces; a hydrosere begins in fresh water; a xerosere grows under dry conditions; and a halosere develops in a salt-marsh.

suckback The state of an apparatus that has been heated, to produce gas in a chemical reaction, and then allowed to cool, drawing water up the DELIVERY TUBE and producing violent boiling.

sucrase An enzyme that breaks down SUCROSE into GLUCOSE and FRUCTOSE.

sucrose ($C_{12}H_{22}O_{10}$) A DISACCHARIDE sugar made up of the MONOSACCHARIDE units GLUCOSE and FRUCTOSE. Sucrose is found in the pith of sugar cane; it can be cane sugar or beet sugar. Sucrose is what is commonly referred to as 'sugar'.

Sudan III test *See* LIPID.

sugar, *saccharide* Any one of a group of CARBOHYDRATES with relatively low RELATIVE MOLECULAR MASS and a typically sweet taste. The term sugar commonly refers to SUCROSE. *See also* MONOSACCHARIDE, DISACCHARIDE, POLYSACCHARIDE.

sulphane Any of the polymeric (*see* POLYMER) forms of hydrogen sulphide, having the general formula H_2S_n.

sulphate A salt containing the sulphate ion, SO_4^{2-}, together with a CATION, either a metal or the ammonium ion, NH_4^+. Sulphates form stable crystals that contain WATER OF CRYSTALLIZATION.

sulphate(IV) *See* SULPHITE.

sulphide Any BINARY COMPOUND containing sulphur and a more electropositive element (*see* ELECTROPOSITIVITY). Non-metallic sulphides are COVALENT and generally unstable. Metallic sulphides are mostly IONIC, though the TRANSITION METAL sulphides have some covalent character (*see* COVALENT BOND) and are insoluble.

sulphite, *sulphate(IV)* A compound containing the sulphite ion, SO_3^{2-}, together with a cation, either a metal or the ammonium ion, NH_4^+. Sulphites are easily oxidized to SULPHATES.

sulphonation In organic chemistry, the introduction of a sulphonic group, HSO_3, to ARENES. For example, the reaction of concentrated sulphuric acid with BENZENE at room temperature gives benzenesulphonic acid.

$$C_6H_5\text{–}H + SO_3 \rightarrow C_6H_5^{+}(H)SO_3^- \xrightarrow{\pm H^+} C_6H_5\text{–}S(=O)_2\text{–}OH$$

benzenesulphonic acid

The reaction is an ELECTROPHILIC SUBSTITUTION in which sulphur trioxide, SO_3, is generated from the sulphuric acid, which serves as the ELECTROPHILE.

Sulphonation reactions are used in the manufacture of detergents and dyes.

sulphonic group The HSO_3 group.

sulphur (S) The element with ATOMIC NUMBER 16; RELATIVE ATOMIC MASS 32.1; melting point 112°C; boiling point 445°C; RELATIVE DENSITY 2.1. Sulphur exists mainly as a yellow non-metallic solid with a distinctive odour, but a red ALLOTROPE also exists that is stable at higher temperatures.

Sulphur burns in air and is soluble in carbon disulphide, but is otherwise fairly unreactive. It forms sulphides with the more reactive metals, but occurs more often in association with oxygen as sulphates. Large amounts of sulphur are used industrially in the manufacture of sulphuric acid.

sulphur dichloride dioxide (SO_2Cl_2) A colourless liquid; melting point –51°C; boiling point 69°C; RELATIVE DENSITY 1.7. It is formed by the reaction of sulphur dioxide on chlorine in the presence of ultraviolet light or iron(III) chloride as a catalyst. Its structure is TETRAHEDRAL, consisting of central sulphur atom forming DOUBLE BONDS with two oxygen atoms and single bonds with two chlorine atoms. It decomposes in water:

$$SO_2Cl_2 + 2H_2O \rightarrow H_2SO_4 + 2HCl$$

Sulphur dichloride dioxide is used as a chlorinating agent in the manufacture of chlorinated hydrocarbons.

sulphur dichloride oxide ($SOCl_2$) A colourless liquid; melting point –100°C; boiling point 76°C. It is formed by the action of phosphorous(V) chloride on sulphur dioxide:

$$SO_2 + PCl_5 \rightarrow SOCl_2 + POCl_3$$

Sulphur dichloride oxide hydrolyses rapidly:

$$SOCl_2 + H_2O \rightarrow SO_2 + 2HCl$$

sulphur dioxide (SO_2) A colourless gas with a distinct odour; melting point –73°C; boiling point –10°C. Sulphur dioxide is made by roasting iron sulphide in air:

$$FeS + O_2 \rightarrow Fe + SO_2$$

It dissolves in water to give sulphurous acid:

$$SO_2 + H_2O \rightarrow H_2SO_3$$

Large amounts of sulphur dioxide are used in the manufacture of sulphuric acid.

sulphuric acid (H_2SO_4) An important MINERAL ACID; melting point 10°C; boiling point 338°C; RELATIVE DENSITY 1.8. Sulphuric acid is manufactured by the CONTACT PROCESS or the LEAD CHAMBER process. Both processes involve oxidizing sulphur dioxide, SO_2, usually obtained by burning sulphur in oxygen, to sulphur trioxide, SO_3, which is then dissolved in water to form sulphuric acid.

Sulphuric acid is a very important chemical FEEDSTOCK, particularly in the fertilizer and paint industries. It is also important as an ELECTROLYTE in LEAD-ACID CELLS. In its concentrated form it has a great affinity for water, which makes it useful as a drying agent, though the heat given off can cause boiling, which means that great care must be taken in handling the concentrated acid. In particular, it should always be diluted by adding acid to water, never water to acid.

sulphurous acid (H_2SO_3) A weak dibasic acid, known only in aqueous solution and from its salts, called sulphites. Sulphurous acid is formed when sulphur dioxide dissolves in water:

$$SO_2 + H_2O \rightarrow H_2SO_3$$

sulphur trioxide (SO_3) A colourless solid; melting point 17°C; boiling point 45°C; relative density 2.0. Sulphur dioxide is obtained by oxidizing sulphur dioxide with oxygen in the presence of a vanadium(V) oxide catalyst:

$$2SO_2 + O_2 \rightarrow 2SO_3$$

Sulphur trioxide dissolves readily in water to give sulphuric acid:

$$SO_3 + H_2O \rightarrow H_2SO_4$$

summing amplifier An AMPLIFIER with two or more inputs and an output equal to the sum of these inputs, each multiplied by a constant.

Sun The star at the centre of the solar system. It has a mean diameter of 1,392,000 km, a mass of 1.9 x 10^{30} kg and lies at a distance of 149,500,000 km from the Earth. It is composed of about 75 per cent hydrogen, 24 per cent helium and about 1 per cent other gases. THERMONUCLEAR REACTIONS take place in the core of the Sun, converting hydrogen into helium and generating vast amounts of energy. Temperatures here are thought to reach 15,000,000°C, but are closer to 6,000°C at the surface. The Sun is classified as a MAIN SEQUENCE STAR. *See also* SOLAR SYSTEM.

superatom *See* BOSE–EINSTEIN CONDENSATION.

supercomputer The most powerful and fastest type of COMPUTER. Supercomputers have hundreds of processors working in parallel, and are generally designed to perform very large numerical calculations for meteorology and other sciences.

superconductivity The disappearance of electrical RESISTANCE, exhibited by some materials, called SUPERCONDUCTORS, at low temperatures. The theory of superconductivity is complex, but it involves electrons forming pairs, called Cooper pairs, with equal and opposite momenta (*see* MOMENTUM). Whenever one member of the pair is scattered, the other receives an equal and opposite momentum change, hence there is no overall change in the rate at which charge is carried through the material.

It has been suggested that the phenomenon of superconductivity could be used to build highly efficient electrical machines. However, so far applications have been limited, owing to the high costs involved in cooling the materials to the low temperatures required. Liquid helium (boiling point 4 K) is often used. Some HIGH-TEMPERATURE SUPERCONDUCTORS are known, which may be cooled with liquid nitrogen (boiling point 77 K), but these materials do not have good mechanical properties.

Another problem is that many electrical machines rely on magnetic fields, and magnetic fields have a tendency to destroy superconductivity. Despite this, superconducting magnets have been used in medical MAGNETIC RESONANCE IMAGING machines and in some PARTICLE ACCELERATORS.

superconductor Any material that exhibits SUPERCONDUCTIVITY. Many metals become superconductors at low enough temperatures, but the temperatures below which superconductivity occurs are often very low, only a few KELVIN. One of the best metallic superconductors is Niobium-tin alloy, which retains its superconductivity up to 22 K in zero magnetic field. *See also* HIGH-TEMPERATURE SUPERCONDUCTOR.

supercooled (*adj.*) Describing a material that has cooled below its melting point, but remained liquid. This effect is usually demonstrated with sodium thiosulphate crystals, $Na_2S_2O_3.5H_2O$, which can be melted and then cooled in a clean test tube to well below their melting point. Adding a single SEED CRYSTAL of sodium thiosulphate will then trigger a rapid CRYSTALLIZATION and the release of large amounts of heat (LATENT HEAT of fusion).

superfluidity The disappearance of all VISCOSITY in liquid helium–4 below 2.2 K. Superfluid helium–4 also displays other unexpected properties, such as the ability to escape from an open vessel by forming a thin film and climbing up the inside of the vessel and down the outside, to collect in droplets at the base of the vessel. Superfluidity is a consequence of a BOSE–EINSTEIN CONDENSATION in the atoms of helium–4 at low temperatures.

superheated The state of a liquid heated above its boiling point without boiling. Such a state can be achieved provided the liquid is clean and the walls of its container smooth, since extra energy is needed to form small bubbles unless CONDENSATION NUCLEI are present. These are normally provided by dirt or imperfections in the walls of the container. When a superheated liquid does boil, it will do so rather violently – an effect called BUMPING.

supernova A sudden brightening of a star as the outer layers are blown off when the core collapses to form a NEUTRON STAR. The brightness of a star may increase by a factor of several million in a period of a few minutes, before decaying over a period of a few weeks.

superoxide An OXIDE formed by the ALKALI METALS, except lithium, containing the ion O_2^-, for example potassium superoxide, KO_2, formed by burning potassium in excess oxygen. All the superoxides are extremely powerful OXIDIZING AGENTS.

superpose (*vb.*) To add together two or more WAVES. *See* INTERFERENCE. *See also* PRINCIPLE OF SUPERPOSITION.

Super Proton Synchrotron *See* SPS.

supersaturated vapour The state of a VAPOUR with a PARTIAL PRESSURE higher than its SATURATED VAPOUR PRESSURE. This is an unstable state, and droplets of liquid will form on any small particles or any irregularities in the surface of the container – these are called CONDENSATION NUCLEI.

supersonic (*adj.*) Faster than the SPEED OF SOUND.

superstring theory A theory of particle physics that postulates that ELEMENTARY PARTICLES are not point-like objects, but extremely small strings in a multidimensional space, where all but four of the dimensions (SPACE-TIME) have collapsed to very small distances. Despite the attractiveness of these theories, there is no experimental evidence to support them.

supplementary angles A pair of angles that add up to 180°.

suppressor T cell *See* T CELL.

suprarenal gland *See* ADRENAL GLAND.

surd An expression containing a square root that is an IRRATIONAL NUMBER, for example $2\sqrt{2}$.

surface tension The force that appears at the surface of a liquid and tends to pull the liquid into spherical droplets. It results from the imbalance of INTERMOLECULAR FORCES acting on a molecule near the surface of a liquid. Such molecules are attracted by all the molecules below it in the liquid, whereas a molecule inside the liquid is attracted by molecules from all sides. The result is physically the same as if the surface were made of an elastic sheet with a constant force across any imaginary line on that surface, proportional to the length this line. The surface tension – which is temperature dependent, falling to zero at the CRITICAL TEMPERATURE of the liquid – is equal to the force per unit length along such a line.

The effect of surface tension allows small objects, such as some insects, to rest on a water surface. Another result is that the pressure inside a bubble is higher than that outside, by an extent which depends on the surface tension of the liquid and the radius of the bubble. It is highest when the bubble is smallest; for similar reasons, a rubber balloon requires the most effort to blow it up when it is small.

For a liquid with surface tension T, in a bubble of radius r, the pressure inside the bubble exceeds that outside by Δp, where

$$\Delta p = 2T/r$$

If the bubble has two surfaces, i.e. is made of a thin film of liquid with gas inside and outside, the difference is then

$$\Delta p = 4T/r$$

surfactant A material, such as a DETERGENT, added to a liquid, usually water, to reduce its SURFACE TENSION, enabling it to wet a solid surface effectively, rather than running off the surface in droplets.

susceptance The reciprocal of REACTANCE. The SI UNIT of susceptance is the SIEMENS.

susceptibility The extent to which a material increases (if FERROMAGNETIC or paramagnetic; *see* PARAMAGNETISM) or decreases (if diamagnetic; *see* DIAMAGNETISM) the strength of a MAGNETIC FIELD in which it is placed.

$$\text{Susceptibility} = \text{relative permeability} - 1$$

See also RELATIVE PERMEABILITY.

suspension A mixture in which small solid particles are suspended in a liquid (or occasionally a gas). Given time, the particles will usually settle to the bottom of the liquid, but they are small enough to have very low TERMINAL VELOCITIES, and so to remain distributed throughout the liquid for long periods.

suture joint An immovable JOINT, such as those between the bones of the skull.

s-wave In SEISMOLOGY, the transverse waves that arrive after the P-WAVES in an EARTHQUAKE. *See also* SEISMIC WAVE.

sweat The secretion from the SWEAT GLANDS.

sweat gland In mammals, a gland located in the SKIN that is responsible for perspiration and therefore heat loss when necessary. Sweat glands are found all over the body of primates, but are more localized in other mammals, for example the feet and face of cats and dogs. There are more sweat glands in male humans than females.

Sweat glands consist of coiled tubes leading via a duct to a pore in the skin surface. They absorb water fluid from the capillaries and transfer it to the skin as sweat (water, salt and urea) to evaporate, which aids heat loss when needed. Sweat is also thought to contain PHEROMONES, which are important in the communication of social messages. Production of sweat is controlled by the AUTONOMIC NERVOUS SYSTEM.

swim bladder An air-filled sac lying between the gut and the spine of a fish that regulates the buoyancy of the fish.

switch In electronics, any device for making or breaking a connection in an electric circuit, or for diverting a current from one part of a circuit to another.

symbiont Any organism living in a state of SYMBIOSIS.

symbiosis A permanent or prolonged close association between two organisms of different species where both benefit from the association. Some definitions also include associations with deleterious effects (*see* PARASITISM). The organisms in association are called symbionts.

Examples of symbiosis occur in the digestive system of animals where PROTOZOANS, bacteria and fungi live and synthesize their own vitamins (K and B), which the host animal can use and provide enzymes essential for digestion of the host's dietary material, and in return receive food and protection. In HERBIVORES these symbionts are essential for the break down of CELLULOSE. They live in the CAECUM, APPENDIX or rumen of RUMINANTS. Ruminants also benefit by using the symbionts as a source of protein.

Another important example occurs in the NITROGEN CYCLE, where nitrogen-fixing bacteria live in nodules on stems, roots or leaves of flowering plants (ANGIOSPERMS) or CONIFERS. LEGUMES, such as beans, peas and clover, have ROOT NODULES where the bacterium *Rhizobium* lives; the bacteria get a source of carbon and the plant is provided with nitrate (independently of soil nitrates). Other examples are

LICHENS and MYCORRHIZAS. *See also* COMMENSALISM, MUTUALISM.

symmetry The property of a system or object remaining unchanged when certain changes are made. In particular, it is the property of certain shapes that remain unchanged under specified transformations. For example, a square has a four-fold symmetry of rotation.

sympathetic nervous system *See* AUTONOMIC NERVOUS SYSTEM.

symplast pathway *See* TRANSLOCATION.

synapse The point at which two NEURONES or a neurone and a muscle meet to transmit a NERVE IMPULSE. The gap between the two cells is called the synaptic cleft and is at least 15 nm in width. The AXON of the transmitting neurone expands at the synapse to form a bulbous synaptic knob, within which are MITOCHONDRIA and synaptic vesicles (released from the GOLGI APPARATUS of nerve cells) containing NEUROTRANSMITTERS. When the nerve impulse reaches the synaptic knob, the neurotransmitter is released and crosses the synaptic cleft to bind to RECEPTOR molecules on the receiving cell's membrane. The receiving cell (after the synapse) is called the post-synaptic neurone and the transmitting cell (before the synapse) is the pre-synaptic neurone.

If the receiving cell is another neurone, then the impulse will be carried from the synaptic cleft of the axon to DENDRITES on the receiving cell. If the receiving cell is a muscle then the release of neurotransmitters will cause it to contract. Most synapses are between axons and dendrites, but axon–axon or dendrite–dendrite synapses can occur. In addition, those axons and dendrites can have synapses with several dendrites and axons, respectively. Less often, the synapses are electrical rather than chemical.

There are a number of drugs that act on the synaptic transmission, either by mimicking the neurotransmitter or affecting its release or receptor binding.

synaptic (*adj.*) Relating to the SYNAPSES.

synchrocyclotron A development of the CYCLOTRON that used a varying frequency for the accelerating electric field to overcome the changing period of particle orbits as they reached RELATIVISTIC speeds. As particle accelerators became larger, the provision of a magnetic field over a large area became increasingly problematic and the synchrocyclotron was superseded by the SYNCHROTRON.

synchronization pulse A signal transmitted, in TELEVISION for example, to keep the transmitting and receiving apparatus in step. In television these are transmitted with the picture information to ensure that the scanning at the receiving end remains in step with the transmitted information.

synchrotron A PARTICLE ACCELERATOR that overcomes the limitations imposed on the CYCLOTRON by the RELATIVISTIC speeds of particles causing a change in orbital period. Unlike the SYNCHROCYCLOTRON, the synchrotron does not require a magnetic field to be maintained over a large area. The synchrotron uses a steadily increasing magnetic field and an increasing frequency of alternating current to accelerate charged particles around a circular evacuated tube, called the beam tube. ELECTRODES placed at various locations around the tube accelerate the particles, whilst ELECTROMAGNETS focus the beam of particles and bend it into a roughly circular path. High energy synchrotrons are generally very large machines as the amount of energy lost by SYNCHROTRON RADIATION decreases rapidly with increasing radius. *See also* COLLIDING BEAM EXPERIMENTS, SPS.

synchrotron radiation A characteristic pattern of ELECTROMAGNETIC RADIATION given off by the charged particles in a SYNCHROTRON, which are accelerating by virtue of their CIRCULAR MOTION. This is an important source of X-rays in medicine and other applications, but it places a limit on the energy that can be achieved by a given size of synchrotron.

syncytium (*pl.* ***syncytia***) Animal tissue formed by the fusion of cells to form a multinucleate mass of PROTOPLASM, for example VOLUNTARY MUSCLE and TROPHOBLAST.

syndiotactic polymer, *syntactic polymer* A POLYMER in which the substituted carbons (e.g. METHYL GROUPS) are arranged alternately above and below the plane of the carbon chain, if it is considered that the carbon atoms all lie in the same plane. One form of POLYVINYL CHLORIDE has this arrangement. *Compare* ATACTIC POLYMER, ISOTACTIC POLYMER.

synergistic (*adj.*) Describing the interaction of two or more substances, organs or organisms to produce an effect greater than the sum of their individual effects. *Compare* ANTAGONISTIC.

synovial fluid A fluid that is found in movable JOINTS in vertebrates. The synovial fluid provides lubrication and nutrients for the cartilage at the end of each bone. It is secreted by the synovial membrane.

synovial joint *See* JOINT.

synovial membrane *See* JOINT.

syntactic polymer *See* SYNDIOTACTIC POLYMER.

synthesis A chemical reaction in which a more complex compound, such as an organic molecule, is manufactured from simpler materials.

syrinx The vocal organ of a bird, located at the lower end of the trachea. *See also* LARYNX.

systems program A piece of software, such as the OPERATING SYSTEM, that controls the performance of a computer system. *Compare* APPLICATIONS PROGRAM.

T

2,2,4-trimethylpentane, *iso-octane* See OCTANE.

tachyon A hypothetical particle travelling faster than the speed of light. This is not forbidden by the SPECIAL THEORY OF RELATIVITY, which only forbids objects from accelerating past the speed of light. Tachyons would have some very distinctive properties if they did exist, but as yet there is no experimental evidence for them.

tangent 1. A line that touches a curve, especially a circle, at one point. The GRADIENT of the line is equal to the gradient of the curve at that point.

2. A function of angle. In a right-angled triangle, the tangent of an angle is equal to the length of the side opposite the angle divided by the side adjacent to the angle. The tangent of any angle is equal to its SINE divided by its COSINE.

tantalum (Ta) The element with ATOMIC NUMBER 73; RELATIVE ATOMIC MASS 180.9; melting point 2,996°C; boiling point 5,427°C; RELATIVE DENSITY 16.7. Tantalum is a TRANSITION METAL, used in some chemical process for its catalytic properties. It is also used in the manufacture of some small electronic components, particularly high quality ELECTROLYTIC CAPACITORS.

tapetum A layer behind the RETINA of the eye in some nocturnal animals, such as cats. It reflects light and improves vision in dim light by providing more opportunities for the light-sensitive cells of the retina to absorb light (hence cat's eyes light up when the light is shone in to them).

tap root The single main ROOT of plants which grows vertically downwards. The tap root is often used for food storage, for example in the carrot.

tar Any of the various dark sticky substances obtained by the DESTRUCTIVE DISTILLATION of organic matter, such as coal, wood or peat, or by the refining of petroleum. Tar is used to cover roads and to stop timber from rotting.

tarnish (*vb.*) Of a metal, particularly silver, to lose the characteristic metallic LUSTRE, due to the formation of compounds, usually OXIDES, by reaction with the atmosphere.

tartrazine, *E102* A DYE used for giving yellow colour to food. It is associated with hyperactivity in children and skin and respiratory problems in those with an allergy to it.

tastebud A chemoreceptor (a RECEPTOR cell that detects chemicals) specialized for taste. In vertebrates, tastebuds are concentrated on the upper surface of the TONGUE.

tau lepton, *tauon* (τ) The heaviest known LEPTON, with a mass over 3,000 times that of the electron, but a HALF-LIFE of only 3×10^{-13} s.

tauon *See* TAU LEPTON.

tautomerism A form of isomerism (*see* ISOMER) in which a material is able to turn from one of its isomers into the other, with an EQUILIBRIUM being reached between the isomers. Such isomers are called tautomers.

taxis (*pl. taxes*) The directional movement of a freely motile organism, or part of one, in response to an external stimulus. Movement towards the stimulus is positive taxis and movement away from a stimulus is negative taxis. Phototaxis is the movement of an organism in response to light, for example green ALGAE swim towards light to increase the rate of PHOTOSYNTHESIS. Chemotaxis (chemical stimulus) is common in many bacteria, which move towards higher concentrations of nutrients. Thermotaxis is movement in response to temperature. *See also* NASTIC MOVEMENT, TROPISM.

taxonomy The study of the CLASSIFICATION of living organisms.

TCA cycle *See* KREBS CYCLE.

T cell, *T lymphocyte* A type of LYMPHOCYTE that is formed in BONE MARROW and passes through the THYMUS (hence *T* cell) before settling in the SPLEEN or a LYMPH NODE.

T cells are involved in CELL-MEDIATED IMMUNITY. They only recognize foreign ANTIGENS in association with cell antigens of the MAJOR HISTOCOMPATIBILITY COMPLEX (MHC). This is called MHC restriction. There are

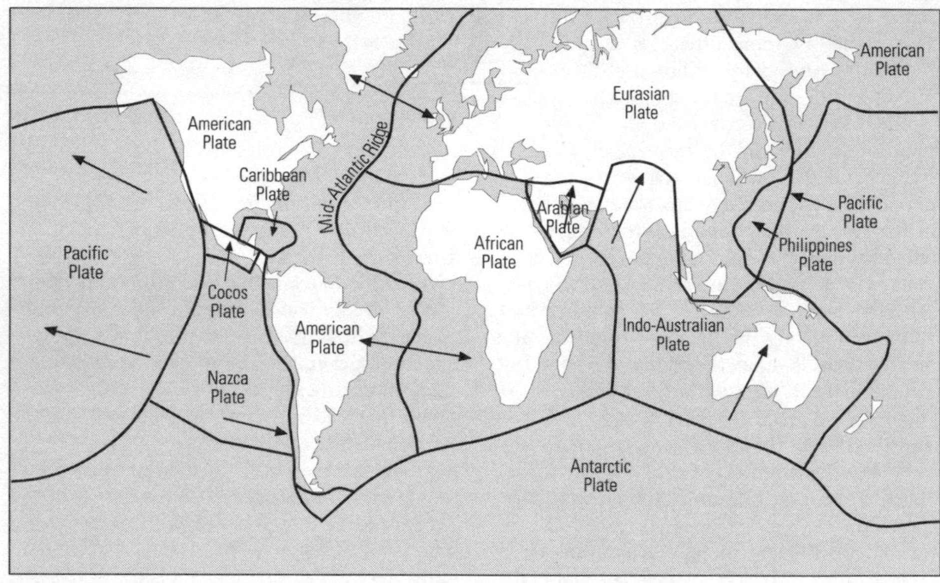

Tectonic plates.

several subsets of T cells: CYTOTOXIC T cells (Tc) that recognize and destroy infected cells; T helper cells (Th) that help B CELLS in their ANTIBODY production; suppressor T cells (Ts) that specifically suppress the IMMUNE RESPONSE; and macrophage-activating cells that produce LYMPHOKINES to activate MACROPHAGES.

Compare B CELLS.

T-cell growth factor *See* GROWTH FACTOR.

tear gland *See* LACHRYMAL GLAND.

technetium (Tc) The element with ATOMIC NUMBER 43, melting point 2,171°C; boiling point 4,876°C. Technetium is a TRANSITION METAL, unusual in having no stable ISOTOPES despite a fairly low atomic number. It is extracted from the NUCLEAR WASTE of uranium fission and some isotopes are used as radioactive tracers in medical diagnosis.

tectonic plate A region of solid rock in the Earth's CRUST. The crust is composed of several such plates which drift around on the surface of the Earth, moving at speeds up to a few centimetres per year. This motion, called continental drift, is driven by convection currents in the MANTLE. The plates effectively float on the mantle, with removal of material from one part of the plate causing the plate to rise up, whilst other plates sink – a process called isostasy.

In some places, such as the mid-Atlantic ridge, plates are moving apart and hot mantle material solidifies. VOLCANOES are found in such areas as MAGMA forces its way to the surface in a semi-molten form called lava. As the magma cools and solidifies, it stores a record of the MAGNETIC FIELD OF THE EARTH, which has reversed many times since the Earth was formed. This enables the rate of spreading to be compared at different places on the Earth. It is believed that these plates originally supported a single large continent called PANGAEA.

tectonics The branch of GEOLOGY concerned with the study of the structure and motion of TECTONIC PLATES.

Teflon The trade name for POLYTETRAFLUOROETHENE.

telecommunications The conveying of information from one place to another, by wire or radio.

telephone A TELECOMMUNICATIONS system for the transmission of voice. The ANALOGUE speech signal is converted into an electrical signal by a microphone in the mouthpiece. This signal is transmitted via a series of EXCHANGES, by wire, fibre optic cables, microwave or satellite, to the receiving telephone, where it is converted back into speech by a loudspeaker in the earpiece. A dialling system and the exchange enables each telephone to be connected to any other telephone in the system. Telephone is an example of a DUPLEX system (two-way communication; in a SIMPLEX system communication can only be achieved in one direction).

Telephone signals are often converted to DIGITAL form if they are to be sent over long distances. In addition to the improved quality and freedom from INTERFERENCE (unwanted signals added to the signal being sent) that can be achieved with digital techniques, the individual digits take less time to send than the speech they represent. This means that several signals can be sent along the same wire, a process called MULTIPLEXING. One form of multiplexing involves sending digital signals representing several different telephone calls in turn down the same wire – this technique is called TIME-DIVISION MULTIPLEXING.

telescope A device that produces a magnified image of a distant object. Telescopes are divided into two classes, reflecting telescopes, which use a CONCAVE mirror to gather the light from the distant object, and refracting telescopes, which use a CONVERGING LENS (the OBJECT LENS). For visual use, the real image produced is then viewed through an eyepiece – a converging lens that acts as a magnifying glass. Most modern large astronomical telescopes are not used directly, however, and photographic film, a CCD or other detector is placed at the FOCUS of the mirror (modern astronomical telescopes are always reflecting telescopes).

To prevent disturbance from light pollution (man-made light sources scattering off particles in the atmosphere) and to minimize REFRACTION from layers of air at different temperatures, large astronomical telescopes are now located on mountain tops or launched into space, as was done with the HUBBLE SPACE TELESCOPE. For use in astronomy, telescopes are built with as large a mirror as possible, this is done both to collect more light, allowing fainter objects to be detected, and to minimize the effects of DIFFRACTION (the WAVEFRONTS of the light entering the telescope are limited by the size of the mirror; the larger the mirror, the less diffraction this will produce).

See also RADIO TELESCOPE.

television A system for the transmission and reception of pictures using RADIO WAVES. Television receivers rely on the CATHODE RAY TUBE. In a television set, voltages are applied to coils around the neck of the tube to produce a scanning pattern called a raster, in which the electron beam moves across the screen from left to right in a series of lines, each line just below the previous one, before returning to the top left-hand corner. As the beam moves, its brightness is controlled by the transmitted signal and a picture is built up on the screen. *See also* COLOUR TELEVISION.

telluride Any compound containing TELLURIUM and a more electropositive element (*see* ELECTROPOSITIVITY). Hydrogen telluride is covalent (*see* COVALENT BOND), but the tellurides of the alkali metals are IONIC, containing the Te^{2-} ion.

tellurium (Te) The element with ATOMIC NUMBER 52; RELATIVE ATOMIC MASS 127.6; melting point 451°C; boiling point 1,390°C; RELATIVE DENSITY 6.2. Tellurium is a METALLOID, used for its semiconducting properties in some electronic devices.

telophase The final stage of MITOSIS or MEIOSIS.

temperature A measure of the hotness or coldness of an object. If two objects are placed in contact with one another, HEAT will flow from the one at the higher temperature to the one at the lower temperature. If they are at the same temperature, there will be no heat flow. The SI unit of temperature is the KELVIN. *See also* TEMPERATURE SCALES, THERMOCOUPLE, THERMOMETER, PYROMETER.

temperature coefficient of resistivity The fractional change in RESISTIVITY for each degree CELSIUS of temperature change. The temperature coefficient of resistance is positive for metals, which have resistances that increase with increasing temperature, but negative for semiconductors, where the resistance decreases with increasing temperature. If there is a change in temperature $\delta\theta$, there will be a change in resistivity $\delta\rho$, given by

$$\delta\rho = \rho\alpha\delta\theta$$

where α is the temperature coefficient of resistivity.

temperature gradient Temperature change divided by the distance over which that change takes place.

temperature scale A system for measuring and comparing TEMPERATURES. A temperature scale defines a particular property, the THERMOMETRIC PROPERTY, which varies with temperature and is to be used for temperature measurements over a certain range. Any scale is also defined by a number of fixed points – temperatures that can be reached precisely and to which temperatures are assigned. Other temperatures can then be measured by first calibrating the THERMOMETER by noting the value of the thermometric property at two fixed points, usually one above and one below the unknown temperature, and then measuring the unknown temperature on the basis that the thermometric property varies in some prescribed way, often linearly, with temperature between the two fixed points.

The CELSIUS scale, for example, takes the freezing and boiling points of water at one atmosphere as fixed points, to which the values 0°C and 100°C are attached. The ABSOLUTE TEMPERATURE scale, or KELVIN scale, is unusual in having only one fixed point, the triple point of water; the other fixed point being effectively ABSOLUTE ZERO. The INTERNATIONAL PRACTICAL TEMPERATURE SCALE, which is used for scientific purposes, has a total of 16 fixed points from 13.8 K to 1,358 K.

See also IDEAL GAS TEMPERATURE SCALE.

temporary hardness HARDNESS in water that can be removed by boiling. It is caused by the presence of calcium (and sometimes magnesium) hydrogencarbonate in the water, which decomposes on heating to give calcium carbonate (which is insoluble):

$$Ca(HCO_3)_2 \rightarrow CaCO_3 + CO_2 + H_2O$$

See HARD WATER.

tendon A band of CONNECTIVE TISSUE consisting of parallel COLLAGEN fibres with FIBROBLASTS in between, attaching muscle to bone. Tendons have great strength but less elasticity than some LIGAMENTS.

tensile (*adj.*) Relating to TENSION.

tensile strain *See* STRAIN.

tensile strength The maximum tensile STRESS a material can withstand before it breaks.

tension A pulling force, tending to increase the length of an object.

tentacle A long, flexible protrusion used by some organisms for feeding, for example in *Hydra* and sea anemone.

tepal In plants, a subdivision of a PERIANTH that is not clearly differentiated into a CALYX and COROLLA.

tera- (T) A prefix indicating that the size of a unit is to be multiplied by 10^{12}. For instance, one Terajoule (TJ) is one million million JOULES.

terbium (Tb) The element with ATOMIC NUMBER 65; RELATIVE ATOMIC MASS 158.9; melting point 1,365°C; boiling point 3,230°C; RELATIVE DENSITY 8.2. Terbium is a LANTHANIDE metal with some uses in electronics as a PHOSPHOR in CATHODE RAY TUBES and in some semiconducting devices.

terephthalic acid *See* 1,4-BENZENEDICARBOXYLIC ACID.

terminal 1. (*electronics*) An electrical contact by which one device can be connected to another.

2. (*computing*) A KEYBOARD and VISUAL DISPLAY UNIT by which a user can communicate with a computer system, particularly where a single computer is provided with more than one such operating position.

terminal velocity The speed at which the FORCE of resistance exerted by the air on a moving body is equal to the weight of the body, or to some other force causing it to accelerate. An object dropped in air will accelerate until it reaches its terminal velocity and then continue to fall at a steady speed.

termination codon *See* STOP CODON.

terrestrial planets Planets that are similar to the Earth in that they are made mostly of rock. MERCURY, the closest to the Sun, is far smaller than the Earth and is too small to have enough gravity to have kept any atmosphere. This means there has been no erosion of its surface, which is heavily cratered, like that of the Moon. VENUS is similar in size to the Earth and has a dense atmosphere of carbon dioxide, which produces a GREENHOUSE EFFECT leading to very high temperatures. After the Earth, the next planet is MARS, which is rather smaller than the Earth and has only a thin atmosphere. The last terrestrial planet is PLUTO, about which little is known except that it is small and being so distant from the Sun it will be very cold.

Tertiary The first PERIOD of the CENOZOIC ERA. It followed the CRETACEOUS period and extended to the QUATERNARY period (that is, it covers the period of time from 65 million to 2 million years ago). The Tertiary period is subdivided into the PALAEOCENE, EOCENE, OLIGOCENE, MIOCENE and PLIOCENE EPOCHS.

During the Tertiary period, modern mammals developed and became the dominant land animals, the continents and climatic regions settled to where they are today and shrubs, grasses and other flowering plants developed.

tervalent *See* TRIVALENT.

Terylene The trade name for a synthetic POLYESTER fibre. It is made by the POLYMERIZATION of GLYCOL and 1,4-BENZENEDICARBOXYLIC ACID (terephthalic acid). Terylene is the most widely produced synthetic fibre and textiles made from these fibres are hard-wearing and wash well.

tesla (T) The SI UNIT of MAGNETIC FIELD. One tesla is equal to the field that will produce a force of one NEWTON on each metre length of a wire that carries a current of one AMPERE at right angles to the field direction.

testa The protective outer coat of a SEED that is usually hard and impermeable. It is formed after fertilization of the OVULE.

testis (*pl.* ***testes***) The male GONAD that produces SPERM and sex HORMONES. In most mammals, including humans, there are a pair of testes (or testicles) that descend from the body cavity shortly before birth and hang below the ABDOMEN in a sac called the SCROTUM. Each testis is divided into lobules containing convoluted SEMINIFEROUS TUBULES that are lined with germinal EPITHELIUM from which sperm develop by a process called spermatogenesis.

Spermatogenesis consists of a series of cell divisions in which spermatogonium divide by MITOSIS to produce primary spermatocytes. These then divide by MEIOSIS to give secondary spermatocytes, and again to give spermatids (which are HAPLOID). Spermatids are protected by the surrounding Sertoli cells and they are modified into spermatozoa. Between the tubules are INTERSTITIAL CELLS, which produce the male hormone TESTOSTERONE.

The seminiferous tubules merge to form the vasa efferentia, which are a group of small ducts through which the sperm pass into the EPIDIDYMIS. Sperm are stored in the epididymis until they gain their motility ready for use in SEXUAL REPRODUCTION.

testosterone In male vertebrates, a STEROID HORMONE produced by the INTERSTITIAL CELLS, of the TESTIS. Testosterone is responsible for the development of secondary sexual characteristics, such as hair growth, deepening of the voice, sexual behaviour and muscle development, and also for stimulating SPERM production. The production of testosterone is controlled by another hormone, the LUTEINIZING HORMONE made by the PITUITARY GLAND. Synthetic testosterone has been used to help muscular development in athletes, although its use has now been banned.

test tube A glass tube, for mixing liquid REAGENTS.

tetrachloromethane *See* CARBON TETRACHLORIDE.

tetragonal (*adj.*) Describing a CRYSTAL structure where the UNIT CELL has all its faces at right angles to one another, with two of the faces being square and the other four rectangular, so there are two different lengths characterizing the shape and size of the unit cell.

tetrahedral (*adj.*) Having the structure of a tetrahedron; that is, a figure with four triangular faces, each side having the same length. In chemistry, the word is applied to molecules or RADICALS which would fill a tetrahedron, with one central atom covalently bonded to an atom at each corner of the tetrahedron. The angle between the bonds being 109.5°. This structure is very common in organic chemistry as the carbon atom forms four single bonds with this structure. Thus methane, CH_4 has a tetrahedral structure. Other molecules, such as chloromethane, $CClH_3$, have similar structures, though the POLAR nature of the carbon–chlorine bond means that the bond angles are no longer equal.

tetrahedron A three-dimensional shape with four faces, each of which is a triangle.

tetrapod Any vertebrate that has evolved from four-legged ancestors. Tetrapods include mammals, birds, reptiles and amphibians. In some cases one pair of limbs has been lost or modified, for example into wings or flippers.

tetravalent (*adj.*) Having a VALENCY of 4.

thalamus Part of the vertebrate FOREBRAIN that is a relay centre for other regions of the brain. The thalamus interprets sensory information and compares it to previously stored information, and sends it to the appropriate area of the CEREBRUM. The thalamus consists of mainly

GREY MATTER and is associated with pain and pleasure.

thalassaemia An inherited blood disorder in which there is a gene defect involving the production of HAEMOGLOBIN. In the HOMOZYGOUS state thalassaemia causes severe anaemia and is usually fatal. It is more often presented in the HETEROZYGOUS state. There are several types of thalassaemia.

thallium (Ti) The element with ATOMIC NUMBER 81; RELATIVE ATOMIC MASS 204.4; melting point 303°C; boiling point 1,460°C; RELATIVE DENSITY 11.9. Thallium's highly toxic nature has been exploited in some pesticides.

thallus A very simple, undivided plant body with no stem, leaves or roots and often thin and flattened. It is typical of some liverworts (*see* BRYOPHYTA).

T helper cell *See* T CELL.

thebaine A highly poisonous white crystalline substance obtained from OPIUM.

theorem A mathematical result derived by the application of the laws of mathematics to a set of AXIOMS.

theory of everything (t.o.e.) The ultimate goal of particle physics, a single theory that explains the FOUR FORCES OF NATURE as aspects of a single force. A good theory of everything would also explain the origin of the masses of the QUARKS and LEPTONS, and why there are three families of these particles. As yet there are no candidate theories which fit all the experimental facts. *See also* GRAND UNIFIED THEORY, STANDARD MODEL, UNIFIED FIELD THEORY.

therm A unit of heat energy, obsolete in science but still sometimes used for commercial purposes. One therm is approximately equal to 1.06×10^8 JOULE.

thermal (*adj.*) Connected with HEAT.

thermal analysis A technique of QUALITATIVE ANALYSIS based on the detection of gases given off, or changes in mass of a solid sample, as it is slowly heated.

thermal conduction The passage of HEAT through matter without any movement of the material itself. FREE ELECTRONS provide an effective mechanism for the transfer of thermal energy from one part of the material to another, thus metals are generally better thermal conductors of heat than non-metals. For this reason saucepans are made with metal bases and plastic handles rather than the reverse. In non-metals, the thermal motion in a hotter section of the material is passed on to cooler areas via intermolecular collisions or, in the case of a solid, by the interatomic bonds. This mechanism is most effective where the material has strong COVALENT BONDS – thus diamond, for example, is a good thermal conductor. The amount of heat conducted through a material depends on the temperature difference between its ends and on the THERMAL RESISTANCE of the material. *See also* THERMAL CONDUCTIVITY.

thermal conductivity The property that measures how well a material conducts heat, independently of the size of the sample being considered. It is the amount of heat flow produced per metre squared of area by a TEMPERATURE GRADIENT of one KELVIN per metre, expressed in $WK^{-1}m^{-1}$. The heat flow, P, through a piece of material with a thermal conductivity k of length l and cross-section A is

$$P = kA\Delta\theta/l$$

where $\Delta\theta$ is the temperature difference between the two ends of the material. *See also* LEE'S DISC, SEARLE'S BAR.

thermal contact resistance An additional THERMAL RESISTANCE that accounts for the fact that actual rates of heat flow are often much smaller than those produced by simple calculations. In a room, for example, the heat flow through the window on a cold day will be less than that suggested by the thermal resistance of the glass because the air just next to the window, within a millimetre or so, will be much cooler than the remainder of the air in the room. *See also* DOUBLE GLAZING.

thermal equilibrium The state of a system where all parts of the system are able to exchange heat energy with one another, but there is no net flow of energy from one part of the system to another.

thermal expansion The increase in length, area or volume of an object as a result of increasing temperature. In broad terms expansion can be explained by saying that molecules move further apart on average as they vibrate more. Not all materials expand on heating, water in the temperature range 0°C to 4°C being an important exception. *See also* EXPANSIVITY.

thermal radiation ELECTROMAGNETIC RADIATION in the INFRARED part of the ELECTROMAGNETIC SPECTRUM. Thermal radiation is given off by all bodies at temperatures above ABSOLUTE ZERO.

See also BLACK-BODY RADIATION, PRÉVOST'S THEORY OF EXCHANGES.

thermal resistance The TEMPERATURE difference between the ends of a piece of material divided by the heat flow. Just as with electrical RESISTANCE, the thermal resistance of a material is proportional to the length of the sample, and inversely proportional to its cross-section. If a temperature difference $\Delta\theta$ causes a heat flow rate of P, the thermal resistance is

$$R = \Delta\theta / P$$

The rules for electrical resistors in SERIES and PARALLEL can also be applied to thermal resistance. A typical case of thermal resistances in series involves heat flow through first one material then another. Thermal resistance in parallel is a useful concept when heat can flow from one location to another via a number of routes – from the inside to the outside of a building through the walls and the windows, for example.

See also THERMAL CONDUCTION, THERMAL CONDUCTIVITY, THERMAL CONTACT RESISTANCE.

thermal runaway A problem that can lead to the self-destruction of JUNCTION TRANSISTORS when operated at high power levels. When a junction transistor becomes warm, more CHARGE CARRIERS are released so more current flows leading to further heating and so on. The circuits that set the working conditions for a transistor before any signal is applied must be designed to compensate for any such potential problems.

thermionic diode A diode that uses the effect of THERMIONIC EMISSION. An electrically heated metal filament (CATHODE) and a metal plate or cylinder (ANODE) are contained in a vacuum. If the metal plate is positive, electrons can flow to it from the heated filament, but if the voltages are reversed so the metal plate is negative, there can be no current flow as it is not hot enough to give off electrons. In electronics, the thermionic diode has now been superseded by the PN JUNCTION DIODE, but a similar device, the X-RAY TUBE, is still used to produce X-RAYS.

thermionic emission The release of electrons from a metal surface when it is heated. The effect of thermionic emission is often demonstrated in a device called a MALTESE CROSS TUBE. Thermionic emission is an ACTIVATION PROCESS with an ACTIVATION ENERGY equal to the WORK FUNCTION of the metal. *See also* CATHODE RAY TUBE, ELECTRON GUN, THERMIONIC VALVE.

thermionic valve Any device in which electrons are emitted from a heated CATHODE by the process THERMIONIC EMISSION, and are attracted to an ANODE at which they are collected. A number of other ELECTRODES may also be present to modify the electron flow. Apart from the CATHODE RAY TUBE, X-RAY TUBE and MAGNETRON, thermionic valves are largely obsolete, having been replaced by solid-state semiconductor devices, which are smaller, more robust and consume less power.

thermistor An electronic device designed to exploit the decrease in resistance of a semiconductor, such as germanium, with increasing temperature. Thermistors are used in electronic circuits that are required to respond to changing temperatures.

Thermit process A reaction used to extract magnesium and chromium from their ores and sometimes also used to produce high temperatures for welding. The oxide of the metal concerned is mixed with finely powdered aluminium and the mixture ignited by setting fire to a strip of magnesium. The aluminium is oxidized and reduces the less reactive metal, for example:

$$Cr_2O_3 + 2Al \rightarrow 2Cr + Al_2O_3$$

thermocouple A pair of junctions between two different metals. If there is a temperature difference between these junctions, the different thermal motion of the electrons in the two metals will cause an ELECTROMOTIVE FORCE by the THERMOELECTRIC EFFECT. This can be used to measure temperature.

thermodynamic equilibrium The state in KINETIC THEORY where individual molecules are exchanging quantities such as energy and MOMENTUM, or reacting chemically, but the total amount of any chemical present, or the total energy, is unchanging. Thus the system can be meaningfully described by quantities such as temperature or the chemical concentration of its constituents.

thermodynamics The study of thermal ENERGY changes and ENTROPY. *See also* ARROW OF TIME, CARNOT ENGINE, FIRST, SECOND and THIRD LAW OF THERMODYNAMICS.

thermoelectric effect, *Seebeck effect* The production of a voltage across a pair of junctions between two different metals. This is a way of converting heat energy directly to electricity, but the EFFICIENCY of energy conversion in the thermoelectric effect is low, and it is usually used as a way of measuring temperature differences rather than as an energy source. Some spacecraft, however, have used a THERMOPILE and heat from a radioactive source to provide electricity. *See also* THERMOCOUPLE.

thermoluminescence LUMINESCENCE of an object when it is heated. This luminescence, which is usually very faint, is caused by electrons becoming trapped within defects in a crystal LATTICE. When heated, lattice vibrations release the electrons. These EXCITED STATES arise as a result of exposure to IONIZING RADIATION, so increase over time. In this way thermoluminescence can be used for the dating of archaeological artefacts, especially pottery.

thermometer Any device for measuring TEMPERATURE. Thermometers are generally classified according to the THERMOMETRIC PROPERTY they exploit. Liquid in glass thermometers are based on the THERMAL EXPANSION of a liquid, usually alcohol or mercury. RESISTANCE THERMOMETERS use the change in resistance of an electric CONDUCTOR, often platinum. The CONSTANT VOLUME GAS THERMOMETER uses the variation in pressure of a fixed mass of gas held in a constant volume.

thermometric property Any property that varies with temperature, such as the volume of a liquid, the pressure of a fixed volume of gas, or the resistance of an electric conductor, and which can be used to measure temperature.

thermonasty The NASTIC MOVEMENT of plants in response to temperature. For example the crocus flower opens at 16°C and closes at temperatures below 16°C.

thermonuclear reaction A CHAIN REACTION in NUCLEAR FISSION maintained by slow moving NEUTRONS. The neutrons are described as thermal because collisions with the MODERATOR in a nuclear reactor give them a distribution of energies which can be characterized by a temperature in a similar way to distribution of energies of the molecules in a gas.

thermopile A BATTERY of THERMOCOUPLES.

thermoplastic, *thermosoftening plastic* Any PLASTIC that can be repeatedly softened on heating and hardened on cooling. In contrast to THERMOSETTING PLASTIC, these thermoplastics do not undergo cross-linking on heating and can therefore be resoftened. Examples include POLY(ETHENE), POLYSTYRENE, POLYVINYL CHLORIDE (PVC).

thermoreceptor A RECEPTOR cell that detects temperature changes. *See* SENSE ORGAN.

thermoset *See* THERMOSETTING PLASTIC.

thermosetting plastic, *thermoset* Any PLASTIC that can be moulded to shape during manufacture but which sets permanently rigid on further heating. This is due to extensive cross-linking that occurs on heating and cannot be reversed by reheating. Examples include PHENOL-FORMALDEHYDE RESINS, EPOXY RESINS, BAKELITE, POLYESTERS, POLYURETHANE and SILICONES.

thermosoftening plastic *See* THERMOPLASTIC.

thermosphere *See* ATMOSPHERE.

thermostat A temperature-controlled switch, usually connected to a heater or REFRIGERATOR for the purpose of maintaining a constant temperature. Most thermostats are based on a pair of contacts, one of which is attached to a BIMETALLIC STRIP, whilst the other can be moved on a screw thread to determine the temperature at which the thermostat operates. The contacts are often magnetized so that they attract one another in order to produce a switch that opens and closes sharply to reduce damage and interference caused by sparks.

thermotaxis *See* TAXIS.

thiamine *See* VITAMIN B.

thigmonasty, *haptonasty* The NASTIC MOVEMENT of plants in response to localized contact, for example, the leaf movements of the Venus flytrap in response to contact with an object.

thigmotropism, *haptotropism* The directional growth of a plant (or part of it) in response to physical contact, for example, tendrils of climbing plants winding around a support. *See* TROPISM.

thin-film interference The INTERFERENCE effect in which light is reflected off the front and back surfaces of a thin layer and interferes when brought to a focus in the eye. The two reflected beams have travelled different distances before reaching the eye and will interfere constructively or destructively (*see* INTERFERENCE) depending on their wavelength and the thickness of the film. The fact that different colours have different wavelengths means that some colours in the white light interfere constructively, others destructively, producing the coloured bands that are seen in soap bubbles and thin layers of oil.

thin-layer chromatography (TLC) A CHROMATOGRAPHY technique widely used for analysing the components in liquid mixtures. The stationary phase is a thin layer of an absorbent solid, such as aluminium oxide, supported on a vertical glass plate.

thiosulphate Any compound containing the thiosulphate ion, $S_2O_3^{2-}$. Sodium thiosulphate is important in the photographic process as a 'fixer', since it dissolves silver HALIDES, the light sensitive component of photographic films.

thiosulphuric acid ($H_2S_2O_3$) An unstable DIBASIC acid, known only in aqueous solution. It is the parent acid of the THIOSULPHATES. Thiosulphuric acid can be formed by the reaction between hydrogen sulphide and sulphuric acid:

$$H_2S + H_2SO_4 \rightarrow H_2S_2O_3 + H_2O$$

thixtropic (*adj.*) Describing the property of some GELS that appear to have a lower viscosity for fast flow rates than for slow ones. This property is exploited in non-drip paint, which will flow more easily at high speeds, as when being applied by a brush, than at low speeds, as in the formation of a drip.

thoracic duct *See* LYMPH VESSEL.

thorax In vertebrates, the chest cavity containing the heart and lungs and protected by the ribcage. In mammals, the thorax is separated from the ABDOMEN by a muscular DIAPHRAGM. In ARTHROPODS the separation is less clear and the thorax represents the body region between the head and the abdomen. In insects, the thorax carries the legs and wings.

The upper area of the thorax is called the PECTORAL GUIDE and is made up of the muscles and bones needed to move the arms and forelimbs.

thorium (Th) The element with ATOMIC NUMBER 90; RELATIVE ATOMIC MASS 232.0; melting point 1,750°C; boiling point 4,790°C; RELATIVE DENSITY 11.7. Thorium is radioactive and produced in nature form the ALPHA DECAY of long-lived ISOTOPES of uranium. Along with uranium and plutonium, it is one of the three FISSILE elements, but this has not been exploited commercially.

threonine An ESSENTIAL AMINO ACID, produced by the HYDROLYSIS of proteins.

threshold The smallest value of energy, or an energy-related quantity such as frequency, that is required to produce a given effect.

threshold of hearing The quietest sound that can be heard. This varies with FREQUENCY and from person to person, but for a person with normal hearing, this is assigned the level 0 dBA. *See also* DECIBEL.

thrombin An enzyme involved in the BLOOD-CLOTTING CASCADE that converts FIBRINOGEN to FIBRIN.

thrombocyte *See* PLATELET.

thrombokinase An enzyme that converts the inactive PROTHROMBIN into THROMBIN, and therefore plays an important role in the clotting of blood. *See* BLOOD, CLOTTING CASCADE.

thrust A pushing or propelling force, as produced by a JET ENGINE, for example.

thulium (Tm) The element with ATOMIC NUMBER 69; RELATIVE ATOMIC MASS 168.9; melting point 1,545°C; boiling point 1,950°C; RELATIVE DENSITY 9.3. Thulium is the least abundant of the LANTHANIDE metals and as such has found few commercial applications.

thunder The sound produced by the sudden heating and consequent rapid expansion of the air when LIGHTNING is formed.

thylakoid One of a number of closed flattened sacs containing photosynthetic pigments such as CHLOROPHYLL. *See also* PHOTOSYNTHESIS, CHLOROPLAST.

thymidine A PYRIMIDINE NUCLEOSIDE, consisting of the organic base THYMINE and the sugar RIBOSE.

thymine An organic base called a PYRIMIDINE that occurs in DNA but not in RNA.

thymosine A HORMONE that stimulates the activity of T CELLS. It is produced in the THYMUS.

thymus An organ found in the upper chest cavity in humans and consisting of primary lymphoid tissue (*see* LYMPHATIC SYSTEM). The thymus is responsible for the maturation of T CELLS, which pass through the thymus before they settle in secondary lymphoid tissue (e.g. SPLEEN and LYMPH NODES).

The thymus reaches full size at puberty and then shrinks, since its role in T cell maturation is completed early in life. Thereafter it acts purely as an endocrine organ (*see* ENDOCRINE GLAND) producing the hormone thymosine, which stimulates the activity of T cells.

thyristor, *silicon controlled rectifier* A SEMICONDUCTOR device that allows current to flow in one direction (from the ANODE to the CATHODE) on the application of a voltage to a third

ELECTRODE, called the gate. Once a current is flowing, it will continue to flow regardless of the gate voltage provided the current remains above a threshold value.

thyroid gland An ENDOCRINE GLAND of vertebrates, located in the neck region, the main role of which is in the regulation of the body's metabolic rate (*see* METABOLISM). In mammals it is a single gland but in amphibians and birds it is paired.

The thyroid gland is under the control of the PITUITARY GLAND, which produces THYROID-STIMULATING HORMONE. In response to this hormone, the thyroid produces thyroxine (T_4) and triiodothyronine (T_3), both of which are hormones derived from the amino acid tyrosine and contain four and three molecules of iodine respectively. Iodine is therefore needed in the diet for the normal functioning of the thyroid. If iodine is limited, T_3 is preferentially formed, otherwise more T_4 is produced. Some of the circulating T_4 is converted to T_3 in the lungs and liver. These hormones stimulate growth and the metabolic rate of cells by increasing the rate at which GLUCOSE is oxidized by cells. A third hormone called CALCITONIN is also produced by the thyroid, which is concerned with regulation of calcium ions in the blood. This is done in conjunction with parathormone produced by the PARATHYROID GLANDS.

An underactive thyroid (hypothyroidism) causes a reduction in an individual's metabolic rate, resulting in a reduced heart and ventilation rate, a lowered body temperature and obesity as well as mental sluggishness. A goitre (swelling in the throat) can occur. In infants the mental retardation is serious and leads to a condition called cretinism. Administration of thyroxine orally eliminates symptoms. An overactive thyroid (hyperthyroidism) leads to an increase in metabolic rate. Treatment for this is destruction of part of the thyroid by surgery or by administration of radioactive iodine.

thyroid-stimulating hormone (TSH), ***thyrotrophin*** A protein produced by the anterior PITUITARY GLAND that stimulates growth of the THYROID GLAND and its production of hormones, such as thyroxine. TSH is itself stimulated by the thyrotrophin-releasing factor (TRF) from the HYPOTHALAMUS.

thyrotrophin *See* THYROID-STIMULATING HORMONE.

thyrotrophin-releasing factor A substance, secreted by the HYPOTHALAMUS, that stimulates the THYROID-STIMULATING HORMONE.

thyroxine A hormone of the THYROID GLAND.

tidal power The generation of electricity on a commercial scale from the tidal rise and fall of water levels. A tidal power station is operating at La Rance in France and one is planned for the Severn Estuary in the UK. Problems arise from the fact that the times of peak generation do not coincide with peaks in demand, the large building costs and the environmental impact on the area flooded.

There are relatively few areas that have sufficiently strong tides or the natural basin needed to build such a scheme, but where they can be constructed they seem economically worthwhile.

tide A change in sea level caused by the different gravitational attraction of the Moon on that part of the Earth which is closest to the Moon at any time compared to that part which is furthest away. In effect one part of the Earth is pulled towards the Moon more strongly than the average whilst the opposite part is left behind. As the Earth rotates, two high and two low tides are seen each day.

Tides are also affected to a lesser extent by the gravitational pull of the Sun. When the effects of the Sun and Moon combine, large tides, called spring tides, are seen. When the Sun and the Moon are at right angles as seen from the Earth, their effects tend to cancel, producing smaller tides, called neap tides.

Tidal effects are observed elsewhere in the SOLAR SYSTEM and are believed to be responsible for the formation of RING SYSTEMS around the GAS GIANTS and for the heating of IO, producing volcanic activity.

timbre The property of a sound that the ear distinguishes in two sounds having the same amplitude and frequency, but produced by two different instruments. This is due to the differing shape of the sound wave, caused by the different HARMONIC content in the sounds. A SINUSOIDAL sound wave will produce a pure sound.

time The continuous passage of existence, originally marked by the rising and setting of the Sun, but now recorded by a wide range of oscillating devices. The SI UNIT of time is the second. The SPECIAL THEORY OF RELATIVITY recognizes that the finite speed of light means that the passage of time is not the same for all observers. *See also* SPACE-TIME, TIME DILATION.

timebase A voltage applied to the X-PLATES of the CATHODE RAY TUBE in a CATHODE RAY OSCILLOSCOPE to move the electron beam across the screen at a steady rate, to enable a graph of voltage against time to be produced.

time constant For any process where a quantity changes exponentially with time, the time constant is the time taken to fall to 1/e times its initial value if decreasing or to rise to e times its initial value if increasing, where e is the EXPONENTIAL constant. Used particularly of circuits containing a RESISTANCE R and a CAPACITANCE C, when the time constant is RC.

time dilation The effect that results in any moving clock running more slowly than when at rest according to the SPECIAL THEORY OF RELATIVITY. If an observer at rest with respect to an object observes a time interval t_0 then an observer moving at a speed v relative to the first will observe time interval t where

$$t = t_0/\sqrt{(1 - v^2/c^2)}$$

where c is the SPEED OF LIGHT. *See also* TWIN PARADOX.

time-division multiplexing A system for sending several pieces of information (such as telephone conversations) down a single link (such as a fibre optic). This is achieved by encoding the information digitally and then sending each set of information very rapidly, one after another.

tin (Sn) The element with ATOMIC NUMBER 50; RELATIVE ATOMIC MASS 118.7; melting point 232°C; boiling point 2,270°C; RELATIVE DENSITY 7.3. Tin is an abundant metal, occurring mostly as tin(IV) oxide, from which it is extracted by reducing the oxide with carbon:

$$SnO_2 + C \rightarrow Sn + CO_2$$

Tin was traditionally alloyed with copper to form bronze, which is much harder than either tin or copper, and more recently it has been used as a corrosion resistant coating to steel in tin cans which are actually made of steel dipped in molten tin.

Chemically, tin is a moderately reactive metal. It reacts with dilute acids to give hydrogen and tin(II) or tin(IV) salts, for example:

$$Sn + 2HCl \rightarrow SnCl_2 + H_2$$

tin chloride Tin(II) chloride, $SnCl_2$, or tin(IV) chloride, $SnCl_4$. Tin (II) chloride is a white solid; melting point 246°C; boiling point 652°C; RELATIVE DENSITY 2.2. It can be formed by reacting tin with hydrochloric acid:

$$Sn + 2HCl \rightarrow SnCl_2 + H_2$$

Tin(IV) chloride is a colourless liquid; melting point –33°C; boiling point 114°C; relative density 2.2. It can be produced by burning tin in chlorine:

$$Sn + 2Cl_2 \rightarrow SnCl_4$$

Tin (IV) chloride is rapidly hydrolysed (*see* HYDROLYSIS) by water:

$$SnCl_4 + 4H_2O \rightarrow Sn(OH)_4 + 4HCl$$

tin(IV) oxide (SnO_2) A white solid; sublimes at 1,850°C; RELATIVE DENSITY 7.0. Tin(IV) oxide is insoluble in water and occurs naturally. It is amphoteric, reacting with acids to form tin salts, and with bases to form STANNATES, for example:

$$SnO_2 + 4HNO_3 \rightarrow Sn(NO_3)_4 + 2H_2O$$

$$SnO_2 + 2NaOH \rightarrow Na_2SnO_3 + H_2O$$

tin sulphide Tin(II) sulphide, SnS, and tin(IV) sulphide, SnS_2. Tin(II) sulphide is a dark grey solid; melting point 882°C; boiling point 1,230°C; RELATIVE DENSITY 5.2. It can be made by heating tin and sulphur together:

$$Sn + S \rightarrow SnS$$

At high temperatures tin(II) sulphide slowly disproportionates to tin(IV) sulphide:

$$2SnS \rightarrow SnS_2 + Sn$$

Tin(IV) sulphide is a golden yellow solid; decomposes on heating; relative density 4.5. It is made by passing hydrogen sulphide through a solution of a soluble tin (IV) salt, for example:

$$Sn(NO_3)_4 + 2H_2S \rightarrow SnS_2 + 4HNO_3$$

tissue A group of CELLS, the same or different, bound together in animals by the EXTRACELLULAR MATRIX, and in plants by the CELL WALLS, that together perform a specific function. Examples include squamous EPITHELIUM (consisting of a single cell type) and MUSCLE (consisting of more than one cell type) in animals, and PARENCHYMA (single cell type) or

XYLEM (more than one cell type) in plants. There are four main types of animal tissues – epithelial, connective, muscular and nervous – and two main types of plant tissues – simple and compound. The study of tissues is called histology. *See also* CONNECTIVE TISSUE, EPITHELIUM, NERVOUS SYSTEM.

tissue culture A technique for maintaining living cells taken from a plant or animal under controlled, sterile conditions in the laboratory. *In vitro* refers to tissue culture outside the plant or organism.

Cells are usually grown in culture dishes in a liquid media that provides nutrients and a balanced pH, and placed in an incubator to maintain a constant temperature, oxygen and carbon dioxide balance and humidity close to that within the organism. Cells can be grown as a monolayer, where they attach to the base of the culture vessel, or as a cell suspension, where no attachment is made and the culture vessel is rotated during the culture period.

A primary culture refers to the original culture established from the tissue of the organism. So that the cells can continue to survive and grow, once they have filled the culture vessel they have to be divided, or subcultured, between more vessels. These are then called secondary cultures. Normal cells grow for a limited number of cell divisions (50–100) but if a culture is treated with a chemical or virus to induce the formation of cancer cells, these are said to be transformed and will divide indefinitely.

Animal tissue culture is used as a research tool for understanding cell functions and interactions better. Tissue culture also has many practical uses, such as in vaccine preparation, production of drugs and in the production of MONOCLONAL ANTIBODIES.

Plant cultures can be used to manufacture useful products such as codeine from poppies for pain relief, but many of these processes are uneconomical. Plant tissue culture is also used to generate plants for agricultural or horticultural use.

See also CONTACT INHIBITION.

tissue fluid, *interstitial fluid* The fluid that bathes cells at the correct PH and salt concentration. Tissue fluid is derived from blood PLASMA by filtration through the capillaries in tissue; it therefore does not have any cells and has a lower protein level than plasma. It acts as a route for DIFFUSION of substances between cells and blood.

Titan The largest satellite of SATURN, notable as one of the few satellites large enough to retain its own atmosphere.

titanium (Ti) The element with ATOMIC NUMBER 22; RELATIVE ATOMIC MASS 47.9; melting point 1,660°C; boiling point 3,280°C; RELATIVE DENSITY 4.5. Titanium's high strength and low density make it an important component of some alloys used in the aerospace industry, but its high cost and difficult engineering properties have prevented its more widespread use.

titanium dioxide *See* TITANIUM(IV) OXIDE.

titanium(IV) oxide, *titanium dioxide* (TiO_2) A white solid; decomposes on heating. Titanium dioxide occurs in nature and is widely used as a white pigment in paint and paper.

titration A technique used in QUANTITATIVE ANALYSIS in which a measured quantity of one reagent is added to another until the chemical reaction between them is complete. The volume of the first reagent is usually measured with a PIPETTE, whilst the second regent is added through a BURETTE until the reaction reaches the end point, which is usually determined by a colour change. If the reaction does not naturally produce a colour change, an INDICATOR is used. *See also* END-POINT, EQUIVALENCE POINT.

T lymphocyte *See* T CELL.

TNT, *trinitrotoluene* ($CH_3C_6H_2(NO_2)_3$) A yellow solid; melting point 81°C. It is prepared from methylbenzene (toluene) by using sulphuric and nitric acids. TNT is a powerful explosive, used as a filling for shells and bombs.

tocopherol *See* VITAMIN E.

tokamak A device for producing energy from NUCLEAR FUSION. In a tokamak, a PLASMA is contained in a torus or ring shape, with the plasma effectively forming the SECONDARY COIL of a TRANSFORMER. Magnetic fields are used to confine the plasma and to induce currents in it, which heat it to the temperatures required for fusion. No tokamak has yet been built that produces a greater energy output than input, but larger machines are proposed, which should be more efficient than the small experimental machines currently operating.

token ring network A form of LOCAL AREA NETWORK. The computers in the network continually pass an electronic signal, or token, from one to another, in order. Any computer on the network can only send data when it holds the token. This prevents collisions occurring when two computers try to send data at the same time. *Compare* ETHERNET.

Tollen's reagent A solution used to distinguish ALDEHYDES from KETONES. The reagent is prepared by mixing sodium hydroxide and silver nitrate to give a brown PRECIPITATE of silver oxide. This silver oxide is then dissolved in aqueous ammonia and when warmed with aldehydes a deposit of silver is left on the test-tube like a mirror. Ketones do not do this.

toluene The common name for METHYLBENZENE.

tomography A MEDICAL IMAGING technique that overcomes the limitations of RADIOGRAPHS due to problems in their interpretation and the reduction in sharpness caused by the scattering of X-rays within the patient. In tomography, the amount of X-rays able to penetrate the patient is measured over a wide range of directions. A computer then reconstructs this information into a measurement of the density of the patients body tissues at each point within the patient. The information is then displayed on a screen as a series of images of slices through the body.

toner Finely powdered ink used in electrostatic printing systems such as the PHOTOCOPIER and LASER PRINTER. The toner is transferred to a piece of paper by a roller and then attached by

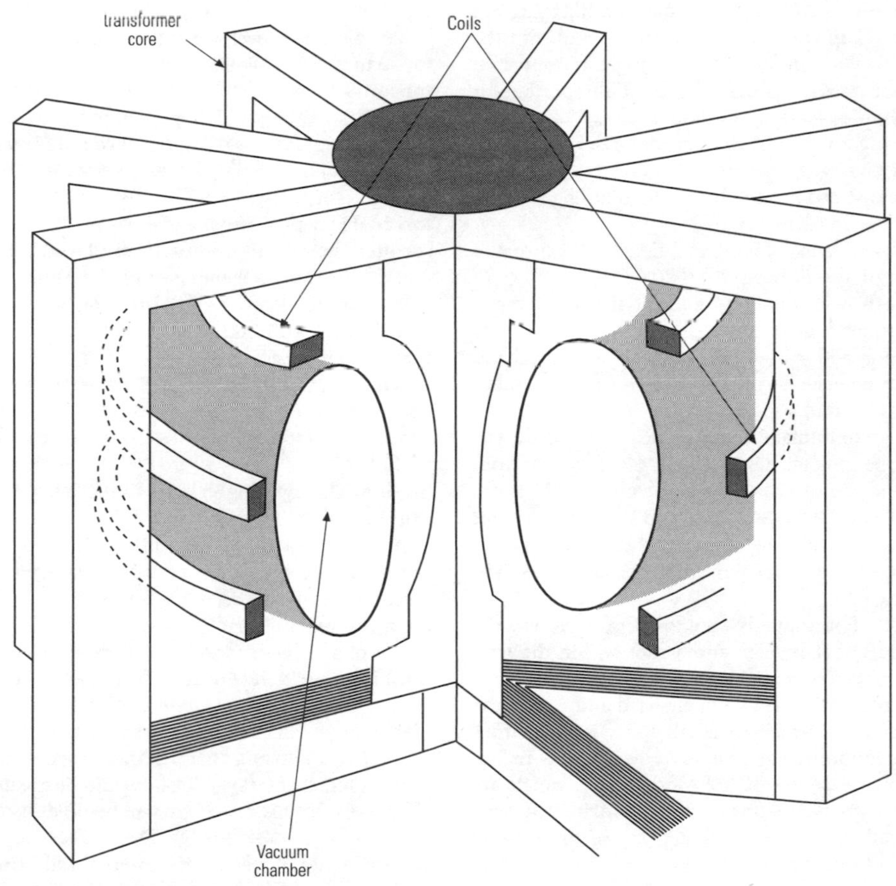

Tokamak.

passing the paper over a heater or heated roller, which melts the toner powder.

tongue In vertebrates, a muscular organ attached to the floor of the mouth and is the SENSE ORGAN for taste. There is a MUCOUS MEMBRANE covering the tongue that contains nerves and tastebuds. These are chemoreceptors (RECEPTOR cells that detect chemicals) involved in detecting taste. Chemoreceptors that detect sweet chemicals are found at the tip of the tongue, those that detect sour chemicals at the sides, those that detect bitter chemicals at the back, and those that detect salt are found all over the tongue. The tastebuds ensure rejection of unsuitable food.

The tongue is also important in assisting chewing and swallowing of food, by directing food to the teeth and then pushing chewed food to the back of the mouth and into the PHARYNX. In humans, the tongue is important for speech; in other animals, it is important for lapping water and grooming.

tonne A unit of MASS. One tonne is equivalent to 1,000 kg.

tonoplast The membrane surrounding the VACUOLE in plant cells.

tonsil A mass of lymphoid tissue in the throat of humans. *See also* LYMPH NODE.

tooth A hard structure in the mouth of vertebrates that is embedded in the bones of the jaws and is used for biting, chewing food and in defence. Dentition is the type and number of teeth in a species.

In humans, there are two sets of teeth. The first milk teeth or deciduous teeth appear from the age of 6 months and consist of 20 teeth. The milk teeth are replaced by the permanent teeth, including a further 12 teeth, from the age of 5 years. There are 32 permanent teeth in total.

The tooth is composed of a crown, the part that is seen, and a root within the jawbone. The outer layer of the crown is called enamel and is extremely hard and covers the dentine, which although softer is still harder than bone. The pulp cavity is a hollow region in the centre of the tooth within which are nerves and blood vessels that provide nutrients and oxygen to the living cells of the dentine and remove waste products. The roots are surrounded by cement to anchor them firmly in their sockets and the gum (soft tissue) that surrounds the base of the tooth.

In some vertebrates, the teeth have a similar shape and differ only in size, but in mammals there are four distinct types of teeth. These are incisors, canines, premolars and molars. The development of these teeth varies between mammals, for example CARNIVORES have well developed canines but in HERBIVORES they may be absent.

Tooth decay is a modern-day problem due mainly to the consumption of sugary foods, which stick to teeth and gums and provide an area for bacteria to thrive.

top In particle physics, a heavy QUARK with charge $+^{2}/_{3}$ of the electron charge. The top quark was long predicted by the STANDARD MODEL of particle physics, and discovered in 1993.

topology The study of shapes and the properties that they retain even when they are distorted.

top-pan balance *See* BALANCE.

topsoil *See* SOIL.

Toricellian vacuum The space above the mercury in a MERCURY BAROMETER. It contains mercury vapour at its SATURATED VAPOUR PRESSURE, which is very low.

toroidal Having the shape of a TORUS.

torque The turning MOMENT produced by a FORCE. Torque is a measure of the ability to produce acceleration in a rotating machine such as an INTERNAL COMBUSTION ENGINE.

torr A unit of PRESSURE used in high-vacuum technology. One torr is equivalent to the pressure produced by 1 mm depth of MERCURY in the Earth's GRAVITATIONAL FIELD (approximately 133 Pa).

torsion The twisting effect produced by two opposing FORCES acting on an object at different points and having different LINES OF ACTION. The size of this effect is measured by the COUPLE produced by the forces.

torus A doughnut shape, formed by bending the ends of a cylinder round until they meet.

total internal reflection The situation where a wave travelling in a dense material meets a boundary with a less dense material at such an ANGLE OF INCIDENCE that the ANGLE OF REFRACTION predicted by SNELL'S LAW would be greater than 90°. In this case there will be no REFRACTION, but only REFLECTION. The smallest angle of incidence at which total internal reflection takes place is called the critical angle. The critical angle for glass is about 42°, which means that totally internally reflecting PRISMS can be made which will change the direction of light by 90° or 180° without any need for a reflective coating.

toxin A poison released by plants, animals and bacteria, that can cause a disease. *See also* ANTITOXIN.

trace element An element that is required in small quantities for certain metabolic functions (*see* in a plant or animal. Most are METABOLISM) metals and iron, copper, zinc and magnesium are amongst the most important.

tracer technique The use of RADIOACTIVITY to determine the path of a particular element in some system. Since radioactive ISOTOPES are chemically no different from stable isotopes of the same element, they can be used to see how a system handles a particular element by introducing a radioactive isotope and monitoring the flow of radioactivity through the system. Examples of the use of tracer techniques include medical diagnosis and the study of the rate of wear in engines.

trachea, *windpipe* In air-breathing vertebrates, the main airway leading into the lungs, extending from the LARYNX. In humans the trachea is about 10 cm long and is strong but flexible due to its reinforcing rings of CARTILAGE along its length. The trachea branches into two tubes called bronchi (*see* BRONCHUS) that carry the air into each of the lungs. The bronchi and trachea have glands secreting a sticky liquid called MUCUS that collects dust and other unwanted particles which are then pushed out towards the mouth for swallowing, aided by CILIA on the walls of both.

In insects, the equivalent system is the tracheal system that consists of small tubes called tracheae opening to the outside at SPIRACLES.

See also BREATHING, RESPIRATORY SYSTEM.

tracheid *See* XYLEM.

tracheole A small branch of the tracheal system in insects (*see* SPIRACLE).

trajectory The path of an object, especially a PROJECTILE or other object moving solely under the influence of gravity.

transactinide element Any element with an ATOMIC NUMBER of 104 and above. Only those with atomic numbers 104, 105 and possibly 106 have been synthesized.

transamination The conversion of one AMINO ACID to another to replace deficient non-essential amino acids. Transamination occurs in the vertebrate LIVER.

transcription Part of the process of PROTEIN SYNTHESIS in living cells, involving the formation of a strand of MESSENGER RNA (mRNA) from a DNA template. The mRNA then carries the information necessary for the actual synthesis of proteins. Transcription involves the action of the enzyme RNA polymerase, which breaks the HYDROGEN BONDS holding the two DNA strands together so that a portion of the DNA strand unwinds to expose the NUCLEOTIDE bases. Complementary RNA nucleotides are then attracted and form BASE PAIRS with the DNA nucleotides. The mRNA moves from the cell nucleus, where it was made, to the CYTOPLASM through the nuclear pores, and towards the RIBOSOMES where it is then translated into proteins (*see* TRANSLATION).

transducer Any device for converting changes a non-electrical signal (pressure, temperature, sound, etc.) to electrical signals, and vice versa. Electronic circuits may then convert these signals to DIGITAL form for processing.

transect A systematic method of sampling the numbers and types of animals and plants in a HABITAT. A line transect consists of a piece of string placed along the ground. Any plant or animal touching or covering the line is recorded. This type of sampling is useful where there is a transition of organisms across an area, for example a sea shore. The height of the line can be varied where this is a major factor in determining the distribution of species. A belt transect is similar to a line transect except that a second line is placed parallel to the first. The species between the lines are systematically recorded a metre at a time, sometimes using a QUADRAT alongside the line transect. *See also* MARK.

transfection 1. In PROKARYOTES, the uptakes of BACTERIOPHAGE DNA by TRANSFORMATION to produce a bacteriophage infection.

2. In EUKARYOTES, a term often used to refer to the uptake of foreign DNA by cultured mammalian or other animal cells. *See also* TRANSFORMATION.

transferase Any one of a group of enzymes that transfer a chemical group from one substance to another, for example phosphorylases transfer PHOSPHATE groups.

transfer RNA (tRNA) A type of RNA, making up 10–15 per cent of the total RNA of a cell, concerned with PROTEIN SYNTHESIS. Transfer RNA is a small single-stranded molecule of about 80 NUCLEOTIDES that forms a clover leaf shape, with one end being the point where a specific amino acid attaches and the other end carrying a specific three base ANTICODON sequence. There are about 20 different types of tRNA, each one combining with a different amino acid and carrying it to MESSENGER RNA (mRNA) where the anticodon sequence matches a three-base codon sequence carried on mRNA. *See also* TRANSLATION.

transformation 1. (*mathematics*) A mapping that shifts all points in a space, or all values of a variable, to new points or values according to a specified procedure. In geometry, a transformation may be a TRANSLATION, ROTATION, REFLECTION, ENLARGEMENT or SHEAR.

2. (*biology*) A change in certain cells (often bacteria) due to their uptake of foreign DNA. This can be as a result of culturing the cells with other killed cells or culture filtrates. The cells acquire characteristics encoded by the foreign DNA and are able to pass these onto their offspring. *See also* TRANSFECTION.

3. (*biology*) An event resulting in the production of a CANCER cell from a normal animal cell. This may be due to infection of the cell or to other factors, such as a chemical stimulus. *See* CARCINOGEN.

transformer A device for changing the voltage of ALTERNATING CURRENT signals and power supplies. A transformer consists of two coils wound on a LAMINATED SOFT iron core. An alternating current in one of these coils (the primary coil) causes a changing magnetic field in the core that induces an alternating current in the other coil (the secondary coil). By altering the ratio of the number of turns in the secondary to the primary (called the TURNS RATIO), the ratio of the secondary to primary voltages can be altered. This ratio is equal to the turns ratio. A transformer with a turns ratio greater than one is called a step-up transformer and will produce an output voltage higher than the input, though at a lower current. Step-down transformers are those with a turns ratio less than one – they produce low voltage outputs but are capable of delivering higher currents. Step-down transformers are needed to generate the high currents required in arc welding, for example.

Transformers are never 100 per cent efficient. Energy losses are divided into two classes – COPPER LOSSES, caused by the electrical RESISTANCE of the transformer coils, and IRON LOSSES, resulting from hysteresis and EDDY CURRENTS in the core.

Transformers are used in electrical power transmission systems. Lower currents are needed at high voltages to transmit a given amount of power, which makes for cheaper cables. Transformers are used to step-up the voltage where the electricity is generated and then to step it down when it reaches the towns where it is used.

If the number of turns in the primary coil of a transformer is n_p and in the secondary n_s, the primary and secondary voltages and currents V_p, V_s and I_p, I_s, then

$$V_s/V_p = n_s/n_p$$

and if the transformer is 100 per cent efficient,

$$V_pI_p = V_sI_s$$

transgenic (*adj.*) Describing an organism that contains foreign GENES, added by GENETIC ENGINEERING. Transgenic mice have become a research tool.

transistor *See* FIELD EFFECT TRANSISTOR, JUNCTION TRANSISTOR.

transition metal An element containing one or two electrons in the outer S-ORBITAL and one to 10 electrons in the outer D-ORBITALS, but no electrons in the next P-ORBITALS. The transition metals show similar chemical properties. They are all metals and often exhibit more than one OXIDATION STATE. They are mostly hard materials and good conductors of heat and electricity. Many form coloured IONS and CO-ORDINATION COMPOUNDS.

transition point For a body moving through a fluid, the point on the surface of the body where the flow of fluid around the body changes from being STREAMLINED to TURBULENT.

translation 1. (*mathematics*) A geometrical operation that maps any point onto another point by moving it by a specified distance in a specified direction. *See also* SHEAR.

2. (*biology*) The stage of PROTEIN SYNTHESIS in which the information carried in the MESSENGER RNA (mRNA) strand formed during TRANSCRIPTION is used to form chains of AMINO ACIDS that are ultimately assembled into PROTEINS.

The order in which amino acids are linked to form POLYPEPTIDES is dictated by the CODONS

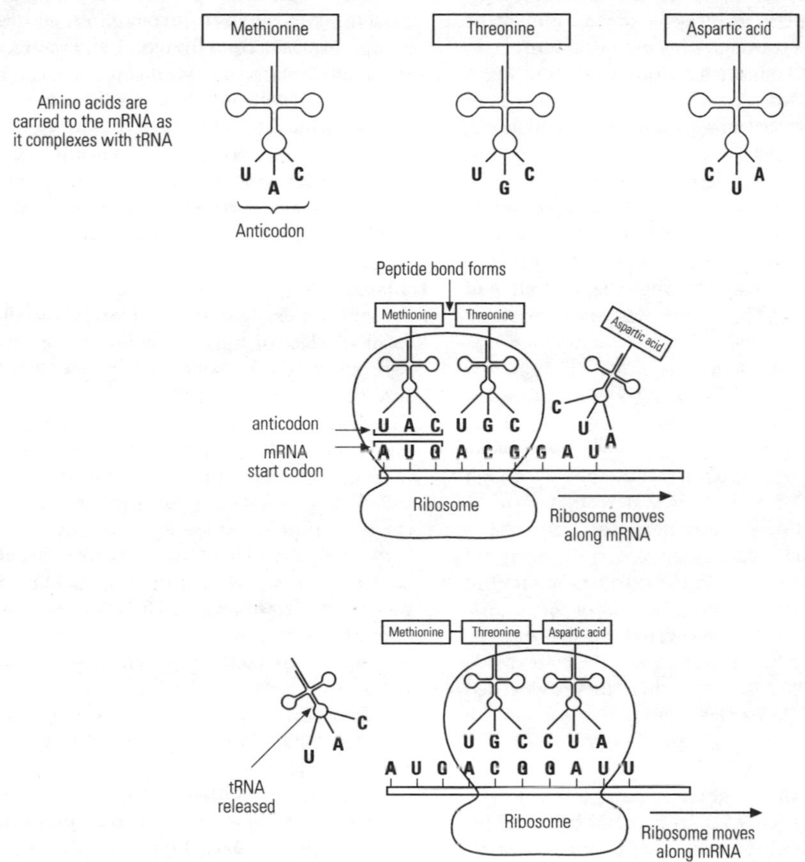

Protein synthesis – translation.

(a triplet of nucleotide bases) on the mRNA. There is always a START CODON, AUG, at one end of the mRNA, which codes for the amino acid methionine. RIBOSOMES attach to the mRNA at this start codon and attract the appropriate complementary anticodon on the TRANSFER RNA (tRNA), which is carrying the amino acid specific to that codon. This continues along the mRNA, with the ribosome moving along holding the mRNA/tRNA complex together until the amino acids have been released and peptide bonds formed between them. When a STOP CODON is reached on the mRNA, the polypeptide chain is released from the mRNA. The tRNA is free to go and get another amino acid. Many polypeptide chains maybe synthesized at the same time by several ribosomes moving along the mRNA one after another. The polypeptides are then assembled into proteins.

translocation In botany, the long-distance transport of water and minerals in a plant. There are three pathways responsible for this movement throughout the plant. (i) The apoplast pathway is the movement of substances in the cell walls of plants. Water enters air spaces between the CELLULOSE fibres of the cell wall of one cell, and as it evaporates through the stomata (*see* STOMA) a tension is created that pulls the water to the next cell wall. This pathway carries the most water. (ii) The symplast pathway is the movement of substances through the cytoplasm of cells, which are connected by tiny strands of cytoplasm called PLASMODESMATA. Water passes from cell to cell along a water potential gradient that results from loss of water by TRANSPIRATION. It is of less importance than the apoplast pathway. (iii) The vacuolar pathway is the movement of water by OSMOSIS along a water potential gradient through the VACUOLE of adjacent cells and through the cell walls and cytoplasm. This contributes least to water movement.

Water and minerals move through the roots and leaves of a plant by these pathways. Water is drawn from the ROOT HAIRS to the XYLEM and upwards to the leaves in the TRANSPIRATION STREAM. Minerals can move through the plant dissolved in water; they can be absorbed by root hairs either passively (DIFFUSION) or by ACTIVE TRANSPORT, and can travel through the xylem to the cells where they are needed. The organic products of PHOTOSYNTHESIS are also translocated throughout the plant from the leaves where they are made through the PHLOEM to where they are needed for growth or storage.

See also CHROMOSOME.

translucent (*adj.*) Describing a material through which light can pass, but with the light being scattered, so that an object cannot be seen clearly through the material. *Compare* OPAQUE, TRANSPARENT.

transmission electron microscope A type of ELECTRON MICROSCOPE where the material to be examined is preserved in a suitable fixative and then embedded in plastic EPOXY RESIN, such as Araldite, so that ultrathin sections can be cut, using an ULTRAMICROTOME. This is necessary because electrons cannot penetrate materials very well. Sections are then stained by various methods to improve their electron scattering ability, often involving HEAVY METALS, and supported on a metal grid that allows electrons to penetrate the section. Electrons are absorbed by some regions of the material (electron dense regions) but penetrate other electron transparent regions to hit the viewing screen and fluoresce. *See also* SCANNING ELECTRON MICROSCOPE, SCANNING TRANSMISSION ELECTRON MICROSCOPE.

transmittance The proportion of incident ELECTROMAGNETIC RADIATION transmitted through a partially TRANSPARENT material, rather than being absorbed or reflected. For a perfectly transparent material the transmittance is 1, for a perfectly OPAQUE material, it is 0.

transmutation The changing of one element into another as a result of RADIOACTIVITY, possibly after bombardment with neutrons in a NUCLEAR REACTOR. Some of the elements produced in this way are not naturally occurring.

transparent (*adj.*) Describing a material through which light, or some other specified form of electromagnetic radiation, can pass without scattering, so that an object can be seen through the material.

transpiration The loss of water from a plant by evaporation. Most water (up to 90 per cent) is lost through pores in the leaves or some stems called STOMATA and the rest through the CUTICLE on the surface of leaves. In woody stems a very small amount of water is lost through LENTICELS. Loss of water from the leaves causes water to be drawn upwards from the roots in a continuous TRANSPIRATION STREAM, although this is not an essential means of obtaining water.

Transpiration seems to be a side-effect of the need to have holes in the leaves for gaseous exchange. The opening and closing of the stomata (and thus the control of transpiration) is determined by the GUARD CELLS that surround them. Internal differences (or adaptions) between plants can affect the rate of transpiration, for example, leaf area, thickness of cuticle and density of stomata (*see* XEROPHYTE, HYDROPHYTE).

See also TRANSLOCATION.

transpiration stream The movement of water in plants by TRANSLOCATION from the ROOTS upwards to the leaves via the XYLEM. Movement is due to the cohesive forces between water molecules (causing them to stick together, *see* COHESION) and the adhesive forces (*see* ADHESION) between the molecules and the wall of the xylem. This creates a tension in the xylem, pulling the water up as some is drawn out of the xylem and across the leaves by TRANSPIRATION. This is called the 'cohesion tension theory'. The movement of water by the transpiration stream is not an essential means of obtaining water, as OSMOSIS would serve this purpose.

transplantation The artificial transfer of a tissue or organ from one animal to another or to a different part of the same animal. A transplantation is usually life-saving.

Most body cells have ANTIGENS on their surface that are unique to the individual (self-antigens). These are coded for by a gene complex called the MAJOR HISTOCOMPATIBILITY COMPLEX (MHC). If skin, for example, is transplanted from one part of the body to another, or between identical twins, these antigens are the same and therefore the transplanted tissue can establish and grow. However, if an organism or tissue is transplanted to another individual, the antigens would be recognized as foreign by the recipient's IMMUNE SYSTEM and rejection of the transplant would occur and it would die. To avoid this, the MHC antigen on the donor's and recipient's cells are matched as much as possible and immunosuppressive drugs are given to reduce the IMMUNE RESPONSE mounted. Despite these precautions, transplants are not always successful.

Kidneys are the organs most successfully transplanted (since 1950), but hearts, lungs, livers, pancreas, bone and bone marrow are also transplanted. The term 'grafting' is used when a small part of an organism is transplanted to another, and usually involves a closer union of tissue than with organ transplants, for example corneal grafting to restore sight to a damaged or diseased eye, and skin grafting. Most transplant material is taken from cadavers (dead bodies).

transport coefficient Any quantity that describes some property of a gas that depends on intermolecular collisions to transfer some quantity from one part of the fluid to another. Examples include THERMAL CONDUCTIVITY, in which energy is transferred from one molecule to another by collisions; VISCOSITY, which relies on molecular collisions to transfer MOMENTUM; and DIFFUSION, where collisions limit the spread of molecules from one place to another. *See also* KNUDSEN REGIME, MEAN FREE PATH.

transposition 1. The act of moving a quantity from one side of a mathematical EQUATION to the other, for example $x + 3 = y$ can have the 3 transposed to give $x = y - 3$.

2. The act of exchanging two items in a SEQUENCE, for example 1, 2, 4, 3 is a transposition of 1, 2, 3, 4.

transuranic element Any element with ATOMIC NUMBER greater than 92 (uranium). All such elements have only radioactive ISOTOPES, with HALF-LIVES that are short compared to the age of the Earth, and so do not occur in nature.

transversal A line that intersects two or more other lines.

transverse wave A WAVE in which the motion of the particles in the wave is at right angles to the direction of propagation. The wave can be thought of as a series of crests and troughs. Examples are waves on a water surface and ELECTROMAGNETIC WAVES. Transverse waves can be further described by their POLARIZATION. *Compare* LONGITUDINAL WAVE.

trapezium A closed shape having four straight sides, two of which are parallel. If the lengths of the parallel sides are a and b, and the distance between them is h, the area A of the trapezium is

$$A = {}^1/_2(a + b)h$$

travelling wave *See* PROGRESSIVE WAVE.

Trematoda A CLASS of the PHYLUM PLATYHELMINTHES (flatworms). The Trematoda are all PARASITES and are of great economic importance. Examples are the liverfluke (*Fasciola hepatica*) of sheep and cattle, with the snail as an intermediate host, and the blood-fluke (*Schistosoma*) of humans, which causes bilharzia or schistosomiasis.

triac A semiconductor device widely used in the control of the power of ALTERNATING CURRENTS. A triac does not conduct until a voltage is applied to an ELECTRODE called the GATE. It will then conduct in either direction, regardless of the gate voltage, until the current through it falls below a THRESHOLD value.

triangle A closed two-dimensional shape having three straight sides. The angles between the three sides add up to 180° and the area of a triangle is equal to one half the length of one side multiplied by the perpendicular distance between that side and the opposite point of the triangle. Triangles are described as equilateral if all the sides have the same length, isosceles if two sides have the same length, and scalene if all three sides have different lengths.

Triassic The first PERIOD of the MESOZOIC ERA. It began about 225 million years ago and lasted about 35 million years until the JURASSIC period. During this period there was a continuation of the desert conditions of the PERMIAN era. Marine animals diversified and reptiles began to dominate on land.

tribology The study of FRICTION and its effects, especially the reduction of friction with lubricating oils.

tricarboxylic acid cycle (TCA cycle) *See* KREBS CYCLE.

trichloroacetaldehyde *See* TRICHLOROETHANAL.

1,1,1-trichloro-2,2-di(4-chloro-phenyl)ethane Systematic name for DDT.

trichloroethanal, *chloral, trichloroacetaldehyde* (CCl_3CHO) A colourless, oily liquid with a pungent odour; boiling point 98°C. It is manufactured by the action of chlorine on ethanol. Useful compounds are formed by the addition of water, such as chloral hydrate, which is a sleep-inducing (or hypnotic) agent. Trichloroethanal is also used in the manufacture of DDT and is decomposed by alkalis to CHLOROFORM.

trichloromethane *See* CHLOROFORM.

trichromatic theory *See* RETINA.

triclinic (*adj.*) Describing a CRYSTAL structure where none of the faces are at right angles to one another. The size and shape of the UNIT CELL is characterized by three lengths and three angles.

tricuspid valve *See* HEART.

triethylamine *See* ETHYLAMINE.

triglyceride *See* LIPID.

trigonal planar A term describing the arrangement of atoms in a molecule or RADICAL where there is a central atom surrounded by three atoms in a plane, with COVALENT BONDS between the central atom and each of the other atoms, the angle between the bonds being 120°. The carbonate ion, CO_3^{2-}, is an example of a trigonal planar structure.

trigonal pyramidal A term describing the arrangement of atoms in a molecule or RADICAL where an atom at the apex of a triangular-based pyramid makes three COVALENT BONDS to atoms at the corners of the base. Ammonia, NH_3, has this structure rather than a TRIGONAL PLANAR structure, due to the presence of a LONE PAIR of electrons in the nitrogen atom.

trigonometry The branch of geometry that deals with the relationships between lengths and angles, particularly those in right-angled triangles, which lead to the SINE, COSINE and TANGENT functions.

trihydric (*adj.*) Describing a compound with three HYDROXYL GROUPS. For example, GLYCEROL is a trihydric ALCOHOL.

triiodomethane, *iodoform* (CHI_3) An ANTISEPTIC that forms pale yellow crystals; melting point 119°C. It has a characteristic 'hospital smell'. It is used to test for the CH_3CO– group in carbonyl compounds (*see* CARBONYL GROUP). *See also* TRIIODOMETHANE TEST.

triiodomethane test, *iodoform test* A useful method of recognizing the CH_3CO– group in carbonyl compounds (*see* CARBONYL GROUP), or groups in other compounds that may be converted to this. Compounds containing the CH_3CO– group give a precipitate of TRIIODOMETHANE when heated in alkaline solution in an excess of iodine. This has a characteristic yellow colour and 'hospital odour'. If alcohols containing the $CH_3CH(OH)$ group are subjected to the same test they are first oxidized to CH_3CO– and subsequently give a positive result in the triiodomethane test. So, for example, ethanol, C_2H_5OH, produces triiodomethane in the test but methanol, CH_3OH, does not. Alcohols usually need heating to react whereas carbonyl compounds may react in the cold.

triiodothyronine A HORMONE of the THYROID GLAND.

trilobite A member of a extinct class of ARTHROPODS, whose fossils are useful in dating rock strata from the PALAEOZOIC ERA. They had a flattened oval body (1–7 cm long), divided into a distinct head, THORAX and segmented tail.

trinitrotoluene *See* TNT.

triose A MONOSACCHARIDE containing three carbon atoms in the molecule.

triple bond A COVALENT BOND in which three sets of orbitals overlap to form a bond stronger than a DOUBLE BOND. In ethyne, C_2H_2, for example, two sp HYBRID ORBITALS overlap to form a SIGMA-BOND whilst two pairs of P-ORBITALS form two PI-BONDS.

triple point The one combination of temperature and pressure at which the solid, liquid and gas PHASES of a substance can exist together in equilibrium. The triple point of water is at a temperature of 0.01°C and a pressure of 600 Pa.

triploblastic (*adj.*) Of an animal, having a body that develops from three GERM LAYERS – the ECTODERM, ENDODERM and MESODERM.

triploid A nucleus, cell or organism that has three sets of CHROMOSOMES.
trisodium phosphate (Na_3PO_4) A white solid, occurring both as the decahydrate $Na_3PO_4.10H_2O$, and the dodecahydrate $Na_3PO_4.12H_2O$. Trisodium phosphate may be formed by neutralizing phosphoric acid with sodium hydroxide:

$$H_3PO_4 + 3NaOH \rightarrow Na_3PO_4 + 3H_2O.$$

Trisodium phosphate is used as a water softener (*see* WATER SOFTENING), precipitating soluble calcium and magnesium salts as insoluble phosphates, for example:

$$3CaCO_3 + 2Na_3PO_4 \rightarrow Ca_3(PO_4)_2 + 3Na_2CO_3$$

tritium The ISOTOPE hydrogen–3. It is widely used in FUSION processes and is unstable, with a HALF-LIFE of 12 years.
trivalent, *tervalent* (*adj.*) Having a VALENCY of 3.
tRNA *See* TRANSFER RNA.
trophic (*adj.*) Relating to NUTRITION.
trophic level The position occupied by a species or group of species in a FOOD CHAIN, for example producers and consumers. *See also* FOOD WEB.
trophoblast In mammals, the outer layer of cells of the BLASTOCYST that develops into the embryonic membrane (CHORION), the VILLI of which invade the wall of the UTERUS and develop into the PLACENTA. In humans, the trophoblast secretes HUMAN CHORIONIC GONADOTROPHIN, which may be detected in the urine of pregnant women and forms the basis of pregnancy tests.
tropical (*adj.*) Relating to the tropics. A tropical AIR MASS is one that has come from equatorial regions.
tropism The directional growth of a plant (or part of it) in response to an external stimulus. The tropism can be positive (growing towards the stimulant) or negative (growing away) and is caused by greater growth on one side of the plant than the other. Light is a major stimulus, the response being PHOTOTROPISM. GEOTROPISM is the response of a plant to gravity; HYDROTROPISM is the response to water; CHEMOTROPISM is the response to a chemical stimulus; and THIGMOTROPISM (haptotropism) is the response to physical contact. *Compare* NASTIC MOVEMENT.
tropomyosin *See* MUSCULAR CONTRACTION.
troponin *See* MUSCULAR CONTRACTION.
tropopause The boundary between the TROPOSPHERE and the STRATOSPHERE. *See* ATMOSPHERE.
troposphere The lowest level of the Earth's ATMOSPHERE, extending up to about 17 km above the Earth's surface. Most of the WEATHER SYSTEMS occur in this layer.
trypsin A PROTEASE enzyme in the vertebrate digestive system that breaks down proteins during digestion. Trypsin is secreted by the PANCREAS in its inactive form as trypsinogen. This is activated in the SMALL INTESTINE by enterokinase, an enzyme secreted by the DUODENUM. Trypsin does not need an acid environment to function.
trypsinogen *See* TRYPSIN.
tryptophan An AMINO ACID produced from proteins by the digestive action of TRYPSIN.
TSH *See* THYROID-STIMULATING HORMONE.
tuber A swollen region of an underground stem (a stem tuber, e.g. the potato) or root (a root tuber, e.g. the dahlia) that is modified for storing food and gives rise to new plants, and so is a structure for VEGETATIVE REPRODUCTION. A tuber lasts for one season only, with new tubers forming the following year in a different place.
tumour, *neoplasm* A swelling or lump caused by an overgrowth of cells in a specific area of the body. Tumours can be malignant or benign. Malignant tumours show unlimited, rapid growth and invade surrounding tissues. They shed cells that can be transported through the blood or LYMPHATIC SYSTEMS to form a secondary tumour (or metastasis) in another part of the body. Benign tumours are not cancerous, are slower growing and non-evasive. They can therefore be surgically removed more easily and do not usually re-occur. Any tumour, even benign, in a difficult site (e.g. the brain) or causing a physical blockage, can be life-threatening. *See also* CARCINOMA, SARCOMA.
tundra A terrestrial BIOME of high latitudes, characterized by treeless expanses and a permanently frozen subsoil.
tuned circuit, *resonant circuit* A circuit containing an INDUCTANCE and a CAPACITANCE. These act in a way analogous to a mass on a spring and have a RESONANT FREQUENCY f given by

$$f = {}^1/_2\pi(LC)^{1/2}$$

where L is the inductance and C the capacitance.

tungsten (W) The element with ATOMIC NUMBER 74; RELATIVE ATOMIC MASS 183.9; melting point 3,410°C; boiling point 5,660°C; RELATIVE DENSITY 19.3. Tungsten has the highest melting point of any metal, which has led to its use in light-bulb filaments. Tungsten carbide, an extremely hard material, is also used to tip drills and other cutting tools.

Turbellaria A CLASS of the PHYLUM PLATYHELMINTHES (flatworms). Unlike the other members of this phylum, which are PARASITES, the Turbellaria are free-living, for example *Planaria.*

turbine A machine for converting the KINETIC ENERGY of a fluid (usually air heated by burning fuel or water). A turbine consists of a rotating set of blades surrounded by fixed blades to channel the fluid flow into the path that produces the largest force on the rotating blades. *See also* GAS TURBINE.

turbocharging A system used in some INTERNAL COMBUSTION ENGINES in which the energy of the expanding exhaust gases is used to drive a small TURBINE which compresses the fuel/air mixture entering the engine, enabling more fuel to be burnt by an engine of a given size, increasing its power output.

turbofan *See* HIGH-BYPASS ENGINE.

turboprop A GAS TURBINE engine used to drive an aircraft propeller.

turbulent (*adj.*) Describing irregular motion in a fluid with sudden unpredictable changes in flow direction over time or from one part of the fluid to the next. *Compare* STREAMLINED. *See also* BOUNDARY LAYER, DRAG, VORTEX.

turgor The rigid condition of a plant cell when it is full of water. Water enters by OSMOSIS, causing the CYTOPLASM and CELL WALL to be pressed together, and the cell is said to be turgid. The pressure exerted by the fluid against the cell wall is called the turgor pressure. Turgor is important in providing support for some plants. *See also* PLASMOLYSIS.

turgor pressure *See* TURGOR.

Turner's syndrome *See* MUTATION.

turning point A point at which a mathematical FUNCTION reaches a local MAXIMUM or MINIMUM; in other words, a point where the gradient of the function is zero, with the sign of the gradient being different on either side of the point.

turns ratio In a TRANSFORMER, the number of turns in the SECONDARY COIL divided by the number of turns in the PRIMARY COIL.

tuyere A metal pipe through which air is blown into a BLAST FURNACE.

twin paradox A RELATIVISTIC PARADOX in which one of a pair of identical twins sets off on a long space journey. When he returns he will have aged less than his twin (TIME DILATION). However, if only relative motion is important, why is it not possible to think of the other twin as being the one who made the journey? The resolution of this paradox lies in careful consideration of what happens when the twin turns round to begin his return journey; up to that point the position of the two twins is symmetrical, but when the travelling twin starts his return journey, he must either jump to a different NON-INERTIAL FRAME OF REFERENCE, or decelerate, making his frame non-inertial. Only when the twins are brought back together is it possible to compare their ages unambiguously.

two-stroke cycle A system of operation used by some smaller PETROL ENGINES, such as those used on small motorcycles. In the two-stroke engine, the space below the cylinder is used to draw in a fresh load of fuel/air mixture as the previous load is leaving the top of the engine. This provides more even power but has disadvantages in that oil must be dissolved in the petrol to provide adequate lubrication of the piston. This oil is partially burnt and expelled in the exhaust.

tympanic membrane *See* EARDRUM.

typhoon *See* HURRICANE.

tyrosine An AMINO ACID present in many proteins. Derivatives of tyrosine include ADRENALINE and NORADRENALINE.

U

UHF (ultra high frequency) RADIO WAVES with wavelength from about 10 cm to 1 m. They are used for many forms of RADIO communication and also for television, where the high rate at which picture information has to be transmitted requires each signal to have a large BANDWIDTH (spread of frequencies). This bandwidth requirement is less serious at high frequencies. As with VHF, only line-of-sight communication is possible.

ultracentrifuge A CENTRIFUGE that operates at very high speeds. It can be used in the laboratory to separate COLLOIDS, submicroscopic particles or particles as small as a NUCLEIC ACID or protein.

ultramicrotome A machine for cutting very thin sections of tissue (embedded in a RESIN such as ARALDITE) for use with a ELECTRON MICROSCOPE. The sections are 20–100 nm thick. A glass or diamond knife is used.

ultrasound SOUND WAVES of too high a frequency for the human ear to hear. The PIEZO-ELECTRIC effect can be used to produce sound waves with frequencies of several megahertz (MHz). Such high frequencies mean that the wavelengths are very short, so narrow beams of sound can be produced without problems from DIFFRACTION. Such ultrasonic beams can then be fired into objects and the reflections studied to give a picture of the structure of the object being studied. *See also* ULTRASOUND IMAGING.

ultrasound imaging A technique for studying the interior of opaque structures, widely used in medicine. ULTRASOUND is especially used in the routine examination of human foetuses. The technique relies on the reflection of ultrasound from boundaries between materials of differing densities. The DOPPLER EFFECT also enables motion, particularly blood flow rates, to be measured by ultrasonic techniques – the shift in frequency of the reflected wave gives a measure of the motion of the reflecting particles.

Whilst ultrasound is non-ionizing, so cannot produce mutations, high levels can produce tissue changes, and more intense ultrasound is used in physiotherapy and to break up kidney stones.

ultra high frequency *See* UHF.

ultraviolet ELECTROMAGNETIC WAVES with wavelengths in the range 4 x 10^{-7}m to about 10^{-9} m. They are produced by the more energetic changes in energy in atomic electrons. Ultraviolet radiation from the Sun is mostly absorbed in the upper layers of the atmosphere (the OZONE LAYER), so relatively little reaches the Earth. That which does reach ground level is responsible for the tanning and burning effect of exposure to sunlight and, with prolonged exposure, is believed to be responsible for skin cancer. Ultraviolet radiation can be detected by PHOTOGRAPHIC FILM and can be made visible by FLUORESCENCE.

umbilical cord The connection between the FOETUS and PLACENTA of placental mammals throughout pregnancy. The umbilical cord has one vein and two arteries that transport oxygen and nutrients to the young and transport waste products away. After birth the cord falls off, leaving a mark called the naval.

uncertainty principle *See* HEISENBERG'S UNCERTAINTY PRINCIPLE.

underdamped (*adj.*) Describing an oscillating system where the DAMPING is less than that required to prevent OSCILLATION. *See also* OVERDAMPED.

unicellular (*adj.*) Describing organisms or their parts that consist of only one cell. *Compare* ACELLULAR, MULTICELLULAR.

unified field theory A theory in particle physics that describes at least two of the FOUR FORCES OF NATURE as being aspects of a single force. The STANDARD MODEL, which brings together the ELECTROMAGNETIC FORCE and WEAK NUCLEAR FORCE is the only truly successful example of such a theory, but physicists have been inspired by this success to apply the idea to the STRONG NUCLEAR FORCE and the gravitational force as well, to create a GRAND UNIFIED THEORY or a THEORY OF EVERYTHING.

unit 1. A fixed size of some quantity, such as length, in terms of which all other examples of that quantity can be expressed. Thus, once the length of one METRE is fixed, all other distances can be measured in terms of the metre. It is possible to fix units for all quantities independently, but in practice once a few units, called base units, have been fixed, other quantities can be defined in terms of these. Speed for example can be measured in terms of metres and seconds, units of time and distance. *See also* C.G.S. UNITS, METRIC SYSTEM, SI UNIT.

2. Relating to the number 1 – a digit immediately to the left of the decimal point, or at the right-hand end of a number with no decimal point, representing the number of 1's in the number, as opposed to the number of 10's for example.

unit cell, *primitive cell,* The smallest part of a crystal LATTICE that is needed to describe the structure of a CRYSTAL. A crystal can be thought of as being made up of repeated unit cells. For example, the unit cell for sodium chloride, NaCl, can be regarded as a cube containing a central sodium ion with half a chlorine ion in the middle of each face and one eighth of a sodium ion at each corner.

universal gas constant *See* MOLAR GAS CONSTANT.

universal gas equation *See* IDEAL GAS EQUATION.

universal indicator A mixture of INDICATORS designed to produce a continuous variation in colour over the whole scale of pH values. Universal indicator can be used to give a quick, but fairly approximate, indication of the degree of acidity or alkalinity in a solution.

universal set The SET that contains all objects that are in any way relevant to a given problem. For example the set of all numbers, or the set of all people in a given population. All other sets are SUBSETS of the universal set.

UNIX An OPERATING SYSTEM designed for MINICOMPUTERS, but increasingly being used on PERSONAL COMPUTERS, MAINFRAMES and SUPERCOMPUTERS. It is designed for systems with more than one user and has the ability to multitask (*see* MULTITASKING).

unnil- A prefix used in giving names to elements with ATOMIC NUMBERS greater than 103. The names used are unnilquadium (104), unnilpentium (105), unnilhexium (106), unnilheptium (107), unniloctium (108), unnilennium (109), unnildecium (110).

unsaturated compound Any organic compound in which the carbon atoms are linked by double or triple COVALENT BONDS. ALKENES, ALKYNES and KETONES are all examples of unsaturated compounds.

BROMINE WATER may be used to to determine whether an organic compound is unsaturated. A red-brown solution of bromine in water is decolorized by unsaturated compounds since they react with the bromine.

unstable equilibrium The situation in which a small displacement from an EQUILIBRIUM state produces forces tending to move the system further from its equilibrium position. A simple example of this is a pencil balanced on its point.

unstriated muscle *See* INVOLUNTARY MUSCLE.

up A FLAVOUR of QUARK, with charge $+^{2}/_{3}$ in units of the electron charge.

upthrust The upward force on an object immersed (totally or partially) in a fluid. *See* FLOTATION.

uracil An organic base called a PYRIMIDINE that occurs in RNA but not in DNA.

uranium (U) The element with ATOMIC NUMBER 92; RELATIVE ATOMIC MASS 238.0; melting point 1,132°C; boiling point 3,818°C; RELATIVE DENSITY 19.1. Uranium is the heaviest and most abundant of the naturally radioactive elements, having several ISOTOPES with HALF-LIVES comparable to the age of the Universe.

Uranium is mined commercially at many sites in the world. The most common isotope, uranium–238 has a half-life of 4.5 x 10^9 years, but the rarer isotope, uranium–235, which makes up just 0.7 per cent of natural uranium, is of greater commercial interest as it is FISSILE. *See also* DEPLETED URANIUM, ENRICHED URANIUM.

uranium(VI) fluoride *See* URANIUM HEXAFLUORIDE.

uranium dioxide *See* URANIUM(IV) HEXAFLUORIDE.

uranium hexafluoride, *uranium(VI) fluoride* (UF_6) A white solid; sublimes at 65°C; RELATIVE DENSITY 5.1. Uranium hexafluoride gas is used in the enrichment of uranium by DIFFUSION. *See also* ENRICHED URANIUM.

uranium(IV) oxide, *uranium dioxide* (UO_2) A black solid; melting point 3,000°C, decomposes on further heating; RELATIVE DENSITY 10.9. Uranium oxide is often used as a fuel in NUCLEAR REACTORS as it has a higher melting point than elemental uranium.

Uranus The seventh planet in order from the Sun, with an orbital radius of 19.2 AU (2.9 billion km). Uranus is a GAS GIANT, with a diameter of 50,000 km (3.9 times that of the Earth) and a mass of 8.7 x 10^{25} kg (14 times that of the Earth). Like the other gas giants it has an atmosphere composed mostly of hydrogen and shows evidence of active weather systems. It has a ring system similar to that of SATURN, but with much finer rings, made of a much blacker material. Uranus is unique in having its axis tilted to lie almost in the plane of its orbit. Fifteen satellites are known, nine of which were discovered following the visit of the VOYAGER spacecraft in 1986. Uranus orbits the sun every 84 years, and rotates on its axis every 17 hours.

urea ($CO(NH_2)_2$) A waste product formed from the break down of ammonia, NH_3, in the mammalian LIVER, which is then excreted in the URINE. Ammonia is itself a waste product derived from the break down of proteins and NUCLEIC ACID but is very toxic and therefore converted in many vertebrates to urea, which is harmless. Many aquatic animals excrete ammonia as their main nitrogenous waste product and are ammoniotelic, in contrast to ureotelic animals, which excrete urea.

Urea is made by liver cells in a cyclic process called the urea or ornithine cycle, which is closely linked to the KREBS CYCLE. Urea is a white solid when purified and has some industrial uses, for example in fertilizers and pharmaceuticals.

See also URINARY SYSTEM.

urea cycle *See* UREA.

ureotelic (*adj.*) Describing an animal that excretes UREA as its main nitrogenous waste product.

ureter A fine tube through which URINE passes from the KIDNEY to the BLADDER. The ureter contains smooth muscle fibres in its wall that contract to assist the movement of urine to the bladder.

urethra A tube in mammals (males and females) that carries URINE from the BLADDER to the outside, the opening of which is a SPHINCTER (constricting muscle) under voluntary control. The

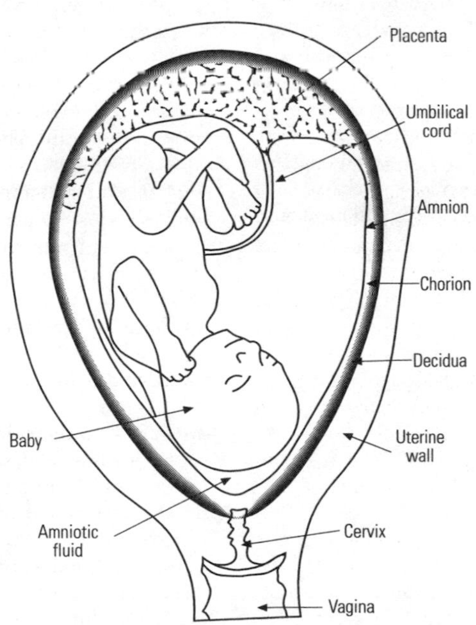

The pregnant human uterus.

urethra in males joins the VAS DEFERENS and also carries SEMEN. *See also* URINARY SYSTEM.

uric acid ($C_5H_4N_4O_3$) A semi-solid nitrogenous waste produced by most land animals that develop in a shell, including reptiles, insects and birds. Uric acid is produced instead of URINE where water is scarce. Humans also produce some uric acid, which if in excess can build up as crystals in joints and tissues, causing gout, or it can form kidney or bladder stones.

uridine A PYRIMIDINE NUCLEOSIDE, consisting of the organic base URACIL and the sugar RIBOSE.

urinary system The system of organs and tubes that removes nitrogenous waste and excess water from the bodies of animals.

In vertebrates, the urinary system consists of a pair of KIDNEYS, two URETERS, a BLADDER and the URETHRA. The kidneys produce URINE, which passes through the ureter, a fine tube extending from each kidney, to the bladder to be stored (up to 0.7 litres in humans) before its discharge. In mammals, urine is then carried to the outside through the urethra, a tube with a SPHINCTER (constricting muscle) at its opening which is under voluntary control.

In other vertebrates (most reptiles, birds, amphibians and many fish), urine drains from the bladder into the CLOACA, a chamber containing all excretory products (digestive and urinary) and into which the reproductive tracts also enter.

See also BOWMAN'S CAPSULE.

urine A fluid made by the KIDNEYS that contains excess water, salt, proteins, some acid and UREA. Reptiles, insects and birds and most land mammals developing in a shell excrete nitrogenous waste as the semi-solid URIC ACID.

urino-genital system The URINARY SYSTEM and REPRODUCTIVE SYSTEM considered together.

uterus, *womb* In female mammals, a muscular organ within which the EMBRYO implants and develops during PREGNANCY. It is located between the BLADDER and RECTUM. The uterus is held in place by ligaments joined to the PELVIS. At the base of the uterus is a ring of muscle called the CERVIX that opens into the VAGINA, which connects the uterus to the outside. The uterus is connected above to the FALLOPIAN TUBES. In humans there is a single uterus of about 5 x 8 cm, but in many mammals there are two uteri joined at the cervix.

The outer wall of the uterus consists of smooth muscle that enables it to expand, to accommodate a growing embryo and contract under hormone stimulation during childbirth. The inner lining of the uterus is called the ENDOMETRIUM, a glandular tissue the structure of which changes in response to hormone stimulation. It is this that is shed at menstruation or thickens during pregnancy to eventually form part of the PLACENTA.

See also MENSTRUAL CYCLE.

utriculus *See* SEMI-CIRCULAR CANAL.

U-value A quantity often used in the study of heat flow, particularly in the design of buildings. The U-value relates to a sheet of material of a specified thickness. It is the heat flow in each square metre of the sheet produced by a one KELVIN temperature difference, expressed in $WK^{-1}m^{-2}$.

V

vaccination, *immunization* The administration of ANTIBODY or a preparation of modified ANTIGEN (the vaccine) to induce specific antibody production and therefore provide artificial IMMUNITY against a particular disease. Edward Jenner (1749–1823) in 1796 was the first English physician to introduce a vaccine, against smallpox. As a result of a world-wide programme of vaccination, smallpox is now eradicated.

A vaccine is usually given orally or by injection and provides immunity lasting from 6 months to 6 years, depending on the disease. Booster injections are often given to maintain the level of protection. Antibodies are given by injection (passive immunity) to fight an established disease or to provide immediate protection to a person at high risk from a particular disease. If antigen is given to induce antibody production (active immunity) then the antigen is modified in one of several ways so that antibodies are made without onset of the disease.

Living micro-organisms can be administered. These have been treated, for example by heating, so that they multiply without causing the disease symptoms. Examples of this include poliomyelitis, rubella (German measles), tuberculosis and measles. Dead micro-organisms can also be used, for example for typhoid, cholera, influenza and whooping cough. Protection against some diseases, for example diphtheria and tetanus, is by injection with a detoxified toxin produced by the PATHOGEN. More recently, artificial antigens have been made by GENETIC ENGINEERING so that they are harmless but immunogenic.

Protection from vaccination is limited when the micro-organism (usually viruses) shows ANTIGENIC VARIATION, for example influenza.

vaccine *See* VACCINATION.

vacuolar pathway *See* TRANSLOCATION.

vacuole A fluid-filled cavity within the cell CYTOPLASM that is bounded by a membrane. In many plant cells a single vacuole takes up most of the volume of the cell and contains cell sap that can be water and other components such as amino acids and sugars as food stores, organic waste materials for later release to the outside, or colour pigments. In animal cells, the vacuoles are smaller but there may be more than one, and they can store food or be PHAGOCYTES.

Contractile vacuoles of single-cell freshwater organisms (e.g. *Amoebae*) are important for osmoregulation (*see* OSMOSIS) because they are able to slowly fill with water and suddenly contract to expel their contents from the cell, so preventing excess water building up following its absorption into the cell by osmosis.

vacuum A region containing no matter of any kind. In particle physics, the true nature of the vacuum state is a matter of some interest. In this context the vacuum is defined as being the lowest energy state, and will contain VIRTUAL PARTICLE/ANTIPARTICLE pairs.

vagina The front passage of female mammals that connects the UTERUS to the outside, into which the PENIS releases its sperm during intercourse and out of which a FOETUS is born. The wall of the vagina contains muscle (and can therefore expand) and secretes a lubricating fluid to neutralize the acidity of the vagina (which would kill the sperm) and allow easy penetration of the erect penis. The vagina opens to the outside through the VULVA.

valence band In the BAND THEORY of solids, the energy band occupied by the VALENCE ELECTRONS. In metals, this band is either only partially full or overlaps with an empty band. In nonmetals it is completely full, so electrons cannot gain energy to conduct heat or electricity. *See also* CONDUCTION BAND

valence electron An electron in the outer SHELL of electrons of an atom, which may be involved in a bonding process. *See also* BAND THEORY, CHEMICAL BOND.

valence shell The outer electron SHELL of an atom, containing the VALENCE ELECTRONS.

valency A number indicating the number of chemical bonds that can be formed by a given element. Many elements show only one valency, for example the valency of the element magnesium is always 2. Other elements, in particular the TRANSITION METALS, can exist in several different valency states. *See also* COVALENCY, ELECTROVALENCY, OXIDATION STATE.

valine An ESSENTIAL AMINO ACID, present in many proteins.

value In mathematics, the particular number that an algebraic quantity represents at a particular moment. For example the value of x is 4.

valve 1. In general, any device that can be closed or opened to allow a fluid to pass through it.

2. ***non-return valve*** A device through which fluid can flow in only one direction. Such a valve generally comprises an opening closed by a flap or ball held in place by a light spring. Fluid pressure on one side of the opening overcomes the force from the spring and opens the valve. Fluid pressure from the other side holds the valve tightly closed.

3. ***safety valve, pressure relief valve*** Similar in structure to a non-return valve, but held firmly closed by a weight or stiff spring. The closing force is only overcome when the fluid pressure exceeds a certain value.

4. An electronic device based on THERMIONIC EMISSION. *See* THERMIONIC VALVE.

vanadium (V) The element with ATOMIC NUMBER 23; RELATIVE ATOMIC MASS 50.9; melting point 1,890°C; boiling point 3,380°C; RELATIVE DENSITY 6.1. Vanadium is a TRANSITION METAL with some useful cataiytic properties. It is also used with chromium in the manufacture of some hard steels. In chemistry, vanadium exists in several OXIDATION STATES, each producing ions of distinctive colours.

vanadium(V) oxide, ***vanadium pentoxide*** (V_2O_5) A white solid that decomposes on heating. Vandium(V) oxide can be formed by burning vanadium metal in oxygen:

$$4V + 5O_2 \rightarrow 2V_2O_5$$

Vandium(V) oxide is widely used for its catalytic properties.

vanadium pentoxide *See* VANADIUM(V) OXIDE.

Van Allen belt One of two doughnut-shaped regions around the Earth in which radiation consisting of high-energy charged particles is trapped by the MAGNETIC FIELD OF THE EARTH. The lower belt, comprising electrons and protons extends from 1,000 to 5,000 km from the Earth's surface, while the upper belt, which contains mostly electrons, is at a distance 15,000 to 25,000 km from the surface.

van de Graaf generator A machine for creating electrostatic charges. The charges are generated by friction in smaller machines and by a beam of charged particles in larger machines. The charges are transferred to an insulating belt, which carries them to the top of the machine, where they are transferred to a large metal dome-shaped ELECTRODE, insulated from its surroundings. *See also* CHARGING BY FRICTION.

van der Waals' bond A very weak bond that holds separate molecules together in molecular solids such as solid carbon dioxide. The bond originates from the VAN DER WAALS' FORCE between neutral molecules. The very weak nature of this bond is reflected in the low melting points of most molecular materials.

van der Waals' equation An EQUATION OF STATE for a REAL GAS. For N MOLES of gas at an ABSOLUTE TEMPERATURE T and pressure p, in a volume V, van der Waals' equation is

$$(p + a/V^2)(V - b) = NRT$$

where R is the UNIVERSAL GAS CONSTANT and a and b are constants that represent the attractive INTERMOLECULAR FORCES and the volume of the gas molecules respectively.

van der Waals' force The attractive force between neutral molecules or atoms. The force is strongest if both molecules are POLAR, but a POLAR MOLECULE will create an induced DIPOLE in a non-polar molecule. There is also a weak induced-dipole to induced-dipole force between non-polar molecules. This arises because at any instant the centres of positive and negative charge may not coincide, even though the molecule may be non-polar on average. These forces are relatively weak and give rise to substances that have melting and boiling points below room temperature.

As with all INTERMOLECULAR FORCES, there is a repulsion at shorter distances, due to the effect of the PAULI EXCLUSION PRINCIPLE on the overlapping electron clouds of the molecules. Van der Waals' forces between non-polar molecules can be described by a POTENTIAL ENERGY with a repulsive core that varies like the inverse twelfth power of the separation and an attractive part, dominant at larger separations, but also falling off with distance, which varies like the inverse sixth power of separation. This is called the LENNARD–JONES 6–12 POTENTIAL.

See also REAL GAS.

vapour The term sometimes used to describe gaseous state of a material below its BOILING POINT.

vapour density A measure of the density of a gas, found by dividing the mass of a sample of the gas by the mass that the same volume of some reference gas (generally hydrogen) would have at the same temperature and pressure.

vapour pressure The PARTIAL PRESSURE exerted by a VAPOUR, especially the vapour found above the surface of a liquid. If the liquid and its vapour are held in a closed container, the vapour pressure will rise until it reaches the SATURATED VAPOUR PRESSURE. *See also* RAOULT'S LAW.

variable An algebraic quantity that may take on a different value at different times. Variables are usually regarded either as independent variables, which can take on any value freely, or dependent variables, whose values are a FUNCTION of an independent variable.

variable capacitor A CAPACITOR whose CAPACITANCE can be changed. This can be constructed by arranging two sets of plates, one fixed and the other moveable, so that the degree of overlap can be altered. *See also* VARICAP DIODE.

variable resistor A RESISTOR whose RESISTANCE can be changed, made from a layer of resistance material with a sliding contact. When just the sliding contact and one end of the resistance are used in a circuit, to control the brightness of a lamp for example, variable resistors are sometimes called rheostats. When used in a POTENTIAL DIVIDER circuit, with a voltage applied between the two fixed ends, a fraction of which is taken off between the sliding contact and one of the fixed contacts, they are called potentiometers.

variation The differences between individuals of the same species, arising as a result of SEXUAL REPRODUCTION. GENOTYPES and PHENOTYPES can show variation, and the variation may be minor or more noticeable. Variations can exist, for example, in colouring, size and behaviour.

There are two types of variation: continuous and discontinuous. Continuous variation is where individuals show a gradation from one extreme to another, such as height, and these characteristics are often controlled by POLYGENES. Discontinuous variation is where there is a number of distinct forms within a population such as blood groups. These characteristics are usually controlled by only one gene and there are no intermediate forms.

The most important cause of variation is genetic as a result of the mixing of two parental genotypes during sexual reproduction and the random distribution of chromosomes during MEIOSIS. Further variation comes from CROSSING-OVER and RECOMBINATION of chromosomes during meiosis, and some variation occurs as a result of MUTATION. Environmental factors can cause variation but these are not passed on. Genetic variation is the basis upon which NATURAL SELECTION can work, and provides species with an opportunity to adapt. It is therefore an evolutionary advantage.

varicap diode A PN JUNCTION DIODE made with a particularly wide junction to enhance its CAPACITANCE when REVERSE BIASED. The capacitance decreases as the reverse bias voltage is increased. Varicap diodes are often used for the electronic control of tuning circuits in radios, etc.

vasa recta *See* LOOP OF HENLE.

vascular bundle The main conducting tissue of VASCULAR PLANTS (for example flowering plants, ferns, mosses and conifers). The vascular bundle extends from the roots of the plant to the leaves and stems. The XYLEM is towards the centre of the bundle and PHLOEM is nearest the epidermis. *See also* CAMBIUM, PROCAMBIUM, TRANSLOCATION.

vascular plant A plant with a VASCULAR BUNDLE for conducting water and sugars. Vascular plants include CONIFERS and ANGIOSPERMS.

vas deferens, *sperm duct* One of a pair of muscular tubes in male vertebrates that carries SPERM (during sexual intercourse) from each testis to the URETHRA and then to the outside. The sperm are carried in a fluid called SEMEN. *See also* EPIDIDYMIS, SEMINAL VESICLE, SEXUAL REPRODUCTION.

vasectomy *See* STERILIZATION.

vas efferens *See* TESTIS.

vasoconstriction A narrowing of the diameter of a BLOOD VESSEL, usually of the ARTERIOLES. Vasoconstriction is under the control of the VASOMOTOR CENTRE.

vasodilation An increase in the diameter of a BLOOD VESSEL, usually of the ARTERIOLES.

Vasodilation is under the control of the VASOMOTOR CENTRE.

vasomotor centre A group of NEURONES in the vertebrate brain that maintain a constant blood pressure by controlling VASOCONSTRICTION (narrowing of blood vessels) and VASODILATION (widening of blood vessels).

vat dye *See* DYES.

VDU *See* VISUAL DISPLAY UNIT.

vector 1. (*mathematics*) A quantity that is described by both its size, or magnitude, and its direction in space. Examples include velocity and force. Vectors can be added by drawing them nose to tail and forming a single vector that joins the ends of the composite vector thus formed. *See also* COMPONENT, RESOLVE.

2. (*biology*) An organism, for example an insect, that acts as a host to a PARASITE and transmits it to another host. *See also* MALARIA.

3. (*biology*) A PLASMID or BACTERIOPHAGE that can carry foreign DNA inserted into its own genetic material and which therefore replicates with it. Plasmid vectors are valuable tools in GENETIC ENGINEERING.

vegetative reproduction A form of ASEXUAL REPRODUCTION in plants in which a new organism develops from structures formed by the parent plant, without the production of spores. Vegetative reproduction is important in the propagation of plants (vegetative propagation). The structures formed are numerous and varied and include BULBS, CORMS, RHIZOMES, TUBERS, STOLONS and RUNNERS. Structures that act as food stores (not stolons and runners) are called perennating organs and allow the plant to survive through the winter.

Examples of the use of vegetative reproduction by humans are grafting and cuttings. Cuttings are sections of the stems, roots or leaves put into soil that produce roots from a node and develop from this into new plants. Grafting involves placing a shoot or bud of one plant into another; they combine to form a plant with some of the advantages of both plants. Grafting is often used to propagate or modify woody plants, for example roses and fruit trees.

vein A vessel in animals with a CIRCULATORY SYSTEM that carries blood from the rest of the body towards the heart. Veins have thin muscular walls with few elastic fibres (a vein cannot alter its diameter as an ARTERY can) and a large LUMEN. All veins (except the PULMONARY vein leaving the lungs and entering the heart) carry deoxygenated blood from the main organs of the body to the heart. All contain valves to ensure the blood flows in one direction only. The blood in veins is not under high pressure and moves slowly and not in pulses, in contrast to the blood in arteries. The main vein entering the heart is the VENA CAVA. Further from the heart small veins are called venules.

The term vein can also refer to other vessels, for example in leaves of plants, that do not contain blood.

velocity The rate of change of displacement of a body. Velocity is the VECTOR quantity related to SPEED; that is, the speed plus the direction of motion.

velocity ratio The factor by which the speed at which the effort force applied to a machine must move is greater than the speed of movement of the LOAD.

$$\text{Velocity ratio} = \frac{\text{distance moved by effort}}{\text{distance moved by load}}$$

velocity selector A mechanism, usually based on electric and magnetic fields, for selecting charged particles moving at a particular speed. *See also* MASS SPECTROMETER.

vena cava Either of the two main VEINS of vertebrates that carries blood to the heart. The anterior vena cava carries deoxygenated blood from the rest of the body to the right side of the heart. The posterior vena cava carries oxygenated blood from the lungs to the left side of the heart.

Venn diagram A diagram showing the relationship between SETS that are SUBSETS of the same UNIVERSAL SET. Individual subsets are represented by circles, and objects that are members

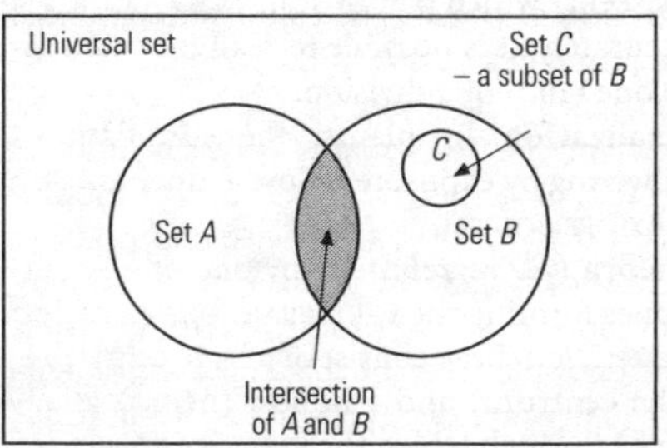

Venn diagram.

of more than one of the sets (called the intersection of the sets) lie in the areas where the circles overlap.

venom A poison released by animals under attack.

ventral (*adj.*) **1.** Of an animal, relating to the front or ABDOMEN.

2. Of a plant, relating to the ANTERIOR or lower surface.

ventral root *See* SPINAL CORD.

ventricle 1. Either one of two chambers in the HEART, with thick muscular walls. Ventricles contract to force blood into the ARTERIES. *See also* ATRIUM, BICUSPID VALVE.

2. Any one of the four main cavities of the vertebrate BRAIN, which contain CEREBROSPINAL FLUID.

venule A small branch of a VEIN.

Venus The second planet in order from the Sun, with an orbital radius of 0.72 AU (108 million km). Venus is similar in size to the Earth, with a diameter of 12,100 km (0.94 times that of the Earth) and a mass of 4.9 x 10^{24} kg (0.81 times that of the Earth). Venus has a dense atmosphere with high levels of carbon dioxide, which have caused an extreme form of the GREENHOUSE EFFECT, leading to high temperatures on the surface. Nothing can be seen of the surface from Earth, due to a constant cover of dense white clouds that are now known to be made of sulphuric acid. The surface of Venus has been mapped using radar and its geology is believed to be similar to that of the other TERRESTRIAL PLANETS. Venus orbits the sun every 225 days and rotates on its axis, from west to east, every 243 days.

vernier A device for fine adjustment or precise measurement. A vernier scale typically consists of a small scale sliding along a main scale. The small scale contains 10 divisions in the space occupied by nine divisions on the main scale. By seeing which pair of scale marks align most accurately, it is possible to read the main scale to one tenth of a division.

vernalization In plants, the stimulation of flowering by exposure to low temperatures. *See* PHOTOPERIODISM.

vertebra (*pl.* ***vertebrae***) Any one of a series of bones forming the VERTEBRAL COLUMN of vertebrates. Vertebrae consist of a large central mass (the centrum) and a hollow (neural canal or arch) through which the SPINAL CORD passes. In humans, there are seven cervical vertebrae (in the neck, with very short ribs), 12 thoracic vertebrae (in the THORAX, bearing the main ribs), five lumbar vertebrae (in the lower back, with no ribs), the sacrum or sacral vertebrae (five fused bones joined to the PELVIS) and the coccyx or caudal vertebrae (four fused vertebrae forming the tailbone). The vertebrae in humans have different shapes according to their position in the body, but in fish they are more similar along the whole length of the vertebral column.

vertebral column, *spine, backbone* The main support of vertebrates connecting the skull, ribs, back muscles and PELVIS and enclosing and protecting the nerve fibres of the SPINAL CORD. The vertebral column is made up of a series of small bones (26 in most mammals) called vertebrae that are linked by LIGAMENTS and have CARTILAGE (intervertebral discs) in-between the bones. The spine is curved to accommodate the larger chest and pelvic regions, and muscles attached to finger-like outgrowths on the vertebrae control the limited movement of the spine.

Vertebrata A major subgroup of the PHYLUM CHORDATA. *See* VERTEBRATE.

vertebrate Any animal with a backbone, skull and well-developed brain from the phylum CHORDATA, including MAMMALS, BIRDS, FISH, AMPHIBIANS and REPTILES. There are about 41,000 species of vertebrates. Vertebrates form the dominant species of land, sea and air, not in numbers but in ecological importance and BIOMASS (because they include most of the larger animals).

vertex A point where two sides of a two-dimensional figure, such as a triangle, meet, or where the edges of a three-dimensional figure meet. In particular, the point furthest from the base of a figure.

very high frequency *See* VHF.

vesicle Any small sac or cavity, especially one filled with fluid, within the CYTOPLASM of a living cell.

vestigial organ An ORGAN whose size and structure has diminished during evolution because of reduced selection pressure (*see* NATURAL SELECTION). An example is the human APPENDIX.

VHF (very high frequency) RADIO WAVES with wavelengths from about 1 m to 10 cm. They are used for many forms of RADIO communication, but because the waves are not reflected by the IONOSPHERE, they can only be used where there is an uninterrupted path from the transmitter to the receiver (this is called line-of-sight communication). This problem can be overcome by transmitting signals from communications satellites or by using RELAY STATIONS.

vibrio Any comma-shaped bacterium. *See* BACTERIA.

villi (*sing.* ***villus***) One of the many finger-like projections from the inner lining of the walls of the SMALL INTESTINE that serve to increase the surface area over which absorption of food can occur. Each villus can be up to 1 mm long in humans. The villi are covered by EPITHELIUM that also has minute projections called MICROVILLI, together forming a BRUSH BORDER, which further increases the area for absorption. Each villus has blood vessels, a small vessel called a lacteal and smooth muscle fibres to allow repeated relaxation and contraction to help food and enzyme mixing.

vinyl chloride, *chloroethene* (CH_2CHCl) A colourless gas; boiling point –14°C. It is used for the manufacture of POLYVINYL CHLORIDE (PVC). Vinyl chloride is made by reacting ethyne with hydrogen chloride over a catalyst. It is CARCINOGENIC.

viroid A small VIRUS consisting of a single strand of NUCLEIC ACID and no protein coat. Viroids cause some important plant diseases and some rare diseases of animals.

virology The study of VIRUSES.

virtual earth A point in a circuit that, though not connected to EARTH, is maintained at a ELECTRIC POTENTIAL close to earth potential. *See also* OPERATIONAL AMPLIFIER.

virtual focus An imaginary point from which light rays appear to have come. *See* FOCUS.

virtual image An optical image where rays of light from a given point on the object only appear to be coming from a corresponding point on the image (as when an object is viewed in a PLANE MIRROR for example). A virtual image cannot be formed on a screen as the light rays do not actually arrive at the point where the image appears to be located. *See also* REAL IMAGE.

virtual particle In QUANTUM THEORY, a particle whose permanent existence is forbidden by the LAW OF CONSERVATION OF ENERGY, but which can exist temporarily according to HEISENBERG'S UNCERTAINTY PRINCIPLE. In particle physics the exchanged GAUGE BOSONS responsible for the forces between particles of matter are virtual particles; they can also exist as real particles, though in the case of the W BOSONS and Z BOSONS, large amounts of energy are required.

virus 1. (*biology*) A small (20–300 nm) infectious particle containing NUCLEIC ACID (DNA or RNA) within a protein shell. Viruses are obligate PARASITES that are unable to multiply except within the living cell of a host. They are smaller than bacteria and cannot be seen through a light microscope.

A mature virus is called a virion and consists of a DNA or RNA core surrounded by a protein coat or capsid, which is sometimes surrounded by host CELL MEMBRANES gained (and maybe modified) on exit from the host cell. The capsid is made of a variable number of capsomere subunits, the symmetry of which is also variable, for example some are spherical (e.g. poliomyelitis), and others are rod-shaped (e.g. tobacco mosaic virus).

Viruses can be classified according to their nucleic acid, which is usually DNA either single-stranded, for example parvovirus, or double-stranded, for example poxvirus, herpesvirus, adenovirus and papovavirus. Some viruses have RNA as their nucleic acid, again either single-stranded, for example the picornavirus (which causes poliomyelitis) and the common cold, or double-stranded, for example reovirus (which causes diarrhoea).

RETROVIRUSES are an important group of RNA viruses that cause AIDS and some human cancers (*see* ONCOGENE). The tobacco mosaic virus (TMV) contains RNA and has been widely studied because of its economic importance. It infects tobacco, potato, tomato, blackcurrant and orchid plants and is highly infectious. Other viruses cause an array of diseases in humans, other animals and plants, for example chicken pox, mumps, measles, rabies, smallpox (although now eradicated), influenza, common cold, Lassa fever, herpes, yellow fever and poliomyelitis. Cancer can also be caused by some DNA viruses (e.g. adenovirus and papovavirus).

VACCINATION has helped to prevent the spread of some viral infections (and has eradicated smallpox), but as viruses mutate continuously it is difficult for the body to develop resistance and vaccination may only be effective against one form. ANTIBIOTICS are ineffective. Viruses can infect different species, causing different symptoms. Some antiviral drugs have been developed but these often affect the host cell too. Interferon is the human body's natural antiviral protein, which has some commercial use but is expensive to produce.

See also BACTERIOPHAGES, PROVIRUS.

2. (*computing*) A piece of self-replicating computer software that can transfer itself from one computer to another without assistance from the user and generally without them being aware of it. Some viruses run a program that is merely annoying; others may cause irreparable damage to a file or the computer system itself.

viscose A type of RAYON, also used to make cellophane.

viscosity A measure of the resistance to flow in a fluid. Forces between molecules in a fluid mean that MOMENTUM given to one part of the fluid tends to be transferred to nearby regions. This creates a force opposing any tendency for different parts of the fluid to move at different speeds, as happens at a boundary with a solid surface for example.

The viscosity of a fluid is measured as the force F per unit area A between two surfaces in the fluid that are moving at unit velocity v relative to one another and are separated by unit distance x.

$$F = \eta A(\mathrm{d}v/\mathrm{d}x)$$

where η is a constant of proportionality, called the coefficient of viscosity. *See also* POISELLE'S EQUATION, THIXTROPIC, SUPERFLUIDITY.

visible light ELECTROMAGNETIC WAVES that can be detected by the human eye. Visible light has wavelengths from about 7×10^{-7} m (red) to 4×10^{-7} m (violet). Visible light is produced by very hot objects (e.g. stars, light-bulb filaments) and also when electrons move from one ENERGY LEVEL to another in an atom. The wavelengths given out depend on which atoms are present and this is an important tool in chemical analysis, particularly in cases where no direct sample can be obtained (for example in looking at the light given out by stars). *See also* COLOUR, LUMINANCE, LUMINOUS FLUX, LUMINOUS INTENSITY, SPECTROMETER, SPECTROSCOPY.

visual display unit (VDU) A television screen based on a CATHODE RAY TUBE, used to display data from a computer.

vital capacity The total amount of air that can be forcibly exhaled. *See* BREATHING.

vitamin Any one of a group of unrelated organic compounds essential in small amounts for normal body growth and metabolism. Vitamins are classified as water-soluble (B, C, H) or fat-soluble (A, D, E, K). Excess water-soluble vitamins are excreted in the urine; fat soluble vitamins can be stored (in the liver in humans) but can build up to lethal concentrations if taken in excess. A normal balanced diet usually provides the vitamin requirements but if there are inadequate levels then deficiency disease results (*see separate entries*).

vitamin A, *retinol* A VITAMIN that is important in skin structure and to form visual pigments. It is found in dairy foods, liver, fruits and vegetables. Lack of vitamin A causes night-blindness (or more severe blindness, xerophthalmia) and dry skin.

vitamin B A complex of B_1 (thiamine), B_2 (riboflavin), B_3 (niacin), B_5 (pantothenic acid), B_6 (pyridoxine) and B_{12} (cyanocobalamin) and biotin and folic acid. They are mostly COENZYMES in cellular RESPIRATION.

Thiamine is found in seeds and grain, and deficiency causes BERIBERI. Riboflavin is a precursor of FAD; niacin forms NAD and NADP; and pantothenic acid ($C_9H_{17}NO_5$) forms part of acetyl coenzyme A (*see* KREBS CYCLE). Lack of niacin causes pellagra, which results in skin lesions, diarrhoea and mental disorders. Pyridoxine ($C_8H_{11}NO_3$) is a coenzyme in amino acid METABOLISM and lack of it causes nervous disorders. Cyanocobalamin is important in synthesis of RNA and is needed for RED BLOOD CELL formation. Lack of vitamin B_{12} (usually made by micro-organisms in the digestive system or found in meat and dairy products) causes PERNICIOUS ANAEMIA.

Biotin (also called vitamin H) is found in yeast and liver and is a coenzyme for certain enzymes that incorporate carbon dioxide into various compounds. Lack of it causes dermatitis (inflammation of the skin). Folic acid is found in liver and green leafy vegetables and is made by intestinal bacteria. It is concerned with nucleoprotein synthesis and red blood cell formation, so lack of it causes anaemia. It is often given to pregnant women to prevent anaemia and neural tube defects (spina bifida).

vitamin C, *ascorbic acid* ($C_6H_8O_6$) A VITAMIN that is found in fresh fruit and vegetables (but is destroyed by soaking or overcooking) and is needed for the synthesis of COLLAGEN. SCURVY results from a deficiency of vitamin C.

vitamin D, *cholecalciserol* A VITAMIN that is found in fatty fish and margarine and made in the skin if exposed to enough sunlight. It is needed for the absorption of calcium and phosphorous and therefore is important for the formation of bones and teeth. Lack of vitamin D causes RICKETS.

vitamin E, *tocopherol* A VITAMIN found in vegetable oil that has an unclear function in humans, but causes sterility in rats.

vitamin K, *phytomenadione* A VITAMIN found in leafy vegetables and liver and synthesized by intestinal bacteria. It is involved with blood clotting and lack of it causes haemorrhaging. It is often given to newborn babies to prevent brain haemorrhage, although its routine use has been questioned. *See also* BLOOD-CLOTTING CASCADE.

vitreous humour A transparent, jelly-like substance found in the vertebrate EYE in a chamber behind the lens. It helps to maintain the shape of the eyeball.

vitriol An obsolete term used to describe SULPHURIC ACID and related compounds, for example oil of vitriol (concentrated sulphuric acid) and blue vitriol (copper(II) sulphate).

vivipary Reproduction in animals where the FOETUS develops and is nourished inside the female. Vivipary involves a PLACENTA in mammals such as humans. Other examples include some reptiles and amphibians. *See also* OVIPARY, OVOVIVIPARY.

voicebox *See* LARYNX.

volatile (*adj.*) Easily EVAPORATED.

volatile memory Temporary computer MEMORY in which data is stored in the memory only as long as the computer is connected to its power supply. Once the computer is switched off, the data is lost.

volcano A geological formation caused by the emergence of semi-molten rock, called LAVA, from areas in the Earth's crust where TECTONIC PLATES are moving apart.

volt (V) The SI UNIT of ELECTRIC POTENTIAL, POTENTIAL DIFFERENCE or ELECTROMOTIVE FORCE. One volt is an energy of one JOULE per COULOMB of CHARGE.

voltage The measure of the amount of ELECTRICAL POTENTIAL ENERGY carried by each unit of charge. The SI UNIT of voltage is the VOLT. *See also* ELECTRIC POTENTIAL, ELECTROMOTIVE FORCE, POTENTIAL DIFFERENCE.

voltaic cell *See* CELL.

voltammeter A container for ELECTROLYSIS experiments with ELECTRODES that can be removed to find their mass. By measuring the change of the mass of the electrodes in an electrolysis experiment, the ELECTROCHEMICAL EQUIVALENT of the element concerned can be found. *See also* HOFMANN VOLTAMMETER.

voltmeter An instrument for measuring VOLTAGE. A voltmeter is connected between the two points between which the POTENTIAL DIFFERENCE is to be measured. Voltmeters should have a very high resistance so that little current flows through them, so as not to affect the circuit to which they are connected. Modern voltmeters are usually based on OPERATIONAL AMPLIFIERS, which have a very high input resistance. The output of the operational amplifier drives a DIGITAL display.

volume 1. The amount of space occupied by an object. For a cube, the volume is equal to the third power of the length of the sides of the cube. For other shapes, the volume can be calculated by imagining them to be built up from a large number of small cubes. Volume is expressed in SI UNITS in metres cubed.

2. The loudness of a sound, related to the amplitude of the sound waves. Sound levels are often measured in DECIBELS.

volumetric analysis A range of techniques of QUANTITATIVE ANALYSIS that involve measuring the volumes of reacting materials. In the gas phase, this involves using graduated vessels attached to MANOMETERS to ensure that the volumes are measured at a known pressure. For liquids, TITRATION is the most common technique. In this, a fixed volume of one reagent has a second reagent added to it from a calibrated vessel, called a BURETTE, until an INDICATOR shows that the reaction is complete.

voluntary muscle, *skeletal muscle, striated muscle, striped muscle* Muscle activated by MOTOR NERVES under voluntary control. *See* MUSCLE.

vortex (*pl.* ***vortices***) A TURBULENT flow characterized by a rotational movement of the fluid. Vortices are produced in aircraft wings, for example, as high pressure air from beneath the wing spills over at the wing tip into the low pressure area above the wing. *See also* DRAG.

Voyager The name given to each of the two unmanned spacecraft that were launched in 1976 to visit the GAS GIANTS. They sent back photographs and other data that greatly increased our knowledge of these planets. The craft took several years to travel from one planet to another, and used the GRAVITY of each planet to accelerate it to the next.

vulcanization A technique used for treating RUBBER to make it harder, stronger and less affected by external temperatures. The technique was discovered by Charles Goodyear (1800–60) in 1839 and involves heating the rubber and chemically combining it with another element, usually sulphur. The rubber is cross-linked by the sulphur forming bridges between the POLYMER chains. Thus the chains are less easily pulled apart on stretching and the rubber has increased elasticity. For special uses, such as tyres, vulcanizing agents other than sulphur are used along with additives to speed up the process. Objects such as tyres can be shaped and vulcanized at the same time by using heated moulds.

vulva The female external genital organs that comprises pairs of outer and inner folds of skin called the labia majora, labia minora and the clitoris (erectile tissue analogous to the male PENIS).

WAN *See* WIDE AREA NETWORK.

warm front A WEATHER SYSTEM in which a body of warm air meets and rises above a layer of cooler air. The arrival of a warm front produces thickening cloud and rain. *See also* COLD FRONT.

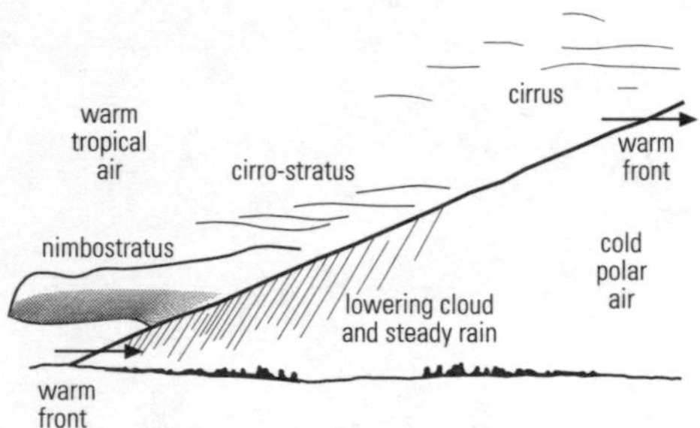

Formation of a warm front.

washing soda *See* SODIUM CARBONATE.

water (H_2O) A colourless, odourless, tasteless liquid, an oxide of hydrogen, which is essential to all living organisms and is the most abundant liquid on Earth. Most water is made up of one atom of hydrogen–1) and two atoms of oxygen–16. A small proportion of water contains other ISOTOPES, such as hydrogen–2 and oxygen–18. When hydrogen–2 (deuterium) is incorporated HEAVY WATER results. Water has the highest SPECIFIC HEAT CAPACITY known, due to the presence of HYDROGEN BONDS.

Water freezes at 0°C and boils at 100°C and has its maximum density at 4°C. One cm^3 of water has a mass of approximately 1 g and forms the unit of RELATIVE DENSITY. Ice is less dense than water and so floats on water, allowing organisms to live beneath it. Most water is found in seas, oceans and rivers, which cover 70 per cent of the earth's surface. The human body contains 60–70 per cent water (about 40 litres) and loss of 8–10 litres can lead to death. *See also* HARD WATER, SOFT WATER, WATER CYCLE.

water cycle The chain of events by which water is re-used in the atmosphere and on the Earth's surface. The condensation of water vapour to produce rain and the evaporation of water from oceans are the key stages in this process, but the absorption of water in soil and its transport to the oceans by rivers are also important, as is the role of RESPIRATION in plants and animals, which absorb water from their surroundings and return it to the atmosphere by evaporation.

water gas The gas formed by passing steam over hot COKE. The steam is reduced (*see* REDUCTION) by the coke, producing a mixture of hydrogen and carbon monoxide, which is used as a fuel in some industrial processes:

$$H_2O + C \rightarrow H_2 + CO$$

The reaction cools the coke and is sometimes used in conjunction with the production of PRODUCER GAS. *See also* BOSCH PROCESS, SEMI-WATER GAS.

water of crystallization Water that forms an important part of the crystal structure in certain salts, particularly NITRATES and SULPHATES. This water can be removed by heating the crystals, which then form a powdery, AMORPHOUS solid. Copper(II) sulphate, $CuSO_4.5H_2O$, forms blue crystals, but on heating water vapour is released and the resulting ANHYDROUS copper sulphate is a white powder. On exposure to atmospheric moisture, water is absorbed and the blue colour returns.

water potential (φ) A measure of the tendency of water in a system to go to its surroundings. The water potential is the difference between the CHEMICAL POTENTIAL of pure water under conditions of STANDARD TEMPERATURE AND PRESSURE and the chemical potential of water in the system.

water softening Any process for converting HARD WATER to SOFT WATER. For some purposes, such as washing clothes, chemicals such as 'Calgon' can be added to the water. Calgon contains sodium phosphate complexes, $(NaPO_3)_6$, which react with the dissolved calcium and magnesium ions to form stable soluble complexes such as $[Ca_2P_6O_{18}]^{2-}$, which prevent the metal ions from reacting with soap.

A more traditional way is to use 'washing soda', sodium carbonate. Sodium and magnesium sulphates and hydrogen carbonates react with this to form insoluble carbonates, removing the metal ions from solution, for example:

$$CaSO_4 + Na_2\,CO_3 \rightarrow CaCO_3 + Na_2SO_4$$

For the softening of water for drinking or washing, ION EXCHANGE methods are used. These involve minerals called PERMUTITES or a modern synthetic equivalent made from fused sand, clay and sodium carbonate.

watt The SI UNIT of POWER. One watt represents WORK done, or ENERGY converted from one form to another at a rate of one JOULE per second.

wattmeter A device for measuring electrical POWER. One form of wattmeter operates on a principle similar to the MOVING-COIL GALVANOMETER, but whilst the voltage of the supply is fed to a moving coil, the current fed to the load flows through an ELECTROMAGNET, which provides the magnetic field in which the coil moves. In this way both current and voltage control the movement of the pointer.

wave A periodic variation in some quantity across a region of space; in particular, a progressive or travelling wave – a motion whereby energy is transmitted through some medium (or empty space in the case of an ELECTROMAGNETIC WAVE) without the particles of the medium moving far from their equilibrium positions.

To support a wave motion, the material through which the wave travels must have INERTIA (so that any given part of the material tends to continue to move after it has been displaced) and a stiffness, tending to restore the material to its original shape. Another consequence of the stiffness of the material is that the displacement of one part of the material tends to cause displacement of nearby regions and in this way a disturbance in one part of the material can spread out, or propagate, through the material.

Waves are divided into two classes: transverse, where the motion is at right angles to the direction in which the wave is travelling; and longitudinal, where the wave motion is to-and-fro along the direction of travel.

Whilst waves can have any shape, SINUSOIDAL waves are the easiest to study, and any more complicated shape can be constructed by adding together sinusoidal waves of various frequencies. In a sinusoidal wave, the wave motion at any fixed point will be SIMPLE HARMONIC MOTION. The AMPLITUDE, PERIOD and FREQUENCY of the wave as a whole are defined

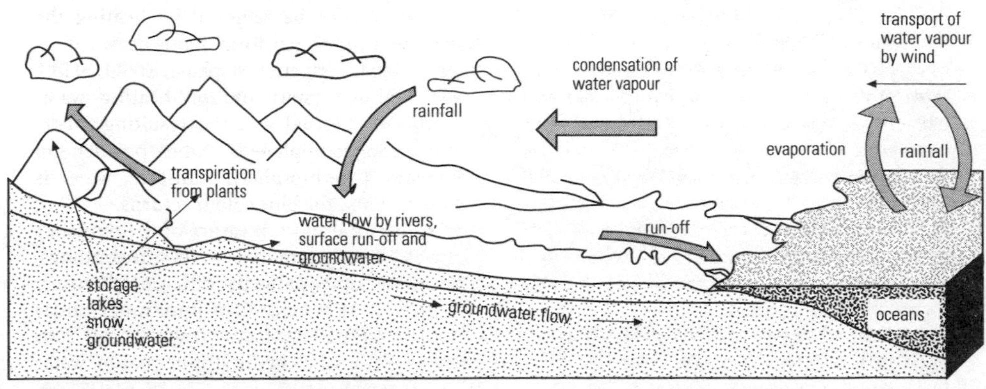

Summary of the water cycle.

as being the same as they are for this simple harmonic motion. Each point along the wave motion performs simple harmonic motion with a phase slightly behind the phase of the previous point. The frequency f of a wave is related to its speed v and WAVELENGTH λ by the equation

$$v = f\lambda$$

See also DIFFRACTION, INTERFERENCE, LONGITUDINAL WAVE, PHASE VELOCITY, REFLECTION, REFRACTION, STANDING WAVE, TRANSVERSE WAVE, WAVEFRONT.

wavefront A line drawn on a diagram that connects all points that are at the same PHASE, such as lines showing the positions of all the crests at a given instant. *See also* HUYGEN'S CONSTRUCTION.

wavefunction In QUANTUM MECHANICS, a PROBABILITY AMPLITUDE, the square of which measures the probability of finding the particle at the point specified or the system in the state described by the particular wavefunction. *See also* SCHRÖDINGER'S EQUATION.

waveguide A metal tube, usually rectangular in cross-section, along which MICROWAVES can travel, being reflected from the inside walls of the tube. Waveguides are used to convey microwaves over relatively short distances, from a transmitter to an AERIAL for example.

wavelength (λ) The closest distance between two points in a WAVE that are moving in PHASE, for example the distance between two adjacent peaks or two adjacent troughs.

wave mechanics The branch of QUANTUM THEORY that deals with the WAVE NATURE OF PARTICLES such as electrons, and the WAVEFUNCTIONS that describe them.

wavemeter A device for measuring the WAVELENGTH of a RADIO WAVE or MICROWAVE, often based on measuring the distance between NODES in a STANDING WAVE.

wave nature of light One aspect of light in the framework of WAVE–PARTICLE DUALITY. The wave nature of light was first demonstrated in 1801 by Thomas Young (1773–1829), in what is now known as YOUNG'S DOUBLE SLIT EXPERIMENT.

wave nature of particles In 1924, Prince Louis de Broglie (1892–1987) suggested that particles would behave like waves with a wavelength now called the DE BROGLIE WAVELENGTH. Three years later this property was confirmed experimentally as ELECTRON DIFFRACTION by Clinton Davisson (1881–1958) and Germer. In electron diffraction experiments, an ELECTRON GUN fires electrons through a thin film of metal or graphite. The regular LATTICE of atoms acts like a DIFFRACTION GRATING for the electron waves and a diffraction pattern is observed on a phosphor screen.

A consequence of the wave nature of electrons was the explanation of the experimental fact that atoms absorb and emit light of certain wavelengths only (*see* SPECTRUM). The explanation for this is that atoms can exist only in certain energy states in much the same way that STANDING WAVES can be produced with only certain frequencies – the permitted states correspond to standing wave patterns of the electrons in the atom. The lowest energy state of any atom is called its ground state. Other, higher energy states are called excited states. In a transition from the ground state to an excited state, an atom will absorb a PHOTON of energy equal to the energy difference between the two states. An atom in an excited state will lose energy by giving off a photon and moving to a lower energy excited state, or to the ground state. In each case, the energy of the photon will be equal to the difference in energy of the two states. The different energy states of different atoms account for the characteristic spectra produced by different elements, which is a useful tool for the analysis of chemical composition, in stars for example.

See also ENERGY LEVEL, HEISENBERG'S UNCERTAINTY PRINCIPLE, NEUTRON DIFFRACTION, SCHRÖDINGER'S EQUATION.

wavenumber The reciprocal of WAVELENGTH, the number of WAVES in one unit of length.

wave-particle duality The name given to the dual behaviour of objects originally thought of as either waves or particles. Thus light, traditionally a wave, can behave as a particle (*see* PHOTON, PHOTOELECTRIC EFFECT), whilst electrons, traditionally thought of as particles, can behave in wave-like ways (*see* ELECTRON DIFFRACTION). *See also* WAVE NATURE OF PARTICLES.

wave power The commercial extraction of energy from waves to generate electricity. Whilst a number of experimental schemes have been tried, there are problems with the high cost of installation and the very large size of any commercial wave power station. The strongest waves are often found far from the places where the most energy is consumed and are also in sites where the installation is most vulnerable to storm damage.

wax A natural solid fatty substance made from ESTERS, FATTY ACIDS or ALCOHOLS. In nature, waxes provide a protective, waterproof covering to many animals and plants. Animal waxes such as beeswax, lanolin and wax from sperm-whale oil are used in cosmetics, ointments and polishes. Mineral waxes are obtained from PETROLEUM and provide a number of products including the soft petroleum jelly used in ointments to the hard wax used for making candles.

W boson, *W particle* The charged particle that, together with the neutral Z BOSON are the GAUGE BOSONS responsible for the WEAK NUCLEAR FORCE. There are two oppositely charged W particles, W^+ and W^-. They have a mass of about 80 times that of the proton.

weak acid An ACID that is not highly IONIZED when dissolved in water, but tends to remain as covalent molecules (*see* COVALENT BOND). Ethanoic acid, CH_3COOH, is an example; only a few ethanoate (CH_3COO^-) ions are formed. Weak acids only react slowly, even with reactive metals. As the supply of ions is used up, the EQUILIBRIUM between ionic and covalent forms is disturbed and more acid DISSOCIATES to the ionic form.

weak intermediate vector boson The W BOSON or Z BOSON.

weak nuclear force The INTERACTION responsible for BETA DECAY. It is mediated by two GAUGE BOSONS called the W BOSON and Z BOSON. The weak force is the only way in which NEUTRINOS can interact with matter. It is also vital in the conversion of hydrogen to helium releasing energy in STARS. *See also* ELECTROWEAK FORCE.

weather systems Regions where AIR MASSES lose water by condensation to form clouds. Water may then fall from the clouds as rain or snow. Weather systems occur when different air masses meet. When TROPICAL MARITIME air meets cooler air, a WARM FRONT is formed. The warm air is less dense, so rises over the cooler air. The result is the formation of cloud as the air cools and becomes SATURATED. This happens first at high altitudes producing wispy clouds composed of ice crystals. These are called cirrus clouds. As the height of the saturated layer moves lower, stratus (layer) clouds are formed and rain falls, or snow if the temperature is below freezing. Sleet is snow that melts as it falls into warmer air.

If cold air runs into warmer air, it tries to force its way underneath, forming a COLD FRONT. The result is a disturbed area in which large heaped (cumulus) clouds form, producing showers. Cumulus clouds can also be formed by uneven solar heating of the ground. Dark areas warm up faster than paler areas and rising air cools and becomes saturated at a certain level, producing flat-bottomed cumulus clouds. As the water vapour condenses in the cloud, the LATENT HEAT released slows the rate of cooling, so the air in the cloud can remain warmer than its surroundings, causing rapid rising currents that reach the STRATOSPHERE. Within such a cloud (called CUMULONIMBUS) water droplets are lifted to great heights and freeze. As they fall, further water freezes on the surface, forming a layered ball of ice, which falls as a hailstone. Collisions between hailstones and water droplets within the cloud produce electrostatic charges, with the top of the cloud positive and the base negative. This charge can become large enough to produce large sparks. The flash of light from the spark is seen as LIGHTNING, whilst the sound is heard as thunder.

The greater HEAT CAPACITY of the sea than the land and the mixing with cooler layers below the surface, which can take place in water but not on dry land, mean that the oceans tend to remain at fairly constant temperatures from one day to the next. The land, on the other hand, heats up during the day and radiates that heat away at night, particularly if the sky is clear. This may set up CONVECTION CURRENTS which cause winds blowing from sea to land by day (sea breezes) and in the opposite direction at night. The cooling of the land on a clear night also leads to condensation which may appear as dew or frost if there is no wind, or as mist or fog if the wind is sufficient to form a layer of cold air just above the cold ground.

See also ANTICYCLONE, ATMOSPHERE, CLIMATE, DEPRESSION.

weber (Wb) The SI UNIT of MAGNETIC FLUX. One weber is the flux produced by a MAGNETIC FIELD of one TESLA over an area of one square metre.

weight The force of GRAVITY on an object.

weightlessness The apparent absence of GRAVITY experienced in an orbiting spacecraft, or any other object falling freely in a GRAVITATIONAL FIELD. *See* FREE FALL.

wet and dry bulb hygrometer A HYGROMETER in which the temperature of the air is measured by one thermometer whilst a second thermometer surrounded by a piece of damp cloth measures a temperature that is lower by an amount called the WET BULB DEPRESSION. The wet bulb depression will be greatest when the air is dry, falling to zero when the air is SATURATED.

wet bulb depression The amount by which the wet bulb temperature is below the dry bulb temperature in a WET AND DRY BULB HYGROMETER. When the wet bulb depression falls to zero, the air is completely saturated with water vapour; that is, the RELATIVE HUMIDITY is 100 per cent.

Wheatstone bridge A circuit containing two pairs of RESISTORS, each pair forming one arm of the bridge, and acting as a POTENTIAL DIVIDER, connected across a single power supply. A GALVANOMETER is connected between the junction of the resistors in each pair, and the bridge is said to be balanced when no current flows through this galvanometer. This will be the case when the POTENTIAL DIFFERENCE across the galvanometer is zero; that is, when the ratios of the two resistance in each potential divider are equal. This balance condition does not depend on the voltage of the supply.

The circuit can be used to compare one resistance with another. The two resistors to be compared form one of the potential dividers, whilst the other takes the form of a POTENTIOMETER, traditionally in the form of a metre length of RESISTANCE WIRE with a sliding contact. The Wheatstone bridge is very sensitive to small changes in resistance, a fact that is exploited in some of its applications (*see* RESISTANCE THERMOMETER, EXPLOSIMETER)

white arsenic *See* ARSENIC OXIDE.

white blood cell, *leucocyte* A type of blood cell made in the white BONE MARROW, the main function of which is defence. White blood cells are larger than RED BLOOD CELLS, are colourless, contain a nucleus and are capable of independent amoeboid (*see* AMOEBA) movement.

In humans, there are about 11,000 white blood cells per millilitre of blood. Their numbers increase (leucocytosis) in response to blood loss, cancer or most infections, or decrease (leucopenia) during starvation, PERNICIOUS ANAEMIA and certain infections. There are two main types of white blood cell: GRANULOCYTES and AGRANULOCYTES. *See also* LYMPHOCYTE, MACROPHAGE, NATURAL KILLER CELL, PHAGOCYTE.

white dwarf A very small hot STAR. As the surface area is small the LUMINOSITY is low. Such stars lie at the bottom left of the HERTZSPRUNG–RUSSELL DIAGRAM. White dwarfs are supported by ELECTRON DEGENERACY PRESSURE – the PAULI EXCLUSION PRINCIPLE places limits on the extent to which the electrons can overlap and the resulting material, similar in many ways to a metal, is able to support itself against further gravitational collapse. Stars of about 0.5 SOLAR MASSES or less simply continue to collapse at the end of their main sequence lifetime. Heavier stars pass through a RED GIANT stage first. *See also* PLANETARY NEBULA.

white lead *See* LEAD CARBONATE HYDROXIDE.

white matter The tissue of the vertebrate BRAIN and SPINAL CORD that consists of AXONS (protected by a MYELIN SHEATH), GLIAL CELLS and blood vessels. It is lighter in appearance than GREY MATTER, due to the myelin sheath. It usually forms an outer layer around the grey matter, although in the cerebral CORTEX of the brain of higher primates it forms an inner layer (*see* CEREBRUM).

white noise NOISE that has a relatively wide range of frequencies.

wide area network (WAN) A data NETWORK that links computers over a large geographical area. Part of the network may use the telephone system.

Wien's law A BLACK BODY will give off the greatest amount of light at a wavelength inversely proportional to the ABSOLUTE TEMPERATURE. The colour of the light from stars can be used as a measure of their surface temperature, with hotter stars appearing more blue whilst cooler stars are orange or red.

Williamson ether synthesis The reaction of HALOGENOALKANES with sodium or potassium ALKOXIDES or PHENOXIDES to give ETHERS. The general reaction is:

$$RHal + NaOR' \rightarrow R\text{-}O\text{-}R' + NaHal$$

where Hal is a HALIDE, and R and R′ are ALKYL GROUPS. *See also* BROMOETHANE.

WIMP 1. (weakly interacting massive particle) In cosmology, one hypothetical form of MISSING MASS in the Universe. *See* COSMOLOGY.

2. (**windows, icons, mice and pointers**) The modern style of information display and control in computers, which relies less on the entry of instructions from a keyboard, and more on the use of a MOUSE to select the desired function from a list on the screen.

window In computer software, a box that appears on the screen containing relevant information or instructions. The box disappears when the information is no longer relevant, and windows relating to different computer functions can be opened or closed by the operator.

windpipe *See* TRACHEA.

wind power The installation of wind-driven electrical generators; one of the more economically effective sources of renewable energy, though only some areas have sufficient wind and enough land available. In Europe, there has been increasing opposition to the visual impact of large 'wind farms'. The costs of installation are low compared to some forms of renewable energy, but no power is produced on still days.

wire wound resistor A RESISTOR made from a coil of RESISTANCE WIRE.

witherite An ore of BARIUM, a mineral form of BARIUM CARBONATE.

womb *See* UTERUS.

woody perennial *See* PERENNIAL PLANT.

woody plant A general term for those plants possessing secondary XYLEM. Woody plants are trees and shrubs, the distinction between these being mostly size. The secondary xylem forms a hard tissue, wood, under the BARK of these plants, which is strengthened by deposits of LIGNIN. The xylem in some trees, for example conifers, does not contain conducting vessels, and tracheids (*see* XYLEM) are used instead to conduct water. These trees, such as pine, provide what is commercially called softwood, in contrast to the hardwood of trees such as oak. Non-woody plants do not undergo secondary growth. *See also* CAMBIUM.

wool A natural protein (KERATIN) fibre mainly obtained from the fleece of sheep. It consists of coiled protein chains joined by HYDROGEN BONDS, which gives it a springy texture.

word *See* BYTE.

word processor A machine or computer program for the manipulation of written text.

work The effect of a FORCE moving through a distance and converting one form of ENERGY to another.

Work done = force × distance moved in the direction of the force

The SI UNIT of work is the JOULE.

work function The minimum amount of energy needed to remove an electron from the surface of a metal in the PHOTOELECTRIC EFFECT.

work hardening The increase in hardness and brittleness of a material (usually a metal) when it undergoes PLASTIC deformation. *See also* ANNEALING.

workstation A powerful desktop computer with strong graphics capabilities, used in engineering (for example, in computer-aided design), scientific research, computer animation, etc. Workstations are generally connected to form a computer NETWORK and most use UNIX as their operating system.

worm A general term for a long, limbless invertebrate. Worms can be from several phyla (*see* PHYLUM). *See* ANNELID, FLATWORM, NEMATODE.

W particle *See* W BOSON.

wrought iron Iron containing less than 0.2 per cent carbon, typically obtained by the PUDDLING process. Wrought iron is easy to work, but too soft to be used in many structural applications. Its use is largely for decorative ironwork.

X

xanthophyll A yellow pigment in plants, of the CAROTENOID group. It functions like CHLOROPHYLL in PHOTOSYNTHESIS.

Xanthophyta A PHYLUM of the KINGDOM PROTOCTISTA that consists of the yellow-green ALGAE.

X-chromosome *See* SEX DETERMINATION.

xenon (Xe) The element with ATOMIC NUMBER 54; RELATIVE ATOMIC MASS 131.3; melting point –112°C; boiling point –107°C. Xenon is a NOBLE GAS, extracted by the FRACTIONAL DISTILLATION from liquid air, where it occurs in minute quantities. A GAS DISCHARGE in xenon has a bright blue-white colour, and is used in flashtubes for photography.

xerophyte A plant that has adapted to living in dry conditions. Xerophytic plants may reduce water loss by TRANSPIRATION by having small leaves (e.g. pine trees) or no leaves (e.g. many cacti), by the orientation of leaves to avoid direct sunlight, or by having a covering of small hairs on the leaf's surface to trap moist air. The STOMATA of xerophytes may be sunken. Many are also able to store water in their leaves or stems (for example, succulents and cacti) and have shallow but extensive root systems to capture surface water. Some of these xerophytic features may be exhibited by other plants where the water supply is limited, for example due to freezing or to high salt concentrations in salt marshes. *See also* HYDROPHYTE.

xerosere A series of plant SUCCESSIONS growing under dry conditions.

x-plates One of the two pairs of plates fitted to a CATHODE RAY TUBE to control the direction of the electron beam. The other pair are the y-plates.

X-ray An ELECTROMAGNETIC WAVE with a wavelength shorter than about 10^{-9} m, produced by the most energetic energy changes of atomic electrons. Such waves are called gamma rays (*see* GAMMA RADIATION) if they are produced by changes within a NUCLEUS. X-rays are a form of IONIZING RADIATION and the shorter wavelengths (hard X-rays) are highly penetrating. They can be detected by photographic film or with a fluorescent screen or by the IONIZATION they produce in a GEIGER COUNTER or other detector of ionizing radiation. The penetrating quality of X-rays has lead to their use in examining the internal structure of various objects, including the human body (*see* MEDICAL IMAGING).

X-ray crystallography The study of the structure of crystals using X-RAY DIFFRACTION. In a simple experiment, the crystal is placed in the path of an X-RAY beam and surrounded by PHOTOGRAPHIC FILM. When the film is developed, spots appear in the directions of CONSTRUCTIVE INTERFERENCE. Since a crystal is a three-dimensional structure, it is possible to identify many crystal planes, or layers of atoms. The result is thus complex, but with modern computational techniques it is possible, even with involved crystal structures, to work back to the structure that produced the pattern.

Some samples cannot be obtained as single crystals, but only as a powder or POLYCRYSTALLINE sample, which effectively contains crystals aligned in all possible directions. The result is that the pattern of spots on the X-ray film becomes a pattern of rings. In the same way that the width of the individual slits in a DIFFRACTION GRATING affects the overall shape of the pattern obtained from the grating, the distribution of the electrons in the atoms affects the overall intensity distribution of the X-rays. It is the electrons that diffract the X-ray beam and thus information can be obtained about the distribution of the electrons in the atoms or molecules from which the crystal is made. It was in this way that the structure of DNA was first uncovered by Crick and Watson in 1953, and similar techniques have more recently been used to unravel the structure of HIGH-TEMPERATURE SUPERCONDUCTORS.

X-ray diffraction The DIFFRACTION of an X-RAY beam off the atoms in a crystal. Because the atoms in a crystalline lattice are arranged regularly, they act rather like a DIFFRACTION GRATING for the X-rays, and strong CONSTRUCTIVE INTERFERENCE is seen in certain directions.

For a beam of X-rays of wavelength λ, striking a crystal at an angle θ to the planes of the crystal, which have a separation d between one plane and the next, constructive interference will occur if

$$2d\sin\theta = n\lambda$$

where n is a whole number. *See also* BRAGG'S LAW.

X-ray tube A device used to produce X-RAYS, similar in structure to the THERMIONIC DIODE but operating at very high voltages (typically 100 kV). At these voltages, electrons gain a great deal of energy from the electric field. This energy is suddenly lost when they hit the metal anode, and some of it is given out as X-rays. A good deal of heat is also produced, so the anode is made of a good thermal conductor, such as copper, and may be cooled by passing water through it.

xylem A compound plant tissue (made of a number of different cell types), the main function of which is to carry water from the roots to other parts of the plant. Xylem consists of PARENCHYMA cells, SCLERENCHYMA fibres, tracheids and conducting vessels. Tracheids are dead, spindle-shaped, overlapping cells with heavily lignified walls (see LIGNIN) and provide strength and support to the plant. They also conduct water in plants with no vessels, such as ferns, mosses and conifers.

The conducting vessels are made up of cells joined end to end whose cross-walls have be broken down to leave long tubes used for carrying water. The vessels are thickened to different extents; protoxylem has rings or spirals of lignin so can still expand, but metaxylem has more lignin arranged in a reticulate pattern. Protoxylem and metaxylem form the primary xylem that is found in non-woody plants. In trees and shrubs further growth occurs and the secondary xylem forms, providing extra support (*see* CAMBIUM). Most flowering plants use vessels for conducting water.

See also TRANSPIRATION, TRANSPIRATION STREAM.

xylene *See* DIMETHYLBENZENE.

Y

yagi A type of radio AERIAL. In a yagi, several elements re-radiate received signals in such a way that they constructively interfere (*see* CONSTRUCTIVE INTERFERENCE) at an element connected to a radio receiver or TV set. Thus a stronger signal is produced than would be achieved by a simple aerial.

Y-chromosome *See* SEX DETERMINATION.

yeast A group of FUNGI of the PHYLUM ASCOMYCOTA that is used by humans as a fermenting agent in baking, brewing and making wines and spirits. Yeasts reproduce asexually by BUDDING or sexually by the formation of a ASCOSPORES developing within a structure called an ASCUS. When growth is rapid, the new daughter cells remain attached to the parent cells forming long chains.

Yeasts produce enzymes that convert starch or sugars to alcohol and carbon dioxide; different yeasts act upon different substrates (for example, cereal grain in the case of beer and grapes in the case of wine). *Saccharomyces cerevisiae* (brewer's yeast) is a yeast commonly used in baking and brewing. Some yeasts are human PATHOGENS, for example *Candida albicans* causes thrush, and some are useful in genetic research.

yield In any reaction, that proportion of the REAGENTS which react to produce the desired product.

yield point For a DUCTILE material, the point beyond which a further small increase in TENSILE STRESS produces a large increase in STRAIN. Beyond the yield point, the material has effectively lost all its strength; it may extend a great deal more before it breaks, but this extra extension will require little more force.

yellow body *See* CORPUS LUTEUM.

yolk A store of food, mostly protein and fat, in the EGGS of many animals. The yolk provides nourishment for the developing EMBRYO. *See also* YOLK SAC.

yolk sac An EXTRAEMBRYONIC MEMBRANE that is important in birds and reptiles, as it contains YOLK to nourish the EMBRYO. In humans, the yolk is less important and combines with the CHORION.

Young's double slit experiment The experiment that first showed the WAVE NATURE OF LIGHT. In this experiment a COHERENT light source (light from a laser, or light diffracted from a single slit) is shone onto a pair of parallel slits. The light from these slits spreads out by DIFFRACTION and as the beams overlap they INTERFERE, producing areas of CONSTRUCTIVE and DESTRUCTIVE INTERFERENCE, which are seen as bright and dark bands called fringes. *See also* FRESNEL'S BIPRISM.

Young's fringes The bright and dark bands seen as a result of constructive and destructive interference of light in YOUNG'S DOUBLE SLIT EXPERIMENT.

Young's modulus A measure of the stiffness of a stretched rod or wire under tension or compression. It is equal to the STRESS of the wire or rod divided by the longitudinal STRAIN.

y-plates One of the two pairs of plates fitted to a CATHODE RAY TUBE to control the direction of the electron beam. The other pair are the x-plates.

ytterbium (Yb) The element with ATOMIC NUMBER 70; RELATIVE ATOMIC MASS 173.0; melting point 819°C; boiling point 1,196°C; RELATIVE DENSITY 4.5. Ytterbium is a LANTHANIDE metal used in some special purpose ALLOYS.

yttrium (Yt) The element with ATOMIC NUMBER 39; RELATIVE ATOMIC MASS 88.9; melting point 1,522°C; boiling point 3,338°C; RELATIVE DENSITY 4.5. Yttrium is a TRANSITION METAL, chemically similar to the LANTHANIDE metals. Yttrium compounds are used in some lasers.

Z

Z boson, *Z particle* The neutral particle that together with the charged W BOSON are the GAUGE BOSONS responsible for the WEAK NUCLEAR FORCE. The Z boson has a mass about 90 times that of the proton.

zeolite Any of a class of minerals based on the aluminosilicate ANION that were traditionally used in WATER SOFTENING and other ION EXCHANGE processes. Zeolites can also absorb other small molecules and are used as 'molecular sieves'. They can also be used in SORPTION PUMPS, improving a vacuum by absorbing gas molecules. Natural zeolites have largely been superseded by synthetic materials.

zidovudine *See* AZT.

Ziegler–Natta catalysts Complex CATALYSTS used in the POLYMERIZATION of ALKENES. The products obtained are denser and tougher than those obtained by traditional high pressure methods of polymerization. Ethene can be polymerized using these catalysts, as can propene, the latter yielding polypropylene. Ziegler-Natta catalysts are prepared by the interaction of an organometallic derivative (*see* ORGANOMETALLIC COMPOUND) and a TRANSITION METAL derivative. For example titanium(IV) chloride and triethylaluminium, $Al(C_2H_5)^3$, are typically combined in a solvent, such as heptane, at low temperatures and pressures.

zinc (Zn) The element with ATOMIC NUMBER 30; RELATIVE ATOMIC MASS 65.4; melting point 290°C; boiling point 732°C; RELATIVE DENSITY 7.1. Zinc is extracted from its ores (mainly zinc sulphide and zinc carbonate) by REDUCTION with carbon, for example:

$$2ZnCO_3 + C \rightarrow 2Zn + 3CO_2$$

Zinc is a fairly reactive TRANSITION METAL, reacting with acids to give hydrogen and zinc salts, for example:

$$Zn + 2HCl \rightarrow ZnCl_2 + H_2$$

Zinc is used as a protective coating on some steel objects (*see* GALVANIZING) and in the manufacture of ZINC-CARBON CELLS. It is also used in the manufacture of brass, an alloy with copper.

zincate Any compound containing the zincate ion, ZnO_2^{2-}. Zincates are formed by the reaction between zinc oxide and strong alkalis, for example:

$$ZnO + 2NaOH \rightarrow Na_2ZnO_2 + H_2O$$

zinc blende *See* BLENDE.

zinc-carbon cell, *dry cell* A common type of electrochemical CELL in which zinc and carbon electrodes are used with an ELECTROLYTE in the form of a paste.

zinc chloride ($ZnCl_2$) A white solid; melting point 290°C; boiling point 732°C; RELATIVE DENSITY 2.9. The ANHYDROUS salt is DELIQUESCENT and can be made by the reaction of zinc with hydrogen chloride:

$$Zn + 2HCl \rightarrow ZnCl_2 + H_2$$

Zinc chloride is used as a flux in some soldering operations.

zinc oxide (ZnO) A white solid (yellow when hot); melting point 1,975°C; decomposes on further heating; RELATIVE DENSITY 5.5. It occurs naturally, and can also be made by heating zinc in air:

$$2Zn + O_2 \rightarrow 2ZnO$$

Zinc oxide is AMPHOTERIC, reacting with acids to form zinc salts, and bases to form ZINCATES, for example:

$$ZnO + 2HCl \rightarrow ZnCl_2 + 2H_2O$$

$$ZnO + 2NaOH \rightarrow Na_2ZnO_2 + 2H_2O$$

Zinc oxide is used as a white pigment and as an antiseptic in some ointments.

zinc sulphate ($ZnSO_4$) A white solid, decomposes on heating. Zinc sulphate usually occurs as the heptahydrate, $ZnSO_4.7H_2O$; RELATIVE DENSITY 1.9. Zinc sulphate can be made by dissolving zinc in sulphuric acid:

$$Zn + H_2SO_4 \rightarrow ZnSO_4 + H_2$$

It is used as a mordant in some dying processes.

zinc sulphide (ZnS) A creamy white solid, which decomposes on heating. Zinc sulphide occurs naturally as zinc blende, and decomposes to zinc on heating:

$$ZnS \rightarrow Zn + S$$

Zinc sulphide is used as a PHOSPHOR in some CATHODE RAY TUBES.

zirconium (Zr) The element with ATOMIC NUMBER 40; RELATIVE ATOMIC MASS 91.2; melting point 1,853°C; boiling point 4,376°C; RELATIVE DENSITY 6.4. Zirconium is a TRANSITION METAL. It is used in the manufacture of FUEL-ROD casings in NUCLEAR REACTORS, where its low neutron absorption coupled with good mechanical properties are an advantage.

zoology The branch of biology that deals with the study of animals. Zoology includes many areas, for example anatomy, physiology, behaviour and evolution.

Zoomastigina A PHYLUM from the KINGDOM PROTOCTISTA that consists of PROTOZOA. The members (flagellates) move by one or more FLAGELLA and reproduce by BINARY FISSION. Flagellates can be free-living or parasitic (*see* PARASITE). An important example is *Trypanosoma*, the parasite that causes sleeping sickness.

zooplankton Animal PLANKTON (small life forms living on the surface of fresh or salt water) that feed on PHYTOPLANKTON (plant plankton) and themselves provide food for larger fishes. Zooplankton are mostly able to move by FLAGELLA.

zoospore A motile SPORE produced in a SPORANGIUM, possessing one or more FLAGELLA, that is present in some ALGAE and FUNGI.

Z particle *See* Z BOSON.

zwitterion An ion with both a positive and a negative charge. For example amino acids in water form zwitterions by the loss of a proton from COOH making it negative, which then goes to the NH_2 group, making it positive.

zygomorphic *See* FLOWER.

Zygomycete A member of the ZYGOMYCOTA PHYLUM of FUNGI.

Zygomycota A PHYLUM of the KINGDOM FUNGI characterized by their absence of septa (partitions) in the HYPHAE and by the production of ZYGOSPORES during SEXUAL REPRODUCTION. Zygomycetes produce a large, branched MYCELIUM. Examples include bread mould (*Rhizopus*) and pin mould (*Mucor*). ASEXUAL REPRODUCTION also occurs by CONIDIA or by SPORES.

zygospore A SPORE produced during SEXUAL REPRODUCTION by members of the PHYLUM ZYGOMYCOTA of FUNGI.

zygote A fertilized OVUM (egg) before it begins cleavage. *See* EMBRYONIC DEVELOPMENT.

APPENDICES

Appendix I: SI units

Base units

Physical quantity	*Name*	*Symbol*	*Definition*
length	metre	m	the length equal to the length of path travelled by light in a vacuum in 1/(299,792,458) seconds
mass	kilogram	kg	the mass equal to that of the international prototype kilogram kept at Sèvres, France
time	second	s	the duration of 9,192,631,770 oscillations of the electromagnetic radiation corresponding to the electron transition between two hyperfine levels of the ground state of the caesium–133 atom
electric current	ampere	A	the constant electric current which, if maintained in two straight parallel conductors of infinite length and negligible cross-section, placed 1 metre apart in a vacuum, would produce a force between these conductors equal to 2 x 10^{-7} metres
thermodynamic temperature	kelvin	K	the fraction 1/273.16 of the thermodynamic temperature of the triple point of water
luminous intensity	candela	cd	the luminous intensity, in a given direction, of a source of monochromatic radiation of frequency 5.4 x 10^{14} Hz and has a radiant intensity in that direction of 1/683 watt per steradian
amount of substance	mole	mol	the amount of substance containing as many atoms (or molecules or ions or electrons) as there are carbon atoms in 12 g of carbon–12.

Supplementary units

plane angle	radian	rad	the plane angle subtended at the centre of a circle by an arc of equal length to the circle radius
solid angle	steradian	sr	the solid angle that encloses a surface on a sphere equal in area to the square of the radius of the sphere

Appendix I/II

Derived SI units

Physical quantity	*Name*	*Symbol*	*SI equivalent*
activity	becquerel	Bq	
dose	gray	Gy	
dose equivalent	sievert	Sv	
electric capacitance	farad	F	$A\ sV^{-1}$
electric charge	coulomb	C	As
electric conductance	siemens	S	
electric potential difference	volt	V	WA^{-1}
electric resistance	ohm	Ω	VA^{-1}
energy	joule	J	Nm
force	newton	N	$kg\ ms^{-2}$
frequency	hertz	Hz	s^{-1}
illuminance	lux	lx	
inductance	henry	H	$V\ sA^{-1}$
luminous flux	lumen	lm	
magnetic flux	weber	Wb	
magnetic flux density	tesla	T	
power	watt	W	$J\ s^{-1}$
pressure	pascal	Pa	Nm^{-2}

Appendix II: SI prefixes

Submultiple	*Prefix*	*Symbol*	*Multiple*	*Prefix*	*Symbol*
10^{-1}	deci	d	10^{1}	deca	da
10^{-2}	centi	c	10^{2}	hecto	h
10^{-3}	milli	m	10^{3}	kilo	k
10^{-6}	micro	μ	10^{6}	mega	M
10^{-9}	nano	n	10^{9}	giga	G
10^{-12}	pico	p	10^{12}	tera	T
10^{-15}	femto	f	10^{15}	peta	P
10^{-18}	atto	a	10^{18}	exa	E

Appendix III: SI conversion factors

Unit name	*Symbol*	*Quantity*	*SI equivalent*	*Unit*
acre		area	0.405	hm^2
ångstrom	Å	length	0.1	nm
astronomical unit	AU	length	0.150	Tm
atomic mass unit	amu	mass	1.661×10^{-27}	kg
bar		pressure	0.1	MPa
barn	b	area	100	fm^2
British thermal unit	btu	energy	1.055	kJ
calorie	cal	energy	4.187	J
cubic foot	cu ft	volume	0.028	m^3
cubic inch	cu in	volume	16.387	cm^3
cubic yard	cu yd	volume	0.765	m^3
curie	Ci	activity	37	GBq
degree (angle)	°	plane angle	$\pi/180$	rad
degree Celsius	°C	temperature	1	K
degree Fahrenheit	°F	temperature	5/9	K
degree Rankine	°R	temperature	5/9	K
dyne	dyn	force	10	μN
electronvolt	eV	energy	0.160	aJ
erg		energy	0.1	μJ
fathom (6 ft)		length	1.829	m
foot	ft	length	30.48	cm
foot per second	$ft\ s^{-1}$	velocity	0.305	$ft\ s^{-1}$
gallon (UK)	gal	volume	4.546	dm^3
gallon (US)	gal	volume	3.785	dm^3
gallon (UK) per mile		consumption	2.825	dm^3km^{-1}
gauss	Gs, G	magnetic flux density	100	μT
hectare	ha	area	1	hm^2
horsepower	hp	power	0.746	kW
inch	in	length	2.54	cm
knot		velocity	1.852	$km\ h^{-1}$
light year	ly	length	9.461×10^{15}	m
litre	l	volume	1	dm^3
Mach number	Ma	velocity	1193.3	$km\ h^{-1}$

SI conversion factors (continued)

Unit name	*Symbol*	*Quantity*	*SI equivalent*	*Unit*
maxwell	Mx	magnetic flux	10	nWb
micron	μ	length	1	μm
mile (nautical)		length	1.852	km
mile (statute)		length	1.609	km
miles per hour	mph	velocity	1,609	km h^{-1}
minute = (1/60)°	′	plane angle	π/10800	rad
oersted	Oe	magnetic field strength	1/(4π)	kA m^{-1}
ounce (avoirdupois)	oz	mass	0.2	g
ounce (troy)		mass	31.103	g
parsec	pc	length	30857	Tm
pint (UK)	pt	volume	0.568	dm^3
pint (US)		volume	0.473	dm^3
pound	lb	mass	0.454	kg
poundal	pdl	force	0.138	N
pound-force	lbf	force	4.448	N
pound-force per inch		pressure	6.895	kPa
pounds per square inch	psi	pressure	6.895 x 10^3	kPa
rad		absorbed dose	0.01	Gy
rem		dose equivalent	0.01	Sv
röntgen	R	exposure	0.258	mC kg^{-1}
second = (1/60)′	″	plane angle	π/648	mrad
solar mass	M	mass	1.989 x 10^{30}	kg
square foot	sq ft	area	9.290	dm^3
square inch	sq in	area	6.452	cm^2
square mile (statute)	sq mi	area	2.590	km^2
square yard	sq yd	area	0.836	m^2
standard atmosphere	atm	pressure	1.101	MPa
stokes	St	viscosity	1	cm^2s^{-1}
therm		energy	0.105	GJ
ton		mass	1.016	Mg
tonne	t	mass	1	Mg
torr, or mm Hg		pressure	0.133	kPa
yard	yd	length	0.914	m

Appendix IV: Common measures

Metric units		Imperial equivalent	Imperial units		Metric equivalent
Length					
	1 millimetre (mm)	0.03937 in		1 inch	2.54 cm
10 mm	1 centimetre (cm)	0.39 in	12 in	1 foot	30.48 cm
10 cm	1 decimetre (dm)	3.94 in	3 ft	1 yard	0.9144 m
100 cm	1 metre (m)	39.37 in	1760 yd	1 mile	1.6093 km
1000 m	1 kilometre (km)	0.62 mi			
Area					
	1 square millimetre	0.0016 sq in		1 square inch	6.45 cm^2
	1 square centimetre	0.155 sq in	144 sq in	1 square foot	0.0929 m^2
100 cm^2	1 square decimetre	15.5 sq in	9 sq ft	1 square yard	0.836 m^2
10,000 cm^2	1 square metre	10.76 sq ft	4,840 sq yd	1 acre	0.405 ha
10,000 m^2	1 hectare	2.47 acres	640 acres	1 square mile	259 ha
Volume					
	1 cubic centimetre	0.016 cu in		1 cubic inch	16.3871 cm^3
1,000 cm^3	1 cubic decimetre	61.024 cu in	1,728 cu in	1 cubic foot	0.028 m^3
1,000 dm^3	1 cubic metre	35.31 cu ft	27 cu ft	1 cubic yard	0.765 m^3
		1.308 cu yd			

Common measures (continued)

Metric units		Imperial equivalent	Imperial units		Metric equivalent
Liquid volume					
	1 litre	1.76 pt		1 pint	0.57 l
100 l	1 hectolitre	22 gal	2 pt	1 quart	1.14 l
			4 qt	1 gallon	4.55 l
Weight					
	1 gram	0.035 oz		1 ounce	28.3495 g
1,000 g	1 kilogram	2.2046 lb	16 oz	1 pound	0.4536 kg
1,000 kg	1 tonne	0.0842 ton cm^2	14 lb	1 stone	6.35 kg
			8 st	1 hundredweight	50.8 kg
			20 cwt	1 ton	1.016 t

Appendix V: Conversion factors

Imperial to metric

from	to	multiply by
Length		
inches	millimetres	25.4
inches	centimetres	2.54
inches	metres	0.0254
hands	millimetres	101.6
links	metres	0.2012
feet	millimetres	304.8
feet	centimetres	30.48
feet	metres	0.3048
yards	metres	0.9144
fathoms	metres	1.829
rods	metres	5.029
chains	metres	20.12
chains	kilometres	0.02012
furlongs	metres	201.2
furlongs	kilometres	0.2012
statute miles	metres	1,609
nautical miles	metres	1,852
statute miles	kilometres	1.609
nautical miles	kilometres	1.852
(statute miles	nautical miles	0.8690
nautical miles	statute miles	1.151)

Metric to Imperial

from	to	multiply by
micrometres	milli-inches	0.03937
millimetres	inches	0.03937
centimetres	inches	0.3937
centimetres	links	0.04971
centimetres	feet	0.03280
metres	inches	39.37
metres	links	4.971
metres	feet	3.281
metres	yards	1.094
metres	fathoms	0.5468
kilometres	chains	49.71
kilometres	furlongs	4.971
kilometres	statute miles	0.6214
kilometres	nautical miles	0.5400

Conversion factors (continued)

Imperial to metric

from	to	multiply by
Area		
square inches	square centimetres	6.452
square inches	square centimetres	6.452
square feet	square centimetres	929.3
square feet	square metres	0.09293
square yards	square metres	0.8361
square yards	ares	0.008361
acres	square metres	4,047
acres	ares	40.47
acres	hectares	0.4047
square miles	hectares	259.0
square miles	square kilometres	2.590
Volume		
cubic inches	cubic centimetres	16.39
cubic inches	cubic decimetres	0.01639
cubic feet	cubic decimetres	28.32
cubic feet	cubic metres	0.02832
cubic yards	cubic metres	0.7646

Metric to Imperial

from	to	multiply by
Area		
square millimetres	square inches	0.00016
square centimetres	square inches	0.1550
square centimetres	square feet	0.00106
square metres	square feet	10.764
square metres	square yards	1.1960
ares	square yards	119.60
ares	acres	0.0247
hectares	acres	2.4710
hectares	square miles	0.0038610
square kilometres	acres	247.105
square kilometres	square miles	0.386102
Volume		
cubic centimetres	cubic inches	0.06102
cubic decimetres	cubic inches	61.02
cubic decimetres	cubic feet	0.03531
cubic metres	cubic feet	35.32
cubic metres	cubic yards	1.308

Conversion factors (continued)

Imperial to metric

from	to	multiply by
Capacity		
UK fluid ounces	millilitres	28.41
US liquid ounces	millilitres	29.57
UK pints	litres	0.5683
US pints	litres	0.4732
UK quarts	litres	1.137
US quarts	litres	0.9464
UK gallons	litres	4.546
US gallons	litres	3.785
(cubic inches	UK fluid ounces	0.5767
cubic inches	US liquid ounces	0.5541
cubic inches	UK pints	0.02884
cubic inches	UK gallons	0.003604
cubic feet	UK gallons	6.2288
cubic feet	US gallons	7.480)
Weight		
ounces (avoirdupois)	grams	28.35
ounces (troy)	grams	31.10
(ounces avoirdupois	ounces troy	0.9115)
pounds	kilograms	0.4536
stones	kilograms	6.350
UK (long) tons	tonnes	1.016

Metric to Imperial

from	to	multiply by
Capacity		
millilitres	UK fluid ounces	0.03520
millilitres	US liquid ounces	0.03381
litres	UK fluid ounces	35.20
litres	US liquid ounces	33.81
litres	UK pints	1.760
litres	US pints	2.113
litres	UK quarts	0.8799
litres	US quarts	1.057
litres	UK gallons	0.2200
litres	US gallons	0.2642
(UK gallons	cubic feet	0.1337
US gallons	cubic feet	0.1605
UK fluid ounces	cubic inches	1.173
UK pints	cubic inches	34.68
UK quarts	cubic inches	69.36
UK gallons	cubic inches	277.4)
Weight		
grams	ounces (avoirdupois)	0.03527
grams	ounces (troy)	0.03215
kilograms	ounces (avoirdupois)	35.27
kilograms	ounces (troy)	32.15
kilograms	pounds	2.205
kilograms	stones	0.1575
tonnes	tons (long)	0.9842

Conversion factors (continued)

Imperial to metric

from	to	multiply by
Velocity		
inches per minute	millimetres per second	0.4233
feet per minute	millimetres per second	5.08
miles per hour	kilometres per hour	1.6093
miles per hour	metres per second	0.4470
international knots	miles per hour	1.5078
international knots	metres per second	0.5144
(international knots	miles per hour	1.1508
miles per hour	international knots	0.8690)
Fuel consumption		
miles per US gallon	kilometres per litre	0.3540
miles per UK gallon	kilometres per litre	0.4251
UK gallons per mile	litres per 100 kilometres	282.5
US gallons per mile	litres per 100 kilometres	235.2
Density and concentration		
ounces per UK gallon	kilograms per cubic metre	6.236
ounces per US gallon	kilograms per cubic metre	7.489
pounds per cubic foot	kilograms per cubic metre	16.02

Metric to Imperial

from	to	multiply by
Velocity		
millimetre per second	inches per minute	2.3622
millimetre per second	feet per minute	0.1969
kilometres per hour	miles per hour	0.6214
kilometres per hour	international knots	0.5400
metres per second	miles per hour	2.2369
metres per second	international knots	1.9438
Fuel consumption		
kilometres per litre	miles per UK gallon	2.825
kilometres per litre	miles per US gallon	2.352
litres per 100 kilometres	UK gallons per mile	0.003540
litres per 100 kilometres	US gallons per mile	0.004251
Density and concentration		
kilograms per cubic metre	ounces per UK gallon	0.1604
kilograms per cubic metre	ounces per US gallon	0.1335
kilograms per cubic metre	pounds per cubic foot	0.06242

Conversion factors (continued)

Imperial to metric

from	to	multiply by
Force		
poundals	newtons	0.1383
ounces-force	newtons	0.2780
pounds-force	newtons	0.1020
(long) tons-force	kilonewtons	9.9640
Pressure		
poundals per square foot	pascals	1.4882
pounds-force per square foot	pascals	47.880
pounds-force per square foot	millibars	0.4788
pounds-force per square inch	millibars	68.948
pounds-force per square foot	kilopascals	6.8948
inches of mercury	millibars	69.947

Metric to Imperial

from	to	multiply by
Force		
newtons	poundals	7.2330
newtons	ounces-force	3.5969
newtons	pounds-force	0.2248
kilonewtons	(long) tons-force	0.1004
Pressure		
pascals	poundals per square foot	0.6720
pascals	pounds-force per square foot	0.0209
millibars	pounds-force per square foot	2.0885
millibars	inches of mercury	0.0295
millibars	pounds-force per square inch	0.0145
kilopascals	pounds-force per square inch	0.1450

Conversion factors (continued)

Imperial to metric

from	to	multiply by
Energy, work, quantity of heat		
calories, 15°C	joules	4.186
calories, international table	joules	4.187
British thermal units	kilojoules	1.055
kilocalories, international table	kilojoules	4.187
kilowatt hours	megajoules	3.6
therms	megajoules	105.5
Power		
UK horsepower	kilowatts	0.7457
Plane angle		
degrees	radians	0.01745

Metric to Imperial

from	to	multiply by
Energy, work, quantity of heat		
joules	calories, 15 C	0.2389
joules	calories, international table	0.2388
kilojoules	British thermal units	0.9478
kilojoules	kilocalories, international table	0.2388
megajoules	kilowatt hours	0.2778
megajoules	therms	0.009478
Power		
kilowatts	UK horsepower	1.341
Plane angle		
radians	degrees	57.30

Appendix VI: Physical concepts in SI units

Concept	*Symbol*	*SI unit*	*SI symbol*	*Defining equation*
Area	A, a	square metre	m^2	$a = l^2$
Volume	V, v	cubic metre	m^3	$V = l^3$
Velocity	v, u	metre/second	ms^{-1}	$v = dl/dt$
Acceleration	a	metre/second^2	ms^{-2}	$a = d^2l/dt^2$
Density	ρ	kilogram/metre^2	kg m^{-2}	$\rho = m/v$
Moment of inertia	I	kilogram metre^2	kg m^2	$I = Mk^2$
Momentum	p	kilogram metre/second	$kgms^{-1}$	$p = mv$
Angular momentum	$I\omega$	kilogram metre^2/second	kgm^2s^{-1}	$I\omega$
Force	F	newton	N	$F = ma$
Torque (moment of force)	T, (M)	newton metre	Nm	
Work (energy, heat)	W, (E)	joule	J	$W = \int F dl$
Kinetic energy	T, (W)	joule	J	$T = {}^1/_2 mv^2$
Potential energy	V	joule	J	$V = \int F dl$
Heat (enthalpy)	Q, (H)	joule	J	$H = U + pV$
Power	P	watt	W	$P = dW/dt$
Pressure (stress)	p, (σ, ϕ)	newton/metre^2	Nm^{-2}	$p = F/A$
Surface tension	γ, (σ)	newton/metre	Nm^{-1}	$\gamma = F/l$
Viscosity, dynamic	η, μ	newton second/metre^2	Nm^{-2}	$F/A = \eta dv/dl$
Viscosity, kinematic	v	metre^2/second	m^2s^{-1}	$v = \eta/\rho$

Physical concepts in SI units (continued)

Concept	*Symbol*	*SI unit*	*SI symbol*	*Defining equation*
Temperature	θ, T	degree Celsius, kelvin	°C, K	$TK = (\theta = 273.15)°C$
Electric charge	Q	coulomb	C	$F = (Q_1 Q_2)/(4\mu\varepsilon_0 r^2)$
Electric potential (potential difference)	V	volt	V	$V = \int_\infty E dl$ ($V_{ab} = -\int^a_b E dl$)
Electric field strength	E	volt/metre	Vm^{-1}	$E = -dV/dl$
Electric resistance	R	ohm	Ω	$R = V/I$
Conductance	G	siemens	S	$G = 1/R$
Electric flux	ψ	coulomb	C	$\psi = Q$
Electric flux density	D	coulomb/metre²	Cm^2	$D = d\psi/dA$
Frequency	f	hertz	Hz	cycles per second
Permittivity	ε	farad/metre	Fm^{-1}	$\varepsilon = D/E$
Relative permittivity	ε_r	farad/metre	Fm^{-1}	$\varepsilon_r = \varepsilon\varepsilon_0$
Magnetic field strength	H	ampere turn/metre	Atm^{-1}	$dH = i\,dl\sin\theta/4\pi r^2$
Magnetic flux	Φ	weber	Wb	$\Phi = -\int e dt$
Magnetic flux density	B	tesla	T	$B = d\Phi/dA$
Permeability	μ	henry/metre	Hm^{-1}	$\mu = B/H$
Relative permeability	μ_r	henry/metre	Hm^{-1}	$\mu_r = \mu/\mu_0$
Mutual inductance	M	henry	H	$e_2 = M di_1/dt$
Self inductance	L	henry	H	$e = L di/dt$
Capacitance	C	farad	F	$C = Q/V$

Physical concepts in SI units (continued)

Concept	*Symbol*	*SI unit*	*SI symbol*	*Defining equation*
Reactance	X	ohm	Ω	$X = \omega L$ or $1/\omega C$
Impedance	Z	ohm	Ω	$X = \sqrt{R^2} = X^2$
Susceptance	B	siemens	S	$B = 1/X$
Admittance	Y	siemens	S	$Y = 1/Z$
Luminous flux	Φ	lumen	lm	lm = cd sr
Illuminance	E	lux	lx	lx = lm m^{-2}

Appendix VII: Fundamental constants

Constant	*Symbol*	*Value*	*Units*
acceleration of free fall	g	9.80665	ms^{-2}
speed of light in vacuum	c	299,792,458	ms^{-1}
elementary charge	e	1.60217733	10^{-19} C
electron rest mass	m_e	9.1093897	10^{-31} kg
permeability of free space	μ_0	$4\pi \times 10^{-7}$	$N A^{-2}$
		= 12.566370614....	$10^{-7} N A^{-2}$
permittivity of free space	ε_0	8.854187187....	$10^{-12} F m^{-1}$
electronic radius	r_e	2.18794092	10^{-15} m
neutron rest mass	m_n	1.6749286	10^{-27} kg
proton rest mass	m_p	1.6726231	10^{-27} kg
Avogadro constant	L, N_A	6.0221367	$10^{23} mol^{-1}$
Boltzmann constant	$k = R/N_A$	1.380658	$10^{-23} JK^{-1}$
Molar volume of ideal gas at STP	V_m	2.241409	$10^{-2} m^3mol^{-1}$
Faraday constant	F	96,485.309	$C mol^{-1}$
Stefan–Boltzmann constant	s	5.67051	$10^{-8} Wm^{-2}K^{-4}$
gas constant	R	8.314510	$J mol^{-1}K^{-1}$
gravitational constant	G	6.67259	$10^{-11} m^3kg^{-1}s^{-2}$
Planck constant	h	6.6260755	$10^{-34}Js$
Bohr radius	r_B	5.29177249	10^{-11} m
Bohr magneton	μ_B	9.2740154	$10^{-24} JT^{-1}$
Nuclear magneton	μ_N	5.0507866	$10^{-27}JT^{-1}$

Appendix VIII: Electromagnetic spectrum

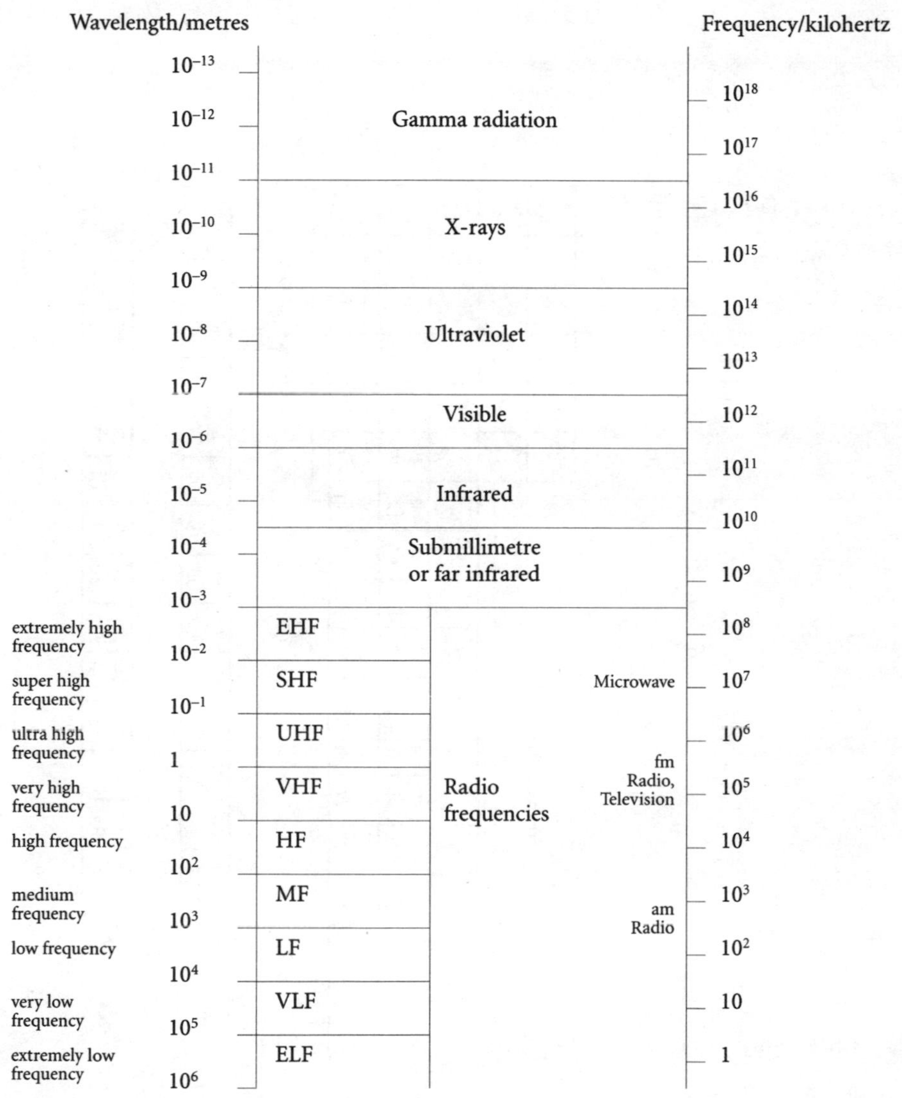

Appendix IX: Periodic table of the elements

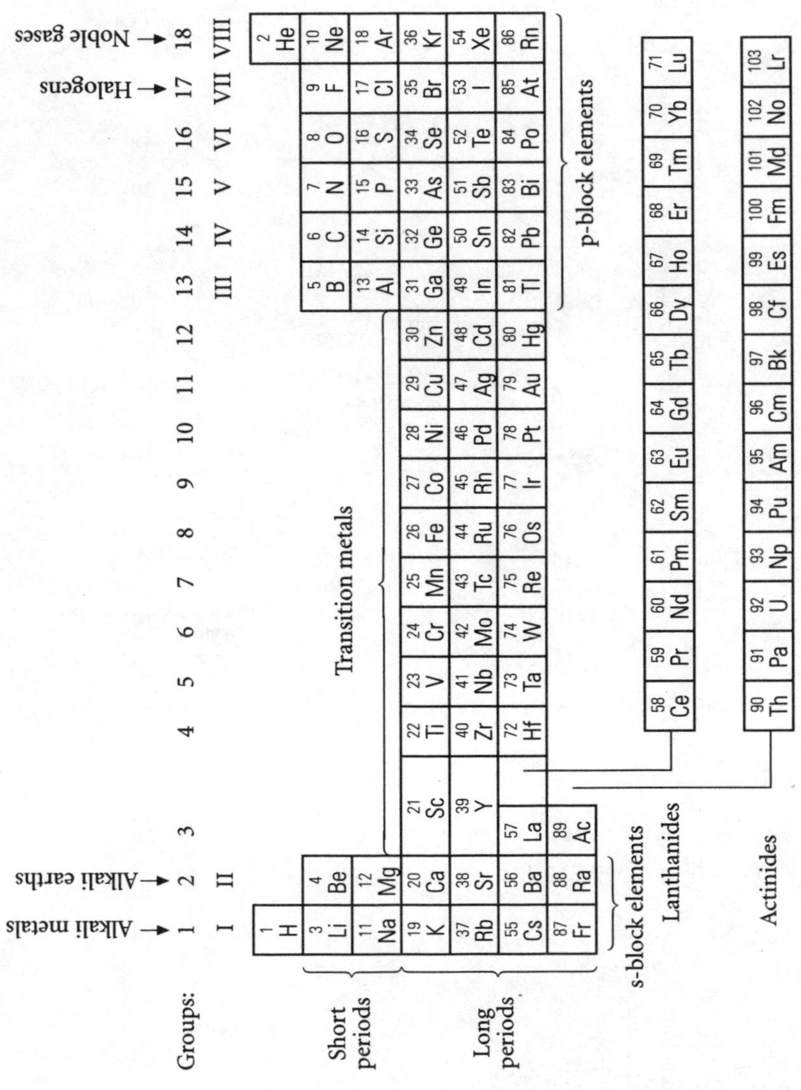

Appendix X: The chemical elements

Name	*Symbol*	*Atomic number*	*Relative atomic mass*	*Melting point °C*	*Boiling point °C*	*Relative density*
Actinium	Ac	89	227*	1,050	3473**	–
Aluminium	Al	13	26.98	660	1,800	2.7
Americium	Am	95	243*	994	2,607	13.7
Antimony	Sb	51	121.8	630	1,750	6.7
Argon	Ar	18	39.95	–189	–185	1.7×10^{-3}
Arsenic	As	33	74.9	–	613	3.9
Astatine	At	85	(210)	302	377	–
Barium	Ba	56	137.3	725	1,640	3.5
Berkelium	Bk	97	247*	–	–	–
Beryllium	Be	4	9.0	1,285	2,970	1.9
Bismuth	Bi	83	208.98	271	1,560	9.8
Boron	B	5	10.8	2030	2,550	2.4
Bromine	Br	35	79.9	–7	58	3.1 (liquid)
Cadmium	Cd	48	112.4	321	765	8.7
Caesium	Cs	55	132.9	28	690	1.9
Calcium	Ca	20	40.1	840	1,484	1.6
Californium	Cf	98	251*	–	–	–
Carbon	C	6	12.0	–	3,500	2.3
Cerium	Ce	58	140.1	798	3,433	6.8
Chlorine	Cl	17	35.5	–101	–35	3.2×10^{-3} (0°C)

The chemical elements (continued)

Name	*Symbol*	*Atomic number*	*Relative atomic mass*	*Melting point °C*	*Boiling point °C*	*Relative density*
Chromium	Cr	24	52.0	1,900	2,640	7.2
Cobalt	Co	27	58.9	1,495	2,870	8.9
Copper	Cu	29	63.5	1,083	2,582	8.9
Curium	Cm	96	247*	–	–	–
Dysprosium	Dy	66	162.5	1,412	2,567	8.6
Einsteinium	Es	99	254*	–	–	–
Erbium	Er	68	167.3	1,529	2,863	9.0
Europium	Eu	63	152.0	852	1,529	5.2
Fermium	Fm	100	(237)	–	–	–
Fluorine	F	9	19.0	–220	–188	1.7×10^{-3} (0°C)
Francium	Fr	87	223*	27	–	–
Gadolinium	Gd	64	157.3	1,312	3,273	7.9
Gallium	Ga	31	69.7	30	2,403	5.9
Germanium	Ge	32	72.6	937	2,830	5.4
Gold	Au	79	197.0	1,064	2,807	19.3
Hafnium	Hf	72	178.5	2,230	4,602	13.3
Helium	He	2	4.0	–269	–268.9	0.2×10^{-3}
Holmium	Ho	67	164.9	1,472	2,700	8.8
Hydrogen	H	1	1.0	–259	–253	0.1×10^{-3}
Indium	In	49	114.8	157	2,080	7.3
Iodine	I	53	126.9	–	183	4.9
Iridium	Ir	77	192.2	2,410	4,130	22.5

The chemical elements (continued)

Name	*Symbol*	*Atomic number*	*Relative atomic mass*	*Melting point °C*	*Boiling point °C*	*Relative density*
Iron	Fe	26	55.8	1,535	2,750	7.9
Krypton	Kr	36	83.8	–157.4	–152	3.5×10^{-3}
Lanthanum	La	57	138.9	918	3464	6.1
Lawrencium	Lw	103	257*	–	–	–
Lead	Pb	82	207.2	328	1,740	11.4
Lithium	Li	3	6.9	180	1,340	0.5
Lutetium	Lu	71	175.0	1,663	3,402	9.8
Magnesium	Mg	12	24.3	651	1,107	1.7
Manganese	Mn	25	54.9	1,244	2,040	7.4
Mendelevium	Md	101	256*	–	–	–
Mercury	Hg	80	200.6	–39	357	13.6
Molybdenum	Mo	42	94.9	2,610	5,560	10.2
Neodymium	Nd	60	144.2	1,016	3,068	7.0
Neon	Ne	10	20.2	–249	–246	0.8×10^{-3}
Neptunium	Np	93	237*	–	–	–
Nickel	Ni	28	58.7	1,450	2,840	8.9
Niobium	Nb	41	92.9	2,468	4,742	8.6
Nitrogen	N	7	14.0	–210	–196	1.2×10^{-3}
Nobelium	No	102	259*	–	–	–
Osmium	Os	76	190.2	3,045	5,027	22.6
Oxygen	O	8	16.0	–214	–183	1.3×10^{-3}
Palladium	Pd	46	106.4	1,551	3,140	12.3

The chemical elements (continued)

Name	*Symbol*	*Atomic number*	*Relative atomic mass*	*Melting point °C*	*Boiling point °C*	*Relative density*
Phosphorus	P	15	31.0	44	280	1.8
Platinum	Pt	78	195.1	1,772	3,800	21.3
Plutonium	Pu	94	244*	641	3,232	–
Polonium	Po	84	210*	254	–	9.2
Potassium	K	19	39.1	64	774	0.9
Praseodymium	Pr	59	140.9	931	3,512	6.8
Promethium	Pm	61	145*	1,042	3,000	7.3
Protactinium	Pa	91	231*	–	–	–
Radium	Ra	88	226*	700	1,140	5.1
Radon	Rn	86	222*	–72	–62	10.0 x 10^{-3} (0°C)
Rhenium	Re	75	186.2	3,180	5,620	22.0
Rhodium	Rh	45	102.9	1,966	3,727	12.4
Rubidium	Rb	37	85.5	38	688	1.5
Ruthenium	Ru	44	101.1	2,310	3,900	12.24
Samarium	Sm	62	150.3	1,075	1,791	7.5
Scandium	Sc	21	45.0	1,540	2,850	3.0
Selenium	Se	34	79.0	217	685	4.8
Silicon	Si	14	28.1	1,410	2,335	2.3
Silver	Ag	47	107.9	962	2,212	10.5
Sodium	Na	11	23.0	98	892	1.0
Strontium	Sr	38	87.62	768	1,300	2.5
Sulphur	S	16	32.1	112	445	2.1

The chemical elements (continued)

Name	*Symbol*	*Atomic number*	*Relative atomic mass*	*Melting point °C*	*Boiling point °C*	*Relative density*
Tantalum	Ta	73	180.9	2,996	5,427	16.7
Technetium	Tc	43	99*	2,171	4,876	11.5
Tellurium	Te	52	127.6	451	1,390	6.2
Terbium	Tb	65	158.9	1,365	3,230	8.2
Thallium	Tl	81	204.4	303	1,460	11.9
Thorium	Th	90	232*	1,750	4,790	11.7
Thulium	Tm	69	168.9	1,545	1,950	9.3
Tin	Sn	50	118.7	232	2,270	7.3
Titanium	Ti	22	47.9	1,660	3,280	4.5
Tungsten	W	74	183.9	3,410	5,660	19.3
Uranium	U	92	238*	1,132	3,818	19.1
Vanadium	V	23	50.9	1,890	3,380	6.1
Xenon	Xe	54	131.3	–112	–107	5.5 x 10–3
Ytterbium	Yb	70	173.0	819	1,196	4.5
Yttrium	Y	39	88.9	1,522	3,338	4.5
Zinc	Zn	30	65.4	290	732	7.1
Zirconium	Zr	40	91.2	1,853	4,376	6.4

*Relative atomic mass of longest-lived isotope

Appendix XI: Elementary particles

Family	*Particle*	*Symbol*	*Mass (MeV)*	*Mean life*
Bosons	photon	*γ	0	stable
Leptons	electron	e	0.5110	stable
	neutrino	ν	0	stable
	muon	μ	106	2.20×10^{-6}
	tau	τ	1784	$<2 \times 10^{-12}$
Baryons	proton	p	938	stable ($>10^{33}$ years)
	neutron	n	939.6	917
	lambda particle	λ	1116	2.6×10^{-10}
Mesons	positive pion	π^{+}	140	2.6×10^{-8}
	neutral pion	π°	135	8.4×10^{-17}
	positive kaon	K^{+}	494	1.24×10^{-8}
	J/psi	J/ψ	3097	10^{-20}

Appendix XII: The Solar System

Planet	*Mean distance from Sun (million km)*	*Mass (Earth masses)*	*Sidereal period*	*Axial rotation (equatorial)*	*Diameter (equatorial) (km)*	*Satellites*	*Atmosphere*
Mercury	57.91	0.054	88 d	58 d 16 h	4,878	0	hydrogen, helium, neon
Venus	108.21	0.815	224.7 d	243 d	12,104	0	carbon dioxide
Earth	149.60	1.000*	365.26 d	23 h 56 m 4 s	12,756	1	nitrogen, oxygen
Mars	227.94	0.107	687 d	24 h 37 m 23 s	6,794	2	carbon, dioxide
Jupiter	778.34	317.89	11.86 y	9 h 50 m 30 s	142,800	16	hydrogen, methane
Saturn	1430	95.14	29.46 y	10 h 14 m	120,000	18	hydrogen, helium
Uranus	2869.6	14.52	164.79 y	16–28 h**	51,000	15	methane, helium, hydrogen
Neptune	4496.7	17.46	164.79 y	18–20 h**	49,500	2	methane, hydrogen
Pluto	5900	0.1 (approx)	247.7 y	6 d 9 h	2,300	1	methane

d = days
y = years
h = hours
km = kilometres
*The mass of the Earth is 5.976 x 10^{24} kilogram
**Different latitudes rotate at different speeds

Appendix XIII: Geological time scale

Era	Period	Epoch	*millions of years ago*
Precambrian	Precambrian		4600
Palaeozoic	Cambrian		570
	Ordovician		500
	Silurian		440
	Devonian		395
	Carboniferous		345
	Permian		280
Mesozoic	Triassic		225
	Jurassic		190
	Cretaceous		136
Cenozoic	Tertiary	Palaeocene	65
		Eocene	
		Oligocene	
		Miocene	
		Pliocene	
	Quaternary	Pleistocene	1.8
		Holocene	

Appendix XIV: Amino acids

Amino acid	*Abbreviation*	*Formula*
arginine	Arg	$H_2N-\underset{\underset{NH}{\parallel}}{C}-NH-CH_2-CH_2-CH_2-\overset{H}{\underset{NH_2}{C}}-COOH$
aspartic acid	Asp	$HOOC-CH_2-\overset{H}{\underset{NH_2}{C}}-COOH$
histidine	His	$HC=C-CH_2-\overset{H}{\underset{NH_2}{C}}-COOH$ (ring: HC–N=CH–NH–C)
isoleucine	Ile	$CH_3-CH_2-\underset{CH_3}{CH}-\overset{H}{\underset{NH_2}{C}}-COOH$
leucine	Leu	$(H_3C)_2CH-CH_2-\overset{H}{\underset{NH_2}{C}}-COOH$
lysine	Lys	$H_2N-CH_2-CH_2-CH_2-CH_2-\overset{H}{\underset{NH_2}{C}}-COOH$

Amino acids (continued)

Amino acid	*Abbreviation*	*Formula*
methionine	Met	$CH_3-S-CH_2-CH_2-\underset{NH_2}{\overset{H}{C}}-COOH$
phenylalanine	Phe	$C_6H_5-CH_2-\underset{NH_2}{\overset{H}{C}}-COOH$
threonine	Thr	$CH_3-\underset{OH}{CH}-\underset{NH_2}{\overset{H}{C}}-COOH$
tryptophan	Trp	$C_6H_4(CH)(C=CH)-CH_2-\underset{NH_2}{\overset{H}{C}}-COOH$
tyrosine	Tyr	$HO-C_6H_4-CH_2-\underset{NH_2}{\overset{H}{C}}-COOH$
valine	Val	$(H_3C)_2CH-\underset{NH_2}{\overset{H}{C}}-COOH$

Appendix XV: Classification of living organisms

Prokaryotae
(*prokaryotes*)

bacteria e.g. *Escherichia coli*	cyanobacteria (blue-green bacteria) e.g. *Nostoc*

Protoctista
(*protoctists*)

Phylum	*Common name*	*Example*
*Rhizopoda	rhizopods	*Amoeba*
*Zoomastigina	flagellates	*Trypanosoma*
*Apicomplexa	sporazoans	*Plasmodium*
*Ciliophora	ciliates	*Paramecium*
Euglenophyta	euglenoid flagellates	*Euglena*
Oomycota	oomycetes	*Phytophthora*
Chlorophyta	green algae	*Chlamydomonas*
Rhodophyta	red algae	*Chondrus*
Phaeophyta	brown algae	*Fucus*

* These 4 phyla together constitute the protozoans

Fungi
(*fungi*)

Phylum	*Common name*	*Example*
Zygomycota	zygomycetes	*Mucor*
Ascomycota	ascomycetes	*Neurospora*
Basidiomycota	basidiomycetes	*Agaricus* (mushroom)

Classification (continued)

Plantae
(*plants*)

Phylum	*Common name*	*Class*	*Common name*	*Example*
Bryophyta	bryophytes	Hepaticae	liverworts	*Pellia*
		Musci	mosses	*Bryum*
Lycopodophyta	club mosses			*Selaginella*
Sphenophyta	horsetails			*Equisetum*
Filicinophyta	ferns			*Pteridium* (bracken)
Coniferophyta	conifers			*Pinus* (Scots pine)
Angiospermophyta	angiosperms (flowering plants)	Monocotyledoneae	monocotyledons	*Triticum* (wheat)
		Dictotyledoneae	dicotyledons	*Ranunculus* (buttercup)

Animalia
(*animals*)

Phylum	*Common name*	*Class*	*Common name*	*Example*
Cnidaria	cnidarians			*Aurelia* (jellyfish)
Platyhelminthes	flatworms	Tubellaria	turbellarians	*Polycelis*
		Trematoda	trematodes or flukes	*Fasciola* (liverfluke)
		Cestoda	cestodes or tapeworms	*Taenia* (tapeworm)
Nematoda	nematodes or roundworms			
Annelida	annalids	Polychatae	polychaetes or marine worms or segmented worms	*Nereis* (ragworm)

Classification: Animalia (continued)

Phylum	*Common name*	*Class*	*Common name*	*Example*
		Oligochaeta	oligochaetes or earthworms	*Lumbricus* (earthworm)
		Hirudinea	leeches	*Hirudu* (medicinal leech)
Mollusca	molluscs	Gastropoda	gastropods	*Helix* (garden snail)
		Pelecypoda	bivalves	*Ostrea* (oyster)
		Cephalopoda	cephalopods	*Sepia* (cuttlefish)
Arthropoda	arthropods	(Superclass)		
		Crustacea	crustaceans	
		Branchiopoda		*Daphnia* (water flea)
		Malacostraca		*Carcinus* (crab)
		Chilopoda	centipedes	*Lithobius* (centipede)
		Diplopoda	millipedes	*Iulus* (millipede)
		Insecta	insects	*Locusta* (locust)
		Arachnida	arachnids	*Scorpio* (scorpion)
Echinodermata	echinoderms	Stelleroidea	star fish and brittle star	*Asterias* (star fish)
		Echinoidea	sea urchins	*Echinocardium* (heart urchin)
Chordata	chordates	Chondrichthyes	cartilaginous fish	*Scyliohinus* (dogfish)
		Osteichthyes	bony fish	*Clupea* (herring)
		Amphibia	amphibians	*Rana* (frog)
		Reptilia	reptiles	*Lacerta* (lizard)
		Aves	birds	*Columba* (pigeon)
		Mammalia	mammals	*Homo* (human)

Appendix XVI: Common differential coefficients and integrals

y	$\frac{dy}{dx}$	$\int y.dx$
x^n	nx^{n-1}	$\frac{1}{n+1}.x^{n+1}$
$\frac{1}{x}$	$\frac{-1}{x^2}$	$\log_e x$
e^{ax}	ae^{ax}	$\frac{1}{a}.e^{ax}$
$\log_a x$	$\frac{1}{x}$	$x(\log_e x^{-1})$
$\cos ax$	$-a.\sin ax$	$\frac{1}{a}.\sin ax$
$\sin ax$	$-a.\cos ax$	$\frac{-1}{a}.\cos ax$
$\tan ax$	$-a.\sec^2 ax$	$\frac{-1}{a}.\log_e \cos ax$
$\cot x$	$-\mathrm{cosec}^2 x$	$\log_e \sin x$
$\sec x$	$\tan x.\sec x$	$\log_e(\sec x + \tan x)$
$\mathrm{cosec}\, x$	$-\cot x.\mathrm{cosec}\, x$	$\log_e(\cos_e x - \cot x)$
$\sin^{-1}\frac{x}{a}$	$\frac{1}{(a^2 - x^2)^{1/2}}$	$x.\sin^{-1}\frac{x}{a} + (a^2 - x^2)^{1/2}$
$\cos^{-1}\frac{x}{a}$	$\frac{-1}{(a^2 - x^2)^{1/2}}$	$x.\cos^{-1}\frac{x}{a} - (a^2 - x^2)^{1/2}$
$\tan^{-1}\frac{x}{a}$	$\frac{a}{a^2 + x^2}$	$x.\tan^{-1}\frac{x}{a} - a\log_e(a^2 - x^2)^{1/2}$

Appendix XVII: Greek alphabet

Α	α	alpha
Β	β	beta
Γ	γ	gamma
Δ	δ	delta
Ε	ε	epsilon
Ζ	ζ	zeta
Η	η	eta
Θ	θ	theta
Ι	ι	iota
Κ	κ	kappa
Λ	λ	lambda
Μ	μ	mu
Ν	ν	nu
Ξ	ξ	xi
Ο	ο	omicron
Π	π	pi
Ρ	ρ	ro
Σ	σ	sigma
Τ	τ	tau
Υ	υ	upsilon
Φ	φ	phi
Χ	χ	chi
Ψ	ψ	psi
Ω	ω	omega

PROSPECT PUBLIC LIBRARY
17 CENTER STREET
PROSPECT CT
$35.00